Lecture Notes in Artificial Intelligence   9651

Subseries of Lecture Notes in Computer Science

More information about this series at http://www.springer.com/series/1244

James Bailey · Latifur Khan
Takashi Washio · Gillian Dobbie
Joshua Zhexue Huang · Ruili Wang (Eds.)

Advances in Knowledge Discovery and Data Mining

20th Pacific-Asia Conference, PAKDD 2016
Auckland, New Zealand, April 19–22, 2016
Proceedings, Part I

 Springer

Editors
James Bailey
The University of Melbourne
Melbourne, VIC
Australia

Latifur Khan
The University of Texas at Dallas
Richardson, TX
USA

Takashi Washio
Osaka University
Osaka
Japan

Gillian Dobbie
University of Auckland
Auckland
New Zealand

Joshua Zhexue Huang
Shenzhen University
Shenzhen
China

Ruili Wang
Massey University
Auckland
New Zealand

ISSN 0302-9743   ISSN 1611-3349 (electronic)
Lecture Notes in Artificial Intelligence
ISBN 978-3-319-31752-6   ISBN 978-3-319-31753-3 (eBook)
DOI 10.1007/978-3-319-31753-3

Library of Congress Control Number: 2016934425

LNCS Sublibrary: SL7 – Artificial Intelligence

Printed on acid-free paper

This Springer imprint is published by Springer Nature
The registered company is Springer International Publishing AG Switzerland

PC Chairs' Preface

PAKDD 2016 is the 20th conference of the Pacific Asia Conference series on Knowledge Discovery and Data Mining. For the first time, the conference is being held in New Zealand. The conference provides a forum for researchers and practitioners to present and discuss new research results and practical applications.

There were 307 papers submitted to PAKDD 2016 and they underwent a rigorous double blind review process. Each paper was reviewed by three Program Committee (PC) members and meta-reviewed by one Senior Program Committee (SPC) member who also conducted discussions with the reviewers. The Program Chairs then considered the recommendations from SPC members, looked into each paper and its reviews, to make final paper selections. At the end, 91 papers were selected for the conference program and proceedings, resulting in an acceptance rate below 30 %, among which 39 papers were assigned as long presentation and 52 papers were assigned as regular presentation. The review process was supported by the Microsoft CMT system.

The conference started with a day of five high-quality workshops and five tutorials. During the next three days, the Technical Program included 19 paper presentation sessions covering various subjects of knowledge discovery and data mining, a data mining contest, and three keynote talks by world-renowned experts.

We would like to thank all the Program Committee members and external reviewers for their hard work to provide timely and comprehensive reviews and recommendations, which were crucial to the final paper selection and production of a high-quality Technical Program. We would also like to express our sincere thanks to Huiping Cao and Jinyan Li together with the individual Workshop Chairs for organizing the workshop program; Hisashi Kashima and Leman Akoglu together with the individual tutorial speakers for arranging the tutorial program; Ruili Wang for compiling all the accepted papers and for working with the Springer team to produce these proceedings.

We hope that participants in the conference in Auckland, as well as subsequent readers of the proceedings, will find the technical program of PAKDD 2016 to be both inspiring and rewarding.

February 2016

James Bailey
Latifur Khan
Takashi Washio

General Chairs' Preface

It is our great pleasure to welcome you to the 20th Conference of the Pacific Asia Conference series on Knowledge Discovery and Data Mining. PAKDD has successfully brought together researchers and developers since 1997, with the purpose of identifying challenging problems facing the development of advanced knowledge discovery. The 20th edition of PAKDD continues this tradition.

We are delighted to present three outstanding keynote speakers: Naren Ramakrishnan from Virginia Tech, Mark Sagar from The University of Auckland, and Svetha Venkatesh from Deakin University.

We are grateful to the many authors who submitted their work to the PAKDD technical program. The Program Committee was led by James Bailey, Latifur Khan and Takashi Washio. A report on the paper selection process appears in the PC Chairs' Preface.

We also thank the other Chairs in the organization team: Muhammad Asif Naeem for running the Contest; David Tse Jung Huang for publicizing to attract submissions and managing the website; Ranjini Swaminathan for handling the registration process and Yun Sing Koh and Ranjini Swaminathan for the local arrangements ensuring the conference runs smoothly.

We are grateful to the sponsors of the conference, Auckland Tourism Events and Economic Development, and BECA, for their generous sponsorship and support, and the PAKDD Steering Committee for its guidance and Best Paper Award, Student Travel Award and Early Career Research Award sponsorship. We would also like to express our gratitude to The University of Auckland for hosting and organizing this conference. Last but not least, our sincere thanks go to all the local team members and volunteer helpers for their hard work to make the event possible. We hope you enjoy PAKDD 2016 and your time in Auckland, New Zealand.

Gillian Dobbie
Joshua Zhexue Huang

Organization

Organizing Committee

General Co-chairs

Gillian Dobbie   University of Auckland, New Zealand
Joshua Zhexue Huang   Shenzhen University, China

Program Committee Co-chairs

James Bailey   The University of Melbourne, Australia
Latifur Khan   University of Texas at Dallas, USA
Takashi Washio   Institute of Scientific and Industrial Research,
                              Osaka University, Japan

Workshop Co-chairs

Huiping Cao   New Mexico State University, USA
Jinyan Li   University of Technology Sydney, Australia

Tutorial Co-chairs

Leman Akoglu   Stony Brook University, USA
Hisashi Kashima   Kyoto University, Japan

Local Arrangements Co-chairs

Yun Sing Koh   University of Auckland, New Zealand
Ranjini Swaminathan   University of Auckland, New Zealand

Proceedings Chair

Ruili Wang   Massey University, New Zealand

Contest Chair

Muhammad Asif Naeem   AUT University, New Zealand

Publicity and Website Chair

David Tse Jung Huang   University of Auckland, New Zealand

Registration Chair

Ranjini Swaminathan   University of Auckland, New Zealand

Steering Committee

Chairs

Tu Bao Ho (Chair)	Japan Advanced Institute of Science and Technology, Japan
Ee-Peng Lim (Co-Chair)	Singapore Management University, Singapore

Treasurer

Graham Williams	Togaware, Australia (see also under Life Members)

Members

Tu Bao Ho	Japan Advanced Institute of Science and Technology, Japan (Member since 2005, Co-Chair 2012–2014, Chair 2015–2017, Life Member since 2013)
Ee-Peng Lim	Singapore Management University, Singapore (Member since 2006, Co-Chair 2015–2017)
Jaideep Srivastava	University of Minnesota, USA (Member since 2006)
Zhi-Hua Zhou	Nanjing University, China (Member since 2007)
Takashi Washio	Institute of Scientific and Industrial Research, Osaka University (Member since 2008)
Thanaruk Theeramunkong	Thammasat University, Thailand (Member since 2009)
P. Krishna Reddy	International Institute of Information Technology, Hyderabad (IIIT-H), India (Member since 2010)
Joshua Z. Huang	Shenzhen Institutes of Advanced Technology, Chinese Academy of Sciences, China (Member since 2011)
Longbing Cao	Advanced Analytics Institute, University of Technology, Sydney (Member since 2013)
Jian Pei	School of Computing Science, Simon Fraser University (Member since 2013)
Myra Spiliopoulou	Information Systems, Otto-von-Guericke-University Magdeburg (Member since 2013)
Vincent S. Tseng	National Cheng Kung University, Taiwan (Member since 2014)

Life Members

Hiroshi Motoda	AFOSR/AOARD and Osaka University, Japan (Member since 1997, Co-Chair 2001–2003, Chair 2004–2006, Life Member since 2006)
Rao Kotagiri	University of Melbourne, Australia (Member since 1997, Co-Chair 2006–2008, Chair 2009–2011, Life Member since 2007, Treasury Co-Sign since 2006)
Huan Liu	Arizona State University, U.S. (Member since 1998, Treasurer 1998–2000, Life Member since 2012)

Ning Zhong   Maebashi Institute of Technology, Japan
                           (Member since 1999, Life member since 2008)
Masaru Kitsuregawa   Tokyo University, Japan (Member since 2000,
                           Life Member since 2008)
David Cheung   University of Hong Kong, China (Member since 2001,
                           Treasurer 2005–2006, Chair 2006–2008, Life Member
                           since 2009)
Graham Williams   Australian National University, Australia (Member since
                           2001, Treasurer since 2006, Co-Chair 2009–2011,
                           Chair 2012–2014, Life Member since 2009)
Ming-Syan Chen   National Taiwan University, Taiwan, ROC (Member since
                           2002, Life Member since 2010)
Kyu-Young Whang   Korea Advanced Institute of Science & Technology, Korea
                           (Member since 2003, Life Member since 2011)
Chengqi Zhang   University of Technology Sydney, Australia (Member
                           since 2004, Life Member since 2012)

Past Members

Hongjun Lu   Hong Kong University of Science and Technology
                           (Member 1997–2005)
Arbee L.P. Chen   National Chengchi University, Taiwan, ROC
                           (Member 2002–2009)
Takao Terano   Tokyo Institute of Technology, Japan
                           (Member 2000–2009)

Program Committee

Senior Program Committee Members

Michael Berthold   University of Konstanz, Germany
Tru Cao   Ho Chi Minh City University of Technology, Vietnam
Ming-Syan Chen   National Taiwan University, Taiwan
Peter Christen   The Australian National University, Australia
Ian Davidson   UC Davis, USA
Guozhu Dong   Wright State University
Bart Goethals   University of Antwerp, Belgium
Xiaohua Hu   Drexel University, USA
Joshua Huang   Shenzhen Institutes of Advanced Technology,
                           Chinese Academy of Sciences, China
George Karypis   University of Minnesota, USA
Ming Li   Nanjing University, China
Jiuyong Li   University of South Australia, Australia
Jinyan Li   University of Technology, Sydney
Chih-Jen Lin   National Taiwan University, Taiwan
Nikos Mamoulis   University of Hong Kong, Hong Kong
Wee Keong Ng   Nanyang Technological University, Singapore

Jian Pei	Simon Fraser University, Canada
Wen-Chih Peng	National Chiao Tung University, Taiwan
Rajeev Raman	University of Leicester, United Kingdom
P. Reddy	International Institute of Information Technology, Hyderabad (IIIT-H), India
Dou Shen	Baidu, China
Kyuseok Shim	Seoul National University, Korea
Myra Spiliopoulou	Otto-von-Guericke-University, Germany
Masashi Sugiyama	The University of Tokyo, Japan
Kai Ming Ting	Federation University, Australia
Hanghang Tong	City University of New York, USA
Vincent S. Tseng	National Cheng Kung University, Taiwan
Koji Tsuda	University of Tokyo, Japan
Wei Wang	University of California at Los Angeles, USA
Haixun Wang	Google, USA
Jianyong Wang	Tsinghua University, China
Xindong Wu	University of Vermont, USA
Xing Xie	Microsoft Research Asia, China
Hui Xiong	Rutgers University, USA
Xifeng Yan	UC Santa Barbara, USA
Jeffrey Yu	The Chinese University of Hong Kong, Hong Kong
Osmar Zaiane	University of Alberta, Canada
Yanchun Zhang	Victoria University, Australia
Min-Ling Zhang	Southeast University, China
Yu Zheng	Microsoft Research Asia, China
Ning Zhong	Maebashi Institute of Technology, Japan
Xiaofang Zhou	The University of Queensland, Australia
Zhi-Hua Zhou	Nanjing University, China

Program Committee Members

Mohammad Al Hasan	Purdue University, USA
Shafiq Alam	University of Auckland, New Zealand
Aijun An	York University, Canada
Gustavo Batista	University of Sao Paulo, Brazil
Chiranjib Bhattachar	Indian Institute of Science, India
Albert Bifet	Universite Paris-Saclay, France
Marut Buranarach	National Electronics and Computer Technology Center, Thailand
Krisztian Buza	Budapest University of Technology and Economics, Hungary
Rui Camacho	Universidade do Porto, Portugal
K. Selcuk Candan	Arizona State University, USA
Jeffrey Chan	RMIT University, Australia
Chia-Hui Chang	National Central University, Taiwan
Muhammad Cheema	Monash University, Australia

Meng Chang Chen   Academia Sinica, Taiwan
Shu-Ching Chen   Florida International University, USA
Songcan Chen   Nanjing University of Aeronautics and Astronautics, China
Zhiyuan Chen   University of Maryland Baltimore County, USA
Yi-Ping Phoebe Chen   La Trobe University, Australia
Zheng Chen   Microsoft Research Asia, China
Chun-Hao Chen   Tamkang University, Taiwan
Enhong Chen   University of Science and Technology of China, China
Ling Chen   University of Technology Sydney, Australia
Jake Chen   Indiana University-Purdue University Indianapolis, USA
Yiu-ming Cheung   Hong Kong Baptist University, Hong Kong
Silvia Chiusano   Politecnico di Torino, Italy
Kun-Ta Chuang   National Cheng Kung University, Taiwan
Bruno Cremilleux   Universite de Caen, France
Alfredo Cuzzocrea   ICAR-CNR and University of Calabria, Italy
Bing Tian Dai   Singapore Management University, Singapore
Dao-Qing Dai   Sun Yat-Sen University, China
Xuan-Hong Dang   UC Santa Barbara, USA
Anne Denton   North Dakota State University, USA
Bolin Ding   Microsoft Research, USA
Wei Ding   University of Massachusetts Boston, USA
Dejing Dou   University of Oregon, USA
Liang Du   Chinese Academy of Sciences, China
Lei Duan   Sichuan University
Christoph Eick   University of Houston
Vladimir Estivill-Castro Griffith University
Philippe Fournier-Viger University of Moncton, Canada
Dragan Gamberger   Rudjer Boskovic Institute, Croatia
Junbin Gao   Charles Sturt University, Australia
Jun Gao   Peking University, China
Yong Guan   Iowa State University, USA
Stephan Gunnemann   TU Munich, Germany
Sunil Gupta   Deakin University
Michael Hahsler   Southern Methodist University, USA
Saotshi Hara   IBM Research Tokyo, Japan
Choochart Haruechaiy   National Electronics and Computer Technology Center,
                        Thailand
Jingrui He   IBM Research, USA
Shoji Hirano   Shimane University, Japan
Jaakko Hollmen   Aalto University, Finland
Tzung-Pei Hong   National University of Kaohsiung, Taiwan
Michael Houle   NII, Japan
Wynne Hsu   National University of Singapore, Singapore
Jun Huan   University of Kansas, USA
Jen-Wei Huang   National Cheng Kung University, Taiwan
Sheng-Jun Huang   NUAA, China

Nam Huynh	Japan Advanced Institute of Science and Technology, Japan
Akihiro Inokuchi	Kwansei Gakuin University
Motoharu Iwata	NTT Communication Science Laboratories, Japan
Sanjay Jain	National University of Singapore, Singapore
Toshihiro Kamishima	National Institute of Advanced Industrial Science and Technology, Japan
Murat Kantarcioglu	University of Texas at Dallas, USA
Hung-Yu Kao	National Cheng Kung University, Taiwan
Yoshinobu Kawahara	Osaka University, Japan
Irena Koprinska	University of Sydney, Australia
Walter Kosters	Universiteit Leiden, Netherlands
Marzena Kryszkiewicz	Warsaw University of Technology, Poland
Satoshi Kurihara	Osaka University, Japan
Hady Lauw	Singapore Management University, Singapore
Wang-Chien Lee	Pennsylvania State University, USA
Yue-Shi Lee	Ming Chuan University, Taiwan
Philippe Lenca	Telecom Bretagne, France
Carson K. Leung	University of Manitoba, Canada
Geng Li	Oracle Corporation, USA
Chun-hung Li	Hong Kong Baptist University, Hong Kong
Zhenhui Li	Pennsylvania State University, USA
Yidong Li	Beijing Jiaotong University, China
Xiaoli Li	Institute for Infocomm Research, Singapore
Wu-Jun Li	Nanjing University, China
Xuelong Li	University of London, UK
Hsuan-Tien Lin	National Taiwan University, Taiwan
Jerry Chun-Wei Lin	Harbin Institute of Technology Shenzhen, China
Xu-Ying Liu	Southeast University, China
Wei Liu	University of Technology Sydney, Australia
Qingshan Liu	NLPR Institute of Automation Chinese Academy of Science, China
Hua Lu	Aalborg University, Denmark
Jun Luo	Hua Wei Noahs Ark Lab, Hong Kong
Shuai Ma	Beihang University, China
Marco Maggini	Universita degli Studi di Siena, Italy
Hiroshi Mamitsuka	Kyoto University, Japan
Giuseppe Manco	Universita' della Calabria, Italy
Florent Masseglia	INRIA, France
Mohammad Mehedy Masud	United Arab Emirates University
Tomoko Matsui	Institute of Statistial Mathematics, Japan
Xiaofeng Meng	Renmin University of China, China
Nguyen Le Minh	JAIST, Japan
Pabitra Mitra	Indian Institute of Technology Kharagpur, India
Yang-Sae Moon	Kangwon National University, Korea

Yasuhiko Morimoto	Hiroshima University, Japan
Emmanuel Mueller	Hasso-Plattner-Institut
J. Nath	Indian Institute of Technology, India
Richi Nayak	Queensland University of Technologies, Australia
Wilfred Ng	Hong Kong University of Science and Technology, Hong Kong
Xuan Vinh Nguyen	University of Melbourne, Australia
Ngoc-Thanh Nguyen	Wroclaw University of Technology, Poland
Hung-Son Nguyen	University of Warsaw, Poland
Duc Dung Nguyen	Institute of Information Technology, Vietnam
Tadashi Nomoto	National Institute of Japanese Literature, Japan
Manabu Okumura	Japan Advanced Institute of Science and Technology, Japan
Salvatore Orlando	University of Venice, Italy
Jia-Yu Pan	Google, USA
Shirui Pan	University of Technology Sydney, Australia
Dhaval Patel	Indian Institute of Technology, Roorkee, India
Dinh Phung	Deakin University, Australia
Vincenzo Piuri	Universita degli Studi di Milano, Italy
Chedy Raissi	INRIA, France
Santu Rana	Deakin University, Australia
Chandan Reddy	Wayne State University, USA
Patricia Riddle	University of Auckland, New Zealand
Hiroshi Sakamoto	Kyushu Institute of Technology, Japan
Yi-Dong Shen	Chinese Academy of Sciences, China
Jialie Shen	Singapore Management University, Singapore
Hong Shen	Adelaide University, Australia
Masashi Shimbo	Nara Institute of Science and Technology, Japan
Andrzej Skowron	University of Warsaw, Poland
Mahito Sugiyama	Osaka University, Japan
Aixin Sun	Nanyang Technological University, Singapore
Yasuo Tabei	Presto, Japan Science and Technology Agency
David Taniar	Monash University, Australia
Xiaohui (Daniel) Tao	The University of Southern Queensland, Australia
Khoat Than	Hanoi University of Science and Technology, Vietnam
Jeffrey Ullman	Stanford University, USA
Ranga Vatsavai	North Carolina University
Kitsana Waiyamai	Kasetsart University, Thailand
Lipo Wang	Nanyang Technological University, Singapore
Jason Wang	New Jersey Science and Technology University, USA
Xin Wang	University of Calgary, Canada
Raymond Chi-Wing Wong	Hong Kong University of Science and Technology, Hong Kong
Xintao Wu	University of Arkansas, USA
Jia Wu	University of Technology Sydney, Australia
Junjie Wu	Beihang University, China

Guandong Xu	University of Technology Sydney, Australia
Takehisa Yairi	University of Tokyo, Japan
De-Nian Yang	Academia Sinica, Taiwan
Min Yao	Zhejiang University, China
Mi-Yen Yeh	Academia Sinica, Taiwan
Tetsuya Yoshida	Nara Womens University
Yang Yu	Nanjing University, China
De-Chuan Zhan	Nanjing University, China
Daoqiang Zhang	Nanjing University of Aeronautics and Astronautics, China
Du Zhang	California State University, USA
Bo Zhang	Tsinghua University, China
Junping Zhang	Fudan University, China
Wenjie Zhang	University of New South Wales, Australia
Ying Zhang	University of New South Wales, Australia
Zhongfei Zhang	Binghamton University, USA
Zili Zhang	Deakin University, Australia
Mengjie Zhang	Victoria University of Wellington, New Zealand
Zhao Zhang	Soochow University, China
Xiuzhen Zhang	RMIT University, Australia
Peixiang Zhao	Florida State University, USA
Shuigeng Zhou	Fudan University, China
Bin Zhou	University of Maryland Baltimore County, USA
Feida Zhu	Singapore Management University, Singapore
Xingquan Zhu	Florida Atlantic University, USA
Arthur Zimek	Ludwig-Maximilians-University Munchen, Germany

Sponsors

Contents – Part I

Feature Extraction and Pattern Mining

Contents – Part II

Anomaly Detection and Clustering

Novel Models and Algorithms

Text Mining and Recommender Systems

Classification

Joint Classification with Heterogeneous Labels Using Random Walk with Dynamic Label Propagation

Yongxin Liao, Shenxi Yuan, Jian Chen, Qingyao Wu[(✉)], and Bin Li

School of Software Engineering,
South China University of Technology, Guangzhou, China
qyw@scut.edu.cn

Abstract. This paper studies a new machine learning strategy called joint classification with heterogeneous labels (JCHL). Unlike traditional supervised learning problems, JCHL uses a single feature space to jointly classify multiple classification tasks with heterogeneous labels. For instance, biologists usually have to label the gene expression images with developmental stages and simultaneously annotate their anatomical terms. We would like to classify the developmental stages and at the same time classify anatomical terms by learning from the gene expression data. Recently, researchers have considered using Preferential random walk (PRW) to build different relations to link heterogeneous labels, thus the heterogeneous label information can be propagated by the instances. On the other hand, it has been shown that learning performance can be significantly enhanced if the dynamic propagation is exploited in PRW. In this paper, we propose a novel algorithm, called random walk with dynamic label propagation (RWDLP), for the JCHL problems. In RWDLP, a joint transition probability graph is constructed to encode the relationships among instances and heterogeneous labels, and we utilize dynamic label propagation in the graph to generate the possible labels for the joint classification tasks with heterogeneous labels. Experimental results have demonstrated the effectiveness of the proposed method.

Keywords: Joint classification · Heterogenerous labels · Random walk · Dynami label propagation

1 Introduction

In machine learning, traditional classification only has a single feature space and a label space [7,17]. An example would be assigning a given email into spam or non-spam classes [1] or assigning a part of speech to each word in a input sentence [12]. But in actual applications, there are quite a few classification scenarios that contain multiple diverse label sets for the training set and it is natural to solve these problems by the joint classification with heterogeneous labels. For example, biologists usually have to label the gene expression images with developmental

J. Bailey et al. (Eds.): PAKDD 2016, Part I, LNAI 9651, pp. 3–13, 2016.
DOI: 10.1007/978-3-319-31753-3_1

stages [6] and anatomical terms annotation of gene expression patterns [20]. Each image is considered as a data instance, and each data instance must be assigned a stage term and one or more anatomical terms. Traditional classification only has one label set, like spam or non-spam classes. Joint classification, however, includes several different label sets, just like parse tree and syntactic structure. Figure 1 shows the difference between traditional classification and the joint classification with heterogeneous labels.

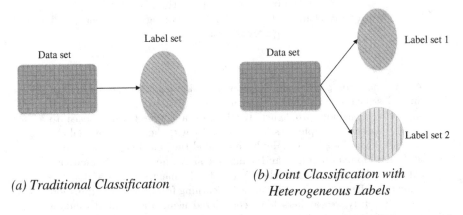

(a) Traditional Classification

(b) Joint Classification with Heterogeneous Labels

Fig. 1. The difference between Traditional Classification and Joint Classification with Heterogeneous Tasks

Random walk is a popular algorithm and has been used in many fields, like economics, biology and computer science. It explains the observed behaviors of many processes and is widely used in classification problems [8,10]. A random walk on a graph can make use of neighbors' information to learn the correct labels in an iterative process. Therefore, Cai *et al.* [2] proposed an graph-based semi-supervised algorithm called preferential random walk (PRW) to solve the JCHL problem. PRW combines the information of both data features with heterogeneous labels and the standard random walk by using data features, and the transition probability matrices of both algorithms are stable.

Recently, another graph-based semi-supervised algorithm was proposed, called dynamic label propagation (DLP) [13], which incorporates the label correlations and instance similarities into a new way of performing label propagation. It was developed on the basis of label propagation (LP) [19] which assumes that nodes connected by edges of large similarity tend to have the same label through information which is propagated within the graph. And it is consistent with random walk. DLP updates the similarity measures dynamically by fusing multi-label/multi-class information. Experimental results have shown that this algorithm is more competitive than those algorithms without the dynamic updating process.

In the light of these previous dicussions, we got inspiration and proposed a graph-based semi-supervised algorithm called RWDLP to solve JCHL problem.

In RWDLP, we also build different relations to link heterogeneous labels by a markov chain random walk, and we employ the dynamic updating process to update the transition probability matrix in each iteration. We develop a new simple way to update the transition probability matrix of random walk, by which the data features and heterogeneous labels could be merged more effectively and produce a better performance. As demonstrated in our experiments, the random walk with dynamic label propagation can successfully deal with JCHL problems.

The rest of the paper is organized as follows. In Sect. 2, we discuss the difference between traditional classification and joint classification, and then we give a brief review of PRW. In Sect. 3, we develop the proposed algorithm. In Sect. 4, we show and discuss the experimental results. In Sect. 5, we give some concluding remarks.

2   Related Works

2.1   Traditional Classification and Joint Classification

Traditional classification has one data feature set and one label set on its input-output mapping functions. In this paper, we study joint classification with heterogeneous lables that has a feature set and multiple diverse label sets. There have been a variety of algorithms proposed to study heterogeneous labels. For example, Jin et al. [5] studied a learning with several views corresponding to different set of class labels. These label sets have a close relation although they are different in the number of labels. However, there are not many studies in joint classification. The main bottleneck of JCHL problem is to deal with multiple tasks in the same time. The solution is to study a new algorithm that can handel the heterogeneous labels, and another idea is to combine the heterogeneous labels into a unified formulation.

Preferential random walk (PRW) proposed by Cai et al. [2] is an algorithm about JCHL problem that combines the heterogeneous labels into an unified formulation. We consider data instances and heterogeneous labels as nodes in a graph, and regard the affinity of data-to-data, label-to-label and data-to-label as edges. We call the graph as Mix-Relevance Graph (MRG). And then the algorithm imagines a random walker which starts from a node (instance) with a known label, and steps to its neighbor nodes with a specific probability given in the transition probability matrix.

2.2   The Construction of MRG

In this subsection, we briefly review the Mix-Relevance Graph in PRW. Throughout this article, we denote a vector as a bold lowercase character $\mathbf{x}$ and a matrix as a bold uppercase character $\mathbf{X}$. If there's no special note, all vectors are column vectors. Specifically, the i-th column vectors of a matrix $\mathbf{X}$ are denoted as $\mathbf{x}_i$. Let $[N : M](M > N)$ denote a set of integers in the range of N to M inclusively, $\mathbf{v}(i)$ denote the i-th entry of a vector $\mathbf{v}$, and $\mathbf{M}(i, j)$ denote the entry at the i-th row and j-th column of a matrix $\mathbf{M}$.

Given a dataset $\mathbf{X}$, there are N data instances and each data instance has M dimensional features, denoted as $\mathbf{X} \in \mathbb{R}^{N \times M}$. On the other hand, there are several output spaces, $\mathbf{Y}_1, \mathbf{Y}_2, ..., \mathbf{Y}_K$, where K is the number of label spaces, and the number of labels in each space is different, denoted as $\mathbf{Y}_1 \in \mathbb{R}^{Q_1}, ..., \mathbf{Y}_K \in \mathbb{R}^{Q_K}$. For simplicity, we write $\mathbf{Y} = [\mathbf{Y}_1, \mathbf{Y}_2, ..., \mathbf{Y}_K]$ We will discuss how to contruct the MRG in the rest of this section.

In dataset $\mathbf{X}$, the number of instances is N. The similarity between $\mathbf{x}_i$ and $\mathbf{x}_j$ in the object feature space can be measured by the affinity $\mathbf{M}_X(i,j)$, while $i \in [1 : N]$ and $j \in [1 : N]$. The affinity $\mathbf{M}_X(i,j)$ can be calculated based on the norm of the difference between their feature vectors $\mathbf{x}_i$ and $\mathbf{x}_j$, while i is not equal to j. In our algorithm, we employ the Gaussian kernel to compute this affinity.

$$\mathbf{M}_X(i,j) = \begin{cases} exp(-\|\mathbf{x}_i - \mathbf{x}_j\|_2 / 2\sigma^2), & i \neq j \\ 0, & otherwise \end{cases} \tag{1}$$

where $\|\mathbf{x}_i - \mathbf{x}_j\|_2$ is the Euclidean distance between the i-th object and the j-th object of dataset $\mathbf{X}$. The parameter σ is regarded as a positive number to control the linkage in the manifold. From $\mathbf{M}_X$, we can construct the transition probability matrix of data instances.

$$\mathbf{S}_X(i,j) = \begin{cases} (1 - \beta_1)\mathbf{M}_X(i,j)/d_i^X, & if d_i^Y > 0 \\ \mathbf{M}_X(i,j)/d_i^X, & otherwise \end{cases} \tag{2}$$

where $d_i^X = \sum_j \mathbf{M}_X(i,j)$, $d_i^Y = \sum_j \mathbf{Y}(i,j)$ and $\beta_1 \in [0,1]$. From $\mathbf{S}_X$, a graph $G_X = (V_X, E_X)$ can be induced, where $V_X = \mathbf{X}$ and $E_X \subseteq V_X \times V_X$. It is clear that $\mathbf{S}_X$ is symmetric and non-negative, therefore G_X is undirected and positively weighted. Since G_X is constructed with the affinity of data points, it is usually called as data graph, such as the left subgraph in Fig. 2. Most of the existing graph-based semi-supervised learning methods [10,16] only make use of the data graph, while RWDLP use both data graph and label graphs.

Now we have the data graph G_X, and then we should build the label graphs. Take label set $\mathbf{Y}_p$ as example, each data instance $\mathbf{x}_i$ belongs to one of Q_p classes $\mathbf{Y}_p = \left\{ \mathbf{y}_1^p, ..., \mathbf{y}_{Q_p}^p \right\}$ represented by $\mathbf{y}_j^p \in \{0,1\}^{Q_p}$, such that $\mathbf{y}_j^p(q) = 1$ if $\mathbf{x}_i$ is classified into class $\mathbf{y}_j^p$, and 0 otherwise. Because there are several heterogenerous label sets, we should build label graph for each label set, and the structure of each label set maybe different. Generally, classification task will be divided into single label task and multiple label task by the number of label on each data point. For this reason, we proposed two strategies to calculate the correlation between labels. Firstly, we compute the affinity of a single label task $\mathbf{Y}_p$ as follow:

$$\mathbf{M}_{Y_p}(i,j) = \left\| \mathbf{y}_i^p - \mathbf{y}_j^p \right\|_F, \tag{3}$$

where $\left\| \mathbf{y}_i^p - \mathbf{y}_j^p \right\|_F$ means Frobenius norm between the i-th label and the j-th label of $\mathbf{Y}_p$. Secondly, for a multiple label task $\mathbf{Y}_q$, we compute the affinity by cosine similarity as follows:

$$\mathbf{M}_{Y_q}(i,j) = \frac{\mathbf{y}_i^q \cdot \mathbf{y}_j^q}{\|\mathbf{y}_i^q\| \cdot \|\mathbf{y}_j^q\|}, \tag{4}$$

where $\|\mathbf{y}_i^q\|$ means absolute value of the i-th label of $\mathbf{Y}_q$. From $\mathbf{M}_{Y_p}$ and $\mathbf{M}_{Y_q}$, we can construct the transition probability matrix of label sets.

$$\mathbf{S}_{Y_k}(i,j) = \begin{cases} (1-\beta_2)\mathbf{M}_{Y_k}(i,j)/d_i^{Y_k}, & if d_i^{Y_k} > 0 \\ \mathbf{M}_{Y_k}(i,j)/d_i^{Y_k}, & otherwise \end{cases} \tag{5}$$

where $\mathbf{Y}_k \subseteq \mathbf{Y}$, $d_i^{Y_k} = \sum_j \mathbf{Y}_k(i,j)$ and $\beta_2 \in [0,1]$. Now, we can also induce label graphs $G_k = (V_k, E_k)$ from $\mathbf{M}_{Y_k}$, where $V_k = \mathbf{Y}_k$ and $E_k \subseteq V_k \times V_k$, just like the right subgraph in Fig. 2. The MRG is mainly composed of data graph and label graph. We have constructed data subgraph and label subgraphs, while these subgraphs are not connected yet. Obviously, the subgraph $G_{XY_p} = (V_X, V_{Y_p}, E_{XY_p})$ connects G_X and G_{Y_p}, whose adjacency matrix is $\mathbf{Y}_p$, where $p \in [1:K]$. Moreover, the subgraph $G_{Y_pY_q} = (V_p, V_q, E_{Y_pY_q})$ connects G_{Y_p} and G_{Y_q}, where $p \in [1:K], q \in [1:K]$ and $p \neq q$. The adjacency matrix of subgraph $G_{Y_pY_q}$ is defined as follow:

$$\mathbf{M}_{Y_pY_q}(i,j) = \frac{\mathbf{y}_i^p \cdot \mathbf{y}_j^q}{\|\mathbf{y}_i^p\| \cdot \|\mathbf{y}_j^q\|}, \tag{6}$$

where $\|\mathbf{y}_i^p\|$ means absolute value of the i-th label of $\mathbf{Y}_p$. From $\mathbf{M}_{Y_pY_q}$, we can construct the transition probability matrix.

$$\mathbf{S}_{Y_pY_q}(i,j) = \begin{cases} \beta_3\mathbf{M}_{Y_pY_q}(i,j)/d_i^{M_{Y_pY_q}}, & if d_i^{M_{Y_pY_q}} > 0 \\ 0, & otherwise \end{cases} \tag{7}$$

where $d_i^{M_{Y_pY_q}} = \sum_j \mathbf{M}_{Y_pY_q}(i,j)$ and $\beta_3 \in [0,1]$.

Lots of graph-based semi-supervised learning methods [19,20] have been proposed in the past, but most of them only use information conveyed by G_X. And PRW fuses data instances and heterogeneous labels information encoded in G. Motivated by PRW [2] and DLP [13], we plan to further develop preferential random walk and dynamic label propagation to measure the relevance among labeled data points and unlabeled data points.

3   Random Walk with Dynamic Label Propagation on MRG

Standard random walk on a graph $\mathbf{G}$ is usually described as a Markov process with transition probability matrix $\mathbf{P} = \mathbf{D}^{-1}\mathbf{G}$, where $\mathbf{D} = diag(d_1, ..., d_n)$ and $d_i = \sum_j \mathbf{G}(i,j)$ are the degree of vertes i. It is clear that $\mathbf{P}^T \neq \mathbf{P}$ and $\sum_j \mathbf{P}(i,j) = 1$. If $\mathbf{G}$ is symmetric, the graph is undirected. If $\mathbf{G}$ is asymmetric, the graph is directed and d_i is the out degree of vertex i. Let $\mathbf{Y}_t$ be the distribution of the random walker at time t. The steady state can be computed by

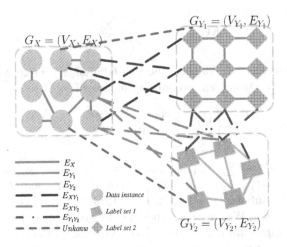

Fig. 2. The Mix-Relevance Graph:solid lines indicate affinity between vertices within in a same subgraph, dashed lines indicates associations between vertices in two different subgraphs

$$\mathbf{Y}_{t+1} = \mathbf{Y}_t \mathbf{P}. \tag{8}$$

It can be seen that the steady state of a standard random walk is just determined by the graph itself. In order to use label information, we propose the RWDLP as follow:

$$\mathbf{Y}_{t+1} = (1 - \alpha)\mathbf{S}_0\mathbf{Y}_t + \alpha\mathbf{Y}_0 \tag{9}$$

where $\mathbf{S}$ is the transition probability matrix constructed in last section, $\mathbf{Y}_0 = [\mathbf{y}_1, \mathbf{y}_2, ..., \mathbf{y}_K]$ contains the given labels of training objects and the unlabels of testing objects at first [15], and $0 \leq \alpha \leq 1$ specifies the importance of initial label information of a data instance, which can affect the ranking of the resulting label. In each iteration, we will update the transition probability matrix by a new label distribution matrix $\mathbf{Y}_{t+1}$, as follow:

$$\mathbf{S}_{t+1} = (1 - \mu)\mathbf{S}_t + \mu\mathbf{Y}_{t+1}\mathbf{Y}_{t+1}^T \tag{10}$$

where μ is the importance of the new label information.

The steady probability distribution $\mathbf{Y}$ can be solved by the iterative method. The overall algorithm is shown in Algorithm 1.

After solving $\mathbf{Y}$ by using Algorithm 1, we predict the heterogenerous labels of data instance in multiple diverse label sets by different methods. For single-label task, we use $y^p(i)$ to represent the stage label of i-th data instance and the $y^p(i)$ is the maximum probability of a label set:

$$y^p(i) = argmax(\mathbf{y}_i^p) \tag{11}$$

where $\mathbf{y}_i^p$ is the i-th row vector of matrix $\mathbf{Y}_p$, and $\mathbf{Y}_p$ is a submatrix of $\mathbf{Y}$, which means the output of p-th task. For multi-label task, we compute a threshold value θ [14] to get the multiple labels $\mathbf{y}^q(i)$ of i-th data instance.

$$\mathbf{y}^q(i) = \begin{cases} 1, & \mathbf{y}_i^q \geqslant \theta \\ 0, & \mathbf{y}_i^q < \theta \end{cases} \qquad (12)$$

where $\mathbf{y}_i^q$ is the row vector of matrix $\mathbf{Y}_q$.

Algorithm 1. Random walk with dynamic label propagation

Input : $\mathbf{S}_0$ and $\mathbf{Y}_0$ is an initial guess stated. α and the tolerance ϵ

Output : $\mathbf{Y}$

Procedure :

1. Set $t = 1$;
2. Compute $\mathbf{Y}_t = (1 - \alpha)\mathbf{S}_{t-1}\mathbf{Y}_{t-1} + \alpha\mathbf{Y}_0$;
3. Compute $\mathbf{S}_t = (1 - \mu)\mathbf{S}_{t-1} + \mu\mathbf{Y}_t\mathbf{Y}_t^T$;
4. Normalize each column of $\mathbf{S}$;
5. If $\|\mathbf{Y}_t - \mathbf{Y}_{t-1}\| < \epsilon$, then stop, Set $\mathbf{Y} = \mathbf{Y}_t$; otherwise set $t = t + 1$ and goto step 2.

---

4 Experiment and Result

Though we joint classifications with heterogenerous labels, we also compare RWDLP's performance with the state-of-the-art classification algorithms and show the performance in this section.

4.1 Data Set

To develop and test our method, we use the Berkeley Drosophila Genome Project (BDGP) gene expression pattern dataset. Recently, a lot of research works have been experienced in the raw data from BDGP. This dataset is widely used to develop and test anatomical annotation methods for Drosophila gene expression pattern images.

The images from BDGP database contain different views, and we just consider three views of images including lateral, dorsal and ventral images in our experiment, because the number of images from other views are not enough. All the images from BDGP have been pre-processed, including aligning and resizing to 128×320 gray images. The SIFT features are extracted from the gray images, and the codebook is made by K-means. The number of clusters is set to 1000, 500 and 250 for lateral, dorsal and ventral images, then we concatenate on the three vectors in one bag. To be specific, let $\mathbf{x}^l \in \mathbb{R}^{1000}$, $\mathbf{x}^d \in \mathbb{R}^{500}$ and $\mathbf{x}^v \in \mathbb{R}^{250}$ denote the bag-of-words vector for images in a bag from lateral, dorsal and ventral view. Therefore, an image bag can be represented as $\mathbf{x} = [\mathbf{x}^l; \mathbf{x}^d; \mathbf{x}^v] \in \mathbb{R}^{1750}$.

Drosophila embryogenesis has 17 stages, which are divided into 6 major ranges, i.e. stages 1–3, 4–6, 7–8, 9–10, 11–12, 13–16 and 17, in the BDGP database [9]. And the total number of anatomical terms is 303, i.e. $foregutAISN$, $maternal$ and so on. We ignore the stage 1–3 and 17 data since the number of anatomical terms is too small. For the same reason, we select 79 anatomical terms in our experiment. At last, there are two classification tasks in a dataset, stage classification and anatomical terms classification. Obviously, stage classification is a single-label/multi-class problem, which has 5 lables, and anatomical terms classification is a multi-label/multi-class problem, which has 79 lables.

We use 5-folds cross-validation and report the average performance of the 5 tails. We use 1-fold for traing and use the remaining 4-folds for testing to imitate actual scenarios in which the known label samples are far less than the unknown label samples. In our experiment, we initialize the testing label with KNN method, because it is simple and intuitive. To be specific, we set $k{=}10$ of KNN and it doesn't matter what it is assigned in our method. There are 5 parameters in our method and it isn't sensitive in certain ranges with good performances. $\beta_1, \beta_2, \beta_3$, which control the jumping among three subgraphs, cannot affect the result much if they are assigned in the range of (0.05,0.5). α is the importance of initial label information of a data instance and it should be assigned in the range of (0.5,0.9). μ is the importance of the new label information and it should be assigned in the range of (0.5,0.9). Since we handle two classifications in the same time, we use accuracy for stage classification to measure the performance between our proposed method and other state-of-the-art methods, and use macro precision, macro recall, macro F1, micro precision, micro recall and micro F1 for anatomical terms classification.

4.2 Compared Methods

Because stage classification is a single-label/multi-class problem, we choose two state-of-the-art methods, SVM and KNN, to compare with our proposed algorithm. The support vector machine (SVM) algorithm constructed by Chang and Lin [3] is one of the most popular methods of single-label problem. We use radial basis function (RBF) kernel and the optimal parameter values for $C = 1$ and $\gamma = 0.9$. This k-nearest neighbor (KNN) method is an unsupervised learning, while SVM is supervised and our method is semi-supervised. We set $k = 50$, however, it doesn't affect the result much. We also predict the anatomical terms for the Drosophila gene expression patterns, which is a multi-label/multi-class problem. It has a different label set compared with stage classification and both of the sets are heterogenerous. Traditionally, the precision, recall and F1 score are the measure of classification performance. But for the multi-label/multi-class classification now, macro and micro average of precision, recall and F1 score are used and suggested by Tomancak, P. et al. [11]. In our experiment, we compare the result of our proposed algorithm with three state-of-the-art multi-label/multi-class methods: local shared subspace(LS) [4], harmonic function (HF) [20] and ML-KNN [18] which is used to do the initialization. LS and HF are proposed to solve the multi-label annotation problem. For these methods, we used the published codes posted on the corresponding author's websites.

4.3 Performance Comparison

We use the average classification accuracy of 5-folds cross-validation to assess the results. The result of stage classification is shown in Fig. 3 on the left.

Our method exhibits the better result on the average prediction accuracy, and is better than other state-of-the-art methods on 4 stages. The average classification accuracy of our method is 85.01 %, while the value of SVM is 80.69 %, and KNN is 77.55 %. The Fig. 3 on the right shows the result of anatomical terms classification results. We can see that our algorithm is better than the other methods on all metrics. From Tables 1 and 2, we can see that our method will have better performance with less labeled data instances, while other algorithms must be given more labeled data instances. On the one hand, our proposed algorithm can joint stage classification and anatomical term classification simultaneously, and the former is single-labe/multi-class classification and the latter is multi-label/multi-class classification. On the other hand, we dynamically update the transition

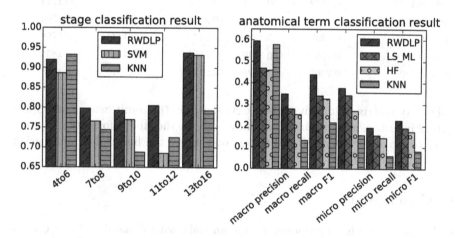

Fig. 3. Stage classification results and anatomical term classification results.

Table 1. The performance of RWDLP and KNN algorithms with different sizes training data

RWDLP	Macro precision	Macro recall	Macro F1	Micro precision	Micro recall	Micro F1
400	0.4713	0.2778	0.3493	0.3027	0.1547	0.1814
500	0.5945	0.3475	0.4383	0.3749	0.1920	0.2252
600	0.5963	0.3486	0.4397	0.3771	0.1932	0.2269
700	0.5960	0.3498	0.4406	0.3766	0.1946	0.2280
800	0.5965	0.3507	0.4414	0.3775	0.1958	0.2292
900	0.5980	0.3522	0.4430	0.3793	0.1975	0.2311
KNN	Macro precision	Macro recall	Macro F1	Micro precision	Micro recall	Micro F1
400	0.6434	0.1077	0.1844	0.1479	0.0476	0.0692
500	0.5783	0.1319	0.2148	0.1568	0.0599	0.0818
600	0.5781	0.1435	0.2299	0.1547	0.0635	0.0869
700	0.5743	0.1530	0.2417	0.1928	0.0698	0.0946
800	0.6183	0.1607	0.2551	0.2339	0.0730	0.0999
900	0.5916	0.1546	0.2451	0.2115	0.0736	0.1017

Table 2. The performance of LS and HF algorithms with different sizes training data

LS	Macro precision	Macro recall	Macro F1	Micro precision	Micro recall	Micro F1
400	0.3678	0.2571	0.2943	0.3329	0.1442	0.1732
500	0.4675	0.2786	0.3394	0.3417	0.1561	0.1899
600	0.5728	0.2905	0.3855	0.3759	0.1604	0.2027
700	0.4933	0.3141	0.3807	0.3768	0.1807	0.2190
800	0.5603	0.3203	0.4076	0.3609	0.1845	0.2262
900	0.5542	0.3191	0.4050	0.3860	0.1873	0.2320
HF	Macro precision	Macro recall	Macro F1	Micro precision	Micro recall	Micro F1
400	0.4516	0.2477	0.3199	0.2554	0.1383	0.1680
500	0.4584	0.2522	0.3254	0.2693	0.1439	0.1728
600	0.4533	0.2499	0.3222	0.2716	0.1447	0.1760
700	0.4576	0.2499	0.3232	0.2775	0.1448	0.1768
800	0.4818	0.2567	0.3349	0.2882	0.1487	0.1834
900	0.4808	0.2663	0.3428	0.2973	0.1570	0.1938

probability matrix in iterative process. As shown in the result, when one work is short of information to do classification, we can use the label information of other works to make the decision, and make good use of the limit label information from the data instances.

5   Conclusions

In this paper, we have proposed a random walk with dynamic label propagation method (RWDLP), which dynamically updates label information by iteration. The experimental results have demonstrated that the proposed algorithm is effective. In the future, we would like to handle more similar tasks. Besides, our method can deal with multi-label classification or multi-instance multi-label classification and just do some improvements.

Acknowledgement. This research was supported by the Guangzhou Key Laboratory of Robotics and Intelligent Software under Grant No. 15180007, and Fundamental Research Funds for the Central Universities under Grant No. D215048w and 2015ZZ029, and National Natural Science Foundation of China (NSFC) under Grant No. 61005061 and 61502177.

References

1. Blanzieri, E., Bryl, A.: A survey of learning-based techniques of email spam filtering. Artif. Intell. Rev. **29**(1), 63–92 (2008)
2. Cai, X., Wang, H., Huang, H., Ding, C.: Joint stage recognition and anatomical annotation of drosophila gene expression patterns. Bioinformatics **28**(12), i16–i24 (2012)

3. Chang, C.-C., Lin, C.-J.: Libsvm: a library for support vector machines. ACM Trans. Intell. Syst. Technol. (TIST) **2**(3), 27 (2011)
4. Ji, S., Tang, L., Yu, S., Ye, J.: Extracting shared subspace for multi-label classification. In: Proceedings of the 14th ACM SIGKDD International Conference on Knowledge Discovery and Data Mining, pp. 381–389. ACM (2008)
5. Jin, X., Zhuang, F., Xiong, H., Du, C., Luo, P., He, Q.: Multi-task multi-view learning for heterogeneous tasks. In: Proceedings of the 23rd ACM International Conference on Conference on Information and Knowledge Management, pp. 441–450. ACM (2014)
6. Kumar, S., Jayaraman, K., Panchanathan, S., Gurunathan, R., Marti-Subirana, A., Newfeld, S.J.: Best: a novel computational approach for comparing gene expression patterns from early stages of drosophila melanogaster development. Genetics **162**(4), 2037–2047 (2002)
7. Lu, Y., Lai, Z., Fan, Z., Cui, J., Zhu, Q.: Manifold discriminant regression learning for image classification. Neurocomputing **166**, 475–486 (2015)
8. Shen, C., Jing, L., Ng, M.K.: Sparse-MIML: a sparsity-based multi-instance multi-learning algorithm. In: Heyden, A., Kahl, F., Olsson, C., Oskarsson, M., Tai, X.-C. (eds.) EMMCVPR 2013. LNCS, vol. 8081, pp. 294–306. Springer, Heidelberg (2013)
9. Tomancak, P., Berman, B.P., Beaton, A., Weiszmann, R., Kwan, E., Hartenstein, V., Celniker, S.E., Rubin, G.M.: Global analysis of patterns of gene expression during drosophila embryogenesis. Genome Biol. **8**(7), R145 (2007)
10. Tong, H., Faloutsos, C., Pan, J.-Y.: Random walk with restart: fast solutions and applications. Knowl. Inf. Syst. **14**(3), 327–346 (2008)
11. Tsoumakas, G., Vlahavas, I.P.: Random k-labelsets: an ensemble method for multilabel classification. In: Kok, J.N., Koronacki, J., Lopez de Mantaras, R., Matwin, S., Mladenič, D., Skowron, A. (eds.) ECML 2007. LNCS (LNAI), vol. 4701, pp. 406–417. Springer, Heidelberg (2007)
12. Voutilainen, A.: Part-of-speech tagging. In: The Oxford handbook of computational linguistics, pp. 219–232 (2003)
13. Wang, B., Tu, Z., Tsotsos, J.K.: Dynamic label propagation for semi-supervised multi-class multi-label classification. In: 2013 IEEE International Conference on Computer Vision (ICCV), pp. 425–432. IEEE (2013)
14. Wang, H., Huang, H., Ding, C.: Image annotation using multi-label correlated green's function. In: 2009 IEEE 12th International Conference on Computer Vision, pp. 2029–2034. IEEE (2009)
15. Wu, Q., Ng, M.K., Ye, Y.: Markov-miml: a markov chain-based multi-instance multi-label learning algorithm. Knowl. Inf. Syst. **37**(1), 83–104 (2013)
16. Wu, Q., Ng, M.K., Ye, Y.: Cotransfer learning using coupled markov chains with restart. IEEE Intell. Syst. **29**(4), 26–33 (2014)
17. Xie, J., Hone, K., Xie, W., Gao, X., Shi, Y., Liu, X.: Extending twin support vector machine classifier for multi-category classification problems. Intell. Data Anal. **17**(4), 649–664 (2013)
18. Zhang, M.-L., Zhou, Z.-H.: Ml-knn: a lazy learning approach to multi-label learning. Pattern Recogn. **40**(7), 2038–2048 (2007)
19. Zhu, X.: Semi-supervised learning literature survey (2005)
20. Zhu, X., Ghahramani, Z., Lafferty, J., et al.: Semi-supervised learning using gaussian fields and harmonic functions. ICML **3**, 912–919 (2003)

Hybrid Sampling with Bagging for Class Imbalance Learning

Yang Lu[1], Yiu-ming Cheung[1(✉)], and Yuan Yan Tang[2]

[1] Department of Computer Science,
Hong Kong Baptist University, Hong Kong, China
{yanglu,ymc}@comp.hkbu.edu.hk
[2] Department of Computer and Information Science,
Faculty of Science and Technology, University of Macau, Macau, China
yytang@umac.mo

Abstract. For class imbalance problem, the integration of sampling and ensemble methods has shown great success among various methods. Nevertheless, as the representatives of sampling methods, undersampling and oversampling cannot outperform each other. That is, undersampling fits some data sets while oversampling fits some other. Besides, the sampling rate also significantly influences the performance of a classifier, while existing methods usually adopt full sampling rate to produce balanced training set. In this paper, we propose a new algorithm that utilizes a new hybrid scheme of undersampling and oversampling with sampling rate selection to preprocess the data in each ensemble iteration. Bagging is adopted as the ensemble framework because the sampling rate selection can benefit from the Out-Of-Bag estimate in bagging. The proposed method features both of undersampling and oversampling, and the specifically selected sampling rate for each data set. The experiments are conducted on 26 data sets from the UCI data repository, in which the proposed method in comparison with the existing counterparts is evaluated by three evaluation metrics. Experiments show that, combined with bagging, the proposed hybrid sampling method significantly outperforms the other state-of-the-art bagging-based methods for class imbalance problem. Meanwhile, the superiority of sampling rate selection is also demonstrated.

Keywords: Class imbalance learning · Hybrid sampling · Sampling method · Ensemble method

1 Introduction

In many classification applications, the problem of learning from imbalanced data is still one of the challenges [22], where the number of data in the minority

---

Y.M. Cheung is the corresponding author. This work was supported by the Faculty Research Grants of Hong Kong Baptist University (HKBU): FRG2/14-15/075, and by the National Natural Science Foundation of China under Grant Number: 61272366.

© Springer International Publishing Switzerland 2016
J. Bailey et al. (Eds.): PAKDD 2016, Part I, LNAI 9651, pp. 14–26, 2016.
DOI: 10.1007/978-3-319-31753-3_2

class is severely under-represented and overwhelmed by the majority class. In this case, the distribution of the classes is skewed. Subsequently, usual classification methods will generate poor results because the distribution is one of the most important factors that affect the performance [7,15,19].

Usually, standard classification algorithms assume that the class distribution is balanced, and the misclassification cost is equal for both classes. However, there exists some cases that the class distribution is skewed and the misclassification cost is extremely unequal. Further, sometimes people focus more on the minority class because it usually contains more information and interest than the majority class. Let us take cancer diagnosis as an example, the number of patients who have cancer is much less than the number of healthy people in regular checkups. It is obvious that the cost for misdiagnosing a healthy person to be sick, which only brings the person mental stress and more payment to further diagnosis, is much less than diagnosing a patient to be health, which may lead to the loss of the patient's life. Therefore, when dealing with imbalanced data, the misclassification cost is one of the most significant factors that affect the process of learning. In addition to the algorithms, evaluation metrics also play important roles in imbalanced learning. Suppose there are 100 cancer patients out of 10,000 people, the normal classifier will tend to predict "healthy", because even all predictions are "healthy", the accuracy of this classifier is still as high as 99 %. Therefore, simply using the accuracy or error rate is not comprehensive enough to measure the performance of a classifier dealing with imbalanced data. Usually, three evaluation metrics for class imbalance problem, i.e. AUC, F1 and G-mean, will be used. In this paper, we focus on the binary classification problem, and following the convention, we treat samples in the minority class as positive and samples in the majority class as negative.

Among various of methods to tackle the imbalance problem, sampling methods have been proved to be effective. Several studies have shown that training on the balanced data set by sampling methods can achieve better overall classification performance than the original imbalanced one [11,21]. Usually, the sampling methods, such as random undersampling or oversampling, are integrated with ensemble methods, such as bagging or boosting, in order to overcome their drawbacks and provide more diversity to the boosted classifier [12].

However, ensemble-based undersampling and oversampling cannot outperform each other, e.g. see a recent survey [12], in which RUSBoost [17] (undersampling based) wins SMOTEBoost [9] (oversampling based) 22 times, draws 4 times and loses 18 times and UnderBagging [1] (undersampling based) wins SMOTEBagging [20] (oversampling based) 18 times, draws 1 time and loses 25 times (shown in Tables XX and XXI in [12]). It can be seen that the results generated by ensemble-based undersampling or oversampling highly depend on the data. In other words, some data has better performance with undersampling, while the other ones with oversampling. Therefore in terms of the sampling process, it is expected that the hybrid of undersampling and oversampling can take advantage of their properties. That is, the hybrid sampling generally outperforms each individual sampling method because undersampling and oversampling are complementary to each other and cure the skewed distribution of class imbalanced data in different extents.

In addition, no matter undersampling, oversampling or even hybrid sampling is adopted, the sampling rate is one of the key factors that affect the performance of a classifier. Most of the sampling based methods, no matter integrated with ensemble methods or not, tend to make the number of data in both classes balanced after sampling, based on the simple assumption that the balanced training set produces the best result. However, producing poor model caused by training on imbalanced data does not imply that the optimal model is produced by training on totally balanced data by sampling. Estabrooks *et al.* [11] shows that the best results from undersampling and oversampling are not always on the balanced case. It means that conducting sampling to achieve the balanced data for training is not guaranteed to be the best solution. Furthermore, the best sampling rate depends on the distribution and complexity of the data set. That is, the best sampling rate of one dataset would be different from one of another dataset. Therefore, it is necessary to select a proper sampling rate for the sampling methods on each data set. To the best of our knowledge, selection of the proper sampling rate has yet to be studied in the literature.

In this paper, we therefore propose a novel method for the class imbalance problem called Hybrid Sampling with Bagging (HSBagging). It adopts a new hybrid scheme that conducts random undersampling in tandem with oversampling technique SMOTE at a certain sampling rate in each bagging iteration. The sampling rate is selected by Out-Of-Bag (OOB) estimate on a specified metric for each data set. To reduce the computational cost, the sampling rate is only estimated in the first several iterations and the averaged estimated sampling rate will be utilized in the rest iterations then. The major advantages of HSBagging are:

- The new hybrid sampling scheme can take advantage of the merits of both undersampling and oversampling.
- Sampling rate selection can effectively select a proper sampling rate which fits the data to achieve best performance.
- The preferred metric can be selected during OOB estimate according to the application requirement.

To validate the effectiveness of the hybrid sampling scheme and sampling rate selection, four experiments are conducted on 26 UCI data sets with statistical significance tests. The experiments show that the proposed HSBagging significantly outperforms individual sampling method with bagging and verify that both hybrid sampling and sampling rate selection contribute to the superiority of HSBagging.

2   Related Work

Over the past years, much work devoted to solve the class imbalance problem has shown great success in the corresponding application domain, in which sampling methods and ensemble methods are two major branches.

Random oversampling and undersampling are two elementary sampling methods to cure imbalance, by randomly replicating data in minority class and discarding data in majority class, respectively. The drawbacks of them are that oversampling will easily cause overfitting and undersampling may discard useful data that leads to information loss. As an improvement to random oversampling, Synthetic Minority Over-sampling TEchnique (SMOTE) [8] synthesizes artificial data in the minority class instead of replication. Borderline-SMOTE [13] and ADASYN [14] improve SMOTE by assuming that the samples close to the borderline are more important, thus synthesize more data there. The idea of combining undersampling and oversampling has been mentioned in [20]. It combines undersampling with oversampling to create a training set with the same number of positive and negative samples. The number of samples in each class after sampling is determined by a predefined re-sampling rate $a\%$.

Ensemble methods such as bagging [4] and boosting [16] cannot solve the imbalanced problem themselves. Usually, they are combined with sampling methods to utilize the diversity provided by sampling to enhance the ensemble classifier. A comprehensive review of ensemble methods for class imbalance problem can be found in [12]. OverBagging [20] and UnderBagging [1] combine random undersampling and oversampling with bagging, respectively. They adopt oversampling or undersampling after bootstrapping the training data to create a balanced training set. As an improvement of OverBagging, SMOTE-Bagging [20] combines SMOTE with random oversampling and the sampling rate of SMOTE increases in every iteration to provide more diversity. As the counterpart of bagging-based methods, SMOTEBoost [8] and RUSBoost [17] are boosting-based. They created balanced training set by SMOTE and random undersampling in each boosting iteration. After sampling applied, the sample weights are normalized. The following steps are the same as Adaboost [16]. IIVotes [3] combines IVotes ensemble [5] and SPIDER [18] data preprocessing to obtain improved balance between the sensitivity and specificity for the minority class.

3   The Proposed Method

Since training in the balanced data set is not guaranteed to produce the best result [11], the proposed HSBagging does not aim to create the balanced training set, but depending on a specified sampling rate p, which is different from the hybrid scheme in [20]. In HSBagging, the minority class is enlarged by p and meanwhile the majority class is shrank by p. Conducting undersampling and oversampling at sampling rate p at the same time can explore the best sampling rate from severe imbalance, slight imbalance to balance or even reversed imbalance (i.e. the minority class becomes majority after sampling). Since each data set tends to have different best sampling rate, it is necessary to estimate the sampling rate during bagging. HSBagging estimates the best sampling rate by Out-Of-Bag (OOB) estimate, which is used to estimate parameters in the bootstrapped set by leaving the samples not selected by bootstrapping as validation set. There are two advantages of using OOB estimate: (1) it acts as

---

Algorithm 1. Hybrid Sampling Bagging

Require: Training set $S = \{(\mathbf{x}_i, y_i)\}$, $i = 1, ..., n$ and $y_i \in \{+1, -1\}$, weak learner L, number of iterations T, number of iterations k for sampling rate estimate, sampling rate selection set I, evaluation metric f_m.

1: **for** $t = 1$ to T **do**
2:     Create a training set B by bootstrapping each class respectively.
3:     Create the OOB set B_o.
4:     **if** $t \leq k$ **then**
5:         **for each** p in I **do**
6:             Create the training set B' by both undersampling and SMOTE set B at sampling rate p.
7:             Learn the classifier $h'_p = L(B')$.
8:         **end for**
9:         $p^*_t = \text{argmax}_{p \in I}\, f_m(h'_p, B_o)$.
10:        $h_t = h'_{p^*_t}$.
11:     **else**
12:        $p^*_t = \frac{1}{k} \sum_{i=1}^{k} p^*_i$.
13:        Create the training set B' by both undersampling and SMOTE set B at sampling rate p^*_t.
14:        Learn the classifier $h_t = L(B')$.
15:     **end if**
16: **end for**
17: **Output:** $H(\mathbf{x}) = sign(\sum_{t=1}^{T} h_t(\mathbf{x}))$

---

validation set, but it needs not separate part of data from training set; (2) the model needs not be trained again on the original training set with the estimated best parameter. Therefore, in HSBagging, the sampling rate is regarded as a parameter to be estimated. The estimate criterion is based on a specified evaluation metric, because commonly used accuracy for classification cannot well assess the class imbalance problem. Usually, it will be computational expensive if the OOB estimate is conducted on every bagging iteration. To save computational cost, we only conduct OOB sampling in the first k iterations, and the rest iterations will use the averaged estimated best sampling rates of the previous iterations.

The proposed HSBagging is shown in Algorithm 1. In each iteration, the training data is bootstrapped on each class, respectively, as shown in Line 2. The bootstrapped training set B keeps the same number of samples for the majority class and minority class as before bootstrapping. The OOB set B_o is then constructed by the samples that are not selected into B. The sampling rate selection is only conducted in the first k iterations in order to save computational cost. In these k iterations, undersampling and SMOTE are used to process B at the same time at sampling rate p in Line 6. The sampling rate $p \in [0, 1]$ is set to each of the values in the set I, in order to find a proper sampling rate for the current data set. For undersampling, it randomly selects $n_{min} + (1 - p)$ $(n_{maj} - n_{min})$ samples from the majority class, and for SMOTE, it synthesizes $p(n_{maj} - n_{min})$ more samples from the minority class and adds them to the original minority class, where n_{maj} and n_{min} represents the number of samples in the majority class and minority class. Therefore, when $p = 0$, the data set B' after sampling is as same as the original data set B and when $p = 1$, the number of samples in the majority class and the minority class gets reversed after sampling. Thus, undersampling and SMOTE are effectively combined. By learning B' by the learner L, a classifier h'_p can be built for the sampling rate p. After that, $f_m(h'_p, B_o)$ estimates the performance of h'_p on the OOB set B_o and metric f_m. The sampling rate p_t^* is set to the p associated with best performance on f_m and $h'_{p_t^*}$ is set to the classifier of the t's iteration h_t. After k iterations of sampling rate selection, the following iterations simply use the averaged value of the first k selected sampling rates to do sampling and train the classifier h_t. At last, each individual classifier is combined into the final boosted classifier $H(x)$.

The computational complexity of HSBagging is $O((T + (k - 1)|I|)\mathcal{L}(n))$, where $|\cdot|$ is the cardinality of a set and $\mathcal{L}(n)$ is the computational cost of the weak learning L with n training samples. Compared with SMOTEBagging [20], although HSBagging costs $(k - 1)|I|$ more iterations to select the sampling rate, it trains only on n samples in each iteration, while SMOTEBagging trains $2n_{maj}$ samples. If the number of iterations T is relatively large and the imbalance problem of the data set is severe, HSBagging will be computational cheaper than SMOTEBagging.

4   Experiments

In this section, we conducted four experiments. Experiment 1 shows the times of best performance on each sampling rate for each data set. It verifies that the sampling rate corresponding to the best performance varies from data to data. Experiment 2 compares the proposed HSBagging with bagging on original imbalanced data set, UnderBagging [1], SMOTEBagging [20] and IIVotes [3]. We denote SMOTE with bagging by full sampling rate ($p = 1$) as SMOTEBagging-1, and SMOTE with bagging by increasing sampling rate in each iteration, which is proposed in [20], as SMOTEBagging-2. Experiment 3 compares HSBagging with those methods on different sampling rates to verify that the superior performance is not only caused by sampling rate selection, but also effected by the hybrid sampling scheme. Experiment 2 and 3 verify that hybrid sampling is significantly better than individual sampling. Finally, Experiment 4 shows the performance of HSBagging on different number k of iterations for sampling rate estimation.

All experiments were conducted on 26 data sets from UCI data repository [2] summarized in Table 1, which cover a wide range of applications and imbalance ratios. The imbalance ratio (IR) is calculated by the number of data in the majority class divided by the number of data in the minority class. All experiments adopted 5-fold cross validation, where 80 % of the samples in each data were used for training and the rest for testing in each fold. The final results were averaged by 10 runs of experiments. The number of iterations T in bagging was set at 10 for all methods, except IIVotes, whose iteration was automatically determined. CART [6] was adopted as the base learner for all bagging-based methods.

Table 1. Information of 26 UCI data sets.

Data set	#Instance	#Attribute	Minority class	Majority class	IR
glass-2	214	9	bwnfp	remainder	1.8
pima	768	8	positive	negative	1.9
vehicle-2	846	18	saab	remainder	2.9
vehicle-1	846	18	opel	remainder	3.0
glass-123vs567	214	9	non-window	remainder	3.2
wpbc	198	33	recur	nonrecur	3.2
vehicle-4	846	18	van	remainder	3.3
haberman	306	3	within-5-year	5-year-or-longer	2.8
cmc	1473	9	long-term	remainder	3.4
ecoli-2	336	7	im	remainder	3.4
car	1728	6	acc	remainder	3.5
wine-quality	6497	11	score 7	remainder	5.0
segment	2310	19	brickface	remainder	6.0
glass-7	214	9	headlamp	remainder	6.4
yeast-4	1484	8	me3	remainder	8.1
ecoli-4	336	7	imU	remainder	8.6
pageblocks	5473	10	remainder	text	8.8
mf-morph	2000	6	class 10	remainder	9.0
mf-zernike	2000	47	class 10	remainder	9.0
cm1	498	21	defects	no-defects	9.2
satimage	6435	36	class 4	remainder	9.3
yeast-5vs347810	1484	8	me2	mit;me3;exc;vac;erl	9.4
abalone	4177	8	class 7	remainder	9.7
balance	625	4	balanced	remainder	11.8
glass-127vs6	214	9	tableware	bwfp;bwnfp;headlamps	19.4
yeast-6	1484	8	me1	remainder	32.7

The number of nearest neighbor for all kNN related methods was set at 5. In the experiments, three evaluation metrics, i.e. AUC, F1 and G-mean, which are commonly adopted as the benchmark assessment metric for class imbalance learning [15], were used to measure the effectiveness of methods.

4.1   Experiment 1: Sampling Rate Verification

Figure 1 shows the number of data sets with the best performance on different sampling rate from 0 to 1. In this experiment, UnderBagging and SMOTEBagging-1 were set to process the data on a specific sampling rate p instead of producing balanced training set. UnderBagging conducted undersampling by discarding $p(n_{maj} - n_{min})$ samples from the majority class while SMOTEBagging-1 conducted SMOTE by synthesizing $p(n_{maj} - n_{min})$ from the

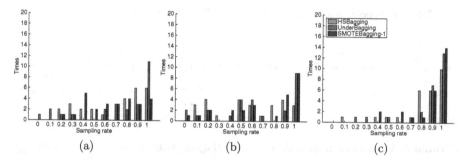

Fig. 1. Number of data sets with best (a) AUC, (b) F1, and (c) G-mean performance generated on different sampling rates.

minority class. HSBagging conducted both undersampling and SMOTE at sampling rate p instead of OOB estimate as described in Algorithm 1. From Fig. 1, it can be observed that the best results of all three methods almost appear on all sampling rates on each evaluation metric. Especially, some best sampling rates of HSBagging occur at sampling rate 0 or 1, which means that the original imbalanced data or the reversed imbalanced data may also be able to generate good results. Furthermore, the sampling rate corresponding to the best performance on different evaluation metrics may also be different, e.g. higher sampling rates generate relatively better results on G-mean. Thus, we can argue that, no matter which sampling method is adopted, selecting a proper sampling rate on a specific metric for each data set is effective and necessary.

4.2 Experiment 2: Comparative Studies

Since CART generates discrete outputs, AUC can only be calculated by the ensemble of CART classifiers and is not available for individual CART classifier. Therefore, we use F1 and G-mean as the metric f_m to select the best sampling rate for HSBagging, denoted as HSBagging-F1 and HSBagging-Gmean, respectively. The number of iterations k for sampling rate estimate is set to 3 and sampling rate selection set $I = \{0, 0.2, 0.4, 0.6, 0.8, 1\}$.

The pairwise comparisons by Wilcoxon signed-rank test [10] is provided to show the statistical significance of the compared methods. It measures the difference between two methods and rank their magnitude among data sets. Greater difference will count more in this evaluation. The sum of ranks of each method is calculated by $R^+ = \sum_{d_i>0} rank(|d_i|) + \frac{1}{2}\sum_{d_i=0} rank(|d_i|)$ and $R^- = \sum_{d_i<0} rank(|d_i|) + \frac{1}{2}\sum_{d_i=0} rank(|d_i|)$ where d_i is the difference of the result of the ith data set. If the significance value N with a certain significance level α is greater than $T = min\{R^+, R^-\}$, the null hypothesis is rejected which indicates one method significantly outperforming the other one. In the following Tables 2, 3 and 4, as well as Table 5 in Sect. 4.4, the method shown in the left upper corner is marked as $+$ and the compared methods are marked as $-$. The sign $(+,-)$ in the T column indicates which method wins more ranks and the symbol $\bullet$ indicates the significance with significance level $\alpha = 0.05$.

Table 2. Wilcoxon signed-rank test for HSBagging-F1 and other methods.

HSBagging-F1 vs.	AUC			F1			G-mean		
	R^+	R^-	T	R^+	R^-	T	R^+	R^-	T
Bagging	335.00	16.00	• 16.00 (+)	331.00	20.00	• 20.00 (+)	341.00	10.00	• 10.00 (+)
UnderBagging	179.00	172.00	172.00(+)	285.00	66.00	• 66.00 (+)	61.00	290.00	• 61.00 (−)
SMOTEBagging-1	205.00	146.00	146.00(+)	279.50	71.50	• 71.50 (+)	264.00	87.00	• 87.00 (+)
SMOTEBagging-2	226.00	125.00	125.00(+)	282.00	69.00	• 69.00 (+)	149.50	201.50	149.50 (−)
IIVotes	350.00	1.00	• 1.00 (+)	350.00	1.00	• 1.00 (+)	328.00	23.00	• 23.00 (+)

Table 3. Wilcoxon signed-rank test for HSBagging-Gmean and other methods.

HSBagging-Gmean vs.	AUC			F1			G-mean		
	R^+	R^-	T	R^+	R^-	T	R^+	R^-	T
Bagging	344.00	7.00	• 7.00 (+)	270.00	81.00	• 81.00 (+)	351.00	0.00	• 0.00 (+)
UnderBagging	221.00	130.00	130.00(+)	296.50	54.50	• 54.50 (+)	177.00	174.00	174.00 (+)
SMOTEBagging-1	269.50	81.50	• 81.50 (+)	210.00	141.00	141.00 (+)	350.00	1.00	• 1.00 (+)
SMOTEBagging-2	257.50	93.50	• 93.50 (+)	171.00	180.00	171.00 (−)	289.00	62.00	• 62.00 (+)
IIVotes	351.00	0.00	• 0.00 (+)	351.00	0.00	• 0.00 (+)	351.00	0.00	• 0.00 (+)

Table 4. Wilcoxon signed-rank test for HSBagging-Gmean and HSBagging-F1.

HSBagging-Gmean vs.	AUC			F1			G-mean		
	R^+	R^-	T	R^+	R^-	T	R^+	R^-	T
HSBagging-F1	276.00	75.00	• 75.00 (+)	117.50	233.50	117.50 (−)	313.00	38.00	• 38.00 (+)

Tables 2 and 3 show the Wilcoxon signed-rank test results of HSBagging-F1 and HSBagging-Gmean compared with the other methods. It can be seen that:

- HSBagging-F1 significantly outperforms all other methods on F1, and HSBagging -Gmean significantly outperforms bagging, SMOTEBagging-1, SMOTEBagging-2 and IIVotes on G-mean.
- Even though the sampling rate is not selected based on AUC, HSBagging-F1 and HSBagging-Gmean also achieve comparable or better performance on AUC. Especially, the performance of HSBagging-Gmean on AUC shows similar significance as its performance on G-mean.
- On G-mean, HSBagging-F1 outperform Bagging, SMOTEBagging-1 and IIVotes, and on F1, HSBagging-Gmean outperform Bagging, UnderBagging and IIVotes.

As a result, it can be observed that no matter the sampling rate is selected on which metric, HSBagging can produce superior results on each metric, especially on its selected metric, i.e. F1 and G-mean.

Table 4 shows the comparison between HSBagging-F1 and HSBagging-Gmean. Both of them have better results on their own selected metrics. However, HSBagging-Gmean significantly outperforms HSBagging-F1 on both AUC and G-mean while HSBagging-F1 is only slightly better than HSBagging-Gmean on F1. Therefore, overall speaking, HSBagging-Gmean performs better than HSBagging-F1.

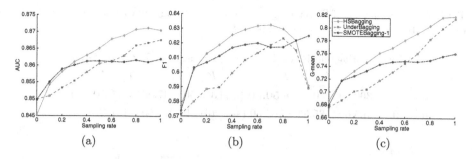

Fig. 2. Average performance of HSBagging, UnderBagging and SMOTEBagging-1 over different sampling rate in terms of (a) AUC, (b) F1, and (c) G-mean.

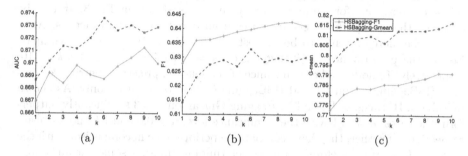

Fig. 3. Average Performance of HSBagging in terms of (a) AUC, (b) F1, and (c) G-mean, respectively, on different number of iterations for sampling rate estimate k.

4.3   Experiment 3: Sampling Rate Comparison

In addition to the sampling rate selection, the new hybrid sampling scheme also plays an important role in terms of the superiority of HSBagging. In this subsection, we show that the effectiveness of the proposed HSBagging depends on not only sampling rate selection, but also the hybrid scheme. The comparison of HSBagging with UnderBagging and SMOTEBagging-1 at different sampling rate is shown in Fig. 2 with the same setting as the experiment in Sect. 4.1. The figures are generated by averaging all 26 UCI data sets. On most of the sampling rates, HSBagging can achieve better results on average than UnderBagging and SMOTEBagging-1. Besides, the best results of HSBagging are better than the best results of UnderBagging and SMOTEBagging among all sampling rates. Figure 2 illustrates that, even sampling rate selection is adopted for UnderBagging and SMOTEBagging, the overall performance cannot be as good as HSBagging. That implies that HSBagging outperforming UnderBagging and SMOTEBagging benefits from not only the choice of a proper sampling rate, but also the hybrid scheme.

Table 5. A comparison of HSBagging with $p = 3$ to HSBagging with $p = 1$ and $p = 10$, respectively, using Wilcoxon signed-rank test.

$p = 3$ vs.	HSBagging-F1 on F1			HSBagging-Gmean on G-mean		
	R^+	R^-	T	R^+	R^-	T
$p = 1$	300.50	50.50	• **50.50** (+)	303.50	47.50	• **47.50** (+)
$p = 10$	165.50	185.50	165.50 (−)	101.50	249.50	101.50 (−)

4.4   Experiment 4: Parameter Selection

The performance of HSBagging-F1 and HSbagging-Gmean over different number k of iterations for sampling rate estimate is shown in Fig. 3. It can be observed that the performance increases from $k = 1$ to 3 for all metrics. After $k = 3$, the increase tends to be modest. Table 5 shows the statistical comparison of the performance of HSBagging on $k = 3$ against $k = 1$ and $k = 10$, respectively. To address the significance of the selected preferred metric, we compare HSBagging-F1 on F1 and HSBagging-Gmean on G-mean only. As shown in Table 5, HSBagging-F1 and HSBagging-Gmean on $k = 3$ significantly outperform the cases on $k = 1$ with the significance level $\alpha = 0.05$ on F1 and G-mean, respectively. Further, they have comparable performance in comparison with the cases on $k = 10$. Therefore, if the longer running time for some certain applications can be tolerated, the selection process is suggested to be conducted in every iteration because it has slightly better performance. Nevertheless, by a rule of thumb, setting $k = 3$ can usually produce significantly better results in comparison with the other bagging-based methods as shown in Tables 2 and 3, meanwhile saving the computational cost compared with $k = 10$.

5   Conclusion

This paper has first investigated the two problems for class imbalance problem. The first is that undersampling and oversampling with ensemble methods have their own irreplaceable property for the imbalanced data. Each of them can only performs well on part of data sets. Second, the sampling rate is crucial to the performance of sampling methods. The sampling rate in regard to the best performance differs from data to data.

A novel method called HSBagging has been proposed to solve the discovered problems. It adopts a new hybrid scheme of undersampling and oversampling integrated with bagging. During the sampling, the sampling rate is selected by OOB estimate on a specified metric. Experiments on 26 UCI data sets have shown that HSBagging can significantly outperform the other related bagging-based methods. The advantages of both the new hybrid sampling scheme and the sampling rate selection are also shown by experiments. Undoubtedly, the hybrid sampling and sampling rate selection are applicable to the other ensemble-based method like boosting as well.

References

1. Barandela, R., Valdovinos, R.M., Sánchez, J.S.: New applications of ensembles of classifiers. Pattern Anal. Appl. **6**(3), 245–256 (2003)
2. Blake, C., Merz, C.J.: UCI repository of machine learning databases (1998)
3. Błaszczyński, J., Deckert, M., Stefanowski, J., Wilk, S.: Integrating selective pre-processing of imbalanced data with ivotes ensemble. In: Szczuka, M., Kryszkiewicz, M., Ramanna, S., Jensen, R., Hu, Q. (eds.) RSCTC 2010. LNCS, vol. 6086, pp. 148–157. Springer, Heidelberg (2010)
4. Breiman, L.: Bagging predictors. Mach. Learn. **24**(2), 123–140 (1996)
5. Breiman, L.: Pasting small votes for classification in large databases and on-line. Mach. Learn. **36**(1–2), 85–103 (1999)
6. Breiman, L., Friedman, J., Stone, C.J., Olshen, R.A.: Classification and Regression Trees. CRC Press, Boca Raton (1984)
7. Chawla, N.V.: Data mining for imbalanced datasets: an overview. In: Maimon, O., Rokach, L. (eds.) Data Mining and Knowledge Discovery Handbook, pp. 853–867. Springer, Heidelberg (2005)
8. Chawla, N.V., Bowyer, K.W., Hall, L.O., Kegelmeyer, W.P.: Smote: synthetic minority over-sampling technique. J. Artif. Intell. Res. **16**(1), 321–357 (2002)
9. Chawla, N.V., Lazarevic, A., Hall, L.O., Bowyer, K.W.: SMOTEBoost: improving prediction of the minority class in boosting. In: Lavrač, N., Gamberger, D., Todorovski, L., Blockeel, H. (eds.) PKDD 2003. LNCS (LNAI), vol. 2838, pp. 107–119. Springer, Heidelberg (2003)
10. Demšar, J.: Statistical comparisons of classifiers over multiple data sets. J. Mach. Learn. Res. **7**, 1–30 (2006)
11. Estabrooks, A., Jo, T., Japkowicz, N.: A multiple resampling method for learning from imbalanced data sets. Comput. Intell. **20**(1), 18–36 (2004)
12. Galar, M., Fernandez, A., Barrenechea, E., Bustince, H., Herrera, F.: A review on ensembles for the class imbalance problem: bagging-, boosting-, and hybrid-based approaches. IEEE Trans. Syst. Man Cybern. Part C Appl. Rev. **42**(4), 463–484 (2012)
13. Han, H., Wang, W.-Y., Mao, B.-H.: Borderline-SMOTE: a new over-sampling method in imbalanced data sets learning. In: Huang, D.-S., Zhang, X.-P., Huang, G.-B. (eds.) ICIC 2005. LNCS, vol. 3644, pp. 878–887. Springer, Heidelberg (2005)
14. He, H., Bai, Y., Garcia, E.A., Li, S.: Adasyn: adaptive synthetic sampling approach for imbalanced learning. In: IEEE International Joint Conference on Neural Networks, 2008. IJCNN 2008. (IEEE World Congress on Computational Intelligence), pp. 1322–1328. IEEE (2008)
15. He, H., Garcia, E.A.: Learning from imbalanced data. IEEE Trans. Knowl. Data Eng. **21**(9), 1263–1284 (2009)
16. Schapire, R.E.: The strength of weak learnability. Mach. Learn. **5**(2), 197–227 (1990)
17. Seiffert, C., Khoshgoftaar, T.M., Van Hulse, J., Napolitano, A.: Rusboost: a hybrid approach to alleviating class imbalance. IEEE Trans. Syst. Man Cybern. Part A Syst. Hum. **40**(1), 185–197 (2010)
18. Stefanowski, J., Wilk, S.: Improving rule based classifiers induced by modlem by selective pre-processing of imbalanced data. In: Proceedings of the RSKD Workshop at ECML/PKDD, Warsaw, pp. 54–65. Citeseer (2007)
19. Sun, Y., Wong, A.K., Kamel, M.S.: Classification of imbalanced data: a review. Int. J. Pattern Recognit. Artif. Intell. **23**(04), 687–719 (2009)

20. Wang, S., Yao, X.: Diversity analysis on imbalanced data sets by using ensemble models. In: IEEE Symposium on Computational Intelligence and Data Mining, 2009. CIDM 2009, pp. 324–331. IEEE (2009)
21. Weiss, G.M., Provosti, F.: The effect of class distribution on classifier learning: an empirical study. Rutgers Univ (2001)
22. Yang, Q., Wu, X.: 10 challenging problems in data mining research. Int. J. Inf. Technol. Decis. Mak. 5(04), 597–604 (2006)

Sparse Adaptive Multi-hyperplane Machine

Khanh Nguyen[1(✉)], Trung Le[1], Vu Nguyen[2], and Dinh Phung[2]

[1] Faculty of Information Technology,
HCMc University of Pedagogy, Ho Chi Minh, Vietnam
khanhndK@hcmup.edu.vn
[2] Pattern Recognition and Data Analytics, Deakin University, Geelong, Australia

Abstract. The Adaptive Multiple-hyperplane Machine (AMM) was recently proposed to deal with large-scale datasets. However, it has no principle to tune the complexity and sparsity levels of the solution. Addressing the sparsity is important to improve learning generalization, prediction accuracy and computational speedup. In this paper, we employ the max-margin principle and sparse approach to propose a new Sparse AMM (SAMM). We solve the new optimization objective function with stochastic gradient descent (SGD). Besides inheriting the good features of SGD-based learning method and the original AMM, our proposed Sparse AMM provides machinery and flexibility to tune the complexity and sparsity of the solution, making it possible to avoid overfitting and underfitting. We validate our approach on several large benchmark datasets. We show that with the ability to control sparsity, the proposed Sparse AMM yields superior classification accuracy to the original AMM while simultaneously achieving computational speedup.

1 Introduction

Max-margin is a powerful principle to construct machine learning algorithms which has been applied to a wide spectrum of areas ranging from kernel method [2], boosting and bagging [6], Bayesian inference [5] to name a few. The margin concept could be flexibly interpreted under the different settings, e.g., the smallest distance from a datum to a decision boundary [2], the smallest absolute decision value [6], the discrepancy between the maximal and runner-up discriminative values [4,10], and the difference between two posteriors [5]. Since the margin of a classifier is reciprocally proportional to its complexity measured by VC-dimension or Flat-dimension, a simple or sparse classifier often induces a large margin and thus offers higher generalization capacity when learning on the general dataset [11]. Therefore, it is desirable to look for classifiers as simple as possible while being able to well present the training set, a philosophy that concurs with Occam's Razor principle.

The max-margin principle has been investigated for multi-class classification problem [4], often in the form of maximizing the discrepancy between two discriminative values: one for the correct label and the other for the runner-up. In [4], a set of hyperplanes, each hyperplane associated with one class, is used for

© Springer International Publishing Switzerland 2016
J. Bailey et al. (Eds.): PAKDD 2016, Part I, LNAI 9651, pp. 27–39, 2016.
DOI: 10.1007/978-3-319-31753-3_3

evaluating the discriminative values. By defining the relevant cost function, the learning problem becomes convex and can be solved analytically. Nonetheless, associating a single hyperplane with one class restricts the representation ability of the model and it therefore cannot well present complex data. To overcome this issue, the work of [1] proposes to associate multiple hyperplanes with each class to increase the representation power of the model. However, this comes with the cost of the learning problem becoming non-convex and only a convergence to a local optimum is guaranteed. This impedes the usage of the aforementioned method to real dataset. In [12], the burden in computation is addressed by applying stochastic gradient descent framework proposed in [8]. The so-called Adaptive Multi-hyperplane Machine (AMM) was proposed for efficiently handling large-scale datasets. AMM has some advantageous features: (1) it has a good representation capacity for learning complex dataset, (2) it is fast and can run online, and (3) the number of hyperplanes associated with each class can be automatically discovered. However, AMM has no principle to tune the complexity and sparsity levels of the solution. The redundant hyperplanes are cut off by an exhaustive pruning weight procedure which heuristically prunes the hyperplanes whose lengths are less than a predefined threshold.

Stochastic gradient descent method [7–9] has recently emerged as building block to develop the fast learning methods for large-scale datasets that can run online and also operate well under the memory budget requirement. In this paper, we leverage the stochastic gradient descent framework with the max-margin principle to propose Sparse Adaptive Multi-hyperplane Machine (SAMM). Besides inheriting several advantages of AMM, with SAMM we can tune the complexity and sparsity levels of the solution to avoid overfitting and underfitting. Our experiment on several large benchmark datasets demonstrates that with the ability to control sparsity, the proposed Sparse AMM yields superior classification accuracy to the original AMM while simultaneously achieving computational speedup.

2   Preliminary

In this section, we present some notations and mathematical tools used throughout the paper. The dot product of two vectors w, x is denoted by $\langle w, x \rangle \triangleq w^\mathsf{T} x$. For any positive number N, the set including the first N positive numbers is defined as $[N] \triangleq \{1, 2, ..., N\}$. Given a logical statement A, $\mathbb{I}_A$ is 1 if A is true and is 0 otherwise. A norm of vector x is denoted by $\|x\|$ and the dual norm is defined as $\|x\|_* \triangleq \sup\limits_{\|w\| \leq 1} \langle w, x \rangle$.

It is known that the dual norm for $\|x\|_2 \triangleq \langle x, x \rangle^{1/2}$ is itself, the dual norm for $\|x\|_p \triangleq \left(\sum_{i=1}^d |x_i|^p \right)^{1/p}$ where $p > 1$ is $\|.\|_q$ where $\frac{1}{p} + \frac{1}{q} = 1$, and the dual norm for $\|x\|_1 \triangleq \sum_{i=1}^d |x_i|$ is $\|x\|_\infty \triangleq \max\limits_{1 \leq i \leq d} |x_i|$.

Given a m-by-n matrix $W = [W_1, W_2, ..., W_n]$ the group norm $\mathbb{L}_{p,q}$ of W is defined as $\|W\|_{p,q} = \|\|W_1\|_p, \|W_2\|_p, ..., \|W_n\|_p\|_q$. The dual norm of a group norm $\|W\|_{p,q}$ is the group norm $\|W\|_{r,s}$ where $\frac{1}{p} + \frac{1}{r} = 1$ and $\frac{1}{q} + \frac{1}{s} = 1$.

We further recall some literature from convex analysis. A set S is convex if for any two vectors $x, y \in S$ and number $\alpha \in [0; 1]$, we have $\alpha x + (1 - \alpha) y \in S$. A function $f : S \to \mathbb{R}$ is convex if $f(\alpha x + (1 - \alpha y)) \le \alpha f(x) + (1 - \alpha) f(y)$ for any $x, y \in S$ and $\alpha \in [0; 1]$.

The sub-gradient of function f at v is denoted by $\partial f(v)$ consisting of vectors λ such that $f(u) - f(v) - \langle \lambda, u - v \rangle \ge 0$ for all $u \in S$.

The Fenchel conjugate of a function $f : S \to \mathbb{R}$ is defined as $f^*(\theta) \triangleq \sup_{w \in S} (\langle w, \theta \rangle - f(w))$.

3 Related Work

In this section, we present the work mostly related to ours. In these work, the max-margin principle embodies as maximizing the discrepancy between to discriminative values, one for the correct label and the other for the runner up. We depart from the original work [4] where the model representation ability is restricted since only a single hyperplane is associated with each class and ends with Adaptive Multi-hyperplane Machine (AMM) [12] where simultaneously multiple hyperplanes are affiliated with each class for raising the representation ability and SGD is applied for speedup.

3.1 Multi-class SVM

Given the training set $\mathfrak{D} = \{(x_n, y_n)\}_{n=1}^N$, where instance $x_n \in \mathbb{R}^D$ is a D-dimensional feature vector and $y_n \in \mathcal{Y} = \{1, ..., M\}$ is the corresponding label of x_n, the goal of multi-class classification problem is to find a decision function $f : \mathbb{R}^D \to \mathcal{Y}$ that can accurately predict label of a new instance.

In multi-class SVM [4], the decision function may be defined by using the discriminative function as $f(x) = \underset{i \in Y}{\arg\max}\, g(i, x_n)$ where $g(i, x) = w_i^\mathsf{T} x$.

Therefore, the predicted label of instance x is the index of the weight vector w_i which maximizes the discriminative value $g(i, x)$. The multi-class problem is now translated into the following optimization problem

$$\min_W P(W) \triangleq \frac{\lambda}{2} \|W\|_2^2 + \frac{1}{N} \sum_{n=1}^N l(W; (x_n, y_n)) \tag{1}$$

where $W = [w_i]_{i=1}^M$, $\|W\|_2^2 = \sum_{i \in \mathcal{Y}} \|w_i\|^2$ and λ is a regularization parameter.

To correctly classify a feature vector x, the corresponding discriminative value $g(i, x)$ of the correct class must be greater than others, i.e., $g(i, x) \ge g(j, x), \forall j \ne i$. The margin of instance (x_n, y_n) is defined as the discrepancy of the maximal and runner-up discriminative values $\rho(x_n, y_n) = g(y_n, x_n) - \max_{i \ne y_n} g(i, x_n)$.

To guarantee the empirical loss is an upper of the empirical error (the number of instances suffering incorrect classifications), the loss function of instance (x_n, y_n) is defined as $l\left(W; (x_n, y_n)\right) = \max\left(0, 1 + \max_{i\in\mathcal{Y}\setminus y_n} g\left(i, x_n\right) - g\left(y_n, x_n\right)\right)$.

In this model, each class is associated with only a single hyperplane and the classifier therefore cannot describe accurately the complex data with several distributions inside.

3.2 Multi-hyperplane Machine

The work of [1] proposed an extended version of multi-class SVM that allows each class in association with an unlimited number of hyperplanes. The discriminative function is redefined as $g(i, x) = \max\limits_{j\in[1...b_i]} (w_{ij}^{\mathsf{T}} x)$ where b_i is the number of hyperplanes associating with the i-th class. The weight matrix W now becomes

$$W = [|w_{1,1}, \ldots w_{1,b_1}|w_{2,1}, \ldots, w_{2,b_2}|, \ldots, |w_{M,1}, \ldots, w_{M,b_M}|].$$

However, the optimization function $P(W)$ is non-convex. To resolve this problem, a latent variable z_n is introduced which specifies the index of the particular hyperplane in y_n-th class being used by the instance (x_n, y_n) to gain its optimal discriminative value. The task of finding the optimal matrix W^* and the latent variables $z = [z_n]_{n=1}^N$ is addressed by solving the optimization problem

1. Given $z = [z_n]_{n=1}^N$, the following optimization problem is solved to find the current optimal matrix W^*

$$\min_W P\left(W|z\right) \triangleq \frac{\lambda}{2}\|W\|^2 + \frac{1}{N}\sum_{n=1}^N l\left(W; (x_n, y_n), z_n\right) \tag{2}$$

where the loss function at (x_n, y_n) is defined as

$$l\left(W; (x_n, y_n); z_n\right) = \max\left(0, 1 + \max_{i\in\mathcal{Y}\setminus y_n} g\left(i, x_n\right) - w_{y_n, z_n}^{\mathsf{T}} x_n\right).$$

2. Given the matrix W^*, we find the current optimal assignment $z = [z_n]_{n=1}^N$ as $z = \operatorname*{argmin}_z P\left(W^*|z\right)$. It means that each latent variable z_n is able to be evaluated as $z_n = \operatorname*{argmin}_k \left(w_{y_n, k}^*\right)^{\mathsf{T}} x_n$.

3.3 Adaptive Multi-hyperplane Machine

An improvement of Multi-Hyperplane Machine, namely Adaptive Multi-hyperplane Machine (AMM), for efficiently handling large-scale dataset was introduced in [12]. AMM is constructed in the spirit of stochastic gradient descent framework proposed in [8]. The weight matrix $W^{(1)}$ is firstly initialized with zero matrix. At each t-th iteration, a random instance (x_t, y_t) is uniformly chosen from the training set $\mathfrak{D}$. The instantaneous objective function associated with t-th instance (x_t, y_t) is defined as $P^{(t)}\left(W|z\right) \triangleq \frac{\lambda}{2}\|W\|_2^2 + l\left(W; (x_t, y_t), z_t\right)$.

The new weight matrix $W^{(t+1)}$ is updated by following the negative direction of sub-gradient $W^{(t+1)} = W^{(t)} - \eta^{(t)}\nabla^{(t)}$ where $\eta^{(t)} = 1/\left(\lambda t\right)$ is the learning rate and the sub-gradient matrix $\nabla^{(t)}$ is computed as

$$\nabla^{(t)} = \left[|\nabla_{1,1}^{(t)}, \ldots, \nabla_{1,b_1}^{(t)}|\nabla_{2,1}^{(t)}, \ldots, \nabla_{2,b_2}^{(t)}|, \ldots, |\nabla_{M,1}^{(t)}, \ldots, \nabla_{M,b_M}^{(t)}|\right]$$

where $\nabla_{i,j}^{(t)} = \nabla_{w_{i,j}^t} P^{(t)}\left(W|z\right)$ is a $D \times 1$ column vector.

Concretely, the sub-gradient matrix $\nabla^{(t)}$ is computed as follows. If the current model predicts correct label of instance x_t, i.e., the loss value $l\left(W; (x_t, y_t), z_t\right) = 0$, then $\nabla_{i,j}^{(t)} = \lambda w_{i,j}^{(t)}$. Otherwise, $\nabla_{i,j}^{(t)}$ is calculated as

$$\nabla_{i,j}^{(t)} = \begin{cases} \lambda w_{i,j}^{(t)} + x_t & \text{if } i = i_t,\, j = j_t \\ \lambda w_{i,j}^{(t)} - x_t & \text{if } i = y_t,\, j = z_t \\ \lambda w_{i,j}^{(t)} & \text{otherwise} \end{cases} \tag{3}$$

where $i_t = \underset{k \in \mathcal{Y} \setminus y_n}{\arg\max}\, g\left(k, x_t\right)$, $j_t = \underset{k}{\arg\max}\left(w_{i_t,k}^{(t)}\right)^{\mathsf{T}} x_t$.

In AMM, the number of hyperplanes which are representing for each class does not require to be prespecified. To each class, a set of non-zero weights and a reserved zero weight are stored and employed. At each iteration, the reserved zero weight of a class can be updated to a non-zero weight and in this case, a new reserved zero weight for this class will be created. The pruning weight procedure is heuristically performed by periodically eliminating the weights whose lengths are less than a predefined threshold. In our viewpoint, subtracting the small-length weights though helps reducing the model size and training time but can impact to the prediction accuracy.

4   Sparse Adaptive Multi-hyperplane Machine (SAMM)

In this section, to encourage the sparsity of the solution, we reformulate the optimization problem of AMM by incorporating the group norm $\|W\|_{2,1}$ associated with the parameter μ. By tuning μ, we can govern the sparsity level of the solution. Concretely, increasing the value of μ leads to a sparser solution in terms of a lower number of hyperplanes per class and non-zero components per hyperplane. Because minimizing $\|W\|_{2,1}$ also inspires as many as possible weights going into zero, the pruning weight procedure therefore is automatically performed in our proposed model without any heuristic or predefined threshold.

4.1   Optimization Problem

In SAMM model, the group norm $\|W\|_{2,2}$ is replaced by the elastic group norm $\Omega\left(W\right) \triangleq \frac{\lambda}{2}\|W\|_{2,2} + \mu\|W\|_{2,1}$. The optimization problem of SAMM becomes

$$\min_{W} P(W) \triangleq \Omega(W) + \frac{1}{N} \sum_{n=1}^{N} l\left(W; (x_n, y_n)\right).$$

In SAMM model, the group norm $\|W\|_{2,1} = \sum_{i=1}^{M} \sum_{j=1}^{b_i} \|w_{i,j}\|_2$ is incorporated. Mathematically, minimizing $\|W\|_{2,1}$ encourages the lengths of the component weights $\|w_{i,j}\|_2$ going to 0 or decreasing to a small amount. It means that the solution is expected to be sparse in terms of a lower number of hyperplanes in use and a smaller number of non-zero elements in each hyperplane. Therefore, the parameter μ in SAMM is used to control the sparsity level of the solution. Furthermore, it can be seen that AMM is a special case of SAMM when $\mu = 0$.

4.2   Optimization Solution

To find the solution of SAMM, we employ two-step alternative approach. In the first step, given the latent variables $z = [z_n]_{n=1}^{N}$, the following optimization is solved to find the current optimal W^*

$$\min_{W} P(W|z) \triangleq \Omega(W) + \frac{1}{N} \sum_{n=1}^{N} l\left(W; (x_n, y_n), z_n\right). \tag{4}$$

In the second step, the latent variables $z = [z_n]_{n=1}^{N}$ are updated as $z_n = \operatorname*{argmax}_{k} \left(w_{y_n,k}^*\right)^{\mathsf{T}} x_n$.

To develop a stochastic gradient descent solution for the optimization problem (4), we base on the primal-dual framework for regularized loss minimization proposed in [7]. Concretely, finding solution of the optimization problem (4) is based on the Theorem 2 in [7] which for completeness we restate its simpler form using our notation.

Theorem 1. *Let f be σ-strongly convex w.r.t $\|.\|$ over a set S. Let $l^{(1)}, l^{(2)}, \dots, l^{(T)}$ be a sequence of convex functions, and L is a positive number such that $\left\|\delta l^{(i)}\left(w^{(i)}\right)\right\|_* \leq L$ for all i. Define $w^{(t)} = \nabla f^*\left(-\frac{1}{t}\sum_{i=1}^{t-1}\delta l^{(i)}\left(w^{(i)}\right)\right)$ then, for any $u \in S$, we have: $\frac{1}{T}\sum_{t=1}^{T}\left(f\left(w^{(t)}\right) + l^{(t)}\left(w^{(t)}\right)\right) \leq \frac{1}{T}\sum_{t=1}^{T}\left(f\left(u\right) + l^{(t)}\left(u\right)\right) + \frac{L^2(1+\log(T))}{2\sigma T}$.*

We define $l^{(t)}(W) = \max\left(0, 1 + \max_{i \in \mathcal{Y}\setminus y_n} g(i, x_t) - w_{y_t, z_t}^{\mathsf{T}} x_t\right)$ and $f(W) \triangleq \Omega(W)$. According to Lemma 3, $f(W)$ is $\frac{\lambda}{2}$-strongly convex to the group norm $\|W\|_{2,2}$ whose duality is $\|W\|_{2,2}^* = \|W\|_{2,2}$. The sub-gradient is able to be computed as $\partial l^{(t)}(W) = \left[\partial l_{1,1}^{(t)}(w_{1,1}), \dots, \partial l_{M,b_M}^{(t)}(w_{M,b_M})\right]$ where $\partial l_{i,j}^{(t)}(w_{i,j})$ is a vector and equal to vector $\mathbf{0}$ if $l^{(t)}(W) = 0$, otherwise it becomes

$$\partial l_{i,j}^{(t)}(w_{i,j}) = \begin{cases} x_t & \text{if } i = i_t, \, j = j_t \\ -x_t & \text{if } i = y_t, \, j = z_t \\ \mathbf{0} & \text{otherwise} \end{cases}$$

where $i_t = \underset{k \in \mathcal{Y} \setminus y_n}{\operatorname{argmax}} g(k, x_t)$, $j_t = \underset{k}{\operatorname{argmax}} \left(w_{i_t,k}^{(t)}\right)^{\mathsf{T}} x_t$.

We note that $\partial l^{(t)}(W)$ is a sparse matrix that has only two non-zero columns when $i = i_t$, $j = j_t$ or $i = y_t$, $j = z_t$. Therefore, we have

$$\left\|\partial l^{(t)}(W)\right\|_{2,2}^2 \leq \|(0, \ldots, x_t, \ldots, -x_t, \ldots, 0)\|_{2,2}^2 \leq 2\|x_t\|_2^2 \leq 2R^2$$

In the above derivation, without loss of generality, we assume that data are bounded in a hypersphere, i.e., $\|x\|_2 \leq R$, $\forall x \in \mathbb{R}^D$.

Relied on Lemma 4, the component $w_{i,j}^{(t+1)}$ in matrix $W^{(t+1)}$ in the update formulation

$$W^{(t+1)} = \nabla f^* \left(-\frac{1}{t+1} \sum_{k=1}^{t} \partial l^{(k)}\left(W^{(k)}\right)\right)$$

is equal to $w_{i,j}^{(t+1)} = \frac{\theta_{i,j}^{(t)} v_{i,j}^{(t)}}{\lambda \left\|\theta_{i,j}^{(t)}\right\|_2}$ where $\theta_{i,j}^{(t)}$ and $v_{i,j}^{(t)}$ are computed as

$$\theta_{i,j}^{(t)} = -\frac{1}{t+1} \sum_{k=1}^{t} \partial l_{i,j}^{(k)}\left(w_{i,j}^{(k)}\right) \text{ and } v_{i,j}^{(t)} = \left|\theta_{i,j}^{(t)} - \mu\right|_+$$

To summarize, we present the pseudo code of SAMM in Algorithm 1.

Algorithm 1. Sparse Adaptive Multi-hyperplane Machine

$W^{(1)} = \mathbf{0}$
$s_{i,j} = \mathbf{0}$ for all $i = 1 \ldots M$, $j = 1 \ldots b_i$
$z^{(0)} = 1$
repeat
    for t = 1 **to** T **do**
        Sampling n_t from $[N]$
        $i_t = \underset{i \in \mathcal{Y} \setminus n_t}{\operatorname{argmax}} g(i, x_{n_t})$
        $j_t = \underset{k}{\operatorname{argmax}} \left(w_{i_t,k}^{\mathsf{T}} x_{n_t}\right)$
        if $\left(l^{(t)}(W^{(t)}) > 0\right)$ $s_{i,j} = s_{i,j} - \mathbb{I}_{[i=i_t; j=j_t]} x_{n_t} + \mathbb{I}_{[i=y_{n_t}; j=z_{n_t}]} x_{n_t}$
                for all $i = 1 \ldots M$, $j = 1 \ldots b_i$
        $\theta_{i,j} = \frac{s_{i,j}}{t+1}$ for all $i = 1 \ldots M$, $j = 1 \ldots b_i$
        $v_{i,j} = \left|\|\theta_{i,j}\|_2 - \mu\right|_+$ for all $i = 1 \ldots M$, $j = 1 \ldots b_i$
        $w_{i,j}^{(t+1)} = \frac{\theta_{i,j}}{\lambda \|\theta_{i,j}\|_2} . v_{i,j}$ for all $i = 1 \ldots M$, $j = 1 \ldots b_i$
    endfor
    Recompute $z^{(r+1)}$
until $z^{(r+1)} \cong z^{(r)}$ or enough epochs

4.3 Generalization Error of SAMM

To investigate the generalization error of SAMM, we refer to Theorem 2 in [12]. Let b_i define the number of using hyperplanes for the i-th class. We have the following generalization error bound.

Theorem 2. *Suppose we are able to correctly classify an i.i.d sampled training set $\mathcal{D}$ using the AMM model*

$$W = [\|w_{1,1}, \ldots w_{1,b_1}\|w_{2,1}, \ldots, w_{2,b_2}\|, \ldots, \|w_{M,1}, \ldots, w_{M,b_M}\|]$$

then we can upper bound the generalization error with probability greater than $1 - \delta$ as

$$\frac{130}{N} \left(\|W\|_{2,2}^2 B \log(4eN) \log(4N) \right) + \log \left(\frac{2(2N)^K}{\delta} \right),$$

where $B = \sum_{i=1}^N b_i + 1 + b_{\max}^2 - b_{\max} - b_{\min}$, $K = \frac{1}{2} \sum_{i=1}^N b_i \sum_{j \neq i} b_j$, $b_{\min} = \min_i \{b_i\}$, and $b_{\max} = \max_i \{b_i\}$.

The above theorem reveals that the generalization error is proportional to $B\|W\|_{2,2}^2$ and the learning method utilizing the set of hyperplanes to characterize each class should offer as sparsest solution as possible in terms of the number hyperplanes for each class and the average number of non-zero components in each hyperplane. This reasons why tuning the sparsity level in SAMM can reduce the generalization error.

5 Experiments

5.1 Experimental Settings

We establish the experiments on 9 benchmark datasets[1]. We make comparison our proposed method SAMM with AMM [12], Pegasos [9] including both linear and kernelized versions (LPegasos and KPegasos), and kernelized LIBSVM (KSVM) [3]. All codes are implemented in C/C++ and the codes of baseline methods are achieved from the corresponding authors. All experiments are performed on the computer with the configuration of core I5 3.2 GHz and 16 GB in RAM.

5.2 Evaluation on Accuracy and Time of the Proposed Method

We run cross validation with 5 folds to select λ for LPegasos, KPegasos, AMM, SAMM, and KSVM and μ for SAMM. The considered ranges are $\lambda \in \{10^{-7}, 10^{-2}\}$ and $\mu \in \{10^{-7}, 10^{-2}\}$. For KSVM and KPegasos, RBF kernel, given by $K(x, x') = e^{-\gamma\|x-x'\|^2}$, is employed. The parameter γ is searched in

---

[1] All datasets can be download at the URL http://www.csie.ntu.edu.tw/~cjlin/libsvmtools/datasets.

the grid $\{2^{-15}, 2^{-13}, \ldots, 2^5\}$. We repeat the experiment for each dataset 5 times and record the corresponding mean and standard deviation.

In Tables 1 and 2, we report the accuracies and training times corresponding to the optimal parameter set. As observed from Table 1, comparing with the linear methods, our proposed SAMM gains superiority accuracies to others on all experimental datasets. The reason is that as compared with AMM, our proposed SAMM can tune the sparsity level of the solution which help boost the generalization capacity of learning method according to Occam's razor principle. Comparing with LPegasos, the fact that SAMM outperforms it is explainable since LPegasos is a linear model that uses only one hyperplane to classify data and is therefore only suitable for linearly separable data in nature. Comparing with the kernelized methods, SAMM also produces the comparable classification accuracies on the experimental datasets.

Regarding the amount of time taken to train, as can be seen from Table 2, the training time of SAMM is always less than AMM and longer than LPegasos. Although the computational complexities of SAMM and AMM are similar, the computational cost in each iteration of SAMM is less than AMM due to the smaller number of using hyperplanes and the sparser non-zero components in each hyperplane itself and thereby makes it faster. The training time of SAMM always exceeds that of LPegasos because of LPegasos's model simplicity. Nonetheless, LPegasos itself cannot be used to accurately learn nonlinearly separable data in nature. Comparing with the kernelized versions, because of the cheaper kernel computation cost, SAMM almost offers the shorter training times. Especially for the large scale datasets namely mnist, webspam, and url, SAMM are much faster than both KSVM and KPegasos. Regarding prediction time, our proposed SAMM always takes less time than AMM (cf. Table 3). This can be explained by the sparser solution in terms of the average number of hyperplanes per class and the average number of non-zero components per hyperplane. By the same reason of the cheaper kernel computation cost, SAMM also provides the shorter prediction times as compared with the kerneilized methods.

We also measure the average number of hyperplanes per class and the average percentage of non-zero components per hyperplane and report them in Table 5. As shown in this table, the average numbers of hyperplanes per class of SAMM and AMM are comparable. This implies that AMM's prune weight procedure operates somehow exhaustively and this may compromise the prediction accuracy of AMM itself. Nonetheless, the average percentage of non-zero components per hyperplane of SAMM is always lower than AMM for all experimental datasets. This observation again confirms our expectation about sparsity level of SAMM.

5.3   Tuning the Sparsity and Its Influence on Performance

To investigate the influence of μ to accuracy, we conduct the experiment where μ is varied and λ is kept fixed. As observed from Table 4, when μ is varied in ascending order, the accuracy is increased at first to its peak and then is gradually decreased. This fact may be partially explained as increasing μ, the

Table 1. Accuracy (in %) comparison.

Dataset	SAMM	AMM	LPegasos	KSVM	KPegasos
a9a	**84.38 ± 0.37**	84.25 ± 0.38	84.27 ± 0.38	84.07 ± 0.26	78.52 ± 6.62
letter	90.35 ± 0.60	68.67 ± 0.60	64.36 ± 0.84	**97.25 ± 0.15**	95.59 ± 1.28
news20	**83.56 ± 0.78**	79.43 ± 0.52	64.36 ± 0.84	83.50 ± 1.10	71.24 ± 5.12
ijcnn1	**98.42 ± 0.26**	90.29 ± 0.28	90.29 ± 0.28	98.25 ± 0.11	95.86 ± 0.23
usps	96.51 ± 0.38	96.40 ± 0.32	95.42 ± 0.42	**98.33 ± 0.22**	95.64 ± 0.11
rcv1_bin	**97.21 ± 0.26**	93.94 ± 0.51	93.96 ± 0.51	95.82 ± 0.17	94.71 ± 0.75
mnist	94.77 ± 0.25	94.66 ± 0.32	93.96 ± 0.51	**99.67 ± 0.45**	96.56 ± 0.56
webspam	97.37 ± 0.15	82.94 ± 0.17	82.98 ± 0.12	**99.2 ± 0.84**	88.69 ± 0.09
url	97.44 ± 0.27	96.95 ± 0.33	96.96 ± 0.33	**98.82 ± 0.55**	98.65 ± 2.3

Table 2. Training time (in seconds) comparison.

Dataset	SAMM	AMM	LPegasos	KSVM	KPegasos
a9a	48.73 ± 3.9	132.84 ± 1.6	20.83 ± 2.6	61.03 ± 0.22	136.70 ± 8.54
letter	61.36 ± 0.9	99.14 ± 4.2	42.31 ± 0.2	4.87 ± 0.04	54.04 ± 0.13
news20	2,472 ± 109	5,138 ± 277	64.36 ± 0.8	222.83 ± 2.91	1604.5 ± 12.8
ijcnn1	92.98 ± 3.4	113.22 ± 8.8	25.16 ± 1.7	197.33 ± 10.72	216.53 ± 11.50
usps	15.65 ± 0.5	48.23 ± 3.7	12.69 ± 0.1	14.95 ± 1.84	21.05 ± 2.84
rcv1_bin	324.94 ± 10.8	749.83 ± 42.8	78.32 ± 7.9	161.60 ± 8.69	214.20 ± 8.60
mnist	145.67 ± 4.5	353.6 ± 41.6	93.96 ± 0.5	1,115.66 ± 39.55	5,590.62 ± 137.8
webspam	890.38 ± 38.8	5,186 ± 85	783.9 ± 11.7	13,293.38 ± 1,838	13,540.3 ± 1,328
url	1,000 ± 184	7,070 ± 715	261.8 ± 2	> 72 h	> 72 h

Table 3. Prediction time (in miliseconds) comparison.

Dataset	SAMM	AMM	Pegasos	KSVM	KPegasos
a9a	8.88 ± 7.9	15.64 ± 0.5	0.64 ± 3.14	12.44 ± 0.10	52.02 ± 4.84
letter	9.92 ± 0.4	17.20 ± 1.4	8.04 ± 1	2.67 ± 0.02	35.05 ± 1.28
news20	629.1 ± 42.6	943.1 ± 76.9	64.36 ± 0.8	42.99 ± 0.70	861.45 ± 12.96
usps	36.24 ± 8.5	93.76 ± 10.9	29.48 ± 7	3.79 ± 0.52	75.07 ± 7.71
ijcnn1	10.32 ± 0.9	13.08 ± 0.5	2.36 ± 0.6	2.61 ± 0.33	140.65 ± 6.84
rcv1_bin	76.88 ± 4.1	98.80 ± 24.5	7.44 ± 7.8	39.47 ± 4.19	117.24 ± 5.71
mnist	334.48 ± 12.5	768.4 ± 83.3	93.96 ± 0.5	354.16 ± 12.13	1,382.35 ± 225.78
webspam	164.60 ± 8.8	626.32 ± 13.4	64.16 ± 2.1	3,742.63 ± 1,030	3,607.05 ± 708
url	4,798±293	19,623 ± 3,141	950 ± 23	32,536 ± 972	34,478 ± 1,126

Table 4. The accuracies (in %) when the parameter μ is varied.

	0	2^{-7}	2^{-6}	2^{-5}	2^{-4}	2^{-3}	2^{-2}	2^{-1}
a9a	83.6559	84.2837	84.2830	84.2161	84.3076	84.2941	84.2370	84.3881
ijcnn1	92.4025	98.3069	98.3937	98.4201	98.3923	98.3501	98.3449	98.3177
rcv1_bin	94.0013	96.1455	97.1712	97.1742	97.1673	97.1930	97.2068	97.1781
webspam	93.1410	97.2012	97.2863	97.3181	97.3288	97.2973	97.3191	97.3709
url	95.4265	96.4540	97.4445	97.4042	97.4010	97.3966	97.3630	97.3752

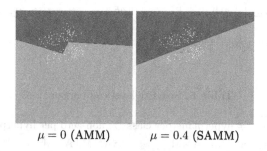

$\mu = 0$ (AMM) $\mu = 0.4$ (SAMM)

Fig. 1. SAMM with $\mu = 0$ (corresponding to AMM) uses 3 hyperplanes to classify data and SAMM with $\mu = 0.4$ uses only 1 hyperplane.

sparsity (simplicity) level of solution is increased as well. It then helps avoid overfitting and brings out an optimal model. However, when μ is increased to a bigger value, the estimated model becomes too simple, and underfitting may consequently happen.

To visually manifest the above reason, we also provide simulation study on $2 - D$ datasets as displayed in Figs. 1 and 2. In Fig. 1, AMM requires 3 hyperplanes to classify the data and the learning seems to be overfitted while SAMM with a tuning of $\mu = 0.4$ uses only 1 hyperplane which is the ideal solution in this case. In Fig. 2, AMM needs 10 hyperplanes to classify the data, but in the meanwhile, SAMM with a tuning of $\mu = 0.04$ requires only 5 hyperplanes which is intuitively a better solution.

6 Conclusion

In this paper, we leverage stochastic gradient descent framework with max-margin principle to propose Sparse Adaptive Multi-hyperplane Machine (SAMM). By incorporating the group norm $\mathbb{L}_{2,1}$ to its model, with SAMM we are able to govern and tune the sparsity level of the solution. It enables SAMM to avoid both overfitting and underfitting. We validate the proposed method on large benchmark datasets. The experimental results show that SAMM can actually tune sparsity level of the solution and consequently yields superior accuracy while simultaneously achieving shorter training time compared with the baselines.

7 Technical Lemmas

Lemma 1. *Given that* $g = \tau f + h$ *where* f *is* σ-*strongly convex w.r.t norm* $\|.\|$ *and* h *is convex,* g *is* $\tau\sigma$-*strongly convex w.r.t norm* $\|.\|$.

Lemma 2. $\frac{1}{2}\|.\|_{p,r}^2$ *is* $\frac{(p-1)(r-1)}{p+r-2}$-*strongly convex w.r.t* $\|.\|_{p,r}$ *provided* $p, r > 1$.

Lemma 3. *The norm* $\Omega(W) = \frac{\lambda}{2}\|W\|_{2,p}^2 + \mu\|W\|_{2,1}$ *is* $\frac{\lambda(p-1)}{p}$-*strongly convex w.r.t the norm* $\|.\|_{2,p}$.

Lemma 4. *Let* $v = \left[\left|\|\theta^1\|_2 - \mu\right|_+, \ldots, \left|\|\theta^F\|_2 - \mu\right|_+\right]$ *then the* j *component of* $\nabla\Omega^*(\theta)$ *is equal to* $\frac{\theta^j}{\lambda\|\theta^j\|_2} \cdot \frac{v_j^{q-1}}{\|v\|_q^{q-2}}$.

Table 5. Sparsity level comparison

Sparsity	# hyperplane per class		% non-zero per hyperplane	
Dataset	SAMM	AMM	SAMM	AMM
a9a	9.6	10	75.66	89.72
letter	4.6	2.3	76.76	100
news20	5.7	4.0	19.37	26.95
usps	4.552	3.1	67.69	99.99
ijcnn1	9.04	5.5	68.11	100
rcv1_bin	9.96	10	15.46	17.82
mnist	5.168	4.1	59.13	69.22
webspam	9.2	10	52.62	49.43
url	8.43	10	2.41	7.44

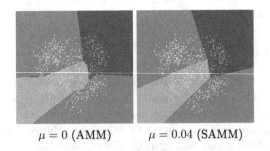

$\mu = 0$ (AMM) $\mu = 0.04$ (SAMM)

Fig. 2. SAMM with $\mu = 0$ (corresponding to AMM) uses 10 hyperplanes to classify data and SAMM with $\mu = 0.04$ uses only 5 hyperplanes.

References

1. Aiolli, F., Sperduti, A.: Multiclass classification with multi-prototype support vector machines. J. Mach. Learn. Res. **6**, 817–850 (2005)
2. Boser, B.E., Guyon, I.M., Vapnik, V.: A training algorithm for optimal margin classifiers. In: Proceedings of the 5th Annual ACM Workshop on Computational Learning Theory, pp. 144–152. ACM Press (1992)
3. Chang, C.-C., Lin, C.-J.: LIBSVM: a library for support vector machines. ACM Trans. Intell. Syst. Technol. **2**(3), 1–27 (2011)
4. Crammer, K., Singer, Y.: On the algorithmic implementation of multiclass kernel-based vector machines. J. Mach. Learn. Res. **2**, 265–292 (2002)
5. Jaakkola, T., Meila, M., Jebara, T.: Maximum entropy discrimination. Technical report, Cambridge, MA, USA (1999)
6. Schapire, R.E., Freund, Y., Bartlett, P., Lee, W.S.: Boosting the margin: a new explanation for the effectiveness of voting methods. Ann. Stat. **26**(5), 1651–1686 (1998)
7. Shalev-Shwartz, S., Kakade, S.M.: Mind the duality gap: logarithmic regret algorithms for online optimization. In: Advances in Neural Information Processing Systems, pp. 1457–1464 (2009)
8. Shalev-Shwartz, S., Singer, Y.: Logarithmic regret algorithms for strongly convex repeated games. The Hebrew University (2007)
9. Shalev-Shwartz, S., Singer, Y., Srebro, N., Cotter, A.: Pegasos: primal estimated sub-gradient solver for SVM. Math. Program. **127**(1), 3–30 (2011)
10. Tsochantaridis, I.: Support vector machine learning for interdependent and structured output spaces. Ph.D. thesis, Providence, RI, USA, AAI3174684 (2005)
11. Vapnik, V.N.: Statistical Learning Theory. Wiley-Interscience, New York (1998)
12. Wang, Z., Djuric, N., Crammer, K., Vucetic, S.: Trading representability for scalability: adaptive multi-hyperplane machine for nonlinear classification. In: Proceedings of the 17th ACM SIGKDD International Conference on Knowledge Discovery and Data Mining, pp. 24–32. ACM, New York (2011)

Exploring Heterogeneous Product Networks for Discovering Collective Marketing Hyping Behavior

Qinzhe Zhang[✉], Qin Zhang, Guodong Long, Peng Zhang, and Chengqi Zhang

The University of Technology Sydney, Sydney, Australia
{Qinzhe.Zhang,Qin.Zhang}@student.uts.edu.au
{Guodong.Long,Peng.Zhang,Chengqi.Zhang}@uts.edu.au

Abstract. Online spam comments often misguide users during online shopping. Existing online spam detection methods rely on semantic clues, behavioral footprints, and relational connections between users in review systems. Although these methods can successfully identify spam activities, evolving fraud strategies can successfully escape from the detection rules by purchasing positive comments from massive random users, i.e., user Cloud. In this paper, we study a new problem, **Collective Marketing Hyping** detection, for spam comments detection generated from the user Cloud. It is defined as detecting a group of marketing hyping products with untrustful marketing promotion behaviour. We propose a new learning model that uses heterogenous product networks extracted from product review systems. Our model aims to mining a group of hyping activities, which differs from existing models that only detect a single product with hyping activities. We show the existence of the Collective Marketing Hyping behavior in real-life networks. Experimental results demonstrate that the product information network can effectively detect fraud intentional product promotions.

1 Introduction

With the booming of online business, user comments on products and online stores are increasingly important for shaping customer decision. These user comments are valuable and represent the opinion and judgement of 'experienced' product users. Sales and profits link with reviews, which entice merchants to hire a group of people to fabricate fake reviews to unjustly hype and denigrate competitors. Therefore, it is essential to detect these fake reviews to restore user confidence on online business.

Previous works focus on capturing important features from user review behavior data, review network data, and store network data [1,2,4–7]. These work can successfully detect the marketing hyping behavior when spammer accounts are manipulated by one or several people with massive false comments. However, applying for an hyping account is costly and it is infeasible to generate a large amount of fake accounts to do marketing hyping. Moreover, existing spam

© Springer International Publishing Switzerland 2016
J. Bailey et al. (Eds.): PAKDD 2016, Part I, LNAI 9651, pp. 40–51, 2016.
DOI: 10.1007/978-3-319-31753-3_4

review approaches are mainly generated by real users on purpose under anomalous comments, and the comments can also have large variety which makes the spam detection even more difficult. The work [12] provided a new perspective to detect the spam comment without any content of reviews and network information, where they define singleton reviewers as the spammers who just reviewed once or very few times. Then, they define the problem as the abnormal detecting problem and utilize statistic method to reveal positive reviews with high rating ascend with singleton users burst in a certain period.

However, existing online review spam detection approaches have obvious limitation in the user Cloud environment. To date, many 3-party platforms provide the service of connecting online sellers with marketing hyping require, where massive random users (user Cloud) conduct real transactions and reviews and then reclaim bonus. In contrast, existing works never take the product network into consider. They only use the user/product network for detection. Because user profile and user behavior can be easily changed or hidden, existing methods can be easily darted by E-commerce Service Providers. In other words, existing works ignore the latent connections in product networks and the connection on product is hard to conceal, especially when the hyping activities has become popular among homogeneous competitors. Moreover, they separately use linguistic clues of deception or relational network information and can only just detect these unjustly activities by a sole product. These shortage not only makes detection rules easily avoid by spam activities, but also leads to inaccurate detection results.

In this work, we study a new problem of detecting a group of marketing hyping users, i.e., **Collective Marketing Hyping** detection. To solve the problem, we need to address the following two challenges:

- How to infer the latent heterogenous product information network? The networks may not be observed directly from the original data sets, to build such networks, we need to infer the relationship matrix among products by their potential connections such as product ID/name, and the store ID/name that sales the similar products. These information are normally latent and require in-depth analysis on the background data.
- How to model the heterogenous product information networks? We need to design a new learning model that can fully combine the power of the relational data and the heterogenous product information networks for discovering collective behaviour among products.

In order to solve the above challenges, we present a new learning model that can use the heterogenous product network information and temporal features to discover a group of marketing hyping activities. As shown in Fig. 1, the product network information is employed to design two product network regularization terms that constraint the matrix factorization based learning function. Specifically, different products within the same store and homogeneous products within different stores are extracted from the original data sets and formulated as the two graph regularization terms in the learning function. Experiments on real-life data show the performance of the proposed method.

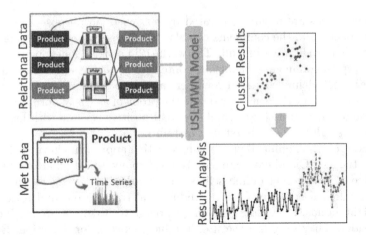

Fig. 1. Our model can cluster products by combining the product network information and the meta data (temporal patterns) extracted from user comments.

The contributions of the paper are summarized as follows,

- We define a new research problem of detecting a group of marketing hyping users, i.e., the **Collective Marketing Hyping** detection problem. Different from existing online spam activity detection problems, we aim to use the latent heterogenous product networks to detect a group of products containing deceitful comments.
- We design a new unsupervised learning model to incorporate both comment patterns (temporal patterns) and online heterogenous product network data for pattern analysis.
- We conduct experiments on real-life data. The results shows that our model can effectively detect the collective marketing hyping activities.

The remainder of this paper is organized as follows. In Sect. 2, we provide an overview of several mainstream approaches for online spam detection and the related work. Section 3 describes the background and methodology of the work. Section 4 presents experimental results on real-life data, followed by the conclusion in Sect. 5.

2 Related Work

In this section, we survey the related research in opinion spam detection and compare them with our present work.

Opinion and Sentimental Mining. There are research works in mining opinions and sentiments behind the rating and review data. McAuley [10] introduced a new dataset of roughly 5 million reviews with multi-dimensional ratings. It is of great challenge to not only model such data but also learning words describe aspects and words refer to sentiment about an aspect simultaneously. After that, they [9] propose statistical models which connect underlying dimensions in rating score to aspects in review text on a large-scale dataset. Besides, a model [11] has been designed and utilized for capturing the evolution of users or community in recommendation system by integrating users experience- as component -into the function of time, when comparing with other method. This model can find similar users even their opinions far apart in a certain time.

However, these works mainly contribute to extracting and summarizing opinions from reviews by utilizing natural language processing and data mining techniques. In other words, they just analyze the latent properties and sentiments on the product or store, but ignore the authenticity of these comments and evaluations. Thus, they cannot tackle the problem of opinion spam unless deriving new features.

Review Spam Detection. Review spam detection techniques are proposed with the evolution of new spammer approaches. Liu [4,5] has studied the trustworthiness of opinions in reviews. It is the first investigation in opinion spam. Three categories of review spam were defined - untruthful opinions, review on brands only and non-reviews. The last two types are relatively easily to be classified by manually labelled. For the first type, it is very hard for labelling by human being. Due to this limitation, only reviews making damage to products and reputation were taken into account and many duplicate or near-duplicated spam reviews (almost certainly) were found. Thus, this no labelled training example problem was solved.

However, the opposite circumstance - reviews which untruthfully promote the product- has not been considered. Furthermore, current spammers can avoid providing massive duplicate text. Li [7] first analyzed several features in terms of spam behaviours, e.g., Content features, Sentiment Features, Product Features, Meta-data features, etc. Then, a two-view semi-supervised method was exploited to identify spam reviews. Feng [2] introduced the notion of natural distribution of opinions and define three types of reviewers: Any-time reviewers, Multi-time reviewers and Single-time Reviewers. Furthermore, these different types of distributional footprints of deceptive reviews were evaluated statistically by NLP techniques, rather than relying on relatively unreliable human judgement. However, one-time reviewers, the most suspicious user type, is more or less stable in the ratio to multi-time reviewers since 2007 in the dataset, which means there may existing spammer groups but cannot be found by their method. Jindal [6] defined several types of expectations based on the original distribution of dataset. It then uses certain unexpectedness measures to rank rules for indicating unusual behaviours as spam activities. However, this study aims to quantify the ratio of abnormal reviews in the on-line e-commercial platforms, rather than detecting the trustiness of the reviews on certain product. Besides, review text understanding, rating is also another element for detecting the spammers. In [1], it mainly exploited models to detect spammed products or product groups by comparing the difference of rating behaviour between suspicious and normal users. The ranking and supervised methods used are high efficient in identifying spammers and the result reveal the spammer have more significant impact on rating, rather than unhelpful reviewers.

Detect Spam Activities by Temporal Feature. Xie [12] abandoned the text data and rating features but just focused on the temporal patterns. This work firstly constructs multidimensional time series refer to singleton reviewers (user who has just written one review) for leveraging correlated anomalies. However, the obvious defect is they assume customers (who really purchased something) will never be spammers and singleton reviewer is surely belongs to the spammer group. They did not consider not only the percentage of genuine buyer purchase several items but just leave one review on one product, but also deceptive users can follow the real purchasing process to give genuine but hired sentiment comments. The work [2] defined the review bursts and use Kernel Density Estimation with several features to detect them. It then proposed a model based on Markov Random Field, and utilize reviewer-reviews-store graph to detect spammers.

3 Methodology

In this section, we first introduce our model in Sect. 3.1, and then describe the algorithm in Sect. 3.2.

Table 1. Dataset statistics

Dataset	Store	Products	Reviews	MRP
Cloth product	10	80	153456	65
Cosmetics product	12	120	223981	82
Electronic product	10	100	183948	76
Food product	10	70	121952	52
Healthy product	10	55	803749	33
Shoes product	10	40	65986	21

3.1 Product Network Regularization

In this section, we first discuss the dataset we collected during our project, and then we will systematically interpret how to model the heterogenous product networks as regularization terms to constrain the matrix factorization learning framework.

In the beginning of 2015, counterfeits crisis on Taobao.com trigger heated debate between Alibaba and the Chinese Consumer Association, shortened for CCA. According to CCA's report, large proportion of fake goods exist in the top 10 Taobao online store of variety industries, including raiment,cosmetics, milk, health etc. Thus, we first search the top 10 stores from these reported industries and then collect at least top 10 best sale products in these stores. In Table 1, we list the collected dataset.

To model the heterogenous information networks, we will discuss the store-Based regularization method and the product-Based regularization approach. In this paper, 'PBR' refers to Product-Based Regularization and 'SBR' represents Store-Based Regularization.

Store-Based Regularization. As mentioned in Sect. 1, online sellers tend to implement untrustful hyping actions regularly, to keep their high rank position. These competitor are more likely to observe each other, rather than directly inform them to united join in. In other words, the collective marketing hyping are an tacit competition among homogeneous stores and products, e.g.,protein powder with same brand in healthy store. They should share similar behaviours when they have similar position in ranking system. Based on this intuition,we impose a Store-based Regularization term to minimize the distance between two same product in different store,

$$R(Store\_based) = \mathbf{VG}_1\mathbf{V}^\top. \tag{1}$$

The Store-Based regularization term makes an postulation that similar ranked products will share similar patterns if they are same merchandise within different stores. Thus, they have very large possibility in the same cluster. However, this may not always be true in the real world. For example, some high rank merchant may make better reputation and profit by regular method while others need untrustful hyping. Hence,a more realistic model should also take other features into consider, e.g., the meta data $\mathbf{G}_1$ is a matrix to represent Store-Based network. Figure 2 describes how to build the network. For every product with the same name and brand within all stores, we set up their connection values as 1; otherwise, 0.

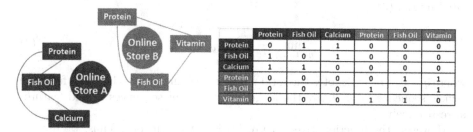

	Protein	Fish Oil	Calcium	Protein	Fish Oil	Vitamin
Protein	0	1	1	0	0	0
Fish Oil	1	0	1	0	0	0
Calcium	1	1	0	0	0	0
Protein	0	0	0	0	1	1
Fish Oil	0	0	0	1	0	1
Vitamin	0	0	0	1	1	0

Fig. 2. The store-based network and matrix $\mathbf{G}_1$.

Product-Based Regularization. The first model we propose imposes a Store-Based regularization term to constrain product similarity among different stores. However, this approach is insensitive to those different products within the same store. This will cause information loss problem, which will result in inaccurate modeling. Hence, in order to tackle this problem, we propose another product regularization term to impose constraints among products within their own store.

Thus, we also introduce second regularization term for minimize the deviation between dissimilar products within same store,

$$R(Product\_based) = \mathbf{V}\mathbf{G}_2\mathbf{V}^\top \tag{2}$$

Specifically, we believe merchants who utilize unfairly technique will hype many of their products, rather than only one type. Therefore, products possess similar sales in certain store should also with similar pattern. In Fig. 3, we describe the Product-Based network and denote the matrix as $\mathbf{G}_2$. In contrast to $\mathbf{G}_1$, we only set up connection value as 1 when products are within the same store; otherwise, 0.

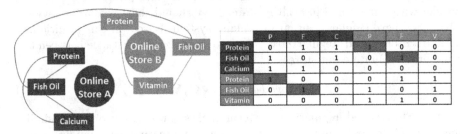

	P	F	C	P	F	V
Protein	0	1	1	1	0	0
Fish Oil	1	0	1	0	1	0
Calcium	1	1	0	0	0	0
Protein	1	0	0	0	1	1
Fish Oil	0	1	0	1	0	1
Vitamin	0	0	0	1	1	0

Fig. 3. The product-based network and matrix $\mathbf{G}_2$.

Feature Representation. Feature transformation can reduce the dimensionality of original online review data [8]. Given a sequence of user comments sequence $T = \{t_1, t_2, ..., t_n\}$ and features $S = \{s_1, s_2, ..., s_k\}$, $\mathbf{X} \in R^{k \times n}$ represent the sequence-transformed data matrix, where $\mathbf{X}_{(\mathbf{s}_i, \mathbf{t}_j)}$ is the distance between each sequential feature s_i and the original data t_j. The distances can be calculated by Eq. (3),

$$\mathbf{X}_{(ij)} = \min_{g=1,...,\bar{q}} \frac{1}{l_i} \sum_{h=1}^{l_i} (\mathbf{t}_{j(g+h-1)} - \mathbf{s}_{i(h)})^2, \tag{3}$$

where $\bar{q} = q_j - l_i + 1$ denotes the total number of segments with length l_i from the review sequence $\mathbf{t}_j$, q_j, l_i are the lengths of comment sequences $\mathbf{t}_j$ and features $\mathbf{s}_i$ respectively.

However, the function given in Eq. (3) is not continuous, Thus, we use the soft minimum function as in Eq. (4) [3]

$$\mathbf{X}_{(ij)} \approx \frac{\sum_{q=1}^{\bar{q}} d_{ijq} \cdot e^{\alpha d_{ijq}}}{\sum_{q=1}^{\bar{q}} e^{\alpha d_{ijq}}}, \tag{4}$$

Review Sequence Similarity. We use $\mathbf{H}_{(ij)}$ to represent the similarity between two temporal features $\mathbf{s}_i$ and $\mathbf{s}_j$, which can be calculated as in Eq. (5),

$$\mathbf{H}_{(ij)} = e^{-\frac{\|d_{ij}\|^2}{\sigma^2}}, \tag{5}$$

Pseudo-Class Label. In this part, we consider unlabeled data training. Note labeled data is a special case of our model where the class label is given a prior, and the iterations on the class label during learning can be removed. Provide that we cluster user review sequence data into c class, the pseudo-class label matrix $\mathbf{V} \in R^{c \times n}$ contains the c labels.

Least Square Minimization. Based on the pseudo-class labels, we wish to minimize the least square error. Let $\mathbf{U} \in R^{k \times c}$ be the classification boundary under the pseudo-class labels, the least square error minimizes the following objective function,

$$\min_{\mathbf{U}} \ \|\mathbf{X} - \mathbf{U}\mathbf{V}\|_F^2 \tag{6}$$

The Learning Model. Based on the above notations, the learning model is a joint optimization problem with respect to variables S, U and V, as in Eq. (7),

$$\min_{\mathbf{W},\mathbf{S},\mathbf{Y}} \frac{1}{2}\|\mathbf{X} - \mathbf{U}\mathbf{V}\|_F^2 + \frac{\lambda_1}{2}\|\mathbf{V}\mathbf{G}_1\mathbf{V}^\top\| + \frac{\lambda_2}{2}\|\mathbf{V}\mathbf{G}_2\mathbf{V}^\top\| + \frac{\lambda_3}{2}\|\mathbf{H(s)}\|_F^2 + \frac{\lambda_4}{2}\|\mathbf{U}\|_F^2, \tag{7}$$

where we aim to minimize the error rate (the first optimization term) under the heterogenous network regularization (the second and third terms). The last two terms are added to avoid overfitting to noisy data.

3.2   The Learning Algorithm

In order to solve the above learning function, we use the coordinate descent algorithm to iteratively solve the three variables as below,

Algorithm
1:   **Input:**
   - The user review sequence data $\mathbf{T}$ with c classes
   - The length and number of sequential features: l_{min}, r, k
   - The number of internal iterations i_{max}
   - The learning rate η
   - Parameters $\lambda_1, \lambda_2, \lambda_3$ and α, σ
2:   **Output:** Sequential feature $\mathbf{S}$ and class label $\mathbf{V}$

3:   **Initialize: $\mathbf{S}_0, \mathbf{V}_0, \mathbf{U}_0$**
4:   **While** Not convergent **do**
5:      **Step 1 :Update V with Fixed U and S:**

$$\mathbf{V}_{ij}^{t+1} = \mathbf{V}_{ij}^t \sqrt{\frac{(X_t^T U_t)_{ij}}{[(\lambda_1 G_1^T + \lambda_2 G_2^T + V_t V_t^T)X_t^T U_t]_{ij}}} \tag{8}$$

6:
   Step 2 :Update U with Fixed V and S:

$$\mathbf{U}_{t+1}^i = \left\{ \begin{matrix} 0 & otherwise \\ (1 - \frac{\lambda_4}{\|(\mathbf{X}_t \mathbf{V}_t^{-1})^i\|})(\mathbf{X}_t \mathbf{V}_t^{-1})^i & if \ \ \|(\mathbf{X}_t \mathbf{V}_t^{-1})^i\| > \lambda_4 \end{matrix} \right\} \tag{9}$$

7:

Step 3 :Update S with Fixed U and V:

$$S^{t+1} = S^t - \alpha[(X_s - UV)\frac{\alpha X_s}{\alpha S} + H_s\frac{\alpha H_s}{\alpha S}]$$ (10)

8:    **end while**
9:    **Output: $S^* = S_{t+1}$; $U^* = U_{t+1}$; $V^* = V_{t+1}$.**

The algorithm iteratively solves the three variables V in Eq. (8), U in Eq. (9) and S in Eq. (10), and eventually converge to the local optimum $S*$, $U*$ and $V*$. Because the objective function given in Eq. (7) is convex, the local solution here is also the global optimum. The algorithm converges fast under the proper parameter settings.

4   Experiments

In this section, we first describe the output of the learning model, and then give a couple of case studies to provide the analysis of the sequential patterns of the problematic products with deceitful user comments. At last, we use human-labeled data to evaluate the results.

4.1   Human Evaluation

In Taobao.com, people purchase items and leave comments. They also have buyer reputation and levels. We invite 20 experienced online buyers who frequently purchase items online to label around 500 products as we collected. Some of these invited users have already participated in online hyping, so they know the rules and strategies of the current hyping techniques.

4.2   Benchmark Methods

To show the strength of the heterogenous product networks in detection, we design three methods for comparisons:

Only with SRB. Intuitively, although two product network information we mentioned above are all very important, we assume that the Store-Based regularization should be more crucial for discovering the collective behaviours. As such promotion is actually an tacit action among all homogeneous sellers. Hence, we modify the objective function in Eq. (7) by setting $\lambda_2 = 0$.

Only with PRB. Similar to above method, we also observe the performance when only use Product-Based regularization in Eq. (7) by setting $\lambda_1 = 0$.

Without Any Regularizations. To verify how significance of the product network information, we take off all the regularization and just consider the features in meta data. Thus, the learning function is in Eq. (7) with $\lambda_1 = 0$ and $\lambda_2 = 0$.

4.3   Experimental Results

We conduct experiments on six datasets. To verify the importance of product network information, we use both network regularizations (Store-Based and Product-Based) in the algorithm, and then use only one of them for comparisons, as given in the benchmark methods.

In Table 2, we show the deviation of the results. It is clear that the accuracy drops significantly without any network regularizations. Furthermore, the precision also drops when just taking one network information into consider. Besides, we found that the Store-Based Regularization is more effective than the Product-Based Regularization. The results explain the latent connections of homogeneous competition among online products in different stores.

Table 2. Comparisons among the benchmark methods.

Dataset	Mean + Std.			
	Our method	SRB only	PRB only	No Reg
Cloth products	**0.9566 ± 0.0082**	0.9156 ± 0.0088	0.9079 ± 0.0059	0.8778 ± 0.0059
Cosmetics products	**0.9498 ± 0.0099**	0.9189 ± 0.0069	0.9083 ± 0.0049	0.8789 ± 0.0058
Electronic products	**0.9506 ± 0.0117**	0.9196 ± 0.0040	0.9096 ± 0.0051	0.8793 ± 0.0064
Food products	**0.9531 ± 0.0131**	0.9205 ± 0.0057	0.9084 ± 0.0037	0.8804 ± 0.0057
Healthy product	**0.9489 ± 0.0126**	0.9161 ± 0.0056	0.9072 ± 0.0053	0.8789 ± 0.0064
Shoes products	**0.9525 ± 0.0110**	0.9182 ± 0.0070	0.9068 ± 0.0047	0.8791 ± 0.0047

4.4   Application: A Case Study

In this part, we mainly focus on analysing several interesting cases in our experimental results.

Collective Marketing Hyping. As we discussed before, online stores tend to hire real people to generate real purchase comments to increase their sales. Thus, the sequential patterns of the comments should be relatively similar to each other given the same certain of a time window.

From the clustering results, we focus on several homogeneous products (they are all healthy products) within different stores with the same class labels, as given in Fig. 3. It is very clear that from May to June, 2015, their comment sequential patterns are highly similar. Furthermore, due to this period is close to the Chinese Mother's day and Father's day, it is very common for customers to purchase healthy products for their parents. Therefore, for ranking in a high position, these Online merchants will pay for hyping activities and these fake comments lead to collective hyping with similar user comment sequential patterns (Fig. 5).

The False Negative Study. There are a few cases that the human label as hyping, but our algorithm categorizes it into Non-Hyping groups. Through our

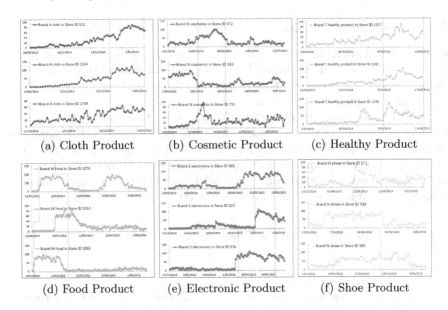

(a) Cloth Product   (b) Cosmetic Product   (c) Healthy Product

(d) Food Product   (e) Electronic Product   (f) Shoe Product

Fig. 4. From all the dataset, we can found obvious evidence of collective marketing hyping activities.

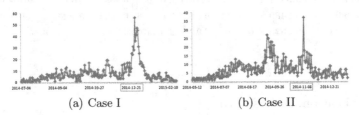

(a) Case I   (b) Case II

Fig. 5. An illustration of the two representative False Negative Cases.

analysis, we can infer that our algorithm is correct. As we can see in Fig. 4, normally, if this problem treats it as an anomaly detection problem, these two stores may be labeled as hyping as they all have a short burst period. However, our method fully considers the overall temporal patterns, not only the burst. Furthermore, we can see that the burst time of these two products are in the special festival In China - the upper one is in the Christmas and New year while another one is close to the 11th of November[1].

Overall, there are few ambiguous results generated, However, by taking the heterogeneous network information into account, we can identify the collective hyping activities correctly,with an average detection accuracy of 90 %.

---

[1] On the holiday, the most popular E-commercial platform in China will appeal the store owner to give special discount or organize group-shopping and flash sale activities, which will significantly increase the store's sale or reviews.

5   Conclusions

In this work we study the problem of collective marketing hyping with fraud intentions generated from a user Cloud. The problem is important and challenging for trustworthy online business. To resolve this problem, we not only use the clustering method to extract user comments' sequential patterns of each single product, but also detect a group of problematic products by using the latent heterogenous product information networks. The experimental results have shown that the heterogenous product networks play an important role in clustering user comments. We also correctly identify the collective marketing hyping behaviour from a real-life large-scale E-commerce platform.

Acknowledgements. This work was supported by Australia ARC Discovery Project (DP140102206) and Australia Linkage Project (LP150100671).

References

1. Fei, G., Mukherjee, A., Liu, B., Hsu, M., Castellanos, M., Ghosh, R.: Exploiting burstiness in reviews for review spammer detection. ICWSM **13**, 175–184 (2013)
2. Feng, S., Xing, L., Gogar, A., Choi, Y.: Distributional footprints of deceptive product reviews (2012)
3. Grabocka, J., Schilling, N., Wistuba, M., Schmidt-Thieme, L.: Learning time-series shapelets. In: KDD 2014, pp. 392–401. ACM (2014)
4. Jindal, N., Liu, B.: Review spam detection. In: WWW 2007, pp. 1189–1190. ACM (2007)
5. Jindal, N., Liu, B.: Opinion spam and analysis. In: WSDM 2008, pp. 219–230. ACM (2008)
6. Jindal, N., Liu, B., Lim, E.-P.: Finding unusual review patterns using unexpected rules, pp. 1549–1552. ACM (2010)
7. Li, F., Huang, M., Yang, Y., Zhu, X.: Learning to identify review spam. In: IJCAI 2011, pp. 2488–2493. AAAI Press (2011)
8. Lines, J., Davis, L.M., Hills, J., Bagnall, A.: A shapelet transform for time series classification. In: KDD 2012, pp. 289–297. ACM (2012)
9. McAuley, J., Leskovec, J.: Hidden factors and hidden topics: understanding rating dimensions with review text. In: RecSys 2013, pp. 165–172. ACM (2013)
10. McAuley, J.J., Leskovec, J., Jurafsky, D.: Learning attitudes and attributes from multi-aspect reviews, abs/1210.3926 (2012)
11. McAuley, J.J., Leskovec, J.: From amateurs to connoisseurs: Modeling the evolution of user expertise through online reviews. In: WWW 2013, pp. 897–908. International World Wide Web Conferences Steering Committee (2013)
12. Xie, S., Wang, G., Lin, S., Yu, P.S.: Review spam detection via temporal pattern discovery, pp. 823–831. ACM (2012)

Optimal Training and Efficient Model Selection for Parameterized Large Margin Learning

Yuxun Zhou[1]([⊠]), Jae Yeon Baek[1], Dan Li[2], and Costas J. Spanos[1]

[1] Department of EECS, UC Berkeley, Berkeley, USA
yxzhou@berkeley.edu
[2] School of EEE, Nanyang Technological University, Singapore, Singapore

Abstract. Recently diverse variations of large margin learning formalism have been proposed to improve the flexibility and the performance of classic discriminative models such as SVM. However, extra difficulties do arise in optimizing non-convex learning objectives and selecting multiple hyperparameters. Observing that many variations of large margin learning could be reformulated as jointly minimizing a parameterized quadratic objective, in this paper we propose a novel optimization framework, namely Parametric Dual sub-Gradient Descent Procedure (PDGDP), that produces a globally optimal training algorithm and an efficient model selection algorithm for two classes of large margin learning variations. The theoretical bases are a series of new results for parametric program, which characterize the unique local and global structure of the dual optimum. The proposed algorithms are evaluated on two representative applications, i.e., the training of latent SVM and the model selection of cost sensitive feature re-scaling SVM. The results show that PDGDP based training and model selection achieves significant improvement over the state-of-the-art approaches.

1 Introduction

Since the first appearance of Support Vector Machines (SVMs) in 1992, many variations of the large margin learning formalism have been extensively studied to improve the model flexibility and the classification performance. To list a few, SVM with latent variables have been proposed in [4,15] for structured learning and object detection when the training labels are not fully observable. The problem of learning large margin convex polyhedron for target detection is considered in [3,18] by combining multiple linear classifiers to implicitly cluster partially labeled samples. Inasmuch as the traditional large margin classifier with hinge loss can be very sensitive to outliers, the authors of [12,13] suggest using ramp loss (truncated hinge loss) with which a "robust" version of SVM is proposed. To cope with the imbalanced cost of type I and type II error, several attempts have been made to formulate the so called cost sensitive SVM [2,9] by introducing extra loss penalty hyperparameters. Moreover, recent works on Automatic Relevance Determination (ARD) kernel method [11] and multiple kernel learning [1,8,17] are all developed within the large margin learning framework and can also be viewed as instances of its variations.

© Springer International Publishing Switzerland 2016
J. Bailey et al. (Eds.): PAKDD 2016, Part I, LNAI 9651, pp. 52–64, 2016.
DOI: 10.1007/978-3-319-31753-3_5

Despite the substantial progress made in adapting large margin learning to different scenarios, two major algorithmic difficulties are widespread. The first one is the problem of optimizing non-convex learning objectives. For example in the case of latent SVM, the incorporation of the "hidden states" induces a concave term in the loss and breaks the convexity of the original formulation. Similarly for robust SVM, the use of ramp loss makes the objective non-convex. In literature, the authors of [3,4] propose Alternating Optimization (AO) to iteratively minimize a subset of decision variables. Another widely applied method is the Concave-Convex Procedure (CCCP) [16]: the objective is written as the sum of a convex part and a concave part, and each time the concave part is approximated with a linear upper bound for minimization [15]. Direct application of Stochastic Gradient Descent (SGD) has also been considered [18]. However, all of these methods are heuristics that only converge to local optimums. In the machine learning context with large scale problems they may lead to severely deteriorated solutions [5].

On the other hand some adaptations of large margin method do preserve the convexity of the learning objective. However another issue arises as these variations usually include a set of additional hyperparameters, which brings about challenging model selection (hyperparameter tuning) problems. For example in cost sensitive SVM and ARD kernel method, extra penalty coefficients and feature re-scaling parameters are introduced to improve the model flexibility, but the choice of these hyperparameters is left to practitioners. For models with a few hyperparameters, classic techniques such as k-fold cross-validation (KCV), leave-one-out cross-validation (LOOCV) [14] could be readily applied. But such exhaustive search method quickly becomes intractable for models with 3 or more such parameters. In [6] the authors propose to implicitly calculate gradients of various error bounds and loss functions and proceed to use these in a gradient-descent algorithm. Yet previous work largely relies on the technique of implicit differentiation, which requires additional smoothness assumptions and is usually done on a case-by-case basis. A more general gradient based approach for choosing different types of multiple hyperparameters is still lacking.

In this work, we propose a novel optimization scheme, namely Parametric Dual sub-Gradient Descent Procedure (PDGDP), to resolve the above two issues in a unified framework. We first observe that many variations of large margin learning objective could be rewritten in a parameterized quadratic form. Then we develop a key property that explicitly relates the optimum to the hyperparameters. Our theoretical result indicates that the optimal dual solution is a piecewise differentiable function of the hyperparameters, with explicit expressions in well-defined critical regions. To solve the training problem, we further show that the optimal objective of the dual is a convex piecewise quadratic function of hidden variables. Hence a sub-gradient descent will converge to global optimality with guarantee. As for the model selection problem, although it is hard to establish the luxury of convexity for generalization cost, its gradient with respect to hyperparameters could be readily obtained based on the explicit expression, and then used for efficient hyperparamter optimization.

This paper is organized as follows. In Sect. 2, we review a general parameterized SVM formulation and show how some typical variations of large margin learning method can be transformed into parameterized problems. In Sect. 3, we derive the explicit form of the dual optimum as a function of hyperparameters under a parametric programming framework. In Sect. 4, we establish the sub-gradient descent algorithm for training and prove its guaranteed convergence. Similarly in Sect. 5 a gradient based model selection algorithm is proposed. Finally, numerical experiments are given in Sect. 6.

2 Large Margin Learning with Multiple Parameters

We denote the training data as $\mathcal{T} = \{(\boldsymbol{x}_1, y_1), \ldots, (\boldsymbol{x}_n, y_n)\}$, with the ith feature vector $\boldsymbol{x}_i \in \mathbb{R}^p$, and its label $y_i \in \{-1, +1\}$[1]. Denote the index sets $\mathcal{I}_+ = \{i | y_i = +1\}$, and $\mathcal{I}_- = \{i | y_i = -1\}$ for the two classes. The primal problem of a general $\boldsymbol{\phi} = [\boldsymbol{s}, \nu, \boldsymbol{\beta}]^T$ parameterized large margin learning reads[2]

$$\min_{w,b,\rho \geq 0, \xi} \quad \mathcal{P}(\boldsymbol{w}; \boldsymbol{\phi}) = \frac{1}{2}\|\boldsymbol{w}\|^2 - \nu\rho + \sum_{i=1}^{n} s_i \xi_i \qquad \text{(Primal)}$$

$$\text{subject to} \quad y_i(\kappa_\beta(\boldsymbol{w}, \boldsymbol{x}_i) + b) \geq \rho - \xi_i; \quad \xi_i \geq 0 \quad \forall i$$

where s_i is a weighting parameter for loss ξ_i. Depending on the context, it can be viewed as a hidden state variable or a hyperparamter for cost sensitive penalty, as will be made clear later. $\kappa_\beta(\cdot)$ is a general kernel function parameterized by $\boldsymbol{\beta}$, which encompasses not only commonly used kernels, but also recent variations such as Gaussian-ARD kernel or linear combination of multiple kernels. The dual of (Primal) can be written as

$$\min_{\boldsymbol{\alpha}} \frac{1}{2} \sum_{i,j} \alpha_i y_i \kappa_\beta(\boldsymbol{x}_i, \boldsymbol{x}_j) y_j \alpha_j \qquad (1)$$

$$\text{subject to} \quad 0 \leq \alpha_i \leq s_i \; \forall i; \quad \boldsymbol{\alpha}^T \boldsymbol{y} = 0; \quad \boldsymbol{1}^T \boldsymbol{\alpha} \geq \nu$$

After a solution of (1) $\boldsymbol{\alpha}^*$ is obtained, we denote the partition of training sample indices as $\{\mathcal{S}_0, \mathcal{S}_b, \mathcal{S}_{ub}\}$, where $\mathcal{S}_0 = \{i \mid \alpha_i^* = 0\}$ denotes non-support vectors, $\mathcal{S}_b = \{i \mid \alpha_i^* = s_i\}$ for bounded support vectors, and $\mathcal{S}_{ub} = \{i \mid 0 < \alpha_i^* < s_i\}$ denotes unbounded support vectors.

For a more compact form, (1) is reformulated into:

$$\min_{\boldsymbol{\alpha}} \quad \mathcal{J}(\boldsymbol{\alpha}; \boldsymbol{\theta}, \boldsymbol{\beta}) = \frac{1}{2}\boldsymbol{\alpha}^T \boldsymbol{Q}_\beta \boldsymbol{\alpha}$$

$$\text{subject to} \begin{cases} \boldsymbol{C}^\alpha \boldsymbol{\alpha} \leq \boldsymbol{C}^\theta \boldsymbol{\theta} + \boldsymbol{C}^0 \\ \boldsymbol{\alpha}^T \boldsymbol{y} = 0, \end{cases} \qquad \text{(Dual)}$$

---

[1] Binary classification is considered here only for ease of notation. The results developed in the following are readily extended to general structured output problems.

[2] ν version, instead of a C-SVM version is used, as it has a more general dual form.

where $(\boldsymbol{Q}_\beta)_{ij} = y_i y_j \kappa_\beta(\boldsymbol{x}_i, \boldsymbol{x}_j)$, and $\boldsymbol{C}^\alpha, \boldsymbol{C}^\theta, \boldsymbol{C}^0$ are constant matrices that encapsulate inequalities of (1). Denote $\Omega(\boldsymbol{\theta})$ as the feasible set for $\boldsymbol{\alpha}$, we will see through examples that the first type of large margin learning variations, such as latent SVM and robust SVM, can be rewritten as jointly optimizing over $\boldsymbol{\alpha}$ and $\boldsymbol{\theta}$ with fixed β, i.e.,

$$\min_{\boldsymbol{\theta} \in \Theta} \min_{\boldsymbol{\alpha} \in \Omega(\boldsymbol{\theta})} \mathcal{J}(\boldsymbol{\alpha}). \tag{J1}$$

Example 1. *Consider the hidden variable SVM proposed in [4] and a ν version in [18]*

$$\min_{\boldsymbol{w}_m, b_m, \rho_m} \frac{1}{2} \sum_{m=1}^{M} ||\boldsymbol{w}_m||^2 - \sum_{m=1}^{M} \nu_m \rho_m + \frac{\gamma}{l} \sum_{i \in I^+} \max_m \{[\rho_m - y_i(\boldsymbol{w}_m \cdot \boldsymbol{x}_i + b_m)]_+\}$$
$$+ \frac{1-\gamma}{l} \sum_{i \in I^-} \min_m \{[\rho_m - y_i(\boldsymbol{w}_m \cdot \boldsymbol{x}_i + b_m)]_+\} \tag{2}$$

The first three terms are still convex, but the last term is not. Introducing hidden state variables $\boldsymbol{s}$ with the trick: $\min_m \{\xi_1, \cdots, \xi_M\} = \min_{\boldsymbol{s} \in \mathbb{S}^M} \sum_{m=1}^{M} s_m \xi_m$, we get $\min_m \{[\rho_m - y_i(\boldsymbol{w}_m \cdot \boldsymbol{x}_i + b_m)]_+\} = \min_{\boldsymbol{s}_i \in \mathbb{S}^M} \sum_{m=1}^{M} s_{im}[\rho_m - y_i(\boldsymbol{w}_m \cdot \boldsymbol{x}_i + b_m)]_+$ where $\mathbb{S}^M$ is the simplex in $\mathbb{R}^M$. Together with the bi-convexity to justify the exchange of minimization orders, the original learning problem is transformed into jointly minimizing

$$\min_{\substack{\boldsymbol{s}_i \in \mathbb{S}^M, \\ i \in I^-}} \min_{\substack{\boldsymbol{w}_m, b_m, \\ \rho_m \geq 0}} \frac{1}{2} \sum_{m=1}^{M} ||\boldsymbol{w}_m||^2 - \sum_{m=1}^{M} \nu_m \rho_m + \frac{\gamma}{l} \sum_{i \in I^+} \max_m \{[\rho_m - y_i(\boldsymbol{w}_m \cdot \boldsymbol{x}_i + b_m)]_+\}$$
$$+ \frac{1-\gamma}{l} \sum_{i \in I^-} \sum_{m=1}^{M} s_{im}[\rho_m - y_i(\boldsymbol{w}_m \cdot \boldsymbol{x}_i + b_m)]_+ \tag{3}$$

By replacing the inner minimization with its Lagrangian dual, we obtain (3) is equivalent to

$$\min_{\boldsymbol{s}_i \in \mathbb{S}^M, \, i \in I^-} \mathcal{J}_d(\boldsymbol{s}) \quad \text{where}$$

$$\mathcal{J}_d(\boldsymbol{s}) = \begin{cases} \min_{\boldsymbol{\alpha}} \frac{1}{2} \sum_{m=1}^{M} \sum_{i,j=1}^{l} \alpha_{im} y_i \kappa_m(\boldsymbol{x}_i, \boldsymbol{x}_j) y_j \alpha_{jm} \\ \text{subject to} \begin{cases} \alpha_{im} \geq 0 \; \forall i, \forall m; \quad \alpha_{im} \leq \frac{1-\gamma}{l} s_{im} \; \forall i \in I^-, \forall m \\ \sum_{m=1}^{M} \alpha_{im} \leq \frac{\gamma}{l} \; \forall i \in I^+; \quad \sum_{i=1}^{l} \alpha_{im} \geq \nu_m \; \forall m \\ \sum_{i=1}^{l} \alpha_{im} y_i = 0 \; \forall m \end{cases} \end{cases} \tag{4}$$

which in a compact matrix form with $\boldsymbol{\theta} = \boldsymbol{s}$ reduces to (J1). As the trick implies, the newly introduced variables can indeed be thought of as indicators for hidden states, and is coherent to the loss weighting parameters in the Primal formulation.

Many other examples, such as robust SVM, hidden structured SVM, polyhedron classifiers, etc., could be processed in a similar way. The inner optimization of the joint form can be viewed as minimizing a quadratic program (the dual) "parameterized" by outer minimization variables $\boldsymbol{\theta}$, i.e. we can regard J1 as

$$\min_{\boldsymbol{\theta} \in \Theta} \ \mathcal{J}(\boldsymbol{\alpha}^*(\boldsymbol{\theta})) \tag{J1'}$$

Seeing that, the key idea of PDGDP is to find the dependence of the optimum $\boldsymbol{\alpha}^*$ of the inner dual on its "parameters" $\boldsymbol{\theta}$, and then proceed to solve the joint optimization.

Now let's consider the model selection problem induced by some other variations of classic SVM. Again we provide a concrete example to facilitate the discussion.

Example 2. *The cost sensitive 2ν-SVM with Gaussian-ARD kernel has the primal*

$$\min_{\boldsymbol{w},b,\rho,\xi} \quad \frac{1}{2}\|\boldsymbol{w}\|_2^2 - \nu\rho + \frac{\gamma}{n}\sum_{i\in\mathcal{I}_+}\xi_i + \frac{1-\gamma}{n}\sum_{i\in\mathcal{I}_-}\xi_i \tag{5}$$

$$subject\ to \quad y_i(\kappa(\boldsymbol{w},\boldsymbol{x}_i)+b) \geq \rho - \xi_i, \quad \xi_i \geq 0 \quad \forall i; \quad \rho \geq 0.$$

with the ARD kernel $\kappa(\boldsymbol{x}_i,\boldsymbol{x}_j) = \exp\left\{-\sum_{k=1}^d \beta_k\left(x_i^k - x_j^k\right)^2\right\}$ the dual reads

$$\max_{\boldsymbol{\alpha}} \quad -\frac{1}{2}\sum_{i=1}^n\sum_{j=1}^n y_iy_j\alpha_i\alpha_j\kappa(\boldsymbol{x}_i,\boldsymbol{x}_j) \tag{6}$$

$$subject\ to \begin{cases} 0 \leq \alpha_i \leq \frac{\gamma}{n} \quad \forall i \in \mathcal{I}_+; \quad 0 \leq \alpha_i \leq \frac{1-\gamma}{n} \quad \forall i \in \mathcal{I}_-; \\ \sum_{i=1}^n \alpha_i y_i = 0; \quad \sum_{i=1}^n \alpha_i \geq \nu \end{cases}$$

The training objective is still convex, and the formulation has advantage of addressing imbalanced cost with penalty coordinator γ as well as reweighting each feature with β_k for better scaling. However this variation incorporates $d+2$ hyperparameters and produces a challenging model selection problem. Once more let's take the "parametric dual" viewpoint: the configuration of hyperparameters and the training data set $\mathcal{T}$ determine the solution $\boldsymbol{\alpha}^*$, with which one can construct a classifier for unseen data. Let $\boldsymbol{\Psi}(\cdot)$ be a function that estimates the generalization cost on a validation data set $\mathcal{Z}$. Given $\mathcal{T}$ and $\mathcal{Z}$, $\boldsymbol{\Psi}$ is a function of the classifier, which depends on hyperparameters $\boldsymbol{\phi}$. Thus the generalization cost has the form $\boldsymbol{\Psi}(\boldsymbol{\alpha}^*(\boldsymbol{\phi}))$, and the problem of model selection becomes

$$\boldsymbol{\phi}^* = \underset{\boldsymbol{\phi}\in\Phi}{\text{argmin}} \quad \boldsymbol{\Psi}_{\mathcal{T},\mathcal{Z}}(\boldsymbol{\alpha}^*(\boldsymbol{\phi})). \tag{J2}$$

We see that for both (J1) and (J2), one can first determine the dependence of optimality $\boldsymbol{\alpha}^*$ on parameters $\boldsymbol{\theta}$ or $\boldsymbol{\phi}$, and then substitute it into the corresponding outer problem for minimization. This constitutes the main idea of PDGDP. The major difficulty, however, is to characterize the dependence as explicit functions. In the subsequent section, we overcome this difficulty by introducing new techniques for parametric optimization.

3   Deriving the Explicit Dependence

In the terminology of operational research and optimization, a problem that depends on multiple parameters is referred to as parametric program, and the task of analyzing the dependence of optimal solution on related parameters is called parametric programming (PP) or Sensitivity Analysis (SA). This section is devoted to solving PP for the Dual with θ and β as parameters.

Note that the Dual is in fact a quadratic PP (QPP). Previously QPP has been addressed in optimization and control community [10], and its special cases have been used in our field for computing regularization path [7]. Nonetheless, for the purpose of this work existing result on QPP is insufficient, because in previous research (1) usually a single parameter is considered, simultaneous variations of multiple parameters, especially those for kernels are not allowed and (2) the so called Linear Independence Constraint Qualification (LICQ) condition [10] is required for the existence of QPP solution, which in our case cannot be satisfied due to the presence of the orthogonal hyperplane property (OHP) constraint $\alpha^T y = 0$. In fact, in the jargon of PP or SA, the problem at hand corresponds to a degenerate case for which existing solution is still lacking. In the subsequent part, by exploiting a sample partition property, we show that the explicit form $\alpha^*(\phi)$ can be obtained under mild conditions. Before we do, it is useful to define some terms.

Definition 1 (Active Constraint). *Assume that a solution of (Dual) has been obtained as $\alpha^*(\phi)$. Then the ith row of the constraint is said to be active at ϕ, if $C_i^\alpha \alpha^*(\phi) = C_i^\theta \theta + C_i^0$, and inactive if $C_i^\alpha \alpha^*(\phi) < C_i^\theta \theta + C_i^0$. The index set of all active inequality constraints i is denoted by $\mathcal{A}$, and all inactive inequality constraints by $\mathcal{A}^C$. We denote $C_\mathcal{A}^\alpha$ as the row selection of matrix C^α, i.e., $C_\mathcal{A}^\alpha$ only contains rows whose index is in $\mathcal{A}$. $C_{\mathcal{A}^C}^\alpha$ is similarly defined.*

Definition 2 (Non-degeneracy by Sample Partition). *We say that a solution of a general large margin learning is Non-degenerate if the set of unbounded support vectors $\mathcal{S}_{ub}$ contains at least one $i \in \mathcal{I}_+$, and at least one $i' \in \mathcal{I}_-$, i.e. both $\mathcal{S}_{ub}^+$ and $\mathcal{S}_{ub}^-$ are non-empty.*

With active set $\mathcal{A}$, we define a matrix that is important in deriving the main result.

$$P_\beta \triangleq C_\mathcal{A}^\alpha \left(\frac{Q_\beta^{-1} y y^T Q_\beta^{-1}}{y^T Q_\beta^{-1} y} - Q_\beta^{-1} \right) C_\mathcal{A}^{\alpha T}. \tag{7}$$

The matrix is symmetric, but in general it is not invertible. A simple example is the case in which training samples only come from one class. The invertibility issue is one of the major difficulties in solving the parametric optimization problem. With a series of lemmas given in the full version of this paper, we prove the following result.

Lemma 1. *If the solution α^* of (Dual) is non-degenerate, then the matrix P_β is strictly symmetric negative definite, hence invertible.*

Consider the mildness of the non-degeneracy requirement: since the unbounded support vectors are essentially the sample points that lie on the decision boundaries and construct the classifiers (including the interception). In order to have meaningful classification in practice this condition is a necessity and is easily satisfied even with just a few training samples. With Lemma 1, we are now able to derive an explicit form for the optimal solution $\boldsymbol{\alpha}^*$ of the Dual as a function of the hyperparameters. We present the main result of this section.

Theorem 1. *Assume that the solution of a general large margin learning is non-degenerate and induces a set of active and inactive constraints $\mathcal{A}$ and $\mathcal{A}^C$, respectively. Then in the **critical region** defined by*

$$\begin{cases} P_\beta^{-1}(C_\mathcal{A}^\theta \boldsymbol{\theta} + C_\mathcal{A}^0) \geq 0 \\ C_{\mathcal{A}^C}^\theta \boldsymbol{\theta} + C_{\mathcal{A}^C}^0 - C_{\mathcal{A}^C}^\alpha T_\beta P_\beta^{-1}(C_\mathcal{A}^\theta \boldsymbol{\theta} + C_\mathcal{A}^0) \geq 0 \end{cases} \tag{8}$$

*the optimal solution $\boldsymbol{\alpha}^*$ of Dual admits a **closed form***

$$\boldsymbol{\alpha}^*(\phi) = T_\beta P_\beta^{-1}(C_\mathcal{A}^\theta \boldsymbol{\theta} + C_\mathcal{A}^0) \tag{9}$$

where $T_\beta \triangleq \left(\dfrac{Q_\beta^{-1} \boldsymbol{y}\boldsymbol{y}^T Q_\beta^{-1}}{\boldsymbol{y}^T Q_\beta^{-1} \boldsymbol{y}} - Q_\beta^{-1} \right) C_\mathcal{A}^{\alpha T}$.

In essence the theorem indicates that each time the inner optimization Dual is solved, full information (closed form solution) in a well-defined neighborhood (critical region) can be retrieved as a function of associated parameters. Besides, the derivation of Theorem 1 only depends on the structure of $C_\mathcal{A}^\alpha, C_{\mathcal{A}^C}^\alpha$ and the PD property of Q_β. Thus, the solution includes many variations of large margin learning that have different forms of $\boldsymbol{\theta}$ and $\kappa_\beta(\cdot)$ as special cases.

4 PDGDP for Training: Global Optimality Guarantee

Recall that the training problem discussed in Sect. 2 is $\min_{\boldsymbol{\theta} \in \Theta} J(\boldsymbol{\alpha}^*(\boldsymbol{\theta}))$ with β fixed. Now we have found the explicit form of $\boldsymbol{\alpha}^*(\boldsymbol{\theta})$, in this section we also characterize the overall geometric structure of the optimality, and show that the optimal objective of the dual, $J(\boldsymbol{\alpha}^*(\boldsymbol{\theta}))$, is globally convex in $\boldsymbol{\theta}$, which guarantees that the PDGDP based training algorithm converge to global optimum, in contrast to existing training methods that only converge to local optimums.

Theorem 2. *Still assuming non-degeneracy, then*

1. *There are finite number of polyhedron critical regions $CR_1, \cdots, CR_{N_r}$ which constitute a **partition** of the feasible set of $\boldsymbol{\theta}$, i.e. each feasible $\boldsymbol{\theta}$ belongs to one and only one critical region.*
2. *The optimal objective $J(\boldsymbol{\alpha}^*(\boldsymbol{\theta}))$ is a globally **convex** Piece-wise Quadratic (PWQ) function of $\boldsymbol{\theta}$, and is almost everywhere differentiable.*

Algorithm 1. PDGDP for Training

Input data $\mathcal{T}$ kernels $\mathcal{K}$
$\boldsymbol{\theta}^1 \leftarrow \text{init}(\mathcal{T})$; $\{\mathcal{Q}, \mathcal{C}\} \leftarrow \text{calMatrix}(\mathcal{S}, \mathcal{K})$
$CR_{explored} \leftarrow \emptyset$, $n \leftarrow itermax$
$\tau \leftarrow stepsize$, $t \leftarrow 1$
while Improved & $t \leq n$ **do**
  if $\boldsymbol{\theta}^t \in \mathcal{R}$ **then**
    $\{\boldsymbol{\alpha}^*, \mathcal{A}\} \leftarrow \text{solDual}(\mathcal{Q}, \mathcal{C})$
    $CR_{new} \leftarrow \text{getRegion}(\boldsymbol{\alpha}^*, \mathcal{A})$
    $CR_{explored} \leftarrow CR_{explored} \cup CR_{new}$
  else
    $\boldsymbol{\alpha}^* \leftarrow \text{thetaInsideR}(\boldsymbol{\theta}^t, \mathcal{A})$
  end if
  $g \leftarrow \text{getGrad}(\boldsymbol{\alpha}^*, \mathcal{A})$
  $\boldsymbol{\theta}^{t+1} \leftarrow \text{Proj}(\boldsymbol{\theta}^t - \tau g)$; $\quad t \leftarrow t+1$
end while
return $\boldsymbol{\alpha}^*, \boldsymbol{\theta}$

The globally convex PWQ structure of $\mathcal{J}(\boldsymbol{\alpha}^*(\boldsymbol{\theta}))$ revealed by part 2 is critical: together with the local explicit solution, the training problem is reduced to minimizing a nonsmooth but convex function in the space $\boldsymbol{\theta} \in \Theta$. Projected subgradient descent method is an immediate choice to search for global optimum. We summarize the PDGDP based training procedure in Algorithm 1. At each step with $\boldsymbol{\alpha}^*$ and $\mathcal{A}$, one can directly compute the critical region boundaries (8), and the gradient:

$$\nabla_{\boldsymbol{\theta}} \mathcal{J}(\boldsymbol{\alpha}^*(\boldsymbol{\theta})) = 2(\boldsymbol{T}_\beta \boldsymbol{P}_\beta^{-1} \boldsymbol{C}_\mathcal{A}^\theta)^T \boldsymbol{Q}_\beta \boldsymbol{\alpha}^*(\boldsymbol{\theta})$$

Since $\boldsymbol{\theta}$ should satisfy simplex constraints, a projection on to that space is needed. By Theorem 1, if $\boldsymbol{\theta}^t$ is in the critical regions that have been explored before, all information could be retrieved in an explicit form and there is no need to solve the inner problem again. However, when the variable goes to a new critical region, a QP solver for Dual has to be invoked for optimal solution $\boldsymbol{\alpha}^*$ and corresponding constraint partition. The following global optimal convergence result is a consequence of Theorem 2:

Theorem 3. *Convergence Guarantee. Let* $\sup_{\boldsymbol{\theta}} \|\boldsymbol{\theta}^1 - \boldsymbol{\theta}\| = B$, *and the Lipschitz constant of* $\mathcal{J}(\boldsymbol{\alpha}^*(\boldsymbol{\theta}))$ *be* G, *then Algorithm I with iteration* n *and optimal step size* $\tau_i = B/G\sqrt{n}$ $\forall i$ *converges to global optimum within* $O(1/\sqrt{n})$. *To be specific, let* $\mathcal{O}^*$ *be the global optimum of the learning objective of J1,* $\mathcal{J}(\boldsymbol{\alpha}^*(\boldsymbol{\theta}_{best}^n)) \triangleq \min\{\mathcal{J}(\boldsymbol{\alpha}^*(\boldsymbol{\theta}^1)), \cdots, \mathcal{J}(\boldsymbol{\alpha}^*(\boldsymbol{\theta}^n))\}$ *then*

$$\mathcal{J}(\boldsymbol{\alpha}^*(\boldsymbol{\theta}_{best}^n)) - \mathcal{O}^* \leq \frac{BG}{\sqrt{n}}.$$

Hence in order to get $\mathcal{J}(\boldsymbol{\alpha}^*(\boldsymbol{\theta}_{best}^n)) - \mathcal{O}^* \leq \epsilon$, the algorithm needs $O(1/\epsilon^2)$ iterations. B is bounded because the feasible set of $\boldsymbol{\theta}$ is simplex. Also as $\mathcal{J}(\boldsymbol{\alpha}^*(\boldsymbol{\theta}))$ is globally convex(hence continuous) and locally quadratic(hence has bounded gradient), G must be bounded as well. The constant step size is optimal in the sense that it minimizes upper bound of the gap. Other choices, such as a diminishing step size could also be used if faster convergence is a concern. Although the inner QP solver is expensive, variety of existing methods for classic SVM dual can be reused for acceleration. In the functions getRegion(), thetaInsiderRegion(), and getGrad(), the computational overhead is mostly matrix inversions. Fortunately, from the proof of the Theorem 1, the involved matrices are either symmetric positive definite or symmetric negative definite hence decomposition methods can be adopted for efficient inversion.

5   PDGDP Based Model Selection Algorithm

In this section, we focus on the second issue of large margin learning variations, i.e. the problem of hyperparameter (model) selection. Different from the training problem discussed in last section, the goal of model selection is to minimize the generalization cost on unseen data $z = (x, y)$. As the true distribution of the data is never known, generalization cost is usually estimated with various validation techniques and error bounds. Let the estimation function be Ψ, recall that the model selection problem reduces to solving $\phi^* = \operatorname{argmin}_{\phi \in \Phi} \Psi_{T,Z}(\alpha^*(\phi))$. The application of PDGDP framework for model selection is now straightforward. To begin with, we first deduce the classifier as an explicit function of hyperparameters, given that α^* has been provided by Theorem 1. Notation wise, we define $d_j = [y_1 \kappa_\beta(x_1, x_j), \ldots, y_n \kappa_\beta(x_n, x_j)]$ and the hyperplane $h(x_j)$

Algorithm 2. PDGDP for Model Selection

$t \leftarrow 0, \ r \in [0.2, 0.8]$
$\phi^0 \leftarrow \nu^0, s^0, \beta^0$
while $||\nabla_\phi \tilde\Psi_{T,Z}(\phi^t)|| \geq \epsilon$ **do**
    $\eta^t \leftarrow \text{initStep}(t)$
    $T, Z \leftarrow \text{randSample}(\mathcal{D})$
    $\alpha^* \leftarrow \text{solDual}(T, \phi^t)$
    $\{\nabla_\phi \tilde\Psi_{T,Z}(\phi^t), \mathcal{R}^t\} \leftarrow \text{getGrad}(Z, \alpha^*)$
    $\tilde\phi^{t+1} = \phi^t - \eta^t \nabla_\phi \tilde\Psi_{T,Z}(\phi^t)$
    while $\tilde\Psi(\tilde\phi^{t+1}) - \tilde\Psi(\tilde\phi^t) > \epsilon'$ **do**
        $\eta^t = r\eta^t$
    end while
    $\phi^{t+1} = \phi^t - \eta^t \nabla_{\phi^t} \tilde\Psi_{T,Z}(\phi^t)$
    $t \leftarrow t + 1$
end while
return ϕ^t

$$h(x_j) = \sum_{i=1}^{n} \alpha_i^* y_i \kappa_\beta(x_i, x_j) + b^* = (d_j^T - d_U^T) T_\beta P_\beta^{-1} (C_A^\theta \theta + C_A^0), \quad (10)$$

then the classifier $f(x_j)$ is merely the sign of $h(x_j)$. Now we have the explicit form of the classifier as a function of the hyperparameters. Consider estimating generalization cost with the empirical cost on the validation data set, i.e., $\hat\Psi_{T,Z}(\phi) = \frac{1}{m} \sum_{(x_j, y_j) \in Z} c_j[1 - y_j \operatorname{sign}(h(x_j))]$. Plugging this in the classifier (10), an explicit form of $\hat\Psi_{T,Z}$ can be obtained as a function of ϕ. To cope with the discontinuity of the sign function, we use a sigmoid function $\tanh(\tau x)$ for approximation. We denote the smoothed empirical generalization cost as $\tilde\Psi_{T,Z}(\phi)$. The following property can be obtained from Theorem 2.

Proposition 1. $\tilde\Psi_{T,Z}(\phi)$ *is almost everywhere differentiable, and the gradient* $\nabla \tilde\Psi_{T,Z}(\phi)$ *is Lipschitz continuous in each critical region.*

Although the objective function to be minimized is non-convex in hyperparameters and it's hard to establish any theoretical guarantee, a stochastic gradient descent scheme is adopted and we justify its effectiveness with empirical studies. The PDGDP based model selection is summarized in Algorithm 2. The involved functions and their computational overhead are very similar to those of Algorithm 1. We also incorporate backtrack line search for a better stepsize by using $r \in [0.2, 0.8]$ to decrease overshot updates. Note that with the form of $\hat{\Psi}_{T,z}$ and the smoothness result in Proposition 1, the algorithm at least converges to local minimums. Similar as the PDGDP for training, the matrix inversions in this algorithm can be efficiently computed with decomposition techniques.

6   Experimental Results

Both of the PDGDP based algorithms were tested for typical large margin learning variations with multiple public datasets. Code, more experimental results and full version with all proofs are available per request.

6.1   PDGDP Training Results

Firstly we test the performance of the PDGDP based training method (Algorithm 1) and compare it with other states-of-the-art methods, including alternating optimization [3], stochastic gradient descent [18], and concave-convex procedure [15]. The public UCI yeast data set is used in this experiment and we focus on the training of latent SVM described in Example 1. For algorithmic comparison purpose we restrict to all linear kernels. Hyperparameters are set with $M = 8$, $\nu_m = 0.01$ and $\gamma = |I^-|/l$. For the initialization of Algorithm 1, a simple K-mean is applied and θ^1 is assigned according to cluster labels. The final value of the objective function, the corresponding testing accuracy, the number of iterations and the time consumed are shown in Table 1. We observer that PDGDP based training achieves a much better objective value, about 45 % lower than the runner-up CCCP. The improved training also leads to 4.62 % increase in testing accuracy. As a global optimization algorithm, it is expected that PDGDP consumes more time than algorithms that only converge to local minimums, as is shown in the last row of the table.

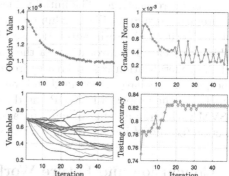

Table 1. Methods comparison

Method	PDGDP	AO	SGD	CCCP
Objective (10^{-5})	**1.09**	3.85	4.27	2.01
Testing Accuracy (%)	**82.77**	74.78	74.27	78.15
# of Iterations	48	12	200	27
Elapsed Time (s)	336	129	**17**	171

Fig. 1. Algorithm 1 Convergence Results

Figure 1 shows the corresponding objective value, gradient norm, latent variable value, and testing accuracy of PDGDP based training in each iteration. Note that for clear presentation only a subset of latent variables with the same

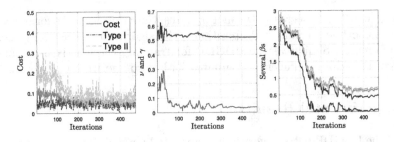

Fig. 2. Convergence result for S500, $c_1 = 1, c_2 = 1$.

initial value is shown. We see that the iterative result is similar to a general subgradient descent for non-smooth convex functions: The learning objective is in general decreasing, with some fluctuations due to non-smoothness between two critical regions. The evolution of gradient norm and latent variables also reflects this character of insider region "exploitation" and beyond region "exploration". The testing accuracy is shown at the bottom right of Fig. 1, which increases correspondingly with the decrease of learning objective.

6.2   PDGDP Model Selection Results

Next we test the proposed model selection method (Algorithm 2) for choosing the hyperparameters of cost sensitive SVM with Gaussian ARD kernel in Example 2. A semi-conductor sensing data set (S500) with 2670 samples and 500 features is used in this experiment. Thus the total number of hyperparameters to choose reaches 502. Figures 2 demonstrate the evolution of generalization cost, values of ν, γ and kernel parameter β in each iteration. The generalization costs were computed with $c_1 = 1, c_2 = 1$, where c_1 and c_2 are cost coefficient for type I and type II error, respectively. The initial values were $\gamma^0 = 0.5, \nu^0 = 0.15, \beta_k^0 = 1 \ \forall k$. The iterative results in Fig. 2 exhibit fluctuations due to the random sampling in the algorithm. More importantly, we observe that the algorithm converges within 350 steps (186 invocations of the quadratic solver solDual() is engaged), and the generalization cost decreases from 0.12 to 0.049. Considering that 502 hyperparameters have to be tuned and the fact that cross validation technique is intractable, the proposed PDGDP based method indeed enables the application of SVM with large number of hyperparameters by providing an efficient model selection technique. For more comparative studies please refer to the full version.

7   Conclusion and Future Work

We highlight our contributions as follows: (1) To the best of our knowledge, the PDGDP based training is the first method that is able to converge to global optimum for this class of non-convex learning problems without resorting to combinatorial search. Both theory and experiment show that it constitutes a promising substitute for existing methods such as AO, CCCP, or SDG. (2) The PDGDP based model selection provides an efficient and unified way to choose high dimensional hyperparameters of different natures, while previous research is limited in terms of both type and number of hyperparameters.

To further improve the efficiency of both algorithms, matrix completion techniques can be applied to approximate large scale problems. Another possible acceleration is to approximately extend critical regions to reduce the invocations of the quadratic solver. Moreover, high-order gradient-based methods with enhanced descent direction are also worth exploring for better convergence.

Acknowledgment. This research is funded by the Republic of Singapore's National Research Foundation through a grant to the Berkeley Education Alliance for Research in Singapore (BEARS) for the Singapore-Berkeley Building Efficiency and Sustainability in the Tropics (SinBerBEST) Program. BEARS has been established by the University of California, Berkeley as a center for intellectual excellence in research and education in Singapore.

References

1. Cortes, C., Mohri, M., Rostamizadeh, A.: Two-stage learning kernel methods. In: ICML(2010)
2. Davenport, M.A.: The 2nu-svm: A cost-sensitive extension of the nu-svm. Technical report, DTIC Document (2005)
3. Dundar, M.M., Wolf, M., Lakare, S., Salganicoff, M., Raykar, V.C.: Polyhedral classifier for target detection: a case study: colorectal cancer. In: ICML (2008)
4. Felzenszwalb, P.F., Girshick, R.B., McAllester, D., Ramanan, D.: Object detection with discriminatively trained part-based models. IEEE Trans. PAMI **32**(9), 1627–1645 (2010)
5. Floudas, C.A.: Nonlinear and Mixed-integer Optimization: Fundamentals and Applications. Oxford University Press, New York (1995)
6. Cawley, G., Talbot, N.: Preventing over-fitting during model selection via bayesian regularisation of the hyper-parameters. J. Mach. Learn. Res. **8**, 841–861 (2007)
7. Giesen, J., Laue, S., Wieschollek, P.: Robust and efficient kernel hyperparameter paths with guarantees. In: Proceedings of the 31st International Conference on Machine Learning (ICML-14), pp. 1296–1304 (2014)
8. Gönen, M., Alpaydın, E.: Multiple kernel learning algorithms. J. Mach. Learn. Res. **12**, 2211–2268 (2011)
9. Masnadi-Shirazi, H., Vasconcelos, N., Iranmehr, A.: Cost-sensitive support vector machines. arXiv preprint.(2012). arxiv:1212.0975
10. Tøndel, P., Johansen, T.A., Bemporad, A.: An algorithm for multi-parametric quadratic programming and explicit MPC solutions. Automatica (2003)
11. Wang, T., Huang, H., Tian, S., Xu, J.: Feature selection for svm via optimization of kernel polarization with gaussian ard kernels. Expert Syst. Appl. **37**(9), 6663–6668 (2010)
12. Wu, Y., Liu, Y.: Robust truncated hinge loss support vector machines. J. Am. Stat. Assoc. **102**(479), 974–983 (2007)
13. Xu, L., Crammer, K., Schuurmans, D.: Robust support vector machine training via convex outlier ablation. In: AAAI, vol. 6, pp. 536–542 (2006)

14. Yong Liu, S.J., Liao, S.: Efficient approximation of cross-validation for kernel methods using bouligand influence function. In: ICML (2014)
15. Yu, C.N.J., Joachims, T.: Learning structural svms with latent variables. In: ICML (2009)
16. Yuille, A.L., Rangarajan, A., Yuille, A.: The concave-convex procedure (cccp). Adv. Neural Inf. Process. Syst. **2**, 1033–1040 (2002)
17. Zhou, Y., Hu, N., Spanos, C.J.: Veto-consensus multiple kernel learning. In: AAAI (2016)
18. Zhou, Y., Jin, B., Spanos, C.J.: Learning convex piecewise linear machine for data-driven optimal control. In: ICMLA (2015)

Locally Weighted Ensemble Learning for Regression

Man Yu, Zongxia Xie, Hong Shi, and Qinghua Hu[✉]

School of Computer Software,
Tianjin University, Tianjin, China
{alexyu,huqinghua}@tju.edu.cn

Abstract. The goal of ensemble regression is to combine a set of regressors in order to improve the predictive accuracy. The key to a successful ensemble regression is to complementally generate base models and elaborately combine their outputs. Traditionally, the weighted average of the outputs is treated as the final prediction. This means each base model plays a constant role in the whole data space. In fact, we know the predictive accuracy of each base model varies across different data spaces. In this paper, we develop a dynamic weighted ensemble method from locality which is called Locally Weighted Ensemble. The weight of each base model varies with sample, which is realized by introducing soft-max function into the objective function. Besides, regularization is also included to make the objective function well-posed. The proposed method is evaluated on several UCI datasets. Compared with single models and other ensemble models, our proposed achieves better performance. From the experiments, we also find that the convergence of Locally Weighted Ensemble is fast.

Keywords: Regression · Locally weighted ensemble · Local learning · Regularization

1 Introduction

Different from single model prediction, ensemble learning trains a set of base models and combines their outputs for producing accurate prediction. And it is robust to data noise [11]. A large number of researches have been proposed for designing effective ensemble system in recent years [7,10,15–17].

In the process of ensemble learning, there exist two tasks: (1) Generating base models; (2) Combining base models to make an accurate prediction [15]. For the first task, base models are built by the same learning algorithm or different learning algorithms. After generating base models, an appropriate integration strategy is required to make an accurate prediction. This makes generalization error of ensemble regression lower [9]. As to integration strategy, lots of algorithms have been proposed [10,16]. These algorithms can be roughly divided into constant weighted methods and dynamic weighted methods [14].

© Springer International Publishing Switzerland 2016
J. Bailey et al. (Eds.): PAKDD 2016, Part I, LNAI 9651, pp. 65–76, 2016.
DOI: 10.1007/978-3-319-31753-3_6

As to constant weighted methods, the coefficients of base models are constant in the whole data space. In 1993, Perrone and Cooper proposed the Basic Ensemble Method. This method calculates the mean of predictions of all base models, and achieves better results than single models [16]. In 2000, Seeger introduced Bayesian Model selection into ensemble strategy and discussed the effectiveness in the context of decision theory [17]. In 2014, Yin *et al.* designed an ensemble measure which focused on both sparsity and diversity of base models [21].

The above ensemble methods learn a unified combination function in the whole data space. However, data distribution is variable. The regression function varies from sample to sample. To deal with this problem, dynamic weighted integration is developed. Mendes-Moreira divided the dynamic weighted methods into selection of similar data and change of integration functions [13]. As to the former, Wooks *et al.* introduced K-nearest neighbor method (KNN) which finds the similar sample in Euclidean Space in 1996 [20]. In addition, the similar data is obtained by other methods, like Discriminant Adaptive Nearest Neighbor [4]. Since 1991, Michal I. Jordan proposed a series of dynamic ensemble methods which are realized by changing integration function. In 1991, mixture of local experts was proposed [7]. Each expert learned to handle a subset of the complete set of training samples. In 1994, the team of Jordan proposed a hierarchical mixture of experts where integration function is different from other integration functions [8]. Li and Hu used classification confidence value to find a subset of base models on each test sample. The final prediction is obtained by weighted voting on the subset [10]. Both the selection of similar data methods and change of integration function methods have achieved good performance. The former is sensitive to dataset. The size of similar data is difficult to obtain. To deal with the problem of multiple objective functions, Bottou and Vapnik proposed local learning algorithm in 1992 [3]. They proved the effectiveness of local learning in solving complex and dynamic distribution problem in theory.

Local learning algorithm contains of a series of methods to obtain the local sample of each test sample, such as KNN, soft-max function and Radial Basic Function network. The objective functions are different from test samples which are built with the local sample [19]. In this paper, we introduce local learning into ensemble regression to deal with the problem of multiple objective function, which is called Locally Weighted Ensemble algorithm (LWE).

LWE is a data-driven method. The weights of base models in different regions are variation continuously which is assigned by soft-max function. However, the introduction of soft-max function makes the objective function ill-posed [7]. [7,8] do not consider this problem. In this paper, we apply L_{21}-regularization, L_F-regularization and Laplacian-regularization to deal with this problem. The experimental assessment is carried on UCI datasets. The experiments show that our method outperforms other methods, such as single methods, constant weighted ensemble methods and dynamic ensemble methods.

The remainder of the paper is organized as follows: Sect. 2 describes two kinds of ensemble integration methods. Section 3 presents the proposed locally weighted ensemble method in detail. Section 4 presents the experiments obtained on UCI datasets. Finally, the conclusions of this paper are drawn in Sect. 5.

2 Related Work

Ensemble integration learns how to combine base models to make a final prediction. During the phase of integration, the discriminative function is

$$f(x) = \sum_{i=1}^{M} w_i \cdot f_i(x), \tag{1}$$

where M is the number of base models, w_i is the weight of the ith base model and $f_i(x)$ is the prediction of ith base model.

2.1 Constant Weighted Ensemble

Yin investigates ensemble method which focuses on both sparsity and diversity [21]. The objective function is

$$
\begin{array}{ll}
f(x) = \sum_{i=1}^{M} \frac{(w^T x_i - y_i)^2}{2} + \alpha \|w\|_1 - \beta \sum_{i=1}^{M} (w^T(x_i)^2 - (w^T x_i)^2) \\
s.t. \qquad w \geq 0
\end{array}
\tag{2}
$$

where α is the control parameter for sparse regularization and β is the parameter for diverse regularization. Sparsity is exposed by the l_1-norm sparsity term. Diversity is expressed by $\sum_{i=1}^{M}(w^T(x_i)^2 - (w^T x_i)^2)$. This is derived from the error-ambiguity decomposition for regression.

2.2 Dynamic Weighted Ensemble

Tsymbal et al. present a dynamic ensemble method through local accuracy [18]. The weight w_i is calculated by

$$w_i(x) = \sum_{j=1}^{k} (\delta(x, x_j)) mr_i(x_j) / \sum_{j=1}^{k} \delta(x, x_j), \tag{3}$$

where k is the size of the neighborhood, $\delta(x, x_j)$ is a distance-based relevance coefficient and $mr_i(x_j)$ is the margin of model i on jth nearest neighbor of x. The margin is defined as

$$mr_i(x) = \begin{cases} 1, & if\, f_i(x) = y(x), \\ -1, & otherwise. \end{cases} \tag{4}$$

The distance-based weight coefficient reflects similarity between two instance

$$\delta(x, x_j) = 1/d_{deom}(x, x_j), \tag{5}$$

where d_{deom} is the heterogeneous Euclidean/overlap metric [6]. The effectiveness of the proposed method is significantly better than constant weighted method. However, the number of nearest neighbors is difficult to determine. In this paper, we introduce a new dynamic integration method.

3   Locally Weighted Ensemble Learning

Figure 1 illustrates its general framework. There are two key steps in this framework. Firstly, a set of base regressors are built by different regression algorithms. In this paper, We think some of the base regressors are not useful for final prediction. The base regressors which Mean Absolute Error is the smallest among the remaining base regressors is added to ensemble learning prior, and the Mean Absolute Error is described in Sect. 4. When the difference of MAE between two iterations is smaller than 0.001, the process of adding base models is stopped. The added regressors are treated as the base models of LWE. Secondly, a weighted function is trained to combine the outputs of base models. Each sample on different base models is assigned with different weights which are realized by soft-max model. Soft-max model is presented as follow

$$w_i(x) = \frac{exp(v_i \cdot x + v_{i0})}{\sum_{j=1}^{M} exp(v_j \cdot x + v_{j0})}, \tag{6}$$

where $x = x_1, x_2, x_3, \cdots, x_n$ is the representation of the input data and M is the number of base models. The parameters in $\{v_{i0}, v_i\}, i = 1, 2, \cdots, M$ should be learned, where i is the ith base model. In the following subsections, we analyze the objective function and the optimizational strategy of our proposed method.

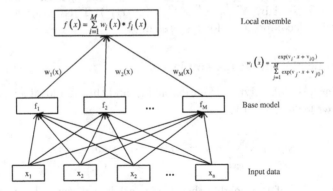

Fig. 1. Architecture of locally weighted ensemble learning:the weight of each base model is not constant, which varies with the input data. We introduce soft-max function $w_j(x)$ to realize this.

3.1   Objective Function of Locally Weighted Ensemble Learning

The objective function of our proposed method is

$$f(x) = \sum_{i=1}^{M} w_i(x) \cdot f_i(x). \tag{7}$$

Regularization is included to make the objective function well-posed [6]. Let $J(w)$ denote the objective function

$$J(w) = \min_{s.t.} \frac{\lambda R_{regular}}{2} + (f(x) - y)^2 \qquad (8)$$
$$s.t. \qquad w(x) \geq 0,$$

where $R_{regular}$ is the regularization term, λ is the regularization factor and y is the real value. The methods of regularization are variable, such as L_1-Norm, L_{21}-Norm, L_2-Norm, L_F-Norm and manifold regularization. In this paper, we introduce.L_{21}-Norm, L_F-Norm and Laplace regularization to our objective function.

A. $L_{2,1}$-Norm Regularization (LWE-$L_{2,1}$). During integration, [21] argues that ensemble some of base models may be better than ensemble all for prediction. This leads to sparse ensemble learning. The used sparse methods are L_0-Norm, L_1-Norm and L_{21}-Norm. For convenient calculation, we introduce $L_{2,1}$-Norm to regularization. Each base model has its own weight value, zero or nonzero. $L_{2,1}$-Norm can select the effective base models [6]. $L_{2,1}$-Norm is written as

$$R_{regular} = ||w(x)||_{21} = \sum_i \sqrt{\sum_j w_{ij}^2} \qquad (9)$$

where w_{ij} is the weight of the i sample and the d-dimensional feature.

B. L_F-Norm Regularization (LWE-L_F). In the matrix norm, $Frobenius$-Norm is a convex function. The introduction of $Frobenius$-Norm makes the objective function strongly convex. If we solve the problem by gradient decent method, the convergence is stable and fast [12]. $Frobenius$-Norm is written as

$$R_{regular} = ||w(x)||_F^2 = \sum_i \sum_j w_{ij}^2. \qquad (10)$$

C. Laplace-Norm Regularization (LWE-L). The solution of an ill-posed problem can be approximated by variational principles, which contains the prior smoothness information [5]. The manifold regularization utilizes the manifold to replace the smoothness, where the manifold is determined by Laplacian. Laplace-Norm keeps the information of spacial structure.

$$R_{regular} = ||w(x)||_L = f(x)Lf(x) \qquad (11)$$

where $L = D - Dis, D_{ii} = \sum_{j=1}^n Dis_{ij}$ and n is the size of samples. Dis_{ij} is the distance between sample i and sample j.

3.2 Optimization of Locally Weighted Ensemble Learning

The base models $f_1, \cdots, f_i, \cdots, f_M$ are trained in advance. During the phase of optimization, these are constant. Therefore, the objective function Eq. 8 is convex. We apply Gradient Descent method to solve it. The introduction of soft-max function makes the constraint condition, $w(x|v) \geq 0$, meet. The gradients of soft-max function are

$$\frac{\partial w_i(x)}{\partial v_m} = w_i(x) \cdot (\delta_m^i - w_m(x)) \cdot x; \tag{12}$$

$$\frac{\partial w_i(x)}{\partial v_{m0}} = w_i(x) \cdot (\delta_m^i - w_m(x)). \tag{13}$$

where w_i is the weight of i_{th} sample and

$$\delta_m^i = \begin{cases} 0, & if \quad m \neq i, \\ 1, & if \quad m = i. \end{cases} \tag{14}$$

The gradients of LWE-$L_{2,1}$, LWE-L_F and LWE-L are presented as follow. For further understanding, we give an example to describe our method.

A. $L_{2,1}$-Norm Regularization (LWE-$L_{2,1}$). The gradients of $L_{2,1}$-Norm regularization are written as

$$\begin{aligned}
\frac{\partial J(w)}{\partial v_m} &= \frac{\lambda \partial \|w(x)\|_{21}}{2\partial v_m} + \sum_{j=1}^n \sum_{i=1}^M 2(f(x_j) - y_j)f_i(x_j)\frac{\partial w_i(x_j)}{\partial v_m} \\
&= \frac{\lambda \partial tr(w^T Dw)}{2\partial v_m} + \sum_{j=1}^n \sum_{i=1}^M 2(f(x_j) - y_j)f_i(x_j)\frac{\partial w_i(x_j)}{\partial v_m} \\
&= \lambda 2Dw\frac{\partial w(x)}{2\partial v_m} + \sum_{j=1}^n \sum_{i=1}^M 2(f(x_j) - y_j)f_i(x_j)\frac{\partial w_i(x_j)}{\partial v_m} \\
&= \sum_{j=1}^n \sum_{i=1}^M (2(f(x_j) - y_j)f_i(x_j) + \lambda Dw)w_i(x_j)(\delta_m^i - w_m(x_j))x_j
\end{aligned} \tag{15}$$

$$\frac{\partial J(w)}{\partial v_{m0}} = \sum_{j=1}^n \sum_{i=1}^M (2(f(x_j) - y_j)f_i(x_j) + \lambda Dw)w_i(x_j)(\delta_m^i - w_m(x_j)) \tag{16}$$

where $D_{ii} = (2\|w_i\|_2)^{-1}$ and n is the size of training set. In each iteration, we get the optimization of w by fixing D. In next iteration, D is updated by the new w.

B. L_F-Norm Regularization (LWE-L_F). The same as $L_{2,1}$-Norm Regularization, the gradients of L_F-Norm regularization are written as

$$\frac{\partial J(w)}{\partial v_m} = \sum_{j=1}^n \sum_{i=1}^M (2(f(x_j) - y_j)f_i(x_j) + \lambda w_i(x_j))w_i(x_j)(\delta_m^i - w_m(x_j))x_j \tag{17}$$

$$\frac{\partial J(w)}{\partial v_{m0}} = \sum_{j=1}^n \sum_{i=1}^M (2(f(x_j) - y_j)f_i(x_j) + \lambda w_i(x_j))w_i(x_j)(\delta_m^i - w_m(x_j)). \tag{18}$$

C. Laplace-Norm Regularization (LWE-L). The gradients of Laplace-Norm regularization are written as

$$\frac{\partial J(w)}{\partial v_m} = \sum_{j=1}^n \sum_{i=1}^M (2(f(x_j) - y_j) + \lambda Lf)f_i(x_j)w_i(x_j)(\delta_m^i - w_m(x_j))x_j \tag{19}$$

$$\frac{\partial J(w)}{\partial v_{m0}} = \sum_{j=1}^n \sum_{i=1}^M (2(f(x_j) - y_j) + \lambda Lf)f_i(x_j)w_i(x_j)(\delta_m^i - w_m(x_j)). \tag{20}$$

D. Example. Assume that there are n samples, a attributes and M base models in training set. They consist in solving the following system

$$\begin{pmatrix} 1 & x_{11} & \cdots & x_{1\alpha} \\ & \vdots & \ddots & \vdots \\ 1 & x_{n1} & \cdots & x_{n\alpha} \end{pmatrix} \begin{pmatrix} v_{01} & \cdots & v_{0M} \\ \vdots & \ddots & \vdots \\ v_{\alpha 1} & \cdots & x_{\alpha M} \end{pmatrix} = \begin{pmatrix} w_{11} & \cdots & w_{1M} \\ \vdots & \ddots & \vdots \\ w_{n1} & \cdots & w_{nM} \end{pmatrix} \qquad (21)$$

The each row of right side of Eq. 21 is the probability of each base model. p_{1M} is the probability of the first sample on Mth base model. The sum of each row is 1. The final prediction is obtained as

$$\begin{pmatrix} w_{11} & \cdots & w_{1M} \\ \vdots & \ddots & \vdots \\ w_{n1} & \cdots & w_{nM} \end{pmatrix} \begin{pmatrix} f_1 \\ \vdots \\ f_M \end{pmatrix} = (f). \qquad (22)$$

3.3 Algorithm of Locally Weighted Ensemble Learning

In this paper, a dynamic ensemble method, which depends on the characteristic of data, is proposed. Different integration methods are applied for different regions. Here we describe the algorithm of LWE.

Local-Weighted Ensemble algorithm (LWE)

```
Input: Training data
Output: Final predictor
Begin
      ensemble generation step
           Build a set of base regressors with all training data;
      ensemble integration step
           Obtain the base model of learning model;
           Calculate descent direction(gradient of objective function);
           Calculate step size with linear search method;
           Update objective function;
           Until convergence
           Ensemble base models
end
```

After analysis, the time complexity of $LWE - L_{2,1}$ is $lO(n^3)$, where l is the number of iteration. The time complexity of $LWE - L_F$ and $LWE - L$ is $lO(n^2)$ and $lO(n^3)$ respectively.

4 Experiments and Analysis

In this section, all of the experimental results under different settings are presented. The UCI datasets that we selected in the experiments are presented in Table 1. For fair comparison, each dataset is randomly split into 2/3 (training data) and 1/3 (testing data) and the regularization parameter is obtained by cross-validation method. In our experiments, seven regression methods, (Least Square (LR), Mat-primal, Support Vector Machine (SVM), Extreme Learning Machine-kernel (ELM), Feed-forward neural network (FNN), Elman Neural Network (ENN), Layer-recurrent neural network (LRN)), are selected as base models. Mean Absolute Error (MAE)and Root Mean Square Error (RMSE) are selected as the criteria [2]

$$MAE = \frac{1}{n_t} \sum_{i=1}^{n_t} |f(x_i) - y_i|; \tag{23}$$

$$RMSE = \sqrt{\frac{1}{n_t} \sum_{i=1}^{n_t} (f(x_i) - y_i)^2}; \tag{24}$$

where n_t is the size of the testing data, x_i is the i_{th} sample and y_i is the i_{th} real value.

4.1   Convergence of Objective Function

Figures 2 and 3 plot the evolution of the objective values of $LWE\text{-}L_{2,1}$, $LWE\text{-}L_F$ and $LWE\text{-}L$. We see that the objective value gets to convergence at the second iteration. This evaluates that our proposed method converges quickly. Sometimes, the convergence is different. For example, on WhiteWine dataset, $LWE\text{-}L_{2,1}$ gets to convergence at the fifth iteration. However, $LWE\text{-}F$ gets to convergence at the second iteration. This evaluates our proposed method is universal.

4.2   Prediction on UCI Datasets

To illustrate the behavior of LWE, we compare with single methods, (LRN, LR, ELM, SVM), constant weighted ensemble method (the weights of base models base on the MSE), dynamic ensemble method [1], AMLE [7] and

Table 1. Description of UCI datasets

Data	Samples	Attributes
Housing	506	14
Forest	517	11
Concrete	1030	9
RedWine	1599	8
Abalone	4177	9
WhiteWine	4898	12
Elevator	16599	19
Physicochemical Properties (Physic)	45730	10

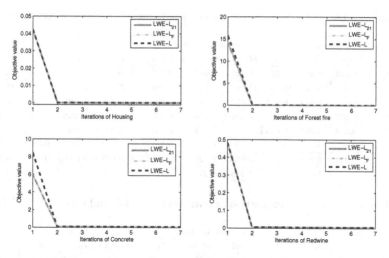

Fig. 2. Convergence of $LWE\text{-}L_{2,1}$, $LWE\text{-}L_F$, $LWE\text{-}L$ on Housing, Forest fire, Concrete, Red-wine datasets

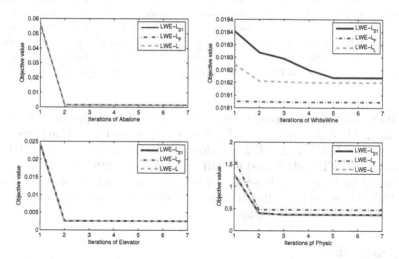

Fig. 3. Convergence of $LWE-L_{2,1}$, $LWE-L_F$, $LWE-L$ on Abalone, White-wine, Elevator, Physic datasets

Adaboost method. For the dynamic ensemble method, we should establish a set of base models for each test point. However there is no prior knowledge about the optimal number of nearest neighbors. We have not calculated the value of dynamic ensemble method on Elevator and Physic.

Tables 2 and 3 present the comparison of MAE and RMSE on different methods respectively. In Tables 2 and 3, we can find the performances of ensemble methods are improved. Dynamic weighted ensemble method is better than constant weighted method. This shows that different regions need different

Table 2. The MAE comparison of ours, single models and ensemble models

Model	Single model			Ensemble methods				LWE		
	LRN	LR	SVM	MSE	Adaboost	Dynamic	AMLE	L_{21}	L_F	L
Housing	0.113	0.279	0.123	0.117	0.105	0.113	0.102	0.101	0.104	**0.097**
Forest	27.40	25.22	15.36	15.41	15.63	19.92	15.511	**15.30**	15.71	15.39
Concrete	6.861	6.835	6.914	6.654	5.912	5.756	5.239	**5.186**	5.240	5.238
RedWine	0.507	0.509	0.520	0.528	0.528	0.502	0.491	**0.485**	0.490	0.485
Ablaone	0.054	0.055	0.053	0.053	0.054	0.052	0.052	**0.051**	0.0522	0.051
WhiteWine	0.552	0.554	0.561	0.515	0.527	0.515	0.512	0.515	**0.504**	0.511
Elevator	0.027	0.032	0.031	**0.024**	0.025	-	0.027	0.025	0.026	0.026
Physic	1.985	1.637	1.575	1.333	1.244	-	1.107	1.108	**1.107**	1.107

Table 3. The RMSE comparison of ours, single models and ensemble models

Model	Single model			Ensemble methods				LWE		
	LRN	LR	SVM	MSE	Adaboost	dynamic	AMLE	L_{21}	L_F	L
Housing	0.179	0.460	0.191	0.178	0.164	0.168	0.156	0.157	0.160	**0.153**
Forest	89.192	64.033	65.153	64.983	65.716	64.528	62.075	62.044	**61.760**	61.864
Concrete	8.657	8.655	8.849	8.854	7.475	7.291	6.723	6.776	**6.701**	6.723
RedWine	0.665	0.671	0.6905	**0.618**	0.681	0.662	0.643	0.635	0.647	0.646
Ablaone	0.071	0.073	0.072	0.070	0.073	0.079	0.070	0.069	0.069	**0.068**
WhiteWine	0.718	0.716	0.725	0.666	0.670	0.661	0.660	0.664	**0.651**	0.658
Elevator	0.335	0.046	0.040	0.034	**0.033**	-	0.037	0.034	0.036	0.036
Physic	1.985	2.436	2.772	2.134	2.065	-	1.975	1.979	**1.970**	1.975

integration methods. Compared with above methods, our proposed methods, LWE-$L_{2,1}$, LWE-L_F and LWE-L, show strong robustness. Compared with Adaboost, the MAE of LWE-L on Housing dataset is decreased by 0.0079. Accordingly, RMSE is decreased by 0.0147. Compared with AMLE, the performance on MAE is similar. However, our proposed method is better than AMLE on RMSE. For example, RMSE is fallen by 4.07 %. This fact confirms the integrating effectiveness of our model.

5   Conclusion

In this paper, a novel dynamic ensemble method is presented, LWEs. Regression methods, such as FNN, LR, ENN, LRN, ELM, mat-primal and SVM, are used to define base regressors. Soft-max function is introduced to assign different integration strategies to different regions. Experiments on eight UCI datasets confirm the effectiveness of LWEs. Some conclusions can be drawn as follow

(1) The distribution of dataset varies from sample to sample. The distribution of dataset is variable. It is not wise to use one global model for prediction.
(2) The different fusion measures must be applied for different regions. A constant weighted ensemble strategy cannot reflect the differences of samples. We introduce the data-driven ensemble method, Locally Weighted Ensemble, to combine the individual models.

(3) The convergence of our proposed method is fast. LWEs only need seconds to deal with 45730 samples.

Acknowledgments. This work is supported by National Program on Key Basic Research Project (973 Program) under Grant 2012CB215201, National Natural Science Foundation of China (NSFC) under Grants 61432011, 61222210, 61170101, and New Century Excellent Talents in University(NCET) under Grant NCET-12-0399.

References

1. Bijalwan, V., Kumar, V., Kumari, P., Pascual, J.: Knn based machine learning approach for text and document mining. Int. J. Database Theor. Appl. **7**(1), 61–70 (2014)
2. Bludszuweit, H., Domínguez-Navarro, J.A., Llombart, A.: Statistical analysis of wind power forecast error. IEEE Trans. Power Syst. **23**(3), 983–991 (2008)
3. Bottou, L., Vapnik, V.: Local learning algorithms. Neural Comput. **4**(6), 888–900 (1992)
4. Didaci, L., Giacinto, G.: Dynamic classifier selection by adaptive k-nearest-neighbourhood rule. In: Roli, F., Kittler, J., Windeatt, T. (eds.) MCS 2004. LNCS, vol. 3077, pp. 174–183. Springer, Heidelberg (2004)
5. Geng, B., Tao, D., Xu, C., Yang, L., Hua, X.S.: Ensemble manifold regularization. IEEE Trans. Pattern Anal. Mach. Intell. **34**(6), 1227–1233 (2012)
6. Haltmeier, M.: Block-sparse analysis regularization of ill-posed problems via l 2, 1-minimization. In: 18th International Conference on Methods and Models in Automation and Robotics, pp. 520–523 (2013)
7. Jacobs, R.A., Jordan, M.I., Nowlan, S.J., Hinton, G.E.: Adaptive mixtures of local experts. Neural Comput. **3**(1), 79–87 (1991)
8. Jordan, M.I., Jacobs, R.A.: Hierarchical mixtures of experts and the em algorithm. Neural Comput. **6**(2), 181–214 (1994)
9. Krogh, A., Vedelsby, J., et al.: Neural network ensembles, cross validation, and active learning. Adv. Neural Inf. Process. Syst. **7**, 231–238 (1995)
10. Li, L., Zou, B., Hu, Q., Wu, X., Yu, D.: Dynamic classifier ensemble using classification confidence. Neurocomputing **99**, 581–591 (2013)
11. Liu, Y., Yao, X.: Ensemble learning via negative correlation. Neural Netw. **12**(10), 1399–1404 (1999)
12. Lo, J.C., Lin, M.L.: Robust H$_\infty$ control for fuzzy systems with frobenius norm-bounded uncertainties. IEEE Trans. Fuzzy Syst. **14**(1), 1–15 (2006)
13. Mendes-Moreira, J., Jorge, A.M., Soares, C., de Sousa, J.F.: Ensemble learning: a study on different variants of the dynamic selection approach. In: Perner, P. (ed.) MLDM 2009. LNCS, vol. 5632, pp. 191–205. Springer, Heidelberg (2009)
14. Mendes-Moreira, J., Soares, C., Jorge, A.M., Sousa, J.F.D.: Ensemble approaches for regression: a survey. ACM Comput. Surv. **45**(1), 10 (2012)
15. Navarro-Arribas, G., Torra, V.: Information fusion in data privacy: a survey. Inf. Fusion **13**(4), 235–244 (2012)
16. Perrone, M.P., Cooper, L.N.: When Networks Disagree: Ensemble Methods for Hybrid Neural Networks. Chapman and Hall, (1993)
17. Seeger, M.: Bayesian model selection for support vector machines, gaussian processes and other kernel classifiers. In: Proceedings of the 13th Annual Conference on Neural Information Processing Systems, pp. 603–609, no. 161324 (2000)

18. Tsymbal, A., Pechenizkiy, M., Cunningham, P., Puuronen, S.: Dynamic integration of classifiers for handling concept drift. Inf. Fusion **9**(1), 56–68 (2008)
19. Vapnik, V., Bottou, L.: Local algorithms for pattern recognition and dependencies estimation. Neural Comput. **5**(6), 893–909 (1993)
20. Woods, K., Bowyer, K., Kegelmeyer Jr., W.P.: Combination of multiple classifiers using local accuracy estimates. In: Computer Vision and Pattern Recognition, pp. 391–396 (1996)
21. Yin, X.C., Huang, K., Yang, C., Hao, H.W.: Convex ensemble learning with sparsity and diversity. Inf. Fusion **20**, 49–59 (2014)

Reliable Confidence Predictions Using Conformal Prediction

Henrik Linusson[1]([⊠]), Ulf Johansson[1], Henrik Boström[2], and Tuve Löfström[1]

[1] Department of Information Technology, University of Borås, Borås, Sweden
{henrik.linusson,ulf.johansson,tuve.lofstrom}@hb.se
[2] Department of Computer and Systems Sciences, Stockholm University,
Kista, Sweden
henrik.bostrom@dsv.su.se

Abstract. Conformal classifiers output confidence prediction regions, i.e., multi-valued predictions that are guaranteed to contain the true output value of each test pattern with some predefined probability. In order to fully utilize the predictions provided by a conformal classifier, it is essential that those predictions are reliable, i.e., that a user is able to assess the quality of the predictions made. Although conformal classifiers are statistically valid by default, the error probability of the prediction regions output are dependent on their size in such a way that smaller, and thus potentially more interesting, predictions are more likely to be incorrect. This paper proposes, and evaluates, a method for producing refined error probability estimates of prediction regions, that takes their size into account. The end result is a binary conformal confidence predictor that is able to provide accurate error probability estimates for those prediction regions containing only a single class label.

1 Introduction

Conformal classifiers [13] are classification models that associate each of their predictions with a measure of confidence; each prediction consists of a set of class labels, and the probability of including the true class label is bounded by a predefined level of confidence. Conformal predictors are automatically valid for any exchangeable sequence of observations, in the sense that the probability of excluding the correct class label is well-calibrated by default.

Apart from validity, the key desideratum for conformal predictors is their *efficiency*, i.e., the size of the prediction regions produced should be kept small, as they limit the number of possible outputs that need to be considered. For conformal classifiers, efficiency can be expressed as a function of the number of class labels included in the prediction regions, given a specific confidence level [12].

H. Linusson—This work was supported by the Swedish Foundation for Strategic Research through the project High-Performance Data Mining for Drug Effect Detection (IIS11-0053) and the Knowledge Foundation through the project Big Data Analytics by Online Ensemble Learning (20120192).

J. Bailey et al. (Eds.): PAKDD 2016, Part I, LNAI 9651, pp. 77–88, 2016.
DOI: 10.1007/978-3-319-31753-3_7

In order to make use of the confidence predictions provided by conformal classifiers, it is necessary that the prediction regions are both small and reliable. The automatic validity of conformal classifiers effectively ensures their reliability for appropriate, i.e., exchangeable, data streams, and much research has been devoted to making conformal classifiers more efficient, see e.g., [2,4–6,8]. However, there is a need for addressing the problem of making predictions that are *simultaneously* small and reliable. The probability of making an incorrect prediction is only valid prior to making said prediction, i.e., we know the probability of the *next* prediction being incorrect. After classifying a sequence of test patterns, however, the *a posteriori* error probability of each particular prediction is dependent on its size; this can easily be seen by noting that an empty prediction region is always incorrect, whereas a prediction region containing all possible outputs is always correct.

This paper proposes a method for utilizing posterior information, i.e., the size of prediction regions produced for a sequence of test patterns, in order to more reliably estimate the error probability of singleton predictions, i.e., predictions containing only a single class label, for binary classification problems.

2   Inductive Conformal Classification

In order to output prediction sets, conformal classifiers combine a *nonconformity function*, which ranks objects based on their apparent strangeness (compared to other observations from the same domain), together with a statistical test that can potentially reject unlikely patterns.

The nonconformity function can be any function on the form $f : X^m \times Y \to R$, but is typically based on a traditional machine learning model according to

$$f[h_Z, (x_i, y_i)] = \Delta[h_Z(x_i), y_i],\qquad(1)$$

where h_Z is a predictive model trained on the problem, Z, and Δ is some function that measures the prediction errors of h_Z. For binary classification problems, a common choice of error function is

$$\Delta[h_Z(x_i), y_i] = 1 - \hat{P}_{h_Z}(y_i \mid x_i).\qquad(2)$$

where $\hat{P}_{h_Z}(y_i \mid x_i)$ is a probability estimate for class y_i when the model h_Z is applied on x_i.

In order to construct an inductive conformal classifier [9,10,13], the following training procedure is used:

1. Divide the training set Z into two disjoint subsets:
   - A *proper training set* Z_t.
   - A *calibration set* Z_c, where $|Z| = q$.
2. Train a classifier h (the underlying model) on Z_t.
3. Let $\{\alpha_1, \ldots, \alpha_q\} = \{f(h, z_i), z_i \in Z_c\}$.

When a new test pattern, x_j, is obtained, its output can be predicted as follows:

1. Fix a significance level $\epsilon \in (0, 1)$.
2. For each class $\tilde{y} \in Y$:
   (a) Tentatively label x_j as $(x_j, \tilde{y})$.
   (b) Let $\alpha_j^{\tilde{y}} = f[h, (x_j, \tilde{y})]$.
   (c) Calculate $p_j^{\tilde{y}}$ as

$$p_j^{\tilde{y}} = \frac{\left|\left\{z_i \in Z_c : \alpha_i > \alpha_j^{\tilde{y}}\right\}\right|}{q + 1} + \theta_j \frac{\left|\left\{z_i \in Z_c : \alpha_i = \alpha_j^{\tilde{y}}\right\}\right| + 1}{q + 1}, \quad (3)$$

   where $\theta_j \sim U[0, 1]$.
   (d) Let $\Gamma_j^{\epsilon} = \left\{\tilde{y} \in Y : p_j^{\tilde{y}} > \epsilon\right\}$.

The resulting prediction set Γ_j^{ϵ} contains the true output y_j with probability $1 - \epsilon$. An error occurs whenever $y_j \notin \Gamma_j^{\epsilon}$, and the expected number of errors made by a conformal classifier is ϵk, where k is the number of test patterns.

3 Conformal Classifier Errors

Conformal predictors are unconditional by default, i.e., while the probability of making an error for an arbitrary test pattern is ϵ, it is possible that errors are distributed unevenly amongst different natural subgroups in the test data, e.g., test patterns with different class labels [7,11,13]. If the output of a test pattern is easily predicted, e.g., because it belongs to the majority class, the probability of making an erroneous prediction on that test pattern might be lower than ϵ, while the opposite might be true for difficult test patterns, e.g., those belonging to the minority class. Hence, we can express the expected number of errors made by a binary conformal classifier as

$$E = \epsilon k = \epsilon_0 k P(c_0) + \epsilon_1 k P(c_1), \quad (4)$$

where ϵ_0 and ϵ_1 are the (unknown) probabilities of making an erroneous prediction for test patterns that belong to class c_0 and c_1 respectively.

Figure 1 illustrates, using the hepatitis data set [1], the (more or less) expected behaviour of an unconditional conformal classifier for binary classification problems where the two classes are of unequal difficulty. The easier (majority) class 'LIVE' shows an error rate below ϵ, while the error rate of the more difficult (minority) class 'DIE' far exceeds ϵ.

3.1 Class-Conditional Conformal Classification

Conditional (or *Mondrian*) conformal classifiers [11,13] effectively let us fix ϵ_0 and ϵ_1 such that $\epsilon = \epsilon_0 = \epsilon_1$ by making the p-values conditional on the class labels of the calibration examples and test patterns. This is accomplished by slightly modifying the p-value equation, so that only calibration examples that

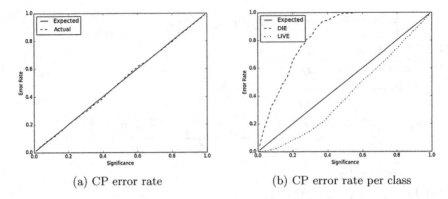

(a) CP error rate   (b) CP error rate per class

Fig. 1. Error rates of a conformal classifier on the hepatitis dataset; (a) overall error rate, i.e., over all test examples; (b) error rates for test examples belonging to the two classes, 'DIE' and 'LIVE', respectively.

share output labels with the test pattern (which is tentatively labeled as $\tilde{y}$) are considered, i.e.,

$$p_j^{\tilde{y}} = \frac{\left|\left\{z_i \in Z_\kappa : \alpha_i > \alpha_j^{\tilde{y}}\right\}\right|}{|Z_\kappa| + 1} + \theta_j \frac{\left|\left\{z_i \in Z_\kappa : \alpha_i = \alpha_j^{\tilde{y}}\right\}\right| + 1}{|Z_\kappa| + 1}, \tag{5}$$

where $Z_\kappa = \{(x_i, y_i) \in Z_c : y_i = \tilde{y}\}$ and $\theta_j \sim U[0, 1]$.

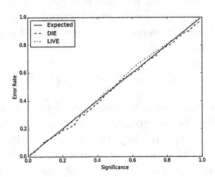

Fig. 2. Error rates of a class-conditional conformal classifier for the two classes, 'DIE' and 'LIVE', on the hepatitis data set.

Figure 2 shows the error rates of a class-conditional conformal classifier for the two classes of the hepatitis dataset. Here, a much more preferable behaviour is observed: the error rate of the 'DIE' and 'LIVE' classes both correspond well to the expected error rate ϵ.

3.2   Utilizing Posterior Information

The overall error probability of a conformal classifier is ϵ, and class-conditional conformal classifiers extend this guarantee to apply to each class individually such that (for a binary classification problem) $\epsilon = \epsilon_0 = \epsilon_1$. This effectively handles the issue of making sure that conformal predictors can provide us with reliable predictions, regardless of class (im)balance. However, we have yet to address the task of making reliable predictions that are also small.

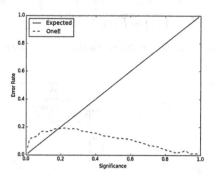

Fig. 3. OneE (error rate on singleton predictions) of a class-conditional conformal classifier on the hepatitis dataset.

For a binary classification problem, the most interesting predictions are, arguably, those containing only a single class label, i.e., the singleton predictions, since empty predictions and double predictions provide us with little actionable information. As illustrated by Fig. 3, conformal classifiers, unfortunately, provide no guarantees regarding the error rate of singleton predictions; as can be seen, for the hepatitis data set, the error rate of singleton predictions (OneE) is substantially greater than ϵ for low values of ϵ.

Hence, we would like some way of expressing the likelihood of a singleton prediction being correct, without requiring knowledge of the true labels of the test patterns. To accomplish this, we are required to slightly shift our point-of-view: rather than guaranteeing the probability of making an erroneous prediction, we need to express the probability of *having made* an erroneous prediction. In the case of a binary classification problem, once k predictions have been made, we can state the expected number of errors as

$$E = \epsilon k = \epsilon k \left(P(e) + P(d) + P(s) \right), \tag{6}$$

where $P(e)$, $P(d)$ and $P(s)$ are the probabilities of making empty, double and singleton predictions respectively. It is clear that we are required to make predictions (at any significance level ϵ) in order to estimate these probabilities, however, we are not required to know the true output labels of the test patterns. Once values for $P(e)$, $P(d)$ and $P(s)$ have been found, we can leverage three pieces of information regarding conformal classifiers and their prediction regions: the overall error rate on the k test patterns is ϵ; double predictions are never erroneous; and, empty predictions are always erroneous. This lets us state the following,

$$\epsilon k = \hat{\epsilon} k P(s) + k P(e) \Rightarrow \hat{\epsilon} = \frac{\epsilon - P(e)}{P(s)}, \tag{7}$$

where $\hat{\epsilon}$ is the expected error rate of the $kP(s)$ singleton predictions made. Alternatively, we can define a smoothed estimate,

$$\hat{\epsilon}_s = \frac{\epsilon}{P(s) + P(e)} \geq \sup\{\epsilon, \hat{\epsilon}\}, \tag{8}$$

where the confidence in a singleton prediction is never allowed to exceed $1 - \epsilon$.

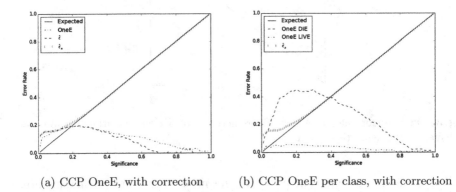

(a) CCP OneE, with correction   (b) CCP OneE per class, with correction

Fig. 4. OneE of a class-conditional conformal classifier on the hepatitis data set, with corrected ($\hat{\epsilon}$, Eq. 7) and smoothed corrected ($\hat{\epsilon}_s$, Eq. 8) singleton error rate estimates: (a) OneE over all test patterns; (b) OneE over test patterns belonging to the 'DIE' and 'LIVE' classes, respectively.

Figure 4 shows, again using the hepatitis data set, that the estimates $\hat{\epsilon}$ and $\hat{\epsilon}_s$ correspond well with the observed error rates on singleton predictions. From Fig. 4a, it is clear that both estimates are better indicators for the OneE scores than the significance level ϵ, however, Fig. 4b displays an obvious issue with both estimates: singleton predictions that indicate that the true class label is 'DIE' are incorrect much more often than expected from both $\hat{\epsilon}$ and $\hat{\epsilon}_s$, while the opposite is true for singleton predictions consisting only of the 'LIVE' class label. Thus, it seems that we have effectively undone the efforts in making sure that the overall error rates are equal for both classes. Indeed, we would ideally want to express a reliable confidence estimate in singleton predictions for each class separately, and thus need to expand on our definition of $\hat{\epsilon}$.

For our binary classification problem, we can write the expected error rate for examples belonging to class c_i as

$$\epsilon_i = P(s_{j \neq i} \mid c_i) + P(e \mid c_i), \tag{9}$$

where, $P(s_{j \neq i} \mid c_i)$ is the probability of (erroneously) making a singleton prediction that does not include the true class c_i, and $P(e \mid c_i)$ is the probability of producing an (automatically incorrect) empty prediction for test patterns belonging to class c_i. From this we can obtain

$$\epsilon_i = P(s_{j \neq i} \mid c_i) + P(e \mid c_i) = \frac{P(c_i \mid s_{j \neq i})P(s_{j \neq i})}{P(c_i)} + P(e \mid c_i) \qquad (10)$$

$$P(c_i \mid s_{j \neq i}) = \frac{P(c_i)\left[\epsilon_i - P(e \mid c_i)\right]}{P(s_{j \neq i})}, \qquad (11)$$

where $P(c_i \mid s_{j \neq i}) = P(c_{i \neq j} \mid s_j)$, i.e., the probability of a prediction region containing only class c_j being erroneous. Unfortunately, this assumes that $P(e \mid c_i)$ is known—something that requires us to obtain the true class labels of our test set—however, if we assume that no empty predictions are made, we can define the estimate

$$P(e) = 0 \Rightarrow P(c_i \mid s_{j \neq i}) = \frac{\epsilon_i P(c_i)}{P(s_{j \neq i})} \geq \frac{P(c_i)\left[\epsilon_i - P(e \mid c_i)\right]}{P(s_{j \neq i})}. \qquad (12)$$

Using our previous notation, we can express the estimate

$$\hat{\epsilon}_j = \frac{\epsilon_{i \neq j} P(c_{i \neq j})}{P(s_j)}, \qquad (13)$$

where $\hat{\epsilon}_j$ is the error probability of a singleton prediction containing only class c_j. It is clear that this is a conservative estimate, since the presence of empty predictions can only decrease the true expected error rate on singleton predictions. We note also that $P(c_{i \neq j})$ can be estimated from the set of calibration examples.

Figure 5, finally, displays the error rates of singleton predictions containing the 'DIE' and 'LIVE' classes, respectively, together with the estimates $\hat{\epsilon}_{DIE}$ and $\hat{\epsilon}_{LIVE}$. In both cases, the true OneE rate is approximately equal to, or lower than, the conservative estimate $\hat{\epsilon}_j$.

4   Experiments

To evaluate the proposed method of obtaining improved error rate estimates of singleton predictions, an experimental evaluation was conducted using 10×10-fold cross-validation on 20 binary classification data sets, obtained from the UCI repository [1] (Table 1). A random forest classifier [3], consisting of 300 trees, was used as the underlying model, and the calibration set size was set to 25 % of the training data for all data sets. Equation 2 was used as the nonconformity function.

Table 2 shows the rate of empty predictions (ZeroC), the rate of singleton predictions (OneC) as well as the error probability of singleton predictions (OneE) of a class-conditional conformal classifier on all 20 data sets at $\epsilon = 0.1$. Error

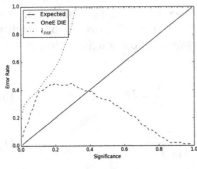
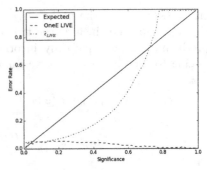

(a) CCP OneE DIE with correction   (b) CCP OneE LIVE with correction

Fig. 5. OneE of a class-conditional conformal classifier on the hepatitis data set, with class-conditional corrected singleton error rate estimates ($\hat{\epsilon}_j$, Eq. 13): (a) OneE rate for predictions containing only the 'DIE' class; (b) OneE rate for predictions containing only the 'LIVE' class.

Table 1. Data sets used in the experiments.

Data set	#Inst.	#Feat.	#C0	#C1	Data set	#Inst.	#Feat.	#C0	#C1
balance-scale	576	5	288	288	hepatitis	155	20	32	123
breast-cancer	286	49	201	85	ionosphere	351	35	126	225
breast-w	699	10	458	241	kr-vs-kp	3196	41	1527	1669
credit-a	690	44	307	383	labor	57	27	20	37
credit-g	1000	62	300	700	liver-disorders	345	7	145	200
diabetes	768	9	500	268	mushroom	8124	122	4208	3916
haberman	306	15	225	81	sick	3772	34	3541	231
heart-c	303	23	165	138	sonar	208	61	111	97
heart-h	294	23	188	106	spambase	4601	58	2788	1813
heart-s	270	14	150	120	tic-tac-toe	958	28	332	626

rates in bold indicate that OneE $> 1.05\epsilon$, i.e., where the one-sided margin of error is greater than 5 %. This error margin is due to the asymptotic validity of conformal predictors—we expect some statistical fluctuations in the observed error rate on a finite data set. For several of the data sets, e.g., breast-cancer, haberman and liver-disorders, the total error probability of singleton predictions ($s_0 \cup s_1$) is much greater than ϵ. This does not appear sufficient, as the singleton predictions would typically be those that are of interest to an analyst. Looking at the error rates of the individual classes, i.e., singleton predictions containing only c_0 (s_0) and singleton predictions containing only c_1 (s_1), the problem is even more pronounced—the error rate of singleton predictions containing a specific class is, for some data sets, several times greater than ϵ. So, while a conformal classifier does indeed provide us with a guarantee on the overall error probability of its predictions (when considering singleton predictions, double predictions

Table 2. Rate of empty predictions (ZeroC), rate of singleton predictions (OneC) and error probability of singleton predictions (OneE) of a class-conditional conformal classifier at $\epsilon = 0.1$.

$\epsilon = 0.1$	$s_0 \cup s_1$			s_0		s_1	
CCP	ZeroC	OneC	OneE	OneC	OneE	OneC	OneE
balance-scale	0.063	0.936	0.042	0.470	0.042	0.467	0.043
breast-cancer	0.000	0.346	**0.292**	0.180	**0.149**	0.166	**0.436**
breast-w	0.085	0.915	0.019	0.588	0.002	0.328	0.048
credit-a	0.004	0.912	**0.107**	0.426	**0.131**	0.486	0.085
credit-g	0.000	0.542	**0.188**	0.200	**0.355**	0.341	0.085
diabetes	0.000	0.616	**0.164**	0.378	0.089	0.238	**0.276**
haberman	0.000	0.374	**0.281**	0.238	0.105	0.136	**0.569**
heart-c	0.000	0.783	**0.127**	0.402	**0.105**	0.381	**0.144**
heart-h	0.000	0.724	**0.132**	0.424	0.064	0.300	**0.216**
heart-s	0.001	0.786	**0.130**	0.408	**0.105**	0.379	**0.149**
hepatitis	0.001	0.614	**0.169**	0.200	**0.379**	0.414	0.047
ionosphere	0.025	0.958	0.069	0.369	**0.120**	0.589	0.034
kr-vs-kp	0.098	0.902	0.001	0.431	0.001	0.471	0.002
labor	0.045	0.679	0.079	0.319	0.095	0.361	0.044
liver-disorders	0.000	0.451	**0.204**	0.239	**0.201**	0.212	**0.187**
mushroom	0.097	0.903	0.000	0.468	0.000	0.436	0.000
sick	0.087	0.913	0.014	0.845	0.001	0.068	**0.174**
sonar	0.002	0.809	**0.116**	0.446	0.097	0.363	**0.118**
spambase	0.064	0.936	0.038	0.560	0.028	0.375	0.054
tic-tac-toe	0.095	0.905	0.001	0.312	0.001	0.593	0.002
mean	0.033	0.750	0.109	0.395	0.103	0.355	0.136
min	0.000	0.346	0.000	0.180	0.000	0.068	0.000
max	0.098	0.958	0.292	0.845	0.379	0.593	0.569

as well as empty predictions), and even though a class-conditional conformal predictor extends this guarantee to each class separately, we cannot state any particular confidence in those prediction regions that would be of most use.

In Table 3, the same singleton error rates are tabulated, together with the exact estimate of singleton error probability $\hat{\epsilon}$ (Eq. 7), the smoothed estimate $\hat{\epsilon}_s$ (Eq. 8) and the class-conditional estimate $\hat{\epsilon}_j$ (Eq. 13). Estimates in bold indicate that OneE $> 1.05\hat{\epsilon}$. For all data sets, the exact estimate $\hat{\epsilon}$ lies close to the empirical error rate of singleton predictions. Although the estimate does exceed the true singleton error rate occasionally, we should expect it to converge with an increasing number of calibration examples and test patterns. The smoothed estimate is automatically conservative whenever the true singleton error rate is

Table 3. Error probabilities of singleton predictions (OneE) of a class-conditional conformal classifier at $\epsilon = 0.1$, together with estimated singleton error probabilities $\hat{\epsilon}$, $\hat{\epsilon}_s$ and $\hat{\epsilon}_j$.

$\epsilon = 0.1$	$s_0 \cup s_1$			s_0		s_1	
CCP	OneE	$\hat{\epsilon}$	$\hat{\epsilon}_s$	OneE	$\hat{\epsilon}_0$	OneE	$\hat{\epsilon}_1$
balance-scale	0.042	**0.039**	0.100	0.042	0.106	0.043	0.107
breast-cancer	0.292	0.289	0.289	0.149	0.165	0.436	0.422
breast-w	0.019	**0.017**	0.100	0.002	0.059	0.048	0.200
credit-a	0.107	0.105	0.109	0.131	0.130	0.085	0.092
credit-g	0.188	0.185	0.185	0.355	0.349	0.085	0.088
diabetes	0.164	0.162	0.162	0.089	0.092	0.276	0.273
haberman	0.281	**0.267**	0.267	0.105	0.111	0.569	0.542
heart-c	0.127	0.128	0.128	0.105	0.113	0.144	0.143
heart-h	0.132	0.138	0.138	0.064	0.085	0.216	0.213
heart-s	0.130	0.126	0.127	0.105	0.109	0.149	0.147
hepatitis	0.169	**0.161**	0.162	0.379	0.397	0.047	0.050
ionosphere	0.069	0.078	0.102	0.120	0.174	0.034	0.061
kr-vs-kp	0.001	0.002	0.100	0.001	0.121	0.002	0.101
labor	0.079	0.080	0.138	0.095	0.204	0.044	0.097
liver-disorders	0.204	0.222	0.222	0.201	0.243	0.187	0.198
mushroom	0.000	0.004	0.100	0.000	0.103	0.000	0.119
sick	0.014	0.015	0.100	0.001	0.007	0.174	1.384
sonar	0.116	0.121	0.123	0.097	0.105	0.118	0.147
spambase	0.038	0.038	0.100	0.028	0.070	0.054	0.162
tic-tac-toe	0.001	0.005	0.100	0.001	0.210	0.002	0.058

lower than the significance level ϵ, and does not substantially underestimate the true singleton error probability for any of the data sets tested on. The class-conditional estimate, $\hat{\epsilon}_j$, is often conservative, in particular for the data sets where the conformal classifier outputs a relatively large number of empty predictions, e.g., balance-scale, breast-w, kr-vs-kp; see Table 2. Again, on the data sets used for evaluation, this estimate never underestimates the singleton error probability substantially; however, for the sick data set in particular, the estimate is extremely conservative on the s_0 predictions (indicating that they are all likely to be incorrect), which is likely a result of the low rate of s_0 predictions (see Table 2).

Overall, it does indeed appear as though these three estimates are better able to more accurately express the true error probability of the singleton predictions than the original significance level ϵ. The smoothed estimate $\hat{\epsilon}_s$ and the class-conditional estimate $\hat{\epsilon}_j$, in particular, tend to overestimate rather than underestimate the true singleton prediction error rate, while the exact estimate $\hat{\epsilon}$ should be expected to converge to the true error probability given enough data.

5 Concluding Remarks

In this paper, a method is proposed for providing well-calibrated error probability estimates for confidence prediction regions from a class-conditional binary conformal classifier. In particular, three estimates are proposed that express the error probability of prediction regions containing only a single class label more accurately than the original significance level, i.e., the acceptable error rate ϵ. The three estimates proposed are: an exact estimate $\hat{\epsilon}$, that expresses the error probability of singleton predictions; a smoothed estimate $\hat{\epsilon}_s$, that expresses the same probability in a conservative manner (it never falls below the original expected error rate ϵ); and, a conservative class-conditional estimate $\hat{\epsilon}_j$, that expresses the error probability of a singleton prediction containing only class c_j. All three estimates are evaluated empirically with good results.

The error probability estimates proposed in this paper do not require knowledge of the true outputs of the test set, however, it is necessary that several predictions are made before the estimates can be calibrated, as they require knowledge of the probabilities of making empty, singleton and double predictions respectively. An alternative approach, left for future work, is to obtain these probabilities from an additional validation set, or, from the calibration set itself. This could, potentially, also allow us to refine the class-conditional estimate, as it would enable us to estimate additional parameters, i.e., the probability of making an empty prediction for a test pattern belonging to a certain class, that are required to express an exact class-conditional estimate rather than a conservative one.

Another interesting direction for future work is to observe the behaviour of the proposed method in an on-line setting. As it stands, the method is best suited for use in a batch prediction setting, due to the requirement of making predictions before calculating the error probability estimates.

Finally, it would be of interest to attempt to extend the proposed method to multi-class problems as well as regression problems.

References

1. Bache, K., Lichman, M.: UCI machine learning repository (2013). http://archive.ics.uci.edu/ml
2. Bhattacharyya, S.: Confidence in predictions from random tree ensembles. Knowl. Inf. Syst. **35**(2), 391–410 (2013)
3. Breiman, L.: Random forests. Mach. Learn. **45**(1), 5–32 (2001)
4. Carlsson, L., Ahlberg, E., Boström, H., Johansson, U., Linusson, H.: Modifications to p-values of conformal predictors. In: Gammerman, A., Vovk, V., Papadopoulos, H. (eds.) SLDS 2015. LNCS, vol. 9047, pp. 251–259. Springer, Heidelberg (2015)

5. Johansson, U., Boström, H., Löfström, T.: Conformal prediction using decision trees. In: 2013 IEEE 13th International Conference on Data Mining (ICDM), pp. 330–339. IEEE (2013)
6. Linusson, H., Johansson, U., Boström, H., Löfström, T.: Efficiency comparison of unstable transductive and inductive conformal classifiers. In: Iliadis, L., Maglogiannis, I., Papadopoulos, H., Sioutas, S., Makris, C. (eds.) Artificial Intelligence Applications and Innovations. IFIP AICT, vol. 437, pp. 261–270. Springer, Heidelberg (2014)
7. Löfström, T., Boström, H., Linusson, H., Johansson, U.: Bias reduction through conditional conformal prediction. Intell. Data Anal. 9(6), 1355–1375 (2015)
8. Löfström, T., Johansson, U., Boström, H.: Effective utilization of data in inductive conformal prediction using ensembles of neural networks. In: The 2013 International Joint Conference on Neural Networks (IJCNN), pp. 1–8. IEEE (2013)
9. Papadopoulos, H.: Inductive conformal prediction: theory and application to neural networks. Tools Artif. Intell. 18(315–330), 2 (2008)
10. Papadopoulos, H., Proedrou, K., Vovk, V., Gammerman, A.J.: Inductive confidence machines for regression. In: Elomaa, T., Mannila, H., Toivonen, H. (eds.) ECML 2002. LNCS (LNAI), vol. 2430, pp. 345–356. Springer, Heidelberg (2002)
11. Vovk, V.: Conditional validity of inductive conformal predictors. Mach. Learn. 92(2–3), 349–376 (2013)
12. Vovk, V., Fedorova, V., Nouretdinov, I., Gammerman, A.: Criteria of efficiency for conformal prediction. Technical report, Royal Holloway University of London, April 2014
13. Vovk, V., Gammerman, A., Shafer, G.: Algorithmic Learning in a Random World. Springer, New York (2006)

Grade Prediction with Course and Student Specific Models

Agoritsa Polyzou[(✉)] and George Karypis

University of Minnesota, Minneapolis, USA
polyz001@umn.edu

Abstract. The accurate estimation of students' grades in future courses is important as it can inform the selection of next term's courses and create personalized degree pathways to facilitate successful and timely graduation. This paper presents future-course grade predictions methods based on sparse linear models and low-rank matrix factorizations that are specific to each course or student-course tuple. These methods identify the predictive subsets of prior courses on a course-by-course basis and better address problems associated with the *not-missing-at-random* nature of the student-course historical grade data. The methods were evaluated on a dataset obtained from the University of Minnesota. This evaluation showed that the course specific models outperformed various competing schemes with the best performing scheme achieving a RMSE across the different courses of 0.632 vs 0.661 for the best competing method.

1 Introduction

Data mining and machine learning approaches are being increasingly used to analyze educational- and learning-related datasets towards understanding how students learn and improving learning outcomes. This has led to the development of various approaches for modeling and predicting the success or failure of students in completing specific tasks in the context of intelligent tutoring systems [9,12,15,16,18,19], building intelligent "early warning systems" that monitor the students' performance during the term [1,3], predicting how well the students will perform by analyzing their activities with the learning management system (e.g., Moodle) [8,11,17], and predicting students' term and final GPA [2,13,14].

Our work focuses on developing methods that utilize historical student-course grade information to accurately estimate how well students will perform (as measured by their grade) on courses that they have not yet taken. Being able to accurately estimate students' grades in future courses is important as it can be used by them (and/or their academic advisers) to identify the appropriate set of courses to take during the next term, and create personalized degree pathways that enable them to successfully and effectively acquire the required knowledge to complete their studies in a timely fashion.

© Springer International Publishing Switzerland 2016
J. Bailey et al. (Eds.): PAKDD 2016, Part I, LNAI 9651, pp. 89–101, 2016.
DOI: 10.1007/978-3-319-31753-3_8

Existing approaches for predicting a student's grade in a future course [4,6,7] rely on neighborhood-based collaborative filtering methods [10]. Despite their relative simplicity, the estimations obtained by these methods are reasonably accurate indicating that there is sufficient information in the historical student-course grade data to make the estimation problem feasible.

In this paper we improve upon these methods by developing various future-course grade prediction methods that utilize approaches based on sparse linear models and low-rank matrix factorizations. These methods rely entirely on the performance that the students achieved in previously taken courses. A unique aspect of many of our methods is that their associated models are either specific to each course or specific to each student-course tuple. This allows them to identify and utilize the relevant information from the prior courses that are associated with the grade for each course and better address problems associated with the *not-missing-at-random* nature of the student-course historical grade data. We experimentally evaluated the performance of our methods on a dataset obtained from the University of Minnesota that contained historical grades that span 12.5 years. Our results showed that the course specific models outperformed various competing schemes and that the best performing scheme, which is based on course-specific regression, achieves a RMSE across the different courses of 0.632 whereas the best competing method achieves an RMSE of 0.661.

The reminder of the paper is organized as follows. Section 2 introduces the notation and definitions used. Section 3 describes the methods developed and Sect. 4 provides information about the experimental design. Section 5 presents an extensive experimental evaluation of the methods and compares them against existing approaches. Finally, Sect. 6 provides some concluding remarks.

2   Definitions and Notations

Throughout the paper, bold lowercase letters will denote column vectors (e.g., $\mathbf{y}$) and bold uppercase letters will denote matrices (e.g., $\mathbf{G}$). Individual elements will be denoted using subscripts (e.g., for a vector y_i, and for a matrix $g_{s,c}$). A single subscript on a matrix will denote its corresponding row. The sets will be represented by calligraphic letters.

The historical student-course grade information will be represented by a sparse matrix $\mathbf{G} \in \mathbb{R}^{n \times m}$, where n and m are the number of students and courses, respectively, and $g_{i,j}$ is the grade in the range of $[0,4]$ that student i achieved in course j. If a student has not taken a course, the corresponding entry will be missing. The course and student whose grades need to be predicted will be called *target course* and *target student*, respectively.

3   Methods

In this section we describe various classes of methods that we developed for predicting the grade that a student will obtain on a course that he/she has not yet taken.

3.1   Course-Specific Regression (CSR)

Undergraduate degree programs are structured in such a way that courses taken by students provide the necessary knowledge and skills for them to do well in future courses. As a result, the performance that a student achieved in a subset of the earlier courses can be used to predict how well he/she will perform in future courses. Motivated by this, we developed a grade prediction method, called *course-specific regression* (CSR) that predicts the grade that a student will achieve in a specific course as a sparse linear combination of the grades that the student obtained in past courses.

In order to estimate the CSR model for course c, we extract from the overall student-course matrix $\mathbf{G}$ the set of rows corresponding to the students that have taken c. For each of these students (rows), we keep only the grades that correspond to courses taken prior to course c. Let $\mathbf{G}^c \in \mathbb{R}^{n_c \times m}$ be the matrix representing that extracted information, where n_c is the number of students that took course c. In addition, let $\mathbf{y}^c \in \mathbb{R}^{n_c}$ be the grades that the students in $\mathbf{G}^c$ obtained in course c (the y_i^c is the grade corresponding the student in the ith row of $\mathbf{G}^c$). Given this, the CSR model $\mathbf{w}^c \in \mathbb{R}_+^m$ for c is estimated as:

$$\underset{\mathbf{w}^c \succeq 0}{\text{minimize}} \ \|\mathbf{y}^c - \mathbb{1}w_0^c - \mathbf{G}^c \mathbf{w}^c\|_2^2 + \lambda_1 \|\mathbf{w}^c\|_2^2 + \lambda_2 \|\mathbf{w}^c\|_1, \tag{1}$$

where w_0^c is a bias term, $\mathbb{1} \in \mathbb{R}^{n_c}$ is a vector of ones and λ_1, λ_2 are regularization parameters to control overfitting and promote sparsity. The model is non-negative because we assume that prior courses can only provide knowledge to future courses. The individual weights of $\mathbf{w}^c$ indicate how much each prior course contributes in the prediction and represent a measure of the importance of the prior course within the context of the estimated model. Using this model, the grade that a student will obtain in course c is estimated as

$$\hat{y}^c = w_0^c + \mathbf{s}^T \mathbf{w}^c, \tag{2}$$

where $\mathbf{s} \in \mathbb{R}^m$ is the vector of the student's grades in the courses he/she has taken so far.

We found that by centering each student's grades around his/hers GPA leads to more accurate predictions (see Sect. 5.1). In this approach, prior to estimating the model using Eq. 1, we first subtract from each $g_{i,j}^c$ grade the GPA of each student (GPA is calculated based on the information in $\mathbf{G}^c$). This centers the data for each student and takes into consideration a notion of student bias as it predicts the performance with respect to the current state of a student. Note that in the case of GPA-centered data, we remove the non-negativity constraint on $\mathbf{w}^c$. We will refer to this model as the CSR-RC (Row Centered) model.

3.2   Student-Specific Regression (SSR)

Depending on the major, the structure of different undergraduate degree programs can be different. Some degree programs have limited flexibility as to the set of courses that a student has to take and at which point in their studies they can take them (i.e., specific semester). Other degree programs are considerably more flexible and are structured around a fairly small number of core courses and a large number of elective courses.

For the latter type of degree programs, a drawback of the CSR method is that it requires the same linear regression model to be applied to all students. However, given that the set of prior courses taken by students in such flexible degree programs can be quite different, a single linear model can fail to capture the various prior course combinations. In fact, there can be cases in which many of the most important courses that were identified by the CSR model were simply not taken by some students, even though these students have acquired the necessary knowledge and skills by taking a different set of courses. To address this limitation, we developed a different method, called *student-specific regression* (SSR), which estimates course-specific linear regression models that are also specific to each student.

The student specific model is derived by creating a student-course specific grade matrix $\mathbf{G}^{s,c}$ for each target student s and each target course c from the $\mathbf{G}^c$ matrix used in CSR method. $\mathbf{G}^{s,c}$ is created in two steps. First, we eliminate from $\mathbf{G}^c$ any grades for courses that were not taken by the target student. Second, we eliminate from $\mathbf{G}^c$ the rows that correspond to students that have not taken a sufficient number of courses that are in common with the target student s. Specifically, if $\mathcal{C}_s$ and $\mathcal{C}_i$ are the set of courses for student s and i respectively, we compute the overlap ratio (OR) $= |\mathcal{C}_s \cap \mathcal{C}_i|/|\mathcal{C}_s|$ and if OR$< t$, then student i is not included in $\mathbf{G}^{s,c}$. The value of t is a parameter of the SSR method and high values ensure that the set of students forming $\mathbf{G}^{s,c}$ have taken many courses in common with s and have followed similar degree plans. Given $\mathbf{G}^{s,c}$, the SSR method proceeds to estimate the model using Eq. 1 (with $\mathbf{G}^{s,c}$ replacing $\mathbf{G}^c$), and uses Eq. 2 for prediction.

3.3   Methods Based on Matrix Factorization

Low rank matrix factorization (MF) approaches have been shown to be very effective for accurately estimating ratings in the context of recommender systems [10]. These approaches can be directly applied to the problem of predicting the grade that a student will achieve on a particular course by treating the student-course grade matrix $\mathbf{G}$ as the user-item rating matrix.

The use of such MF-based approaches for grade prediction is postulated on the fact that there is a low dimensional latent feature space that can jointly represent both students and courses. Given the nature of the domain, this latent space can correspond to the space of knowledge components. Each course vector is the set of components associated with a course and each student vector represents the student's level of knowledge across these knowledge components.

By applying the common approaches of MF-based rating prediction to the problem of grade prediction, the grade that student i will obtain on course j is estimated to be

$$\hat{g}_{i,j} = \mu + sb_i + cb_j + \mathbf{p}_i \mathbf{q}_j{}^T, \tag{3}$$

where μ is a global bias term, sb_i and cb_j are the student and course bias terms, respectively, and p_i and q_j are the latent representations for student i and course j, respectively. The parameters of the MF method ($\mu, \mathbf{sb} \in \mathbb{R}^n, \mathbf{cb} \in \mathbb{R}^m, \mathbf{P} \in \mathbb{R}^{n \times l}$, and $\mathbf{Q} \in \mathbb{R}^{n \times l}$) are estimated following a matrix completion approach that considers only the observed entries in $\mathbf{G}$ as

$$\underset{\mu, \mathbf{sb}, \mathbf{cb}, \mathbf{P}, \mathbf{Q}}{\text{minimize}} \sum_{g_{i,j} \in \mathbf{G}} \left(g_{i,j} - \mu - sb_i - cb_j - \mathbf{p}_i \mathbf{q}_j^T\right)^2 + \lambda(\|\mathbf{P}\|_F^2 + \|\mathbf{Q}\|_F^2$$

$$+ \|\mathbf{sb}\|_2^2 + \|\mathbf{cb}\|_2^2), \quad (4)$$

where λ is a regularization parameter and l is the dimensionality of the latent space, which is a parameter to this method.

The accurate recovery of the low rank model (when such a model exists) from a set of partial observations depends on having a sufficient number of observed entries, and on these entries be randomly sampled from the entries of the target matrix $\mathbf{G}$ [5]. However, in the context of student grade data, the set of courses that students take is not a random subset of the courses being offered as they need to satisfy their degree program requirements. As a result, such an MF approach may lead to suboptimal prediction performance.

In order to address this problem we developed a *course specific matrix factorization* (CSMF) approach that estimates an MF model for each course by utilizing a course specific subset of the data that is denser (in terms of the number of observed entries and the dimensions of the matrix). As a result, it contains a larger number of random by sampled subsets of sufficient size.

Given a course c and a set of students $\mathcal{S}^c$ for which we need to estimate their grade for c (i.e., the students in $\mathcal{S}^c$ have not taken this course yet), the data that CSMF utilizes are the following: (i) the students and grades of the $\mathbf{G}^c$ matrix and $\mathbf{y}^c$ vector of the CSR method (Sect. 3.1), (ii) the students in $\mathcal{S}^c$ and their grades. This data is used to form a matrix $\mathbf{X}^c \in \mathbb{R}^{(n_c + n_t) \times (m_c + 1)}$, where n_c is the number of students in $\mathbf{G}^c$, $n_t = |\mathcal{S}^c|$, and m_c is the number of distinct courses that have at least one grade in $\mathbf{G}^c$ or $\mathcal{S}^c$. The values stored in $\mathbf{X}^c$ are the grades that exist in $\mathbf{G}^c$ and $\mathcal{S}^c$. The last column of $\mathbf{X}^c$ stores the grades $\mathbf{y}^c$ for the course c that were obtained from the students in $\mathbf{G}^c$. Thus, $\mathbf{X}^c$ contains all the prior grades associated with the students who have already taken course c and the students for which we need to have their grade on c predicted. Matrix $\mathbf{X}^c$ is then used in place of matrix $\mathbf{G}$ in Eq. 4 to estimate the parameters of the CSMF method, which are then used to predict the missing entries of the last column of $\mathbf{X}^c$, which are the grades that need to be predicted.

4   Experimental Design

4.1   Dataset

The student-course-grade dataset that we used in our experiments was obtained from the University of Minnesota which has a very flexible degree program.

Table 1. Statistics for course-specific datasets.

Prior courses	5	7	9
Average number of students in training set	270	232	212
Average number of students in test set	22	21	20
Average number of prior courses	141	141	145
Average number of grades	3,872	3,663	3,663
Courses predicted	92	90	80
Grades predicted	2,088	1,959	1,666

It contains the students that have been part of the Computer Science and Engineering (CSE) and Electrical and Computer Engineering (ECE) programs from Fall of 2002 to Spring of 2014. Both of these degree programs are part of the College of Science & Engineering (CS&E) in which students have to take a common set of core science courses during the first 2–3 semesters. We removed from the dataset any courses that are not part of those offered by CS&E departments, as these correspond to various liberal arts and physical education courses, which are taken by few students and in general do not count towards degree requirements. Furthermore, we eliminated any courses that were taken as pass/fail. The initial grades were in the A–F scale, which was converted to the 4–0 scale using the standard letter-grade to GPA conversion. The resulting dataset consists of 2,949 students, 2,556 different courses, and 76,748 student-course grades.

We used this dataset to assess the performance of the different methods for the task of predicting the grades that the students will obtain in the last semester (i.e., the most recent semester for which we have data). For this reason, the dataset was further split into two parts, one containing the students that are still *active*, i.e., have taken courses in the last semester (D_{active}) and one that contains the remaining students ($D_{inactive}$). D_{active} contains 876 students, 19,089 grades, out of which 3,427 grades are for the 475 distinct classes taken in the last semester. $D_{inactive}$ contains 2,073 students and 57,659 grades.

These datasets were used to derive various training and testing datasets for the different methods that we developed. Specifically, for the CSR method we extracted the course specific training and testing datasets as follows. For each course c that was offered in the last semester, we extracted course-specific training and testing sets ($D_{train}^{c,\geq k}$ and $D_{test}^{c,\geq k}$) by selecting from $D_{inactive}$ and D_{active}, respectively, the students that have taken c, and prior to taken c, they also took at least k other courses. The reason that these datasets were parametrized with respect to k is because we wanted to assess how the methods perform when different amount of historical student performance information is available. In our experiments we used k in the set $\{5, 7, 9\}$. That information will create the grade matrix $\mathbf{G}^c$, where $g_{i,j}^c$ is the grade of the ith student on the jth course from the training set $D_{train}^{c,\geq k}$. Table 1 shows various statistics about the various course-specific datasets for different values of k.

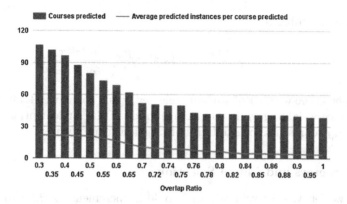

Fig. 1. Statistics of the datasets used in SSR w.r.t overlap ratio.

For the CSMF method, the training dataset for course c was obtained by combining $D_{train}^{c,\geq k}$ and $D_{test}^{c,\geq k}$ into a single matrix after removing the grades that the target students achieved in course c.

For the MF method, the matrix is constructed as the union of the sets $D_{train}^{c,\geq k}$ and $D_{test}^{c,\geq k}$ for every course to be predicted after removing the grades that the active students achieved in the courses we want to predict. We formulated the dataset in this way in order to provide the same information for training and testing to all our models.

In the SSR, the grade matrix $\mathbf{G}^{s,c}$ is created by selecting from $D_{train}^{c,\geq k}$ the set of courses that were also taken by student s and the set of students whose OR with s is at least t. Figure 1 shows some statistics about these datasets as a function of t.

Finally, we did not consider the models that have less than 20 students in their corresponding dataset, as we consider them to have too few training instances for reliable estimation.

4.2 Competing Methods

In our experiments, we compared our methods with the following competing approaches.

1. **BiasOnly.** We only took into consideration local and global bias to predict the students' grades. These biases were estimated using Eq. 4 when $l = 0$.
2. **Student-Based Collaborative Filtering (SBCF).** This method implements the approach described in [4]. For a target course c, every student i is represented by a vector formed with his/hers grades in courses taken prior to c. The vector of a target student s is compared against the vectors of the other students that have taken course c with the Pearson's correlation coefficient. We select the students with positive similarity to perform grade prediction

for s in c according to:

$$\hat{g}_{s,c} = \bar{g}_s + \frac{\min(r, nbr)}{r} \frac{\sum_{i=1}^{nbr}(g_{i,c} - \bar{g}_i)sim_{s,i}}{\sum_{i=1}^{nbr} sim_{s,i}}, \qquad (5)$$

where nbr is the number of students selected, r is a confidence lower limit for significance weighting, $\bar{g}_i$ is the average grade of the student prior taking c, and $sim_{s,i}$ represents the similarity of target student s with i.

4.3 Parameters and Model Selection

For CSR, we let λ_1 take values from 0 to 40 in increments of 2.5 and λ_2 from 0 to 50 in increments of 2.5. For SSR, we let λ_1 take values from 0 to 10 in increments of 1 and λ_2 from 0 to 14 in increments of 2. For MF and CSMF, we let λ take values from 0 to 6 in increments of 0.05. For SSR, the range of the tested values for overlap ratio is 0.3 to 1, in increments of 0.04. For MF and CSMF methods we tested the number of latent dimensions with the values 2, 5 and 8.

As we could not use cross validation for the SSR, we did not apply it for any regression model, in order to be fair with our comparisons. The best models are selected based on their performance on the test set. For MF based approaches, we used the semester before the target semester to estimate and select the best parameters.

4.4 Evaluation Methodology and Performance Metrics

We evaluated the performance of the different approaches by using them to predict the grades for the last semester in our dataset using the data from the previous semester for training.

We assessed the performance using the root mean square error (RMSE) between the actual grades and the predicted ones. Since the courses whose grades are predicted have different number of students, we computed two RMSE-based metrics. The first is the overall RMSE in which all the grades across the different courses were pooled together, and the second is the average RMSE obtained by averaging the RMSE values for each course. We will denote the first by RMSE and the second as AvgRMSE.

5 Experimental Results

5.1 Course-Specific Regression

Table 2 shows the performance achieved by the CSR and CSR-RC models when trained using the three different datasets discussed in Sect. 4.1. These results show that among the two models, CSR-RC, which operates on the GPA-centered grades leads to considerably lower errors both in terms of RMSE and AvgRMSE.

Table 2. The performance achieved by Linear Course-Specific Regression.

	RMSE			AvgRMSE		
Prior courses	5	7	9	5	7	9
CSR	0.751	0.761	0.779	0.757	0.785	0.762
CSR-RC	0.634	0.632	0.632	0.585	0.579	0.543

The performance of the models trained on the different datasets were evaluated on the $D_{test}^{\geq 9}$ test set, which is the common subset among their respective test sets.

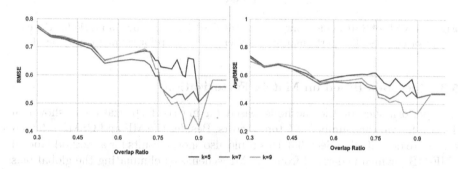

Fig. 2. The performance achieved by the SSR model w.r.t. overlap ratio.

In terms of the sensitivity of their performance on the amount of historical information that was available when estimating these models (i.e., the minimum number of prior courses), we can see that for CSR-RC, the RMSE performance of the models does not change significantly; though the AvgRMSE performance improves when going from five to nine prior courses. This indicates that training sets with more number of prior courses tend to help smaller courses.

5.2   Student-Specific Regression

As one of the parameters for this problem was the overlap ratio between the courses of the target student and other students, Fig. 2 presents the behavior of the model's RMSE (left) and AvgRMSE (right) as we vary the overlap ratio for $D_{test}^{c, \geq 5}(k = 5), D_{test}^{c, \geq 7}(k = 7)$ and $D_{test}^{c, \geq 9}(k = 9)$. When the overlap ratio is increased, the selected students have more courses in common with the target user and that results to better performance. In order to compare the performance of SSR against CSR-RC, Fig. 3 shows the RMSE of the best CSR-RC and SSR models. The RMSE values were computed as the subsets of the test set that was predicted by both models. If the overlap ratio is more than 0.8, then SSR is more accurate. However, the capability of this method to predict courses is very low, i.e., we can predict 50 % less courses than the CSR model for $k = 9$ when the overlap ratio is more than 0.8, because there are not as many students that had followed the same degree plan as the selected student.

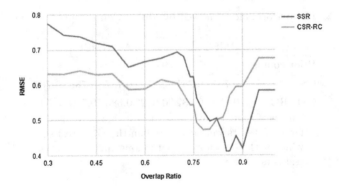

Fig. 3. RMSE of SSR model compared to the CSR-RC w.r.t. overlap ratio for the case of 9 prior courses. The performance for other choices of prior courses is very similar.

5.3 Methods Based on Matrix Factorization

The performance of the methods based on matrix factorization is shown in Table 3 for various number of latent factors. Besides the MF and CSMF schemes that were described in Sect. 3.3, this table also shows results for a method labeled "MF-GB", which is derived from the MF scheme by eliminating the global bias term (μ) of Eq. 4. These results show that CSMF leads to lower RMSE values when there are more than nine prior courses per student, which confirms that by building matrix factorization models on smaller but denser course-specific sub-matrices, we can derive low-rank models that lead to more accurate matrix completion. Even for the case with more than five prior courses, if we focus on denser models, the majority of courses are predicted better by CSMF* than by the best model, MF-GB. In terms of the number of latent factors, we can see that in most cases, the best performance is achieved with small number of latent factors. This should not be surprising, as the average number of grades per student is low, which does not support a large number of latent factors.

5.4 Comparison with other methods

Table 4 compares the performance of the baseline approaches described in Sect. 4.2 (BiasOnly and SBCF) with the best-performing course-specific regression method (CSR-RC), and the best CSMF method (two latent factors). In addition, the results labeled "CSMF*" correspond to those obtained by CSMF in which the best-performing number of latent factors for each course can be different and was selected based on their performance on the validation set (10 % of the training data). CSR-RC and CSMF lead to RMSE and AvgRMSE values that are substantially better than either BiasOnly or SBCF. In terms of the methods that we developed, we see that CSR-RC consistently outperforms CSMF, suggesting that sparse linear regression methods are better than those based on matrix factorization for this setting. Finally, comparing the performance of CSMF* against CSMF, we see that even though the former achieved

Table 3. The performance achieved by the methods based on matrix factorization model w.r.t. the number of prior courses and the number of latent factors.

Prior courses	Latent Factors		MF	MF-GB	CSMF	CSMF*
5	2	RMSE	0.662	0.661	0.683	0.676
	5		0.666	0.667	0.682	0.682
	8		0.667	0.672	0.679	0.676
	2	AvgRMSE	0.597	0.581	0.648	0.645
	5		0.603	0.569	0.643	0.647
	8		0.604	0.596	0.645	0.644
7	2	RMSE	0.667	0.671	0.684	0.679
	5		0.673	0.675	0.680	0.677
	8		0.676	0.681	0.681	0.676
	2	AvgRMSE	0.590	0.598	0.641	0.643
	5		0.603	0.607	0.638	0.640
	8		0.604	0.610	0.637	0.640
9	2	RMSE	0.675	0.684	0.683	0.671
	5		0.677	0.687	0.676	0.672
	8		0.681	0.692	0.677	0.674
	2	AvgRMSE	0.581	0.600	0.653	0.648
	5		0.582	0.607	0.645	0.646
	8		0.579	0.599	0.648	0.647

Table 4. Comparison of the performance achieved from our methods with the competing approaches.

	RMSE			AvgRMSE		
Prior courses	5	7	9	5	7	9
BiasOnly			0.728			0.687
SBCF			0.677			0.675
CSR-RC	0.634	0.632	0.632	0.585	0.579	0.543
CSMF	0.679	0.680	0.676	0.645	0.638	0.645
CSMF*	0.676	0.676	0.671	0.644	0.640	0.648

The performance of the models trained on the different datasets were evaluated on the $D_{test}^{\geq 9}$ test set, which is the common subset among their respective test sets.

better performance, the difference is not very large, which suggests that CSMF's performance is more consistent across its different model parameters.

6   Conclusions

In this paper, we presented two course-specific approaches based on linear regression and matrix factorization that perform better than existing approaches based on traditional methods. This suggests that focusing on a course specific subset of the data can result in more accurate predictions. A student-course specific approach was also developed but its accuracy in grade prediction is limited by the diverse nature of degree plans. The course-specific regression was the one with the best results compared to any other method tested.

Aknowledgements. This work was supported in part by NSF (IIS-0905220, OCI-1048018, CNS-1162405, IIS-1247632, IIP-1414153, IIS-1447788) and the Digital Technology Center at the University of Minnesota. Access to research and computing facilities was provided by the Digital Technology Center and the Minnesota Supercomputing Institute. http://www.msi.umn.edu.

References

1. Starfish: Earlyalert. http://www.starfishsolutions.com/home/student-success-solutions/. Accessed 4 October 2015
2. Al-Barrak, M.A., Al-Razgan, M.: Predicting students final GPA using decision trees: a case study. Int. J. Inf. Educ. Technol. **6**(7), 528 (2016)
3. Arnold, K.E., Pistilli, M.D.: Course signals at purdue: using learning analytics to increase student success. In: Proceedings of the 2nd International Conference on Learning Analytics and Knowledge, pp. 267–270. ACM (2012)
4. Bydžovská, H.: Are collaborative filtering methods suitable for student performance prediction? In: Pereira, F., Machado, P., Costa, E., Cardoso, A. (eds.) EPIA 2015. LNCS, vol. 9273, pp. 425–430. Springer, Heidelberg (2015)
5. Chen, Y., Bhojanapalli, S., Sanghavi, S., Ward, R.: Coherent matrix completion. arXiv preprint (2013). arXiv:1306.2979
6. Denley, T.: Course recommendation system and method. http://www.google.com/patents/US20130011821. Accessed 4 October 2015
7. Denley, T.: Austin peay state university: degree compass. EDUCAUSE Review Online (2012). http://www.educause.edu/ero/article/austin-peay-state-university-degree-compass
8. Elbadrawy, A., Studham, R.S., Karypis, G.: Collaborative multi-regression models for predicting students' performance in course activities. In: Proceedings of the Fifth International Conference on Learning Analytics And Knowledge, pp. 103–107. ACM (2015)
9. Hwang, C.S., Su, Y.C.: Unified clustering locality preserving matrix factorization for student performance prediction. IAENG Int. J. Comput. Sci. **42**(3), 245–253 (2015)
10. Kantor, P.B., Rokach, L., Ricci, F., Shapira, B.: Recommender Systems Handbook. Springer, New York (2011)

11. Luo, J., Sorour, E., Goda, K., Mine, T.: Predicting student grade based on free-style comments using word2vec and ANN by considering prediction results obtained in consecutive lessons, pp. 396–399, June 2015
12. McKay, T., Miller, K., Tritz, J.: What to do with actionable intelligence: E 2 coach as an intervention engine. In: Proceedings of the 2nd International Conference on Learning Analytics and Knowledge, pp. 88–91. ACM (2012)
13. Ogunde, A., Ajibade, D.: A data mining system for predicting university students? graduation grades using ID3 decision tree algorithm. J. Comput. Sci. Inf. Technol. **2**(1), 21–46 (2014)
14. Osmanbegović, E., Suljić, M.: Data mining approach for predicting student performance. Econ. Rev. **10**(1), 3–12 (2012)
15. Pardos, Z.A., Heffernan, N.T.: Using HMMs and bagged decision trees to leverage rich features of user and skill from an intelligent tutoring system dataset. J. Mach. Learn. Res. W & CP (2010)
16. Romero, C., Ventura, S., Espejo, P.G., Hervás, C.: Data mining algorithms to classify students. In: Educational Data Mining 2008 (2008)
17. Sorour, S.E., Mine, T., Goda, K., Hirokawa, S.: A predictive model to evaluate student performance. J. Inf. Process. **23**(2), 192–201 (2015)
18. Thai-Nghe, N., Drumond, L., Horváth, T., Schmidt-Thieme, L.: Using factorization machines for student modeling. In: UMAP Workshops (2012)
19. Toscher, A., Jahrer, M.: Collaborative filtering applied to educational data mining. In: KDD Cup (2010)

Flexible Transfer Learning Framework for Bayesian Optimisation

Tinu Theckel Joy[✉], Santu Rana, Sunil Kumar Gupta, and Svetha Venkatesh

Centre for Pattern Recognition and Data Analytics,
Deakin University, Geelong 3216, Australia
{ttheckel,santu.rana,sunil.gupta,svetha.venkatesh}@deakin.edu.au

Abstract. Bayesian optimisation is an efficient technique to optimise functions that are expensive to compute. In this paper, we propose a novel framework to transfer knowledge from a completed source optimisation task to a new target task in order to overcome the cold start problem. We model source data as noisy observations of the target function. The level of noise is computed from the data in a Bayesian setting. This enables flexible knowledge transfer across tasks with differing relatedness, addressing a limitation of the existing methods. We evaluate on the task of tuning hyperparameters of two machine learning algorithms. Treating a fraction of the whole training data as source and the whole as the target task, we show that our method finds the best hyperparameters in the least amount of time compared to both the state-of-art and no transfer method.

1 Introduction

Whether it is the design of new products in manufacturing, or tuning hyperparameters in machine learning algorithms, it is expensive to search for the best solution exhaustively because these functions are expensive to evaluate. Bayesian optimisation offers powerful solutions in this space [1]. It is efficient in terms of the number of function evaluations required, and is powerful to model objective functions without knowing its form [2]. Bayesian optimisation has been successfully applied in many different fields including learning optimal robot mechanics [3], sequential experimental design [4], optimal sensor placement [5], etc.

Recently, it has found popularity in tuning hyperparameters for machine learning algorithms [6,7]. A problem arises in the case of "cold start", when a new tuning task is tackled. In initial trials, many bad set of hyperparameters may be recommended before a good region is found. When data is large and model is complex, tuning hyperparameters can be excruciatingly long. Reducing the time to optimally tune remains an important problem to solve.

One solution is to induce transfer learning by leveraging the data from previous tasks. Bardenet et al. [8] build a model from past experience by biasing search in a new problem towards the part of the hyperparameter space where optimal hyper-parameters can be found. Incorporating a surrogate based ranking method, they can collaboratively optimise similar objective functions.

J. Bailey et al. (Eds.): PAKDD 2016, Part I, LNAI 9651, pp. 102–114, 2016.
DOI: 10.1007/978-3-319-31753-3_9

Yogatama and Mann [9] use a Bayesian optimisation setting to transfer knowledge from one dataset to the next by using a Gaussian process to model deviations from the per dataset mean. However, both assume that the transfer occurs where the source (previous dataset) and the target (current dataset) tasks are highly related e.g. [8] assumes strong similarity in terms of ranking behavior and [9] assumes strong similarity in the deviations from the respective means. *Therefore, transfer learning approach for Bayesian optimisation that can handle different relatedness among the source and the target tasks in a principled manner is still an open problem.*

Addressing this problem we propose an alternate framework to transfer knowledge across tasks. Intuitively, we do this by modeling the source data as noisy observations of the target function. We achieve this through the modification of the kernel of the Gaussian process, adding more noise variance to source observations. We start by assuming that the source and target functions lie within some envelope of each other. The width of the envelope is determined by the noise variance - smaller noise variance imply that the source observations provide a strong prior knowledge of the target function, larger noise variance imply that the source does not influence the target function. Former is required when tasks are related and the latter is desirable when the tasks are unrelated. We estimate the appropriate noise variance for the target and a source task from the data in a Bayesian setting.

We apply our algorithm for hyperparameter tuning using a novel setting. We sample a small fraction of the data, treating it as a source. This source dataset is then evaluated exhaustively on a number of different hyperparameters. Since the number of data in the source is small, the cost of exhaustive evaluation is low. This knowledge is now used to tune the hyperparameters of the original dataset. We experiment on three benchmark classification datasets for finding the best hyperparameters for two machine learning approaches - elastic net and support vector machine with RBF kernel. In all the experiments, our method is able to find the best hyperparameters in the least amount of time (considering time taken by both source sampling and the target optimisation) over both the current state-of-the-art and the usual tuning algorithm without transfer learning.

In short our contributions are:

– Proposal of a new transfer learning algorithm for Bayesian optimisation in the most general setting that includes source and target tasks with different similarity. This is achieved by modeling the source as a noisy observation of the target function and automatically estimating the noise variance from data.
– Proposal of a novel setting for tuning hyperparameters that exploits the proposed transfer learning framework to improve efficiency. Using a small fraction of the training dataset as a source task, we accelerate the hyperparameter tuning for the whole training set, which is modeled as the target task.
– Evaluation of the proposed algorithm for the best hyperparameter search for two machine learning algorithms on three benchmark classification datasets. Our method is around 6 times faster than methods that tune hyperparameters without transfer learning and around 3 times faster than the state-of-art transfer learning for Bayesian optimisation [9].

2   Preliminaries

2.1   Gaussian Process

Gaussian processes (GPs) [10] are a way of specifying prior distributions over the space of smooth functions. The properties of the Gaussian distribution allow us to compute the predictive means and variances in closed form. It is specified by its mean function, $\mu(x)$ and covariance function, $k(x, x')$. A sample from a Gaussian process is a function as,

$$f(x) \sim \mathcal{GP}(\mu(x), k(x, x')). \tag{1}$$

where the value $f(x)$ at an arbitrary x is a Gaussian distributed random variable specified by a mean and a variance. Without any loss in generality, the prior mean function can be assumed to be a zero function making the Gaussian process fully defined by the covariance function. A popular choice of covariance function is squared exponential function,

$$k(\mathbf{x}, \mathbf{x}') = exp(-\frac{1}{2}||\mathbf{x} - \mathbf{x}'||^2) \tag{2}$$

Other choice of covariance functions include linear kernel, Matérn kernel etc.

Let us assume that we have data points $\mathbf{x}_{1:p}$ and say that the function values corresponding to those data points are sampled from the prior Gaussian process with mean zero and the covariance function $k(\mathbf{x}_i, \mathbf{x}_j)$. Let us denote $\mathbf{y}_{1:p} = f(\mathbf{X}_{1:p})$ as the function values corresponding to the data points $\mathbf{X}_{1:p}$. The function values $\mathbf{y}_{1:p}$ jointly follow a multivariate Gaussian distribution $\mathbf{y}_{1:p} \sim \mathcal{N}(\mathbf{0}, \mathbf{K})$, where

$$\mathbf{K} = \begin{bmatrix} k(\mathbf{x}_1, \mathbf{x}_1) \dots k(\mathbf{x}_1, \mathbf{x}_p) \\ \vdots \quad \ddots \quad \vdots \\ k(\mathbf{x}_p, \mathbf{x}_1) \dots k(\mathbf{x}_p, \mathbf{x}_p) \end{bmatrix} \tag{3}$$

is called the kernel matrix. For a new data point $\mathbf{x}_{p+1}$, let the function value be $y_{p+1} = f(\mathbf{x}_{p+1})$. Then, by the properties of Gaussian process, $\mathbf{y}_{1:p}$ and y_{p+1} are jointly Gaussian as,

$$\begin{bmatrix} \mathbf{y}_{1:p} \\ y_{p+1} \end{bmatrix} \sim \mathcal{N}(\mathbf{0}, \begin{bmatrix} \mathbf{K} & \mathbf{k} \\ \mathbf{k}^T & k(\mathbf{x}_{p+1}, \mathbf{x}_{p+1}) \end{bmatrix}) \tag{4}$$

where $\mathbf{k} = [k(\mathbf{x}_1, \mathbf{x}_{p+1}) \, k(\mathbf{x}_2, \mathbf{x}_{p+1}) \, \dots \, k(\mathbf{x}_p, \mathbf{x}_{p+1})]$. Using Sherman-Morrison-Woodburry [10] formula, the predictive distribution of the function value at a new location $(\mathbf{x}_{p+1})$ can be written as,

$$P(y_{p+1}|\mathbf{X}_{1:p}, \mathbf{y}_{1:p}) \sim \mathcal{N}(\mu_p(\mathbf{x}_{p+1}), \sigma_p^2(\mathbf{x}_{p+1})) \tag{5}$$

where the predicted mean and the variance is given by $\mu_p(\mathbf{x}_{p+1}) = \mathbf{k}^T \mathbf{K}^{-1} \mathbf{y}_{1:p}$ and $\sigma(\mathbf{x}_{p+1}) = k(\mathbf{x}_{p+1}, \mathbf{x}_{p+1}) - \mathbf{k}^T \mathbf{K}^{-1} \mathbf{y}_{1:p}$ respectively. If the observation is a noisy estimate of the actual function value i.e. $y = f(x) + \eta$, where $\eta \sim \mathcal{N}(0, \sigma_{noise}^2)$ the modified predictive distribution becomes

$$P(y_{p+1}|\mathbf{X}_{1:p}, \mathbf{y}_{1:p}) \sim \mathcal{N}(\mu_p(\mathbf{x}_{p+1}), \sigma_p^2(\mathbf{x}_{p+1}) + \sigma_{noise}^2) \qquad (6)$$

where $\mu_p(\mathbf{x}_{p+1}) = \mathbf{k}^T[\mathbf{K} + \sigma_{noise}^2\mathbf{I}]^{-1}\mathbf{y}_{1:p}$ and $\sigma_p^2(\mathbf{x}_{p+1}) = k(\mathbf{x}_{p+1}, \mathbf{x}_{p+1}) - \mathbf{k}^T[\mathbf{K} + \sigma_{noise}^2\mathbf{I}]^{-1}\mathbf{y}_{1:p}$ respectively.

2.2 Bayesian Optimisation

Bayesian optimisation is an efficient method for the global optimisation of costly objective functions [2]. It is especially used in situations where one does not have access to the function form. The user only has the access to the noisy evaluations. Examples in machine learning include tuning hyperparameters of a machine learning model, where the function that relates the choice of hyperparameters to the model performance is unknown and can be very complex.

The unknown function is modeled using a Gaussian process. Bayesian optimisation then employs a simple strategy where it makes use of a surrogate utility function, which is easy to evaluate. The surrogate utility function is called *acquisition function*. The role of the acquisition function is to guide us to reach the optimum of the underlying function. Essentially, acquisition functions are defined in such a way that a high value of the acquisition function corresponds to the potential high value of the underlying function when the optimisation problem is a maxima problem. The new point (*e.g.* new hyperparameter setting) to evaluate next, is then obtained by maximizing the acquisition function.

Acquisition Functions. Acquisition function can be defined either using improvement based criteria or using confidence based criteria. Improvement based criteria such as Probability of Improvement (PI) [11] or Expected Improvement (EI) [12] results in maximizing the probability of improvement over the current best or the improvement in the expected sense. Confidence based criteria such GP-UCB [13] use the upper confidence bound of the GP predictive distribution as an acquisition function. Sometime, a mix of them can also be used as the acquisition function [4]. In this paper, we use EI as the criteria for its usefulness and simplicity. A brief description of EI is provided below.

Expected Improvement (EI). Let us assume that our optimisation problem is optimizing $\arg\max_{\mathbf{x}} f(\mathbf{x})$ and the current best is at $\mathbf{x}^+ = \arg\max_{\mathbf{x}_i \in \mathbf{X}_{1:p}} f(\mathbf{x}_i)$. The improvement function is defined as,

$$I(\mathbf{x}) = \max\{0, f(\mathbf{x}) - f(\mathbf{x}^+)\} \qquad (7)$$

The acquisition function is then defined on the expected value of $I(\mathbf{x})$ [2] as,

$$\arg\max_{\mathbf{x}} \mathbb{E}(I(\mathbf{x})|D_{1:p}) \qquad (8)$$

Algorithm 1. Generic Bayesian Optimisation Algorithm

1: **Input:** The initial observation $D \equiv \{\mathbf{X}_{1:p}, \mathbf{y}_{1:p}\}$.
2: **Output:** $\{x_n, y_n\}_{n=1}^{T}$
3: **for** $n = 1, 2, ..T$
4:          Find $x_n = \arg \max_{\mathbf{x}} \mathbb{E}(I(\mathbf{x})|D)$ of (9).
5:          Evaluate the objective function: $y_n = f(x_n)$.
6:          Augment the observation set $D = D \cup (x_n, y_n)$ and update the GP.
7: **end for**

where $D_{1:p} \equiv \{\mathbf{x}_{1:p}, \mathbf{y}_{1:p}\}$. The analytic form of $E(I(\mathbf{x}))$ can be obtained as [12],

$$\mathbb{E}(I(\mathbf{x})) = \begin{cases} (\mu(\mathbf{x}) - f(\mathbf{x}^+))\Phi(z) + \sigma(\mathbf{x})\phi(z) & \text{if } \sigma(\mathbf{x}) > 0 \\ 0 & \text{if } \sigma(\mathbf{x}) = 0 \end{cases} \quad (9)$$

where $z = (\mu(\mathbf{x}) - f(\mathbf{x}^+))/\sigma(\mathbf{x})$. $\Phi(.)$ and $\phi(.)$ are the CDF and PDF of a standard normal distribution respectively.

The generic Bayesian optimisation algorithm is presented in Algorithm 1.

3   Proposed Method

The generic Bayesian optimisation algorithm suffers from "cold start" problem i.e. at the beginning it may take many trials before it reaches a good region. To improve the efficiency, we propose a novel Bayesian optimisation framework using transfer learning. We elaborate our framework in the context of hyperparameter optimisation.

Let us denote the source observations as $\{\mathbf{x}_i^s, y_i^s\}_{i=1}^{N^s}$, where $\mathbf{x}_i^s$ denotes the hyperparameter setting, y_i^s is the performance of the model built using the hyperparameters $\mathbf{x}_i^s$ and N^s is the size of source observations. The source observations are generated either on a grid or at random hyperparameter settings. No optimisation is performed at this stage. We assume that the source and the target function lies within a close proximity with each other since they only differ in the amount of training data; source models the mapping from hyperparameter setting to the model performance on a small subset of the whole training data, whereas target function maps the same for the whole training data. We model the difference between the source and target. We use source data to provide us with a rough guideline about the target function, $f^t(.)$. To accomplish this, we model source observations as a noisy measurement of the target function as,

$$y_i^s = f^t(\mathbf{x}_i^s) + \epsilon_i^s, \forall i = 1, \dots, N^s \quad (10)$$

where $\epsilon^s \sim \mathcal{N}(0, \sigma_s^2)$ is a random noise. This implies that the source function values lies within $3\sigma_s$ ball of the target function values with a probability close to 1.

Let us denote the observations from the target task as $\{\mathbf{x}_i^t, y_i^t\}_{i=1}^{N^t}$, where N^t is the number of target observations so far. We combine data from both the source and target and create a combined observation set: $\mathbf{X} = \{\mathbf{X}^s, \mathbf{X}^t\}$ and

$\mathbf{y} = \{\mathbf{y}^s, \mathbf{y}^t\}$. The target GP is built using the combined observation. The kernel matrix for the combined data is computed and then it is updated to incorporate the noise of the source as,

$$\mathbf{K} = \mathbf{K} + \begin{bmatrix} \sigma_s^2 I_{N^s \times N^s} & \mathbf{0} \\ \mathbf{0}^T & \sigma_t^2 I_{N^t \times N^t} \end{bmatrix} \qquad (11)$$

where σ_s^2 models the closeness between the source and the target function and σ_t^2 is the measurement noise for the target. The value of the σ_s^2 reflects our belief on how close the source and target functions are. If they are thought to be very close then we should set σ_s^2 small and large otherwise. The value of σ_s^2 can greatly effect the efficiency of the Bayesian optimisation. In the next section, we will provide a principled way to estimate its value from the target observations.

Estimating the Source Noise Variance (σ_s^2). We estimate the source noise variance from the data by placing an inverse gamma distribution with parameters α_0 and β_0 as the prior distribution as,

$$\sigma_s^2 \sim InvGamma(\alpha_0, \beta_0) \qquad (12)$$

We start with a wide prior and then update the posterior from the observation of output value of the target (y^t) and the source (y^s) for the same hyperparameter setting. We use the evaluated target value at the recommended settings and use the source predicted value ($\hat{y}^s$) at those settings. The source function is modeled with a Gaussian process. Since the inverse gamma is a conjugate prior to the variance, the posterior is also an inverse gamma distribution with updated parameters α_n and β_n as,

$$P(\sigma_s^2 / \{y_i^s - \hat{y}_i^s\}_{i=1}^{N^t}) \sim InvGamma(\alpha_n, \beta_n) \qquad (13)$$

Assuming the mean of the difference to be zero, the parameters α_n and β_n is updated as follows,

$$\alpha_n = \alpha_0 + n/2 \qquad (14)$$

$$\beta_n = \beta_0 + \frac{\sum\limits_{i=1}^{N^t} (y_i^t - \hat{y}_i^s)^2}{2} \qquad (15)$$

We use the mode of the posterior distribution as the value of source noise variance and it is given by,

$$\sigma_s^2 = \frac{\beta_n}{\alpha_n + 1} \qquad (16)$$

The kernel matrix is recomputed following (11) and the Bayesian optimisation is sequentially performed using this kernel matrix. The proposed algorithm is illustrated in Algorithm 2.

Algorithm 2. Proposed Transfer Learning Algorithm.

1: **Input**: Source Observations: $\{\mathbf{X}^s, \mathbf{y}^s\}$, Target Observations: $\{\mathbf{X}^t, \mathbf{y}^t\}$,
     Combined Observation set: $\mathbf{X} = \{\mathbf{X}^s, \mathbf{X}^t\}$, $\mathbf{y} = \{\mathbf{y}^s, \mathbf{y}^t\}$, $D= \{\mathbf{X}, \mathbf{y}\}$.
2: **Initial Settings:** Fit a GP at the source points $\{\mathbf{X}^s, \mathbf{y}^s\}$, Fit a GP at the combined
     Observation Set D, Compute σ_s^2 and update $\mathbf{K}$ using (16) and (11).
3: **Output**:$\{x_n, y_n\}_{n=1}^T$.
4: **for** $n = 1, 2, ..T$
5:     Find $x_n = \arg\max_{\mathbf{x}} \mathbb{E}(I(\mathbf{x})|D)$ of (9).
6:     Evaluate the target function: $y_n = f^t(x_n)$.
8:     Compute $\hat{y}_n^s$ at x_n using the GP.
9:     Update α_n and β_n using (14) and (15).
10:    Update source noise variance σ_s^2 and $\mathbf{K}$ using (16) and (11).
11:    Augment the observation set $D = D \cup (x_n, y_n)$ and Update the GP.
12: **end for**

---

4   Experiments

4.1   Experimental Setup

We conduct experiments on both synthetic and real world datasets. Through two different synthetic datasets, we create two transfer learning situations by varying the similarity between the source and the target functions and analyse the behavior of the proposed algorithm in those cases. For the experiment with synthetic data, we do not tune hyperparameters of any classifier, rather we assume that the source and the target functions are known and the task is to reach the maximum of the target function. Experiments with real world dataset are performed to evaluate our algorithm with respect to the baselines on the efficiency of tuning hyperparameters for two classification algorithms. For the experiment on tuning hyperparameters, a fraction of the training data is treated as the source and whole as the target task.

We compare the proposed method with the following baselines:

- **Efficient-BO** [9]: In this transfer learning approach, a common function for source and target is learnt where the common function is represented as deviations from their respective means.
- **Generic-BO**: Algorithm 1 is used only on the target dataset.

We evaluate based on both the number of iteration taken and total time taken to reach the maximum performance. All timings reported in the experiments, are computed for programs running on a workstation with 3.4 GHz quad-core processor having 8 GB RAM.

4.2   Experiment with Synthetic Data

We generate two synthetic datasets: in synthetic dataset-I, we create a a target function that is highly similar to the source function and in synthetic dataset-II, we create the target function to be very different from the source function.

The source function is always fixed in both the scenarios whilst the target function is varied. The source function is a 2-variate normal probability distribution function with mean at [0,0] and covariance matrix as $I_{2\times2}$. The target functions are also modeled by another 2-variate normal probability distribution function with the same covariance matrix but at different means. For synthetic dataset-I, the target task mean is set at $(0.1, 0.1)$ which is very close to the source function mean. The two functions are shown in Fig. 1a. For synthetic dataset-II, the target task mean is set at $(1.5, 1.5)$ which is far from the source mean. The two functions for this scenario is shown in Fig. 1b. For both the scenarios source functions are represented with 25 data points sampled randomly between $[-3, 6]$ along both the dimensions.

Figure 1c plots the percentage of the maximum value reached as a function of the iteration for the proposed method and the baselines for synthetic dataset-I. All the three methods start from the same position, but our proposed method is able to gain faster reaching 80 % of the maximum value within 7 and reaching at the maximum value by 22. In comparison, Efficient-BO reaches 80 % of the maximum value after 10, although finally reaching at the maximum around the same time as the source.

The generic Bayesian optimisation without any help from source function knowledge, reaches 80 % of the maximum value only after 15 iterations and not reaching the maximum even after 30 iterations. Figure 1e shows the noise variance estimate after each iteration for the proposed method. The variance starts with a high prior and decreases fast as the two functions are very close to each other.

Figure 1d plots the percentage of the maximum value reached as a function of the iteration for the proposed method and the baselines for synthetic dataset-II. We see that, even if all of them start from the same position, our proposed method is able to gain faster reaching close to the maximum value by 15th iteration. In comparison, the generic Bayesian optimisation reaches to a similar value only after 28th iteration. In this case, since the source and the target functions are quite different, Efficient-BO fails to reach beyond 60 % of the maximum with 35 iterations. Figure 1f shows the noise variance estimate after each iteration for the proposed method. The variance starts at the same position as it started for synthetic dataset-I, but instead of decreasing, the variance increases with the iteration as the the source and the target functions are quite different for this scenario.

4.3   Experiment with Real World Datasets

We experiment with three real world datasets for tuning hyperparameters for two classification algorithms: Elastic net and Support Vector Machines (SVM) with radial basis function (RBF) kernel. All three datasets are benchmark datasets used in [14]. A brief description of the datasets are provided in Table 1. For elastic net, the hyperparameters are the L_1 and L_2 penalty weights. The bounds for both of them are chosen to be within $[10^{-5}, 10^{-1}]$. For SVM with RBF kernel, the hyperparameters are the cost parameter (C) of the SVM formulation and the

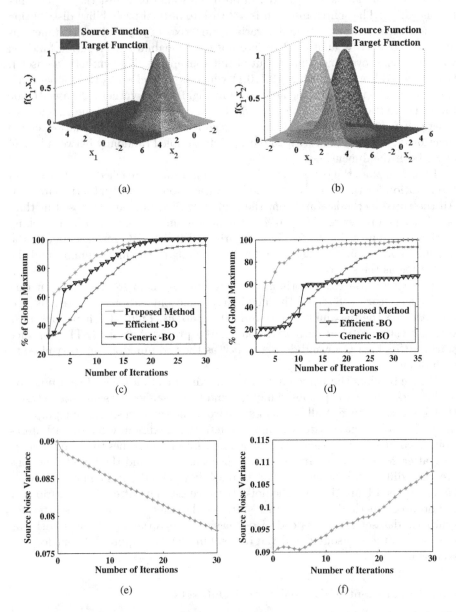

Fig. 1. Synthetic dataset-I (left column) and Synthetic dataset-II (right column): (a, b) the source (blue) and the target functions (red), (c, d) percentage of the maximum value achieved as a function of number of iterations for the three different methods and (e, f) estimated source noise variance after each iteration for the proposed method (Color figure online).

width of the RBF kernel. We choose $10^{-3}$to 10^3 and $10^{-5}$to 10^0 as bounds for C and the width of the kernel, respectively. As the ranges are high, we perform Bayesian optimisation on the logarithmic of the values of the hyperparameters. For each dataset, 5 separate training datasets are created by randomly sampling 70 % of the data for training. Average results over these training set are reported.

Table 1. Datasets used in the experiments.

Dataset	Number of data points	Number of features
Liver disorders	345	6
Heart Diseases	270	13
Breast Cancer	683	10

The results for the hyperparameter tuning of the machine learning algorithms on the different datasets are presented in Fig. 2. The results show that the proposed method achieves the maximum accuracy in the least number of evaluations compared to both the methods. The transfer learning based Efficient-BO follows closely but have never been able to reach to the optimal hyperparameters faster than the proposed method. The Generic-BO without any transfer learning performs the slowest and mostly not being able to reach to the best within 30 iterations.

In Table 2, we present the actual time taken to find the best hyperparameters in CPU seconds. On all the datasets and for both the classification algorithms, our proposed method reaches to the optimal hyperparameters the quickest. It is around 6 times faster than the no transfer method and around 3 times faster than the Efficient-BO [9]. This clearly demonstrate the usefulness of our proposed method for tuning hyperparameters.

Table 2. Time in CPU seconds to reach the best hyperparameter when 40 % of the whole training data is used as source.

Datasets	Elastic Net			SVM with RBF Kernel		
	Proposed Method	Efficient-BO	Generic-BO	Propsed Method	Efficient-BO	Generic-BO
Liver Disorders	**4.8**	12.8	30.3	**6.7**	2 3.3	54.4
Heart Diseases	**3.9**	11.3	27.8	**5.9**	19.3	49.7
Breast Cancer	**6.4**	13.8	33.8	**8.3**	23.7	62.9

In Fig. 3, we plot the total time taken to tune the hyperparameters by our proposed method with respect to the size of the source data. Plots for all three datasets are shown. When a smaller percentage of data is used as source, the source to target difference may be higher leading to more optimisation iterations for target. This amounts to higher computational demand. Increasing source percentage implies larger computational demand for source but decreasing computational requirement for target. In other extreme, a large source means time taken for evaluating source itself is very high. This leads to a nice 'U' shaped

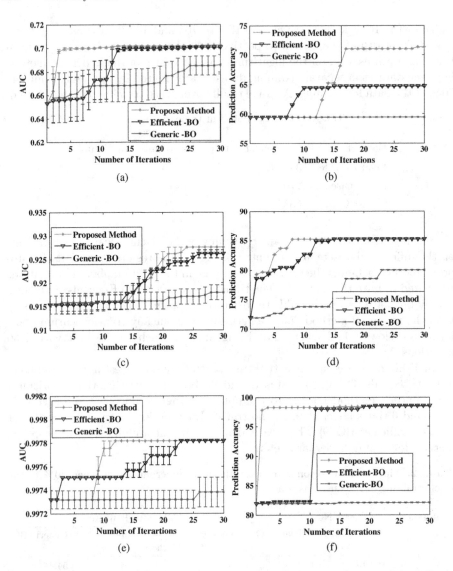

Fig. 2. Hyperparameter tuning experiment on three datasets at three rows (top: liver disorder, middle: heart disease and bottom: breast cancer): current best performance as a function of the iteration for three methods for Elastic Net (left column) SVM with RBF Kernel (right column).

efficiency curve for our proposed algorithm. The optimal lies in the middle and at around 40 % for all three datasets for both classifiers. For much larger data it may be possible to reach to the maximum even with a smaller fraction of data but from our experience 20–40 % is a good starting point.

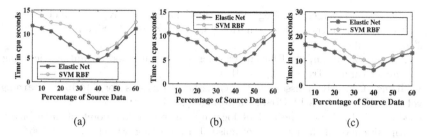

Fig. 3. Time taken to reach to the best hyperparameter vs source size as a percentage of the whole data: (a) Liver disorders, (b) Heart Diseases and (c) Breast cancer Datasets.

5   Conclusion

In this paper, we proposed a novel transfer learning framework for Bayesian optimisation. We model source task as a noisy observation of the target function and use the source observations to avoid the cold start problem for target task optimisation. The noise variance is estimated from data based on the data in a Bayesian setting. This enabled us to address the limitations of the existing methods that only work when tasks are closely related. The proposed method performs around 6 times faster than the generic Bayesian optimisation method and around 3 times faster than the current state-of-art on the task of tuning hyperparameters.

References

1. Snoek, J., Larochelle, H., Adams, R.P.: Practical Bayesian optimization of machine learning algorithms. In: Advances in neural information processing systems, pp. 2951–2959 (2012)
2. Mockus, J.: Application of Bayesian approach to numerical methods of global and stochastic optimization. J. Global Optim. **4**(4), 347–365 (1994)
3. Lizotte, D.J., Wang, T., Bowling, M.H., Schuurmans, D.: Automatic gait optimization with Gaussian process regression. IJCAI **7**, 944–949 (2007)
4. Brochu, E., Cora, V.M., De Freitas, N.: A tutorial on bayesian optimization of expensive cost functions, with application to active user modeling and hierarchical reinforcement learning (2010). arXiv preprint arXiv:1012.2599
5. Garnett, R., Osborne, M.A., Roberts, S.J.: Bayesian optimization for sensor set selection. In: Proceedings of the 9th ACM/IEEE International Conference on Information Processing in Sensor Networks, pp. 209–219. ACM (2010)
6. Bergstra, J., Bardenet, R., Kégl, B., Bengio, Y.: Implementations of algorithms for hyper-parameter optimization. In: NIPS Workshop on Bayesian Optimization, p. 29 (2011)
7. Thornton, C., Hutter, F., Hoos, H.H., Leyton-Brown, K.: Auto-weka: combined selection and hyperparameter optimization of classification algorithms. In: Proceedings of the 19th ACM SIGKDD International Conference on Knowledge Discovery and Data Mining, pp. 847–855. ACM (2013)

8. Bardenet, R., Brendel, M., Kégl, B., et al.: Collaborative hyperparameter tuning. In: Proceedings of the 30th International Conference on Machine Learning (ICML 2013), pp. 199–207 (2013)
9. Yogatama, D., Mann, G.: Efficient transfer learning method for automatic hyperparameter tuning. Transfer 1, 1 (2014)
10. Williams, C.K., Rasmussen, C.E.: Gaussian Processes for Machine Learning, vol. 2, 3rd edn. The MIT Press, Cambridge (2006)
11. Kushner, H.J.: A new method of locating the maximum point of an arbitrary multipeak curve in the presence of noise. J. Fluids Eng. 86(1), 97–106 (1964)
12. Mockus, J., Tiesis, V., Zilinskas, A.: The application of bayesian methods for seeking the extremum. Towards Global Optim. 2(117–129), 2 (1978)
13. Srinivas, N., Krause, A., Kakade, S., Seeger, M.: Gaussian process optimization in the bandit setting: no regret and experimental design. In: Proceedings of International Conference on Machine Learning (ICML) (2010)
14. Chang, C.-C., Lin, C.-J.: Libsvm: a library for support vector machines. ACM Trans. Intell. Syst. Technol. (TIST) 2(3), 27 (2011)

A Simple Unlearning Framework for Online Learning Under Concept Drifts

Sheng-Chi You and Hsuan-Tien Lin[(✉)]

Department of Computer Science and Information Engineering,
National Taiwan University, No.1, Sec. 4, Roosevelt Road, Taipei, Taiwan
{r02922068,htlin}@csie.ntu.edu.tw

Abstract. Real-world online learning applications often face data coming from changing target functions or distributions. Such changes, called the concept drift, degrade the performance of traditional online learning algorithms. Thus, many existing works focus on detecting concept drift based on statistical evidence. Other works use sliding window or similar mechanisms to select the data that closely reflect current concept. Nevertheless, few works study how the detection and selection techniques can be combined to improve the learning performance. We propose a novel framework on top of existing online learning algorithms to improve the learning performance under concept drifts. The framework detects the possible concept drift by checking whether forgetting some older data may be helpful, and then conduct forgetting through a step called unlearning. The framework effectively results in a dynamic sliding window that selects some data flexibly for different kinds of concept drifts. We design concrete approaches from the framework based on three popular online learning algorithms. Empirical results show that the framework consistently improves those algorithms on ten synthetic data sets and two real-world data sets.

Keywords: Online learning · Concept drift

1 Introduction

Online learning is a machine learning setup where the learning algorithm needs to learn from and make predictions on streaming data efficiently and effectively [3,4,9]. The setup enjoys many potential applications, such as predicting the weather, customer preferences, or stock prices [16].

Traditional online learning algorithms, such as the passive-aggressive algorithm (PA) [3], the confidence weighted algorithm (CW) [4] and the adaptive regularization of weight algorithm (AROW) [9], are designed under the assumption that the target function to be learned is fixed. In many applications, however, change in the underlying environment can result in change of the target function (concept) as time goes by. That is, the concept can be drifting [17] instead of fixed. For example, the best popular-cloth predictor (concept) is consistently drifting as the fashion trend evolves [10]. The drifting concept possesses

© Springer International Publishing Switzerland 2016
J. Bailey et al. (Eds.): PAKDD 2016, Part I, LNAI 9651, pp. 115–126, 2016.
DOI: 10.1007/978-3-319-31753-3_10

difficulty for traditional online learning algorithms and are studied by two families of works. One family of works focuses on the detection of concept drift from the data stream [1,5,12,15]. Those works generally conduct statistical analysis on the data distribution and set up an alert threshold to reliably detect concept drift. The other family tries to construct learning models from selected instances of the data stream, with the hope that such instances match the drifting concept better [2,13]. The simplest approach of this family is to use a sliding window to capture the newest instances for learning [13]. While the two kinds both deal with concept-drifting data, it is not fully clear on how they could be combined to improve the learning performance and will be the main focus of this work.

In particular, we propose a framework on top of existing online learning algorithms to improve the learning performance under concept drifts. The framework detects the possible concept drift by checking whether forgetting some older data may be helpful, where the detection is motivated by the confidence terms used in modern online learning algorithms. Then, it conducts forgetting by unlearning older data from the current model. By greedily repeating the detection and unlearning steps along with online learning, the framework effectively results in a dynamic sliding window that can suit different concept drifts. We design concrete approaches of the framework based on PA [3], AROW [9] and CW [4]. Our empirical results demonstrate that the framework can reach better accuracy on artificial and real-world data. The results justify the usefulness of the framework.

The paper is organized as follows. Section 2 establishes the setup and lists related online learning algorithms. Section 3 introduces the proposed framework. Section 4 discusses the experimental results and Sect. 5 concludes the paper.

2   Preliminaries

In this paper, we consider the online learning problem for binary classification. In each round of this problem, the learning algorithm observes a coming instance and predicts its label to be $+1$ or -1. After the prediction, the true label is revealed and the algorithm can then take the new instance-label pair to improve its internal prediction model. The goal of the algorithm is to make as few prediction errors as possible.

We shall denote the instance-label pair in round t as $(\mathbf{x}_t, y_t)$, where $t \in \{1, 2, \ldots, T\}$. Each $\mathbf{x}_t \in \mathbb{R}^n$ represents the instance (feature vector) and $y_t \in \{+1, -1\}$ indicates the label. The prediction in the t-th round is denoted as $\hat{y}_t$, and the error refers to the zero-one loss $\ell_{01}(y_t, \hat{y}_t)$, which is 1 if and only if $y_t \neq \hat{y}_t$, and 0 otherwise.

In this work, we consider the linear model for online learning, where some linear weight vector $\mathbf{w}_t \in \mathbb{R}^n$ is maintained within round t and $\hat{y}_t = \text{sign}(\mathbf{w}_t \cdot \mathbf{x}_t)$ with $\cdot$ denoting an inner product. The linear model generally enjoys efficiency in online learning and is often the focus of study in many online learning works [3,4,9,14]. For the linear model, improving would then mean updating from $\mathbf{w}_t$ to $\mathbf{w}_{t+1}$, and we denote the difference as $\Delta\mathbf{w}_t = \mathbf{w}_{t+1} - \mathbf{w}_t$. The core of different online learning algorithms is then to design a proper update function

---

Algorithm 1. The linear model for online learning

1: initialize $\mathbf{w}_1 \leftarrow (0, 0, ..., 0)$
2: **for** $t = 1$ **to** T **do**
3:     receive instance $\mathbf{x}_t \in R^n$ and predict $\hat{y}_t \leftarrow \text{sign}(\mathbf{w}_t \cdot \mathbf{x}_t)$
4:     receive label: $y_t \in \{-1, +1\}$
5:     $\Delta\mathbf{w}_t \leftarrow \text{UPDATE}(\mathbf{w}_t, \mathbf{x}_t, y_t)$ and $\mathbf{w}_{t+1} \leftarrow \mathbf{w}_t + \Delta\mathbf{w}_t$
6: **end for**

UPDATE$(\mathbf{w}_t, \mathbf{x}_t, y_t)$ that calculates $\Delta\mathbf{w}_t$. The details steps of the linear model for online learning is shown in Algorithm 1, where we assume $\mathbf{w}_1$ to be the zero vector for simplicity.

One of the most popular algorithms for online learning with the linear model is the Passive-Aggressive algorithm (PA) [3]. PA calculates the signed margin of the labeled instance by $y_t(\mathbf{w}_t \cdot \mathbf{x}_t)$, which indicates how confident the prediction $\hat{y}_t = \text{sign}(\mathbf{w}_t \cdot \mathbf{x}_t)$ was. PA then aims to adjust the weights $\mathbf{w}_t$ to the closest $\mathbf{w}_{t+1}$ (passive) in terms of the Euclidean distance, such that the hinge loss $\ell_h(\mathbf{w}; (y_t, \mathbf{x}_t)) = \max(0, 1 - y_t(\mathbf{w} \cdot \mathbf{x}_t))$ is decreased to $\ell_h(\mathbf{w}_{t+1}; (y_t, \mathbf{x}_t)) = 0$ (aggressive). The aim leads to the following UPDATE$(\mathbf{w}_t, \mathbf{x}_t, y_t)$ for PA:

$$\Delta\mathbf{w}_t = \frac{\ell_h(\mathbf{w}_t; (y_t, \mathbf{x}_t))}{\|\mathbf{x}_t\|^2} y_t \mathbf{x}_t. \tag{1}$$

The Confidence weighted (CW) algorithm [4] is extended from PA. Instead of considering a single weight vector $\mathbf{w}_t$, the algorithm considers the weight *distribution*, modeled as a Gaussian distribution with mean $\mathbf{w}_t$ and covariance Σ_t. During each UPDATE for CW, both $\mathbf{w}_t$ and Σ_t are taken into account, and updated to $\mathbf{w}_{t+1}$ and Σ_{t+1}. The updating step adjusts $(\mathbf{w}_t, \Sigma_t)$ to the closest $(\mathbf{w}_{t+1}, \Sigma_{t+1})$ (passive) in terms of the KL divergence, such that the probabilistic zero-one loss under the new Gaussian distribution is smaller than some $(1 - \eta)$ (aggressive).

An extension of CW is called adaptive regularization of weight (AROW) [9], which improves CW by including more regularization. In particular, the updating step of AROW solves an *unconstrained* optimization problem that calculates $(\mathbf{w}_{t+1}, \Sigma_{t+1})$ by

$$\operatorname*{argmin}_{\mathbf{w}, \Sigma} D_{\text{KL}}(\mathcal{N}(\mathbf{w}, \Sigma) \| \mathcal{N}(\mathbf{w}_t, \Sigma_t)) + \lambda_1 \ell_h^2(\mathbf{w}; (y_t, \mathbf{x}_t)) + \lambda_2 \mathbf{x}_t^T \Sigma \mathbf{x}_t. \tag{2}$$

The first term is exactly the KL divergence that the passive part of CW considers; the second term embeds the aggressive part of CW with the squared hinge loss (similar to PA); the third term represents the confidence on $\mathbf{x}_t$ that should generally grow as more instances have been observed. In particular, the confidence term represents how different $\mathbf{x}_t$ is from the current estimate of Σ. The confidence term acts as a regularization term to make the learning algorithm more robust. In this work, we set the parameters λ_1 and λ_2 by $\lambda_1 = \lambda_2 = 1/(2\gamma)$ as the original paper suggests [9].

One special property of the three algorithms above, which is also shared by many algorithms for the linear model of online learning, is that $\Delta\mathbf{w}_t$ is a scaled version of $y_t\mathbf{x}_t$, as can be seen in (1) for PA. Then, by having $\mathbf{w}_1$ as the zero vector, each $\mathbf{w}_t$ is simply a *linear combination* of the previous data $y_1\mathbf{x}_1, y_2\mathbf{x}_2, \cdots, y_{t-1}\mathbf{x}_{t-1}$. We will use this property later for designing our framework.

The three representative algorithms introduced above do not specifically focus on concept-drifting data. For example, when concept drift happens, being passive like the algorithms do may easily lead to slow adaptation to the latest concept. Next, we illustrate more on what we mean by concept drift in online learning. [16] defines concept drift to mean the change of "property" within the data. Some major types of concept drifts that will be considered here are abrupt concept drift, gradual concept drift and virtual concept drift. The first two entail the change of the relation between instances and labels. Denote the relation as the ideal target function f such that $y_t = f(\mathbf{x}_t) +$ noise, *abrupt concept drift* means that the ideal target function can change from f to a very different one like $(-f)$ at some round t_1, and *gradual concept drift* means f is slowly changed to some different f' between rounds t_1 and t_2.

Virtual concept drift, unlike the other two, is generally a consequence of the change of some hidden context within the data [6]. The change effectively causes the distribution of $\mathbf{x}_t$ to vary. While the target function that characterizes the relation between $\mathbf{x}_t$ and y_t may stay the same for virtual concept drift, the change of distribution places different importance on different parts of the feature space for the algorithm to digest.

Two families of methods in the literature focus on dealing with concept-drifting data for online learning. One family [1,5,12,15] is about drift detection based on different statistical property of the data. [5] proposes the drift detection method (DDM) that tracks the trend of the zero-one loss to calculate the drift level. When the drift level reaches an alert threshold, the method claims to detect the concept drift and resets the internal model. While the idea of DDM is simple, it generally cannot detect gradual concept drift effectively. [1] thus proposes the early drift detection method (EDDM) to cope with gradual concept drift, where the distribution of errors instead of the trend is estimated for detection. Some other popular detection criteria include the estimated accuracy difference between an all-data model and a recent-data model [12], and the estimated performance difference between models built from different chunks of data [15]. Generally, similar to [5], after detecting the concept drift, the methods above reset the internal model. That is, all knowledge about the data received before detection are effectively forgotten. Nevertheless, forgetting all data before the detection may not be the best strategy for gradual concept drift (where the earlier data may be somewhat helpful) and virtual concept drift (where the earlier data still hint the target function).

The other family [2,13] makes the internal model adaptive to the concept drift by training the model with selected instances only. The selected instances are often within a sliding window, which matches the fact that the latest instances should best reflect the current concept. Most of the state-of-the-art methods consider dynamic sliding windows. For instance, [13] takes the leave-one-out error estimate of the support vector machine to design a method that computes the best dynamic sliding window for minimizing the leave-one-out error. [2] proposes a general dynamic sliding window method by maintaining a sliding window such that the "head" and "tail" sub-windows are of little statistical difference. The sliding-window methods naturally trace concept drifts well, especially gradual concept drifts. Nevertheless, calculating a good dynamic sliding window is often computationally intensive. It is thus difficult to apply the methods within this family to real-world online learning scenario where efficiency is highly demanded.

In summary, drift-detection methods are usually simple and efficient, but resetting the internal model may not lead to the best learning performance under concept drifts; sliding-windows methods are usually effective, but are at the expense of computation. We aim to design a different framework for better online learning performance under the concept drift. Our framework will include a simple detection scheme and directly exploits the detection scheme to efficiently determine a dynamic sliding window. In addition, the framework can be flexibly coupled with existing online learning algorithms with linear models.

3 Unlearning Framework

The idea of our proposed unlearning framework is simple. Between steps 5 and 6 of Algorithm 1, we add a procedure UNLEARNINGTEST to check if forgetting some older instance can be beneficial for learning. In particular, the decision of "beneficial" is done by comparing a regularized objective function before and after the forgetting, where the regularized objective function mimics that being used by AROW. If forgetting is beneficial, a new $\mathbf{w}'_{t+1}$ (and its accompanying Σ'_{t+1} in the case of CW or AROW) replaces the original $\mathbf{w}_{t+1}$. There are then two issues in describing the framework concretely: what the regularized objective function and unlearning step are, and which "older" instance to check? We will clarify the issues in the next subsections.

3.1 Unlearning Test

Denote $(\mathbf{x}_k, y_k)$, $k \in \{1, 2, \cdots, t-1\}$ as the selected instance for UNLEARN-INGTEST. Recall that in round t, each $\mathbf{w}_t$ is simply a linear combination of the previous data $y_1\mathbf{x}_1, y_2\mathbf{x}_2, \cdots, y_{t-1}\mathbf{x}_{t-1}$. That is, every old instance has its (possibly 0) footprint within $\mathbf{w}_{t+1}$ if we record $\Delta\mathbf{w}_k$ along with the online learning process. Then, one straightforward step to unlearn $(\mathbf{x}_k, y_k)$ is to remove it from $\mathbf{w}_{t+1}$. That is,

$$\mathbf{w}'_{t+1} \leftarrow \mathbf{w}_{t+1} - \Delta\mathbf{w}_k.$$

The Σ'_{t+1} accompanying $\mathbf{w}_{t+1}$ can also be calculated similarly by recording $\Delta\Sigma_k$ along with the online learning process.

Now, $\mathbf{w}'_{t+1}$ represents the weight vector after removing some older instance, and $\mathbf{w}_{t+1}$ represents the original weight vector. Our task is to pick the better one for online learning with concept drift. A simple idea is to just compare their loss, such as the squared hinge loss used by AROW. That is, unlearning is conducted if and only if

$$\ell_h^2(\mathbf{w}'_{t+1}; (\mathbf{x}_t, y_t)) \leq \ell_h^2(\mathbf{w}_{t+1}; (\mathbf{x}_t, y_t)).$$

We can even make the condition more strict by inserting a parameter $\alpha \leq 1.0$ that controls the demanded reduction of loss from the original weight vector. That is, unlearning is conducted if and only if

$$\ell_h^2(\mathbf{w}'_{t+1}; (\mathbf{x}_t, y_t)) \leq \alpha \ell_h^2(\mathbf{w}_{t+1}; (\mathbf{x}_t, y_t)).$$

Then, $\alpha = 0.0$ makes unlearning happen only if $\mathbf{w}'_{t+1}$ is fully correct on $(\mathbf{x}_t, y_t)$ in terms of the hinge loss, and the original online learning algorithms are *as if* using $\alpha < 0$.

In our study, we find that only using ℓ_h^2 as the decision objective makes the unlearning procedure rather unstable. Motivated by AROW, we thus decide to add two terms to the decision objective. One is the confidence term used by AROW, and the other is the usual squared length of $\mathbf{w}$. The first term regularizes against unwanted update of Σ, much like AROW does. The second term regularizes against unwanted update of $\mathbf{w}$ to a long vector, much like the usual ridge regression regularization. That is, given $(\mathbf{x}_t, y_t)$, the framework considers

$$\text{obj}(\mathbf{w}, \Sigma) = \ell_h^2(\mathbf{w}; (\mathbf{x}_t, y_t)) + \beta \mathbf{x}_t^T \Sigma \mathbf{x}_t + \gamma \|\mathbf{w}\|^2 \tag{3}$$

and conduct unlearning if and only if $\text{obj}(\mathbf{w}'_{t+1}, \Sigma'_{t+1}) \leq \alpha \text{obj}(\mathbf{w}_{t+1}, \Sigma_{t+1})$. The parameters β and γ balances the influence of each term.

The final missing component is how to specify β and γ. To avoid making the framework overly complicated, we only consider using those parameters to balance the numerical range of the terms. In particular, we let β be the average of

$$\frac{1}{2} \left(\frac{\ell_h^2(\mathbf{w}_{\tau+1}, \mathbf{x}_\tau, y_\tau)}{\mathbf{x}_\tau^T \Sigma_{\tau+1} \mathbf{x}_\tau} + \frac{\ell_h^2(\mathbf{w}'_{\tau+1}, \mathbf{x}_\tau, y_\tau)}{\mathbf{x}_\tau^T \Sigma'_{\tau+1} \mathbf{x}_\tau} \right). \tag{4}$$

for $\tau \in \{1, 2, \ldots, t\}$ so $\beta \mathbf{x}_t^T \Sigma \mathbf{x}_t$ can be of a similar numerical range to ℓ_h^2. Similarly, we let γ be the average of

$$\frac{1}{2} \left(\frac{\ell_h^2(\mathbf{w}_{\tau+1}, \mathbf{x}_\tau, y_\tau)}{\|\mathbf{w}_{\tau+1}\|^2} + \frac{\ell_h^2(\mathbf{w}'_{\tau+1}, \mathbf{x}_\tau, y_\tau)}{\|\mathbf{w}'_{\tau+1}\|^2} \right). \tag{5}$$

The details of UNLEARNINGTEST is listed in Algorithm 2.

3.2   Instance for Unlearning Test

Unlearning is completed by the unlearning test at a certain selected instance $(\mathbf{x}_k, y_k)$. But how to determine the k from all previous processed instances? We proposed three possible unlearning strategies to deciding the instance $(\mathbf{x}_k, y_k)$.

Algorithm 2. Unlearning test for some instance $(\mathbf{x}_k, y_k)$

1: input parameter: $\alpha \in [0.0, 1.0]$
2: **procedure** UNLEARNINGTEST($\mathbf{w}_{t+1}, \Sigma_{t+1}, \mathbf{x}_k, y_k$)
3:      $\Delta\mathbf{w}_k, \Delta\Sigma_k \leftarrow$ UPDATEHISTORY($\mathbf{x}_k, y_k$)   ▷ previous updated status on $(\mathbf{x}_k, y_k)$
4:      $\mathbf{w}'_{t+1} \leftarrow \mathbf{w}_{t+1} - \Delta\mathbf{w}_k, \quad \Sigma'_{t+1} \leftarrow \Sigma_{t+1} - \Delta\Sigma_k$
5:      set β, γ as the average of (4) and (5), respectively
6:      **if** obj($\mathbf{w}'_{t+1}, \Sigma'_{t+1}$) $\leq \alpha$obj($\mathbf{w}_{t+1}, \Sigma_{t+1}$) **then**   ▷ see (3)
7:          **return** $\mathbf{w}'_{t+1}, \Sigma'_{t+1}$
8:      **else**
9:          **return** $\mathbf{w}_{t+1}, \Sigma_{t+1}$
10:     **end if**
11: **end procedure**

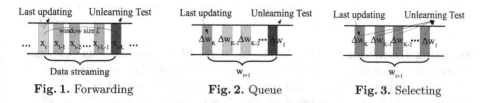

Fig. 1. Forwarding   **Fig. 2.** Queue   **Fig. 3.** Selecting

Forwarding-Removing: Traditional sliding window technique tries to maintain a window that keeps the recent accessed examples, and drops the oldest instance according to some set of rules [2]. Here, the unlearning test is substituted for the rules. Forward-removing considers $(\mathbf{x}_{t-L}, y_{t-L})$ subject to a fixed window size L as the as the selected instance for unlearning test. The strategy is illustrated by Fig. 1, where the older instances are at the right-hand-side of the data stream. After updating on $\mathbf{x}_t$ is done, the unlearning test examines the red instance $\mathbf{x}_{t-L}$.

With some studies on parameter $L = \{1, 10, 100, 1000\}$, $L = 100$ is sufficiently stable and will be used to demonstrate this strategy in Sect. 4.

Queue-Removing: Instead of considering the instance that is L rounds away, this strategy selects the oldest one within the current model $\mathbf{w}_{t+1}$. Recall that the current model $\mathbf{w}_{t+1}$ is a combination of some updated parts $\Delta\mathbf{w}_i$ on previous updated instance $(\mathbf{x}_i, y_i)$. We record those $\Delta\mathbf{w}_i$ like a data list, as illustrated in Fig. 2.

$$\mathbf{w}_{t+1} = \sum_{i=1}^{K} \Delta\mathbf{w}_i = \sum_{i=1}^{K} \tau_i \mathbf{x}_i y_i \quad \text{where } \tau_i \neq 0. \tag{6}$$

Take $\mathbf{w}_{t+1}$ as a queue, unlearning test will be executed at the red updated part $\Delta\mathbf{w}_1$, which is the oldest updated instance in model. As $(\mathbf{x}_i, y_i)$ are added and removed from $\mathbf{w}_t$, the size of the queue can change dynamically, resulting in a dynamic sliding window effectively.

Table 1. The properties of the ten data sets

Data set	Properties		
	Features	Drift type	Drifting details
SINE1	2 real	Abrupt	Reversed wave: $y = sin(x)$
SINE2	2 real	Abrupt	Reversed wave: $0.5 + 0.3sin(3\pi x)$
SINIRREL1	2 real + 2 irrelevant	Abrupt	Same as SINE1 function
SINIRREL2	2 real + 2 irrelevant	Abrupt	Same as SINE2 function
MIXED	2 real + 2 boolean	Abrupt	Reversed 1 function with 1 boolean condition
STAGGER	3 boolean	Abrupt	Switching between 3 boolean conditions
GAUSS	2 real	Virtual	Switching between 2 distributions
CIRCLES	2 real	Gradual	Switching between 4 circles [5]
LINES	2 real	Gradual	changing line functions: shift and rotate
MULTILINES	4–15 real	Gradual	changing hyperplanes: $\Sigma_i^d \mathbf{w_i x_i} = w_0$ [8]

Selecting-Removing: Above strategies both select one particular instance under different structure. However, those strategies neither consider all candidates in their window nor find out the best unlearned weight $\mathbf{w}'_{t+1}$ for current instance $(\mathbf{x}_t, y_t)$. Illustrated by Fig. 3, Selecting-removing will test all K instances and take the instance that can decrease obj the most as the instance to be unlearned.

4   Empirical Evaluation

We take these three unlearning strategies in Sect. 3 with PA [3], AROW [9] and CW [4]. In those algorithms, we set $a = 1.0$, $\phi = 0.0001$ in CW and $r = 0.1$ in AROW. The parameter α in unlearning test is individually selected from $\{0.1, 0.2, \ldots, 0.9\}$ due to the different properties on these algorithms.

All ten synthetic data sets contain different concept drifts described in Table 1. The first eight data sets are used by [5]. Because most of them are about abrupt concept drift, we construct two more data sets, LINES and MULTI-LINES, whose drifting type is gradual. The target function of LINES is changed by shifting and rotating gradually in 2D, and MULTILINES is a d dimensional version defined in [8].

Previous works [1,5] assume every concept contains a fixed number of instances, and examine on small size data sets. Here we construct these artificial data sets with three differences to make the data sets more realistic. First, the number and the timing of concept drifts are randomly assigned and all drift events are recorded so that we could simulate a perfect drifting detection, Concept-removing, which resets $\mathbf{w}_{t+1}$ immediately after a concept drift happens. We take Concept-removing as an upper bound benchmark for using the ideal drifting information. Second, at least 1,000,000 instances are generated in each data set for the robustness. Finally, we inject noise made by flipping binary labels under different probabilities to check the robustness of the proposed framework. All artificial data sets are generated under different flipping level within $\{0.00, 0.05, \cdots, 0.30\}$.

Table 2. Ranking all unlearning strategies under three types of drifting data

strategy \ drifting type	Abrupt	Gradual	Virtual
None	4.031 ± 0.347	3.047 ± 0.402	3.333 ± 0.890
Forwarding-removing	4.325 ± 0.308	3.809 ± 0.471	5.476 ± 0.534
Queue-removing	$\mathbf{3.373 \pm 0.235}$	$\mathbf{2.984 \pm 0.387}$	$\mathbf{2.190 \pm 0.499}$
Selecting-removing	3.769 ± 0.254	3.666 ± 0.517	3.714 ± 0.730
EDDM	$\mathbf{3.309 \pm 0.264}$	3.174 ± 0.385	3.000 ± 0.427
Concept-removing	1.269 ± 0.138	3.809 ± 0.501	2.666 ± 0.930

For each data, a simple second-order polynomial transform is applied to improve the accuracy. Two evaluation criteria are considered, *ranking performance* and *cumulative classification accuracy*. A smaller average rank (along with standard deviation) indicates that an higher classification accuracy performed among compared methods.

4.1 Results and Discussion

In addition to the three proposed strategies within the framework, and the ideal Concept-removing strategy, we also compare the proposed framework with EDDM [1]. Our experimental results are summarized in following tables with different control variables. Table 2 compares all unlearning strategies under three kinds of concept-drifting data. Table 3 compares the relation between different unlearning strategies and each online learning algorithm individually. Table 4 evaluates the influence on the best unlearning strategy with different noise level. The individual accuracy performances for each data set are recorded in Table 5.

Table 2 makes comparison by different kinds of concept-drifting data. The ideal Concept-removing strategy performs very well for abrupt drifting and virtual drifting, as expected. But the immediate resetting cannot work for gradual drifting data, and the ideal detection is not realistic anyway. Our proposed framework, on the other hand, performs well on all kinds of data when using Queue-removing.

Table 3 is evaluated under individual learning algorithms. On the strategy side, Queue-removing preforms the best ranking on average in four unlearning strategies. Note that Selecting-removing is worse than Queue-removing, which indicates that overly searching for the "best" instance to unlearn is not necessary. On the algorithm sides, a significant ranking gap between Concept-removing and the others is presented in AROW. All four unlearning strategies show the smaller ranking than original AROW. For the other two algorithms, only Queue-removing and EDDM gets smaller ranking on PA. But almost all unlearning approaches do not have great advantage in CW. The cause of non-improving is their individual updating rules, which does not consider confidence term in PA and squared hinge-loss in CW.

Table 3. Ranking all unlearning strategies under each learning algorithm

algorithm / strategy	PA	AROW	CW	Average
None	3.257 ± 0.325	5.642 ± 0.336	2.100 ± 0.215	3.666 ± 0.263
Forwarding-removing	5.128 ± 0.352	5.371 ± 0.239	2.357 ± 0.236	4.285 ± 0.245
Queue-removing	$\mathbf{3.100 \pm 0.348}$	3.485 ± 0.335	2.528 ± 0.284	$\mathbf{3.138 \pm 0.195}$
Selecting-removing	4.228 ± 0.352	4.457 ± 0.411	2.514 ± 0.235	3.733 ± 0.228
EDDM	3.342 ± 0.394	$\mathbf{2.200 \pm 0.152}$	4.171 ± 0.279	3.238 ± 0.200
Concept-removing	2.242 ± 0.467	1.242 ± 0.140	3.028 ± 0.483	2.171 ± 0.248

Table 4. Ranking three main unlearning strategies under different bias data sets

Unlearning strategy	Noise level					
	0.05	0.10	0.15	0.20	0.25	0.30
None	2.22 ± 0.17	2.16 ± 0.16	2.11 ± 0.17	2.18 ± 0.16	2.06 ± 0.16	2.02 ± 0.16
Queue-removing	1.90 ± 0.08	1.97 ± 0.11	1.97 ± 0.11	1.97 ± 0.11	2.01 ± 0.12	1.97 ± 0.14
Concept-removing	1.28 ± 0.12	1.38 ± 0.14	1.36 ± 0.13	1.34 ± 0.13	1.40 ± 0.14	1.44 ± 0.14

We study Queue-removing more in Table 4, which shows the ranking performance under different noise levels. From lowest to highest bias, Concept-removing is still the best in three strategies but Queue-removing shows its effectiveness in all noise levels. When the noise becomes larger, Queue-removing is closer to the ideal Concept-removing strategy.

Table 5 explains whether unlearning framework reflects the significant difference from original algorithms. We conducted the t-test experiment by its cumulative classification accuracy at each data set 30 times for all artificial data and directly evaluated two real data, MNIST[1] and ELEC2[2]. For two real data, we directly compare the accuracy performance with EDDM and Queue-removing.

The t-test is evaluated in three different strategies. Queue-removing shows better accuracy than no-unlearning, and those p-value(N-Q) are mostly smaller than 0.01, which indicates the performance gap is significant enough. Concept-removing reveal the upper bound accuracy and the nearly 0 on p-value(Q-C) comparing with the Queue-removing in all data sets except for CIRCLES.

MNIST [11] is a handwritten digits data. Although it is not a concept-drifting data, we test whether our unlearning framework will deteriorate the classifying performance. We use one versus one to evaluate 45 binary classifications for those digits under online learning scenario. To handle all classifications quickly, we scale each image by 25 % and take its pixel as feature. Because MNIST data does not contain significant drifting and the nearly same accuracies are presented, it implies our unlearning framework can work well in the normal data set.

ELEC2 [7] is the collection of the electricity price. Those prices are affected by demand and supply of the market, and the labels identify the changing prices

---

[1] Handwritten digits: http://yann.lecun.com/exdb/mnist.

[2] Electricity price data: http://www.inescporto.pt/~jgama/ales/ales.html.

Table 5. Cumulative accuracy and t-test on ten artificial and two real-world data

Properties	Average accuracy among three algorithms			P-value	
Strategy	None	Queue-removing	Concept-removing	N-Q	Q-C
SINE1	0.6696 ± 0.0232	**0.6816 ± 0.0244**	0.7541 ± 0.0267	0.0352	0.0000
SINE2	0.6373 ± 0.0161	**0.6422 ± 0.0154**	0.6984 ± 0.0166	0.0117	0.0000
SINIRREL1	0.6819 ± 0.0212	**0.7202 ± 0.0199**	0.7687 ± 0.0215	0.0000	0.0000
SINIRREL2	0.6395 ± 0.0181	**0.6660 ± 0.0175**	0.7071 ± 0.0169	0.0000	0.0000
MIXED	0.6792 ± 0.0220	**0.6938 ± 0.0214**	0.7469 ± 0.0211	0.0011	0.0000
STAGGER	0.7476 ± 0.0219	**0.7517 ± 0.0216**	0.7996 ± 0.0223	0.0001	0.0000
GAUSS	0.6452 ± 0.0189	**0.6676 ± 0.0188**	0.6871 ± 0.0182	0.0001	0.0000
CIRCLES	0.7179 ± 0.0194	**0.7262 ± 0.0208**	0.7244 ± 0.0217	0.0000	0.5950
LINES	0.7557 ± 0.0244	**0.7783 ± 0.0246**	0.7970 ± 0.0230	0.0002	0.0002
MULTILINES	0.7687 ± 0.0222	0.7566 ± 0.0240	0.7865 ± 0.0239	0.0002	0.0000
Strategy	None	Queue-removing	EDDM	N-Q	Q-C
MNIST	0.9774 ± 0.0032	0.9774 ± 0.0032	0.9758 ± 0.0033	NA	NA
ELEC2	0.8423 ± 0.0765	**0.8742 ± 0.0575**	0.8342 ± 0.1312	NA	NA

related to a moving average. It is a widely used for concept-drifting. We predict the current price rises or falls by its all 8 features. The result shows that Queue-removing preforms better than no-unlearning and EDDM.

5   Conclusion

We present an unlearning framework on top of PA-based online algorithms to improve the learning performance under different kinds of concept-drifting data. This framework is simple yet effective. In particular, the queue-removing strategy, which is the best-performing one, results in a dynamic sliding window on the *mistaken* data and dynamically unlearns based on a simple unlearning test as the drift detection. Future work includes more sophisticated ways to balance between loss and regularization for the unlearning test.

Acknowledgment. The work arises from the Master's thesis of the first author [18]. We thank Profs. Yuh-Jye Lee, Shou-De Lin, the anonymous reviewers and the members of the NTU Computational Learning Lab for valuable suggestions. This work is partially supported by the Ministry of Science and Technology of Taiwan (MOST 103-2221-E-002-148-MY3) and the Asian Office of Aerospace Research and Development (AOARD FA2386-15-1-4012).

References

1. Baena-García, M., del Campo-Ávila, J., Fidalgo, R., Bifet, A., Gavaldà, R., Morales-Bueno, R.: Early drift detection method. In: International Workshop on Knowledge Discovery from Data Streams, pp. 77–86 (2006)

2. Bifet, A., Gavalda, R.: Learning from time-changing data with adaptive windowing. In: SIAM International Conference on Data Mining, pp. 443–448 (2007)
3. Crammer, K., Dekel, O., Keshet, J., Shalev-Shwartz, S., Singer, Y.: Online passive-aggressive algorithms. J. Mach. Learn. Res. **7**, 551–585 (2006)
4. Dredze, M., Crammer, K., Pereira, F.: Confidence-weighted linear classification. In: Proceedings of the 25th International Conference on Machine Learning, pp. 264–271 (2008)
5. Gama, J., Medas, P., Castillo, G., Rodrigues, P.: Learning with drift detection. In: Bazzan, A.L.C., Labidi, S. (eds.) SBIA 2004. LNCS (LNAI), vol. 3171, pp. 286–295. Springer, Heidelberg (2004)
6. Gama, J., Žliobaitė, I., Bifet, A., Pechenizkiy, M., Bouchachia, A.: A survey on concept drift adaptation. ACM Comput. Surv. (CSUR) **46**(4), 1–37 (2014)
7. Harries, M., Wales, N.S.: Splice-2 comparative evaluation: electricity pricing (1999)
8. Hulten, G., Spencer, L., Domingos, P.: Mining time-changing data streams. In: Proceedings of the Seventh ACM SIGKDD International Conference on Knowledge Discovery and Data Mining, pp. 97–106 (2001)
9. Crammer, K., Alex Kulesza, M.D.: Adaptive regularization of weight vectors. Mach. Learn. **91**, 155–187 (2013)
10. Kolter, J.Z., Maloof, M.A.: Dynamic weighted majority: an ensemble method for drifting concepts. J. Mach. Learn. Res. **8**, 2755–2790 (2007)
11. LeCun, Y., Cortes, C.: MNIST handwritten digit database. AT&T Labs (2010)
12. Nishida, K., Yamauchi, K.: Detecting concept drift using statistical testing. In: Corruble, V., Takeda, M., Suzuki, E. (eds.) DS 2007. LNCS (LNAI), vol. 4755, pp. 264–269. Springer, Heidelberg (2007)
13. Ralf, K., Joachims, T.: Detecting concept drift with support vector machines. In: Proceedings of the Seventeenth International Conference on Machine Learning, pp. 487–494 (2000)
14. Shalev-Shwartz, S.: Online learning and online convex optimization. Found. Trends Mach. Learn. **4**, 107–194 (2012)
15. Sobhani, P., Beigy, H.: New drift detection method for data streams. Adapt. Intell. Syst. **6943**, 88–97 (2011)
16. Tsymbal, A.: The problem of concept drift: definitions and related work. Technical report, Computer Science Department, Trinity College Dublin (2004)
17. Widmer, G., Kubat, M.: Learning in the presence of concept drift and hidden contexts. Mach. Learn. **23**, 69–101 (1996)
18. You, S.C.: Dynamic unlearning for online learning on concept-drifting data. Masters thesis, National Taiwan University (2015)

User-Guided Large Attributed Graph Clustering with Multiple Sparse Annotations

Jianping Cao[1], Senzhang Wang[2(✉)], Fengcai Qiao[1], Hui Wang[1], Feiyue Wang[1], and Philip S. Yu[3]

[1] College of Information Systems and Management,
National University of Defense Technology, Changsha, Hunan, China
[2] State Key Laboratory of Software Development Environment,
Beihang University, Beijing, China
szwang@buaa.edu.cn
[3] Department of Computer Science,
University of Illinois at Chicago, Chicago, IL, USA

Abstract. One of the key challenges in large attributed graph clustering is how to select representative attributes. Previous studies introduce user-guided clustering methods by letting a user select samples based on his/her knowledge. However, due to knowledge limitation, a single user may only pick out the samples that s/he is familiar with while ignore the others, such that the selected samples are often biased. We propose a framework to address this issue which allows multiple individuals to select samples for a specific clustering. With wider knowledge coming from multiple users, the selected samples can be more relevant to the target cluster. The challenges of this study are two-folds. Firstly, as user selected samples are usually sparse and the graph can be large, it is non-trivial to effectively combine the different annotations given by the multiple users. Secondly, it is also difficult to design a scalable approach to cluster large graphs with millions of nodes. We propose the approach *CGMA* (*C*lustering *G*raphs with *M*ultiple *A*nnotations) to address these challenges. *CGMA* is able to combine the crowd's consensus opinions in an unbiased way, and conducts an effective clustering with low time complexity. We show the effectiveness and efficiency of the proposed approach on real-world graphs, by comparing with existing attributed graph clustering approaches.

Keywords: User-guided · Large attributed graph · Clustering · Sparse

1 Introduction

With the coming of big data era, the clustering of large attributed graphs has drawn a lot of research attention. One of the major challenges in this field is the attribute selection, which has not been fully explored. Recent research addressed

---

J. Cao—This work is supported by NSFC grants: 71331008, 61105124 and 61303017.

J. Bailey et al. (Eds.): PAKDD 2016, Part I, LNAI 9651, pp. 127–138, 2016.
DOI: 10.1007/978-3-319-31753-3_11

this problem either by using the properties of datasets (e.g. the data density, the topology) [1,2] or by applying user preference to guide clustering [7]. As user-guided clustering is more interpretable and flexible, it attracts more research attentions recently [4–7]. Different from conventional unsupervised clustering methods, user-guided clustering is semi-supervised which allows a user to select a small amount of samples for a particular cluster based on his/her preference. However, existing user-guided clustering assumes there is only one user to annotate the preferred samples. The clustering results largely rely on the labeled samples given by the user. Thus the clustering results may largely rely on the user selected samples based on his/her knowledge on the graph. A potential issue is that the clustering can be biased to the user's preference or knowledge.

In this paper, we propose a framework for user-guided large attributed graph clustering which allows multiple users to annotate their preferred samples independently. An individual may only have partial knowledge about the target clusters, and multiple annotators provide us an effective way to reveal the common knowledge toward a specific issue. Here we borrow the idea of crowdsourcing [18,24] for user-guided clustering with multiple annotations. The general idea can be illustrated by Fig. 1. Suppose three annotators answer such a common question, e.g., "who are the data mining researchers?" As depicted in Fig. 1, each of the annotators gives his/her own annotation based on their own knowledge independently. However, since the data is too big, the annotations are sparse and may hardly overlap [23]. If we only use one of the annotations, we may get the clusters of "IBM Ph.D students," "computer science professors," or "PAKDD authors." However, if we combine the annotations together, we may find out the clusters of "data mining researchers" through the shared conferences of "data mining."

There are two major challenges for introducing multiple annotators to annotate the samples for cluster analysis. (1) It is challenging to combine the annotations of different users due to the fact that the annotations are not only sparse but also overlap very little. Thus it is hard to apply conventional techniques like majority voting to combine them in a straightforward manner. (2) It is also challenging to address the scalability issue of the proposed approach. Under the background of big data, the graph scale can be extremely large and the attribute dimensions can be very high, developing a scalable algorithm is becoming critically important.

To address the above mentioned challenges, we propose an approach for **C**lustering **G**raphs with **M**ultiple **A**nnotations (*CGMA*). A basic assumption here is that each annotator may label the samples of the preferred clusters based on only a few of the sample attributes instead of all of them [7]. With such an assumption, we map each annotation to the attribute space to obtain the weight vector denoting the relevance of attributes. In this way, the problem of combining sparse annotations is transformed into combining the weight vectors corresponding to the annotators in the common attribute space. Once the combined weight vector is obtained, we use it re-weigh the entire network to obtain a pure seed set which we used it for further clustering. The target cluster will be obtained by expanding these seed sets using a local partitioning method. The contributions of this paper can be addressed as follows,

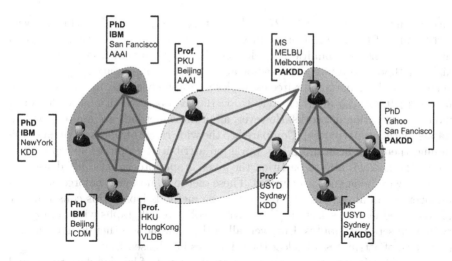

Fig. 1. A toy example of the studied problem. Different annotations with little overlapping are given. Each annotation contains several objects sharing a few focused attributes within it. But the sharing attributes among different annotations may not be the same. The aim of this paper is to identify the target cluster that complies to the multiple annotations as much as possible.

- We introduce a novel problem of user-guided clustering in large attributed networks with multiple annotations. Different from previous user-preference guided clustering, which is often biased, using multiple annotations can alleviate the bias. To the best of our knowledge, this is the first paper applying multiple annotations for graph clustering.
- We propose a two-step clustering approach $CGMA$ to address the proposed problem. $CGMA$ combines multiple annotations in an unbiased way, and it also amplifies the sparse annotations by re-sampling and expansion process. The proposed approach has near-linear time complexity.
- We conduct a series of experiments on various large networks to examine $CGMA$. The experimental results show the effectiveness and efficiency of our method.

The rest of this paper is organized as follows. Section 2 will introduce the related work of this research. Section 3 gives the details of the $CGMA$ algorithm. Next, we will show the experimental results of $CGMA$ on real networks compared with some competitive baselines in Sect. 4. Finally, Sect. 5 concludes the paper.

2 Related Work

Clustering of homogeneous graphs can be sorted into two groups, the plain graph clustering and the attributed graph clustering. Traditional methods mostly target at plain graphs, and they have been well studied in literatures, for example, the partitioning methods $METIS$ [27] and spectral clustering [14] aim to find a k-way partitioning of the graph. Community detection methods [16] cluster the graph into variable size communities, which is significantly different from partitioning-based methods. Autopart, cross-associations [4], and information theoretic co-clustering [13] are parameter-free examples to graph clustering methods. Several methods [19,20] also allow clusters to overlap as observed in real-world social and communication networks. However, all of these methods are limited to plain graphs (without attributes). Compared to the wide range of works on plain graph mining, there has been much less works on attributed graphs. The representative methods [2,11] aim to partition the given graph into structurally dense and attribute wise homogeneous clusters. These methods, however, enforce attribute homogeneity in all attributes. Recently some methods loosen this constraint by unsupervised feature selection [1] to extract cohesive subgraphs with homogeneity in a subset of attributes. However, all of these methods either do not perform a selection of attributes, or select the attributes in a biased way.

Semi-supervised clustering applies a small amount of labeled data to aid and bias the clustering of unlabeled data [8]. There are various kinds of methods for semi-supervised clustering considering user-given pairwise constraints like 'must-link' and 'cannot-link' [10]. It is also known as constraint-based clustering where the constraints are often strict to follow [12]. However, most of these methods are based on vector data, thus they are not applicable to graphs with attributes. Methods on seeded community mining [19,22] find communities around (user-given) seed nodes. However, those methods find structural communities on plain graphs and neither are applicable to attributed graphs, nor enable user guidance on attributes. Our proposed method has two advantages compared with above mentioned methods. First, we apply user-given example sets to automatically infer the possible combination of representative attributes. Second, the constraints of traditional semi-supervised clusterings are hard, while the constraints given by different users are soft, causing the combination problem to be addressed in this study.

3   Method CGMA

In this section, we will present the framework of $CGMA$ to address the problem of using multiple annotations to guide attributed graph clustering. First of all, we give the formulation of our problem. *Given a large attributed graph* $G(V, E, F)$ *with* $|V| = n$ *nodes and* $|E| = m$ *edges, where each node is associated with* $|F| = d$ *attributes, we target to extract cluster C from G with the guidance of K users. Each user independently labels the samples based on his/her own knowledge. The samples annotated by the k-th user are denoted as U^k. For each set U^k, we assume that nodes inside it are similar to each other, and they are dissimilar to the nodes outside the set.*

3.1   Framework

The proposed approach $CGMA$ combines the annotations first in an unbiased way to obtain the guidance information. Then, a local clustering method is applied to cluster the graph with the guidance of combined annotations. Thus, $CGMA$ addresses the problem in two phases, the annotations combination and cluster extraction.

Annotations Combination. Since the annotations are sparse labels with little overlaps, straightforward methods like majority voting may not effectively capture the relations among the annotations. In this paper, we combine the annotations through each one's inferred weights in relevance to the feature space. Here are two major steps. The first step is mapping the annotations to the attribute space to facilitate measuring the similarity of the annotations. For different annotators, the attributes they think are essential to a particular cluster may be different due to their biased knowledge. Our first goal is to infer the attribute weights of $U^k (k \in \{1, \cdots, K\})$ that make the example nodes as similar to each other as possible. The similarity between two nodes can be measured by the (inverse) Mahalanobis distance: the distance between two nodes with feature vectors f_i and f_j is $(f_i - f_j)^T A^k (f_i - f_j)$. To ensure it as a metric, we set the weight matrix A^k as a positive definite matrix [3], and it denotes the attribute weight that is relevant to annotator $k's$ preference.

The process of learning A^k from annotation U^k is known as the distance metric learning problem [3]. The essence is to minimize the distance among the nodes in U^k. The optimal A^k can be obtained by solving the following convex optimization problem.

$$\min_{A^k} \sum_{(i,j) \in P_C^k} (f_i - f_j)^T A^k (f_i - f_j) - \gamma log(\sum_{(i,j) \in P_D^k} \sqrt{(f_i - f_j)^T A^k (f_i - f_j)}) \qquad (1)$$

Here, P_C^k and P_D^k denote the similar and dissimilar set of the k-th annotation, respectively. Following [7], we consider the annotated node pairs as similar set, and the un-annotated node pairs as dissimilar set. The un-annotated pairs are randomly selected from the edges of un-annotated part. To emphasize the difference between similar and dissimilar set, we set $|P_D^k| = d|P_C^k|$ by over-sampling of P_D^k [17]. According to [3], the above objective function is convex and enables efficient, local-minima-free algorithms to solve it, especially for a diagonal solution.

The second step in this phase is to combine the attribute weights A^k of each sample set. Since A^k is a diagonal matrix, we assign attribute vector $\beta^k = diag(A^k)(k \in 1, \cdots, K)$ and combine the vectors β^k according to its importance [15]. Each weight vector β^k can be viewed as a point in a d-dimensional Euclidean space, where the distances d_{ij} $(i, j \in 1, \cdots, K)$ between β^i and β^j be measured by Euclidean distance. For each point β^k, we first compute its local density ρ_k as its importance. Here, the local density of β^k refers to the number of points within a distance d_c to it (Eq. 2).

$$\rho_k = \sum_{l=1, l \neq k}^{K} \chi(d_{kl} - d_c) \qquad (2)$$

where $\chi(x) = 1$ if $x < 0$ and $\chi(x) = 0$ otherwise, and d_c is a distance threshold. The algorithm is only sensitive to the relative magnitude of ρ_k in different points.

Algorithm 1. Combination: The Combining of Annotations

Input: example annotations $U^1, \cdots, U^K$
Output: combined attribute weights vector β
1: //Computing attribute weights vectors
2: **for all** U^k **do**
3:     Similar pairs $P_S^k = \emptyset$, Dissimilar pairs $P_D^k = \emptyset$
4:     **for all** $u \in U^k$, $v \in U^k$ **do**
5:         $P_S^k = P_S^k \cup (u, v)$
6:     **end for**
7:     **repeat**
8:         Random sample u from set $V \backslash U^k$
9:         Random sample v from set $V \backslash U^k$
10:         $P_D^k = P_D^k \cup (u, v)$
11:     **until** $d|P_S|$ dissimilar pairs are generated, $d = |F|$
12:     Oversample from P_S such that $|P_S| = |P_D|$
13:     Solve objective function in Eq. (1) for diagonal A^k
14:     $\beta^k = diag(A^k)$
15: **end for**
16: //Combining the attribute weights vectors
17: **for all** β^k **do**
18:     Compute ρ_k by Eq. (2)
19: **end for**
20: Calculate $\beta = norm(\sum_k \rho_k \beta^k)$
21: **return** combined attribute weights vector β

Thus the results of analysis are robust with respect to the choice of d_c [15]. Finally, we get the combination β of weight vectors $\beta^1, \cdots, \beta^K$, according to each vector's importance ρ_k.

We give the details of combining the annotation results in Algorithm 1. The step of inferring the attribute weights of an annotation is illustrated in $A1$ Lines 2–15, and the combination of the attribute vectors is shown in $A1$ Lines 17–22. In our setting, all pairs of example nodes in U^k constitute P_S^k ($A1$ Line 3). We create P_D^k by randomly drawing pairs of nodes that do not belong to user k's example set ($A1$ Lines 7–11). Note that if $\rho_k = 0$ ($A1$ Line 18), β^k will have no contribution to the combined vector β. That denotes user k's opinion will be ignored ($A1$ Line 20). In the last step, we get a combined β, and then we normalize it.

Cluster Extraction. We use the information of combined annotations to extract the target cluster from the graph. Since a global clustering method would be time-consuming and can not scale well to large graph, we apply a seed-set-expansion algorithm to identify clusters locally with lower computational complexity.

There are two major problems in this step. First, how to extract the seed set samples of the cluster based on the different annotations from multiple annotators and the combined attribute weight vector? Second, how to develop a local partitioning method so that the expansion of seed sets will be scalable to large graphs? Therefore, we explain our algorithm focusing on two parts, the identification of seed set S of the cluster and its expansion rules.

In the process of identifying a pure seed set, we first apply the combined vector β to re-weigh the entire graph, then select the edges with high weight (similarity) to shape the seed set. We call this process as "re-sampling", which aims to enrich the samples space for the expansion process. Specifically, we firstly measure the weights of all the edges. Then, we assign the edges with high weights over threshold w_r as seeds. Simply, we assign a linear interpolation as w_r over the weights of samples, $w_r = \lambda w_{max} + (1 - \lambda) w_{min}$, where λ is a parameter falls in $[0, 1]$. w_{max} and w_{min} represent the maximum and minimum value of the example edges weighted by β, respectively. Algorithm 2 details the process of finding the pure seed set by re-sampling.

Next, we expand the seed set S to the target clusters C through a series of strict rules. Following the expansion process in [19], the expansion process carefully adds new nodes to each component of S. In this paper, we apply conductance [19] to measure the quality of a cluster as it accounts for both the cut size and the total volume/density retained within the cluster. The weighted conductance $\phi^{(w)}(S, G)$ of a set of nodes in graph $G(V, E, F)$ is defined as follows,

$$\phi^{(w)}(S, G) = \frac{W_{cut}(S)}{W_{vol}(S)} = \frac{\sum_{(i,j) \in E, i \in S, j \in V \setminus S} w_{ij}}{\sum_{(i,j) \in S} \sum_{(i,j) \in E} w_{ij}} \tag{3}$$

Here, $W_{cut}(S)$ and $W_{vol}(S)$ are the total weight of cut edges and within edges of S, respectively. The lower the conductance of a cluster is, the better the quality of the cluster is with few cross-cut edges and large within-density.

In each step, the expansion process selects all the nodes in the margin of a component, and adds the ones that will decrease the conductance of the cluster. The process will simultaneously kick out the (nodes) edges within a cluster that will decrease the conductance of the cluster. The process continues until there is no node changing that would decrease the quality of a component. Due the page limitation, we do not illustrate the algorithm, please refer [15] for more details.

3.2 Complexity Analysis

(1) The combination of annotations. Since every annotator provides the same amount of examples, we take $U^k (k \in 1, \cdots K)$ as an example to analyze the time complexity of this step. First, we create similar and dissimilar node pairs which we use to infer the attribute weights. Since the optimization objective in Eq. 1 is convex and we aim to find a diagonal solution, local-optima-free gradient descent techniques will take $O(d/\epsilon^2)$ for an ϵ−approximate answer [26]. The clustering process of combination is not time consuming because the number of annotators is significantly small to the data scale, $K << n$. According to [15], we calculate that the complexity is $O(K^2)$. Therefore, the total computational complexity of the first part is $O(Kd/\epsilon^2) + O(K^2)$.

Algorithm 2. Find seed set by re-sampling

Input: attributed graph $G(V, E, F)$, combined weight vector β, annotations
$\quad U^1, \cdots, U^K$

Output: seed set S for expansion

1: re-weigh edges by β getting edge re-weight $w(u, v)$
2: **for all** $(u, v) \in E$ **do**
3: $\quad w(u, v) = 1/(\sqrt{(f_u - f_v)^T diag(\beta)(f_u - f_v)} + \epsilon)$
4: **end for**
5: seed node set $V' = \emptyset$
6: $w_{max}(w_{min}) = max(min)\{w(u, v)|u, v \in \{U^1 \cup \cdots \cup U^K\}\}$
7: **if** $w(u, v) > w_r = \lambda w_{max} + (1 - \lambda)w_{min}$ **then**
8: $\quad$ seed nodes $V' = V' \cup \{u, v\}$
9: **end if**
10: build seed set graphs $g(V', E', F)$ where
11: $\forall u, v \in V', (u, v) \in E, w(u, v) \geq w'\text{iff}(u, v) \in E'$
12: seed set $S \leftarrow G(V', E', F)$
13: **return** seed set S

---

(2) The finding and expansion of seed set. Since β is supposed to be sparse with only a few non-zero entries for focused attributes, the multiplicative factor becomes effectively constant yielding a complexity of $O(m)$. In the process of expansion, we enlist all the non-member neighbors as the candidate set and evaluate their weighted Δ conductance. As discussed above, the complexity is $\sum_{n \in S} d(n)$. Since $S \subseteq V$, it is equivalent $O(m)$. As we add one node at each iteration, the total complexity becomes $O(|S|m)$ where $|S|$ is the node scale of the seed set, and $|S| << n$.

To sum up, the complexity of the two phases comes to $O(Kd/\epsilon^2 + K^2 + |S|m)$. It is critically low comparing to the large scale of graphs.

4   Experiments

In order to evaluate the clustering quality and scalability of $CGMA$, we compare it with two representative graph clustering techniques $METIS$ [27] and $FocusCO$ [7] on real-world datasets. $METIS$ is a classical graph partitioning algorithm which expects the number of clusters as input. $FocusCO$ is a local clustering approach proposed recently using the guidance of a single user.

To introduce $CGMA$ clearly, we conduct our experiments on the "four-area" dataset, a co-authorship network of computer science researchers. The attributes of the authors are the conferences in which they have published papers in the areas of database (DB), data mining (DM), machine learning (ML), and information retrieval (IR). We use multiple annotations from 50 persons, each person gives 20 sample nodes in responding to the same question in one experiment. For the convenience of study, our problem is identifying the researchers belonging to the four areas, respectively. The ground truth clusters of an area consists of all the researchers whoever published at least one paper in the area.

Since the re-sampling process affects the final clusters significantly, we conduct a parameter study of λ, and we show the F1-score of the clustering results with different settings of λ. As shown in Fig. 2, the F1-score of the final clustering results is not linearly related to λ. One can see that without the re-sampling step the F1-score of the final results in each of experiments is critically low, about or less than half of the value when $\lambda = 0.4$. The F1-score of the final clustering results presents that the re-sampling properly will improve clustering performance significantly.

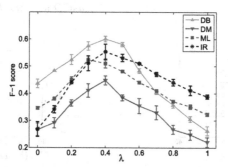

Fig. 2. The λ effects on the clustering results, 50 annotations in each of the experiment.

Accuracy. We compare the cluster results with *METIS* and *FocusCO*. Here we set the clusters number of *METIS* as four, which performs the best on this dataset. As shown in Fig. 3, one can see that F1-score of our method is significantly higher than that of the two baselines, which shows the superior performance of the proposed method. The experimental results show that our method significantly outperforms the other two methods.

Stability. We also examine the stability of the proposed *CGMA*. Although we have different annotations as inputs, they are annotated under the same question. Therefore, the annotations are all theoretically related a common clustering. We use the normalized mutual information (NMI) to evaluate the stability of the proposed clustering approach. Here, we use average NMI between each pair of clustering results to indicate the stability of a method. Higher NMI implies a more stable clustering result. As shown in Fig. 5, the proposed approach *CGMA* gets more stable results than other two methods. With the increasing of λ, the stability of *CGMA* improves significantly.

Scalability. We select five subsets of "four-area" with the size from 100 to 27200. Each dataset scale is three times larger than its previous one. Then we conduct extensive experiments on these datasets. Note that the annotation volumes change with the scale of experimental graphs. Larger scale of the graph needs more annotations. For each dataset, we run the experiments for ten times and average the results. The experimental results are shown in Fig. 5. Note that for *CGMA*, the extraction of cluster can be performed in parallel, thus the

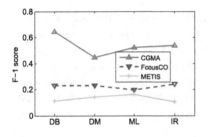

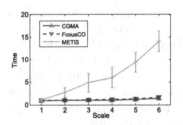

Fig. 3. The accuracy of *CGMA*, *FocusCO* and *METIS*.

Fig. 4. The scalability of *CGMA*, *FocusCO* and *METIS*.

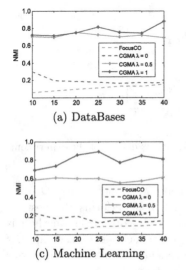

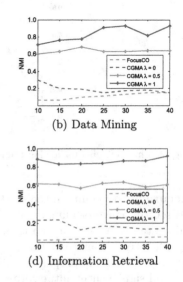

Fig. 5. The comparison of average NMI value in the 4 clusterings. The horizontal axis in all sub-figures represents the number of annotations we randomly selected.

computing time can be significantly reduced. As the figure shown, the running time of *METIS* increases with the increasing of graph scale. However, the running time of *CGMA* and *FocusCO* is stable. In such case, the running time of *CGMA* is also comparable to *METIS* and *FocusCO* (Fig. 4).

To further examine the scalability of *CGMA*, we conduct more experiments on two different types of real world attributed networks. The first is crawled from *PolBlogs*, a citation network among a collection of online blogs that discuss political issues. The attributes of *PolBlogs* are the keywords in the blogs. The second dataset is crawled from *Twitter*, and it is a following network with a collection of discussed topics. The attributes are the keywords in their posts. Statistics of the datasets are given in also given in Table 1. The average running time (total) and their standard deviations are in Table 1. The running time of *CGMA* is the total running time including the annotation combination and

Table 1. Comparisons on the Scalability of $CGMA$

Dataset	$|V|$	$|E|$	$|F|$	$|C|$	Running time (sec)
PolBlog	362	1288	44839	10	0.4772 ± 0.0591 (**CGMA**)
					0.8772 ± 0.0839 (**METIS**)
					3.0561 ± 0.0471 (**FocusCO**)
Twitter	14078	44619	17839	10	1.2135 ± 0.0322 (**CGMA**)
					1.9425 ± 0.0381 (**METIS**)
					6.8772 ± 0.0491 (**FocusCO**)

clustering extractions steps. As it shows, the running time demonstrates the efficiency of our approach. It only takes less than 2 s to cluster the Twitter dataset with more than 14 thousand nodes, which shows CGMA can be scalable to very large graphs. The experimental results prove that $CGMA$ is a scalable approach that can deal with various datasets.

5 Conclusions

In this work, we introduced a novel problem of finding clusters with multi-example sets in large attributed graphs. The challenge here is how to combine them in an unbiased way in order to conduct a clustering. To address these challenges, we proposed $CGMA$ in this paper which has two major phases: (1) combining the various example sets, (2) re-sampling the seed sets and expanding them to find a batch of densely connected clusters. Extensive experiments are conducted to examine the $CGMA$, and the experimental results showed that the proposed approach outperforms baseline methods.

References

1. Tang, J., Liu, H.: Unsupervised feature selection for linked social media data. In: SIGKDD (2012)
2. Akoglu, L., Tong, H., Meeder, B., Faloutsos, C.: PICS: parameter-free identification of cohesive subgroups in large attributed graphs. In: SDM (2012)
3. Xing, E.P., Jordan, M.I., Russell, S., et al.: Distance metric learning with application to clustering with side-information. Adv. Neural Inf. Process. Syst. **15**, 505–512 (2002)
4. Yin, X., Han, J., Yu, P.S.: Cross-relational clustering with user's guidance. In: SIGKDD (2005)
5. Yin, X., Han, J., Yu, P.S.: CrossClus: user-guided multi-relational clustering. In: SIGKDD (2007)
6. Sun, Y., Norick, B., Han, J., et al.: Integrating meta-path selection with user-guided object clustering in heterogeneous information networks. In: SIGKDD (2012)
7. Perozzi, B., Akoglu, L., Iglesias Snchez, P., et al.: Focused clustering and outlier detection in large attributed graphs. In: SIGKDD (2014)

8. Basu, S., Banerjee, A., Mooney, R.: Semi-supervised clustering by seeding. In: ICML (2002)
9. Sánchez, P.I., Muller, E., Laforet, F., et al.: Statistical selection of congruent subspaces for mining attributed graphs. In: ICDM (2013)
10. Chapelle, O., Schölkopf, B., Zien, A., et al.: Semi-Supervised Learning. MIT Press, Cambridge (2006)
11. Zhou, Y., Cheng, H., Yu, J.X., Zhou, Y., Cheng, H., Yu, J.X.: Graph clustering based on structural, attribute similarities. J. VLDB **2**(1), 718–729 (2009)
12. Han, J., Kamber, M., Pei, J.: Data Mining: Concepts and Techniques. Elsevier, Amsterdam (2011)
13. Dhillon, I.S., Mallela, S., Modha, D.S.: Information-theoretic co-clustering. In: SIGKDD (2003)
14. Ng, A.Y., Jordan, M.I., et al.: On spectral clustering: analysis and an algorithm. In: NIPS (2002)
15. Rodriguez, A., Laio, A.: Clustering by fast search and find of density peaks. Science **344**, 1492–1496 (2014)
16. Flake, G.W., Lawrence, S., Giles, C.L.: Efficient identification of web communities. In: SIGKDD (2000)
17. Wang, S., Li, Z., Chao, W.-H., Cao, Q.: Applying adaptive over-sampling technique based on data density and cost-sensitive SVM to imbalanced learning. In: IJCNN (2012)
18. Zhou, D., Liu, Q., Platt, J.C., Meek, C.: Aggregating ordinal labels from crowds by minimax conditional entropy. In: ICML (2014)
19. Andersen, R., Chung, F., Lang, K.: Local graph partitioning using pagerank vectors. In: IEEE SFCS (2006)
20. Yang, J., Leskovec, J.: Overlapping community detection at scale: a nonnegative matrix factorization approach. In: WSDM (2013)
21. Tong, H., Lin, C.-Y.: Non-negative residual matrix factorization with application to graph anomaly detection. In: SDM (2011)
22. Gleich, D.F., Seshadhri, C.: Vertex neighborhoods, low conductance cuts, and good seeds for local community methods. In: SIGKDD (2012)
23. Wang, S., Xie, S., Zhang, X., Li, Z., Philip, S.Y., Xinyu, S.: Future influence ranking of scientific literature. In: SDM (2014)
24. Ruvolo, P., Whitehill, J., Movellan, J.R.: Exploiting commonality and interaction effects in crowdsourcing tasks using latent factor models. In: NIPS (2013)
25. Zhou, D., Basu, S., Mao, Y., Platt, J.C.: Learning from the wisdom of crowds by minimax entropy. In: NIPS (2012)
26. Boyd, S., Vandenberghe, L.: Convex Optimization. Cambridge University Press, Cambridge (2004)
27. Karypis, G., Kumar, V.: Multilevel algorithms for multi-constraint graph partitioning. In: ACM/IEEE Conference on Supercomputing (1998)

Early-Stage Event Prediction
for Longitudinal Data

Mahtab J. Fard[1]([✉]), Sanjay Chawla[2,3], and Chandan K. Reddy[1]

[1] Computer Science Department, Wayne State University, Detroit, MI 48202, USA
mahtab.jahanbanifard@wayne.edu, reddy@cs.wayne.edu
[2] Qatar Computing Research Institute, HBKU, Ar-rayyan, Qatar
schawla@qf.org.qa
[3] University of Sydney, Sydney, NSW, Australia
sanjay.chawla@sydney.edu.au

Abstract. Predicting event occurrence at an early stage in longitudinal studies is an important problem which has high practical value. As opposed to the standard classification and regression problems where a domain expert can provide the labels for the data in a reasonably short period of time, training data in such longitudinal studies must be obtained only by waiting for the occurrence of sufficient number of events. The main objective of this work is to predict the event occurrence in the future for a particular subject in the study using the data collected at the initial stages of a longitudinal study. In this paper, we propose a novel Early Stage Prediction (ESP) framework for building event prediction models which are trained at early stages of longitudinal studies. More specifically, we develop two probabilistic algorithms based on Naive Bayes and Tree-Augmented Naive Bayes (TAN), called ESP-NB and ESP-TAN, respectively, for early stage event prediction by modifying the posterior probability of event occurrence using different extrapolations that are based on Weibull and Lognormal distributions. The proposed framework is evaluated using a wide range of synthetic and real-world benchmark datasets. Our extensive set of experiments show that the proposed ESP framework is able to more accurately predict future event occurrences using only a limited amount of training data compared to the other alternative approaches.

Keywords: Prediction · Regression · Longitudinal data · Survival analysis

1 Introduction

Developing effective prediction models to estimate the outcome of a particular event of interest is a critical challenge in various application domains such as healthcare, reliability, engineering, etc. [12]. In longitudinal studies, event prediction is an important area of research where the goal is to predict the event occurrence during a specific time period of interest [9]. Obtaining training data

© Springer International Publishing Switzerland 2016
J. Bailey et al. (Eds.): PAKDD 2016, Part I, LNAI 9651, pp. 139–151, 2016.
DOI: 10.1007/978-3-319-31753-3_12

for such a time-to-event problem is a daunting task. As opposed to the standard supervised learning problems where a domain expert can provide labels in a reasonable amount of time, training data for longitudinal studies must be obtained only by waiting for the occurrence of sufficient number of events. Therefore, the ability to leverage only a limited amount of available information at early stages of longitudinal studies to forecast the event occurrence at future time points is an important and challenging research task.

Let us consider an illustrative example shown in Fig. 1. In this example, a longitudinal study is conducted on 5 subjects and the information for event occurrence until time t_c is recorded, where only subjects B and E have experienced the event. The goal of our paper is to predict the event occurrence by the time t_f where t_f is much greater than t_c. It can be seen that, except subjects B and E, all the remaining subjects are considered to be censored at t_c (marked by red 'x') and the event will occur for subject A within the time period t_f. This scenario is applicable for many real-world applications where it is critical to obtain early stage time-to-event predictions. For example, in the healthcare domain, let us say that there is a new treatment option (or drug) which is available and one would like to study the effect of such a treatment on a particular group of patients in order to understand the efficacy of the treatment. This patient group is monitored over a period of time and an event here corresponds to the patient being hospitalized (or occurrence of death) because the treatment has failed. The effectiveness of this treatment must be estimated as early as possible when there are only a few hospitalized patients.

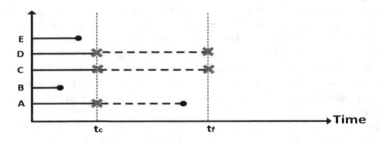

Fig. 1. An illustration to demonstrate the problem of early stage event prediction for time t_f using the information of event occurrence until time t_c.

This practical problem clearly emphasizes the need to build algorithms that can effectively predict events using the training data that contains only the event information at an early stage of a longitudinal study. It should be noted that the previous research in the field of statistics mainly focuses on the prediction of survivability up to a certain specific time point. Predicting events at future timepoints using the available information at the initial phases of the study remains to be a relatively unexplored area of research. Thus, in this paper, we develop prediction models using the data collected at earlier time points in longitudinal studies. More specifically, the contributions of this paper are as follows:

- Propose an **E**arly **S**tage **P**rediction (ESP) framework which estimates the probability of event occurrence for a future timepoint using different extrapolation techniques.
- Develop a probabilistic algorithms based on Naive Bayes and Tree-Augmented Naive Bayes (TAN), called ESP-NB and ESP-TAN, respectively, for early-stage event prediction by modifying the posterior probability of event occurrence.
- Evaluate the proposed algorithms using several synthetic and real-world benchmark datasets.

This paper is organized as follows. In Sect. 2, we present a summary of existing works on using survival analysis and machine learning methods for longitudinal data. In Sect. 3, we explain the problem formulation and describe two probabilistic classifiers, namely, Naive Bayes and Tree-Augmented Naive Bayes. In Sect. 4, we introduce the proposed extrapolation methods and then explain our novel Early Stage Prediction (ESP) framework based on Naive Bayes and TAN algorithms. In Sect. 5, the results of the proposed methods along with those of the competing algorithms on various synthetic and real-world datasets are presented. In the last section, we conclude our paper with a summary of the main results of the proposed work.

2   Related Work

Survival analysis is a subfield of statistics where a wide range of techniques have been proposed to model time-to-event data (e.g., failure, death, admission to hospital, emergence of disease, etc.) [13]. For such a time-to-event prediction problem, there have also been many attempts using different machine learning methodologies that were modified and applied to this problem [19,21]. On the other hand, longitudinal data cannot be modeled solely using traditional classification or regression approaches since certain observations have event status and the rest have an unknown status up until that specific time of study.

Several machine learning approaches have been adapted to handle the concept of censoring in survival data [15]. Modifications of decision trees [8,17], artificial neural networks [6] and support vector machines [11,18] represent some of the works on this topic. Another popular choice in the predictive modeling literature is the Bayesian approach. However, there was only a little work in the literature using Bayesian methods for survival data [1,14,20].

The work that is being developed in this paper is significantly different from the above mentioned algorithms since none of the existing works perform forecasting of event occurrence at future points in the context of survival data. They basically use the training data that is collected at the same time point as the test data. The basic idea of the proposed model is to develop Naive Bayes and its extension Tree-Augmented Naive Bayes (TAN), to build a predictive probabilistic model which will allow us to adapt the prior probability of events for forecasting the event occurrence at different points of time in the future. It is important to note that discriminative models are not suitable for the forecasting framework due to the lack of the prior probability component.

3   Preliminaries

The aim of our work is to address the following question: "when will a subject in longitudinal study experience an event?" The fundamental challenge here is to determine which subject in the study will experience the event at a certain timepoint based on event occurrence information that is available only until prior points of time (usually much earlier than the timepoint used during estimation). Before describing the details of the proposed model, we formalize the problem and transform it to a binary classification task. Then, we describe two well-known probabilistic classification approaches, namely, Naive Bayes and Tree-Augmented Naive Bayes (TAN). Table 1 describes the notations used in this paper.

Table 1. Notations used in this paper

Name	Description
n	Number of samples
m	Number of features
$\mathbf{x}$	$n \times m$ data matrix
T	$n \times 1$ vector of event times
C	$n \times 1$ vector of last follow-up times
O	$n \times 1$ vector of observed time which is $min(T, C)$
δ	$n \times 1$ binary vector of censored status
t_c	Specified time until which information is available
t_f	Desired time at which the forecast of future events is made
$y_i(t)$	Event status for subject i at time t

3.1   Problem Formulation

Let us consider a longitudinal study where the data about n independent subjects are available. Let the feature vector for sample i be represented by $\mathbf{x_i} = \langle x_{i1}, ..., x_{im} \rangle$ where x_{ij} is the jth feature for subject i. For each subject i, we can define T_i as the event time, and C_i as the last follow-up time or censoring time (the time after which the subject has left the study). For all the subjects $i = \{1, ..., n\}$, O_i denotes the observed time which is defined as $min(T_i, C_i)$. Then, the event status can be defined as $\delta_i = \mathbf{I}\{T_i \leq C_i\}$. Thus, a longitudinal dataset can be represented as $(\mathbf{x_i}, T_i, \delta_i)$ where $\mathbf{x_i} \in \mathbf{R}^m$, $T_i \in \mathbf{R}^+$, $\delta_i \in \{0, 1\}$.

It should be noted that we only have the information for few events until the time t_c. Our aim is to predict the event status at time t_f where $t_f > t_c$. Let us define $y_i(t_c)$ as event status for subject i at time t_c. We consider t_c to be less than the observation time since we aim to forecast the event occurrence at early stage of the study. Suppose, among n subjects in the study, only $n(t_c)$ will experience the event at time t_c. For each subject i we can define

$$y_i(t_c) = \begin{cases} 1 & \text{if } O_i \leq t_c \text{ and } \delta_i = 1, \\ 0 & \text{otherwise} \end{cases}$$

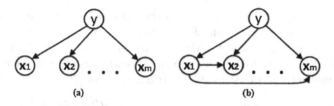

Fig. 2. An illustration of the basic structure of (a) Naive Bayes and (b) TAN classifier.

In this transformed formulation, given the training data $(x_i, y_i(t_c))$, we can build a binary classifier using $y_i(t_c)$ as the class label. If $y_i(t_c) = 1$, then the event has occurred for subject i and if $y_i(t_c) = 0$, the event has not occurred. It should be noted that a new classifier will have to be built to estimate the probability of event occurrence at t_f based on the training data that is available at t_c.

3.2 Naive Bayes Method

Naive Bayes is a well-known probabilistic model in the machine learning domain. Assume we have a training set in Fig. 1 where the event occurrence information is available up to time t_c. Based on the binary classification transformation explained above, using Naive Bayes algorithm, the event probability can be estimated as follows:

$$P\big(y(t_c) = 1 \mid \mathbf{x}, t \le t_c\big) = \frac{P\big(y(t_c) = 1, t \le t_c\big) \prod_{j=1}^{m} P\big(\mathbf{x_j} \mid y(t_c) = 1\big)}{P(\mathbf{x}, t \le t_c)} \quad (1)$$

The first component of the numerator is the prior probability of the event occurrence at time t_c. The second component is a conditional probability distribution which can be estimated as follows:

$$P\big(\mathbf{x_j} \mid y(t_c) = 1\big) = \frac{\sum_{i=1}^{n} \big(y_i(t_c) = 1, x_{ij} = \mathbf{x_j}\big)}{\sum_{i=1}^{n} \big(y_i(t_c) = 1\big)} \quad (2)$$

where x_{ij} is the value of attribute j for subject i. Thus, it is a natural estimate for the likelihood function in Naive Bayes to count the number of times that event occurred at time t_c in conjunction with j^{th} attribute that takes a value of x_j. Then we count the number of times the event occurred at time t_c in total and finally take the ratio of these two terms. This formula is valid for discrete attributes; However, it can be easily adapted for continues variables as well [10].

3.3 Tree-Augmented Naive Bayes Method

A prominent extension of Naive Bayes is the Tree-Augmented Naive Bayes (TAN) where the independence assumption between the attributes is relaxed [7]. The TAN algorithm imposes a tree structure on the Naive Bayes model by restricting the interaction between the variables to a single level. This method allows every attribute x_j to depend upon the class as well as at most one other attribute, $x_p(j)$, called the parent of x_j. Illustration of the basic structure of the dependency in Naive Bayes and TAN is shown in Fig. 2. Given the training set $(x, y(t_c))$, firstly the tree for the TAN model should be constructed based on the conditional mutual information between two attributes [7].

$$I(x_j, x_k \mid y(t_c)) = \sum_{x_j, x_k, y(t_c)} P(x_j, x_k, y(t_c)) \frac{P(x_j, x_k \mid y(t_c))}{P(x_j \mid y(t_c)) P(x_k \mid y(t_c))} \qquad (3)$$

Then, a complete undirected graph in which the vertices correspond to the attributes x_j is constructed. Using Eq. (3), the weight of all the edges can be computed. A maximum weighted spanning tree is built and finally, an undirected tree is transformed into a directed one by randomly choosing a root variable and setting the direction of all the edges outward from the root. After the construction of the tree, the conditional probability of each attribute on its parent and the class label is calculated and stored. Hence, the probability of event at time t_c, can be defined as follows:

$$P(y(t_c) = 1 \mid x, t \leq t_c) = \frac{P(y(t_c) = 1, t \leq t_c) \prod_{j=1}^{m} P(x_j \mid y(t_c) = 1, x_p(j))}{P(x, t \leq t_c)}$$

$$(4)$$

The numerator consists of two components; the prior probability of the event occurrence at time t_c and the conditional probability distributions which can be estimated using the maximum likelihood estimation (MLE).

4   The Proposed ESP Framework

In this section, we describe the proposed **E**arly **S**tage **P**rediction (ESP) framework. First, we describe our proposed prior probability extrapolation based method using different distributions and then we will introduce ESP-NB and ESP-TAN algorithms which utilize the extrapolation method.

4.1   Prior Probability Extrapolation

In order to predict event occurrence in longitudinal data, we develop a technique that can estimate the ratio of event occurrence beyond the original observation range or in other words, compute the *extrapolation for prior probability of event occurrence*. This extrapolation approach will be based on Weibull and Lognormal distributions which are used widely in the literature for modeling the time-to-event data [3,16]. We will integrate such extrapolated values later with the proposed learning algorithms in order to make predictions at future timepoints.

Weibull: We estimate the shape and scale parameters, α_{t_c} and β_{t_c}, in Weibull distribution, by fitting the distribution to data obtained until t_c and then making the following extrapolation

$$p(t_f) = \frac{t_f^{\alpha-1}}{\beta^\alpha} exp\big(-(t_f/\beta)^\alpha \big) \qquad (5)$$

Lognormal: We can also assume that the time to event follows a log-normal distribution, and then we can estimate μ_{t_c} and σ_{t_c}, mean and standard deviation of log-normal distribution, from the training data. The extrapolation is given as follows:

$$p(t_f) = \frac{1}{\sqrt{2\pi}\sigma_{t_c}t_f}exp-\left(log(t_f)-\mu_{t_c}\right)^2/2\sigma_{t_c}^2. \tag{6}$$

4.2 The ESP Algorithm

We will now describe the ESP Algorithm which consists of two phases. In the first phase, the conditional probability distribution is estimated using training data which is obtained until time t_c (see Sects. 3.2 and 3.3). In the second phase, we extrapolate the prior probability of event occurrence for time t_f which is beyond the observed time using different extrapolation techniques as follows:

$$P\left(y(t_f) = 1, t \leq t_f\right) = p(t_f) \tag{7}$$

It should be noted that the Eq. (7) can be estimated using Eqs. (5) and (6). Thus, the posterior probability for event occurrences at time t_f can be estimated as:

ESP-NB:

$$P\left(y(t_f) = 1 \mid \mathbf{x}, t \leq t_f\right) = \frac{p(t_f) \prod_{j=1}^{m} P\left(\mathbf{x_j} \mid y(t_c) = 1\right)}{P(\mathbf{x}, t \leq t_f)}. \tag{8}$$

ESP-TAN:

$$P\left(y(t_f) = 1 \mid \mathbf{x}, t \leq t_f\right) = \frac{p(t_f) \prod_{j=1}^{m} P\left(\mathbf{x_j} \mid y(t_c) = 1, \mathbf{x_p}(j)\right)}{P(\mathbf{x}, t \leq t_f)}. \tag{9}$$

Algorithm 1 outlines the proposed ESP method. In the first phase (lines 1–4), for each attribute j, the algorithm estimates the conditional probability using the data available until time t_c. In the second phase, a probabilistic model is built to predict the event occurrence at t_f. In lines 5–7, the prior probability for event occurrence at time t_f is estimated using different extrapolation techniques. Then, in lines 8–12, for each subject i, we adapt the posterior probability of event occurrence at time t_f. The time complexity of the ESP algorithm follows the time complexity of the learning method that is chosen. It should be noted that the complexity of the extrapolation component is a constant and does not depend on either m or n. Hence, for ESP-NB, the overall complexity is $O(mn)$ and for ESP-TAN, it is $O(m^2n)$, where n is the total number of subjects and m is the number of features in the dataset.

5   Experimental Results

In this section, we will describe the datasets that are used for evaluating the proposed methods along with the comparisons of the proposed algorithms with various baseline prediction methods.

Algorithm 1. Early Stage Prediction (ESP) Framework

Require: Training data $D(t_c) = (\mathbf{x}, y(t_c), T)$, t_f
Output: Probability of event at time t_f
Phase 1: Conditional probability estimation at t_c
1. **for** $j = 1, ..., m$
2.   Naive Bayes: $P(\mathbf{x_j} \mid y(t_c) = 1)$ (Eq. (2))
3.   TAN: $P(\mathbf{x_j} \mid y(t_c) = 1, x_p(j))$ (Eq. (3))
4. **end**
Phase 2: Predict probability of event occurrence at t_f
5. Estimate $P(y(t_f) = 1, t \leq t_f)$
6.   Weibull: $t_f{}^{\alpha-1}/\beta^\alpha exp(-(t_f/\beta)^\alpha)$ (Eq. (5))
7.   Lognormal: $1/\sqrt{2\pi}\sigma_{t_c} t_f exp-(log(t_f)-\mu_{t_c})^2/2\sigma_{t_c}^2$ (Eq. (6))
8. **for** $i = 1, ..., n$
9.   Estimate $P(y_i(t_f) = 1 \mid \mathbf{x}_i, t \leq t_f)$
10.   ESP-NB: Eq. (8)
11.   ESP-TAN: Eq. (9)
12. **end**
13: return $P(y(t_f) = 1 \mid \mathbf{x}, t \leq t_f)$

5.1   Dataset Description

We evaluated the performance of the models using both synthetic and real-world survival datasets which are summarized in Table 2.

Synthetic Datasets: We generated synthetic dataset in which the feature vectors $\mathbf{x}$ are generated based on a normal distribution $N(0,1)$. Covariate coefficient vector β is generated based on a uniform distribution $Unif(0,1)$. Thus, T can be generated using the method described in [2]. Given the observed covariates $\mathbf{x}_i$ for observation i, the failure time can be generated by

$$T_i = -\left(\frac{log(Unif(0,1))}{\lambda \dot{e}xp(\beta'\mathbf{x}_i)}\right)^\nu \tag{10}$$

In our experiments, we set $\lambda = 0.01$ and $\nu = 2$.

Real-World Survival Datasets: Several real-world survival benchmark datasets were used in our experiments. We used primary biliary cirrhosis (PBC), breast and colon cancer datasets (available in the survival data repository[1]) which are widely used in evaluating longitudinal studies. We also used Framingham heart study dataset which is publicly available [4]. In addition, we also used two in-house proprietary datasets. One is the electronic health record (EHR) data from heart failure patients collected at the Henry Ford Health System in Detroit, Michigan. This data contains patient's clinical information such as procedures, medications, lab results and demographics and the goal here is to predict the number of days for the next readmission after the patient is discharged from

---
[1] http://cran.rproject.org/web/packages/survival/.

Table 2. Number of features, instances and events. T_{50} and T_{100} corresponds to the time taken for the occurrence of 50 % and 100 % of the events, respectively.

Dataset	#Features	#Instances	#Events	T_{50}	T_{100}
Syn1	5	100	50	1014	3808
Syn2	20	1000	602	943	7723
Breast	8	673	298	646	2659
Colon	13	888	445	394	3329
PBC	17	276	110	1191	4456
Framingham	16	5209	1990	1991	5029
EHR	77	4417	3479	50	4172
Kickstarter	54	4175	1961	21	60

the hospital. Another dataset was obtained from Kickstarter, a popular crowd-funding platform. Each project has been tracked for a specific period of time. If the project reaches the desired funding goal within deadline date then it is considered to be a success (or event occurred). On the other hand, the project is considered to be censored if it fails to reach its goal within the deadline date.

5.2 Performance Evaluation

The performance of the proposed models is measured using following metrics,

- *AUC* is the area under the receiver operating characteristic (ROC) curve. The curve is generated by plotting the true positive rate (TPR) against the false positive rate (FPR) by varying the threshold value.
- *F-measure* is defined as a harmonic mean of precision and recall. A high value of *F*-measure indicates that both precision and recall are reasonably high.

$$F - measure = \frac{2 \times Precision \times Recall}{Precision + Recall}.$$

Implementation Details: The proposed ESP-NB and ESP-TAN methods are implemented using *e1071* package available in the R programming language [5]. The same package used for comparison results from Naive Bayes and TAN classification model. The *coxph* model in the survival package is employed to train the Cox model. The source code of the proposed algorithms in R programming environment is available at http://dmkd.cs.wayne.edu/codes/ESP.

5.3 Results and Discussion

For performance benchmarking, we compare the proposed ESP-NB and ESP-TAN algorithms using Weibull and Lognormal distributions as extrapolation techniques with Cox regression, Naive Bayes (NB) and Tree-Augmented Naive Bayes (TAN) classification methods which are trained at time when only 50 %

Table 3. Comparison of AUC values for Cox, NB and TAN with proposed ESP-NB and ESP-TAN methods using Weibull (W) and Lognormal (L) extrapolation methods (with standard deviation values).

Data	AUC						
	Cox	NB	TAN	ESP-NB(W)	ESP-NB(L)	ESP-TAN(W)	ESP-TAN(L)
Syn1	0.697	0.702	0.713	0.865	0.841	**0.865**	0.849
	(0.004)	(0.007)	(0.002)	(0.003)	(0.003)	**(0.001)**	(0.001)
Syn2	0.703	0.699	0.705	0.818	0.811	**0.821**	0.817
	(0.003)	(0.009)	(0.005)	(0.002)	(0.003)	**(0.002)**	(0.002)
Breast	0.612	0.621	0.632	0.655	0.633	**0.662**	0.635
	(0.011)	(0.009)	(0.004)	(0.001)	(0.003)	**(0.007)**	(0.005)
Colon	0.601	0.615	0.617	0.621	0.617	**0.627**	0.619
	(0.024)	(0.011)	(0.014)	(0.013)	(0.014)	**(0.009)**	(0.011)
PBC	0.665	0.643	0.679	0.765	0.761	**0.768**	0.763
	(0.009)	(0.003)	(0.01)	(0.001)	(0.004)	**(0.003)**	(0.001)
Framingham	0.863	0.945	0.953	0.953	0.959	0.961	**0.971**
	(0.006)	(0.002)	(0.005)	(0.007)	(0.003)	(0.004)	**(0.002)**
EHR	0.612	0.633	0.638	0.654	0.624	**0.649**	0.628
	(0.022)	(0.019)	(0.025)	(0.018)	(0.021)	**(0.011)**	(0.026)
Kickstarter	0.761	0.811	0.816	0.821	0.825	0.822	**0.831**
	(0.018)	(0.022)	(0.025)	(0.024)	(0.023)	(0.019)	**(0.018)**

of events have occurred and the event prediction is done at the end of study. Tables 3 and 4 summarize the comparison result in AUC and F-measure evaluation metrics, respectively. We used stratified 10-fold cross-validation and average values (along with the standard deviations) of the results on all the ten folds are being reported. For all of the datasets, our results evidently show that the proposed ESP-based methods using either Weibull or lognormal distribution will provide significantly better prediction results compared to the other methods. The choice of the optimal distribution will depend on the nature of the dataset being considered, in particular, the distribution that the event occurrence follows. Furthermore, ESP-NB build on independence assumption between the attributes which does not hold in many survival applications. Thus, the introduced ESP-TAN relaxed the independence assumption which leads to improved AUC and F-measure values in almost all of the results.

The results clearly show that our models can obtain practically useful results using the data collected at an early stage of the study. This is due to the fact that classification methods do not have the ability to predict the event occurrence for a time beyond the observation time. Also, in the Cox regression model, the baseline hazard is undefined after the observation time t_c. Thus, from our experiments, we can conclude that the proposed framework is able to obtain practically useful results at the initial phases of a longitudinal study and can provide good insights about the event occurrence by the end of the study.

In Fig. 3, we present the prediction performance of different methods by varying the percentage of event occurrence information that is available to train the model for the PBC dataset. For example, 20 % on the x-axis corresponds to the

Table 4. Comparison of F-measure values for Cox, NB and TAN with proposed ESP-NB and ESP-TAN methods using Weibull (W) and Lognormal (L) extrapolation methods (with standard deviation values).

Data	F-measure						
	Cox	NB	TAN	ESP-NB(W)	ESP-NB(L)	ESP-TAN(W)	ESP-TAN(L)
Syn1	0.632	0.753	0.762	0.775	0.771	**0.785**	0.785
	(0.023)	(0.021)	(0.026)	(0.021)	(0.022)	**(0.019)**	(0.023)
Syn2	0.629	0.638	0.647	0.764	0.763	**0.777**	0.769
	(0.025)	(0.034)	(0.023)	(0.025)	(0.029)	**(0.02)**	(0.021)
Breast	0.628	0.543	0.555	0.712	0.653	**0.723**	0.679
	(0.031)	(0.053)	(0.034)	(0.039)	(0.042)	**(0.039)**	(0.039)
Colon	0.496	0.523	0.529	0.619	0.606	**0.626**	0.623
	(0.163)	(0.169)	(0.184)	(0.145)	(0.151)	**(0.148)**	(0.15)
PBC	0.603	0.529	0.535	0.709	0.664	**0.715**	0.698
	(0.141)	(0.121)	(0.11)	(0.11)	(0.109)	**(0.098)**	(0.114)
Framingham	0.755	0.787	0.798	0.865	0.873	0.894	**0.905**
	(0.079)	(0.085)	(0.073)	(0.073)	(0.093)	(0.069)	**(0.056)**
EHR	0.672	0.616	0.623	0.781	0.750	**0.798**	0.781
	(0.125)	(0.156)	(0.198)	(0.126)	(0.206)	**(0.16)**	(0.12)
Kickstarter	0.672	0.713	0.719	0.747	0.742	0.762	**0.775**
	(0.084)	(0.058)	(0.067)	(0.034)	(0.054)	(0.048)	**(0.032)**

training data obtained when only 20 % of the events have occurred and prediction of the event occurrences was made for the end of the study period. From this plot we can see that the AUC values improve when there is more information on the event occurrence in the training data. For all the cases, our proposed ESP framework gives better prediction performance compared to other techniques. Furthermore, it should be noted that the improvements of the proposed methods are more significant over the baseline methods when there is only a limited amount (20 % or 40 %) of training data. Also, when 100 % of the training data is available, the performance of the proposed methods will converge to that of the standard Naive Bayes and TAN methods since the prior probabilities in both scenarios will be the same and fitting a distribution will not have any impact when evaluated at the end of the study. The proposed prediction framework is an extremely useful tool for domains where one has to wait for a significant period of time to collect sufficient amount of training data. The practical implication of this result is the fact that using the proposed models, one can obtain an approximate result and gain insights about the problem within the early stage of the study. Thus, it is not needed to wait until the end of the study to obtain the model performance. Also, we can observe that, in many real-world datasets, 50 % of the events typically occur within 25 % of the total study time. Such an early stage model building is an extremely useful tool for domains where one has to wait for longer time periods to collect the required training data.

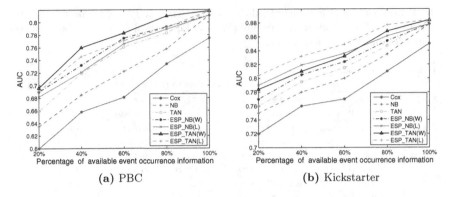

Fig. 3. AUC values of different methods obtained by varying the percentage of event occurrence information for the PBC and Kickstarter dataset (Color figure online).

6 Conclusion

In many real-world application domains, it is important to be able to forecast the occurrence of future events by only using the data collected at early stages in longitudinal studies. In this paper, we developed event prediction algorithms by extending Bayesian methods through fitting a statistical distribution to time-to-event data with fewer available events at the early stages. This enables us to have a reliable prediction of event occurrence for future time points. Our extensive experiments using both synthetic and real datasets demonstrate that the proposed ESP-based algorithms are more effective than Cox model and other classification methods in forecasting events at future time points. Also, we investigated different kinds of extrapolation approaches by fitting various distributions such as Weibull and log-normal. Though motivated by biomedical and healthcare application scenarios (primarily for estimating survival), the proposed algorithms are also applicable to various other domains where one needs to predict event occurrences at early stage of analysis when there are only a relatively fewer set of events that have occurred until a certain time point.

Acknowledgements. This work was supported in part by the National Science Foundation grants IIS-1527827 and IIS-1231742.

References

1. Bandyopadhyay, S., Wolfson, J., Vock, D.M., Vazquez-Benitez, G., Adomavicius, G., Elidrisi, M., Johnson, P.E., O'Connor, P.J.: Data mining for censored time-to-event data: a bayesian network model for predicting cardiovascular risk from electronic health record data. Data Min. Knowl. Disc. **29**(4), 1033–1069 (2015)
2. Bender, R., Augustin, T., Blettner, M.: Generating survival times to simulate Cox proportional hazards models. Stat. Med. **25**, 1978–1979 (2006)

3. Carroll, K.J.: On the use and utility of the Weibull model in the analysis of survival data. Control. Clin. Trials **24**(6), 682–701 (2003)
4. Dawber, T.R., Kannel, W.B., Lyell, L.P.: An approach to longitudinal studies in a community: the Framingham study. Ann. N.Y. Acad. Sci. **107**(2), 539–556 (1963)
5. Dimitriadou, E., Hornik, K., Leisch, F., Meyer, D., Weingessel, A., Leisch, M.F.: Package e1071. R Software package (2009). http://cran.rproject.org/web/packages/e1071/index.html
6. Donovan, M.J., Donovan, M.J., Hamann, S., Clayton, M., et al.: Systems pathology approach for the prediction of prostate cancer progression after radical prostatectomy. J. Clin. Oncol.: Off. J. Am. Soc. Clin. Oncol. **26**(24), 3923–3929 (2008)
7. Friedman, N., Geiger, D., Goldszmidt, M.: Bayesian network classifiers. Mach. Learn. **29**(2–3), 131–163 (1997)
8. Gordon, L., Plshen, R.: Tree-structured survival analysis. Cancer Treat Rep. **69**(10), 1065–1074 (1985)
9. Hosmer, D.W., Lemeshow, S.: Applied Survival Analysis: Regression Modeling of Time to Event Data. Wiley, New York (1999)
10. John, G.H., Langley, P.: Estimating continuous distributions in Bayesian classifiers. In: Proceedings of the 11th Conference on Uncertainty in Artificial Intelligence, pp. 338–345. Morgan Kaufmann Publishers Inc. (1995)
11. Khan, F.M., Zubek, V.B.: Support vector regression for censored data (SVRc): a novel tool for survival analysis. In: 8th IEEE International Conference on Data Mining, pp. 863–868 (2008)
12. Lavrac, N.: Selected techniques for data mining in medicine. Artif. Intell. Med. **16**, 3–23 (1999)
13. Lee, E.T., Wang, J.: Statistical Methods for Survival Data Analysis, vol. 476. Wiley, New York (2003)
14. Lucas, P.J.F., van der Gaag, L.C., Abu-Hanna, A.: Bayesian networks in biomedicine and health-care. Artif. Intell. Med. **30**(3), 201–214 (2004)
15. Reddy, C.K., Li, Y.: A review of clinical prediction models. In: Reddy, C.K., Aggarwal, C.C. (eds.) Healthcare Data Analytics. Chapman and Hall/CRC Press, Boca Raton (2015)
16. Royston, P.: The lognormal distribution as a model for survival time in cancer, with an emphasis on prognostic factors. Stat. Neerl. **55**(1), 89–104 (2001)
17. Segal, M.R.: Regression trees for censored data. Biometrics **44**(1), 35–47 (1988)
18. Shiao, H.-T., Cherkassky, V.: Learning using privileged information (LUPI) for modeling survival data. In: 2014 International Joint Conference on Neural Networks (IJCNN), pp. 1042–1049, July 2014
19. Štajduhar, I., Dalbelo-Bašić, B.: Uncensoring censored data for machine learning: a likelihood-based approach. Expert Syst. Appl. **39**(8), 7226–7234 (2012)
20. Wolfson, J., Bandyopadhyay, S., Elidrisi, M., Vazquez-Benitez, G., Vock, D.M., Musgrove, D., Adomavicius, G., Johnson, P.E., O'Connor, P.J.: A Naive Bayes machine learning approach to risk prediction using censored, time-to-event data. Stat. Med. **34**(21), 2941–2957 (2015)
21. Zupan, B., DemšAr, J., Kattan, M.W., Beck, J.R., Bratko, I.: Machine learning for survival analysis: a case study on recurrence of prostate cancer. Artif. Intell. Med. **20**(1), 59–75 (2000)

Toxicity Prediction in Cancer Using Multiple Instance Learning in a Multi-task Framework

Cheng Li[1(✉)], Sunil Gupta[1], Santu Rana[1], Wei Luo[1], Svetha Venkatesh[1], David Ashely[2], and Dinh Phung[1]

[1] Center for Pattern Recognition and Data Analytic,
Deakin University, Geelong, Australia
`cheng.l@deakin.edu.au`
[2] School of Medicine, Deakin University, Geelong, Australia

Abstract. Treatments of cancer cause severe side effects called toxicities. Reduction of such effects is crucial in cancer care. To impact care, we need to predict toxicities at fortnightly intervals. This toxicity data differs from traditional time series data as toxicities can be caused by one treatment on a given day alone, and thus it is necessary to consider the effect of the singular data vector causing toxicity. We model the data before prediction points using the multiple instance learning, where each bag is composed of multiple instances associated with daily treatments and patient-specific attributes, such as chemotherapy, radiotherapy, age and cancer types. We then formulate a Bayesian multi-task framework to enhance toxicity prediction at each prediction point. The use of the prior allows factors to be shared across task predictors. Our proposed method simultaneously captures the heterogeneity of daily treatments and performs toxicity prediction at different prediction points. Our method was evaluated on a real-word dataset of more than 2000 cancer patients and had achieved a better prediction accuracy in terms of AUC than the state-of-art baselines.

1 Introduction

Cancer kills. It is one of leading causes of morbidity and mortality worldwide. In 2012 alone there were 14 million new cases and 8.2 million cancer-related deaths [1]. Effective treatment and care remains a dominant health concern.

One expects to get better when treated. A crucial problem is that cancer treatments themselves make you sick. They cause severe adverse events, called toxicities, such as anemia and neutropenia. These toxicities greatly impair patient care and must be reduced in clinical. But oncologists are unable to predict which patient will suffer toxicities. This is because the causes that underlie patient reactions to treatments is still poorly understood. Any predictive system

---

Electronic supplementary material The online version of this chapter (doi:10.1007/978-3-319-31753-3_13) contains supplementary material, which is available to authorized users.

J. Bailey et al. (Eds.): PAKDD 2016, Part I, LNAI 9651, pp. 152–164, 2016.
DOI: 10.1007/978-3-319-31753-3_13

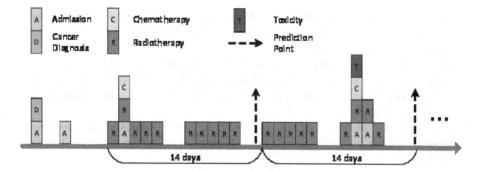

Fig. 1. Treatment timeline, wherein different treatments may be given in one day. A brace denotes a fortnight. Prediction points are placed fortnightly.

would be invaluable because it will provide objective data to oncologists to adjust treatments, thus improving patient care.

We address this important problem and find solutions to provide objective decision support to oncologists. The data is particularly interesting. A patient identified with cancer undergoes several bouts of treatments that can include both radiotherapy and chemotherapy (see Fig. 1). The problem is to predict toxicities continually - fortnightly, after the first treatment. The data before the prediction point can be used to predict future toxicities. This data differs from traditional time series data in the two ways: toxicity can be caused by one treatment on a given day, but treatments on other days may not cause adverse effects. Thus the toxicity outcome is not correlated with all the inputs. Further, if we aggregate all data in between prediction points, the effect of the singular data vector causing toxicity will be diminished. Thus a new approach is required for this data to solve this important problem.

We cast the problem as a multiple-instance problem. We deliver a solution through a novel approach integrating the multi-instance learning into a multi-task learning framework. In the multi-instance learning, each data point is a bag of feature vectors, or instances. In our case, an instance consists of daily treatments and patient-related variables (see Table 1). A bag has several instances, but only one label - it is positive if any instance in the bag is positive and negative, otherwise. The multi-instance formulation models the impact of daily treatments on toxicity. The multi-task part of our framework induces a sharing of model parameters across different prediction points, exploiting their statistical similarities. This is realized by modeling these multiple model parameters through a common subspace, which can be obtained by combining the Indian Buffet Process (IBP) prior and factor analysis. The proposed model is called the factorial multi-task learning with multiple instance learning (FMT-MIL). We derive an efficient Gibbs sampling for this model. We evaluate our model on a synthetic and a real-world cancer dataset of more than 2000 patients. The experimental results show that our FMT-MIL outperforms the baselines - the individual multi-instance algorithm [2] and the simple multi-task use of multi-instance learning [3]. Our contributions are:

- The first formulation of toxicity prediction in cancer through a multi-instance framework. The method captures the heterogeneity of daily treatments to model the data uniqueness.
- The proposal of a multi-task framework for toxicity prediction across prediction points. Our model can automatically learn the number of factors and transfer knowledge across prediction points through a common subspace shared by predictors;
- The validation on a real-world dataset. Our model improves the prediction accuracy as much as 10 %, which validates the use of our approach for toxicity prediction.

2 Related Work and Background Knowledge

2.1 Toxicity Prediction

Toxicity prediction provides treatment-decision support for oncologists. However, toxicity prediction is a difficult task since toxicities are determined by complex interactions between treatment and patient-related variables. For example, older adults more likely suffer from toxicity from cytotoxic treatments than younger patients due to the reduction of organ functionality [4]. In addition, oncologists may consider a patient's ability to tolerate particular treatments or the effect of previous treatments to adjust treatment schemes, which lead to increasing difficulties in toxicity prediction.

Research has performed toxicity prediction using statistical methods [5,6]. For example, Hurria et al. [5] identified risk factors of chemotherapy toxicity based on p-value and then applied a multivariate logistic regression model that incorporates identified factors to compute the probability of toxicity occurring. Kim et al. [6] studied factors in a logistic regression which cause radiation pneumonitis in lung cancer. However, these methods are not able to capture the interactions between factors and hence they perform limited predictive power in clinical practice [7].

Some studies [7–9] use machine learning and data mining approaches for toxicity prediction. Gulliford et al. [8] applied the artificial neural network (ANN) to predict biological outcomes by learning the relationship between treatments and effects. EI Naqa et al. [7] proposed a modified SVM kernel method to model the nonlinear relationship between factors, which can be generalized to unseen data. Pella et al. [9] implemented large scale optimization methods and traditional classification techniques to predict acute toxicity. All these studies do not consider the impact of daily treatments on toxicity. Our proposed approach not only captures the heterogeneity of daily treatments but also performs toxicity prediction at different prediction points.

2.2 The Multi-instance Learning

In a standard logistic regression classifier, each data point is represented by a feature vector and has a label. The posterior probability of the label is modeled using a sigmoid function, that is, $p(y = 1 \mid x) = \sigma(\omega^T x)$ and $p(y = 0 \mid x) = 1 - p(y = 0 \mid x) = 1 - \sigma(\omega^T x)$, where $\sigma(z) = 1/(1 + e^{-z})$ is a sigmoid function and ω is a classifier weight vector. Multiple instance learning (MIL) assumes that each data point can be represented by a bag of feature vectors (or instances). A bag has only one label which means that the instances in a bag share a label. Note that instances often do not have explicit labels [10].

Let $\mathcal{D} = \{x_i, y_i\}_{i=1}^N$ denote N training data points, where $x_i = \{x_{ij} \in \mathbb{R}^d\}_{j=1}^{M_i}$ is a bag of instances and y_i is the shared label. For a binary classification problem $y_i \in \{0, 1\}$, if all instances in a bag are negative, the bag is negative and hence the posterior probability of a negative bag is defined as $p(y_i = 0 \mid x_i) = \prod_{j=1}^{M_i}[1 - \sigma(\omega^T x_{ij})]$. Similarly, the posterior probability of a positive bag is $p(y_i = 1 \mid x_i) = 1 - p(y_i = 0 \mid x_i) = 1 - \prod_{j=1}^{M_i}[1 - \sigma(\omega^T x_{ij})]$.

The goal of MIL is to learn a classifier ω to predict labels of unseen bags. If every bag consists of only one instance, the MIL will be reduced to a standard logistic classifier. ω can be estimated by maximizing the likelihood using the AnyBoost framework [11] or the MAP framework [3].

MIL has received increasing attentions in many real applications. It was first used in drug activity detection [2]. Later MIL has been widely used in computer vision, such as scene classification [12] and object detection [11]. The Bayesian version of the MIL has been proposed in [3]. We employ MIL to model the generation of toxicity with the assumption that a treatment interval is a bag, wherein each instance is a feature vector constructed from daily treatments and patient-specific attributes.

2.3   Multi-task Learning Using Nonparametric Factor Analysis

Multi-task learning jointly models multiple tasks to improve the performance of some tasks using the knowledge from other tasks. A lot of experimental work have shown the advantages of the multi-task learning compared to the single task learning [13, 14].

Gupta et al. [15] developed a flexible factorial multi-task learning framework, where task predictors are assumed to lie in a subspace and can be decomposed as

$$\omega_{N \times D} = F_{N \times K} \phi_{K \times D} + E_{N \times D} \tag{1}$$

where $\omega_{N \times D}$ is a matrix consisting of N data points lying in D-dimensional Euclidean space; $\phi_{K \times D}$ is a subspace consisting of the basis factors in the transformed subspace; $F_{N \times K}$ is the representation matrix of $\omega_{N \times D}$ in the subspace ϕ and $E_{N \times D}$ is the offset error. We further decompose $F_{N \times K} = z_{N \times K} \odot h_{N \times K}$ so that $h_{N \times K}$ is the actual subspace representation of ω in ϕ and $z_{N \times K}$ is a binary matrix indicating the presence or absence of the factors. The number of factors K may vary from applications to applications. In this model, the K can be inferred automatically due to the introduction of the Indian buffet process (IBP) [17] as a nonparametric prior.

The Indian buffet process (IBP) [17] is a Bayesian nonparametric prior over a binary matrix $z_{N \times K}$. The binary matrix consists of a finite row and

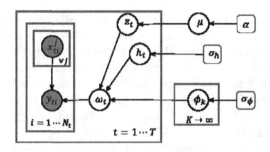

Fig. 2. A directed graphical representation for the FMT-MIL

an unbounded number of columns. Each row can be interpreted as an object and each column can be taken as a feature. For each feature k $(k = 1, \cdots, K)$, let μ_k be a prior probability over the presence of the feature k in an object i (which we denote as z_{ik}). A prior Beta$(\frac{\alpha}{K}, 1)$ is further placed on μ_k, where α is the strength parameter of the IBP. The generative model of z is described as: $\mu_k \sim$ Beta$(\frac{\alpha}{K}, 1)$, $z_{ik} \mid \mu_k \sim$ Bernoulli(μ_k).

The stick-breaking IBP construction has been proposed in [17]. Specifically, let $\boldsymbol{\mu} = \{\mu_{(1)}, \mu_{(2)}, \cdots, \mu_{(K)}\}$ be a decreasing order of $\{\mu_1, \mu_2, \cdots, \mu_K\}$. Teh et al. [17] shows the following construction when $K \to \infty$, i.e., $\nu_{(k)} \overset{\text{i.i.d}}{\sim}$ Beta$(\alpha, 1)$, $\mu_{(k)} = \nu_{(k)}\mu_{(k-1)} = \prod_{l=1}^{K} \nu_{(k)}$. We denote the construction as $\boldsymbol{\mu} \sim$ stickIBP(α). The densities of the subsequent $\mu_{(k)}$'s have been derived as

$$p(\mu_{(k+1)} \mid \mu_{(k)}) = \alpha \mu_{(k)}^{-\alpha} \mu_{(k+1)}^{\alpha-1} \mathbb{I}(0 \le \mu_{(k+1)} \le \mu_{(k)}) \tag{2}$$

where $\mathbb{I}(Q)$ equals to 1 if Q is true and 0, otherwise. Note that $\mu_{(k)}$'s have a Markov structure, with $\mu_{(k+1)}$ only conditionally dependent on its previous one $\mu_{(k)}$. The conditional probability of z_{ik} is presented

$$p(z_{ik} = 1 \mid \boldsymbol{z}_{-ik}, \alpha, \boldsymbol{x}) \propto p(z_{ik} = 1 \mid \boldsymbol{z}_{-ik}, \alpha)p(x_i \mid \boldsymbol{z}_{-ik}, z_{ik} = 1, \boldsymbol{x}_{-i})$$
$$\propto \frac{\mu_{(k)}}{\mu*}p(x_i \mid \boldsymbol{z}_{-ik}, z_{ik} = 1, \boldsymbol{x}_{-i}) \tag{3}$$

where $\boldsymbol{z}_{-ik} = \boldsymbol{z} \backslash z_{ik}$ denotes others excluding z_{ik}; the $\mu*$ is chosen to be the length of the stick for the last active feature [17]. We have proposed a similar factorial multi-task learning framework that has incorporated the MIL paradigm. Our model differs from the one in [3], which simply assumes that all weight vectors share the same Gaussian prior and thus cannot capture the dynamic process between tasks.

3   The Proposed Framework

3.1   Model Description

In this section, we propose a model incorporating the multi-instance learning into a multi-task framework. Our goal is to improve the prediction performance of different tasks. In our model, the multi-instance learning is utilized to model the classification problem in which each data point consists of a bag of feature vectors. We then perform the multi-task learning with the assumption that all task predictors lie in a subspace. The subspace can be modeled by combining the IBP prior and factor analysis. We refer our model as the factorial multi-task learning with the multi-instance learning (FMT-MIL).

Suppose that we have T tasks. In each task t ($t \in 1, \cdots, T$), the training data consist of N_t bags $D_t = \{(\boldsymbol{x}_{ti}, y_{ti})\}_{i=1}^{N_t}$, where $\boldsymbol{x}_{ti}$ is a bag of instances and y_{ti} is the label of the bag $\boldsymbol{x}_{ti}$. The task predictor of the task t is denoted as ω_t. As discussed in Sect. 2.3, we handle $\boldsymbol{\omega}$ by defining $\omega_{T \times D} = (\boldsymbol{z}_{T \times K} \odot \boldsymbol{h}_{T \times K}) \boldsymbol{\phi}_{K \times D}$, where $\boldsymbol{\phi} = \{\phi_k\}_{k=1}^{K}$ is the subspace; $\boldsymbol{z} = \{z_t\}_{t=1}^{T}$ is a binary matrix indicating the presence or absence of the factors; $\boldsymbol{h} = \{h_t\}_{t=1}^{T}$ is the actual subspace representation of $\boldsymbol{\omega}$ in $\boldsymbol{\phi}$; and K is number of factors. We introduce the sticking IBP [17] as the prior of $\boldsymbol{z}$. Based on the prior of $\boldsymbol{z}$ and the likelihood of $\boldsymbol{\omega}$, we can infer the K_t of each task and the final K. The graphical representation of the FMT-MIL model is shown in Fig. 2 and its generative process is described as

$$\boldsymbol{\mu} \sim \text{stickIBP}(\alpha) \qquad h_t \sim \mathcal{N}(0, \sigma_h^2 \mathbf{I}), t = 1 \cdots T \tag{4}$$

$$\boldsymbol{z} \sim \text{Bernoulli}(\boldsymbol{\mu}) \qquad \omega_t \sim \mathcal{N}(\boldsymbol{\phi}(z_t \odot h_t), \sigma_t^2 \mathbf{I}), t = 1 \cdots T \tag{5}$$

$$\boldsymbol{\phi} \sim \mathcal{N}(0, \sigma_\phi^2 \mathbf{I}) \qquad y_{ti} \sim \text{MIL}(\boldsymbol{x}_{ti}, \omega_{ti}), t = 1 \cdots T \tag{6}$$

where α is the strength parameter of the IBP; $\mathcal{N}$ is a Gaussian distribution; and σ_h, σ_ϕ, σ_t are hyperparameters. $y_{ti} \sim \text{MIL}(\boldsymbol{x}_{ti}, \omega_t)$ is the multi-instance learning and can be specified by

$$p_{ti} = p(y_{ti} = 1 \mid \boldsymbol{x}_{ti}, \omega_{ti}) = 1 - \prod_{j=1}^{M_{ti}}(1 - \sigma(\omega_t^T x_{ti}^j)) \tag{7}$$

$$p(y_{ti} = 0 \mid \boldsymbol{x}_t, \omega_t) = \prod_{j=1}^{M_{ti}}(1 - \sigma(\omega_t^T x_{ti}^j)) \tag{8}$$

where σ is the sigmoid function and M_{ti} is the number of instances in bag i of task t.

In our **toxicity prediction** application, each fortnightly prediction point is a task. Each instance is a feature vector constructed from daily treatments and patient-specific attributes. The label is positive if the toxicities have been searched in 28 days after the prediction point and negative, otherwise. We aim to obtain the classifier weight vector $\boldsymbol{\omega}$ to predict toxicity for unseen patients.

3.2   Model Inference

The closed form of the proposed model is intractable. We use the Gibbs sampling for the model inference. In our model, we need to update the main latent variables $\{\boldsymbol{\omega}, \boldsymbol{z}, \boldsymbol{h}, \boldsymbol{\phi}\}$.

Sampling ω. The ω can be sampled independently for each task. We sample ω_t from the following conditional distribution

$$p(\omega_t \mid h_t, \phi, \boldsymbol{y}_t) \propto p(\omega_t \mid h_t, \phi) p(\boldsymbol{y}_t \mid \omega_t, \boldsymbol{x}_t) \propto e^{-R_t/2\sigma_t^2} \left[\prod_{i=1}^{N_t} s_{ti}^{y_{ti}} (1 - s_{ti})^{1-y_{yi}} \right]$$

where $R_t = \|\omega_t - (z_t \odot h_t)\phi\|_2^2$ and $s_{ti} = p_{ti} = p(y_{ti} = 1)$. The expression above is intractable as it cannot be reduced to a standard distribution. We employ the Laplace approximation to estimate the posterior of ω_t. The Laplace approximation aims to find the mode of the posterior distribution and fits a Gaussian whose mean lies at the mode. The mode of the posterior can be computed by maximizing the posterior distribution $p(\omega_t \mid \cdots)$, i.e., $\nabla_{\omega_t} \ln p(\omega_t \mid \cdots) = 0$. We use the Newton-Raphson update in our implementation to obtain the optimization solution (denoted as $\omega_t^{Laplace}$). The co-variance matrix of the approximate Gaussian is obtained by computing the negative of the inverse of the Hessian of the log posterior, i.e., $\Sigma_{\omega_t}^{Laplace} = -\text{inv}(\nabla_{\omega_t}^2 \ln p(\omega_t \mid \cdots) \mid \omega_t = \omega_t^{Laplace})$. Therefore, the new ω_t could be sampled from $\mathcal{N}(\omega_t^{Laplace}, \Sigma_{\omega_t}^{Laplace})$. The first and the second derivatives of log posterior is computed as following

$$\nabla_{\omega_t} \ln p(\omega_t \mid \cdots) = \sum_{i=1}^{N_t} [y_{ti}\beta_{ti} - (1-y_{ti})] \sum_{j=1}^{M_{ti}} x_{ti}^j \sigma(\omega_t^T x_{ti}^j) - \frac{1}{\sigma_t^2}(\omega_t - (z_t \odot h_t)\phi)) \quad (9)$$

$$\nabla_{\omega_t}^2 \ln p(\omega_t \mid \cdots) = \sum_{i=1}^{N_t} [y_{ti}\beta_{ti} - (1-y_{ti})] \sum_{j=1}^{M_{ti}} x_{ti}^j (x_{ti}^j)^T \sigma(\omega_t^T x_{ti}^j) \sigma(-\omega_t^T x_{ti}^j) - \frac{\mathbf{I}}{\sigma_t^2}$$

$$- \sum_{i=1}^{N_t} y_{ti}\beta_{ti}(\beta_{ti} + 1) \left[\sum_{j=1}^{M_{ti}} x_{ti}^j \sigma(\omega_t^T x_{ti}^j) \right] \left[\sum_{j=1}^{M_{ti}} x_{ti}^j \sigma(\omega_t^T x_{ti}^j) \right]^T \quad (10)$$

where $\beta_{ti} = (1 - p_{ti})/p_{ti}$ and N_t is the number of data points in the task t.

Sampling z. The z is a $T \times K$ binary matrix indicating which subspace bases are used to generate the task predictors. z_{tk} can be sampled from

$$p(z_{tk} = 1 \mid \boldsymbol{z}_{-tk}, \alpha, \omega_t) \propto p(z_{tk} = 1 \mid \boldsymbol{z}_{-tk}, \alpha) p(\omega_t \mid \phi, z_{tk} = 1, \boldsymbol{z}_{-tk}, h_t) \quad (11)$$

The first term in the right hand is the prior of the stick-breaking IBP, which could be obtained from the part of Eq. (3). The second term in the right hand is the likelihood of ω_t given all factors and other variables. Finally, $p(z_{tk} \mid \boldsymbol{z}_{-tk}, \alpha, \omega_t)$ is a Bernoulli distribution.

Sampling ϕ. The variable ϕ is sampled from the following conditional distribution

$$p(\phi \mid \sigma_\phi, \omega_{1:T}, z_{1:T}, h_{1:T}) \propto p(\phi \mid \sigma_\phi^2) p(\omega_{1:T} \mid \phi, z_{1:T}, h_{1:T}) \quad (12)$$

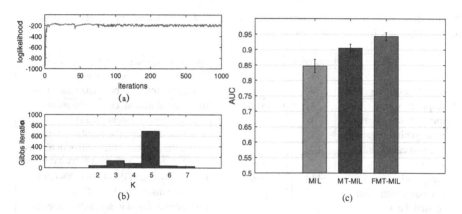

Fig. 3. (a) Convergence (loglikelihood) of the FMT-MIL; (b) Posterior over number of factors for the FMT-MIL; (c) AUC comparison of different algorithms;

From the Eq. (5) and Eq. (6), we know that the above conditional distribution is a normal distribution. We further expand it and derive it as $p(\phi \mid \cdots) \sim \mathcal{N}(\mu_\phi, S_\phi)$, where $\mu_\phi \triangleq S_\phi(\sum_t \frac{(z_t \odot h_t)^T \omega_t}{\sigma_t^2})$ and $S_\phi^{-1} = \sum_t \frac{(z_t \odot h_t)^T (z_t \odot h_t)}{\sigma_t^2} + \frac{\mathbf{I}_K}{\sigma_\phi^2}$. The detailed derivation is provided in Appendix.

Sampling h. The posterior of h can be sampled using the conditional distribution below

$$p(h_t \mid \omega_t, \phi) \propto p(h_t \mid \sigma_t^2)p(\omega_t \mid \phi, h_t) \tag{13}$$

Similar to the sampling of ϕ, we derive the posterior as a normal distribution $p(h_t \mid \cdots) \sim \mathcal{N}(\mu_{h_t}, S_{h_t})$, where $\mu_{h_t} \triangleq S_{h_t} \frac{D_{zt} \phi \omega_t^T}{\sigma_t^2}$ and $S_{h_t}^{-1} = \sum_t \frac{D_{zt} \phi \phi^T D_{zt}}{\sigma_t^2} + \frac{\mathbf{I}_K}{\sigma_h^2}$ and $D_{zt} = \text{diag}(z_t)$.

Sampling Hyperparamters. We can assume that σ_h is sampled from a Gamma distribution. Due to the conjugate property of the Gamma distribution and the Gaussian distribution, the conditional probability of σ_h is an inverse-gamma distribution, which is tractable. The updates of other hyperparameters σ_t and σ_ϕ are similar to σ_h.

4   Experiments

4.1   Synthetic Data

We construct a synthetic data consisting of 20 tasks. The number of factors (i.e., K) is 5. The hyperparameters $\{\alpha, \sigma_h, \sigma_t, \sigma_\phi\}$ are 1 and each task predictor (i.e., ω_t) is generated following Eq. (5). We also generate 100 data points in each task and each data point consists of 1~4 instances, which are randomly sampled from a standard Gaussian distribution with the dimensionality of 10. The labels of data points then could be generated by using Eqs. (7) and (8). We perform our experiment comparison for the following algorithms:

Table 1. Left: Number of cancer patients for different cancer types. Right: Features and labels used in learning toxicity predictors

cancer	patients	basic	age, gender, marital_statues, cancer types, tumor size, cancer stage, treatment_intent_type, metastasis_flag
breast	447		
digestive organs	390		
respiratory and intrathoracic organs	341	treatment	treatment duration, chemo duration, chemo intervals (mean and std.), radiation duration, the number of used drugs, the number of toxicities in history
male genital organs	185		
lip, oral cavity and pharynx	142		
lymphoid, haematopoietic and related tissue	113	toxicity (ICD10 code)	blood diseases: D60~D64,D70~D77 nervous system: G50~G59,G90~G99 digestive disorders: K00~K14 disorders of the skin and subcutaneous tissue: L55~L59
others	383		

- **FMT-MIL.** The proposed factorial multi-task learning with multi-instance learning.
- **MT-MIL.** The multi-task learning joint with the Bayesian multiple instance learning without feature selection [3]. All predictors share a single prior.
- **MIL.** The Bayesian multiple instance learning without feature selection [3]. Tasks are independent.

In our algorithm running, we initialized all the hyperparameters $\{\alpha, \sigma_h, \sigma_t, \sigma_\phi\}$ to 1. We randomly split the data into training and test sets such that each of them contains roughly half of the entire data set. We run 1000 Gibbs iteration and show the convergence in Fig. 3(a, b). Our algorithm can converge during 200 iterations and return the true number of factors K during 1000 iterations. The AUC from the test data (seen in Fig. 3(c)) shows that our algorithm outperforms the baselines.

4.2 Real Data Description

There is no public cancer toxicity dataset available. Our dataset is from a regional hospital. The data are in form of EMRs (electronic medical records), which include diagnosis codes, demographic information and treatment procedures. We use all patients who have been given both chemotherapy and radiotherapy since their toxicity information is comparatively rich (seen in Fig. 5). Table 1 shows groupings of cancer patients in the dataset and the variables and labels for learning toxicity predictors. We combine the patient-specific diagnosis, demographic and treatment properties as the feature sources. We only focus on the toxicities specified by ICD-10 codes [18]. The blood diseases, nervous system and digestive disorders are usually caused by chemotherapy and the disorders of the skin and subcutaneous tissue are often caused by radiotherapy.

We further illustrate how to construct features in Fig. 4. The prediction points are placed in every fortnight after the first treatment. We use the data before

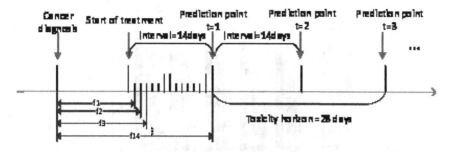

Fig. 4. Feature construction: the prediction points are set every fortnight. Feature vectors $\{f1, f2, \cdots, f_{14}\}$ form a bag. The toxicity horizon is 28-days.

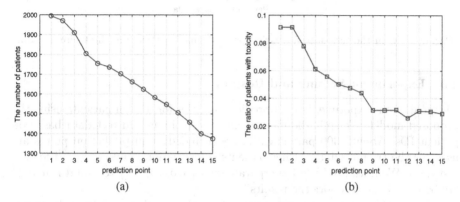

Fig. 5. (a) The number of patients vs. prediction point; (b) The ratio of patients with toxicity vs. prediction point.

the prediction point, including basic and treatment data, to construct a bag of instances. An instance is a feature vector which is extracted in between cancer diagnosis to daily treatments (e.g. $f_1, f_2, \cdots, f_{14}$). If the treatments have not been given in the day, we skip this day. Thus, for a patient, there are a maximum of 14 instances in a bag. The bag is labeled positively if toxicities have been searched in toxicity horizon (28 days in our experiments) and is labeled negatively otherwise. We also extract a feature vector at each prediction point used for a standard logistic regression (LR) classifier. The LR algorithm does not consider the influence of daily treatments on toxicity.

We show the patients for 15 prediction points in Fig. 5. The number of patients is decreasing along the prediction points as patients discharge after weeks of treatments. In the first 5 prediction points (*i.e.* 10 weeks), the toxicity ratios are relatively high since oncologists may adopt intensive treatments at the initial stage. From the sixth to the tenth prediction point, the toxicity ratios decrease obviously. Possible reasons are that oncologists adjust treatments to patient care or patients have adapted the treatments. The ratios remains stable after the tenth prediction point. We perform our experiments on the first 10 prediction points.

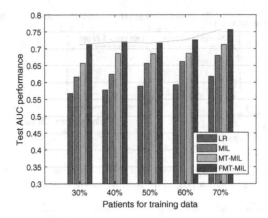

Fig. 6. AUC of the test data for varied proportions of patients in the training data.

4.3   Experiment Setting and Results

The goal of our experiments is to predict toxicities for unseen patients using the proposed model. We split the dataset into the training and test data based on patient IDs. About 1300 patients who are included in all prediction points are used for the training data. Other patients in prediction points are used for the test data. We initialized the hyperparameters $\{\alpha, \sigma_h, \sigma_t, \sigma_\phi\}$ to 1 and run 1000 Gibbs iterations to report the results.

We use different proportions of patients in the training data to evaluate our FMT-MIL and the baselines. The average AUC of the test data are shown in Fig. 6. As the number of the training patients increases, the performance of all models rises. Moreover, our model performs much better than the baselines when

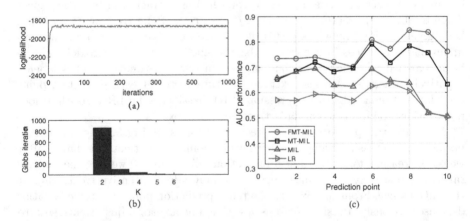

Fig. 7. (a) Convergence (loglikelihood) of the FMT-MIL; (b) Posterior over number of factors for the FMT-MIL; (c) AUC comparison of different algorithms (FMT-MIL, MT-MIL and MIL) at different prediction points;

a small number of training data is given. For example, when 30 % patients of the training data are used for training the algorithms, the AUC of our model increases by as much as 10 % compared to the MT-MIL, 15 % compared to the MIL and 30 % compared to the LR. This is because our model joints the multi-instance learning and the muli-task learning, enhancing the generalized performance of individual tasks. It is noted that the LR performs worse than the MIL, which validates our use of the MIL for toxicity prediction.

We further give insights into each task. We perform the experiment using 70 % of the training data. Our model can converge at about 200 iterations (seen in Fig. 7(a)). The number of factors computed from our model centers at 2 (seen in Fig. 7(b)). The AUC results are presented in Fig. 7(c). Our algorithm performs best in all cases. The AUC gradually increases between the 1∼8 prediction points and then begins to decrease thereafter. The trend is significant in the MIL algorithm since the positive samples reduce dramatically around the 9∼10 prediction points (seen in Fig. 5(b)). Individual tasks in the MIL algorithm cannot share knowledge while tasks in the FMT-MIL and the MT-MIL algorithms share the common subspace. This has validated our use of the multi-task learning for the multi-point prediction.

5 Conclusion

We have proposed a multi-task framework incorporating the multiple instance learning for toxicity prediction. The prediction points occur fortnightly after the first treatment. We treat daily treatments before the prediction points and patient-specific attributes to be multi-instance data. We further combine all prediction points to perform a multi-task framework. The factors in a shared subspace can be inferred automatically. We derive an efficient Gibbs sampling for our model. We evaluate our model on a synthetic dataset and a real-world dataset. The experiment results show that our model outperforms the state of the art, with the prediction accuracy increased by as much as 10 %.

Toxicities can be caused by many factors in clinical treatment. No single approach can handle toxicity prediction perfectly. This paper has explored the problem using a machine learning approach. Our approach not only captures the influence of daily treatments but also performs the multi-point prediction simultaneously. Our current data do not include the pathology test and physician's assessments, which may possibly improve the prediction accuracy if available.

References

1. Torre, L.A., Bray, R.L., Siegel, F., Jemal, J., Ferlay, A., Lortet-Tieulent, J.: Global cancer statistics, 2012. CA Cancer J. Clin. **65**(2), 87–108 (2015)
2. Dietterich, T.G., Lathrop, R.H., Lozano-Pérez, T.: Solving the multiple instance problem with axis-parallel rectangles. Artif. Intell. **89**(1–2), 31–71 (1997)
3. Raykar, V.C., Krishnapuram, B., Bi, J., et al.: Bayesian multiple instance learning: automatic feature selection and inductive transfer. In: ICML, pp. 808–815 (2008)

4. Hurria, A., Lichtman, S.M.: Pharmacokinetics of chemotherapy in the older patient. Cancer Control **14**(1), 32–43 (2007)
5. Hurria, A., Togawa, K., et al.: Predicting chemotherapy toxicity in older adults with cancer: a prospective multicenter study. J. Clin. Oncol. **29**(25), 3457–3465 (2011)
6. Kim, M., Lee, J., Ha, B., et al.: Factors predicting radiation pneumonitis in locally advanced non-small cell lung cancer. Radiat. Oncol. J. **29**(3), 181–190 (2011)
7. Ei Naqa, I., Bradley, J.D., Lindsay, P.E., et al.: Predicting radiotherapy outcomes using statistical learning techniques. Phys. Med. Biol. **54**(18), 9–30 (2009)
8. Gulliford, S.L., Webb, S., et al.: Use of artificial neural networks to predict biological outcomes for patients receiving radical radiotherapy of the prostate. Radiother. Oncol. **71**(1), 3–12 (2004)
9. Pella, A., Cambria, R., et al.: Use of machine learning methods for prediction of acute toxicity in organs at risk following prostate radiotherapy. Med. Phys. **38**(6), 2859–2867 (2011)
10. Foulds, J.R., Frank, E.: A review of multi-instance learning assumptions. Knowl. Eng. Rev. **25**(1), 1–25 (2010)
11. Viola, P., Platt, J.C., Zhang, C.: Multiple instance boosting for object detection. In: NIPS, pp. 1419–1426 (2006)
12. Maron, O., Ratan, A.L.: Multiple-instance learning for natural scene classification. In: ICML, pp. 341–349, San Francisco, CA, USA (1998)
13. Caruana, R.: Multitask learning. Mach. Learn. **28**(1), 41–75 (1997)
14. Evgeniou, T., Pontil, M.: Regularized multi-task learning. In: KDD, pp. 109–117 (2004)
15. Gupta, S., Phung, D., Venkatesh, S.: Factorial multi-task learning: a Bayesian nonparametric approach. In: ICML, vol. 28, pp. 657–665 (2013)
16. Griffiths, T.L., Ghahramani, Z.: The indian buffet process: an introduction and review. J. Mach. Learn. Res. **12**, 1185–1224 (2011)
17. Teh, Y.W., Görür, D., Ghahramani, Z.: Stick-breaking construction for the indian buffet process. In: AISTATS, pp. 556–563. MIT Press, Cambridge, MA, USA (2007)
18. World Health Organization. International Statistical Classification of Diseases and Related Health Problems, ICD 2010. World Health Organization, 2010 edn. (2012)

Shot Boundary Detection Using Multi-instance Incremental and Decremental One-Class Support Vector Machine

Hanhe Lin[(✉)], Jeremiah D. Deng, and Brendon J. Woodford

Department of Information Science,
University of Otago, PO Box 56, Dunedin 9054, New Zealand
{hanhe.lin,jeremiah.deng,brendon.woodford}@otago.ac.nz

Abstract. This paper presents a novel framework to detect shot boundaries based on the One-Class Support Vector Machine (OCSVM). Instead of comparing the difference between pair-wise consecutive frames at a specific time, we measure the divergence between two OCSVM classifiers, which are learnt from two contextual sets, i.e., immediate past set and immediate future set. To speed up the processing procedure, the two OCSVM classifiers are updated in an online fashion by our proposed multi-instance incremental and decremental one-class support vector machine algorithm. Our approach, which inherits the advantages of OCSVM, is robust to noises such as abrupt illumination changes and large object or camera movements, and capable of detecting gradual transitions as well. Experimental results on some benchmark datasets compare favorably with the state-of-the-art methods.

Keywords: Support vector machine · One-class · Kernel method · Online learning · Shot boundary detection

1 Introduction

A video shot, which represents a continuous action in time and space, is composed of a series of related, consecutive frames taken contiguously by a single camera [1]. Both pre-edited and unedited video footages may contain shots, and partitioning a video into shots is the fundamental prerequisite for further video content analysis, editing, browsing and retrieval applications. During the last decade, various approaches [2–4] for shot boundary detection (SBD), have been proposed. The easiest approach is to analyze the difference between two successive frames [2]. This is straightforward to implement and is effective to detect abrupt changes. However, it is sensitive to noises, e.g. flashlight frames, and it cannot detect gradual transitions because the pair-wise difference is rather small. Automatic thresholding is employed in [5], where changes on optical flows of frames within a sliding window are thresholded by a value that is equal to the median plus two times of the standard deviation. In [3], the strength of using graph partition for SBD is discussed. To deal with varying characteristics

© Springer International Publishing Switzerland 2016
J. Bailey et al. (Eds.): PAKDD 2016, Part I, LNAI 9651, pp. 165–176, 2016.
DOI: 10.1007/978-3-319-31753-3_14

of videos which challenge threshold setting, a support vector machine (SVM) based approach is adopted to treat the SBD as a classification problem. This approach however has two deficiencies: firstly the availability of ground-truth data frame-by-frame for training, and secondly, the costly training time because of the use of SVMs. Recently, [4] adopts a candidate segment and singular value decomposition to cut down the processing time.

In this paper, inspired by previous work [6], we propose a novel framework to detect shot boundaries on the basis of One-Class Support Vector Machine (OCSVM) [7]. Our approach hence differs from [3], as it uses an online learning OCSVM that operates in an unsupervised manner. Since OCSVMs work to incorporate as many data points inside as possible but leave out outliers, the basic idea of our approach is to use two frame sets, an Immediate Past Set (IPS) and an Immediate Future Set (IFS), to train two OCSVMs as the summary of the two frame sets. We then examine the divergence between the two OCSVMs, and significant divergence will indicate the shot boundary.

The main contributions of our paper are as follows:

- We propose a novel learning method, referred to as Multi-instance Incremental and Decremental One-Class Support Vector Machine (MID-OCSVM), to update OCSVM classifier in an online fashion, requiring very low computational cost.
- We present a unified framework to detect different types of shot boundary rather than adopting a set of classifiers for each specific shot boundary type. This simplifies the computational complexity of the approach while maintaining a high accuracy.
- Instead of comparing the difference between successive frames, we define a function to measure the divergence between IPS and IFS. This approach contributes to robustness to noises such as abrupt illumination changes and large movements.

The rest of the paper is organized as follows. Details of the proposed computational framework are presented in Sect. 2. Section 3 reports the experimental settings and the results obtained on the TRECVID 2007 SBD datasets. We conclude the paper in Sect. 4, also discussing some possible future work.

2   Computational Framework

2.1   Overview

The flowchart of our approach is illustrated in Fig. 1. Given the IPS and IFS with fixed length m at time t, we first extract corresponding feature descriptor for each frame in each set, where an OCSVM classifier is trained respectively. We measure the divergence between the two sets based on the concept of OCSVM. From t to $t + 1$, we add a new frame and remove the oldest one in each set while updating the classifier using the proposed MID-OCSVM algorithm. Shot boundaries are detected based on the divergence output.

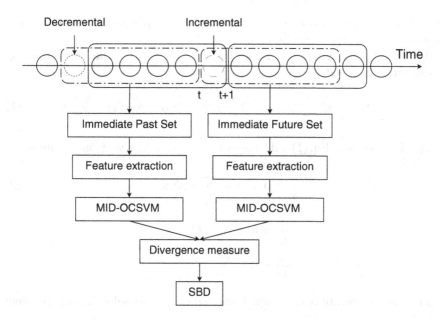

Fig. 1. The flowchart of our approach (Color figure online).

2.2 Feature Extraction

Following [3], we adopt a blocked-based histogram as our feature representation. Specifically, a frame is divided into $2^l \times 2^l$ blocks, where l is the pyramid level, $l = 0, 1, \ldots, n$. In each block a 48-bin (16 bins for each channel in the RGB space) colour histogram is calculated, and the frame is represented by concatenating colour histograms from all blocks. We chose this feature for several reasons. First, it is simple to implement and has little computational cost. Secondly, it provides a trade-off between sensitivity and invariance. Note that we have also investigated other features, e.g., pixel-based [2], but the performance is worse than that of the blocked-based histogram, so we will not report the results due to the space limit.

2.3 OCSVM

Given a set of training data $X = \{\mathbf{x}_1, \ldots, \mathbf{x}_n\}$, OCSVM aims to find an optimal separating function $f(\mathbf{x}) = \mathbf{w} \cdot \Phi(\mathbf{x}) - \rho$, which contains most of the training data in a compact region. Here Φ is a feature map that transforms $\mathbf{x}$ from the input space X to the feature space $\mathcal{F}$.

To acquire the parameters $\mathbf{w}$ and ρ, one can solve the following quadratic programming problem:

$$\min_{\mathbf{w}, \xi, \rho} : \frac{1}{2}\|\mathbf{w}\|^2 + C\sum_i \xi_i - \rho$$

$$s.t. : \mathbf{w} \cdot \Phi(\mathbf{x}_i) \geq \rho - \xi_i, \ \xi_i \geq 0,$$

where ξ_i are slack variables, and C is a regularization parameter, controlling the trade-off between structure and empirical risks.

By introducing Lagrange multipliers $\alpha_i, \beta_i \geq 0$, the corresponding Lagrangian is formulated as:

$$L = \frac{1}{2}\|\mathbf{w}\|^2 + C\sum_i \xi_i - \rho - \sum_i \alpha_i(\mathbf{w} \cdot \Phi(\mathbf{x}_i) - \rho + \xi_i) - \sum_i \beta_i\xi_i. \quad (1)$$

The derivatives of Eq. (1) with respect to the variables $\mathbf{w}$, ξ, and ρ are set to zero, giving:

$$\frac{\partial L}{\partial \mathbf{w}} = 0 \Rightarrow \mathbf{w} = \sum_i \alpha_i\Phi(\mathbf{x}_i), \quad (2)$$

$$\frac{\partial L}{\partial \xi} = 0 \Rightarrow \alpha_i = C - \beta_i,$$

$$\frac{\partial L}{\partial \rho} = 0 \Rightarrow \sum_i \alpha_i = 1.$$

Substituting the above three equations into Eq. (1), we solve the dual problem instead:

$$\min_{\alpha} : \frac{1}{2}\sum_{ij} \alpha_i\alpha_j k(\mathbf{x}_i, \mathbf{x}_j)$$
$$s.t. : 0 \leq \alpha_i \leq C, \ \sum_i \alpha_i = 1, \quad (3)$$

where $k(\mathbf{x}_i, \mathbf{x}_j) = \Phi(\mathbf{x}_i) \cdot \Phi(\mathbf{x}_j)$ is a kernel function that measures the similarity between the two examples $\mathbf{x}_i$ and $\mathbf{x}_j$, with maximum similarity 1 and no similarity 0. Correspondingly, the separating function is rewritten as:

$$f(\mathbf{x}) = \sum_i \alpha_i k(\mathbf{x}_i, \mathbf{x}) - \rho.$$

Considering the histogram-based representation of our feature descriptor, we adopt the Histogram Intersection kernel [8] as the similarity measure between two inputs:

$$k(\mathbf{x}_i, \mathbf{x}_j) = \sum_b \min(\mathbf{x}_i^b, \mathbf{x}_j^b), \quad (4)$$

where b indicates the corresponding bin in $\mathbf{x}_i$ and $\mathbf{x}_j$.

2.4   MID-OCSVM

From time t to $t+1$, each set has to add a new frame feature data and discard the oldest one to learn a new OCSVM classifier for divergence measure. It is very time-consuming to train a classifier in the batch mode whenever a new input comes. Therefore, we propose MID-OCSVM by extending previous work [9–11].

Karush-Kuhn-Tucker Conditions. To elaborate our approach, we rewrite the dual problem Eq. (3) as a saddle-point formulation:

$$\max_{\rho} \min_{0 \leq \alpha_i \leq C} : W = \frac{1}{2} \sum_{ij} \alpha_i \alpha_j k(\mathbf{x}_i, \mathbf{x}_j) - \rho(\sum_i \alpha_i - 1).$$

The first-order conditions on W reduce to the Karush-Kuhn-Tucker (KKT) conditions:

$$g_i = \frac{\partial W}{\partial \alpha_i} = \sum_j \alpha_j k(\mathbf{x}_i, \mathbf{x}_j) - \rho$$

$$\implies f(\mathbf{x}_i) \begin{cases} \geq 0, & \text{if } \alpha_i = 0; \\ = 0, & \text{if } 0 < \alpha_i < C; \\ \leq 0, & \text{if } \alpha_i = C; \end{cases} \tag{5}$$

$$\frac{\partial W}{\partial \rho} = \sum_j \alpha_j - 1 = 0. \tag{6}$$

On the basis of Eq. (5), the training data X is divided into three subsets: margin support vectors S, error support vectors E, and the remaining set O. In parallel, their corresponding indexes set are given as: $I_S = \{i : \mathbf{x}_i \in X, 0 < \alpha_i < C\}$, $I_E = \{i : \mathbf{x}_i \in X, \alpha_i = C\}$, $I_O = \{i : \mathbf{x}_i \in X, \alpha_i = 0\}$. In the following, we will abbreviate $k(\mathbf{x}_i, \mathbf{x}_j)$ to k_{ij}. For two subsets S and O, k_{SO} denotes the kernel matrix, whose rows are indexed by S, and the columns are indexed by O.

Derivation. As depicted in Fig. 1, suppose we add a new arriving data $\mathbf{x}_a$ (red dashed circle) and remove the obsolete data $\mathbf{x}_r$ (blue dotted circle) simultaneously. We first remove $\mathbf{x}_r$ from X (i.e., $X \leftarrow X \setminus \mathbf{x}_r$), and the coefficient of $\mathbf{x}_a$ is initialized as 0 (i.e. $\alpha_a = 0$). If $g_a > 0$, we append $\mathbf{x}_a$ directly to O because it already satisfies the KKT conditions. Likewise, we discard $\mathbf{x}_r$ directly if its corresponding coefficient a_r equals to 0. For $\mathbf{x}_a$ and $\mathbf{x}_r$ having $g_i \leq 0$, the KKT conditions are to be kept:

$$\Delta g_i = k_{ia} \Delta \alpha_a + k_{ir} \Delta \alpha_r + \sum_{j \in I_S} k_{ij} \Delta \alpha_j + \Delta \rho, \quad \forall i \in I_X \cup a, \tag{7}$$

$$0 = \Delta \alpha_a + \Delta \alpha_r + \sum_{j \in I_S} \Delta \alpha_j.$$

For all margin support vectors S, $g_i \equiv 0, \forall i \in I_S$. The above equations can be re-written in matrix notations:

$$\underbrace{\begin{bmatrix} 0 & 1 \\ 1 & k_{SS} \end{bmatrix}}_{K} \begin{bmatrix} \Delta \rho \\ \Delta \alpha_S \end{bmatrix} = - \begin{bmatrix} 1 & 1 \\ k_{Sa} & k_{Sr} \end{bmatrix} \begin{bmatrix} \Delta \alpha_a \\ \Delta \alpha_r \end{bmatrix}. \tag{8}$$

The same as in [11], the change directions of $\Delta \alpha_A$ and $\Delta \alpha_R$ are given as:

$$\Delta \alpha_a = \eta(C\mathbf{1} - \alpha_a),$$
$$\Delta \alpha_r = -\eta \alpha_r, \tag{9}$$

where η is a step length. Together with Eq. (8), we can write:

$$\begin{bmatrix} \Delta\rho \\ \Delta\alpha_S \end{bmatrix} = \eta\Phi, \tag{10}$$

where

$$\Phi = \begin{bmatrix} \phi_\rho \\ \phi_S \end{bmatrix} = -\underbrace{K^{-1}}_{Q} \begin{bmatrix} 1 & 1 \\ k_{Sa} & k_{Sr} \end{bmatrix} \begin{bmatrix} C - \alpha_a \\ -\alpha_r \end{bmatrix}. \tag{11}$$

Substituting (9) and (11) into (7):

$$\Delta g_i = \eta\Psi_i, \tag{12}$$

where

$$\Psi_i = k_{ia}(C - \alpha_a) - k_{ir}\alpha_r + \begin{bmatrix} 1 & k_{iS} \end{bmatrix}\Phi, \quad \forall i \notin I_S. \tag{13}$$

Online Update. The ideal situation is that η equals to 1, so adding data $\mathbf{x}_a$ becomes an error support vector and α_r becomes zero. However, we cannot obtain the new OCSVM state directly in most situations since in Eqs. (10) and (12) the composition of the sets S, E and O changes relative to the change of $\Delta\alpha_S$ and Δg_i. Therefore, as shown in Fig. 2, we have identified the following conditions:

1. g_a reaches zero, corresponding to $\mathbf{x}_a$ joining S. The largest step is computed as:

$$\eta^a = \min\frac{-g_a}{\Psi_a}.$$

2. g_i in E becomes zero, equivalent to $\mathbf{x}_i$ transferring from E to S. The most likely occurred constrain η^E equals to finding the minimal increment:

$$\eta^E = \min\frac{-g_i}{\Psi_i}, \quad \forall i \in I_E \cap \Psi_i > 0.$$

3. g_i in O becomes zero, equivalent to $\mathbf{x}_i$ transferring from O to S. The most likely occurred step is computed as:

$$\eta^O = \min\frac{-g_i}{\Psi_i}, \quad \forall i \in I_O \cap \Psi_i < 0.$$

4. $\mathbf{x}_i$ in S reaches a bound, α_i with equality 0 is equivalent to transferring $\mathbf{x}_i$ from S to O, and equality C from S to E. The most likely increment equals:

$$\eta^S = \min\frac{\Delta\alpha_i^S}{\phi_i}, \quad \forall i \in I_S,$$

where

$$\Delta\alpha_i^S = \begin{cases} C - \alpha_i, & \text{if } \phi_i > 0; \\ -\alpha_i, & \text{if } \phi_i < 0. \end{cases}$$

The largest possible step length η is determined as:

$$\eta = \min(\eta^a, \eta^E, \eta^O, \eta^S, 1).$$

Once obtaining η, we can update ρ, α_i, and g_i through Eqs. (9), (10) and (12). The procedures are repeated until η becomes 1, where all the data, i.e., X and $\mathbf{x}_a$, satisfy KKT conditions and α_r reaches 0.

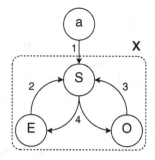

Fig. 2. The corresponding composition change with the change of $\Delta\alpha_S$ and Δg_i.

Recursive Update of Q. It is time-consuming if we compute the inverse matrix Q whenever the set S has changed. Fortunately, by applying the Sherman-Morrison-Woodbury formula [12] for block matrix inversion, we can update the matrix Q in an efficient way. More specifically, let $\hat{Q}$ be the enlarged inverse matrix, when a data $\mathbf{x}_i$ transfers to S, the inversed kernel matrix expands to:

$$\hat{Q} = \begin{bmatrix} 0 & 1 & 1 \\ 1 & k_{SS} & k_{iS}^T \\ 1 & k_{iS} & k_{ii} \end{bmatrix}^{-1} = \begin{bmatrix} K & V^T \\ V & k_{ii} \end{bmatrix}^{-1}, \tag{14}$$

where

$$V = \begin{bmatrix} 1 & k_{iS} \end{bmatrix}.$$

Using the Sherman-Morrison-Woodbury formula, the update rule from Q to $\hat{Q}$ is computed as:

$$\hat{Q} = \begin{bmatrix} Q & 0 \\ 0 & 0 \end{bmatrix} + \frac{1}{\zeta} \begin{bmatrix} QV^TVQ & -QV^T \\ -VQ & 1 \end{bmatrix}, \ i \notin I_S,$$

where $\zeta = k_{ii} - VQV^T$.

Similarly, to remove a data $\mathbf{x}_i$ from the set S, the update rule shrinking the original inverse matrix $\hat{Q}$ to the reduced inverse matrix Q is calculated as:

$$Q = \hat{Q}_{SS} - \hat{Q}_{ii}^{-1}\hat{Q}_{iS}^T\hat{Q}_{iS}, \ i \notin I_S.$$

2.5   OCSVM Divergence

To measure the divergence between two OCSVM classifiers, we first analyse the representation of OCSVM in the feature space $\mathcal{F}$. Using $\mathbf{v}_i$ to denote $\Phi(\mathbf{x}_i)$, for any $\mathbf{v}_i$ we have $||\mathbf{v}_i|| = k(\mathbf{x}_i, \mathbf{x}_i) = 1$. In other words, the training set X are all mapped on a hypersphere S with radius $r = 1$. The OCSVM in $\mathcal{F}$ corresponds to find the optimal hyperplane $\mathbf{w}$ that most mapped training set $\mathbf{v}_i$ have $\mathbf{w} \cdot \mathbf{v}_i - \rho > 0$. Figure 3(a) illustrates the OCSVM classifiers in $\mathcal{F}$, forming a segment cut by the hyperplane $\mathbf{w}$ on the hypersphere S.

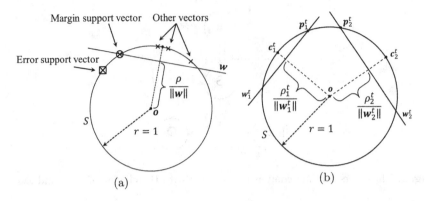

Fig. 3. (a) In the feature space $\mathcal{F}$, OCSVM aims to find the smallest segment while enclosing the mapped training set $\mathbf{v}_i$ as many as possible, i.e., maximizing the margin $\rho/\|\mathbf{w}\|$. (b) Two OCSVM classifiers in $\mathcal{F}$, where $\mathbf{w}_1^t$ and $\mathbf{w}_2^t$ are the optimal hyperplanes of the immediate past set B_1^t and the immediate future set B_2^t. This situation corresponds to a shot boundary as both segments get separated from each other.

Let $B_1^t = \{\mathbf{x}_{t-m}, \mathbf{x}_{t-m+1}, \ldots, \mathbf{x}_t\}$ and $B_2^t = \{\mathbf{x}_{t+1}, \mathbf{x}_{t+2}, \ldots, \mathbf{x}_{t+m}\}$ be the IPS and IFS at time t, where the length of both sets are m. If there is a shot boundary, their segments in $\mathcal{F}$ should be different from each other. In other words, the size as well as the location of the two segments are different from each other. To this end, we define a divergence function D in the following.

Let $\mathbf{c}_1^t$ and $\mathbf{c}_2^t$ be centre points of segments learnt from B_1^t and B_2^t respectively, and $\mathbf{p}_1^t$ and $\mathbf{p}_2^t$ be arbitrary points residing on the boundary of their corresponding segments, as shown in Fig. 3, the divergence function D between B_1^t and B_2^t is given as:

$$D(B_1^t, B_2^t) = \frac{\overparen{\mathbf{c}_1^t \mathbf{c}_2^t}}{\overparen{\mathbf{c}_1^t \mathbf{p}_1^t} + \overparen{\mathbf{c}_2^t \mathbf{p}_2^t}}, \tag{15}$$

where $\overparen{\mathbf{c}_1^t \mathbf{c}_2^t}$ is the arc distance from $\mathbf{c}_1^t$ to $\mathbf{c}_2^t$, and $\overparen{\mathbf{c}_1^t \mathbf{p}_1^t}$ $(\overparen{\mathbf{c}_2^t \mathbf{p}_2^t})$ is the arc distance from $\mathbf{c}_1^t$ $(\mathbf{c}_2^t)$ to $\mathbf{p}_1^t$ $(\mathbf{p}_2^t)$. Equation (15) indicates the divergence is large if two segments are well separated, whereas it is small for strongly overlapped segments.

However, we cannot calculate Eq. (15) directly in feature space because we do not know the explicitly representation of feature map $\Phi(\mathbf{x})$. Therefore, we have to transform D from $\mathcal{F}$ into X. Specifically, for any two points $\mathbf{a}$ and $\mathbf{b}$ lying on an arbitrary sphere, the arc distance is given by:

$$\overparen{\mathbf{ab}} = r\theta, \tag{16}$$

where r is the radius, and θ is the central angle between $\mathbf{a}$ and $\mathbf{b}$.

Meanwhile, the dot product between vector $\mathbf{a}$ and vector $\mathbf{b}$ is given by:

$$\mathbf{a} \cdot \mathbf{b} = \|\mathbf{a}\|\|\mathbf{b}\| \cos\theta, \tag{17}$$

where $||\mathbf{a}|| = ||\mathbf{b}|| = r$.

Combining Eqs. (16) and (17) together with r equals to 1, we have:

$$\widehat{\mathbf{ab}} = \arccos(\mathbf{a} \cdot \mathbf{b}). \tag{18}$$

Therefore, computing $\widehat{\mathbf{ab}}$ is simplified to find coordinates of $\mathbf{a}$ and $\mathbf{b}$ that reside on the sphere.

As seen in Fig. 3(b), based on the concept of OCSVM, we know line $\mathbf{oc}_1^t$ ($\mathbf{oc}_2^t$) is perpendicular to $\mathbf{w}_1^t$ ($\mathbf{w}_2^t$), so we can replace $\mathbf{c}_1^t$ ($\mathbf{c}_2^t$) with $\mathbf{w}_1^t/||\mathbf{w}_1^t||$ ($\mathbf{w}_2^t/||\mathbf{w}_2^t||$) after some geometric computations. The arc distance $\widehat{\mathbf{c}_1^t \mathbf{c}_2^t}$ therefore is calculated as:

$$\widehat{\mathbf{c}_1^t \mathbf{c}_2^t} = \arccos \left(\frac{\mathbf{w}_1^t \cdot \mathbf{w}_2^t}{||\mathbf{w}_1^t|| ||\mathbf{w}_2^t||} \right). \tag{19}$$

Substitute Eq. (2) into Eq. (19), we have:

$$\widehat{\mathbf{c}_1^t \mathbf{c}_2^t} = \arccos \left(\frac{(\boldsymbol{\alpha}_1^t)^T k_{12}^t \boldsymbol{\alpha}_2^t}{\sqrt{(\boldsymbol{\alpha}_1^t)^T k_{11}^t \boldsymbol{\alpha}_1^t} \sqrt{(\boldsymbol{\alpha}_2^t)^T k_{22}^t \boldsymbol{\alpha}_2^t}} \right), \tag{20}$$

where $\boldsymbol{\alpha}_1^t$ and $\boldsymbol{\alpha}_2^t$ are the coefficient sets of the OCSVM classifier learnt from B_1^t and B_2^t respectively.

Similarly, the arc distance $\widehat{\mathbf{c}_1^t \mathbf{p}_1^t}$ ($\widehat{\mathbf{c}_2^t \mathbf{p}_2^t}$) is given as:

$$\widehat{\mathbf{c}_i^t \mathbf{p}_i^t} = \arccos \left(\frac{\rho_i^t}{\sqrt{(\boldsymbol{\alpha}_i^t)^T k_{ii}^t \boldsymbol{\alpha}_i^t}} \right), \quad i = 1, 2. \tag{21}$$

Compared with the traditional approach of comparing features of successive frames directly, assessing the divergence between two OCSVMs trained on frame sets have two potential advantages. First of all, it is more robust to noises such as flashlight frames, as they will be regarded as error support vectors and will not affect the OCSVM classifiers. Second, it is effective to detect gradual transitions, e.g., dissolve, and wipe etc. Even though the difference between two consecutive frames is not significant in these gradual transitions, the divergence between two sets is supposed to large as the overall distribution will be quite different.

3   Experimental Results

3.1   Setup

We have carried out experiments on TREC Video Retrieval Evaluation (TRECVID) 2007 SBD dataset[1]. The TRECVID is an annually worldwide benchmarking activity, whose goal is to encourage research on content-based information retrieval in digital video. The TRECVID 2007 SBD dataset contains 17 video sequences of $637,805$ frames, where $2,320$ shots are annotated.

---

[1] http://trecvid.nist.gov/trecvid.data.html.

Among them, 90 % of the shots are hard cuts, and the rest are gradual transitions. Contents of the dataset are diverse, covering a wide range from news reports to archived grayscale videos.

For comparison purposes, three criteria are selected to evaluate the SBD performance, i.e., recall, precision and F_1, given as:

$$recall = \frac{true\ positive}{true\ positive + false\ positive},$$

$$precision = \frac{true\ positive}{true\ positive + false\ negative},$$

$$F_1 = 2 \cdot \frac{recall \cdot precision}{recall + precision},$$

where "true positive" and "false positive" correspond to numbers of correctly and falsely detected shot boundaries respectively, and "false negative" is number of missed shot boundaries. The process of performance evaluation is as follows: we first compute the divergence output for each video sequence, then we tune the threshold of divergence to identify shot boundaries. Following the measurement of TRECVID, a correct shot boundary detection is defined as at least one frame overlap between the detected transition and the annotated transition.

3.2 Performance Evaluation

To obtain the optimal performance, we first evaluate the impact of parameters on SBD performance. Three parameters have to be evaluated, namely, parameter C of OCSVM, set length m, and pyramid level l. The optimal settings are acquired by fine-tuning one parameter while fixing the rest of them. The impact of C on performance is displayed in Fig. 4(a), where m and l are set as 20 and 2 randomly. It shows that the OCSVM classifier with C of 0.2 achieves the best performance. With C of 0.2 and l of 2, Fig. 4(b) illustrates the impact of m, where 20 attains the best recall and precision. Figure 4(c) shows the influence of different l, where C and m are set as 0.2 and 20. It is improved with l increasing from 0 to 3, and saturates when l enlarges from 3 to 4. Therefore, we report our final results and compare with the start-of-the-art approaches using the aforementioned optimal settings, i.e., $C = 0.2$, $m = 20$, and $l = 3$.

Table 1. Performance comparison on TRECVID 2007 SBD dataset.

Method	Recall (%)	Precision (%)	F_1(%)	Running time (seconds)
Mühling et al. [13]	93.1	90.7	91.9	7000
Zhao et al. [14]	91.3	90.0	90.6	-
Ren et al. [15]	94.1	91.9	93.0	5185
Kawai et al. [16]	90.5	94.4	92.4	1697
MID-OCSVM	**90.7**	**93.4**	**92.0**	**5102**

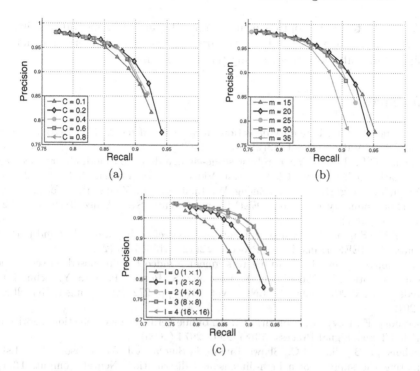

Fig. 4. Parameter tuning on the TRECVID 2007 SBD dataset. (a) Different settings of C in OCSVM; (b) Different set length m; (c) Different pyramid levels l.

Our final SBD results are reported in Table 1, achieving **92.0 %** on F_1 performance. Compared with the published results in TRECVID 2007[2], it is comparable to [15] (93.0 %) and [16] (92.4 %), better than [13] (91.9 %) and [14] (90.6 %).

Using the platform of MATLAB R2014a on a desktop computer with 2.7GHZ Intel Core i5 and 8GB RAM, the overall running time our approach is 5102 seconds, faster than most of the approaches except [16]. Specifically, the speed has increased from 92 frames/second using batch mode OCSVM to 125 frames/second using MID-OCSVM, where the processing time for each OCSVM classifier training has decreased from 5.4×10^{-3} second to 2.5×10^{-3} second.

4   Conclusion and Future Work

In this paper we present a novel approach to address the problem of shot boundary detection. Using online OCSVMs, a unified framework to detect all types of boundaries is proposed. We reduce the computational cost through a multi-instance incremental and decremental learning algorithm. By inheriting the properties of OCSVM, our method is robust to noises while effective to gradual transitions. Experimental results on a challenging benchmark dataset exhibit the competitive performance of our approach compared with the state-of-the-art.

---

[2] http://www-nlpir.nist.gov/projects/tvpubs/tv.pubs.7.org.html.

We intend to further optimize the incremental and decremental algorithmic design in our future work. As there are various spatial and temporal information within video, multiple kernel learning [17] is also considered.

References

1. Hanjalic, A.: Shot-boundary detection: unraveled and resolved? IEEE Trans. Circuits Syst. Video Technol. **12**(2), 90–105 (2002)
2. Huang, C.L., Liao, B.Y.: A robust scene-change detection method for video segmentation. IEEE Trans. Circuits Syst. Video Technol. **11**(12), 1281–1288 (2001)
3. Yuan, J., Wang, H., Xiao, L., Zheng, W., Li, J., Lin, F., Zhang, B.: A formal study of shot boundary detection. IEEE Trans. Circuits Syst. Video Technol. **17**(2), 168–186 (2007)
4. Lu, Z.M., Shi, Y.: Fast video shot boundary detection based on SVD and pattern matching. IEEE Trans. Image Process. **22**(12), 5136–5145 (2013)
5. Kowdle, A., Chen, T.: Learning to segment a video to clips based on scene and camera motion. In: Fitzgibbon, A., Lazebnik, S., Perona, P., Sato, Y., Schmid, C. (eds.) ECCV 2012, Part III. LNCS, vol. 7574, pp. 272–286. Springer, Heidelberg (2012)
6. Desobry, F., Davy, M., Doncarli, C.: An online kernel change detection algorithm. IEEE Trans. Signal Process. **53**(8), 2961–2974 (2005)
7. Schölkopf, B., Platt, J.C., Shawe-Taylor, J., Smola, A.J., Williamson, R.C.: Estimating the support of a high-dimensional distribution. Neural Comput. **13**(7), 1443–1471 (2001)
8. Swain, M.J., Ballard, D.H.: Color indexing. Int. J. Comput. Vis. **7**(1), 11–32 (1991)
9. Cauwenberghs, G., Poggio, T.: Incremental and decremental support vector machine learning. Adv. Neural Inf. Process. Syst. **13**, 409–415 (2001)
10. Laskov, P., Gehl, C., Krüger, S., Müller, K.R.: Incremental support vector learning: analysis, implementation and applications. J.Mach. Learn. Res. **7**, 1909–1936 (2006)
11. Karasuyama, M., Takeuchi, I.: Multiple incremental decremental learning of support vector machines. In: Advances in Neural Information Processing Systems, pp. 907–915 (2009)
12. Golub, G.H., Van Loan, C.F.: Matrix Computations, vol. 3. JHU Press, Baltimore (2012)
13. Mühling, M., Ewerth, R., Stadelmann, T., Zöfel, C., Shi, B., Freisleben, B.: University of Marburg at TRECVID 2007: Shot Boundary Detection and High Level Feature Extraction. In: TRECVID (2007)
14. Zhao, Z.C., Zeng, X., Liu, T., Cai, A.N.: BUPT at TRECVID 2007: Shot Boundary Detection. In: TRECVID (2007)
15. Ren, J., Jiang, J., Chen, J.: Determination of Shot Boundary in MPEG videos for TRECVID 2007. In: TRECVID (2007)
16. Kawai, Y., Sumiyoshi, H., Yagi, N.: Shot Boundary Detection at TRECVID 2007. In: TRECVID (2007)
17. Gönen, M., Alpaydın, E.: Multiple kernel learning algorithms. J. Mach. Learn. Res. **12**, 2211–2268 (2011)

Will I Win Your Favor? Predicting the Success of Altruistic Requests

Hsun-Ping Hsieh[1], Rui Yan[2], and Cheng-Te Li[3(✉)]

[1] National Cheng Kung University, Tainan, Taiwan
sandoh714@gmail.com
[2] Baidu Inc., Beijing, China
yanrui02@baidu.com
[3] Academia Sinica, Taipei, Taiwan
ctli@citi.sinica.edu.tw

Abstract. As those in need increasingly ask for favors in online social services, having a technique to accurately predict whether their requests will be successful can instantaneously help them better formulating the requests. This paper aims to boost the accuracy of predicting the success of altruistic requests, by following the similar setting of the state-of-the-art work ADJ [1]. While ADJ has an unsatisfying prediction accuracy and requires a large set of training data, we develop a novel request success prediction model, termed *Graph-based Predictor for Request Success* (GPRS). Our GPRS model is featured by learning the correlation between success or not and the set of features extracted in the request, together with a label propagation-based optimization mechanism. Besides, in addition to the textual, social, and temporal features proposed by ADJ, we further propose three effective features, including centrality, role, and topic features, to capture how users interact in the history and how different topics affect the success of requests. Experiments conducted on the requests in the "Random Acts of Pizza" community of *Reddit.com* show GPRS can lead to around 0.81 and 0.68 AUC scores using sufficient and limited training data respectively, which significantly outperform ADJ by 0.14 and 0.08 respectively.

Keywords: Social media · Altruistic requests · Success prediction

1 Introduction

With the maturity of World Wide Web, online services provide various functions for social good. Those in need can use social communities, such as *DonorsChoose.org*, *flyingv.cc*, and *Reddit.com*, to request donations or any help. Recent studies are studying why and how a posted request can get accepted and become successful, because understanding the hidden factors driving the requests to be satisfied by givers can be a great benefit for not only those in need to write their requests but also more people to help promote such kind of requests [16,21]. Existing work had identified some key factors, including the scale of the request

© Springer International Publishing Switzerland 2016
J. Bailey et al. (Eds.): PAKDD 2016, Part I, LNAI 9651, pp. 177–188, 2016.
DOI: 10.1007/978-3-319-31753-3_15

(e.g., a simple question vs. a big financial need) [22], whether the giver receives in return [21], and the social interaction between the receiver and the giver [8]. Furthermore, to uncover how the linguistic factor solely influences the success of requests, the ADJ work [1] focuses on *altruistic* requests, in which the giver receive *no* rewards. Based on the qualitative analysis of linguistic factors on decision making [26], ADJ shows that the linguistic presentation, including narrative structure, politeness, evidentiality, and reciprocity, has strong correlation to the success of requests. They also demonstrate the predictability of the success of requests by treating the measured scores of these factors as feature values.

This paper aims to boost the performance of predicting the success of altruistic requests via following the similar setting and data used by ADJ [1]. While what ADJ mainly contributes is developing a series of textual features to characterize the linguistic presentation of altruistic requests, study their correlation with request success, and use the logistic regression model to test the predictability, we need to point out four aspects of insufficiency of ADJ. First, the prediction accuracy is not satisfied (the AUC score is only 0.67). Second, ADJ does not study the importance of features, but in practice different requests can resort to or concentrate on various factors to seek for the success. Third, ADJ uses a large set of requests to build the predictive model (70 % for training and 30 % for testing). However, in real-world applications, we might not have many request data with the labels of success and unsuccess for training. Fourth, the features considered in ADJ cannot model how users interact with each other affect the success, and more importantly how the topics of the requests have impact on the success. We believe user interactions and topic information also play a deterministic role in the success prediction of altruistic requests.

We think a highly accurate predictive model can bring practical advices for requesters to optimize their presentation in a *real-time* manner when asking for favors. To have a more powerful method with limited training data to accurately predict the success of altruistic requests, we devise a novel model, *Graph-based Predictor for Request Success* (termed *GPRS*). In addition to the three features (i.e., textual, social, and temporal) proposed by ADJ, we further propose three additional features, including centrality, role, and topic. Our GPRS model is designed to jointly learn the feature weights and predict the success of query altruistic requests. We evaluate the effectiveness of our GPRS model using the dataset provided by ADJ, i.e., "Random Acts of Pizza" (RAOP)[1] in *Reddit.com*, an online community established for giving away free pizza to strangers that ask for one. The results exhibit GPRS is able to not only significantly outperform ADJ with the AUC score up to 0.8, but also produce a satisfying accuracy (AUC score = 0.68) using only 20 % request data for training.

2   Dataset and Features

2.1   Data

We use the Pizza Request Dataset[2] compiled by ADJ [1], which is the entire collection of the Random Acts of Pizza Subreddit (RAOP) from December 8,

---

[1] www.reddit.com/r/Random%5fActs%5fOf%5fPizza.

[2] http://cs.stanford.edu/%7ealthoff/raop-dataset/.

2010 to September 29, 2013, in *Reddit.com*. Totally there are 5728 altruistic pizza requests (24.6 % success rate) and 1.87M relevant posts by RAOP users (for computing user features).

2.2 Features

We expand the list of features proposed by ADJ to develop a novel prediction model. We first describe these features considered in ADJ, including textual, social, and temporal. These features mainly models how language usages, social reputation, and the request time/date affect the success of requests. Then we give the details about another three of our proposed novel features, including centrality, role, and topic. Our three features are designed to further capture how user interactions in *Reddit.com* via comments and the topic information of the requests have impact on the success, and thus can be considered as the important and informative complements of the previous three features.

Textual Features. There are six categories of textual features, which are designed to capture whether or not people will help the requesters based on the textual contents of the requests.

1. **Narrative.** The narrative of a request can significantly determine the success of that request [16,21]. We follow ADJ to use the relevant vocabularies from "Linguistic Inquiry and Word Count (LIWC)" [25] to characterize five different narratives (Money, Job, Student, Family, Craving) (please refer to ADJ for the detailed vocabularies), and compute the word counts of each narrative to be the features.
2. **Politeness.** Expressing in a polite manner can leave positive impression to the potential receivers. Some qualitative studies have shown that people have higher potential to help polite requesters [6]. We take advantage of 20 politeness strategies developed in the computational politeness model [12]. We consider that each of these politeness strategies is a binary feature value that indicates whether or not such strategy is used in the request text. Stanford Dependency Parser [11], together with Regular expression and the lexicons of each politeness strategy, is used to extract the features.
3. **Evidentiality.** To make the requests more convincing, the requesters can provide images to show the evidence of what they state and their need. For example, the pictures about the house status, the screenshot of bankbook, and the proof of disabled or unemployment. By using regular expression, we compute whether or not a request contains an image URL and the number of image URLs as features.
4. **Reciprocity.** People tend to help if they received help themselves [28]. We extract the binary feature that indicates whether or not the past posts of the request contain any phrases like "pay it forward," "pay it back," or "return the favor."
5. **Sentiment.** People is more likely to help those delivering positive mood [14]. Therefore we use the Stanford CoreNLP Package[3] to estimate whether the

---

[3] http://nlp.stanford.edu/software/.

sentiment of a request text exceeds the average fraction of positive sentiment sentences, and compute the counts of lexicons of positive and negative vocabularies from LIWC. In addition, we also consider the emoticons as another set of sentiment features.

6. **Length.** Longer requests can reflect more efforts of the need and be more successful. We use the word count of the request as the feature.

Social Features. Existing study showed that the inter-evaluation between people have positive correlation with the success of a request, especially user status and user similarity [2]. (1) **Status:** The *karma point* of a user measured by her activity in *Reddit.com* is used as the feature. (2) **Similarity:** People tend to help those who resemble them [10]. We use intersection size and the Jaccard similarity to compute the similarity between the requester and the other users on RAOP based on the set of Subreddits of users.

Temporal Features. We measure the specific season, month, workday or weekend, weekday, day of the month, and hour of the day of the request, as well as the number of months since the beginning of RAOP community to be the temporal features.

Centrality Features. We would like to investigate how the extent of user interaction in *Reddit.com* affects the success of a request. Users in *Reddit.com* are allowed to interact with one another through commenting others' posts. We construct an directed weighted *interaction graph* to represent their interaction behaviors, in which each node is a user and each directed edge refers to a comment from the user of the comment to the recipient. Each edge is weighted by the reciprocal of the number of comments from one user to another. Lower weight values mean frequent/stronger interaction. Then to characterize how the requester interact with others, based on the constructed interaction graph, we calculate several structural **centrality** measures, including *in-degree, out-degree, clustering coefficient, closeness, betweenness, eigenvector, PageRank,* and *HITS* scores (hub and authority), as the feature values.

Role Features. We also measure the **role** of interaction of the requester among other users in terms of *network communities* (a community refers to a set of nodes that are densely connected internally and loosely connected externally in a graph). A user who is exposed to less communities can belong to the minority that needs help [13]. But a user with connections to more communities is also capable of reaching more information, and thus has a higher possibility to earn resources [23]. Therefore, we aim to quantify how the role of a requester who plays among communities affects the success of his/her altruistic request. We detect communities in the constructed interaction graph using *Louvain's algorithm* [5]. Then based on some social theories, we measure four social roles as the features.

1. **Structural Hole.** *Structural hole theory* [7] suggests that nodes act as an intermediary between groups have higher potential to access more information and earn the favor. We measure the extent of being structural holes for nodes using the number of overlapping communities [19]. Nodes overlapped by more communities tend to act as structural holes.

2. **Structural Diversity.** A person participating in fewer diverse social contexts has been validated to have lower probability to obtain his/her material and mental need [27]. We score the structural diversity by computing the number of disjoint connected components in the neighboring induced subgraph of a node. Higher scores mean higher structural diversity.

3. **Bridging Effect.** Some people act as the role of transmitting information, instead of giving or receiving help. Since the bridging nodes in a graph have been proved effective in distributing information [24], we adopt their proposed measure, *rawComm*, to quantify the bridging effect a node involves.

4. **Group Core.** Individuals with more friends who connect to each other, i.e., higher *Triangle Participation Ratio* (TPR, i.e., the number of triangles involved) and thus are closer to the core of a group, have higher potential to receive both information and help [20]. Nodes with higher TRP values tend to have higher social visibility and thus have higher potential to receive help [3]. We compute the *Triangle Participation Ratio* as the feature values.

Topic Features. We assume that the success of altruistic requests may depend on which topic that a request belongs to. Topics like bereavement are more likely to evoke public sympathy and earn the favor than other topics like unemployment. Therefore, we aim at extract the hidden topic of a request text, and treat the hidden topics as the features. Based on the textual content, we consider three kinds of manners to model the hidden topics.

1. **Bag-Of-Word** (BOW) is one of the simplest but useful features in many natural language tasks. We consider the words with NN (noun), NR (proper noun), VV (verb), VA (adjective), and AD (adverb) for the BOW features. Word counts in a document are treated as the feature values.

2. **N-gram** is to estimate the likelihood of a sentence by conditional probability. Here we use N-gram to capture whether or not there are some important terms that can determine the success of the requests. By eliminating stop words, Bi-gram and Tri-gram are considered. For BOW, Bi-gram, and Tri-gram features, we perform feature selection (dimension reduction) to identify the important and discriminative features. The technique of *Recursive Feature Elimination with Cross Validation* [17] is used to select the best number and set of features.

3. **LDA Hidden Topic.** We exploit the *Latent Dirichlet Allocation* (LDA) [4], which is a well-known topic modeling technique, to derive the hidden topic for each request. In LDA, a document-word matrix D is decomposed to a topic-hidden matrix H and a hidden-word matrix W, i.e., $D = H \times W$, in which a parameter n is used to determine the number of hidden topic categories. Since each row vector in H (denoted by H_v) can be regarded as a request document v's hidden topic, which is represented by the distribution over hidden topic categories, we use H_v as the feature values. In addition, for a request, we propose to estimate its degree of interestingness by users in terms of hidden topics. We append the user-word matrix X to D in a row-wise manner, and obtain a combined matrix M, i.e., $M = [D; X]$. LDA is applied

again for matrix decomposition: $M = H' \times W'$. We compute and sum up the Cosine similarity score $s(v)$ between the request v's hidden topic vector H'_v and each user i's hidden topic vector H'_i, i.e., $s(v) = \sum_{i \in X} cos(H'_v, H'_i)$. The obtained score $s(v)$ is treated as the feature that characterizes the degree of interestingness for the request v over all the users.

3   The Proposed GPRS Model

We devise a novel model, *Graph-based Predictor for Request Success* (**GPRS**), to predict whether or not a request will be successful (i.e., a binary success label, 0 or 1). GPRS consists of two stages: constructing a *Request Graph*, and *Propagation-based Optimization*. The basic idea lies in representing the feature similarity-based correlation between requests in a graph structure, and jointly learning the feature weights and computing the success labels of unseen requests by spreading the probabilities of success labels in the request graph such that those requests with higher similarity with each other have the same success label (0 or 1).

3.1   Constructing Request Graph

A *Request Graph* (RG) $G = (\mathcal{V}, E)$ is devised to model the feature-based correlation between request nodes, in which $\mathcal{V}$ is the set of nodes and $\mathcal{V} = V \cup U$, where V and U are the node sets of training and testing requests respectively. The construction of RG has two parts: (1) each testing request node $u \in U$ is connected to the top-k_1 similar training node $v \in V$, and (2) each testing request $u \in U$ is connected to the top-k_2 similar testing requests $u' \in U(u' \neq u)$ in terms of features, where k_1 and k_2 are determined by a parameter $\lambda \in [0, 1]$: $k_1 = \lambda \cdot |V|$ and $k_2 = \lambda \cdot |U|$. We will show how λ affects the effectiveness and efficiency our model. On the other hand, each request $x \in \mathcal{V}$ is associated with two probabilities, $P_{s_x}(x)$, corresponding to its success label $s_x = 0$ or $s_x = 1$. $P_1(x)$ and $P_0(x)$ are the probabilities that the request x is successful or not respectively. For each training request $v \in V$ whose success label is 1 (i.e., $s_v = 1$), we always fix $P_1(v) = 1$ and $P_0(v) = 0$; $P_0(v) = 0$ and $P_0(v) = 1$ if $s_v = 0$. We also initialize $P_1(u) = 0$ and $P_0(u) = 0$ for each testing request $u \in U$.

Each edge in RG is associated as a weight that represents the feature-based correlation between requests. Given a certain feature $\mathcal{F}_d$, the *feature-based request correlation* $frc_{\mathcal{F}_d}(x, y)$ between nodes x and y, $(x, y) \in E$, can be derived from their feature difference $frc_{\mathcal{F}_d}(x, y) = \Delta \mathcal{F}_d(x, y)$, where $\Delta \mathcal{F}_d$ is their feature difference, defined by $\Delta \mathcal{F}_d = \|\mathbf{f_d}(x) - \mathbf{f_d}(y)\|$. Given a list of features $F = \{\mathcal{F}_d\}$ ($d = 1, ..., m$, m is the number of features), we compute *feature-aware request correlation* value $frc(x, y)$ via the weighted sum of their correlation $frc_{\mathcal{F}_d}$, given by:

$$frc(x, y) = \exp(-\sum_{d=1}^{m} \pi_d^2 \times frc_{\mathcal{F}_d}(x, y)), \tag{1}$$

where π_d is the weight of feature $\mathcal{F}_d$. The combined correlation is considered as the edge weight $w_{x,y} = frc(x, y)$ for edge $(x, y) \in E$ in RG.

3.2 Propagation-Based Optimization

The idea is two-fold. First, the probabilities of training request $v \in V$, $P_0(v)$ and $P_1(v)$, are used to infer the probabilities of testing request $u \in U$, $P_0(u)$ and $P_1(u)$. Second, $P_0(u)$ and $P_1(u)$ are inferred from both u's neighboring training and testing requests in RG, expressed by $P_s(u) = \frac{1}{deg_u} \sum_{(u,x) \in E} w_{u,x} \cdot P_s(x)$, where deg_u is the degree of node u in RG. Putting these together, we seek for an optimal set of edge weights $\mathcal{W}$ in RG, where edge weights can be further determined by feature weights π_d, such that after inference, the testing request and its neighboring requests that possess similar features tend to have close probabilities, which lead to the same success labels. When the iterative propagation process is finalized, we can choose the success label $s_u^\star$ with the higher probability to be the predicted result, given by $s_u^\star = argmax_{s_u}\{P_{s_u}(u)\}, s_u = 0, 1$.

Recall that edge weights are obtained by the weighted sum over $frc_{\mathcal{F}_d}(x, y)$ of features $\mathcal{F}_d$ with feature weights $\{\pi_d\}$. That said, the determination of feature weights first influences the edge weights, and then edges weights take effect on the inference of success labels for testing requests. Hence, our ultimate goal is to learn a set of feature weights from training and testing requests in RG.

We propose a heuristic objective for learning $\{\pi_d\}$: minimizing the *average entropy of success probabilities* $H(P^U)$ for testing requests $u \in U$:

$$H(P^U) = \frac{1}{|U|} \times - \sum_{s=0,1} P_s(u) \log P_s(u) + (1 - P_s(u)) \log(1 - P_s(u)), \quad (2)$$

where $|U|$ is the number of testing requests in RG. Our idea is that assigning the testing requests $u \in U$ the success probabilities $P_0(u)$ and $P_1(u)$ that produce the lower entropy values can make the inference be less uncertain and higher confidence. Therefore, we take advantage of the minimization of $H(P^U)$ so that through Eq. (2) the inferred success probabilities of each testing request tends to be squeezed and constrained at a success label $c_u^\star$ which possesses the highest probability.

We design a *mutually reinforced* flow to iteratively minimize $H(P^U)$ during the propagation process in RG: the learned feature weights π_d triggers an update of edge weights $w_{x,y}$ that update the success probabilities $P_1(x)$ and $P_0(x)$ for every testing request $x \in U$, which further determine their average success probability entropy $H(P^U)$ to be minimized in Eq. (2). This flow proceeds iteratively till the convergence is reached. To enable this flow, we exploit the technique of *gradient descent* on π_d to obtain an updated set of feature weights $w_{x,y}$ that minimizes $H(P^U)$. The gradient $\frac{\partial w_{x,y}}{\partial \pi_d}$ can be derived by computing $\frac{\partial H(P^U)}{\partial \pi_d}$:

$$\frac{\partial H(P^U)}{\partial \pi_d} = \frac{1}{|U|} \sum_{x \in U} \log \frac{1 - P(x)}{P(x)} \frac{\partial P(x)}{\partial \pi_d}. \quad (3)$$

Using the chain rule of differentiation, we can have the final gradient as:

$$\frac{\partial w_{x,y}}{\partial \pi_d} = 2 \cdot w_{x,y} \cdot frc_{\mathcal{F}_d}(x, y) \cdot \pi_d. \quad (4)$$

In short, in each iteration, we update the feature weights $\pi_d = \pi_d - 2 \cdot w_{x,y} \cdot frc_{\mathcal{F}_d}(x, y) \cdot \pi_d$. Then a new set of edge weights $w_{x,y}$ can be derived using the updated π_d via Eq. (1). Based on new $w_{x,y}$, we can generate the new success probabilities of testing requests $P_1(x)$ and $P_0(x)$ from x's neighbors in RG. Then the average success probability entropy $H(P^U)$ is updated accordingly via Eq. (3). The iterative updating procedure will continue and be terminated till $H(P^U)$ converges. Finally, using the derived success probabilities of each testing request $x \in U$, we can find its predicted success label $s_x^\star$. Note that we can prove the convergence by establishing a reduction from the graph-based label propagation [29]. However, due to the page limit, we skip the proof.

4   Evaluation

The experiment consists of four parts. First, we aim to show whether the proposed GPRS model can perform better than ADJ [1] using the same volume of training requests. Second, we will show GPRS can still work well using a small set of training requests. Third, we will present the effectiveness of each feature category, i.e., textual, social, temporal, centrality, role, and topic. Fourth, the time efficiency of GPRS by varying the parameter λ will be reported.

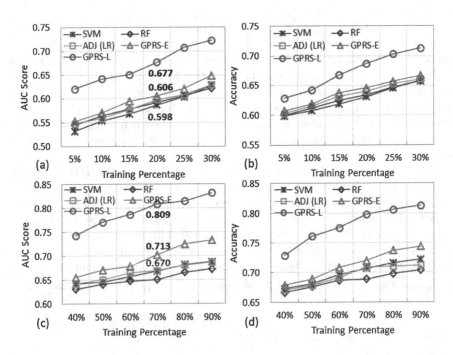

Fig. 1. AUC and accuracy by varying the percentage of training data (Color figure online).

Settings. We vary the percentage of requests for training data using two scales: 40 %–90 % and 5 %–30 % to evaluate the first two parts. We set $\lambda = 0.1$ by default, and vary λ (it determines the size of the request graph) to show the time efficiency of GPRS-W using 70 % requests as the training data. In addition, to further know whether learning feature weights π_d can benefit the prediction accuracy, we divide our GPRS into two versions: equally assigned (**GPRS-E**) and automatically learned (**GPRS-L**) from the validation set. We have two evaluation metrics. The first one follows ADJ to use the *Area Under Receiver-Operating Characteristic (ROC)* Curve (AUC). Second, we define an accuracy measure $acc = \frac{\#hits}{|U|}$, where $\#hits$ is the number of correctly predicted testing requests and $|U|$ is the number of testing requests. We randomly select the corresponding percentages of training and testing data, and repeat the experiment up to 100 times. The average AUC score and the average accuracy are reported.

Competitors. In addition to L_1-penalized logistic regression (LR) model [15] used by ADJ, we further compare our GPRS model with several typical supervised learning methods (treated as baseline models), including Support Vector Machine (SVM) [9], and Random Forest (RF) [18].

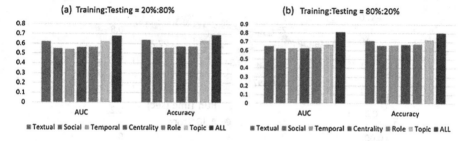

Fig. 2. AUC and accuracy of each feature category using our GPRS-L model under training:testing $= 80\,\%{:}20\,\%$ and training:testing $= 20\,\%{:}80\,\%$.

The results under different training percentages are shown in Fig. 1. We can find that both GPRS-L and GPRS-E outperform ADJ and other competitors in every case, especially for GPRS-L whose AUC and accuracy scores are significantly higher than ADJ. Since ADJ uses solely 70 % training percentage to have 0.67 AUC, we especially highlight the results at 70 % for GPRS and show that both AUC and accuracy of GPRS-L are up to 0.8. It is also worthwhile to notice that GPRS-L can still achieve 0.67 AUC using only 20 % requests for training, and exhibits the prediction power of GPRS-L when tackling rare training data. Besides, the result that GPRS-L beats GPRS-E validates the efficacy of learning feature weights.

The resulting effectiveness of each category of feature using our GPRS-L model is shown in Fig. 2. We can find that our proposed three feature categories (i.e., centrality, role, and topic) lead to the competitive performance, comparing with the three feature categories proposed by ADJ (i.e., textual, social, and temporal). Among our proposed three feature categories, the topic feature can generate a bit more accurate prediction results than the textual feature, which is the best among the three features used in ADJ. Such outcome demonstrates the topic of the request can be more deterministic than the language usage (e.g., politeness and narrative), the provided evidence, and the conveyed sentiment, which are modelled by the textual feature. In addition, it is worthwhile to notice that the performance of the topic feature will be better (compared to the textual feature) if more training data are used (e.g., the performance of the topic feature under 80 % training is better than that under 20 % training data). We think the reason could be the hidden topics can be learned only if there are sufficient training data. Finally, combing all of these six feature categories can generate the best results.

The results of time efficiency (in second) and the AUC score using our GPRS-L model under different λ values are reported in Fig. 3. There is a trade-off between accuracy and efficiency. Higher λ values that add more edges in the request graph lead to higher AUC scores, but cost more time to run the GPRS-W model. Nevertheless, we find $\lambda = 0.1$ is a good choice since it balances the prediction accuracy (up to 0.8 AUC score) and run time (only 3 s), and suggest $\lambda = 0.1$ for real usages.

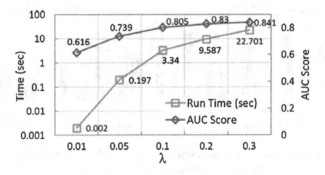

Fig. 3. AUC and run time by varying λ under the proposed GPRS-L model (Color figure online).

5   Conclusion

The contribution of our GPRS model is four-fold. First, our GPRS model is able to significantly boost the accuracy of predicting the success of altruistic requests. Second, it is capable of tacking the problem of inadequate training data while keeping the accuracy. Third, three additional features (i.e., centrality, role, and topic) are proposed and evaluated to be effective in predicting the success of requests. Fourth, the technique that jointly learns feature weights and predicts labels can be served as a novel framework to solve other NLP tasks like sentiment detection and POS tagging.

Acknowledgments. This work was sponsored by Ministry of Science and Technology of Taiwan under grant 104-2221-E-001-027-MY2. This work is also supported by Multidisciplinary Health Cloud Research Program: Technology Development and Application of Big Health Data, Academia Sinica, Taipei, Taiwan under grant MP10212-0318.

References

1. Althoff, T., Danescu-Niculescu-Mizil, C., Jurafsky, D.: How to ask for a favor: a case study on the success of altruistic requests. In: Proceedings of AAAI International Conference on Web and Social Media, ICWSM (2014)
2. Anderson, A., Huttenlocher, D., Kleinberg, J., Leskovec, J.: Effects of user similarity in social media. In: Proceedings of the 5th ACM International Conference on Web Search and Data Mining, WSDM, pp. 703–712 (2012)
3. Backstrom, L., Huttenlocher, D., Kleinberg, J., Lan, X.: Group formation in large social networks: membership, growth, and evolution. In: Proceedings of the ACM SIGKDD International Conference on Knowledge Discovery and Data Mining, KDD, pp. 44–54 (2006)
4. Blei, D.M., Ng, A.Y., Jordan, M.I.: Latent Dirichlet allocation. J. Mach. Learn. Res. **3**, 993–1022 (2003)
5. Blondel, V.D., Guillaume, J.-L., Lambiotte, R., Lefebvre, E.: Fast unfolding of communities in large networks. J. Stat. Mech.: Theor. Exp. **2008**(10), P1000 (2008)
6. Brown, P., Levinson, S.C.: Universals in language use: politeness phenomena. Questions and Politeness: Strategies in Social Interaction, pp. 56–311. Cambridge University Press, Cambridge (1978)
7. Burt, R.S., Holes, S.: The Social Structure of Competition. Harvard University Press, Cambridge
8. Ceyhan, S., Shi, X., Leskovec, J.: Dynamics of bidding in a P2P lending service: effects of herding and predicting loan success. In: Proceedings of the 20th International Conference on World Wide Web, WWW, pp. 547–556 (2011)
9. Chang, C.-C., Lin, C.-J.: LIBSVM: a library for support vector machines. ACM Trans. Intell. Syst. Technol. (TIST) **2**(3), 1–27 (2011)
10. Chierco, S., Rosa, C., Kayson, W.A.: Effects of location, appearance, and monetary value on altruistic behavior. Psychol. Rep. **51**(1), 199–202 (1982)
11. de Marneffe, M.-C., MacCartney, B., Manning, C.D.: Generating typed dependency parses from phrase structure parses. In: Proceedings of the Language Resources and Evaluation Conference, LREC, pp. 449–454 (2006)
12. Danescu-Niculescu-Mizil, C., Sudhof, M., Jurafsky, D., Leskovec, J., Potts, C.: A computational approach to politeness with application to social factors. In: Proceedings of the 51st Annual Meeting of the Association for Computational Linguistics, ACL, pp. 250–259 (2013)
13. Fagan, J., Meares, T.L.: Punishment, deterrence and social control: the paradox of punishment in minority communities. Ohio State J. Crim. Law **6**, 173 (2008)
14. Forgas, J.P.: Asking nicely? The effects of mood on responding to more or less polite requests. Pers. Soc. Psychol. Bull. **24**(2), 173–185 (1998)
15. Friedman, J., Hastie, T., Tibshirani, R.: Regularization paths for generalized linear models via coordinate descent. J. Stat. Softw. **33**(1), 1–22 (2010)
16. Greenberg, M.D., Pardo, B., Hariharan, K., Gerber, E.: Crowdfunding support tools: predicting success & failure. In: CHI 2013 Extended Abstracts on Human Factors in Computing Systems, pp. 1815–1820 (2013)

17. Guyon, I., Weston, J., Barnhill, S., Vapnik, V.: Gene selection for cancer classification using support vector machines. Mach. Learn. **46**(1–3), 389–422 (2002)
18. Liaw, A., Wiener, M.: Classification and regression by randomForest. R News **2**(3), 18–22 (2002)
19. Lou, T., Tang, J.: Mining structural hole spanners through information diffusion in social networks. In: Proceedings of the ACM International Conference on World Wide Web, WWW, pp. 825–836 (2013)
20. Malliaros, F.D., Vazirgiannis, M.: To stay or not to stay: modeling engagement dynamics in social graphs. In: Proceedings of ACM International Conference on Information and Knowledge Management, CIKM, pp. 469–478 (2013)
21. Mitra, T., Gilber, E.: The language that gets people to give people to give: phrases that predict success on kickstarter. In: Proceedings of the 17th ACM Conference on Computer Supported Cooperative Work and Social Computing, CSCW, pp. 49–61 (2014)
22. Mollick, E.: The dynamics of crowdfunding: an exploratory study. J. Bus. Ventur. **29**(1), 1–16 (2014)
23. Pearce, J., Witten, K., Hiscock, R., Blakely, T.: Are socially disadvantaged neighbourhoods deprived of health-related community resources? Int. J. Epidemiol. **36**, 348–355 (2007)
24. Scripps, J., Tan, P.-N., Esfahanian, A.-H.: Exploration of link structure and community-based node roles in network analysis. In: Proceedings of the IEEE International Conference on Data Mining, ICDM, pp. 649–654 (2007)
25. Tausczik, Y.R., Pennebaker, J.W.: The psychological meaning of words: LIWC and computerized text analysis methods. J. Lang. Soc. Psychol. **29**(1), 24–54 (2010)
26. Teevan, J., Morris, M.R., Panovich, K.: Factors affecting response quantity, quality, and speed for questions asked via social network status messages. In: Proceedings of the AAAI International Conference on Web and Social Media, ICWSM (2011)
27. Ugander, J., Backstrom, L., Marlow, C., Kleinberg, J.: Structural diversity in social contagion. Proc. Nat. Acad. Sci. U.S.A. (PNAS) **109**(16), 5962–5966 (2012)
28. Willer, R., Flynn, F.J., Feinberg, M., Mensching, O., de Mello Ferreira, V.R., Bianchi, A.M., Choshen-Hillel, S., Weisel, O., Peng, K., Fetchenhauer, D.: Do people pay it forward? Gratitude fosters generalized reciprocity. Technical report (2013)
29. Zhou, D., Bousquet, O., Lal, T.N., Weston, J., Scholkopf, B.: Learning with local and global consistency. In: Proceedings of Advances in Neural Information Processing Systems, NIPS 2003, pp. 321–328 (2003)

Feature Extraction and Pattern Mining

Unsupervised and Semi-supervised Dimensionality Reduction with Self-Organizing Incremental Neural Network and Graph Similarity Constraints

Zhiyang Xiang[1,2], Zhu Xiao[1,2], Yourong Huang[1], Dong Wang[1(✉)], Bin Fu[1], and Wenjie Chen[3]

[1] College of Computer Science and Electronics Engineering,
Hunan University, Changsha, China
{z_xiang,zhxiao,h_y_r,wangd,fubin}@hnu.edu.cn
[2] State Key Laboratory of Integrated Services Networks,
Xidian University, Xian, China
[3] Business College, Central South University
of Forestry and Technology, Changsha, China
wendychen711@126.com

Abstract. The complexity of optimizations in semi-supervised dimensionality reduction methods has limited their usage. In this paper, an unsupervised and semi-supervised nonlinear dimensionality reduction method that aims at lower space complexity is proposed. First, a positive and negative competitive learning strategy is introduced to the single layered Self-Organizing Incremental Neural Network (SOINN) to process partially labeled datasets. Then, we formulate the dimensionality reduction of SOINN weight vectors as a quadratic programming problem with graph similarities calculated from previous step as constraints. Finally, an approximation of distances between newly arrived samples and the SOINN weight vectors is proposed to complete the dimensionality reduction task. Experiments are carried out on two artificial datasets and the NSL-KDD dataset comparing with Isomap, Transductive Support Vector Machine etc. The results show that the proposed method is effective in dimensionality reduction and an efficient alternate transductive learner.

Keywords: Dimensionality reduction · Self-Organizing Incremental Neural Network · Graph cut · Semi-supervised learning

1 Introduction

Dimensionality reduction or referred as embedding is an important technique for data mining and pattern recognition applications. It is often employed in high dimensional data processing to enable data visualization and or as a metric learning method. In real world applications it is always involved in two types of tasks. First, it can construct and find out the most informative dimensions, thus

© Springer International Publishing Switzerland 2016
J. Bailey et al. (Eds.): PAKDD 2016, Part I, LNAI 9651, pp. 191–202, 2016.
DOI: 10.1007/978-3-319-31753-3_16

to enable machine learning techniques that only perform well on low dimensional techniques. Second, it can function as a metric learning process to augment distance based supervised learning and semi-supervised learning (SSL). Depending on the availability of labeled information, dimensionality reduction can be categorized into supervised, unsupervised and semi-supervised ones.

There are linear and nonlinear dimensionality reduction methods. Multi-Dimensional Scaling (MDS) and Fisher Discriminant Analysis (FDA) are examples of the linear methods; and Isomap [17], semi-definite programming (SDP) based embeddings [20] are examples of nonlinear methods. The major difference between them is that nonlinear methods employs more computationally costly methods such as graph similarity and SDP to construct the kernel matrix. Linear methods such as classical MDS are efficient but are ineffective to learn the nonlinear embedding of data from high dimensional spaces to lower ones. On the other hand, the nonlinear methods are often less efficient because the graph similarity calculation, SDP and other nonlinear optimizations are hard to solve, especially when applying to real world large scale problems.

Semi-supervised dimensionality reduction are often constructed by manipulations of the kernel matrix from their supervised or unsupervised counterparts. Such examples can be seen in [2,3,21]. The manipulation can come either as a form of pairwise constraints or approximations like graph kernels. An optimization problem would be difficult if its number of variables are of the same magnitude of the number of samples in training datasets. This has limited their usage in real world applications. Graph construction methods are often slow, too [6].

There is a recent trend to solve transfer domain learning with dimensionality reduction, which has an impact on SSL research because SSL is a special form of transfer domain learning [11,12]. Semi-supervised dimensionality reduction is equivalent to an embedding task on two domains, one of which is the data feature domain and the other is the augmented domain with labels as extra features. Transfer domain embedding methods such as [10,19] relies on SDP to find the kernel matrix. Unfortunately, SDPs are often hard to solve. An SDP with several thousand variables may be difficult for a mainstream personal computer to solve, but real world applications will easily reach this magnitude.

Clustering is frequently employed to reduce the problem size. Competitive learning neural networks such as Self-Organizing Maps [8], Growing Neural Gas (GNG) [5] and Self-Organizing Incremental Neural Network (SOINN) [15] are widely used as data reduction techniques, because of their incremental and topology learning abilities. However, their applicability as feature reduction techniques are not fully explored. Moreover, the SSL designs of SOINN are not as thorough as GNG. In [1,9] the consensus and heuristical online label inferring extensions to GNG are studied. However, the two extensions assumed that the labeled information is so little that it can not affect the data distribution learned by GNG.

In this paper we propose a semi-supervised dimensionality reduction method based on semi-supervised SOINN and linear dimensionality reduction. Our main contributions are as follows.

1. A nonlinear dimensionality reduction framework with graph similarity from SOINN as constraints is proposed.
2. Semi-supervised extensions to SOINN in the form of positive and negative competitive learning is proposed.
3. With the semi-supervised SOINN, the kernel constructed for dimensionality reduction is simplified. As a result, the efficiency of dimensionality reduction is improved.

The rest of the paper is organized as follows. Section 2 gives the notations and preliminaries used in later sections. Section 3 introduces the framework of our work and the semi-supervised extension to SOINN. Then in Sect. 4 are the experiments and in Sect. 5 are the conclusions.

2   Preliminaries

In this section, theories and algorithms our work is based upon are introduced. More specifically, the linear dimensionality reduction and the single layered SOINN are described.

2.1   Linear Dimensionality Reduction

Suppose $\mathbf{P} = \{p_{ij}\}$ is the dissimilarity matrix of a training dataset. According to [7,18], the linear dimensionality reduction can be described as follows. First, calculate a matrix $\mathbf{H}$ as [7,18]

$$(\mathbf{H})_{ij} = -\frac{|p_{ij}|^2}{2} + \frac{\sum_{m=1}^{l} |p_{mj}|^2 + \sum_{n=1}^{l} |p_{in}|^2}{2l} - \frac{\sum_{m,n=1}^{l} |p_{mn}|^2}{2l^2} \tag{1}$$

Where l is the desired dimension count after dimensionality reduction. Then calculate eigendecomposition as $\mathbf{P} = \mathbf{U}\mathbf{D}\mathbf{U}^T$, where $\mathbf{D}$ is a diagonal matrix with eigenvalues of $\mathbf{H}$ along the diagonal and only the non-zero l eigenvalues are selected. If all the eigenvalues are non-negative, then the dimensionality reduction is achieved by $\mathbf{X} = \mathbf{D}^{0.5}\mathbf{U}^T$, where the columns of $\mathbf{X}$ are the data items after dimensionality reduction. If there are negative eigenvalues, then $\mathbf{X} = (\mathbf{MD})^{0.5}\mathbf{U}^T$, where the $n \times n$ matrix $\mathbf{M} = diag(\mathbf{I}_{n+}, -\mathbf{I}_{n-})$ with $n = n^+ + n^-$ where the pair (n^+, n^-) is called the signature of the pseudo-Euclidean space [7,18]. The distance metric is altered as [18]

$$\delta_{ij} = \sqrt{(X_i - X_j)^T \mathbf{M}(X_i - X_j)} \tag{2}$$

For a new data item, its dimensionality reduction can be calculated by two steps. First, from the distances of the new data item to existing data items $P = \{p_i\}$ calculate a vector $H = \{h_i\}$ [18]

$$h_i = -\frac{|p_i|^2}{2} + \frac{\sum_{m=1}^{l} |p_m|^2 + |p_i|^2}{2l} - \frac{\sum_{m} |p_m|^2}{2l^2} \tag{3}$$

Then the dimensionality reduction is accomplished as

$$X = (\mathbf{MD})^{0.5}\mathbf{U}H \tag{4}$$

where $H = [h_1, h_2, \cdots, h_i, \cdots]^T$.

2.2   The Single Layered SOINN

The single layered SOINN [15] can be divided into three phases. First, on an arrival of a data item, it decides whether to insert a new vector into the set $\mathbf{W}$. Second, update the weight vectors. Third, discover the topology structure by label propagation.

In step 1, each weight vector is said to control a spherical area in the data space. The sphere is centered at the position of the vector itself, and the diameter is the vector to its farthest topological neighbor by metric of Euclidean distance or the smallest distance to its neighbors if it has no topological neighbor. When a new sample V arrives and it is not controlled by its nearest neighbor W_s or second nearest neighbor W_t in $\mathbf{W}$, then this vector is added to $\mathbf{W}$.

In step 2, if there is no edge between W_s and W_t, add one. Then if in step 1 the new sample is not added to $\mathbf{W}$, update W_s and its topological neighbors towards the new sample as

$$W_s = W_s + \frac{1}{c_s}(V - W_s) \tag{5}$$

$$W_k = W_k + \frac{\phi}{c_s}(V - W_k), \forall k \in neighbor(s) \tag{6}$$

where c_s is the winning times of W_s. ϕ is a constant and in [15] it is set to 0.01.

Step 3 is only invoked one time every λ times of sample arrival. It assumes the convexity of data and try to divide graph learned into trivial graphs in the original SOINN algorithms. Since the purpose of our method is not for clustering under convexity assumption, this step is not inherited in our work. The step 3 is not introduced and not used in the semi-supervised extension which is to be introduced in later sections.

3   Dimensionality Reduction with Semi-supervised SOINN

In this section, the details of the proposed method is given. First, the problem formation and algorithm framework are elaborated. Second, the semi-supervised extension to SOINN is explained. The key idea is a knowledge reuse framework, where the learning result from SOINN is employed in defining the similarity of samples in the same space and under the same distribution as the training dataset (including the labeled and unlabeled). In the SOINN learned data representation, weight vectors that are similar to each other is more likely to be linked by more paths in the graph [15].

3.1 Problem Formation and Algorithm Framework

We approach dimensionality reduction by three steps. First, train semi-supervised SOINN on the input data. Second, calculate the dimensionality reduction of SOINN weight vectors. Third, calculate the dimensionality reduction of newly arrived data by their similarity to the SOINN weight vectors.

The major problem of a dimensionality reduction method is to construct the kernel matrix. We use the SOINN trained graph as a graph kernel for further construction of a kernel matrix. In SOINN the more similar the two weight vectors are, the higher the probability that the two vectors are linked by more paths on the graph. Similarity between graph vertices can be calculated by the Edmonds-Karp algorithm [4]. However, graph similarity defined by graph cuts are often used for classification purposes. Since the variation of the similarity values can be highly skewed and possibly can not be embedded to an Euclidean space. If the graph similarities defined by graph cut are directly used in dimensionality reduction, the eigendecomposition of graph similarity matrix may generate many negative eigenvalues. As a result, in our work we try to preserve the distance matrix $\mathbf{P}$ calculated from the original data, and use the graph similarity $\mathbf{G} = \{g_{ij}\}$ calculated by Edmonds-Karp algorithm on the SOINN graph as constraints. The graph similarity $\mathbf{G}$ can be calculated from edge weights of

$$\text{flow}_{ij} = g(\|W_i - W_j\|) \tag{7}$$

where $g(x)$ is a decreasing and positive valued function. This means that when connected by a single edge, the closer the two vector is, the higher the graph similarity. The specific choice of $g(x)$ is irrelevant as long as it is decreasing.

Then the problem of embedding weight vectors can be described as

$$\min_{\mathbf{P}'} \sum_i \sum_j (p'_{ij} - p_{ij})^2 \tag{8}$$

$$s.t. \, \forall g_{ij}, g_{mn} \in \mathbf{G}, if \, g_{ij} \leq g_{mn}, p'_{ij} \geq p'_{mn}$$

Put the elements of $\mathbf{G}$ in the upper triangle (and excluding the diagonal) in a vector $S = \{s_k\}$ (standing for similarity) and sort in decreasing order; and construct two vectors $D = \{d_k\}$ (standing for distance), $K = \{k_k\}$ for each element $s_k = g_{ij}, d_k = p_{ij}, k_k = p'_{ij}$. Since the diagonal of $\mathbf{P}, \mathbf{P}'$ must be zero, the optimization described by Eq. (8) is equivalent to

$$\min_K \sum_i (k_i - d_i)^2 \tag{9}$$

$$s.t. \, 0 \leq k_1 \leq k_2 \cdots \leq k_j \cdots$$

which is a quadratic programming problem

$$\min_K \frac{1}{2} K^T K - D^T K \tag{10}$$

$$s.t. \, 0 \leq k_1 \leq k_2 \cdots \leq k_j \cdots$$

This can be solved by a standard solver. This approach is similar to Structure Preserving Embedding (SPM) [14] which is based on SDP. However, the proposed approach tries to preserve the dissimilarity in the Minkovsky space instead of the Euclidean space.

After acquiring the kernel matrix $\mathbf{P}'$ through K. The dimensionality reduction of weight vectors $\mathbf{W}$ from SOINN can be accomplished by linear dimensionality reduction. However, to calculate the dimensionality reduction of data items other than $\mathbf{W}$ with Eq. (3), the distance from this new data item V to each weight vector in $\mathbf{W}$ must be acquired. The distances can not simply be Euclidean distances $\|V - W_i\|$, since the distances between weight vectors in $\mathbf{W}$ is changed in the new data space. We design a distance that emphasizes the local similarity as

$$t_i = k\left(\frac{1/\|V - W_i\|}{\sum_{W_j \in \mathbf{W}} 1/\|V - W_j\|}\right) \tag{11}$$

where $k(x)$ is a kernel function, and it should be increasing and positive valued. This design is to ensure two properties. First, if a data item is close to W_i, its distance to other weight vectors will be similar to those in the kernel matrix calculated from Eq. (8). Second, the dimensionality reduction will try to preserve the smoothness of data distribution in the original space. Finally, the distance vector is constructed as

$$\hat{P} = \mathbf{P}'T \tag{12}$$

where $\mathbf{P}'$ is the kernel matrix calculated by quadratic programming and $T = [t_1, t_2, \cdots, t_i, \cdots]^T$. The dimensionality reduction of a data item V can then be calculated by Eqs. (3) and (4).

Besides, it is beneficial to add another add-hoc element to handle the trivial groups which are separated from the main cluster of weight vectors. One common way is to connect each trivial graph to its nearest trivial graph. This is to preserve the order of similarities between the trivial graphs. This strategy can be altered to fit real world problems.

3.2   Semi-supervised Extension to the Single Layered SOINN

Labeled information can augment the learning result of competitive learning neural network. In this paper we propose a positive negative competitive learning strategy that is different from previous extensions in two ways.

1. Robust to noise in labeled information.
2. Avoid label conflicts of weight vectors when learning with large labeled datasets.

The basic idea of positive and negative competitive learning is that when the label of the winner weight vector is contradicting with the current sample, the winner weight vector and its topological neighbors are moved away from the current sample. The implementation details are as follows.

Step 1 is the same as the unsupervised learning step while the major difference is in step 2. In step 2, the compatibility of labels between the new input V and the winner W_s must be checked first. If the label of V is the same as W_s or that V is unlabeled, step 2 will proceed the same as in the unsupervised learning, which is the positive competitive learning. If they are conflicting, W_s will be pushed away from V. But the scale of the push can not simply be $\frac{1}{c_s}(V - W_s)$, because the input V might be a noise so far way from W_s that W_s would be pushed away too much. To solve this problem we assume that the input V is highly likely to be consistent to the second winner W_t. Then with the help of W_t, we have

$$\omega = min(\frac{||V - W_s||^\varsigma}{||V - W_s||^\varsigma + ||V - W_t||^\varsigma}, \frac{||V - W_t||^\varsigma}{||V - W_s||^\varsigma + ||V - W_t||^\varsigma}) \qquad (13)$$

$$W_s = W_s - \frac{\omega}{c_s}(V - W_s) \qquad (14)$$

$$W_k = W_k - \frac{\omega\phi}{c_s}(V - W_k), \forall k \in neighbor(s) \qquad (15)$$

where ς is a user defined parameter. This updating strategy is under the assumption that the closer V is to W_s, the less credential it is. We call this the negative competitive learning. Finally, the complete algorithm of semi-supervised dimensionality reduction with SOINN is illustrated in Algorithm 1.

4   Experiments

First, experiments are carried out on artificial datasets to demonstrate the unsupervised and semi-supervised learning abilities of the proposed method. Then the evaluations are moved to the NSL-KDD [16] dataset. The reason for choosing this dataset is that the test dataset is from a different distribution of the training dataset, and handling such concept drift is one of the primary goals of SSL algorithm designs.

4.1   Artificial Datasets

The first dataset (illustrated in Fig. 1) is unlabeled, on which the dimensionality reduction abilities of the proposed method and PCA, Isomap, LLE [13] are compared. The second dataset is partially labeled as illustrated in Fig. 2. The second dataset is to demonstrate the SSL property of the proposed method. Experiment results of the unsupervised dimensionality reduction is in Fig. 3; and results on the second dataset is in Fig. 4.

It is shown in the unsupervised experiments that the proposed method is able to perform dimensionality reduction as PCA and Isomap, and the results are smooth as expected. The proposed method separates the two rings better than PCA and the topology of data is better preserved than Isomap and LLE.

The results on the semi-supervised dataset show that the after processing by the proposed method, the data items with different labels are redistributed such that they are more easily to be separated linearly. Moreover, the in the proposed method the differently labeled data are separated by a low density area.

Algorithm 1. Semi-Supervised SOINN for Dimensionality Reduction (SOINN-DR)

Input: Sequence $\{V\}$ and the labels $\{l\}$, max_age, λ, dimension d, SSL parameter ζ.
Output: Data after dimensionality reduction.

1: Initialize set $\mathbf{W}$ with 2 samples randomly drawn from input dataset.
2: **while** $\{V\}$ is not empty **do**
3:    Draw a new sample V_i from $\{V\}$ and find its nearest and second nearest neighbors (s and t).
4:    **if** Labels of V_i and W_s do not conflict **then**
5:      **if** Need to insert neuron **then**
6:        Add a new weight vector with weight V_i to $\mathbf{W}$, and set winning times $c_k = 0$.
7:        Add a new edge between s and t if none exists, and set edge weight to 0.
8:      **else**
9:        Update $\mathbf{W}$ with positive competitive learning. Increase edges ages connecting to s, and the winning times c_s by 1.
10:       If V_i is labeled and W_s is unlabeled, label the later with the label of the former; and if this cause contradictions of labels between s and its topology neighbors, remove edges causing the contradictions.
11:     **end if**
12:   **else**
13:     Update $\mathbf{W}$ with negative competitive learning. Increase edges ages connecting to s, and the winning times c_s by ω from equation (13).
14:   **end if**
15:   Delete edges with age larger than max_age.
16:   **if** number of input samples divides λ **then**
17:     Remove neurons with less than 1 neighbors.
18:   **end if**
19: **end while**
20: Eliminate trivial graphs by connecting weight vectors in different graphs that are nearest to each other.
21: Calculate embeddings of $\mathbf{W}$ by equation (1) and eigendecompositions.
22: **while** There is a query about data item V' **do**
23:   Calculate its dimensionality reduction with equation (12), equation (3), and equation (4), and output.
24: **end while**

4.2 The Intrusion Detection Dataset

In this experiment the 34 numerical values of NSL-KDD intrusion detection dataset are used. The SSL learning ability of the proposed method is evaluated qualitatively and compared with other SSL methods such as label propagation, TSVM. To compare with the unsupervised and supervised learning methods, support vector machine (SVM) and Isomap are included as well. The 1 % of the training dataset is drawn to form the labeled dataset; and the complete test dataset is selected as the unlabeled dataset. The training of SSL methods are on the mixture of the labeled and unlabeled datasets and the evaluation results are on the unlabeled datasets. In this experiment we link all the trivial

Fig. 1. Data are located on 2 interlocking rings.

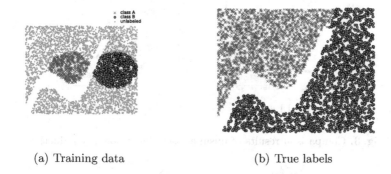

(a) Training data   (b) True labels

Fig. 2. A dataset to test the SSL abilities. Two classes are separated by a low density area.

Table 1. Comparison results (%) on the NSL-KDD dataset

	SVM	SVM-PCA	SVM-LLE	SVM-Isomap	TSVM	SVM-SOINN-DR
Precision	82.8	79.9	20.1	85.5	82.6	**86.3**
Recall	74.7	74.3	44.9	83.5	80.8	**86.2**
F-measure	74.1	74.0	27.8	83.6	80.8	**86.2**

graphs generated by semi-supervised SOINN to a weight vector with a confirmed intrusion label, because samples far from the main body of data are more likely to be intrusions. The parameters are selected by grid search combining with cross validation. The selected parameters of the proposed method are $max\_age = 500$, $\lambda = 350$, $d = 15$, $\zeta = 0.6$. The experiment results are listed in Table 1.

It is shown that the proposed method gives the best results. It is easy for SSL methods to outperform the supervised and unsupervised methods because of the concept drift. Besides, the proposed method is able to finish much faster than TSVM (which takes hours per training). This is because the transductive kernel construction is time consuming and the dataset is large in size (with over 20000 samples). In summery, the proposed method is an accurate and efficient transductive learner alternative for TSVM.

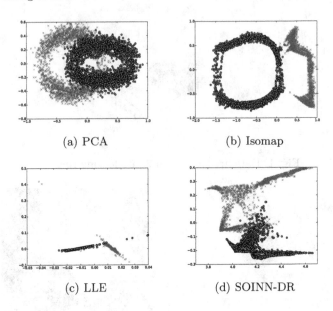

(a) PCA   (b) Isomap

(c) LLE   (d) SOINN-DR

Fig. 3. Comparison results of unsupervised dimensionality reduction.

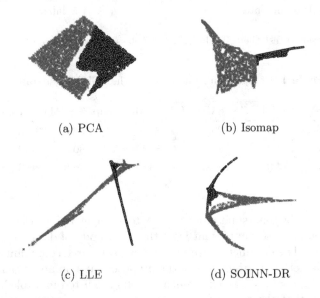

(a) PCA   (b) Isomap

(c) LLE   (d) SOINN-DR

Fig. 4. Comparison results of semi-supervised embedding.

5   Conclusions and Future Works

In this paper, we employ the competitive learning neural network SOINN to solve
the semi-supervised dimensionality reduction problem. A semi-supervised extension to SOINN is introduced to process the mixture of labeled and unlabeled

information. Then, weight vectors from SOINN are embedded to a lower dimensional space by a quadratic programming optimization, where the distances between weight vectors are preserved as much as possible and the graph similarities defined by the graph kernel from SOINN are satisfied as constraints. Then a similarity approximation is designed to process the remaining samples. The experimental results show that the proposed method is capable of unsupervised and semi-supervised dimensionality reduction, and it is an accurate and efficient transductive learner alternative for TSVM.

There are a few techniques proposed in this paper that can be employed in future works. First, the semi-supervised extension to SOINN can be employed in other competitive learning neural networks such as GNG, Growing Cell Structures etc., with some modifications. Second, the proposed method still require add-hoc designs when dealing with real world applications. It is worth investigating a substitute for semi-supervised SOINN. Third, the algorithm frameworks generalization to transfer learning is possible since SSL is a special case of transfer domain learning.

Acknowledgment. This work was partly supported by National Natural Science Foundations of China (No. 61301148, No. 61272061 and No. 71403299), the fundamental research funds for the central universities of China (No. 531107040263, 531107040276), the Research Funds for the Doctoral Program of Higher Education of China (No. 20120161120019 and No. 20130161110002), Hunan Natural Science Foundation of China (No. 14JJ7023).

References

1. Beyer, O., Cimiano, P.: Online semi-supervised growing neural gas. Int. J. Neural Syst. **22**(05), 1250023 (2012)
2. Cai, X., Wei, J., Wen, G., Yu, Z.: Local and global preserving semisupervised dimensionality reduction based on random subspace for cancer classification. IEEE J. Biomed. Health Inform. **18**(2), 500–507 (2014)
3. Cai, X., Wen, G., Wei, J., Yu, Z.: Relative manifold based semi-supervised dimensionality reduction. Frontiers Comput. Sci. **8**(6), 923–932 (2014)
4. Edmonds, J., Karp, R.M.: Theoretical improvements in algorithmic efficiency for network flow problems. J. ACM (JACM) **19**(2), 248–264 (1972)
5. Fritzke, B., et al.: A growing neural gas network learns topologies. Adv. Neural Inf. Process. Syst. **7**, 625–632 (1995)
6. Gomory, R.E., Hu, T.C.: Multi-terminal network flows. J. Soc. Ind. Appl. Math. **9**(4), 551–570 (1961)
7. Graepel, T., Herbrich, R., Bollmann-Sdorra, P., Obermayer, K.: Classification on pairwise proximity data. Adv. Neural Inf. Process. Syst. 438–444 (1999)
8. Kohonen, T.: The self-organizing map. Neurocomputing **21**(1), 1–6 (1998)
9. Maximo, V.R., Quiles, M.G., Nascimento, M.C.: A consensus-based semi-supervised growing neural gas. In: 2014 International Joint Conference on Neural Networks (IJCNN), pp. 2019–2026. IEEE (2014)
10. Pan, S.J., Kwok, J.T., Yang, Q.: Transfer learning via dimensionality reduction. In: AAAI, vol. 8, pp. 677–682 (2008)

11. Pan, S.J., Yang, Q.: A survey on transfer learning. IEEE Trans. Knowl. Data Eng. **22**(10), 1345–1359 (2010)
12. Raina, R., Battle, A., Lee, H., Packer, B., Ng, A.Y.: Self-taught learning: transfer learning from unlabeled data. In: Proceedings of the 24th International Conference on Machine Learning, pp. 759–766. ACM (2007)
13. Roweis, S.T., Saul, L.K.: Nonlinear dimensionality reduction by locally linear embedding. Science **290**(5500), 2323–2326 (2000)
14. Shaw, B., Jebara, T.: Structure preserving embedding. In: Proceedings of the 26th Annual International Conference on Machine Learning, pp. 937–944. ACM (2009)
15. Shen, F., Yu, H., Sakurai, K., Hasegawa, O.: An incremental online semi-supervised active learning algorithm based on self-organizing incremental neural network. Neural Comput. Appl. **20**(7), 1061–1074 (2011)
16. Tavallaee, M., Bagheri, E., Lu, W., Ghorbani, A.A.: A detailed analysis of the KDD CUP 99 data set. In: 2009 Proceedings of the Second IEEE Symposium on Computational Intelligence for Security and Defence Applications (2009)
17. Tenenbaum, J.B., de Silva, V., Langford, J.C.: A global geometric framework for nonlinear dimensionality reduction. Science **290**(5500), 2319–2323 (2000)
18. Torgerson, W.S.: Theory and methods of scaling (1958)
19. Wang, J.: Maximum variance unfolding. In: Wang, J. (ed.) Geometric Structure of High-Dimensional Data and Dimensionality Reduction, pp. 181–202. Springer, Heidelberg (2011)
20. Weinberger, K.Q., Packer, B.D., Saul, L.K.: Nonlinear dimensionality reduction by semidefinite programming and kernel matrix factorization. In: Proceedings of the Tenth International Workshop on Artificial Intelligence and Statistics, pp. 381–388. Citeseer (2005)
21. Yang, X., Fu, H., Zha, H., Barlow, J.: Semi-supervised nonlinear dimensionality reduction. In: Proceedings of the 23rd International Conference on Machine Learning, pp. 1065–1072. ACM(2006)

Cross-View Feature Hashing for Image Retrieval

Wei Wu[1(✉)], Bin Li[1,2], Ling Chen[1], and Chengqi Zhang[1]

[1] Centre for Quantum Computation and Intelligent Systems,
University of Technology Sydney, Ultimo, NSW 2007, Australia
william.third.wu@gmail.com, {ling.chen,chengqi.zhang}@uts.edu.au
[2] Machine Learning Research Group, National ICT Australia,
Eveleigh, NSW 2015, Australia
bin.li@nicta.com.au

Abstract. Traditional cross-view information retrieval mainly rests on correlating two sets of features in different views. However, features in different views usually have different physical interpretations. It may be inappropriate to map multiple views of data onto a shared feature space and directly compare them. In this paper, we propose a simple yet effective Cross-View Feature Hashing (CVFH) algorithm via a "partition and match" approach. The feature space for each view is bi-partitioned multiple times using B hash functions and the resulting binary codes for all the views can thus be represented in a compatible B-bit Hamming space. To ensure that hashed feature space is effective for supporting generic machine learning and information retrieval functionalities, the hash functions are learned to satisfy two criteria: (1) the neighbors in the original feature spaces should be also close in the Hamming space; and (2) the binary codes for multiple views of the same sample should be similar in the shared Hamming space. We apply CVFH to cross-view image retrieval. The experimental results show that CVFH can outperform the Canonical Component Analysis (CCA) based cross-view method.

1 Introduction

As data collection channels and means become diverse, many real-world information retrieval tasks can involve multiple views for the same samples collected from different information sources. In these cases, people may have some examples in one view while intend to query the database in another view, which we refer to cross-view information retrieval. There are many real-world examples: In image retrieval, text captions [2,8] or point-of-interest features [17,23] can be used to query images which are stored in color features. Such cross-view scenarios are inherently different from tag-based image retrieval, where the images in the database are associated with tags thus the underlying retrieval task is indeed text matching. In document retrieval, a document written in French can be used to query documents written in English [12,20]. Other cross-view scenarios include audio-visual clustering [4], image-url clustering [14], image-video retrieval [23], rating-attribute recommendation [6].

© Springer International Publishing Switzerland 2016
J. Bailey et al. (Eds.): PAKDD 2016, Part I, LNAI 9651, pp. 203–214, 2016.
DOI: 10.1007/978-3-319-31753-3_17

Cross-language text retrieval is one of the earliest applications involving cross-view learning, where different translations of the same document can be viewed as multiple views. An early approach to cross-view learning was based on Latent Semantic Indexing (LSI) [12] until Canonical Component Analysis (CCA) was introduced [8,20]. CCA is able to find two projections to maximize the correlation between two sets of variables. A recent example [24] applies an improved version of CCA to learning query and image similarities. By applying CCA to cross-view learning, the data in two different feature spaces can be projected to a common low-dimensional space and the projected features thus become in the same representation for similarity comparison. However, directly comparing two sets of data, which have different physical interpretations, is unreasonable, although they have been linearly projected to the "same" space. In fact, such "same" space only means that the dimensionality is identical for two data sets; while the interpretation of each dimension of this "same" space is still different for two data sets. Due to this limitation, CCA shows less advantage in cross-view information retrieval [17].

Spectral embedding [1] can be used as a compatible representation for different views of data since the embedded data are derived from affinity graphs, which have the same physical interpretation for different views. Some works have been proposed to solve multi-view learning problems based on spectral embedding [14,23]. A limitation is that spectral embedding can be hardly extended to out-of-sample setting. The linear version [9] of spectral embedding may be applied to address this limitation by using a linear projection. However, linear projection will induce the same problem like that in CCA since linear projection involves comparison of the original different views of data.

Recently, "learning to hash" becomes an active research topic in information retrieval [10,13,15,19,22]. Different from classical Locality-Sensitive Hashing (LSH) techniques used in databases [3,7], where hash functions are designed manually, learning to hash aims to learn good hash functions from training data to preserve locality better while using fewer bits. By hashing, the data in the original feature spaces are encoded into binary codes. Some existing works have considered learning to hash in cross-view scenarios: Relational-aware heterogeneous hashing [16] and heterogeneous translated hashing [21] project data from different views to Hamming spaces with different lengths. In contrast, the idea of this work is motivated from the following observation:

Hashing indeed performs bi-partitions on the original feature spaces for B times using B hash functions, and the resulting B-bit binary codes of the original data can be interpreted as indices of the cells. If we can match the partitions of original feature spaces for different views, we can compare similarity of the binary codes from different views using Hamming distance.

Motivated by the existing "learning to hash" research, we propose a simple yet effective Cross-View Feature Hashing (CVFH) algorithm via a "partition and match" approach. The feature space for each view is bi-partitioned multiple times using B hash functions and the resulting binary codes for all the views can thus be represented in a compatible B-bit Hamming space. More specifically, we first bi-partition multiple graphs of different views using Normalized Cuts (NCut) [18], based on a combined graph Laplacian of multiple views to preserve a consensus locality. The consensus locality preserving balances the geometry structures of multiple views. The NCut results (B binary label sets for the nodes on the graphs) are used for supervised hash function training. Since the hash functions for different views are all learned on the same labels, the partitions of different feature spaces can be matched. For test, our method hashes features in different views and uses the resulting binary codes of one view to query another view. We apply CVFH to cross-view image retrieval on the NUS-WIDE-LITE image data. The experimental results show that CVFH can clearly outperform the CCA-based cross-view method.

The remainder of the paper is organized as follows: We first formulate the cross-view information retrieval problem in Sect. 2. Then we introduce a preliminary cross-view information retrieval method based on CCA and spectral hashing in Sect. 3. Our CVFH algorithm is presented in Sect. 4. The experimental results are reported in Sect. 5 and the paper is concluded in Sect. 6.

2 Problem Formulation

Suppose there are S data sources. A sample $\mathbf{x}_n$ is generated from these data sources that combine S views of heterogeneous features, i.e. $\mathbf{x}_n = \{\mathbf{x}_n^{(1)}, \ldots, \mathbf{x}_n^{(S)}\}$. The data for the s-th view is represented by a matrix $\mathbf{X}^{(s)} = [\mathbf{x}_1^{(s)} \ldots \mathbf{x}_N^{(s)}] \in \mathbb{R}^{D^{(s)} \times N}$, where N is the number of samples and $D^{(s)}$ is the dimension of the feature space of the s-th view.

Our goal is to learn S sets of hash functions to map the multi-view data $\{\mathbf{x}_n^{(1)}, \ldots, \mathbf{x}_n^{(S)}\}$ from their original heterogenous feature spaces, $\mathbb{R}^{D^{(1)}}, \ldots, \mathbb{R}^{D^{(S)}}$, onto a common Hamming space $\{-1, 1\}^B$. For the s-th view, there are B hash functions $h_1^{(s)}, \ldots, h_B^{(s)}$, where $h_b^{(s)} : \mathbb{R}^{D^{(s)}} \mapsto \{-1, 1\}$, for $b = 1, \ldots, B$. By the following feature hashing

$$\mathbf{y}_n^{(s)} = [h_1^{(s)}(\mathbf{x}_n^{(s)}) \ldots h_B^{(s)}(\mathbf{x}_n^{(s)})]^\top \tag{1}$$

the mapped B-bit binary data for all the data views, $\{\mathbf{y}_n^{(1)}, \ldots, \mathbf{y}_n^{(S)}\} \in \{-1, 1\}^B$, become comparable in the common Hamming space. Similarly, we let a sample in the mapped Hamming space $\mathbf{y}_n = \{\mathbf{y}_n^{(1)}, \ldots, \mathbf{y}_n^{(S)}\}$, and $\mathbf{Y}^{(s)} = [\mathbf{y}_1^{(s)} \ldots \mathbf{y}_N^{(s)}] \in \{-1, 1\}^{B \times N}$.

The mapping in (1) can be viewed as an encoding process such that the data in the original heterogenous feature spaces, $\mathbb{R}^{D^{(1)}}, \ldots, \mathbb{R}^{D^{(S)}}$, which cannot be directly compared with one another, are encoded in the same representations, $\{-1, 1\}^B$, which are compatible across S views of data.

The S sets of hash functions are learned to satisfy the following two criteria:

- The neighbors in the original feature spaces should be also close in the Hamming space; and
- The binary codes for multiple views of the same sample should be similar in the Hamming space.

3   Preliminary and A Baseline Approach

In this section, we introduce a baseline approach to cross-view information retrieval by combining CCA and spectral hashing. This baseline approach only maximizes the correlation but does not preserve locality between two sets of features in different views.

3.1   Canonical Correlation Analysis

Canonical Correlation Analysis (CCA) was proposed by Hotelling in 1936 and is a frequently used method in multivariate analysis. It was introduced to multi-view learning by Shawe-Taylor and his colleagues with a number of studies [8,20]. The aim of CCA is to find basis vectors, $\mathbf{v}^{(s)}$ and $\mathbf{v}^{(t)}$, for two sets of variables such that the correlation between the projections of the two sets of variables onto these basis vectors is mutually maximized. Take the data in two views, $\mathbf{X}^{(s)}$ and $\mathbf{X}^{(t)}$, for example, the objective of CCA [8] is to

$$\max_{\mathbf{v}^{(s)},\mathbf{v}^{(t)}} \left[\mathbf{v}^{(s)}\right]^{\top} \mathbf{\Sigma}^{(s,t)} \mathbf{v}^{(t)} \tag{2}$$

$$\text{s.t.} \quad \left[\mathbf{v}^{(s)}\right]^{\top} \mathbf{\Sigma}^{(s,s)} \mathbf{v}^{(s)} = 1 \tag{3}$$

$$\left[\mathbf{v}^{(t)}\right]^{\top} \mathbf{\Sigma}^{(t,t)} \mathbf{v}^{(t)} = 1 \tag{4}$$

where $\mathbf{\Sigma}^{(s,s)}$ and $\mathbf{\Sigma}^{(t,t)}$ are within-sets covariance matrices while $\mathbf{\Sigma}^{(s,t)}$ is between-sets covariance matrix, which are computed from $\mathbf{X}^{(s)}$ and $\mathbf{X}^{(t)}$. The optimal basis vectors can be obtained by solving an eigen decomposition problem.

The data in the original feature spaces then can be projected onto a shared B-dimensional space by using the basis vectors with top B eigenvalues, i.e., $[\mathbf{V}^{(s)}]^{\top}\mathbf{X}^{(s)}$ and $[\mathbf{V}^{(t)}]^{\top}\mathbf{X}^{(t)}$, where the columns of $\mathbf{V}^{(s)}$ and $\mathbf{V}^{(t)}$ are B eigenvectors. The projections from two data views can thus be compared by computing a distance. A large number of cross-view learning methods are based on this approach [2,4,8,17,20].

3.2   Spectral Hashing

Spectral Hashing (SH) [22] is a recently proposed hashing algorithm based on the well developed graph theory in machine learning. It reveals the fact that graph bi-partition is equivalent to the locality-sensitive criterion used in Locality-Sensitive Hashing (LSH) [7]. Based on this observation, spectral hashing resorts to bi-partitioning a graph using Laplacian eigenmaps [1] and uses the signs of a set of eigenvectors[1] as the hashing codes. The objective of spectral hashing [22] is to

---

[1] The signs of eigenvectors of a graph Laplacian are used as bi-partition labels in spectral clustering.

$$\min_{\mathbf{Y}^{(s)}} \frac{1}{2} \sum_{n,m} \mathbf{W}_{nm}^{(s)} ||\mathbf{y}_n^{(s)} - \mathbf{y}_m^{(s)}||^2$$

$$= trace(\mathbf{Y}^{(s)} \mathbf{L}^{(s)} [\mathbf{Y}^{(s)}]^\top) \tag{5}$$

$$\text{s.t.} \quad \mathbf{Y}^{(s)} \in \{-1, 1\}^{B \times N} \tag{6}$$

$$\mathbf{Y}^{(s)} \mathbf{1} = \mathbf{0} \tag{7}$$

$$\mathbf{Y}^{(s)} [\mathbf{Y}^{(s)}]^\top = N\mathbf{I} \tag{8}$$

where $\mathbf{W}_{nm}^{(s)}$ denotes the similarity between $\mathbf{x}_n^{(s)}$ and $\mathbf{x}_m^{(s)}$, $\mathbf{L}^{(s)} = \text{diag}(\mathbf{W}^{(s)} \mathbf{1}) - \mathbf{W}^{(s)}$ is the Laplacian matrix. This is an integer programming known as NP-hard. Spectral hashing relaxes the problem by removing the constraint $\mathbf{Y}^{(s)} \in \{-1, 1\}^{B \times N}$ and it reduces to a spectral embedding problem.

The obtained eigenvectors with B smallest eigenvalues (except the last one) can be used to encode the training samples while the out-of-sample test remains a problem. The authors in [22] adopt a set of data independent eigenfunctions as the hash functions while the authors in [13] extend the eigenvectors to eigenfunctions using Nyström method. As a result, the test samples also can be encoded into binary codes.

3.3 CCA+SH Baseline Algorithm

We combine the advantages of the above two techniques (i.e., CCA for cross-view and SH for hashing) to introduce a baseline algorithm for cross-view information retrieval:

1. Use CCA to compute two projection matrices, $\mathbf{V}^{(s)}$ and $\mathbf{V}^{(t)}$, which can maximize the correlation between $\mathbf{X}^{(s)}$ and $\mathbf{X}^{(t)}$. Project $\mathbf{X}^{(s)}$ and $\mathbf{X}^{(t)}$ onto the shared B-dimensional common space and obtain $[\mathbf{V}^{(s)}]^\top \mathbf{X}^{(s)}$ and $[\mathbf{V}^{(t)}]^\top \mathbf{X}^{(t)}$.
2. Use spectral hashing to hash $[\mathbf{V}^{(s)}]^\top \mathbf{X}^{(s)}$ and $[\mathbf{V}^{(t)}]^\top \mathbf{X}^{(t)}$ and obtain two sets of binary codes, $\mathbf{Y}^{(s)}$ and $\mathbf{Y}^{(t)}$, for the two views of the data set.
3. Use a code in one view (e.g., $\mathbf{y}_n^{(s)}$) to query the database in the other view (e.g., $\mathbf{Y}^{(t)}$) based on Hamming distance.

This baseline approach will be compared to the proposed CVFH algorithm in the experiments. Note that this algorithm only can be applied to two-view setting since it is based on CCA.

4 CVFH: Cross-View Feature Hashing

4.1 Objective

The objective of CVFH is twofold: (1) For each view, the distance of any pair of mapped data, $\mathbf{y}_n^{(s)}$ and $\mathbf{y}_m^{(s)}$, in the Hamming space should be proportional to the distance of their preimages, $\mathbf{x}_n^{(s)}$ and $\mathbf{x}_m^{(s)}$, in the original feature space. In other words, CVFH is a *locality-preserving* mapping. (2) For all the views, the mapped data of different views for the same sample, $\mathbf{y}_n^{(1)}, \ldots, \mathbf{y}_n^{(S)}$, should be as close as possible in the Hamming space. In other words, CVFH is a *cross-view adaptation* mapping.

Fortunately, we have found a way for our problem to satisfy the above two criteria. Since the objective of spectral hashing can satisfy the first one, we also adopt the idea of graph bi-partition for feature hashing. For the second one, as we argued in Sect. 1, since linear projection like CCA is not interpretable for comparing samples based on distance, we will resort to another way to bridge different views.

4.2 "Bi-Partition and Match" Strategy

As we have highlighted in Sect. 1, the essence of partition-based hashing approach is to bi-partition the original feature spaces for B times using B hash functions and then index the obtained cells with binary codes. Thus, the Hamming spaces after feature hashing for different views actually have the same interpretation, that is the cell indices of the original feature spaces. Thus, as long as we can match the partitions of the original feature spaces for different views, we can compute similarity of the binary codes from different views in Hamming distance.

To match the partitions of the original feature spaces means that, in the corresponding cells of multi-view original feature spaces, the samples should be the same ones in different views. For example, in Fig. 1, after bi-partitioning the two original spaces, the samples of two views fall in each cell (e.g., cell '11') are the same data points (e.g., the red ones). If the multi-view data in all the partitioned cells are the same ones, we say that the multi-view original spaces are completely matched. Of course, on a real-world data set, we can hardly achieve a complete matching.

To achieve a good matching between different views, we adopt a combined graph Laplacian of multiple views to preserve a consensus locality, i.e., $\sum_s \mathbf{L}^{(s)}$. Since we expect the resulting binary codes are as similar as possible for the multi-view data of the same sample, we let the eigenvectors for all views be identical as $\mathbf{Y}$. We solve the following problem

$$\min_{\mathbf{Y}} trace(\mathbf{Y} \sum_s \mathbf{L}^{(s)} \mathbf{Y}^\top) \tag{9}$$

$$\text{s.t.} \quad \mathbf{Y} \in \{-1, 1\}^{B \times N} \tag{10}$$

$$\mathbf{Y1} = \mathbf{0} \tag{11}$$

$$\mathbf{YY}^\top = N\mathbf{I} \tag{12}$$

This problem is very similar to the one in spectral hashing. To achieve more balanced graph cuts, we adopt Normalized Cuts (NCut) [18] to bi-partition the combined graph. The consensus locality preserving is able to balance the geometry structures of multiple views. Then, we can select B sets of eigenvectors with smallest eigenvalues (except the last one) to extract their signs as the binary codes for the training samples. Note that the multi-view data of the same sample in all views will get the same binary codes. These binary codes will be used as supervision information to learn hash functions for different views.

4.3   Hash Functions

We will learn S sets of hash functions, based on the supervision information of NCut results (B binary label sets for the nodes on the consensus graph), for S views and each set has B hash functions. Formally, to learn a hash function $h_b^{(s)}$ for the b-th bit in the s-th view, we perform supervised learning to train a binary classifier on $\mathbf{X}^{(s)}$ with labels $\mathbf{y}_b$ (the b-th row in $\mathbf{Y}$). Since all the hash functions for different views are learned on the same labels, the cells of different feature spaces partitioned by the learned hash functions can be matched.

Any binary classifier can be used as hash functions for CVFH. It can also be naturally extended to nonlinear functions by using kernels. For simplicity, we adopt the following form for the linear case

$$h_b^{(s)}(\mathbf{x}_n^{(s)}) = \text{sign}([\mathbf{w}_b^{(s)}]^\top \mathbf{x}_n^{(s)}) \tag{13}$$

where $\mathbf{w}_b^{(s)} = \mathbf{X}^{(s)}\mathbf{y}_b$, and adopt the following one for the nonlinear case

$$h_b^{(s)}(\mathbf{x}_n^{(s)}) = \text{sign}(\sum_m \mathbf{y}_{b,m}\kappa(\mathbf{x}_m^{(s)}, \mathbf{x}_n^{(s)})) \tag{14}$$

where κ is a kernel function, such as the Gaussian kernel used in our experiments. Note that no optimization is required for the above two forms of hash functions.

For test, a sample $\mathbf{x}_n^{(s)}$ is first hashed into a B-bit binary code using $h_1^{(s)}, \ldots, h_B^{(s)}$, and the obtained binary code can be used for query in the database in other views.

4.4   CVFH Algorithm

We summarize the CVFH algorithm as follows:

1. Use NCut to perform bi-partitions on the consensus graph Laplacian, $\sum_s \mathbf{L}^{(s)}$, of S views. Extract the signs of the obtained eigenvectors $\mathbf{Y}$ with smallest B eigenvalues (except the last one).
2. Perform supervised learning to learn a hash function $h_b^{(s)}$ for the b-th bit in the s-th view, on the data set $\mathbf{X}^{(s)}$ with labels $\mathbf{y}_b$, for $s = 1, \ldots, S, b = 1, \ldots, B$.
3. Hash a test sample $\mathbf{x}_n^{(s)}$ using $h_1^{(s)}, \ldots, h_B^{(s)}$ into a B-bit binary code and use the code to query the database in the other views based on Hamming distance.

Since CVFH dose not involve additional optimization problems compared to spectral hashing, the computational complexity is as same as that for spectral hashing.

5   Experiments

In this section, we report our experimental results obtained from both synthetic and real-word data sets. In the first part of the experiments, we use a toy data

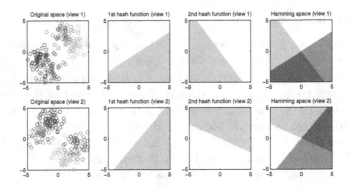

Fig. 1. Cross-view feature hashing on a toy data set where row 1 and row 2 represent view 1 and view 2, respectively. After two bi-partitions on the original feature spaces (the 1st column) using two hash functions (the 2nd and the 3rd columns, two gray scales denote '0' and '1', respectively), the data of two views are matched in the Hamming space (the 4th column, four gray scales denote '00', '01', '10', and '11', respectively) (Color figure online).

set to intuitively demonstrate the idea of CVFH for matching different views of data in the shared Hamming space. In the second part of the experiments, we use the NUS-WIDE-LITE[2] image data set as our testbed to validate the effectiveness of the CVFH algorithm for cross-view information retrieval.

5.1   Results on Toy Data

We generate the toy data as follows: (1) Let the column vectors in

$$\begin{bmatrix} -1 & -3 & 1 & 3 \\ -3 & -1 & 3 & 1 \\ 3 & -1 & -3 & 1 \\ -1 & 3 & 1 & -3 \end{bmatrix}$$

be the means of four Gaussians. (2) Let a 4×4 identity matrix be the covariance matrices for the four Gaussians. (3) Draw 50 samples from each Gaussian and obtain 200 samples in total, represented by a 4×200 matrix $\mathbf{X}$, one column for one sample. (4) Let the first two dimensions be the first view, $\mathbf{X}^{(1)}$, and the rest two dimensions be the second view, $\mathbf{X}^{(2)}$, i.e., $\mathbf{X} = \begin{bmatrix} \mathbf{X}^{(1)} \\ \mathbf{X}^{(2)} \end{bmatrix}$. The generated samples are plotted in Fig. 1, where the upper row is for the first view and the bottom row for the second view, and four colors are used to indicate the samples from four different Gaussians.

We perform CVFH on the two views of the toy data using the linear form of hash functions (13) and the results are shown in Fig. 1. The first hash functions

---

[2] http://lms.comp.nus.edu.sg/research/NUS-WIDE.htm.

for both views are illustrated in the second column and the second hash functions in the third column. One can clearly find that the hash functions bi-partition the feature spaces into two parts. By integrating the hashing results of both the first and the second hash functions, we finally partition the two feature spaces into four cells (2-bit Hamming space). One can see that the four cells in both views can be completely matched (e.g., the cells of '11' in both views comprises the red samples), although the samples in two views have different distributions in the original feature spaces.

From the illustration in Fig. 1, it is clear that the essence of CVFH is to bi-partition the original multi-view feature spaces by preserving the cross-view locality, and match the cells in both views after B bi-partitions. Through this approach, the samples in different views can be finally mapped into a common Hamming space with an interpretable physical meaning.

5.2 Results on NUS-WIDE-LITE Image Data

The NUS-WIDE-LITE image data set comprises 55,615 images, half of which (27,807 images) are used for training and the rest (27,808 images) for testing. Some sample images are shown in Fig. 2. Each image is tagged (or annotated) with one or multiple concepts. There are 81 concepts (organized in a hierarchical structure) in total for all images in the data set. Examples of concepts include Animal, Person, Sports, Dancing etc. Each image is represented in five sets of low-level features, including (1) 64-D color histogram (CH), (2) 144-D color correlogram (CORR), (3) 225-D block-wise color moments (CM), (4) 73-D edge direction histogram (EDH), and (5) 128-D wavelet texture (WV). We concatenate the three color-related features as the first view and concatenate the other two texture-related features as the second view, and obtain the data for two views as follows,

1. First View (CH+CORR+CM = 433 dimensions): a 433×27807 matrix $\mathbf{X}_{train}^{(1)}$ for training and a 433×27808 matrix $\mathbf{X}_{test}^{(1)}$ for testing.
2. Second View (EDH+WV = 201 dimensions): a 201×27807 matrix $\mathbf{X}_{train}^{(2)}$ for training and a 201×27808 matrix $\mathbf{X}_{test}^{(2)}$ for testing.

Fig. 2. Examples of the NUS-WIDE-LITE image data set.

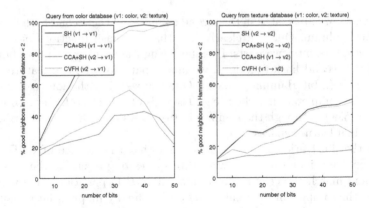

Fig. 3. Performance comparison on the NUS-WIDE data set. "Good neighbors" are defined as pairs of data in the original feature space whose distances are in the top 5th percentile (same setting as that in [10,22]). The performance of SH and PCA+SH are single-view results and they are plotted as the upper bound for cross-view results. In both "View 2 → View 1" (left panel) and "View 1 → View 2" (right panel) settings, the proposed CVFH clearly outperforms the CCA-based baseline method, especially in the second setting (right panel), where CVFH even approaches the upper bound.

We first perform CVFH on the training data set with two views to learn hash functions and apply them to the test data set. We then obtain the binary codes of the test data set for two views, represented in two $B \times 27808$ binary matrices, $\mathbf{Y}_{test}^{(1)}$ and $\mathbf{Y}_{test}^{(2)}$, respectively. We investigate the impact of the number of bits B for the retrieval performance by changing its value in $\{5, 10, 15, \dots, 50\}$.

We follow the same performance evaluation method used in [10,22] by counting the retrieved "good" neighbors in Hamming space that the distance between the queried example and the retrieved one is smaller than 2. The "good" neighbors are defined as pairs of data in the original feature space whose distances are in the top 5th percentile. We plot the cross-view image retrieval results in Fig. 3. The performances of Spectral Hashing (SH) [22] and PCA+SH are not cross-view but single-view results and they are plotted as the upper bound for cross-view results. That is, the closer to the performance of SH and PCA+SH, the better the cross-view retrieval is, by approaching the performance of querying and searching in the same view. The CCA+SH approach is used as the cross-view baseline introduced in Sect. 3. The Matlab implementation for SH used in our experiments is downloaded from the authors' homepage[3] and for the PCA and CCA methods, we use the built-in functions in Matlab.

The left panel shows the results of "using examples in View 2 (texture) to query from the database in View 1 (color)" for CCA+SH and CVFH; while the right panel shows the results of "using examples in View 1 (color) to query from the database in View 2 (texture)" for CCA+SH and CVFH. The performance is averaged over 10 times and, for each time, we randomly select 100 examples in

---

[3] http://www.cs.huji.ac.il/yweiss/SpectralHashing/.

one view of the test set to query the test data in the other view. From Fig. 3, we can find that the proposed CVFH algorithm can clearly outperform the baseline method CCA+SH, especially in the second setting (right panel), where CVFH even approaches the upper bound.

Apart from the promising performance, we also observe a phenomenon that the performance of CVFH tends to decline at 30~40 bits after continuous increasing. A possible explanation is that, with the increasing number of hash functions, the feature spaces of both views are partitioned into more and more cells, and the feature spaces can be hardly matched in such a fine granularity.

6 Conclusion and Future Work

In this paper, we propose a simple yet effective Cross-View Feature Hashing (CVFH) algorithm via a "partition and match" approach. We argue that existing approaches for multi-view learning, such as Canonical Component Analysis (CCA), mainly project different feature spaces to a common low-dimensional space for computing similarities. But directly comparing two sets of features, which have different physical interpretations, may be unreasonable, although they have been linearly projected to the "same" space. To address the problem, we propose a "partition and match" cross-view feature hashing algorithm. By bi-partitioning multi-view feature spaces into cells based on a consensus locality-preserving graph cut, CVFH can match the original feature spaces for different views in a compatible Hamming space. In doing so, the multi-view data from different feature spaces can be directly compared for computing similarities, with an interpretable physical meaning. Experimental results on both synthetic and real-world data sets validated the effectiveness of the proposed CVFH algorithm.

In the future work, we will extend the "partition and match" strategy to match hierarchically partitioned spaces for recursive hashing [5,11].

References

1. Belkin, M., Niyogi, P.: Laplacian eignemaps and spectral techniques for embedding and clustering. In: NIPS, pp. 585–591 (2002)
2. Blaschko, M.B., Lampert, C.H.: Correlational spectral clustering. In: CVPR, pp. 1–8 (2008)
3. Broder, A., Charikar, M., Frieze, A., Mitzenmacher, M.: Min-wise independent permutations. In: STOC, pp. 327–336 (1998)
4. Chaudhuri, K., Kakade, S.M., Livescu, K., Sridharan, K.: Multi-view clustering via canonical correlation analysis. In: ICML, pp. 129–136 (2009)
5. Chi, L., Li, B., Zhu, X.: Context-preserving hashing for fast text classification. In: SDM, pp. 100–108 (2013)
6. Fu, B., Xu, G., Cao, L., Wang, Z., Wu, Z.: Coupling multiple views of relations for recommendation. In: Cao, T., Lim, E.-P., Zhou, Z.-H., Ho, T.-B., Cheung, D., Motoda, H. (eds.) PAKDD 2015. LNCS, vol. 9078, pp. 732–743. Springer, Heidelberg (2015)

214 W. Wu et al.

7. Gionis, A., Indyk, P., Motwani, R.: Similarity search in high dimensions via hashing. In: VLDB, pp. 518–529 (1999)
8. Hardoon, D.R., Szedmak, S., Shawe-Taylor, J.: Canonical correlation analysis: an overview with application to learning methods. Neural Comput. **16**(12), 2639–2664 (2004)
9. He, X., Niyogi, P.: Locality preserving projections. In: NIPS, pp. 585–591 (2004)
10. Kulis, B., Darrell, T.: Learning to hash with binary reconstructive embeddings. In: NIPS, pp. 1042–1050 (2010)
11. Li, B., Zhu, X., Chi, L., Zhang, C.: Nested subtree hash kernels for large-scale graph classification over streams. In: ICDM, pp. 399–408 (2012)
12. Littman, M., Dumais, S.T., Landauer, T.K.: Automatic cross-language information retrieval using latent semantic indexing. In: Grefenstette, G. (ed.) Cross-Language Information Retrieval, Chapter 5, pp. 51–62. Springer, New York (1998)
13. Liu, W., Wang, J., Kumar, S., Chang, S.F.: Hashing with graphs. In: ICML, pp. 1–8 (2011)
14. Long, B., Yu, P.S., Zhang, Z.: A general model for multiple view unsupervised learning. In: SDM, pp. 822–833 (2008)
15. Marukatat, S., Sinthupinyo, W.: Improved spectral hashing. In: Huang, J.Z., Cao, L., Srivastava, J. (eds.) PAKDD 2011, Part I. LNCS, vol. 6634, pp. 160–170. Springer, Heidelberg (2011)
16. Ou, M., Cui, P., Wang, F., Wang, J., Zhu, W., Yang, S.: Comparing apples to oranges: a scalable solution with heterogeneous hashing. In: KDD, pp. 230–238. ACM (2013)
17. Quadrianto, N., Lampert, C.H.: Learning multi-view neighborhood preserving projections. In: ICML, pp. 425–432 (2011)
18. Shi, J., Malik, J.: Normalized cuts and image segmentation. IEEE Trans. Pattern Anal. Mach. Intell. **22**(8), 888–905 (2000)
19. Tokui, S., Sato, I., Nakagawa, H.: Locally optimized hashing for nearest neighbor search. In: Cao, T., Lim, E.-P., Zhou, Z.-H., Ho, T.-B., Cheung, D., Motoda, H. (eds.) PAKDD 2015. LNCS, vol. 9078, pp. 498–509. Springer, Heidelberg (2015)
20. Vinokourov, A., Shawe-Taylor, J., Cristianini, N.: Inferring a semantic representation of text via cross-language correlation analysis. In: NIPS, pp. 1473–1480 (2003)
21. Wei, Y., Song, Y., Zhen, Y., Liu, B., Yang, Q.: Scalable heterogeneous translated hashing. In: KDD, pp. 791–800. ACM (2014)
22. Weiss, Y., Torralba, A., Fergus, R.: Spectral hashing. In: NIPS, pp. 1753–1760 (2009)
23. Xia, T., Tao, D., Mei, T., Zhang, Y.: Multiview spectral embedding. IEEE Trans. Syst. Man Cybern.-Part B Cybern. **40**(6), 1438–1446 (2010)
24. Yao, T., Mei, T., Ngo, C.W.: Learning query and image similarities with ranking canonical correlation analysis. In: ICCV, pp. 28–36 (2015)

Towards Automatic Generation of Metafeatures

Fábio Pinto$^{(\boxtimes)}$, Carlos Soares, and João Mendes-Moreira

INESC TEC/Faculdade de Engenharia, Universidade do Porto,
Rua Dr. Roberto Frias s/n, 4200-465 Porto, Portugal
fhpinto@inesctec.pt, {csoares,jmoreira}@fe.up.pt

Abstract. The selection of metafeatures for metalearning (MtL) is often an *ad hoc* process. The lack of a proper motivation for the choice of a metafeature rather than others is questionable and may originate a loss of valuable information for a given problem (e.g., use of *class entropy* and not *attribute entropy*). We present a framework to systematically generate metafeatures in the context of MtL. This framework decomposes a metafeature into three components: meta-function, object and post-processing. The automatic generation of metafeatures is triggered by the selection of a meta-function used to systematically generate metafeatures from all possible combinations of object and post-processing alternatives. We executed experiments by addressing the problem of algorithm selection in classification datasets. Results show that the sets of systematic metafeatures generated from our framework are more informative than the non-systematic ones and the set regarded as state-of-the-art.

Keywords: Metalearning · Systematic metafeatures · Algorithm selection · Classification

1 Introduction

A central task in the data mining process is the selection and training of a learning algorithm on a dataset. Given the number of learning algorithms available, this task can become very time consuming, especially if the data analyst does not have the necessary experience to focus on the most promising ones. Therefore, there is a need for systems that automate this process and guide the data analyst in the search for a learning algorithm that better suits a given dataset [1]. Such systems must reduce the amount of time for model development without significant loss of model performance when compared to the best learning algorithm. Metalearning (MtL) is one approach that can be used to address this need. Brazdil et al. defined MtL as the study of principled methods that exploit meta-knowledge to obtain efficient models and solutions by adapting machine learning and data mining processes [2].

Although the MtL literature proposes many metafeatures of different types for a wide range of problems (e.g., statistics and landmarks), most of those metafeatures are developed in a *ad hoc* way. For instance, some papers report the use of the entropy function applied to the target atribute in classification

© Springer International Publishing Switzerland 2016
J. Bailey et al. (Eds.): PAKDD 2016, Part I, LNAI 9651, pp. 215–226, 2016.
DOI: 10.1007/978-3-319-31753-3_18

problems, i.e. the *class entropy* metafeature, but only a few use metafeatures based on the application of the same function to independent attributes, i.e. *attribute entropy* [3]. Very often, there is no justification for such options. We claim that the literature lacks an unifying framework to categorize and develop metafeatures. Therefore, this paper proposes one such framework to support the systematic generation of metafeatures for MtL problems.

Our proposal is a framework that decomposes a metafeature into three fundamental components: meta-function, object and post-processing functions. A meta-function (e.g., entropy) is applied to an object (e.g., set of independent variables) and the result is post-processed (e.g., average value), resulting in a metafeature (e.g., average attribute entropy). This decomposition enables the systematic generation of sets of metafeatures by applying the meta-function to all possible objects and process the result with all the possible post-processing functions.

In the experiments described in this paper, we use three meta-functions to generate systematic metafeatures: entropy, mutual information and correlation. We compare our approach with state-of-the-art metafeatures: the set of simple, statistical and information-theoretic metafeatures proposed by Brazdil et al. [3], landmarkers [4] and the pairwise meta-rules proposed by Sun and Pfahringer [5]. We address the problem of selecting the best algorithm for a classification dataset.

This paper is organized as follows. Section 2 describes the state-of-the-art in MtL regarding applications and metafeatures. In Sect. 3 we present the framework that supports systematic generation of metafeatures and we use it to decompose several metafeatures proposed in the literature. In Sect. 4 we present the results of the experiments that we carried out. Finally, Sect. 5 presents the conclusions and indicates some directions for future work.

2   Metalearning

Figure 1 illustrates a common MtL framework for algorithm recommendation, step by step. The process starts with a collection of datasets and learning algorithms. For each of those datasets, we extract metafeatures that describe their characteristics (A). Then, each algorithm is tested on each dataset and its performance is estimated (B). The metafeatures and the estimates of performance are stored as metadata. The process continues by applying a learning algorithm (a meta-learner) that induces a meta-model that relates the values of the metafeatures with the best algorithm for each dataset (C). Given the metafeatures of a new dataset (D), this meta-model is used to recommend one or more algorithms for that dataset (E).

As in any other ML task, the success of the application depends on the ability to include informative (meta) features in the data. The literature clearly groups metafeatures into three types: (1) simple, statistical and information-theoretic (2) model-based and (3) landmarkers [2]. In the first group we can find the *number of examples* of the dataset, *correlation between numeric attributes* or

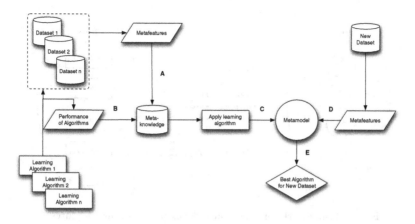

Fig. 1. Metalearning framework for algorithm recommendation.

class entropy, to name a few. Application of these kinds of metafeatures provides not only informative metafeatures but also interpretable knowledge about the problems [3]. The model-based ones [6] capture some characteristics of a model generated by applying a learning algorithm to a dataset, e.g., the *number of leaf nodes of a decision tree*. Finally, landmarkers [4] are generated by making a quick performance estimate of a simple learning algorithm in the dataset. For instance, the predictive performance of a Decision Stump.

The main focus of MtL research has been the problem of algorithm recommendation and is most commonly applied to classification problems. Brazdil et al. [3] proposed an approach that provides recommendations in the form of rankings of learning algorithms. They used simple, statistical and information-theoretic metafeatures. Sun and Pfahringer [5] extended the work of Brazdil et al. with two main contributions: the pairwise meta-rules (PMR), a higher-level type of metafeatures generated by comparing the performance of individual base learners in a one-against-one manner; and a new meta-learner for ranking algorithms. Besides PMR, they characterized datasets mainly with landmarkers.

However, MtL has also been used in other applications: time series forecasting [7], parameter tuning [8], data streams [9,10] and others [2]. This lead to a large set of metafeatures proposed in the literature for very different problems. It is common to find discrepancies between the use of a function such as entropy or mutual information to measure exclusively a specific object. Often, it is mandatory to adapt the set of metafeatures to the problem domain. For instance, metafeatures that characterize the target feature in regression cannot be used directly in classification. We believe that it would be useful to decompose all these metafeatures into a common framework. Furthermore, such framework must also help the MtL user to systematically develop new metafeatures.

3 Systematic Generation of Metafeatures

In this section, we propose a framework to enable a systematized and standardized development of metafeatures for MtL problems. The framework decomposes

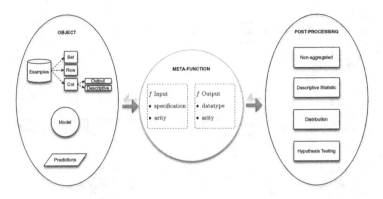

Fig. 2. Framework for systematic development of metafeatures.

a metafeature into three fundamental components: meta-function, object and post-processing. Figure 2 illustrates the framework.

The key component of the framework is the meta-function, f. This component is selected by the user according to his/her knowledge of the MtL problem. Although we acknowledge that this choice may be based on an *ad hoc* decision, the interest of a meta-function for a MtL problem should be well motivated. For example, entropy is a concept used in several Machine Learning algorithms, including decision trees [11]. Therefore, it is expected to be useful to better understand the learning behavior of those algorithms. Furthermore, this decision is made at a more abstract level than the typical design of metafeatures and is, thus, easier. For instance, it is indisputable that the concept of entropy is important for learning decision trees and that it is likely that some of the metafeatures that can be based on this function contain useful information on the behavior of tree learning algorithms. It is less clear that *class entropy* is useful and *attribute entropy* is not, or vice-versa. Finally, given the choice of a meta-function, the methodology generates metafeatures that characterize all components of the data with which it is compatible. This makes sure that the metafeatures that contain useful information, if any, will be generated.

The formal definition of f is given in Definition 1.

Definition 1. *A meta-function f is defined as*

$$f : X \to Y. \tag{1}$$

X is formed according to the data specification and the arity suitable to f input and the data specification and the arity of the O elements that are available for the problem. Given O as a type of object about a learning problem, namely, a set of examples, a row, an output column, a descriptive column, a model or a set of predictions, X is any set composed of those six object; and Y is the set of all possible results of applying f to X.

The selected meta-function has intrinsic characteristics that affect the metafeatures that can be generated from it. Particularly, the *data specification* and the *arity* of meta-function's input, and the *datatype* and *arity* of the meta-function's output. *Arity* corresponds to the number of inputs of a function. For instance, the arities of entropy and mutual information are, respectively, 1 and 2. Regarding the *data specification*, we use the data mining ontology proposed by Panov et al. [12] to guide the characterization. We take into account the role of the element in the context of the learning problem, i.e., if the column corresponds to *descriptive* data or *output* data. For instance, entropy is a meta-function that allows *discrete descriptive data, boolean descriptive data, discrete output data* and *boolean output data* as input. The arity of the objects X need to be equal to 1 in order to suit the meta-function's input. These characteristics are used to identify which of the available objects X can be used to generate metafeatures.

Let us present some examples of objects. The *examples* type can be detailed into a *set* of examples (e.g., all the examples of the dataset, a bootstrap sample of the examples, a specific subset, etc.), a single *row* or a *column*. The *column* can also be detailed into a column that represents an *output* variable or a column that represents a *descriptive* one. For the *model* type, we refer to all the information that can be measured regarding its induction and final output variable (e.g., the value of a parameter or a characteristic of the model - number of leaf nodes of a decision tree or the number of support vectors of a SVM). Finally, the *predictions* type considers all the information that can be extracted from the output of a predictive model when used to predict. For instance, to compute landmarkers, it is necessary two types of objects: an *output* column and a set of *predictions*. We give more detailed examples in Sect. 3.1.

The post-processing function p concerns the aggregation of the meta-function output, Y. Formally, p is defined in Definition 2.

Definition 2. *A post-processing function p is defined as*

$$p : Y \to MF \qquad (2)$$

where MF is the set of all possible metafeatures. The datatype and arity of Y defines implicitly the post-processing function p that can be used to form a metafeature.

For the meta-function's output Y, we need to characterize its *datatype* (real, discrete, boolean, etc.) and *arity*. Same thing for each post-processing function p. If the characteristics of Y match with the ones from p, a metafeature is formed. For instance, if the meta-function entropy is applied to n discrete descriptive attributes, the *datatype* of Y is real and the *arity* is n. This allows to generate metafeatures using post-processing functions such as *mean, maximum value, standard deviation, histogram bins*, etc.

Our framework splits p into four groups: *non-aggregated, descriptive statistic, distribution* and *hypothesis testing*. The *non-aggregated* alternative uses the meta-function output in its raw state directly as metafeature(s). In some MtL problems it might be useful not to aggregate the information. This is particularly frequent in MtL applications such as time series or data streams where the data has the same morphology [9] or when the MtL algorithm is relational [13].

For instance, instead of computing the mean of the correlation between pairs of numerical attributes, one could use the correlation between all pairs of numerical attributes. It can also be the case that Y does not need aggregation and, therefore, the non-aggregated post-processing function is applied.

The *descriptive statistic* case is perhaps the most common approach to aggregate information and generate a metafeature. This can be accomplished by using the *mean, maximum, minimum, standard deviation, mode*, etc. However, such aggregation can cause loss of valuable information.

Another option is the *distribution* alternative. It captures a representation of the output provided by the meta-function by characterizing its distribution. This fine-grained aggregation can be achieved through the use of *histograms* with a fixed number of bins as proposed in [14]. In this case, each bin is used as metafeature, providing a description of the distribution of the meta-function's output.

Finally, the *hypothesis testing* subcomponent. Here, the output Y provided by the meta-function f is used to test an assumption. For instance, it can test whether the values of Y follow a normal distribution. The output of this test (it can be a p-value or a nominal variable) is used as metafeature.

From Definitions 1 and 2, a metafeature $mf \in MF$ can be defined as

$$mf = p(f(x)) \tag{3}$$

where $x \in X$.

3.1 Decomposing Metafeatures

A first test to the validity of the proposed framework is to check if existing metafeatures could be the result of its use. We show examples from three types of metafeatures: simple, statistical and information-theoretic; model-based and landmarkers.

Figure 3 illustrates the decomposition of five metafeatures. For instance, the *absolute mean correlation between numeric attributes* is very similar to the *correlation between numeric attributes* (used in data streams applications [9]) except for the post-processing alternative. In this case, the application domain makes it feasible and potentially more informative to not aggregate the correlation values. The framework decomposes the computation of the metafeatures in detail. Furthermore, it allows the comparison between two or more metafeatures.

Still regarding Fig. 3, the decomposition of the two last metafeatures shows that is possible to use the framework for more complex metafeatures. The *number of nodes of a decision tree* is an example of a model-based metafeature. The object component is the decision tree model, the meta-function is count and the post-processing option is non-aggregated. Peng et al. [6] propose several model-based metafeatures (for decision trees models) of this kind. Finally, we also show an example of a landmarker. The *decision stump landmarker*, uses as object a set of predictions and the output column of the dataset. In the example given, the meta-function is individual accuracy. As post-processing function, the most

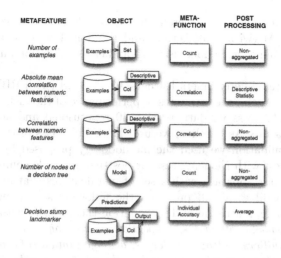

Fig. 3. Metafeatures decomposed using our framework.

common one is average. However, it could be used another, such as histogram bins. This kind of flexibility is one of the advantages of our framework.

4 Experiments

The experiments that we present in this Section aim to answer two questions.

- (1) Is the proposed framework able to develop sets of systematic metafeatures that are consistently more informative than non-systematic sets?
- (2) Is the set of systematic metafeatures computed with the proposed framework more informative than the state-of-the-art metafeatures?

For both questions, we executed experiments by addressing the problem of selecting the best algorithm in a set with a classification perspective [4, 14].

4.1 Experimental Setup

For the first question, we carried out a set of experiments with the goal of providing a proof of concept of our framework. By testing whether the systematic generation of metafeatures increases the amount of information of a set of metafeatures, we show that our framework can be useful and help MtL users to avoid an *ad hoc* selection of metafeatures. For the second question, we executed experiments in which we compare the set of metafeatures generated by our framework with a set of metafeatures regarded as state-of-the-art. All the experiments were executed on 58 UCI classification datasets [15]. The selection of the datasets was done randomly. To speed up the experiments, we limited the number of instances in larger datasets to a maximum of 5000 instances with stratified random sampling.

Six classification algorithms were tested as base learners: NaiveBayes, k-NN, C5.0, CART, SVM (with RBF kernel) and Random Forest. The estimates of algorithm performance were done using 10-fold cross validation and accuracy as error measure.[1]

We tested three different meta-learners: C5.0, SVM (with RBF kernel) and Random Forests. Again, the estimates of performance of the meta-learners were done using 10-fold cross validation (with 30 repetitions) and accuracy is the performance measure. As baseline, we use the default class of the training set. For statistical validation we used the methodology proposed by Demšar [16]: Friedman rank test with Nemenyi test for post-hoc multiple comparisons.

We compare our approach with a set of metafeatures that are widely used in the field. Brazdil et al. [3] proposed the following set of metafeatures of the so called simple, statistical and information-theoretic: *number of examples, proportion of symbolic attributes, proportion of missing values, proportion of attributes with outliers, entropy of classes, average mutual information of class and attributes* and *canonical correlation of the most discriminating single linear combination of numeric attributes and the class distribution*. To this set of simple, statistical and information-theoretic metafeatures we also added the *absolute average correlation between numeric attributes*. Finally, we included two landmarkers: *Decision Stump sub-sampling landmarker* and a *Naive Bayes sub-sampling landmarker*. We call this set of 10 metafeatures the *Traditional* one.

In order to provide a fair comparison, we designed metafeatures based as much as possible on the concepts involved in the *Traditional* metafeatures. Therefore, the set of objects consists on the dataset and on two sets of predictions: obtained with Naive Bayes and Decision Stump. As post-processing functions p, we used: mean, weighted mean, standard deviation, variance, minimum, maximum and histogram bins. We generated four sets of systematic metafeatures by selecting three meta-functions: entropy, mutual information and correlation, resulting in 20, 36 and 19 metafeatures, respectively. The fourth set includes *all* the metafeatures from the three previous sets (75 in total). The choice of the meta-functions was not done randomly. We used meta-functions that are used in the *Traditional* set with a non-systematic approach. Our approach is to use the very same meta-functions together with our framework to generate systematic sets of metafeatures.

One of the disadvantages of using our framework to generate systematic metafeatures is the curse of dimensionality. The number of metafeatures generated is usually very high. Since MtL applications don't have a large number of examples, this can be emphasized by our approach. So, we rely on two feature selection algorithms to tackle this problem: ReliefF [17] and correlation feature selection (CFS).

4.2   Systematized vs Unsystematized

The Critical Difference (CD) diagrams generated with the first set of experiments is presented in Fig. 4. We set $\alpha = 0.05$ for all experiments.

---

[1] The estimates of performance showed that NaiveBayes was better in 4 datasets, k-NN in 9, C5.0 in 23, CART in 2, SVM in 14 and Random Forest in 6.

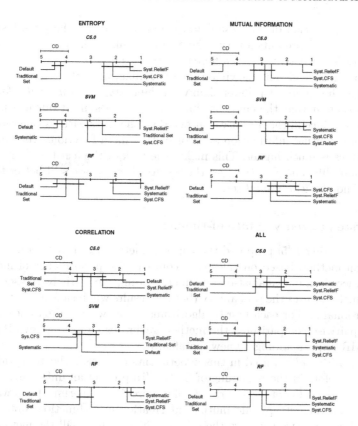

Fig. 4. Critical Difference diagrams for the meta-functions: *entropy, mutual information, correlation* and *all*.

The *Traditional* set in those experiments changes according to the meta-function which is being compared, to maximize the fairness of the comparison, as discussed earlier. For instance, the *Traditional* set used against the systematic metafeatures of the entropy meta-function consists of the following metafeatures: *entropy of classes, average entropy of symbolic attributes* and the two landmarkers mentioned above. Similarly, if the meta-function for generating systematic metafeatures is correlation, the *Traditional* set consists of the *absolute average correlation between numeric attributes* and, again, the two landmarkers.

Overall, the metafeatures generated from our framework present superior performance. This result is consistent regarding the meta-function and the meta-learner. However, it is noticeable that the results obtained with the correlation function are worst than in the other cases. It is probably related to the fact that this meta-function cannot be applied to the target variable, as this is a nominal variable and the input to the function must be numerical.

The set of metafeatures generated both with the meta-functions entropy and mutual information present good results when compared with the *Traditional* set. The combination of the *Systematic* metafeatures with the feature selection algorithm ReliefF presents a very good average rank in almost all CD diagrams. This result is consistent across different meta-learners.

Although this is not shown on the CD diagrams, the average accuracy obtained with the *all* set is lower than with the *entropy* and *mutual information* set. This suggest that the curse of dimensionality does affect our methodology, as expected. Since in the *all* set we gathered all the metafeatures generated from the entropy, mutual information and correlation meta-functions, the number of metafeatures is much higher. This makes the task of the feature selection algorithms more difficult. Nevertheless, the results obtained are still better than the baseline and the *Traditional* set.

4.3   Systematized vs State-of-the-art

Sun and Pfahringer [5] proposed the pairwise meta-rules (PMR), a metafeature generation method based on rules that compare the performance of individual base learners in a one-against-one manner. Adding the PMR to the sets of systematic metafeatures the probability that the results will be affected by the curse of dimensionality. For each pair of algorithms (since we test 6 base-learners, we have 15 pairwise comparisons) the method generates on average two PMR. So, using PMR implies adding 30 new metafeatures.

The *Traditional* set used in this experiments comprises the 10 metafeatures mentioned before in the beginning of Sect. 4. To prevent an unfair advantage of our approach, we do not use the results of the previous experiments, in which the sets based on the entropy and mutual information meta-functions obtained the best results. Thus, we compare the all set, which includes all the metafeatures, with the state of the art approaches. We added the respective PMR to the *Traditional* and *Syst.ReliefF* sets, forming *Traditional + PMR* and *Syst.ReliefF + PMR*. We compared the four sets of metafeatures with the same meta-learners used previously. We use ReliefF as feature selection algorithm.

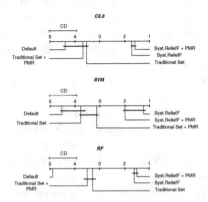

Fig. 5. Critical Difference diagrams of systematic metafeatures vs state-of-the-art in the classification experiments.

Figure 5 presents the results of the experiments. Comparing the sets of metafeatures, it is noticeable that those generated using our framework present a superior predictive performance in comparison with the *Traditional* sets. This difference is statistically significant. Also, the result is consistent across different meta-learners.

Regarding the addition of the PMR, the gain both on the *Traditional* or on the *Syst.ReliefF* set is not statistically significant. Overall, the sets *Syst.ReliefF* and *Syst.ReliefF + PMR* are the most informative across all meta-learners.

5   Conclusion and Future Work

This paper proposes a generic framework to develop metafeatures for MtL problems. This framework is a step towards the automatic generation of metafeatures. The framework is structured in such a way that the systematic generation of metafeatures is triggered through the selection of a meta-function. Then, given the objects and post-processing functions that are available, the framework outputs a set of metafeatures generated according to the characteristics (e.g., domain of the inputs) of the selected meta-function. The process can be repeated with several meta-functions. The selection of the meta-function is crucial and it should be chosen intuitively according to the MtL application.

Our experiments aim to answer two questions: (1) are the systematic sets of metafeatures better than the non-systematic ones? (2) are the systematic sets generated with the framework better than the state-of-the-art? In the first set of experiments, we found that the systematic metafeatures generated are consistently more informative than the non-systematic ones. In the second set of experiments, we found that the systematic sets are also more informative than state-of-the-art methods such as PMR.

As for future work, we plan to use this framework to generate systematic metafeatures for different MtL problems that we have been working on, particularly, MtL for pruning of bagging ensembles and dynamic integration of models [18].

Acknowledgments. This research has received funding from the ECSEL Joint Undertaking, the framework programme for research and innovation horizon 2020 (2014–2020) under grant agreement no. 662189-MANTIS-2014-1.

References

1. Serban, F., Vanschoren, J., Kietz, J.U., Bernstein, A.: A survey of intelligent assistants for data analysis. ACM Comput. Surv. (CSUR) **45**(3), 31 (2013)
2. Brazdil, P., Carrier, C.G., Soares, C., Vilalta, R.: Metalearning: Applications to Data Mining. Springer, Heidelberg (2008)
3. Brazdil, P.B., Soares, C., Da Costa, J.P.: Ranking learning algorithms: using IBL and meta-learning on accuracy and time results. Mach. Learn. **50**(3), 251–277 (2003)

4. Pfahringer, B., Bensusan, H., Giraud-Carrier, C.: Tell me who can learn you and I can tell you who you are: landmarking various learning algorithms. In: International Conference on Machine Learning, pp. 743–750 (2000)
5. Sun, Q., Pfahringer, B.: Pairwise meta-rules for better meta-learning-based algorithm ranking. Mach. Learn. **93**(1), 141–161 (2013)
6. Peng, Y.H., Flach, P.A., Soares, C., Brazdil, P.B.: Improved dataset characterisation for meta-learning. In: Lange, S., Satoh, K., Smith, C.H. (eds.) DS 2002. LNCS, vol. 2534, pp. 141–152. Springer, Heidelberg (2002)
7. Prudêncio, R.B., Ludermir, T.B.: Meta-learning approaches to selecting time series models. Neurocomputing **61**, 121–137 (2004)
8. Reif, M., Shafait, F., Dengel, A.: Meta-learning for evolutionary parameter optimization of classifiers. Mach. Learn. **87**(3), 357–380 (2012)
9. Rossi, A.L.D., de Leon Ferreira, A.C.P., Soares, C., De Souza, B.F.: Metastream: a meta-learning based method for periodic algorithm selection in time-changing data. Neurocomputing **127**, 52–64 (2014)
10. van Rijn, J.N., Holmes, G., Pfahringer, B., Vanschoren, J.: Algorithm selection on data streams. In: Džeroski, S., Panov, P., Kocev, D., Todorovski, L. (eds.) DS 2014. LNCS, vol. 8777, pp. 325–336. Springer, Heidelberg (2014)
11. Quinlan, J.R.: Induction of decision trees. Mach. Learn. **1**(1), 81–106 (1986)
12. Panov, P., Soldatova, L., Džeroski, S.: Ontology of core data mining entities. Data Min. Knowl. Disc. **28**(5–6), 1222–1265 (2014)
13. Getoor, L., Mihalkova, L.: Learning statistical models from relational data. In: ACM SIGMOD International Conference on Management of Data, pp. 1195–1198. ACM (2011)
14. Kalousis, A., Theoharis, T.: Noemon: design, implementation and performance results of an intelligent assistant for classifier selection. Intell. Data Anal. **3**(5), 319–337 (1999)
15. Lichman, M.: UCI Machine Learning Repository. University of California, Irvine, School of Information and Computer Sciences (2013). http://archive.ics.uci.edu/ml
16. Demšar, J.: Statistical comparisons of classifiers over multiple data sets. J. Mach. Learn. Res. **7**, 1–30 (2006)
17. Robnik-Šikonja, M., Kononenko, I.: Theoretical and empirical analysis of ReliefF and RReliefF. Mach. Learn. **53**(1–2), 23–69 (2003)
18. Pinto, F., Soares, C., Mendes-Moreira, J.: Pruning bagging ensembles with metalearning. In: Schwenker, F., Roli, F., Kittler, J. (eds.) MCS 2015. LNCS, vol. 9132, pp. 64–75. Springer, Heidelberg (2015)

Hash Learning with Convolutional Neural Networks for Semantic Based Image Retrieval

Jinma Guo, Shifeng Zhang, and Jianmin Li[(⊠)]

State Key Lab of Intelligent Technology and Systems,
Tsinghua National Lab for Information Science and Technology,
Department of Computer Science and Technology, Tsinghua University,
Beijing 100084, China
guojinma@gmail.com, zsffq999@163.com, lijianmin@mail.tsinghua.edu.cn

Abstract. Hashing is an effective method of approximate nearest neighbor search (ANN) for the massive web images. In this paper, we propose a method that combines convolutional neural networks (CNN) with hash learning, where the features learned by the former are beneficial to the latter. By introducing a new loss layer and a new hash layer, the proposed method can learn the hash functions that preserve the semantic information and at the same time satisfy the desirable independent properties of hashing. Experiments show that our method outperforms the state-of-the-art methods by a large margin on image retrieval. And the comparisons with baseline models show the effectiveness of our proposed layers.

Keywords: Hashing · Convolutional Neural Network · Image retrieval

1 Introduction

The amount of web data, images especially, is growing rapidly. How to retrieve images that meet users' requirements from this extremely tremendous data with efficient storage and computation has attracted extensive attentions from academia and industry [3].

Exhaustive nearest neighbor search is intractable. Approximate nearest neighbor search (ANN) can return satisfactory results within logarithmic ($O(log(n))$) or even constant ($O(1)$) time by organizing data with structures that keep the distance metric. Especially, hashing-based methods [5, 14–16, 19, 24, 26] with lookup tables consume only constant time on a query. The compact codes of hashing can also bring down the demand of storage, and the bitwise operations needed for a query make hashing competent even in the case of exhaustive ranking.

Conventional hashing methods usually take low-level features as input and use shallow models to generate the hash codes. However, the hand-crafted features are fixed and not learnable for further improvements. Recently CNNH [24] gains a great performance boost via deep model to learn hash codes. But this method breaks the learning process into two separate stages. Firstly, pseudo hash codes are learned from images' labels. Then the codes are fixed and used to train a convolutional neural network (CNN) model for later prediction. But some information

© Springer International Publishing Switzerland 2016
J. Bailey et al. (Eds.): PAKDD 2016, Part I, LNAI 9651, pp. 227–238, 2016.
DOI: 10.1007/978-3-319-31753-3_19

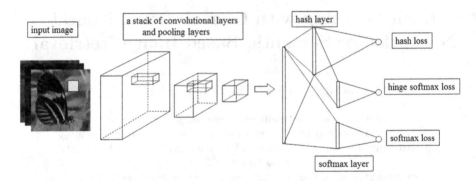

Fig. 1. The architecture of our proposed model in the training stage to learn hash functions. The input size of the model is fixed, and hash layer is fully connected with the prior. When training is done, the two softmax layers can be simply dropped and the outputs of the hash layer are binarized as hash codes.

will be lost in the first stage. Although Lai *et al.* [16] and Zhao *et al.* [26] propose one-stage ranking-based hashing methods respectively, both of which take only ranking as the supervisory information, they do not use the classes information.

In this paper, we propose a novel model that learn deep features and hash functions at the same time. As shown in Fig. 1, the model consists of three parts, which are a stack of convolutional layers, one softmax loss layer for classification, a new proposed hash layer and hinge softmax loss layer for hash code learning. Of the above three parts, the first is used to learn semantic-preserving features, the second is used to encourage the model to learn discriminant features from class labels, while the third part will learn more hashing-like codes. When training is done, the three loss layers will be dropped away and outputs of the hash layer are binarized with 0 to generate the final hash codes. The proposed model is an end-to-end system where feature extraction and hashing are combined.

The specific contributions of our work are as follows:

(1) we learn hash functions via CNN in the form of multi loss layers
(2) we introduce the hinge softmax loss layer and a hash layer into hash learning
(3) as far as we know, our results on the experimental datasets outperform the state-of-the-art.

The remaining is organized as follows: related works are briefly reviewed in Sect. 2. And the methodology of our work is described in Sect. 3. The experiments and discussions are presented in Sects. 4 and 5. Finally, we conclude the whole paper in Sect. 6.

2   Related Work

To generate n-bit code, hashing methods need n hash functions the k_{th} of which generally takes the following form:

$$h_k(x) = \begin{cases} 1 & f_k(x) \geq b_k \\ 0 & f_k(x) < b_k \end{cases} \qquad (1)$$

where x is the representation of a data sample, f_k is the hashing function, and b_k is the corresponding threshold. Based on the method to get f_k, hashing can be divided into data-independent methods and data-dependent (or learning-based) methods, of which the latter attempts to capture the inherent distribution of the task domain by learning. And learning-based hashing can be classified into unsupervised and supervised methods according to using annotation information or not.

A typical category of Locality Sensitive Hashing (LSH) [5] uses random projection to construct hash functions. The property of LSH, that samples within small Hamming distance in hash space are more likely to be near in their feature space, makes it very attractive. But the metrics are asymptotically preserved with increasing code length. Thus, to achieve high precision, LSH-related methods require large hash tables.

Unsupervised methods use only unlabeled data as training set, among which are methods such as Kernelized LSH [15], Semantic Hashing [12,20] and Spectral Hashing [23].

Supervised hashing utilizes human-annotated similarities or labels to get satisfying codes. Supervised Hashing with Kernels (KSH) [19] uses kernel-based model to minimize the Hamming distances of learned hash codes between similar data samples while maximize the distances between dissimilar ones at the same time. Binary Reconstruction Embedding (BRE) [14] learns hash functions by minimizing the differences between original distances of any two samples and the corresponding Hamming distances in hashing space. While initially proposed as unsupervised hashing, BRE can be easily extended to a supervised one by setting similar pairs with distance 0 and dissimilar pairs with distance 1.

These methods are kind of shallow and usually leverage some feature extraction algorithms to get the image representations. But the relationships between samples in semantic space are not maintained in low-level feature apace. And even combined with high-level features, the conventional hashing methods are very likely to perform no better than an end-to-end system which learns the feature extractor and hash functions together [26].

On the other hand, explosive interests in computer vision have been attracted to CNN [13] since 2012. Its remarkable successes in kinds of tasks such as object recognition [13,17,22], detection [13,22], image parsing [4] and video classification [10] have narrowed the gap between machine and human vision by a large step.

It has been suggested that the features in deep layers learned from ImageNet possess great capability to represent visual content of images, and can be used for different tasks, such as scene parsing [2], detection [6] and image retrieval [1]. Neural codes [1] uses activations of a fully-connected layer from an ImageNet-pretrained CNN as descriptors of the input image. And then Euclidean distances are computed to measure similarities. When retrained with datasets related to the query field, the retrieval performance can be comparable with the state-of-the-art.

CNNH and CNNH+ [24] take raw image as input, but divide the learning process into two different stages. In the first stage, similarity/dissimilarity matrix is decomposed to get the pseudo binary codes for training images. In the second stage, the raw image pixels and the corresponding binary codes (CNNH+ together with their one-hot binary labels) are fed to a CNN whose objective is to minimize the error between outputs and the target binary codes. But the decomposition stage would bring about extra errors. And the pseudo codes are fixed once the first stage is done, thus are not tunable for further improvement.

Lai et al. [16] proposes a deep neural network model to learn the hash functions. And the model's input is in the form of triplet, i.e., (I, I^+, I^-), meaning image I is more similar with I^+ than with I^-. The sub-networks for each element image of the triplet share parameters with each other. And the triplet ranking loss function is:

$$loss(x, x^+, x^-) = \max(0, ||x - x^+||_2^2 - ||x - x^-||_2^2 + 1)$$
$$s.t. \ x, x^+, x^- \in [0, 1]^k \tag{2}$$

where x, x^+ and x^- are the sub-networks' outputs of I, I^+, I^- respectively, and k is the length of x, x^- and x^+. Zhao et al. [26] takes a similar method, but the loss of every triplet is assigned with a weight which is defined by the numbers of shared labels between query image and two result images.

In spite of the above method, [18, 21, 25] assume that good hash codes should also be easily classified by a linear classifier, and take this target into hash learning. In addition to the classification task, [21, 25] penalize the output of embedding functions to make it close to -1 and 1 as much as possible, while [25] also requires the mean of each function to be 0.

3 Methodology

Given a set of class labels $\mathcal{Y} = \{1, \ldots, C\}$ and an image dataset $\mathcal{I} = \{I_1, I_2, \ldots I_N\}$ where each image is associated with one label y_n, our goal is to learn k hash functions which is used to encode images into k-bit hash codes. When using the codes for retrieval, the images sharing same label with the query image will be ranked on top of the result list.

In this paper, we propose a CNN architecture to learn the semantic-preserving hash functions, as shown in Fig. 1. The input image first goes through a stack of convolutional and pooling layers and then arrives at the concatenation layer from where the model branches into two separate paths. Of the two paths, one is the original softmax loss layer and the other is our hash layer and hinge softmax layer.

Normally, suppose that x^l is the output of the l-th layer of a CNN. Then if l-th layer is a softmax layer which is used to predict a vector p of which the c-th element is the probability of class c, the formulation is given by

$$p_c = \frac{\exp(w_c^T x^{l-1})}{\sum_{c'} \exp(w_{c'}^T x^{l-1})} \qquad (c = 1, 2, \ldots, C) \tag{3}$$

where w_c is the weights related with class c, and x^{l-1} is the output of the prior layer.

3.1   Hash Layer

The hash layer is a fully connected layer which has no nonlinear function inserted. By using fully connection, the hash layer can learn global semantic representations of the input image. And the hash layer's output will be penalized by the following formulation:

$$\text{hash\_loss}(\mathbf{x}) = \sum_{i=1}^{l_n} ||1 - \text{abs}(g(\mathbf{x}_i))||^2 \tag{4}$$

where l_n is the number of neurons in the hash layer. And $g(x_i)$ will take the following form:

$$g(\mathbf{x}_i) = \begin{cases} -1 & \mathbf{x}_i \leq -1 \\ 1 & \mathbf{x}_i \geq 1 \\ \mathbf{x}_i & otherwise \end{cases} \tag{5}$$

This loss can encourage the neurons to generate outputs that distributes less around 0, which will be used as the threshold to binarize the outputs and get the hash codes.

3.2   Hinge Softmax Loss

In hash learning, the target is more a rank problem than classification. It is sufficient to make prediction of ground truth label p_y larger than the rest, while the traditional softmax loss $loss(y, p) = -log(p_y)$ can be too harsh. So we define a loss modified from softmax, which takes the following form:

$$\text{hinge\_softmax\_loss}(y, p) = \begin{cases} 0 & p_y \geq \max(p_{\tilde{y}}) + m \\ -log(p_y) & p_y < \max(p_{\tilde{y}}) + m \end{cases} \tag{6}$$

where y is the ground truth label, $\tilde{y}$ is the rest of label set, p is the prediction possibilities for every class and m is the slack that controls when the model should be penalized.

By this formulation, for those samples that have been classified correctly by a slack larger than m, the loss will be forced to be 0 and thus back propagate no changes to the learnable parameters. Otherwise, the semantic representations learned by CNN can be not so good, thus the penalization term will be taken. So by this setting, the features of our hash layer can be semantically correct.

When m is set to be 1, because all the prediction possibilities are between 0 and 1, the hinge softmax loss will only execute the lower part and thus becomes conventional softmax loss. And when m is no greater than 0, the loss can be easily stuck in a local minimum, for example when the probabilities of all classes are equal to a certain value.

3.3   The Model

Traditional softmax loss can help in training the networks more discriminant. To combine classification task with hash learning, we branch out from the layer prior to the softmax layer and add a hash layer together with a hinge softmax layer. By this way, the softmax loss in our model performs as an auxiliary classifier like [17].

The overall loss function is given as:

$$\min_{W} L_1(y, p) + \alpha L_2(\mathsf{x}_h) + \beta L_3(y, p) + \lambda ||W||^2 \qquad (7)$$

where W is all the parameters that are to be learned in the network, L_1 is the hinge softmax loss, L_2 is the loss of our proposed hash layer, L_3 the softmax loss for classification and the last term is the weight decay, α and β are hyperparameters.

By our loss function, the representations learned by the proposed hash layer can preserve the semantic information and be more hashing-like.

Compactness is an important property of hash codes. The binary code generated by every function should be independent with each other and the information carried by the binary bit should be maximized. Compared with [16], we take advantage of dropout's [8] capability to prevent co-adaption where a feature detector is only helpful in the presence of several other specific feature detectors. With a dropout layer inserted between the hash layer and hinge softmax loss layer, the neurons of hash layer can be independent of each other.

3.4   Hash Codes

The networks are trained by stochastic gradient descent. When training is done, the two softmax layers can be simply dropped and use the rest architecture to generate the hash codes. When a new image comes, it is first filtered by the model so as to being encoded into a k-dimension vector, and then the vector is binarized into the final hash codes according to Eq. (1) with all b_k set to be 0.

4   Experiments

4.1   Experimental Settings

We compare the proposed model with one data-independent method LSH [5], and four supervised methods BRE [14], KSH [19], CNNH [24] and [16] on two widely used benchmark datasets, i.e., the CIFAR-10 dataset[1] [11,16] and the Street View House Number (SVHN) dataset[2]. And we will call the model of [16] TRCNNH for short.

For fair comparison, we sample 1000 images from each dataset as query set and another 5000 from the rest for training like [16]. For LSH, all the data except

---

[1] http://www.cs.toronto.edu/~kriz/cifar.html.

[2] http://ufldl.stanford.edu/housenumbers/.

the query set are training set. The 5000 keeping-label samples serve as training set for the other methods. And following [16], images are represented by 512-dim GIST features for non-CNN methods.

As for our proposed method, the 5000 labeled samples in each datasets are divided into training set and validation set by which to find the most suitable architecture and tune the hyper-parameters. Then all the 5000 samples are used to retrain the model from scratch.

The description of an architecture is given in the following way: $3 \times 32 \times 32$-32C5P2-MP3S2-32C5P0-D0.5-SL10 represents a CNN with inputs of 3 channel of 32×32 pixels, a convolutional layer with 32 filters whose size is 5×5 and 2 paddings around the input maps, a max pooling layer (AP for average pooling) of 3×3 size and stride 2, a 32-filters convolutional layer whose kernel size is 5×5 and have 0 padding around the input maps, a dropout layer whose dropout ratio is 0.5, and finally a softmax loss layer with 10 classes. And our hash layer is always connected with the last hidden layer. Rectifier Linear Unit (ReLU) is used as the nonlinear transformation neurons for all convolutional layers.

Hash lookup and Hamming ranking are two widely used methods to conduct search with hashing [16,24]. Hash lookup constructs a lookup table with radius r in advance, and all the samples within the radius will be returned as results, thus can decrease the query time to a constant value. However, the number of results returned will dramatically decrease with the code's length increases. On the other hand, Hamming ranking will traverse the dataset all through at a new coming.

We evaluate the performances on three metrics, i.e. Precision curves within Hamming radius 2, Precision-Recall curves and Precision curves with respect to different returned number with ranking.

Our models are implemented with Caffe [9], an open-source CNN framework. On both datasets, our networks are trained with stochastic gradient descent. The momentum is set to 0.9. Weight decay coefficients for convolutional layers and fully connected layers are separately set to 0.001 and 0.25. For the hyper-parameters in Eq. (7), α and β are decided by validation and fixed at 0.1, while γ decreases from 0.3 to 0 in the whole training stage. The margin in the hinge softmax loss layer is set to 0.1 in our experiments, which won't hurt too much, and can back propagate at the same time.

Table 1. MAP of Hamming ranking w.r.t different number of bits on two datasets.

Method code length	CIFAR10(MAP)				SVHN(MAP)			
	12 bits	24 bits	32 bits	48 bits	12 bits	24 bits	32 bits	48 bits
Ours	**0.611**	**0.632**	**0.645**	**0.641**	**0.911**	**0.931**	**0.934**	**0.942**
TRCNNH [16]	0.552	0.566	0.558	0.581	0.899	0.914	0.925	0.923
CNNH [24]	0.484	0.476	0.472	0.489	0.897	0.903	0.904	0.896
KSH [19]	0.311	0.348	0.353	0.366	0.576	0.631	0.658	0.662
BRE [14]	0.150	0.172	0.174	0.176	0.156	0.168	0.169	0.180
LSH [5]	0.106	0.119	0.121	0.124	0.132	0.143	0.128	0.151

The implementations of BRE [14] and KSH[3] [19] are provided by their authors. For LSH, the projections are randomly sampled from a Gaussian distribution with zero-mean and identity covariance to construct the hash tables. And the results of CNNH and TRCNNH are obtained from [16].

4.2   CIFAR-10

The CIFAR-10 dataset is an images collection containing 60,000 color images of 32×32 pixels. All the samples are evenly labeled with 10 mutually exclusive classes, ranging from airplane to bird.

For this dataset, the architecture we choose is a six-conv-layer model. The main branch of the model is $3 \times 32 \times 32$-32C3P1-32C3P1-MP3S2-D0.5-32C3P1-32C3P1-MP3S2-D0.5-64C3P1-64C3P1-AP3S2-D0.5-SL10. For the convenience of later quotation, we name the third pooling layer "Pool3".

The MAP of Hamming ranking can be seen on Table 1 and the performance curves are shown in Fig. 2. For all the four experimented code length, i.e. 12, 24, 32 and 48, the MAP of our model surpass the state-of-the-art by more than 5.9

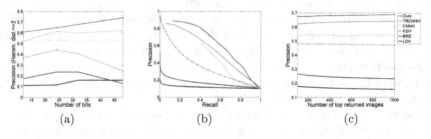

(a)   (b)   (c)

Fig. 2. The results on CIFAR-10. (a) Precision curves of hash lookup within Hamming radius 2. (b) Precision-recall curves of Hamming raking with code's length of 48 bits. (c) Precision curves with respect to number of returned images of Hamming raking with code's length of 48 bits.

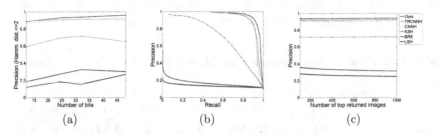

(a)   (b)   (c)

Fig. 3. The results on SVHN. (a) Precision curves of hash lookup within Hamming radius 2. (b) Precision-recall curves of Hamming raking with code's length of 48 bits. (c) Precision curves with respect to number of returned images of Hamming raking with code's length of 48 bits.

---

[3] http://www.ee.columbia.edu/ln/dvmm/downloads/WeiKSHCode/dlform.htm.

percent. Figure 2(a) witnesses that precision of images within 2 bits Hamming distance returned by our model increases along with the length of hash code. The LSH reported in this paper used random projection which relies on the input feature to maintain the similarity. But because of the semantic gap between low-level feature and content, LSH's performances are not stable. While BRE and KSH perform better than LSH, they are still worse than CNN based methods. And in Fig. 2(b) the precision-recall curve of our model encloses more spaces than the others, which is consistent with Table 1. Then Fig. 2(c) demonstrates our model's best performance on Hamming ranking, from 64 % to 69 %, nearly by 8 % beyond TRCNNH.

Comparison Between Different Features. On CIFAR-10, we conduct some other experiments to compare. One is image retrieval using Euclidean distance between corresponding features of query and dataset's images as similarity, and the features are right the outputs of hash layer (the "L2_fc" method). Another two experiments we undertake are using KSH and LSH with outputs of "Pool3" as descriptors of an input image (the "Pool3_feat"). LSH is also conducted with 512-dimension Gist feature as image descriptions. The network used in this part is the 32-bit network. The results are shown in Table 2.

Surprisingly, we find that L2_fc is not the best, which may be resulted from the denoising property of binarizing. KSH learns hash functions on top of the features extracted from "Pool3", right the preceding layer of our hash layer. The retrieval precision of "Pool3_feat + KSH" is inferior to our method by a small margin. And the last two LSH-based experiments show that the features learned by CNN are indeed better than Gist.

Comparison with Baseline. We evaluate the performance of our proposed hash layer and hinge softmax loss layer by comparing with three baseline models on CIFAR-10. The first baseline model is just like our model but without the hash loss. The second baseline model is similar with [18], i.e. without the auxiliary path and the hash loss. In addition to the second baseline, the third baseline model replaces the hinge softmax loss with traditional softmax loss. The parameters of the preceding convolutional layers are kept the same. The results are presented in Table 3. Our proposed method performs best on three

Table 2. Ranking precisions of different methods at 500 and 1500 on CIFAR-10

Precision \ Top N Method	500	1500
Ours	**68.52%**	**69.51%**
L2_fc	68.30%	66.17%
Pool3_feat + KSH	68.40%	68.88%
Pool3_feat + LSH	58.84%	53.96%
Gist_512 + LSH	16.39%	15.21%

Table 3. Comparison of the proposed model with baseline on CIFAR-10.

Method code length	CIFAR10(MAP)			
	12 bits	24 bits	32 bits	48 bits
Ours proposed model	0.611	**0.632**	**0.645**	**0.641**
Ours without hash loss	0.607	0.621	0.633	0.641
Ours with only hinge_softmax	**0.617**	0.624	0.633	0.638
CNN with only softmax	0.594	0.629	0.619	0.636

of the four tested code lengths. The third baseline model without any of our modifications performs worst except 24 bits and fluctuate with the code length.

4.3 SVHN

The SVHN dataset consists of 630, 420 color house number images collected from Google Street View images. The data is provided in two formats and the second is used in our experiments. Each data sample is of size 32×32 pixels and annotated with one label from 1 to 10. And the dataset is preprocessed with local contrast normalization, following Goodfellow *et al.* [7].

On this dataset, the main structure is similar to the CIFAR-10 model: $3 \times 32 \times 32$-128C3P1-128C3P1-MP2S2-D0.5-128C3P1-128C3P1-MP2S2-D0.5-128C3P1-128C3P1-AP2S2-D0.5-SL10.

Similar with the results on CIFAR-10, the MAP values related with the four code lengths of our model on CIFAR10 outperform the state-of-the art. Figure 3(a) shows the precision of returns with hash lookup of radius 2. When using 12-bit codes for retrieval, the precision of our model is just comparable with TRCNNH [16]. But with the increment of code length, the gap between our model and TRCNNH is getting larger. On all the three metrics, our model is the best one.

5   Discussion

Experiments have shown that the precision of our proposed model on both SVHN and CIFAR-10 within Hamming radius of 2 improves with more hash bits, which is consistent with our intuition.

Although KSH uses kernel functions for hashing and our hash layer uses linear functions, the KSH with our CNN-feature performs slightly inferior than our method. The hash layer can be considered as learning-based linear projections, whose inputs are also learnable. During training, the weights and biases in CNN are adjusted to render the filters expressive enough so as to allow the simple classifier on the top perform well. And the raw pixels are at the same time transformed into a description space where similarity relationships are correlated with Euclidean distance.

Unlike CNNH and TRCNNH both of which need the amount of annotations to be square or even cubic of the number of training images, our method trains the CNN with single image whose label indicates the class. So more training data will not burden it too much.

6 Conclusion

As a method of ANN search, interests of many researchers and companies have been attracted by hashing. We propose a new method which can obtain the binary hash code of a given image just by binarizing the outputs of our hash layer, and achieves the best result on both SVHN and CIFAR-10.

In consideration of the fact that all the models use only 5000 samples to train, we can expect an improvement of performance with a larger training set. Besides, a large amount of unlabeled data remains untouched, which can be another key element to enhancement.

Acknowledgments. This work was supported by the National Basic Research Program (973 Program) of China (Nos. 2012CB316301 and 2013CB329403), and the National Natural Science Foundation of China (No. 61332007).

References

1. Babenko, A., Slesarev, A., Chigorin, A., Lempitsky, V.: Neural codes for image retrieval. In: Fleet, D., Pajdla, T., Schiele, B., Tuytelaars, T. (eds.) ECCV 2014, Part I. LNCS, vol. 8689, pp. 584–599. Springer, Heidelberg (2014)
2. Chen, L.-C., Papandreou, G., Kokkinos, I., Murphy, K., Yuille, A.L.: Semantic image segmentation with deep convolutional nets, fully connected CRFs (2014). arXiv preprint arXiv:1412.7062
3. Datta, R., Joshi, D., Li, J., Wang, J.Z.: Image retrieval: ideas, influences, and trends of the new age. ACM Comput. Surv. (CSUR) 40(2), 5 (2008)
4. Farabet, C., Couprie, C., Najman, L., Lecun, Y.: Learning hierarchical features for scene labeling. IEEE Trans. Pattern Anal. Mach. Intell. 35(8), 1915–1929 (2013)
5. Gionis, A., et al.: Similarity search in high dimensions via hashing. In: Proceedings of 25th International Conference on Very Large Data Bases VLDB 1999, September 7–10, Edinburgh, Scotland, UK, pp. 518–529 (1999)
6. Girshick, R., Donahue, J., Darrell, T., Malik, J.: Rich feature hierarchies for accurate object detection and semantic segmentation. In: Proceedings of the IEEE Conference on Computer Vision and Pattern Recognition, pp. 580–587. IEEE (2014)
7. Goodfellow, I., Warde-farley, D., Mirza, M., Courville, A., Bengio, Y.: Maxout networks. In: Proceedings of the 30th International Conference on Machine Learning (ICML 2013), pp. 1319–1327 (2013)
8. Hinton, G.E., Srivastava, N., Krizhevsky, A., Sutskever, I., Salakhutdinov, R.R.: Improving neural networks by preventing co-adaptation of feature detectors (2012). arXiv preprint arXiv:1207.0580
9. Jia, Y., Shelhamer, E., Donahue, J., Karayev, S., Long, J., Girshick, R., Guadarrama, S., Darrell, T.: Caffe: convolutional architecture for fast feature embedding. In: Proceedings of the ACM International Conference on Multimedia, pp. 675–678. ACM (2014)

10. Karpathy, A., Toderici, G., Shetty, S., Leung, T., Sukthankar, R., Fei-Fei, L.: Large-scale video classification with convolutional neural networks. In: Proceedings of the IEEE Conference on Computer Vision and Pattern Recognition, pp. 1725–1732. IEEE (2014)
11. Krizhevsky, A., Hinton, G.E.: Learning multiple layers of features from tiny images (2009)
12. Krizhevsky, A., Hinton, G.E.: Using very deep autoencoders for content-based image retrieval. In: Proceedings of the European Symposium on Artificial Neural Networks. Citeseer (2011)
13. Krizhevsky, A., Sutskever, I., Hinton, G.E.: Imagenet classification with deep convolutional neural networks. In: Advances in Neural Information Processing Systems 25, pp. 1097–1105 (2012)
14. Kulis, B., Darrell, T.: Learning to hash with binary reconstructive embeddings. In: Advances in Neural Information Processing Systems 22, pp. 1042–1050 (2009)
15. Kulis, B., Grauman, K.: Kernelized locality-sensitive hashing for scalable image search. In: IEEE 12th International Conference on Computer Vision, pp. 2130–2137. IEEE (2009)
16. Lai, H., Pan, Y., Liu, Y., Yan, S.: Simultaneous feature learning and hash coding with deep neural networks. In: Proceedings of the IEEE Conference on Computer Vision and Pattern Recognition (2015)
17. Lee, C.-Y., Xie, S., Gallagher, P., Zhang, Z., Tu, Z.: Deeply-supervised nets. In: Proceedings of the International Conference on Artificial Intelligence and Statistics, pp. 562–570 (2015)
18. Lin, K., Yang, H.-F., Hsiao, J.-H., Chen, C.-S.: Deep learning of binary hash codes for fast image retrieval. In: Proceedings of the IEEE Conference on Computer Vision and Pattern Recognition Workshops, pp. 27–35 (2015)
19. Liu, W., Wang, J., Ji, R., Jiang, Y.-G., Chang, S.-F.: Supervised hashing with kernels. In: Proceedings of the IEEE Conference on Computer Vision and Pattern Recognition, pp. 2074–2081. IEEE (2012)
20. Salakhutdinov, R., Hinton, G.: Semantic hashing. Int. J. Approximate Reasoning **50**, 969–978 (2009)
21. Shen, F., Shen, C., Liu, W., Shen, H.T.: Supervised discrete hashing. In: Proceedings of the IEEE Conference on Computer Vision and Pattern Recognition. IEEE (2015)
22. Simonyan, K., Zisserman, A.: Very deep convolutional networks for large-scale image recognition (2014). arXiv preprint arXiv:1409.1556
23. Weiss, Y., Torralba, A., Fergus, R.: Spectral hashing. In: Advances in Neural Information Processing Systems 21, pp. 1753–1760 (2008)
24. Xia, R., Pan, Y., Lai, H., Liu, C., Yan, S.: Supervised hashing for image retrieval via image representation learning. In: Proceedings of the AAAI Conference on Artificial Intellignece, pp. 2156–2162 (2014)
25. Yang, H.-F., Lin, K., Chen, C.-S.: Supervised learning of semantics-preserving hashing via deep neural networks for large-scale image search (2015). arXiv preprint arXiv:1507.00101
26. Zhao, F., Huang, Y., Wang, L., Tan, T.: Deep semantic ranking based hashing for multi-label image retrieval. In: Proceedings of the IEEE Conference on Computer Vision and Pattern Recognition, pp. 1556–1564 (2015)

Bayesian Group Feature Selection for Support Vector Learning Machines

Changde Du[1,2,4], Changying Du[1,3,4(✉)], Shandian Zhe[5], Ali Luo[2], Qing He[1], and Guoping Long[3]

[1] Key Lab of Intelligent Information Processing of Chinese Academy of Sciences (CAS), Institute of Computing Technology, CAS, Beijing 100190, China
cddu@nao.cas.cn, {ducy,heq}@ics.ict.ac.cn
[2] Key Laboratory of Optical Astronomy, National Astronomical Observatories, Chinese Academy of Sciences, Beijing 100012, China
lal@bao.ac.cn
[3] Laboratory of Parallel Software and Computational Science, Institute of Software, Chinese Academy of Sciences, Beijing 100190, China
[4] University of Chinese Academy of Sciences, Beijing 100049, China
[5] Department of Computer Science, Purdue University, West Lafayette, IN 47907, USA

Abstract. Group Feature Selection (GFS) has proven to be useful in improving the interpretability and prediction performance of learned model parameters in many machine learning and data mining applications. Existing GFS models were mainly based on square loss and logistic loss for regression and classification, leaving the ϵ-insensitive loss and the hinge loss popularized by Support Vector Learning (SVL) machines still unexplored. In this paper, we present a Bayesian GFS framework for SVL machines based on the pseudo likelihood and data augmentation idea. With Bayesian inference, our method can circumvent the cross-validation for regularization parameters. Specifically, we apply the mean field variational method in an augmented space to derive the posterior distribution of model parameters and hyper-parameters for Bayesian estimation. Both regression and classification experiments conducted on synthetic and real-world data sets demonstrate that our proposed approach outperforms a number of competitors.

Keywords: Group feature selection · Support vector machine · Support vector regression · Variational bayesian inference · Data augmentation

1 Introduction

Feature selection which involves choosing an optimal subset of raw input features such that the subset does not contain any irrelevant or redundant features plays an important role in many applications. However, individual features may not reveal the structural information among raw inputs, thus we are more interested in finding some important feature groups instead of individual features in

© Springer International Publishing Switzerland 2016
J. Bailey et al. (Eds.): PAKDD 2016, Part I, LNAI 9651, pp. 239–252, 2016.
DOI: 10.1007/978-3-319-31753-3_20

some scenarios. Group Feature Selection (GFS) is a technique of making use of structural information among features to select an optimal subset of relevant features in a grouped manner. The advantages of GFS can be summarized as improving the prediction performance, reducing training and utilization time and enhancing the interpretability of learned parameters. So far, GFS has been successfully applied in a number of domains, such as multi-sensor data fusion [6], birth weight prediction [23], gene finding [11], waveband selection [16], etc.

As the most popular method employed for GFS, Group-Lasso [23] is an important extension of lasso [19]. It consists in estimating a linear regression model by minimizing the square loss function evaluated on training data, under a series of constraints which enforce sparsity at the group level. Besides square loss, a logistic loss has also been considered to address classification problems with group sparsity [11]. Though Group-Lasso models are useful, their main problem concerns the lack of meaningful variance estimates for model coefficients because these methods only provide a point estimate for model parameters. To overcome such problems, Bayesian Group-Lasso is proposed in [13] as a full Bayesian treatment of the Group-Lasso. In [7], the authors used a generalized spike-and-slab prior to encourage group sparsity in linear regression.

These existing GFS models were mainly based on square loss and logistic loss for regression and classification analysis, leaving the ϵ-insensitive loss and the hinge loss popularized by Support Vector Learning (SVL) machines still unexplored. SVL machines, such as Support Vector Machine (SVM) and Support Vector Regression (SVR) [4] are widely used in various machine learning applications owing to their arguably good generalization performance and the merit of only using support vectors in decision function. Nevertheless, their performance can be seriously affected when the input data are very high dimensional, with many non-informative or noisy features. Such a situation could be alleviated if the structural information among features is exploited, e.g., by combining SVL with GFS to select the most relevant feature groups.

In this paper, based on the pseudo likelihood and data augmentation idea [12,25], we propose a new Bayesian GFS framework for SVL machines. To the best of our knowledge, this is the first effort to integrate GFS into SVL with Bayesian learning, which allows us to circumvent the time consuming cross-validation for regularization parameters. Specifically, our new framework employs a group Automatic Relevance Determination (ARD) [20] prior to select the most relevant feature groups, and a pseudo likelihood term for Bayesian SVL. With the data augmentation idea, we re-express the pseudo likelihood into different forms for different SVL tasks. To derive the posterior distribution of model parameters and hyper-parameters for Bayesian estimation, we perform mean field variational inference in the augmented variable space. Finally, both regression and classification experiments conducted on synthetic and real-world data sets demonstrate that our proposed approach outperforms other state-of-the-art GFS approaches and the direct SVR and SVM learning.

2   Related Work

Existing GFS methods can be divided into two kinds. One kind are deterministic methods addressing the optimization problem directly, such as the Group-Lasso for logistic regression [11], the Group-Lasso for generalized linear models [14], the Group-Lasso with overlap between groups [9], the sparse Group-Lasso [15], the online learning algorithms for GFS [21,22], etc. Another kind are Bayesian inference based approaches which can apply different likelihoods and priors on the model conveniently. The Bayesian Group-Lasso [13] imposes multivariate Laplace priors on separate groups, with a Monte Calo sampling scheme for inference. The Variational Relevant Group Selector (VRGS) [17], as an extension of the Relevance Vector Machine (RVM) [20], is similar to sparse Group-Lasso, which have sparse effects both on groups and on individual features. In [7], the authors used a generalized spike-and-slab prior to encourage group sparsity with Expectation Propagation (EP) inference. In [2], Babacan presented a general class of multivariate priors for group sparse modeling and developed Bayesian inference methods via variational Bayesian approximation.

All methods mentioned above were based on square loss and logistic loss due to their convenient form and tractable solution. The problem of them is their more risk of overfitting compared to the ϵ-insensitive loss and the hinge loss popularized by SVL machines when only small training data is available.

There are also some sparse learning methods for SVMs [18,24]. But unlike GFS models, these methods typically didn't consider the structural information among features, thus can only identify relevant features rather than feature groups.

3   Bayesian GFS for SVL Machines

In this section, we first review the basic group sparse model, and then present our proposed framework along with the learning models. Fast variational inference procedures are developed to infer the model parameters and hyper-parameters in Bayesian manner. In the sequel, suppose we have a data matrix $\mathbf{X} \in \mathbb{R}^{d \times N}$ consisting of N observations $\{\mathbf{x}_i\}_{i=1}^N$ in d-dimensional feature space, and a $N \times 1$ response vector $\mathbf{y}$, with y_i denoting the response of the i-th observation.

3.1   Group Sparse Model

Group sparse modeling is a natural generalization of the traditional sparse modeling methods. In group sparse model, the sparsity constraint is imposed on groups instead of the individual features. It effectively models the structural properties of the feature vector, such that dependencies among features are taken into account. A general optimization formulation for group sparse model is

$$\min_{\mathbf{w}} \sum_{i=1}^{N} \ell(y_i, \mathbf{x}_i; \mathbf{w}) + \rho \|\mathbf{w}\|_{1,2}, \tag{1}$$

where $\mathbf{w}$ is a column vector of coefficients, $\ell(y_i, \mathbf{x}_i; \mathbf{w})$ denotes the loss of model $\mathbf{w}$ on data $(\mathbf{x}_i, y_i)$, $\rho \geq 0$ is the regularization parameter controlling the strength of the enforced sparsity over groups, and $\|\mathbf{w}\|_{1,2} = \sum_{g=1}^{G} \|\mathbf{w}_g\|_2$, where G is the number of groups, and $\mathbf{w}_g$ denotes the coefficients of g-th group. Assume that the groups are not overlapping, and the size of g-th group is denoted by d_g, such that $\sum_{g=1}^{G} d_g = d$. Obviously, the traditional l_1 norm based formulation is a special case of this formulation (when $d_g = 1, \forall g$).

In previous work, typically the square loss was used for regression problems, i.e., $\ell(y_i, \mathbf{x}_i; \mathbf{w}) = (y_i - \mathbf{w}^T \mathbf{x}_i)^2$, and the logistic loss was used for classification problems, i.e., $\ell(y_i, \mathbf{x}_i; \mathbf{w}) = \log(1 + \exp(-y_i(\mathbf{w}^T \mathbf{x}_i + c)))$, where c is the intercept. Compared with these losses, the ϵ-insensitive loss and hinge loss are usually more attractive due to their better generalization performance and the merit of only using support vectors in decision function. On the other hand, the performance of traditional SVL machines can be seriously affected when the input data contains a lot of non-informative or noisy features. Therefore, it is meaningful to exploit the structural information among features by combining SVL with group sparse model.

In addition, an important issue in the group sparse model described above is to choose the regularization parameter ρ. Usually, the time consuming cross-validation is employed to select an optimal value for ρ. This problem can be circumvented by Bayesian inference, as shown in the following subsection.

3.2   The Proposed Framework

In this subsection, we will present a Bayesian Group Feature Selection (GFS) framework for SVL machines based on the pseudo likelihood and data augmentation idea. The Bayesian modeling of Eq. 1 requires the definition of a joint distribution of all observed and latent variables. Typically, this joint distribution includes the prior distributions over the latent variables, and the likelihood of latent variables on the observed variables.

To impose group structure constraints on $\mathbf{w}$, each group $(\mathbf{w}_g)$ could have a multivariate Gaussian prior controlled by a distinct hyper-parameter z_g ($g = 1, \ldots, G$). We assume a multivariate variance mixture of Gaussian prior over the coefficient vector $\mathbf{w}$,

$$p(\mathbf{w}|\mathbf{z}) = \prod_{g=1}^{G} \mathcal{N}(\mathbf{w}_g | \mathbf{0}_{d_g}, z_g^{-1} \mathbf{I}_{d_g}),$$

where z_g is the inverse of the prior variances for each component of $\mathbf{w}_g$, that is, all components of $\mathbf{w}_g$ have the same prior variances. Furthermore, we select a conjugate prior for z_g by choosing a Gamma distribution,

$$p(\mathbf{z}) = \prod_{g=1}^{G} \Gamma(z_g | \alpha_g, \beta_g),$$

where α_g and β_g are the shape and rate parameter of Gamma distribution respectively.

Note that the hierarchical priors on $\mathbf{w}_g$ and z_g are motivated by Automatic Relevance Determination (ARD). This group ARD formulation can easily obtain the sparsity at the group level by considering a different hyper-parameter z_g for each group of coefficients. A variety of distributions over $\mathbf{w}$ can be represented in this fashion by different selections of the hyper-prior distribution $p(z_g^{-1})$ [2].

With the group ARD prior for GFS, we still need to define a likelihood function to link SVL with the above Bayesian model. To this end, it is necessary to utilize the loss function of SVL. However, the ϵ-insensitive loss and hinge loss do not lend themselves to a convenient description of a likelihood function. To overcome this situation, we have to transform the loss function of SVL into a Gaussian pseudo likelihood, which has the form

$$p(\mathbf{y}|\mathbf{w}, \mathbf{X}) = \prod_{i=1}^{N} \exp[-2 \cdot \ell(y_i, \mathbf{x}_i; \mathbf{w})]. \tag{2}$$

The difference between our pseudo likelihood and an actual likelihood is that the former is unnormalized with respect to $\mathbf{y}$. By defining $\ell(y_i, \mathbf{x}_i; \mathbf{w})$ with ϵ-insensitive loss and hinge loss respectively, this general framework can handle both regression and classification problems.

Note that the hyper-parameters $\mathbf{z}$ introduced to control the group structure of $\mathbf{w}$ is directly related to the regularization parameters in traditional Support Vector Regression (SVR) and Support Vector Machine (SVM). Since $\mathbf{z}$ can be derived automatically by Bayesian inference, our framework can circumvent the cross-validation procedure for regularization parameters. In the following, we will present the regression and classification models (BGFS-SVR and BGFS-SVM) with linear assumption.

BGFS-SVR. For regression model ($y_i \in \mathbb{R}$, $i = 1, \ldots, N$), we employ the ϵ-insensitive loss to re-express Eq. 2, which implies a pseudo likelihood of the form

$$p(\mathbf{y}|\mathbf{w}, \mathbf{X}) = \prod_{i=1}^{N} \exp[-2 \cdot \ell(y_i, \mathbf{x}_i; \mathbf{w})] = \prod_{i=1}^{N} \exp[-2 \max(|\mathbf{w}^T \mathbf{x}_i - y_i| - \epsilon, 0)],$$

where ϵ is a margin of tolerance. It has been shown that $p(\mathbf{y}|\mathbf{w}, \mathbf{X})$ admits a dual scale mixture of normals representation by data augmentation [25], such that

$$p(\mathbf{y}|\mathbf{w}, \mathbf{X}) = \prod_{i=1}^{N} \int_0^\infty \exp[\frac{(\lambda_i + \mathbf{w}^T \mathbf{x}_i - y_i - \epsilon)^2}{-2\lambda_i}] \frac{d\lambda_i}{\sqrt{2\pi\lambda_i}} \cdot \int_0^\infty \exp[\frac{(\theta_i - \mathbf{w}^T \mathbf{x}_i + y_i - \epsilon)^2}{-2\theta_i}] \frac{d\theta_i}{\sqrt{2\pi\theta_i}},$$

where λ_i and θ_i ($i = 1, \ldots, N$) are the augmented variables introduced to deal with the max function. Let $\boldsymbol{\lambda} = [\lambda_1, \ldots, \lambda_N]^T$, $\boldsymbol{\theta} = [\theta_1, \ldots, \theta_N]^T$, then the unnormalized joint distribution of $\mathbf{y}$, $\boldsymbol{\lambda}$ and $\boldsymbol{\theta}$ can be expressed as

$$p(\mathbf{y}, \boldsymbol{\lambda}, \boldsymbol{\theta}|\mathbf{w}, \mathbf{X}) = \prod_{i=1}^{N} \frac{1}{\sqrt{2\pi\lambda_i}} \exp[\frac{(\lambda_i + \mathbf{w}^T \mathbf{x}_i - y_i - \epsilon)^2}{-2\lambda_i}] \cdot \frac{1}{\sqrt{2\pi\theta_i}} \exp[\frac{(\theta_i - \mathbf{w}^T \mathbf{x}_i + y_i - \epsilon)^2}{-2\theta_i}].$$

This allows us to regard the pseudo posterior distribution as the marginal of the augmented pseudo posterior distribution which has the form

$$p(\mathbf{w}, \mathbf{z}, \boldsymbol{\lambda}, \boldsymbol{\theta}|\mathbf{y}, \mathbf{X}) \propto p(\mathbf{y}, \boldsymbol{\lambda}, \boldsymbol{\theta}|\mathbf{X}, \mathbf{w})p(\mathbf{w}|\mathbf{z})p(\mathbf{z}).$$

Directly solving for this augmented pseudo posterior is intractable, and previous approximate inference for probabilistic ϵ-insensitive loss relys on Gibbs sampling [25], which is inefficient for large data sets. Here we employ the mean field variational inference method, which has attractive computational properties along with good estimation performance. Specifically, we assume there are a family of fully factorized but free-form variational distributions $q(\mathbf{w}, \mathbf{z}, \boldsymbol{\lambda}, \boldsymbol{\theta}) = q(\mathbf{w})q(\mathbf{z})q(\boldsymbol{\lambda})q(\boldsymbol{\theta})$, and our goal is to minimize the Kullback-Leibler divergence $\mathrm{KL}(q(\mathbf{w}, \mathbf{z}, \boldsymbol{\lambda}, \boldsymbol{\theta}) \| p(\mathbf{w}, \mathbf{z}, \boldsymbol{\lambda}, \boldsymbol{\theta} | \mathbf{X}, \mathbf{y}))$ between the approximating distribution and the target posterior. To this end, we first initialize the moments of all factor distributions $q(\mathbf{w})$, $q(\mathbf{z})$, $q(\boldsymbol{\lambda})$ and $q(\boldsymbol{\theta})$ appropriately and then iteratively optimize each of the factors in turn using the current estimates for all of the other factors. Convergence is guaranteed because the KL divergence is convex with respect to each of the factors. It can be shown that when keeping all other factors fixed, the optimal distribution $q^*(\mathbf{w})$ satisfies

$$q^*(\mathbf{w}) \propto \exp\langle \ln p(\mathbf{w}, \mathbf{z}, \boldsymbol{\lambda}, \boldsymbol{\theta}, \mathbf{X}, \mathbf{y}) \rangle_{-\mathbf{w}}, \tag{3}$$

where $\langle \cdot \rangle_{-\mathbf{w}}$ denotes the expectation of the term inside the angled brackets with respect to $q(\mathbf{w}, \mathbf{z}, \boldsymbol{\lambda}, \boldsymbol{\theta})$ over all variables except for $\mathbf{w}$. Here $p(\mathbf{w}, \mathbf{z}, \boldsymbol{\lambda}, \boldsymbol{\theta}, \mathbf{X}, \mathbf{y})$ is the joint distribution of all observed and latent variables with the form

$$p(\mathbf{w}, \mathbf{z}, \boldsymbol{\lambda}, \boldsymbol{\theta}, \mathbf{X}, \mathbf{y}) = p(\mathbf{y}, \boldsymbol{\lambda}, \boldsymbol{\theta} | \mathbf{w}, \mathbf{X})p(\mathbf{w}|\mathbf{z})p(\mathbf{X})p(\mathbf{z}).$$

Plugging all involved quantities into Eq. 3, we can further get:

$$q^*(\mathbf{w}) = \mathcal{N}\left(\mathbf{w}|\langle\mathbf{w}\rangle, (\mathbf{X}\Lambda_{\boldsymbol{\lambda}\boldsymbol{\theta}}\mathbf{X}^T + \Lambda_{\mathbf{z}})^{-1}\right),$$
$$\langle\mathbf{w}\rangle = \boldsymbol{\Sigma}_{\mathbf{w}} \cdot \{\mathbf{X}[\mathbf{y} \odot (\langle 1 \oslash \boldsymbol{\lambda}\rangle + \langle 1 \oslash \boldsymbol{\theta}\rangle) + \epsilon(\langle 1 \oslash \boldsymbol{\lambda}\rangle - \langle 1 \oslash \boldsymbol{\theta}\rangle)]\},$$

where $\Lambda_{\boldsymbol{\lambda}\boldsymbol{\theta}} = \mathrm{diag}((\langle\lambda_i^{-1}\rangle + \langle\theta_i^{-1}\rangle))$, i.e., the diagonal elements in matrix $\Lambda_{\boldsymbol{\lambda}\boldsymbol{\theta}}$ are $((\langle\lambda_i^{-1}\rangle + \langle\theta_i^{-1}\rangle))$ $(i = 1, \dots, N)$, while the other elements are all zero; $\Lambda_{\mathbf{z}} = \mathrm{diag}((\langle z_g\rangle))$, i.e., every $\langle z_g\rangle$ repeats d_g $(g = 1, \dots, G)$ times in the diagonal position of matrix $\Lambda_{\mathbf{z}}$; $\odot$ and $\oslash$ denote element-wise multiplication and division, respectively. Similarly, we can get the optimal distributions $q^*(\mathbf{z})$, $q^*(\boldsymbol{\lambda})$ and $q^*(\boldsymbol{\theta})$ as:

$$q^*(\mathbf{z}) = \prod_{g=1}^{G} \Gamma\left(z_g | \alpha_g + d_g/2, \beta_g + \langle\mathbf{w}_g^T\mathbf{w}_g\rangle/2\right),$$
$$q^*(\boldsymbol{\lambda}) = \prod_{i=1}^{N} \mathcal{GIG}\left(\lambda_i | 1/2, 1, \langle(\mathbf{w}^T\mathbf{x}_i - y_i - \epsilon)^2\rangle\right),$$
$$q^*(\boldsymbol{\theta}) = \prod_{i=1}^{N} \mathcal{GIG}\left(\theta_i | 1/2, 1, \langle(y_i - \mathbf{w}^T\mathbf{x}_i - \epsilon)^2\rangle\right),$$

where $\mathcal{GIG}(\cdot)$ denotes the generalized inverse Gaussian distribution.

After the above variational inference procedure converges, the predicted response for a new data point $\mathbf{x}_{new}$ can be computed by $y_{new} = \langle\mathbf{w}\rangle^T\mathbf{x}_{new}$. It should be noted that once the model hyper-parameters $\boldsymbol{\alpha} \in \mathbb{R}^{G \times 1}$ and $\boldsymbol{\beta} \in \mathbb{R}^{G \times 1}$ have been tuned carefully, BGFS-SVR can infer its model parameters automatically via Bayesian inference. Compared with the non-Bayesian group sparse models whose regularization parameters need to be re-tuned for different data sets, BGFS-SVR can apply the chosen hyper-parameters to all data sets. Therefore our approach provides the user a simple way to select the most relevant feature groups.

BGFS-SVM. For binary classification model $(y_i \in \{+1, -1\}, \; i = 1, \ldots, N)$, we utilize hinge loss to re-express Eq. 2 such that it has the following form

$$p(\mathbf{y}|\mathbf{w}, \mathbf{X}) = \prod_{i=1}^{N} \exp[-2 \cdot \ell(y_i, \mathbf{x}_i; \mathbf{w})] = \prod_{i=1}^{N} \exp[-2 \max(1 - y_i \mathbf{w}^T \mathbf{x}_i, 0)].$$

To deal with the max function, we still need to transform $p(\mathbf{y}|\mathbf{w}, \mathbf{X})$ into a tractable formulation. Fortunately, the Gaussian pseudo likelihood $p(\mathbf{y}|\mathbf{w}, \mathbf{X})$ can be re-expressed as a location-scale mixture of normals by data augmentation [12], such that

$$p(\mathbf{y}|\mathbf{w}, \mathbf{X}) = \prod_{i=1}^{N} \int_{0}^{\infty} 1/\sqrt{2\pi\lambda_i} \cdot \exp[-(\lambda_i + 1 - y_i \mathbf{w}^T \mathbf{x}_i)^2/(2\lambda_i)] \, d\lambda_i,$$

where λ_i $(i = 1, \ldots, N)$ are the augmented variables. This data augmentation idea provides an elegant way to incorporate max-margin principle into Bayesian learning.

Let $\boldsymbol{\lambda} = [\lambda_1, \ldots, \lambda_N]^T$, then the joint distribution of $\mathbf{y}$ and $\boldsymbol{\lambda}$ can be expressed as

$$p(\mathbf{y}, \boldsymbol{\lambda}|\mathbf{w}, \mathbf{X}) = \prod_{i=1}^{N} 1/\sqrt{2\pi\lambda_i} \cdot \exp[-(\lambda_i + 1 - y_i \mathbf{w}^T \mathbf{x}_i)^2/(2\lambda_i)],$$

and the augmented pseudo posterior distribution satisfies

$$p(\mathbf{w}, \mathbf{z}, \boldsymbol{\lambda}|\mathbf{X}, \mathbf{y}) \propto p(\mathbf{y}, \boldsymbol{\lambda}|\mathbf{w}, \mathbf{X})p(\mathbf{w}|\mathbf{z})p(\mathbf{z}).$$

Since exact inference for this augmented pseudo posterior is intractable, here we also use the mean field variational inference to approximate $p(\mathbf{w}, \mathbf{z}, \boldsymbol{\lambda}|\mathbf{X}, \mathbf{y})$. The derivation procedure is similar as that in BGFS-SVR, so we only provide the optimal distributions for the latent and augmented variables:

$$q^*(\mathbf{w}) = \mathcal{N}\left(\mathbf{w}|\Sigma_{\mathbf{w}} \cdot \{\mathbf{X}[\mathbf{y} \odot (1 + \langle 1 \oslash \boldsymbol{\lambda}\rangle)]\}, (\mathbf{X}\Lambda_{\boldsymbol{\lambda}}\mathbf{X}^T + \Lambda_{\mathbf{z}})^{-1}\right),$$

$$q^*(\mathbf{z}) = \prod_{g=1}^{G} \Gamma\left(z_g|\alpha_g + d_g/2, \beta_g + \langle \mathbf{w}_g^T \mathbf{w}_g \rangle/2\right),$$

$$q^*(\boldsymbol{\lambda}) = \prod_{i=1}^{N} \mathcal{GIG}\left(\lambda_i|1/2, 1, \langle(1 - y_i \mathbf{w}^T \mathbf{x}_i)^2\rangle\right),$$

where $\Lambda_{\boldsymbol{\lambda}} = \mathrm{diag}(\langle \lambda_i^{-1}\rangle)$.

After this procedure converges, we can predict the label y_{new} $(y_{new} \in \{+1, -1\})$ for a new data point $\mathbf{x}_{new}$ as $y_{new} = \mathrm{sgn}(\langle \mathbf{w}\rangle^T \mathbf{x}_{new})$, where $\mathrm{sgn}(\cdot)$ denotes the signum function.

Note that the hyper-parameter $\mathbf{z}$ in the proposed framework plays a similar role as the regularization parameter ρ in group sparse models and SVL machines, but $\mathbf{z}$ is more flexible. Since each group of $\mathbf{w}$ is controlled by distinct z_g, $\mathbf{z}$ can be seen as a generalization of ρ. With Bayesian inference, our models infer $\mathbf{z}$ automatically from the data such that the time-consuming cross-validation can be circumvented.

3.3 Computational Complexity

The computational complexities of BGFS-SVR and BGFS-SVM are the same. For each iteration of our variational inference on training data, we need $O(d^3 + dN)$ computation, where $O(d^3)$ is spent on the inversion of covariance matrix $\Sigma_{\mathbf{w}}$. Given d, our method scales linearly in the number of data samples, making it suitable for large-scale data set.

4   Experiments

To evaluate our models, we conduct a series of experiments on both synthetic and real-world data sets. On each data set, we perform 100 independent trials for our algorithm and various competitors, and the averaged results are reported. The parameters and hyper-parameters of all algorithms were carefully tuned on our data. More specifically, the hyper-parameters of the proposed models and the other Bayesian competitors were tuned manually. For Bayesian methods, once the hyper-parameters have been tuned carefully, they can be used to infer the model parameters automatically from the data. So, in our experiments the chosen hyper-parameters were applied to all data sets for regression and classification, respectively. On the other hand, the regularization parameters of the non-Bayesian methods must be re-tuned for different data sets, so we obtained them by conducting a 5-fold cross-validation on each training data set.

4.1   Regression

We compare the proposed BGFS-SVR with the following algorithms: SVR, Group-Lasso (G-Lasso) [23], Bayesian Group-Lasso (BG-Lasso) [13], Bayesian Group-Sparse model with Jeffreys prior (Jeffreys) [2], a model also assumes the group ARD prior but is based on square loss (G-ARD). We use the LIBLINEAR package [5] for SVR, and the SLEP package [10] for G-Lasso. For BGFS-SVR, we set the hyper-parameters $\alpha_g = 10^3$, $\beta_g = 10^{-3}$ $(g = 1, \ldots, G)$, the maximum number of iterations T = 100, and we vary the tolerance parameter ϵ in $\{10^{-5}, 10^{-4}, \ldots, 1\}$. For BG-Lasso we assign a Gamma prior with shape and scale hyper-parameters $k = 10^{-6}$ and $\theta = 10^{-6}$ on a and then integrate out a, where a is the parameter of multivariate Laplace distribution. For G-ARD, we set the hyper-parameters $\alpha_g = 1$ and $\beta_g = 10^{-5}$ $(g = 1, \ldots, G)$. For SVR, we vary the regularization parameter C in $\{e^{-8}, e^{-6}, \ldots, e^8\}$ and the tolerance parameter ϵ in $\{10^{-5}, 10^{-4}, \ldots, 1\}$. Finally, for G-Lasso we vary the regularization parameter ρ in $\{10^{-4}, 10^{-3}, \ldots, 10^3\}$.

Synthetic Data. The performance of BGFS-SVR is first tested on a sparse signal reconstruction problem. We generate the synthetic data similar as in [2]. More precisely, we assume the length of weight vector (signal) $\mathbf{w}$ is $d = 500$ and fix the group size to 20 $(d_g = 20, \forall g)$, such that the number of groups is $G = 25$. We assign the 20 adjacent components into one group. From the 25 groups,

Table 1. Average relative reconstruction error on the synthetic regression data.

N/d	G-Lasso	BG-Lasso	Jeffreys	G-ARD	SVR	BGFS-SVR
0.10	0.96 ± 0.04	0.96 ± 0.03	0.95 ± 0.05	0.95 ± 0.03	0.98 ± 0.01	$\mathbf{0.93 \pm 0.04}$
0.15	0.85 ± 0.05	0.87 ± 0.05	0.84 ± 0.06	0.84 ± 0.06	0.92 ± 0.01	$\mathbf{0.74 \pm 0.06}$
0.20	0.68 ± 0.06	0.70 ± 0.08	0.68 ± 0.08	0.67 ± 0.08	0.89 ± 0.02	$\mathbf{0.63 \pm 0.07}$

only 5 randomly chosen groups contain components that generated from the standard Gaussian distribution with zero mean and unit standard deviation, and the remaining groups of components are zeros. The $d \times N$ measurement matrix $\mathbf{X}$ is generated by drawing its elements from a standard Gaussian distribution and then normalizing the rows to have unit l_2-norm. The $N \times 1$ observations vector $\mathbf{y}$ is generated by $\mathbf{y} = \mathbf{X}^T \mathbf{w} + \mathbf{n}$, where $\mathbf{n} = [n_1, \ldots, n_N]^T \sim \mathcal{N}(\mathbf{0}, \sigma_n^2 \mathbf{I})$ are Gaussian noise with $\sigma_n = 10^{-2}$.

Given $\mathbf{w}$, $\mathbf{X}$ and $\mathbf{y}$, we use the Relative Reconstruction Error (RRE) $\|\hat{\mathbf{w}} - \mathbf{w}\|_2 / \|\mathbf{w}\|_2$ to evaluate the reconstruction performance of different methods. Here $\hat{\mathbf{w}}$ and $\mathbf{w}$ denote the estimated signal and the true signal, respectively. The experiments are independently repeated 100 times, i.e. in every time we have different realizations of the signal $\mathbf{w}$, the measurement matrix $\mathbf{X}$ and the observations vector $\mathbf{y}$.

The average RREs of all methods with varying N/d ratios are presented in Table 1, from which we can get two observations. First, the proposed BGFS-SVR model performs better than those square loss based group sparse modeling approaches. Second, the baseline method SVR obtains a significantly worse reconstruction error which is due to the fact that SVR cannot make use of the grouping information.

Gas Sensor Array Data. In order to evaluate the performance of BGFS-SVR in real-world applications, we apply it to the Gas Sensor Array (GSA) data, which is available in the UCI repository. This data set contains $13,910$ instances from 16 chemical sensors exposed to 6 gases at different concentration levels. Each instance vector contains the 128 features extracted from 16 sensors. We regard those features from the same sensor as a group such that the number of groups is $G = 16$, and the group size is $d_g = 8$ for $g = 1, \ldots, G$. We utilize all 3600 instances gathered in the 36-th month (600 instances of each class), and the goal is to predict the concentration levels as accurate as possible. First, the data set is normalized so that each row of the data matrix $\mathbf{X} \in \mathbb{R}^{d \times N}$ has unit l_2 norm. Then, we randomly select training instances from each class with the size in $\{100, 150, 200, 250, 300, 350, 400\}$, and the rest instances are used as test set. We do experiments for each class, then the averaged root-mean-square errors (RMSE) over these six regression problems are shown in Fig. 1, and the averaged numbers of selected groups are listed in Table 2.

We can observe that BGFS-SVR obviously outperforms other approaches in term of the RMSE, while keeping the number of selected groups comparable with

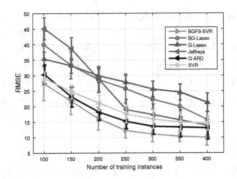

Fig. 1. Average RMSE of various regression methods on GSA data.

Table 2. Average number of selected groups for various regression methods on GSA data.

#Training data	G-Lasso	BG-Lasso	Jeffreys	G-ARD	SVR	BGFS-SVR
100	13.1 ± 1.4	14.2 ± 0.9	14.6 ± 1.1	13.9 ± 1.7	16.0 ± 0	12.4 ± 1.3
400	14.1 ± 1.0	15.9 ± 0.2	15.7 ± 0.5	15.4 ± 1.3	16.0 ± 0	15.2 ± 0.7

Table 3. Average predictive accuracy and F1 score on the synthetic classification data.

#Training data	Accuracy (%)				F1 (%)			
	G-Lasso	SG-Lasso	SVM	BGFS-SVM	G-Lasso	SG-Lasso	SVM	BGFS-SVM
100	59.9 ± 6.9	61.1 ± 6.0	56.3 ± 5.7	$\mathbf{64.3 \pm 7.8}$	49.2 ± 6.3	54.5 ± 5.5	52.3 ± 5.2	$\mathbf{57.7 \pm 8.2}$
500	72.4 ± 4.8	76.2 ± 5.0	77.6 ± 5.1	$\mathbf{78.0 \pm 6.1}$	70.2 ± 4.2	73.4 ± 4.5	74.5 ± 4.2	$\mathbf{76.0 \pm 5.8}$
1500	93.6 ± 3.1	95.0 ± 2.3	94.9 ± 2.4	$\mathbf{95.3 \pm 2.3}$	92.2 ± 3.5	93.7 ± 2.8	93.2 ± 3.3	$\mathbf{94.2 \pm 2.7}$

those of other approaches. The advantage of BGFS-SVR over SVR demonstrates the benefit of considering grouping information of features and only retaining useful feature groups for model learning and prediction. The performance differences between BGFS-SVR and those square loss based Group Feature Selection (GFS) approaches verify again that combining SVR and GFS is meaningful.

4.2 Classification

We compare the proposed BGFS-SVM with the following algorithms: SVM, Group-Lasso with logistic loss (G-Lasso) [11], Sparse Group-Lasso with logistic loss (SG-Lasso) [15]. We use the LIBLINEAR package for SVM, and the SLEP package for G-Lasso and SG-Lasso. For BGFS-SVM, we set $\alpha_g = 10^{-3}$, $\beta_g = 10^{-3}$ $(g = 1, \ldots, G)$ and T = 100. For SVM, we vary the regularization parameter C over the grid $\{e^{-8}, e^{-6}, \ldots, e^8\}$. For G-Lasso and SG-Lasso, we vary the regularization parameter ρ in $\{10^{-4}, 10^{-3}, \ldots, 10^3\}$. In the experiments, we increased the size of training data from a small value until it is large enough such that most algorithms perform well. This is to study the performance change of each method as training data increase.

Synthetic Data. We generate the synthetic data set similar to [11,22,23]. Before generating the data, we first generate a true model vector $\mathbf{w}$ consisting of ten blocks of ten dimensions, i.e., $\mathbf{w} \in \mathbb{R}^{100}$, $G = 10$, $d_g = 10$ for $g = 1, \ldots, G$. The numbers of non-zero weights in the first six blocks are $10, 8, 6, 4, 2, 1$ respectively, with their values chosen at random from $\{+1, -1\}$. All other elements in the first six blocks and the remaining four blocks are set to be zero. Then, we generate N data points $\mathbf{x}_i$, $i = 1, \ldots, N$ by $\mathbf{x}_i = L\mathbf{v}_i$, where $\mathbf{v}_i \sim \mathcal{N}(\mathbf{0}, \mathbf{I}_d)$ and L is the Cholesky decomposition of the correlation matrix Σ. The (i,j)-th entry in the g-th group of Σ is $\Sigma_{i,j}^g = 0.2^{|i-j|}$, and entries involves different groups are all zero. We finally get the response y_i by $y_i = \text{sgn}(\mathbf{w}^T \mathbf{x}_i + \sigma_o)$, where σ_o is a Gaussian noise with standard deviation 4.

We randomly generate the training data with size in $\{100, 500, 1500\}$, and the goal is to predict the weights $\mathbf{w}_i \in \{+1, -1, 0\}$, $i = 1, \ldots, 100$. Let the component $\hat{\mathbf{w}}_i = 0$ if $|\hat{\mathbf{w}}_i| < 0.01$, and then regard $\text{sgn}(\hat{\mathbf{w}})$ as the final estimated weight vector. Table 3 lists the averaged results in term of accuracy and the F1 score, which aim to verify whether the learned weight has the same sign as the true model weight. We calculate the F1 scores on the tasks of $+1$ vs.$\{-1, 0\}$, -1 vs.$\{+1, 0\}$, and 0 vs.$\{+1, -1\}$ and average these three F1 scores. A larger F1 indicates a better estimation.

Table 3 shows that the accuracies and F1 scores of all algorithms increase with the number of training instances. Among them, BGFS-SVM gets the best results, and the performance difference is especially prominent when the number of training instances is small. Furthermore, compared with G-Lasso and SG-Lasso methods which are based on logistic loss, BGFS-SVM is based on the popular maximum margin principle, thus tend to have a good generalization performance when the number of training instances is small. Finally, BGFS-SVM is more powerful than SVM, because it can make use of the structure information among features and exclude the redundant and non-informative features better. This is in accord with the conclusion in regression analysis, i.e., we can recover the true weights more accurate by combining both the advantages of group sparse modeling and the maximum margin modeling.

Real-World Data. The GSA data set has been described in regression experiment. We utilize all instances gathered in the 36-th month and our goal is to discriminate six different analytes regardless of their concentration. The Smartphones data set [1] built from the recordings of 30 subjects performing activities of daily living has been used for human activity recognition. It contains totally 10299 instances from 6 categories. Feature vectors in Smartphones data set are obtained by calculating variables from both the time and frequency domain, and all features can be grouped into 18 groups according to the variable types. The USPS data set [8] contains totally 9298 handwritten digits from 10 categories (i.e., 0, 1, ..., 9), and each digit is represented as a 16 by 16 matrix. Inspired by the fact that people can recognize a digit even when it loses some columns of pixels, we group these 256 pixels into 16 columns in the experiments.

Table 4. Average predictive accuracy (%) on three real-world classification data sets.

Data set	Training data	G-Lasso	SG-Lasso	SVM	BGFS-SVM
GSA	50	79.8 ± 4.4	83.0 ± 4.8	83.3 ± 2.6	**84.3 ± 1.7**
	100	84.3 ± 3.6	85.3 ± 4.0	87.6 ± 2.1	**92.0 ± 1.6**
	200	86.4 ± 3.5	88.3 ± 2.7	96.7 ± 0.6	**98.1 ± 0.8**
Smartphones	50	57.5 ± 3.5	58.2 ± 3.5	**83.2 ± 1.2**	82.1 ± 1.6
	100	71.3 ± 4.1	72.3 ± 2.8	90.0 ± 0.7	**92.6 ± 0.8**
	200	73.3 ± 2.7	74.1 ± 2.8	93.5 ± 0.4	**95.6 ± 0.3**
USPS	50	69.0 ± 1.3	71.5 ± 1.5	85.7 ± 0.6	**86.6 ± 1.2**
	100	69.6 ± 1.4	72.2 ± 0.9	89.4 ± 0.4	**90.7 ± 0.4**
	200	72.7 ± 1.2	74.1 ± 1.2	91.5 ± 0.3	**92.5 ± 0.3**

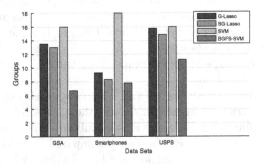

Fig. 2. Average number of selected groups for various classification methods.

For multi-class classification, we choose the method proposed in [3] for SVM, and *one vs. rest* strategy for BGFS-SVM, G-Lasso and SG-Lasso. All data sets are normalized so that each row of data matrix $\mathbf{X} \in \mathbb{R}^{d \times N}$ has unit l_2 norm. For each data set, we randomly select training instances from each category with the size in $\{50, 100, 200\}$, and the rest instances are used as test set. The averaged predictive accuracies of various methods are reported in Table 4, and the averaged numbers of selected groups when the number of training instances in each category is 200 are shown in Fig. 2 (we calculate the average number of selected groups over all binary classification sub-problems).

Several observations can be drawn from Table 4 and Fig. 2. First, the BGFS-SVM and SVM prominently outperform G-Lasso and SG-Lasso approaches on all three data sets. This can be related to the fact that the maximum margin principle generally yields better generalization ability. Second, the numbers of selected groups of our approach are obviously less than those of other approaches, which indicates that our model is more effective in identifying and selecting the most relevant groups for classification. Finally, although SVM obtains a comparable performance to our BGFS-SVM, it cannot identify and select the most relevant feature groups which can improve the interpretability of the learned model parameters and reduce the complexity of computing.

5   Conclusion

We have presented a new Bayesian Group Feature Selection (GFS) framework for Support Vector Learning (SVL) machines based on the pseudo likelihood and data augmentation idea. Compared with traditional GFS models, our SVL based approach generally yields better generalization performance, and it can circumvent the cross-validation for regularization parameters with Bayesian inference. Extensive experimental results demonstrated that our proposed models are superior to several existing GFS models and the direct SVL models. There still exist some future work, e.g., to impose sparsity restriction on both the groups and the individual features, to take into account the group overlapping, and to estimate the group partition automatically when it is unknown.

Acknowledgments. This work was supported by the National Natural Science Foundation of China (No. 9154610306, 61573335, 61473273, 61473274, 11390371, 11233004), National Key Basic Research Program of China (Grant No. 2014CB845700), National High-tech R&D Program of China (863 Program) (No. 2014AA015105), Guangdong provincial science and technology plan projects (No. 2015 B010109005).

References

1. Anguita, D., Ghio, A., Oneto, L., Parra, X., Reyes-Ortiz, J.L.: A public domain dataset for human activity recognition using smartphones. In: ESANN (2013)
2. Babacan, S.D., Nakajima, S., Do, M.N.: Bayesian group-sparse modeling and variational inference. IEEE Trans. Sig. Process **62**(11), 2906–2921 (2014)
3. Crammer, K., Singer, Y.: On the algorithmic implementation of multiclass kernel-based vector machines. J. Mach. Learn. Res. **2**, 265–292 (2002)
4. Drucker, H., Burges, C.J., Kaufman, L., Smola, A., Vapnik, V.: Support vector regression machines. In: NIPS, pp. 155–161 (1997)
5. Fan, R.E., Chang, K.W., Hsieh, C.J., Wang, X.R., Lin, C.J.: Liblinear: a library for large linear classification. J. Mach. Learn. Res. **9**, 1871–1874 (2008)
6. Hall, D.L., McMullen, S.A.: Mathematical Techniques in Multisensor Data Fusion. Artech House, Norwood (2004)
7. Hernández-Lobato, D., Hernández-Lobato, J.M., Dupont, P.: Generalized spike-and-slab priors for bayesian group feature selection using expectation propagation. J. Mach. Learn. Res. **14**(1), 1891–1945 (2013)
8. Hull, J.: A database for handwritten text recognition research. IEEE Trans. PAMI **16**(5), 550–554 (1994)
9. Jacob, L., Obozinski, G., Vert, J.P.: Group lasso with overlap and graph lasso. In: ICML, pp. 433–440 (2009)
10. Liu, J., Ji, S., Ye, J.: Slep: Sparse Learning with Efficient Projections. Arizona State University, Tempe (2009)
11. Meier, L., Van De Geer, S., Bühlmann, P.: The group lasso for logistic regression. J. R. Stat. Soc. Ser. B (Stat. Methodol.) **70**(1), 53–71 (2008)
12. Polson, N.G., Scott, S.L.: Data augmentation for support vector machines. Bayesian Anal. **6**(1), 1–23 (2011)
13. Raman, S., Fuchs, T.J., Wild, P.J., Dahl, E., Roth, V.: The bayesian group-lasso for analyzing contingency tables. In: ICML, pp. 881–888 (2009)

14. Roth, V., Fischer, B.: The group-lasso for generalized linear models: uniqueness of solutions and efficient algorithms. In: ICML, pp. 848–855 (2008)

15. Simon, N., Friedman, J., Hastie, T., Tibshirani, R.: A sparse-group lasso. J. Comput. Graph. Stat. **22**(2), 231–245 (2013)

16. Subrahmanya, N., Shin, Y.C.: Sparse multiple kernel learning for signal processing applications. IEEE Trans. PAMI **32**(5), 788–798 (2010)

17. Subrahmanya, N., Shin, Y.C.: A variational bayesian framework for group feature selection. Int. J. Mach. Learn. Cybern. **4**(6), 609–619 (2013)

18. Tan, M., Wang, L., Tsang, I.W.: Learning sparse svm for feature selection on very high dimensional datasets. In: ICML, pp. 1047–1054 (2010)

19. Tibshirani, R.: Regression shrinkage and selection via the lasso. J. R. Stat. Soc. Ser. B (Methodol.) **58**(1), 267–288 (1996)

20. Tipping, M.E.: Sparse bayesian learning and the relevance vector machine. J. Mach. Learn. Res. **1**, 211–244 (2001)

21. Wang, J., Zhao, Z.Q., Hu, X., Cheung, Y.M., Wang, M., Wu, X.: Online group feature selection. In: IJCAI, pp. 1757–1763 (2013)

22. Yang, H., Xu, Z., King, I., Lyu, M.R.: Online learning for group lasso. In: ICML, pp. 1191–1198 (2010)

23. Yuan, M., Lin, Y.: Model selection and estimation in regression with grouped variables. J. R. Stat. Soc. Series B **68**(1), 49–67 (2006)

24. Zhu, J., Rosset, S., Hastie, T., Tibshirani, R.: 1-norm support vector machines. In: NIPS, pp. 49–56 (2004)

25. Zhu, J., Chen, N., Perkins, H., Zhang, B.: Gibbs max-margin topic models with data augmentation. J. Mach. Learn. Res. **15**, 1073–1110 (2014)

Active Distance-Based Clustering
Using K-Medoids

Amin Aghaee[(✉)], Mehrdad Ghadiri[(✉)], and Mahdieh Soleymani Baghshah

Department of Computer Engineering, Sharif University of Technology, Tehran, Iran
{aghaee,ghadiri}@ce.sharif.edu, soleymani@sharif.edu

Abstract. k-medoids algorithm is a partitional, centroid-based clustering algorithm which uses pairwise distances of data points and tries to directly decompose the dataset with n points into a set of k disjoint clusters. However, k-medoids itself requires all distances between data points that are not so easy to get in many applications. In this paper, we introduce a new method which requires only a small proportion of the whole set of distances and makes an effort to estimate an upper-bound for unknown distances using the inquired ones. This algorithm makes use of the triangle inequality to calculate an upper-bound estimation of the unknown distances. Our method is built upon a recursive approach to cluster objects and to choose some points actively from each bunch of data and acquire the distances between these prominent points from oracle. Experimental results show that the proposed method using only a small subset of the distances can find proper clustering on many real-world and synthetic datasets.

Keywords: Active k-medoids · Active clustering · Distance-based clustering · Centroid-based clustering

1 Introduction

As the production of data is expanding at an astonishing rate and the era of big data is coming, organizing data via assigning items into groups is inevitable. Data clustering algorithms try to find clusters of objects in such a way that the objects in the same cluster are more similar to each other than to those in other clusters. Nowadays clustering algorithms are widely used in data mining tasks.

There are different categorization for clustering algorithms, e.g., these algorithms can be categorized into Density-based, Centroid-based and Distribution-based methods. In centroid-based clustering methods, each cluster is shown by a central object. This object which can be a member of the dataset denotes a prototype of the whole cluster. When these algorithms are appointed to find K clusters, they usually find K central objects and assign each element to the nearest centroid. As they go on, they attempt to decrease the energy and total error of clusters by finding better central elements. *K-medoids* and *K-means* are the

---

A. Aghaee and M. Ghadiri—Equally Contributed Authors.

© Springer International Publishing Switzerland 2016
J. Bailey et al. (Eds.): PAKDD 2016, Part I, LNAI 9651, pp. 253–264, 2016.
DOI: 10.1007/978-3-319-31753-3_21

two most popular centroid-based algorithms. Although, they both partition the data into groups such that the sum of the squared distances of the data points to the nearest center to them is minimized, they have different assumptions about centroids. Indeed, the k-medoids algorithm chooses the centroids only from the data points and so these centroids are members of the whole dataset while k-means algorithm can select the centroids from the whole input space.

In [3], some usages and applications of k-medoids algorithm are discussed. According to this study, in resource allocation problems, when a company wants to open some branches in a city, in such a way that the average distance from each residential block to the closest branch is intended to be minimized, the k-medoids algorithm is a proper option. Additionally, in mobile computing context, it is an issue to save communication cost when devices need to choose super-nodes among each other, which should have minimum average distances to all devices and k-medoids can solve this problem. Furthermore, as reported by [3], medoid queries also arise in the sensor networks and many other fields.

Active learning is a machine learning field that have bee attended specially in the last decade. Until now, many active methods for supervised learning that intend to select more informative samples to be labeled have been proposed. Active unsupervised methods have also been received attention recently. In an unsupervised learning manner, finding the similarities or distances of samples from each other may be difficult or infeasible. For example, the sequence similarity of proteins [23] or similarity of face images [7] which needs to be obtained from human as an oracle, may be difficult to be responded. The active version of some of the well known clustering algorithms have been recently presented in [14, 26].

In this paper, we propose the active k-medoid algorithm that inquires a subset of pairwise distances to find the clustering of data. We use a bottom-up approach to find more informative subset of the distances to be inquired. Extensive experiments on several data sets show that our algorithm usually needs a few percentage of the pairwise distances to cluster data properly.

In the rest of this paper, we first discuss about the works that have been done in the field of active clustering in Sect. 2. In Sect. 3, we introduce our algorithm. The result of experiments on different datasets have been presented in Sect. 4. At last, we discuss about some aspects of our algorithm and conclude the paper in Sect. 5.

2   Related Work

Active learning is a machine learning paradigm that endeavors to do learning with asking labels of a few number of samples which are more important in the final result of learning. Indeed, most of supervised learning algorithms need a large amount of labeled samples and gathering these labeled samples may need unreasonable amount of time and effort. Thus, active learning tries to ask labels for more important samples where important samples may be interpreted as most informative ones, most uncertain ones, or the ones that have a large effect in the results [18]. The active clustering problem has been recently received much attention. Until now, the active version of some well known clustering methods has been proposed in [14,26]. In the active clustering problem, a query is a pair of data whose similarity must be determined. The purpose of the active learning approach is reducing the number of required queries via active selection of them instead of random selection [21].

The existing active clustering methods can be categorized into constraint-based and distance-based ones [24]. In the most of the constraint-based methods, must-link and cannot-link constraints on pairs of data points indicating these pairs must be in the same cluster or different clusters are inquired. Some constraint-based methods for active clustering have been proposed in [6,7,11,24–26,28]. In distance-based methods, the response to a query on a pair of data points is the distance of that pair according to an objective function. Distance-based methods for active clustering have been recently attended in [10,13,14,19,23,27].

In [14], an algorithm for active DBSCAN clustering is presented. In this algorithm, the distances that have not been queried are estimated with a lower bound. A score indicating the amount and the probability of changes in the estimated distances by asking a query is used to select queries. Moreover, an updating technique is introduced in [14] that update clustering after a query.

In [19,27], distance-based algorithms are presented for active spectral clustering in which a perturbation theory approach is used to select queries. A constraint-based algorithm has also been presented in [26] for active spectral clustering that uses an approach based on maximum expected error reduction to select queries.

An active clustering method for k-median clustering has also been proposed in [23]. This method selects some points as the landmarks and ask the distances between these landmarks and all the other data points as queries. Finally, k-median clustering is done using these distances.

3 Proposed Method

In this section, we introduce the proposed active k-medoids clustering. We assume that our algorithm intends to partition n samples into k different clusters. As mentioned above, many clustering algorithms such as K-medoids, PAM [12], and some other distance-based methods, calculate an $n \times n$ distance matrix at first and perform the clustering algorithm on this distance matrix. We show the distance matrix by D where d_{ij} denotes the distance between the ith sample and the jth one.

We introduce a method to estimate unknown distances during an active clustering process. In a metric space, a satisfying and eminent upper-bound estimation for any distance metric can be obtained by the triangle inequality. For example, when we know the exact distances between d_{ax}, d_{xy} and d_{yb}, we can determine the upper-bound estimation for d_{ab} as:

$$d_{ab} \leqslant d_{ax} + d_{xy} + d_{yb} \tag{1}$$

We find an upper-bound estimation of the distances using the triangle inequality and the known distances asked from an oracle already. Therefore, we have

$$\forall i, j, 1 \leqslant i, j \leqslant n : D(i, j) \leqslant D_e(i, j) \tag{2}$$

where D_e shows the estimated distances.

First, upper-bound estimations for all distances are infinity and we update these distances by asking some of them and make better estimations for the other unknown distances using the triangle inequality and new distances taken from the oracle. The update will be done by replacing exact values for the asked distances and getting the minimum of the old and the new upper-bound estimation for unknown distances. By asking the landmark distances, we intend to take a better estimation of distances required for the k-medoids algorithm.

Consider some data points which are partitioned to m groups where the distances within each group are known or estimated. However, the distances between data points from different groups are unknown. Our goal is to estimate these unknown distances instead of asking them. In such situation, we can choose t finite points from each group and ask the distances between these mt points between different groups and estimate the other distances using these asked distances. The number of these distances is $\binom{m}{2}t^2$. Figure 1 gives an intuition about this distance estimation method. We want to estimate the distance between a and b and the distances between those points that are connected by solid lines and dotted lines are known.

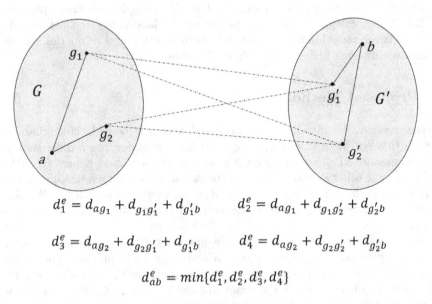

$$d_1^e = d_{ag_1} + d_{g_1 g_1'} + d_{g_1' b} \qquad d_2^e = d_{ag_1} + d_{g_1 g_2'} + d_{g_2' b}$$

$$d_3^e = d_{ag_2} + d_{g_2 g_1'} + d_{g_1' b} \qquad d_4^e = d_{ag_2} + d_{g_2 g_2'} + d_{g_2' b}$$

$$d_{ab}^e = min\{d_1^e, d_2^e, d_3^e, d_4^e\}$$

Fig. 1. Upper-bound distance estimation between a, b. Superscript e determines that the distance is an estimation.

Such estimation algorithm can be done in $O(n^2 t^2)$, when n is the number of data points. If we have $t \ll n$, the time complexity of the algorithm will be $O(n^2)$.

Based on the estimation method used for unknown distances, we present an active k-medoids algorithm. The approach is based on partitioning the data points into some groups by a hierarchical manner. In the other words, we partition data points into b groups and partition each of these groups to b groups and so on until we get to a threshold like t_h for the number of data points in each partition. In this level, we ask all the distances within each group among all its data points and choose t points from each group (to ask their distances) and using the explained algorithm for estimating distances in a bottom-up approach until we get to the highest level. After that, we cluster the data using the k-medoids algorithm on the estimated distances. According to these explanation, it seems that choosing t points in each group is a critical step and choosing a bad point can lead to unfavourable estimations. For this purpose, consider a group of data points like G which its inner distances are known or estimated. In order to choosing t points, we perform k-medoids algorithm on G and find clusters and medoids for this group. Then, we choose medoids and s random points from each cluster as the points whose distances are needed to be asked. Therefore, the number of the chosen data points in G will be $t = k(s+1)$. It is obvious that a greater s will lead to more accurate estimations.

Algorithm 1 present the pseudo code of the proposed method. This function clusters n data points into k different categories and return clusters of data. Here, b shows the branching factor which is used to partition data points to b different groups with the same size. The partitioning algorithm will perform for each group recursively. There is also a threshold t_h which clarify the minimum size of a group of data points. Clearly, $t_h \geq k(s+1)$ since we need to choose at least $k(s+1)$ points in each group. It is also noteworthy that if $n \leq 2t_h$, we need to ask all distance pairs since these data points cannot be partitioned.

Figure 2 shows an example workflow for Algorithm 1 for 1600 data points with branching factor 2 and threshold 400.

Now we calculate the complexity of our Algorithm 1. It makes a tree with the branching factor b and the threshold t_h for the number of the data points in the leaves of the tree. Therefore, the height of this tree is $\lceil \log_b (n/t_h) \rceil$. The number of nodes in the ith level of the tree is b^i and each node of the ith level has n/b^i data points. According to [20], the time complexity of the k-medoids algorithm is $O(kn^2)$ for n data points and k clusters in each iteration. Consider p as the maximum number of iterations used in the k-medoids algorithm, then the time complexity in each node in the ith level of the tree is $O((n^2/b^{2i})kp)$. Therefore, the overall complexity of Algorithm 1 is

$$O\left(\sum_{i=1}^{\lfloor \log_b (n/t_h) \rfloor + 1} (\frac{n^2}{b^{2i}} kp) b^i \right) = O(n^2 kp). \qquad (3)$$

A major factor that measures the quality of an active clustering algorithm, is the number of distances that the algorithm demands from the oracle. The number of the asked queries in the internal nodes of the tree is $O(b^2(k(s+1))^2)$ and in the leaves is $O(nt_h)$. Since, in the proposed method, there are n/t_h

---

Algorithm 1. Active k-medoids

INPUT: D_e, n, k, b, t_h ▷ distance estimation, #data, #clusters, branch factor, threshold

OUTPUT: $C_1, \ldots, C_k$

1: **procedure** ACTIVEKMEDOIDS(D_e, n, k, b, t_h)
2:    **if** $n \le 2t_h$ **then**
3:       Update D_e by querying all distances
4:       $C_1, \ldots, C_k \leftarrow kmedoids(D_e, k)$ ▷ kmedoids function is a regular k-medoids
5:       **return**
6:    **end if**
7:    Partition data to b equal size groups, like $G_1, \ldots, G_b$
8:    **for** i **from** 1 **to** b **do**
9:       $T_1, \ldots, T_k \leftarrow ActiveKmedoids(D_e(G_i), |G_i|, k, b, t_h)$ ▷ $D_e(G_i)$ is the part of the estimated distance matrix corresponding to G_i
10:       $G_i^c \leftarrow$ medoids of $T_1, \ldots, T_k$ and s random points from each of them.
11:    **end for**
12:    Update D_e by querying distances between all those pairs that one of them is in G_i^c and the other is in G_j^c.
13:    Update D_e by the triangle inequality and new inquired distances.
14:    $C_1, \ldots, C_k \leftarrow kmedoids(D_e, k)$
15: **end procedure**

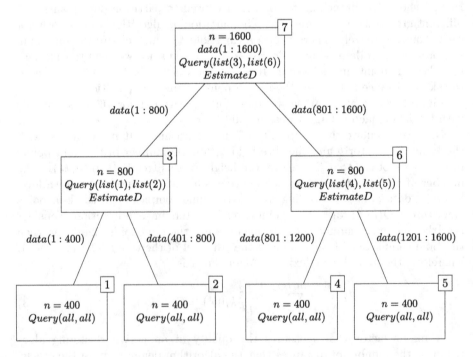

Fig. 2. ActiveKmedoids workflow for 1600 points ($b = 2$, $t_h = 400$)

leaves and each of them has t_h data points, the ratio of the asked distances to all of the distances is almost equal to

$$(b^2(k(s+1))^2 \frac{b^{\lfloor \log_b (n/t_h) \rfloor} - 1}{b-1} + nt_h)/\binom{n}{2} \tag{4}$$

where $(b^{\lfloor \log_b (n/t_h) \rfloor} - 1)/(b-1)$ shows the approximate number of the internal nodes in the tree. This ration can approximately be simplified to the following formula for large values of n:

$$\frac{2b^2(k(s+1))^2}{n(b-1)t_h} + \frac{2t_h}{n} \tag{5}$$

In order to improve the accuracy of the estimated distances, we can increase the s value. However, it may raise running time of calculating upper-bound estimations and the number of queries that should be asked.

To have a good baseline for comparison, we introduce an algorithm based on random selection of the distance queries, called *Random-Rival* and compare results of our algorithm with those of this algorithm. In Sect. 4, we show that asking queries using our method is much better than asking them randomly that is the aim of any active clustering algorithm.

Random-Rival(RR) algorithm, considers data points as the vertices of a weighted graph in which the weight of each edge shows the distance between its endpoints. Firstly, RR asks some distances randomly and then estimates all unknown distances using the triangle inequality and *Floyd-Warshall* [9, p. 693] algorithm. In the other words, we find shortest path distances of all pairs using the available distances in the aforementioned graph. Since our distance function is a distance metric, the length of these shortest paths is an upper-bound estimation of the true distances. Finally, RR clusters data runs the k-medoids algorithm on the estimated distances. Since Floyd-Warshall worst case time complexity is $O(n^3)$ [9, p. 695], we can consider the runtime complexity of RR as $O(n^3)$ at worst. Algorithm 2 shows the pseudo code of the RR algorithm.

Algorithm 2. Random Rival

INPUT: n, k, B ▷ #data, #clusters, budget
OUTPUT: $C_1, \ldots, C_k$
  1: **procedure** RANDOMRIVAL(n, k, B)
  2:      $D \leftarrow n \times n$ infinity matrix
  3:      $(x_1, y_1), \ldots, (x_B, y_B) \leftarrow$ random pairs such that $1 \leqslant x_i < y_i \leq n$
  4:      $\forall i, 1 \leqslant i \leqslant B$: update $D(x_i, y_i)$ by querying distances
  5:      $D_e \leftarrow FloydWarshall(D)$
  6:      $C_1, \ldots, C_k \leftarrow kmedoids(D_e, k)$
  7: **end procedure**

---

4 Empirical Results

In this section, we show the results of our algorithm on some synthesized and real world datasets. General information about these datasets are presented in

Table 1. Most of them are real world datasets, but some are synthesized which are marked with letter s in Table 1. NORM-10 [4] contains 10000 data points having 20 features. This dataset has been generated by choosing 10 real centers uniformly at random from the hypercube of side length 50. Then, for each of the real centers, 1000 points from a Gaussian distribution of variance one centered at the corresponding point is generated. We converted samples in NEC_animal [15] and ALOI200 [1] dataset into 32×32 grayscale images. ALOI [1] (Object Viewpoint version) has 1000 classes but we use only 200 classes of it.

Table 1. General information about datasets

Data	#Samples	#Features	#Classes	#Distances	Ref.
Vary-density(s)	150	2	3	11175	[2]
Seeds	210	7	3	21945	[5]
Mouse(s)	500	2	4	124750	[2]
Fisheriris	150	4	3	11175	[5]
Data_2000(s)	2000	2	5	1999000	[8]
Trace	200	275	4	19900	[29]
Multi-features	2000	649	10	1999000	[5]
TwoDiamonds(s)	800	2	2	319600	[22]
EngyTime(s)	4096	2	2	8386560	[22]
COIL100	7200	1024	100	25916400	[16]
NORM10(s)	10000	20	10	49995000	[4]
NEC_animal	4371	1024	60	9550635	[15]
ALOI200	14400	1024	200	103672800	[1]

Although active version of some clustering algorithms like DBSCAN and spectral clustering have been introduced in [14, 26], these clustering algorithms are substantially different from the k-medoids algorithm. For example, DBSCAN and spectral clustering methods can find clusters of different shapes while k-medoids cannot. Thus, we cannot compare results of our active k-medoid with those of active DBSCAN and active spectral clustering methods. One way to evaluate an active clustering method that asks distances is to compare it with a clustering method that asks a random subset of distances. We compare our method with the Random-Rival algorithm introduced in Sect. 3 that tries to use a random subset of distances to estimate the whole distance matrix (using shortest paths on the graph of data points and the triangle inequality). It must be mentioned that in both the proposed active k-medoid and the Random-Rival algorithm, the clustering algorithm that is run on the obtained distance matrix will be k-medoids. One of the most common measures for comparison of clustering algorithms is *normalized mutual information* (NMI) [17]. This measure shows the agreement of the two assignments, ignoring permutations. In the other words, NMI for the clustering obtained by an algorithm shows the agreement between the obtained grouping by this algorithm and the ground truth grouping of data.

We run our algorithm with $s = 1, 3$ and branching factor $b = 2$ over all the datasets. Threshold t_h is set to the minimum possible value which is equal to the number of classes for each dataset. Greater value of s, branching factor b, or threshold t_h can improve NMI score for some datasets. However, it would also increases the number of queries which is usually quite unsatisfactory. We also run Random-Rival over these datasets which requires a specified proportion of distances starting from 5 % to 10 % (with the step 5 %).

Results of our method with the parameters mentioned in the previous paragraph are shown in Table 2. For each algorithm and each dataset, NMI score and the ratio of the asked distances are presented in the table cells. For the Random-Rival algorithm, the results for the proportion of distances (between 5 % and 100 %) that is the first place where the number of inquired distances is greater than or equal to the inquired ones in our method are reported.

Table 2. NMI results of the methods. Numbers in parenthesis show percent of the asked distances.

Data	RR	$s = 1$	$s = 3$
Vary-density	70.8 (10.0 %)	95.0 (9.7 %)	96.6 (10.5 %)
Seeds	55.7 (10.0 %)	90.3 (7.1 %)	89.5 (7.3 %)
Mouse	58.1 (5.0 %)	75.5 (4.1 %)	73.6 (4.3 %)
Fisheriris	65.3 (10.0 %)	85.6 (9.6 %)	89.3 (10.2 %)
Data_2000	88.5 (5.0 %)	77.1 (1.9 %)	78.3 (1.9 %)
Trace	45.7 (10.0 %)	51.2 (9.6 %)	52.4 (10.4 %)
Multi-features	46.9 (5.0 %)	78.7 (2.8 %)	77.6 (2.8 %)
TwoDiamonds	97.5 (5.0 %)	100.0 (1.8 %)	100.0 (1.8 %)
EngyTime	71.2 (5.0 %)	99.9 (1.6)	99.6 (1.6 %)
COIL100	42.3 (10.0 %)	75.6 (7.6 %)	75.6 (8.0 %)
NORM10	95.5 (5.0 %)	94.3 (1.6 %)	94.6 (1.6 %)
NEC_animal	23.6 (10.0 %)	66.6 (7.6 %)	67.3 (7.9 %)
ALOI200	48.0 (10.0 %)	79.6 (7.7 %)	79.7 (8.0 %)

Moreover, results of the Random-Rival algorithm which uses 0 percent of distances up to 100 % of them is presented for some datasets in Fig. 3. According to Fig. 3, RR shows an ascending trend by asking extra distances and it gets close to the maximum value by asking about 20 % of distances. Although it sounds to be a good algorithm, the proposed active k-medoids algorithm gets better NMI values with asking fewer number of distances. Moreover, the time complexity of our active k-medoid is also better. These results state the power of our algorithms which find accurate clusters by asking only a small subset of distances.

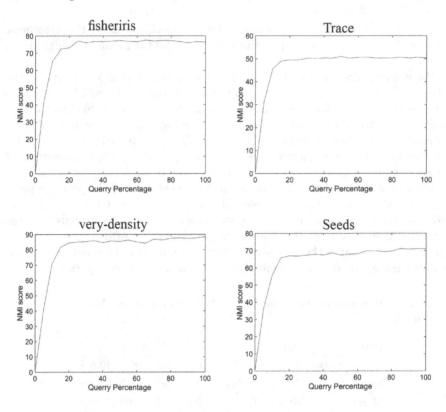

Fig. 3. Random-Rival over four datasets.

5 Conclusion

In this paper, we introduce an innovative active distance-based clustering method. Its goal is to cluster n points from a metric dataset into k clusters by using lowest number of distances that is possible. We design a recursive model that makes a tree and split data with a branching factor b unless the number of objects is less than a threshold t_h. Then, it actively selects and ask some pairwise similarities from and oracle. After that it tries to make an upper-bound estimation for unknown distances utilizing triangular inequality. Eventually, it clusters data with a simple k-medoids algorithm. We run our algorithm over some synthesized and real world datasets. In order to show privilege of our method and to compare the results, we also introduce an algorithm which randomly selects pairwise distances and estimates unknown ones using Floyd-Warshall algorithm.

References

1. Amsterdam library of object images (aloi) (2004). http://aloi.science.uva.nl/
2. Example data sets for elki (2013). http://elki.dbs.ifi.lmu.de/wiki/DataSets

3. Arbelaez, A., Quesada, L.: Parallelising the k-meds clustering problem using space-partitioning. In: Proceedings of the Sixth Annual Symposium on Combinatorial Search, SOCS, Leavenworth, Washington, USA, 11–13 July 2013
4. Arthur, D., Vassilvitskii, S.: k-means++: the advantages of careful seeding. In: Proceedings of the Eighteenth Annual ACM-SIAM Symposium on Discrete Algorithms, SODA, pp. 1027–1035, New Orleans, Louisiana, USA, 7–9 January 2007
5. Asuncion, A., Newman, D.: UCI machine learning repository datasets (2007). https://archive.ics.uci.edu/ml/datasets.html
6. Basu, S., Banerjee, A., Mooney, R.J.: Active semi-supervision for pairwise constrained clustering. In: Proceedings of the Fourth SIAM International Conference on Data Mining, pp. 333–344, Lake Buena Vista, Florida, USA, 22–24 April 2004
7. Biswas, A., Jacobs, D.W.: Active image clustering with pairwise constraints from humans. Int. J. Comput. Vis. **108**(1–2), 133–147 (2014)
8. Chen, M.: Synthesized dataset for k-medoids. http://www.mathworks.com/matlabcentral/fileexchange/28898-k-medoids/
9. Cormen, T.H., Leiserson, C.E., Rivest, R.L., Stein, C.: Introduction to Algorithms, 3rd edn. The MIT Press, Cambridge (2009)
10. Eriksson, B., Dasarathy, G., Singh, A., Nowak, R.D.: Active clustering: robust and efficient hierarchical clustering using adaptively selected similarities. In: Proceedings of the Fourteenth International Conference on Artificial Intelligence and Statistics, AISTATS, pp. 260–268, Fort Lauderdale, USA, 11–13 April 2011
11. Grira, N., Crucianu, M., Boujemaa, N.: Active semi-supervised fuzzy clustering. Pattern Recogn. **41**(5), 1834–1844 (2008)
12. Kaufman, L., Rousseeuw, P.J.: Finding Groups in Data: An Introduction to Cluster Analysis. Wiley, Hoboken (1990)
13. Krishnamurthy, A., Balakrishnan, S., Xu, M., Singh, A.: Efficient active algorithms for hierarchical clustering. In: Proceedings of the 29th International Conference on Machine Learning, ICML, Edinburgh, Scotland, UK, June 26-July 1, 2012
14. Mai, S.T., He, X., Hubig, N., Plant, C., Böhm, C.: Active density-based clustering. In: IEEE 13th International Conference on Data Mining, pp. 508–517, Dallas, TX, USA, 7–10 December 2013
15. Mobahi, H., Collobert, R., Weston, J.: Deep learning from temporal coherence in video. In: Proceedings of the 26th Annual International Conference on Machine Learning, pp. 737–744, ICML, Montreal, Quebec, Canada, 14–18 June 2009
16. Nayar, S., Nene, S.A., Murase, H.: Columbia object image library (coil 100). Department of Computer Science, Columbia University, Technical report, CUCS-006-96 (1996)
17. Nguyen, X.V., Epps, J., Bailey, J.: Information theoretic measures for clusterings comparison: is a correction for chance necessary? In: Proceedings of the 26th Annual International Conference on Machine Learning, ICML, pp. 1073–1080, Montreal, Quebec, Canada, 14–18 June 2009
18. Settles, B.: Active learning literature survey. Univ. Wis. Madison **52**(55–66), 11 (2010)
19. Shamir, O., Tishby, N.: Spectral clustering on a budget. In: Proceedings of the Fourteenth International Conference on Artificial Intelligence and Statistics, AISTATS, pp. 661–669, Fort Lauderdale, USA, 11–13 April 2011
20. Singh, S.S., Chauhan, N.: K-means v/s k-medoids: a comparative study. In: National Conference on Recent Trends in Engineering and Technology, vol. 13 (2011)
21. Tong, S., Koller, D.: Support vector machine active learning with applications to text classification. J. Mach. Learn. Res. **2**, 45–66 (2001)

22. Ultsch, A.: Fundamental clustering problems suite (fcps). Technical report, University of Marburg (2005)
23. Voevodski, K., Balcan, M., Röglin, H., Teng, S., Xia, Y.: Active clustering of biological sequences. J. Mach. Learn. Res. **13**, 203–225 (2012)
24. Vu, V., Labroche, N., Bouchon-Meunier, B.: Improving constrained clustering with active query selection. Pattern Recogn. **45**(4), 1749–1758 (2012)
25. Wagstaff, K., Cardie, C., Rogers, S., Schrödl, S.: Constrained k-means clustering with background knowledge. In: Proceedings of the Eighteenth International Conference on Machine Learning (ICML), Williams College, pp. 577–584, Williamstown, MA, USA, June 28–July 1, 2001
26. Wang, X., Davidson, I.: Active spectral clustering. In: ICDM 2010, The 10th IEEE International Conference on Data Mining, pp. 561–568, Sydney, Australia, 14–17 December 2010
27. Wauthier, F.L., Jojic, N., Jordan, M.I.: Active spectral clustering via iterative uncertainty reduction. In: The 18th ACM SIGKDD International Conference on Knowledge Discovery and Data Mining, KDD 2012, pp. 1339–1347, Beijing, China, 12–16 August 2012
28. Xiong, C., Johnson, D.M., Corso, J.J.: Active clustering with model-based uncertainty reduction. CoRR, abs/1402.1783 (2014)
29. Chen, Y., Keogh, E., Batista, G.: UCR time series classification archive (2015). http://www.cs.ucr.edu/~eamonn/time_series_data/

Analyzing Similarities of Datasets Using a Pattern Set Kernel

A. Ibrahim[1(✉)], P.S. Sastry[1], and Shivakumar Sastry[2]

[1] Indian Institute of Science, Bangalore, India
{ibrahim,sastry}@ee.iisc.ernet.in
[2] University of Akron, Akron, USA
ssastry@uakron.edu

Abstract. In the area of pattern discovery, there is much interest in discovering small sets of patterns that characterize the data well. In such scenarios, when data is represented by a small set of characterizing patterns, an interesting problem is the comparison of datasets, by comparing the respective representative sets of patterns. In this paper, we propose a novel kernel function for measuring similarities between two sets of patterns, which is based on evaluating the structural similarities between the patterns in the two sets, weighted using their relative frequencies in the data. We define the kernel for injective serial episodes and itemsets. We also present an efficient algorithm for computing this kernel. We demonstrate the effectiveness of our kernel on classification scenarios and for change detection using sequential datasets and transaction databases.

1 Introduction

One of the reasons for the effectiveness of frequent pattern methods is that some of the frequently occurring patterns can capture crucial aspects of the underlying semantics of the data. There exist many techniques to characterize transaction as well as sequence data with a small representative set of patterns [7,9,13,15,16]. An interesting question that follows is the quantification of the similarity between two sets of patterns. Such a similarity measure can, in many ways, allow us to compare different data sets by comparing the characterizing subsets of patterns representing them. This would be useful, e.g., in change detection and classification. In this work, we address this problem of quantifying similarity between two sets of patterns, where patterns could be itemsets or serial episodes.[1]

One way of using the representative set of patterns is to employ what is known as the Bag of Words (BoW) representation [3,9]. For example, in a classification application, we can first discover a good representative subset of patterns (called a dictionary) from the training data. Then any given data instance can be represented as a feature vector whose dimension is same as the size of the dictionary. Each component of the feature vector specifies presence or absence (or number

---

[1] A preliminary version of this paper was presented as a poster at 2nd IKDD Conference on Data Sciences, CoDS 2015 [6].

© Springer International Publishing Switzerland 2016
J. Bailey et al. (Eds.): PAKDD 2016, Part I, LNAI 9651, pp. 265–276, 2016.
DOI: 10.1007/978-3-319-31753-3_22

of occurrences) of the corresponding dictionary pattern in the data instance. The feature vectors can then be used to find similarities between different data instances.

The BoW representation treats different patterns in the dictionary as independent features and it cannot take into account any similarities between different patterns in the dictionary. When comparing datasets, patterns representing different datasets may be quite similar without being exactly the same. Hence, what is desirable is a measure that gives weightage to structural similarity between patterns such as the number of shared subpatterns. In addition to structural similarity between patterns, we also need to consider their relative importance to data. A pattern occurring in a dataset with a high frequency may be more important as opposed to one with low frequency.

There exists a host of methods for comparing different types of data. Most of the methods are based on (probability) model comparison or comparison of sets of patterns representing the data. [8] maps vectors belonging to a set, to a Hilbert space and fits a Gaussian distribution to the set using Kernel PCA. The kernel between two sets of vectors is then defined as the Bhattacharya's measure of affinity between Gaussians. A similar kernel, where a Gaussian Mixture is fitted over a Hilbert space is proposed in [12]. Similarity measures for structured data like graphs, consider the structural similarity of patterns [4]. Even though subgraph isomorphism is NP-Hard, there exists efficient graph kernels which look at simpler substructures [4]. Two such polynomially computable graph kernels are proposed in [12]. For string comparison, string kernels based on the counts of shared substrings have been proposed in [11].

However, in the context of episodes or itemset patterns, similarity measures based on comparison of patterns, usually do not take into account structural similarity of patterns; rather they look at the collective similarity such as the amount of data the collection of patterns share [17]. Similarity measures considering the amount of compression achieved by different sets of patterns in representing the data have also been considered for transaction data [14,15].

In this paper, we propose a kernel, called the *Pattern Set Kernel*, for comparing two *sets of patterns*. We define this kernel for serial episodes and itemsets. We first define what we call a *Pattern Kernel*, which is a measure of similarity between two patterns. The measure depends on the extent of subpatterns shared by the two patterns along with frequencies of the patterns. Pattern Set Kernel is then defined using the Pattern Kernel. Even though, our pattern kernel is based on the number of common subpatterns, which could be exponential, we present efficient algorithms for calculating the kernel; complexity of which grows only as the product of lengths of the two patterns. We demonstrate the effectiveness of this new measure of similarity through extensive empirical studies.

The rest of the paper is organized as follows. Section 2 presents the Pattern Kernel for serial episodes and itemsets. We present the Pattern Set Kernel in Sect. 3. Section 4 gives the simulation results showing the effectiveness of our kernel and we conclude the paper in Sect. 5.

2 Pattern Kernel

In this section, we present the kernel for a pair of patterns. We first define the pattern kernel for injective serial episodes, and then extend it for itemsets.

2.1 Episode Kernel: Pattern Kernel for Injective Serial Episodes

In the episodes framework, the data, referred to as an *event sequence*, is denoted by $\mathbb{D} = \langle(E_1, t_1), (E_2, t_2), \ldots, (E_n, t_n)\rangle$, where n is the number of events. In each tuple (E_i, t_i) (called an *event*), E_i denotes the event-type and t_i denotes the time of occurrence of the event. $E_i \in \mathcal{E}$, a finite alphabet, and $t_i \leq t_{i+1}$, $1 \leq i < n$.

A serial episode is a totally ordered set of event types. A k-node serial episode, α, is denoted as $(e_1 \rightarrow e_2 \rightarrow \cdots \rightarrow e_k)$ where $e_i \in \mathcal{E}$, $\forall i$. An episode β is a subepisode of α, denoted as $\beta \preceq \alpha$, if there exists integers $i_1, i_2, \ldots, i_m$ with $m \leq k$ and $1 \leq i_1 < i_2 < \cdots < i_m \leq k$, such that $\beta = (e_{i_1} \rightarrow e_{i_2} \rightarrow \cdots \rightarrow e_{i_m})$. An episode $\alpha = (e_1 \rightarrow e_2 \rightarrow \cdots \rightarrow e_k)$ is called an *injective serial episode* if $e_i \neq e_j, \forall i, j, i \neq j$. In this paper, we call injective serial episodes as serial episodes (or just episodes). We denote the frequency of α (in a given data sequence) as $fr(\alpha)^2$. $\mathcal{E}^+$ denotes the set of all serial episodes of size 1 or more.

The Episode Kernel, $K_{EK}(\alpha, \beta)$, to compare two serial episodes, α and β, is defined as $K_{EK}(\alpha, \beta) = \phi(\alpha)^T \phi(\beta)$ where ϕ is a function that maps each serial episode to a vector in the feature space $\Re^{|\mathcal{E}^+|}$. We use $\gamma \in \mathcal{E}^+$ to index the components of ϕ. The mapping ϕ for an episode α is given by

$$\phi_\gamma(\alpha) = \begin{cases} fr(\alpha), & \text{if } \gamma \preceq \alpha \\ 0, & \text{otherwise} \end{cases}$$

Note that since $K_{EK}(\alpha, \beta) = \phi(\alpha)^T \phi(\beta)$, those coordinates γ for which either $\phi_\gamma(\alpha)$ or $\phi_\gamma(\beta)$ are zeros do not contribute to the kernel. Thus we obtain

$$K_{EK}(\alpha, \beta) = \sum_{\gamma : \gamma \preceq \alpha, \gamma \preceq \beta} fr(\alpha) \cdot fr(\beta) = fr(\alpha) \cdot fr(\beta) \cdot |\{\gamma : \gamma \preceq \alpha, \gamma \preceq \beta\}| \quad (1)$$

Also note that the size of the set, $\{\gamma : \gamma \preceq \alpha, \gamma \preceq \beta\}$ in Eq. (1) could be exponential in the size of alphabet. For instance, if $\alpha = \beta$, with α, being an M-node episode, then $|\{\gamma : \gamma \preceq \alpha, \gamma \preceq \beta\}| = 2^M - 1$. However, the size of this set can be efficiently calculated in $O(MN)$ steps, where M and N are lengths of α and β respectively, as described below.

Efficient Computation of K_{EK}. We rewrite the episode kernel as $K_{EK}(\alpha, \beta) = fr(\alpha) \cdot fr(\beta) \cdot K_1(\alpha, \beta)$ where $K_1(\alpha, \beta) = |\{\gamma : \gamma \preceq \alpha, \gamma \preceq \beta\}|$.

---

[2] There are various definitions of frequency proposed for episodes [1]. We are not imposing any condition on what frequency we are considering, and hence $fr(\alpha)$ could be any measure of relative significance of episode α in the data.

For an n node serial episode $\alpha = (\alpha[1] \to \alpha[2] \to \cdots \to \alpha[n])$, the *i-node prefix subepisode* of α is defined as the serial episode $(\alpha[1] \to \alpha[2] \to \cdots \to \alpha[i])$, for $i \leq n$. We denote the i-node prefix subepisode of α by $\alpha_{[1..i]}$.

Let α and β be two serial episodes of length M and N respectively. We assign $K_1(X, Y) = 0$, if $X = \varnothing$ or $Y = \varnothing$, where $\varnothing$ denotes empty or null episode whose size is zero. The iterative algorithm for computing $K_1(\alpha, \beta)$ is as follows

for $i = 1..M$ **do**

  for $j = 1..N$ **do**

$$K_1(\alpha_{[1..i]}, \beta_{[1..j]}) = \begin{cases} 2 \times K_1(\alpha_{[1..(i-1)]}, \beta_{[1..(j-1)]}) & +1 \text{ if } \alpha[i] = \beta[j] \\ K_1(\alpha_{[1..i]}, \beta_{[1..(j-1)]}) & + \text{ if } \alpha[i] \neq \beta[j] \\ K_1(\alpha_{[1..(i-1)]}, \beta_{[1..j]}) & - \\ K_1(\alpha_{[1..(i-1)]}, \beta_{[1..(j-1)]}) & \end{cases}$$

In the algorithm above, both $\alpha_{[1..0]}$ and $\beta_{[1..0]}$ denote the null episode and $K_1(\alpha_{[1..M]}, \beta_{[1..N]})$ gives the value of $K_1(\alpha, \beta)$. It is easily seen that the complexity of the above algorithm is $O(MN)$.

Proof of Correctness for Calculating K_1. We first note that the episodes we consider are injective episodes. Thus, each event-type occurs in an episode at most once. Suppose we want to find $K_1(\alpha_{[1..i]}, \beta_{[1..j]})$. We have two conditions based on the values of $\alpha[i]$ and $\beta[j]$.

1. Let $\alpha[i] = \beta[j] = A$. Let $\mathcal{A}$ be the set of all subepisodes common to $\alpha_{[1..(i-1)]}$ and $\beta_{[1..(j-1)]}$. Episodes in $\mathcal{A}$ are also common subepisodes to $\alpha_{[1..i]}$ and $\beta_{[1..j]}$. For every episode $\gamma \in \mathcal{A}$, $(\gamma \to A)$ is a subepisode common to $\alpha_{[1..i]}$ and $\beta_{[1..j]}$. The one-node episode A is an obvious common subepisode for $\alpha_{[1..i]}$ and $\beta_{[1..j]}$ and no other subepisode is common to both $\alpha_{[1..i]}$ and $\beta_{[1..j]}$. Thus $K_1(\alpha_{[1..i]}, \beta_{[1..j]}) = 2|\mathcal{A}| + 1 = 2 \times K_1(\alpha_{[1..(i-1)]}, \beta_{[1..(j-1)]}) + 1$.

2. Let $\alpha[i] \neq \beta[j]$. Any subepisode γ common to $\alpha_{[1..i]}$ and $\beta_{[1..j]}$ should belong to any one of the three *mutually exclusive* sets
   (a) $\mathcal{A} = \{\gamma \in \mathcal{E}^+ : \gamma \preceq \alpha_{[1..i]}, \gamma \preceq \beta_{[1..(j-1)]}, \alpha[i] \text{ is the last event of } \gamma\}$.
   (b) $\mathcal{B} = \{\gamma \in \mathcal{E}^+ : \gamma \preceq \alpha_{[1..(i-1)]}, \gamma \preceq \beta_{[1..j]}, \beta[j] \text{ is the last event of } \gamma\}$.
   (c) $\mathcal{C} = \{\gamma \in \mathcal{E}^+ : \gamma \preceq \alpha_{[1..(i-1)]}, \gamma \preceq \beta_{[1..(j-1)]}\}$
   It is easy to see that neither $\alpha[i]$ nor $\beta[j]$ is the last event for $\gamma \in \mathcal{C}$. Episodes in $\mathcal{A}$ are formed by aligning $\alpha[i]$ (which forms the last event of the episodes) with some element $\beta[k], k < j$, such that $\beta[k] = \alpha[i]$. Similarly, episodes in $\mathcal{B}$ are formed by aligning $\beta[j]$ with some event $\alpha[g], g < i$ such that $\alpha[g] = \beta[j]$. Noting that $K_1(\alpha_{[1..i]}, \beta_{[1..(j-1)]}) = |\mathcal{A} \cup \mathcal{C}|$ and $K_1(\alpha_{[1..(i-1)]}, \beta_{[1..j]}) = |\mathcal{B} \cup \mathcal{C}|$ and that $\mathcal{A}, \mathcal{B}, \mathcal{C}$ are mutually exclusive, we have

$$\begin{aligned} K_1(\alpha_{[1..i]}, \beta_{[1..j]}) &= |\mathcal{A} \cup \mathcal{B} \cup \mathcal{C}| \\ &= |\mathcal{A} \cup \mathcal{C}| + |\mathcal{B} \cup \mathcal{C}| - |\mathcal{C}| \\ &= K_1(\alpha_{[1..i]}, \beta_{[1..(j-1)]}) + K_1(\alpha_{[1..(i-1)]}, \beta_{[1..j]}) \\ &\quad - K_1(\alpha_{[1..(i-1)]}, \beta_{[1..(j-1)]}) \end{aligned}$$

This completes the proof of correctness of the algorithm.

2.2   Itemset Kernel: Pattern Kernel for Itemsets

In this section, we define Pattern Kernel for itemsets. Itemsets are patterns obtained from transaction data. Let $\mathcal{I} = \{i_1, i_2, \ldots, i_d\}$ be a set of items. An itemset is a subset of $\mathcal{I}$. Let $\mathcal{D} = \{t_1, t_2, \ldots, t_N\}$ be the set of all transactions, where $t_i \subseteq \mathcal{I}, \forall i$. Suppose we give an ordering for the items in $\mathcal{I}$ and order the items in the itemsets based on that order. Then an itemset $\{i_{j_1} i_{j_2} \ldots i_{j_M}\}$ would correspond to the unique injective[3] serial episode $(i_{j_1} \to i_{j_2} \to \cdots \to i_{j_M})$. The itemset kernel is exactly episode kernel on these injective serial episodes.

Given two itemsets α and β. There exists a unique largest common subset for α and β, which is $\alpha \cap \beta$. Then any common subset of α and β is a subset of $\alpha \cap \beta$ and hence the number of shared itemsets between α and β is $2^{|\alpha \cap \beta|} - 1$. Hence, itemset kernel, denoted as K_{IK} for two itemsets α and β is $K_{IK}(\alpha, \beta) = fr(\alpha) \cdot fr(\beta) \cdot |\{\gamma : \gamma \preceq \alpha, \gamma \preceq \beta\}| = fr(\alpha) \cdot fr(\beta) \cdot (2^{|\alpha \cap \beta|} - 1)$.

3   Pattern Set Kernel

Now, using the pattern kernel, we define the Pattern Set Kernel for sets of patterns. We denote the pattern kernel (for both the pattern types) as K_{PK}.

Definition 1. *The Pattern Set Kernel, denoted as K_{PSK}, between two sets of patterns $\mathcal{F}_1$ and $\mathcal{F}_2$ is defined as $K_{PSK}(\mathcal{F}_1, \mathcal{F}_2) = \Phi(\mathcal{F}_1)^T \Phi(\mathcal{F}_2)$, where Φ is function that maps each set of patterns $\mathcal{F}_i$ to a vector in feature space $\Re^{|\mathcal{E}^+|}$, where $\mathcal{E}^+$ denotes the set of all patterns of size one or more.*

Based on the context, $\mathcal{E}^+$ could represent the set of serial episodes or itemsets of size one or more. For each $\mathcal{F}$, the γ coordinate of Φ is given by

$$\Phi_\gamma(\mathcal{F}) = \sum_{\substack{\alpha: \\ \alpha \in \mathcal{F}, \gamma \preceq \alpha}} fr(\alpha)$$

We now show how K_{PSK} can be computed using the underlying pattern kernel.

Proposition 1. $K_{PSK}(\mathcal{F}_1, \mathcal{F}_2) = \sum_{\alpha \in \mathcal{F}_1, \beta \in \mathcal{F}_2} K_{PK}(\alpha, \beta)$.

Proof.

$$
\begin{aligned}
K_{PSK}(\mathcal{F}_1, \mathcal{F}_2) &= \sum_{\gamma \in \mathcal{E}^+} \left(\sum_{\substack{\alpha: \\ \alpha \in \mathcal{F}_1, \gamma \preceq \alpha}} fr(\alpha) \right) \left(\sum_{\substack{\beta: \\ \beta \in \mathcal{F}_2, \gamma \preceq \beta}} fr(\beta) \right) \\
&= \sum_{\alpha \in \mathcal{F}_1} \sum_{\beta \in \mathcal{F}_2} \sum_{\substack{\gamma: \\ \gamma \preceq \alpha, \gamma \preceq \beta}} fr(\alpha) fr(\beta) \\
&= \sum_{\alpha \in \mathcal{F}_1} \sum_{\beta \in \mathcal{F}_2} K_{PK}(\alpha, \beta)
\end{aligned}
$$

---

[3] Injective because itemsets, by definition, do not have repetitive items.

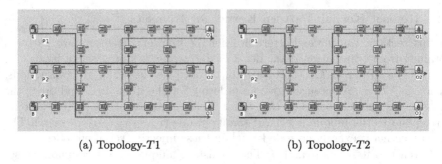

(a) Topology-$T1$ (b) Topology-$T2$

Fig. 1. Two example topologies of CCS.

Pattern Set Kernel between two sets of patterns is thus the summation of Pattern Kernels of pairs of patterns from $\mathcal{F}_1$ and $\mathcal{F}_2$.

3.1   Complexity for Finding the Pattern Set Kernel

Let $|\mathcal{F}|$ denote the number of patterns in set $\mathcal{F}$. Then, based on Proposition 1, $K_{PSK}(\mathcal{F}_1, \mathcal{F}_2)$ computation takes $|\mathcal{F}_1| \times |\mathcal{F}_2|$ summations (of K_{PK}). In the case of episodes, assuming that the maximum size of the episodes in $\mathcal{F}_1$ and $\mathcal{F}_2$ as M and N respectively, each Episode Kernel (K_{EK}) calculation would cost $O(MN)$. For itemsets, the itemset kernel (K_{IK}) computation would cost only $O(M+N)$. Hence the total cost of computing $K_{PSK}(\mathcal{F}_1, \mathcal{F}_2)$ is $O(|\mathcal{F}_1||\mathcal{F}_2|MN)$ for episodes and $O(|\mathcal{F}_1||\mathcal{F}_2|(M+N))$ for itemsets.

4   Simulations

In this section, we show the effectiveness of Pattern Set Kernel (PSK), where the patterns are either itemsets or episodes. We show the effectiveness of PSK, by comparing data from different sources. We also show the effectiveness of PSK for change detection and classification. All algorithms are implemented in Matlab and experiments were executed single threaded on an Intel i7 4-core processor with 16 GB memory running over linux OS.

We define a measure of similarity between two sets of patterns as

$$sim\text{-}score(\mathcal{F}_1, \mathcal{F}_2) = \frac{\Phi(\mathcal{F}_1)^T \Phi(\mathcal{F}_2)}{\sqrt{\Phi(\mathcal{F}_1)^T \Phi(\mathcal{F}_1)}\sqrt{\Phi(\mathcal{F}_2)^T \Phi(\mathcal{F}_2)}} = \frac{K(\mathcal{F}_1, \mathcal{F}_2)}{\sqrt{K(\mathcal{F}_1, \mathcal{F}_1)}\sqrt{K(\mathcal{F}_2, \mathcal{F}_2)}}$$

which is the usual cosine similarity measure between two vectors, normally used with kernels. We also define a corresponding distance metric between two sets of patterns as $dist(\mathcal{F}_1, \mathcal{F}_2) = 1 - (sim\text{-}score(\mathcal{F}_1, \mathcal{F}_2))$.

4.1   Measuring Similarity Between Sequences

We first consider sequence data, where patterns are injective serial episodes. Of various techniques for summarizing sequences using serial episodes [7,9,13],

we use the method CSC-2 [7], which retrieves a representative set of injective serial episodes from any sequence data. The algorithm does not need any user specified parameter such as frequency threshold. We consider the Coupled Conveyor
Systems (CCS) [2] for our first set of sequence data. CCS are reconfigurable conveyor systems for moving material/packages from input sources to output desti-
nations. They are built with units called Segments and Turns that each operate
autonomously. A Segment moves packages over a predetermined length over its
belt. A Turn is a unit that can serve as merger or splitter for package flow. The
system can be configured to different topologies that determine different paths
over which packages move. There are sensors at each turn and segment that record
events of packages moving through them with time stamps. Data mining over such
data streams is useful for remote monitoring and visualization of such systems.
The data we consider here is obtained through a detailed discrete event simulator
of CCS [2]. (See [7] for more details on this application).

We generate data streams from two topologies, $T1$ and $T2$, shown in Fig. 1a
and b. The topologies share the same subpaths, but the actual paths are different
in different topologies. For our simulation, for each topology, we generate two
sets of sequences, corresponding to two different input rates (assumed Poisson)
of 0.2 and 0.8, by running the simulator for a period of 5000 s each. Each such
stream is cut into 5 disjoint chunks of time span 1000 s.

We obtain a characterizing set of serial episodes (along with their frequencies normalized to the chunk size) for each chunk using CSC-2 algorithm [7]
and compare these sets using the Pattern Set Kernel. Three sets of experiments
were conducted, comparing (a) chunks from the same sequences (same topology
and same input rates) (b) chunks from different sequences generated using the
same topology, but different input rates and (c) chunks from different sequences
generated from different topologies, but with the same input rate. Some representative samples of these results are shown in Tables 1, 2 and 3. Each chunk is
referred in the tables using the notation, T_i-$\langle rate\rangle$-$\langle chunknumber\rangle$. As can be
seen, chunks from the same topology and same rate have very high similarity
scores (greater than 0.9), chunks from same topology but different rates have
somewhat lower scores (greater than 0.55) and chunks from different topologies
have much lower similarity scores (less than 0.38). The results show the PSK
based similarity measure is able to capture the difference in characteristics of
the datasets.

Table 1. *sim-score* between **Table 2.** *sim-score* between **Table 3.** *sim-score* between
T_1-0.20-i (row) vs T_1-0.20-j T_1-0.20-i (row) vs T_1-0.80-j T_1-0.20-i (row) vs T_2-0.20-j
(col).   (col).   (col).

i,j	1	2	3	4	5
1	1	0.95	0.92	0.92	0.97
2	0.95	1	0.97	0.95	0.96
3	0.92	0.97	1	0.97	0.92
4	0.92	0.95	0.97	1	0.91
5	0.97	0.96	0.92	0.91	1

i,j	1	2	3	4	5
1	0.88	0.70	0.90	0.55	0.90
2	0.85	0.66	0.91	0.58	0.91
3	0.82	0.65	0.95	0.59	0.94
4	0.85	0.67	0.97	0.62	0.96
5	0.86	0.67	0.86	0.59	0.86

i,j	1	2	3	4	5
1	0.32	0.31	0.33	0.33	0.33
2	0.32	0.33	0.33	0.35	0.33
3	0.32	0.33	0.33	0.34	0.32
4	0.33	0.31	0.34	0.34	0.34
5	0.34	0.33	0.35	0.37	0.35

4.2   Pattern Set Kernels for Classification

Next we present results to show the effectiveness of PSK as a distance metric for classification. We show results on sequences from the CCS problem and a few benchmark sequence datasets in the domain of sign languages [9]. We compare performance of three classifiers. Our first classifier is a PSK based *K-Nearest Neighbors* (KNN) classifier, where distance metric used is *dist*, introduced earlier. We denote this classifier as PSK-KNN. The second classifier is also a KNN classifier, but has a Bag of Word (BoW) representation of data sequences using selected episodes of all classes as dictionary. The third classifier we use is linear Support Vector Machine (SVM) classifier, where again each sample is represented using BoW representation.

For data sequences in CCS problem, we compare performance of the three classifiers on predicting which topology a given data chunk comes from. The training data consists of data chunks generated from different topologies and we use CSC-2 algorithm for selecting a representative set of serial episodes for each class. Test data chunks are also characterized by episodes discovered from them.

For this classification problem, we generated three datasets. The first dataset consisted of sequences from Topologies $T1$ and $T2$ (Fig. 1). The data samples for the two classes are sequences of chunk size 50 s generated from the respective topologies. The second data consisted of sequences from the same topology $T1$, but with different input rates, 0.5 and 0.8. The two classes correspond to the two rates. The third data is similar to the first, except that chunk sizes have been halved to 25 s. For KNN classifier, we tried different values of K. For SVM classifier, we varied the error cost parameter 'C' over a range of values. The results we present are the best among all parameter values. We show the classification results on CCS data in the upper part of Table 4. We see that PSK based KNN classifier outperforms other classifiers for all three datasets.

The other datasets we consider contain labeled small sequences, corresponding to different actions in sign language and have been used earlier as benchmark sequence datasets [9]. Extracting characterizing serial episodes from very small sequences is not generally possible. Hence, we generate new datasets, where individual sequences corresponding to a class are generated by randomly selecting 10 sequences (of the original dataset) of the same class and concatenating them.

Table 4. Classification accuracy for CCS and sign language datasets.

Datasets	PSK-KNN	KNN	SVM
$T1$-$T2$-50s	100	96.5	95
$T1$-$T1$-Diff-rates	100	97.5	98
$T1$-$T2$-25s	95	84	74.5
aslbu	98.5	97.1	99.3
aslgt	71.6	64.3	60.8
auslan2	78	77	74
context	100	98.3	98.3
pioneer	100	100	100

Table 5. Precision-Recall values for change detection for different streams.

	$T1$vs$T2$	$T1$vs$T3$	$T2$vs$T3$
Precision	0.9	1	1
Recall	0.85	0.91	0.94

We experimented with five such benchmark datasets all of which are multi-class. (The SVM classifier for the multi-class scenario is implemented using the one versus rest approach). The results are shown in the lower part of Table 4. We see that, except for 'aslbu' dataset (where it achieves slightly lower accuracy), the PSK-KNN achieves higher accuracy as compared to other classifiers.

4.3 Change Detection in Streaming Data from Conveyor Systems

In this section, we use the PSK based distance metric to detect changes in streams obtained from different topologies of the conveyor system. We generate sequences from three topologies, $T1$, $T2$ and $T3$ ($T3$ is another topology, very similar to $T1$ and $T2$ such that they share many subpaths, yet have different sets of actual paths). For the change detection experiment, we generate data streams consisting of a random mixture of data chunks from two different topologies. For this, we adopt the following scheme. We randomly choose a number r between 5 and 10, and select the next r chunks, alternatively from one of the two topologies. We continue this process until we get 100 chunks which forms the data stream for change detection. Each chunk is compared with the previous one using the metric, $dist$, to see how dissimilar they are. Results are shown in Fig. 2. Vertical lines correspond the actual points of change in the stream. It is easy to see that the $dist$ metric using PSK is very effective in detecting points of change.

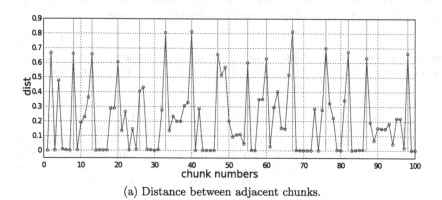

(a) Distance between adjacent chunks.

Fig. 2. Change detection: $T1$ vs. $T2$ with random breaks

We also show how PSK based $dist$ metric performs for predicting changes, using suitable thresholds on distances. We generated 10 instances of random streams for $T1 - T2$, $T1 - T3$ and $T2 - T3$. The thresholds are calculated from training data as follows. We take the 20th percentile of $dist$ values corresponding to change points in the training data as the *distance threshold*. We also calculate the sequence of 'derivatives', which are changes in successive distances. We then calculate $derivative(i) - derivative(i + 1)$. As is easy to deduce, at points of change, these correspond to a fall from a high value of derivative (at the point of change) to a low value of derivative (at the subsequent point). We again take the 20th percentile value of the fall (in training data) as *fall threshold*. For prediction, we report a change, when distance and fall of a point is above 90 % of both thresholds. We report average precision and recall values in Table 5. We see that PSK based distance measure achieves high precision and recall values.

4.4   Measuring Similarity Between Transaction Data

In this section, we show the effectiveness of Pattern Set Kernel, when patterns are itemsets. We use the Krimp algorithm [16] for mining itemsets from transaction datasets. Krimp is an MDL based lossless compression algorithm, which outputs a set of itemsets, that summarizes transaction data well. The dataset we consider is 2-class mushroom transaction dataset [10]. Mushroom dataset consists of 8124 transactions, each having 22 categorical attributes. Transactions belonging to each class are equally divided into 5 chunks of transaction data. The classes are denoted by $c1$ and $c2$ respectively. Since Krimp initially mines the set of frequent itemsets before finding the best summarizing subset of itemsets, we specify the frequency threshold as 12 % (for mining) of the total number of transactions in each chunk. Each chunk is referred using the format $\langle class\rangle$-$\langle chunknumber\rangle$.

Tables 6, 7 and 8 show the results on the different chunks. We see that *sim-scores* are higher between chunks belonging to the same class. The similarity scores between the chunks of different classes are extremely low, thus showing the effectiveness of our kernel based similarity score.

Table 6. *sim-score* betw-een $c1$-i (row) vs. $c1$-j (col).

i, j	1	2	3	4	5
1	1	0.81	0.81	0.85	0.97
2	0.81	1	0.82	0.81	0.79
3	0.81	0.82	1	0.80	0.84
4	0.85	0.81	0.80	1	0.83
5	0.97	0.79	0.84	0.83	1

Table 7. *sim-score* betw-een $c2$-i (row) vs. $c2$-j (col).

i, j	1	2	3	4	5
1	0.029	0.020	0.018	0.019	0.020
2	0.017	0.018	0.016	0.017	0.019
3	0.017	0.018	0.016	0.017	0.018
4	0.017	0.018	0.016	0.017	0.018
5	0.018	0.019	0.017	0.018	0.020

Table 8. *sim-score* betw-een $c1$-i (row) vs. $c2$-j (col).

i, j	1	2	3	4	5
1	0.029	0.020	0.018	0.019	0.020
2	0.017	0.018	0.016	0.017	0.019
3	0.017	0.018	0.016	0.017	0.018
4	0.017	0.018	0.016	0.017	0.018
5	0.018	0.019	0.017	0.018	0.020

Remark 1. We would like to point out that we have not compared our similarity measure with any other known similarity measures for comparing datasets. Even though there is a sequence kernel for comparing sequences [11], it is computationally inefficient for long sequences. The rest of the similarity measures are not directly comparable with the measures we proposed here. Thus, although methods for comparing sequences such as strings are known, as per our knowledge, this is the first work, wherein structural comparison of sets of patterns is considered for analyzing similarities between sequential or transaction data.

5 Conclusion

The goal of frequent pattern mining is to gain useful information about data. Recently, many algorithms have been proposed for characterizing or representing data using a small subset of (frequent) patterns [7,9,13,15,16]. In such scenarios, when data is represented by a set of 'characteristic' patterns, a natural question is that of comparing sets of patterns representing the datasets in order to gain insights into the similarity of different datasets. In this paper, we have looked at this problem of pattern set comparison. We proposed a Pattern Kernel for quantifying similarity between pairs of patterns and then used it to define a Pattern Set Kernel for comparing sets of patterns. Pattern Kernel was defined for injective serial episodes and itemsets. The Pattern Kernel value depends on number of common subpatterns shared by the two patterns. We also presented efficient algorithms for calculating these kernels.

The effectiveness of the Pattern Set Kernel was shown for different sequence datasets as well as for a 2-class transaction dataset. We defined a similarity score and a distance measure using Pattern Set Kernel and used it for scenarios involving direct comparison of datasets, change detection and classification. On many sequence datasets as well as on the transaction dataset, it is seen that our kernel-based similarity measure is very effective in capturing similarities/differences between data.

Sequential patterns constitute another type of patterns that are used for analyzing sequence data [18]. While we did not consider these in this paper, our pattern kernel can be defined for sequential patterns also [5]. In the case of sequential patterns we need to assume a form of injectiveness that is somewhat restricted, to be able to compute the kernel efficiently. We would be further exploring this and other issues of extending our pattern kernel to all types of frequent patterns in our future work. The field of pattern mining has opened-up a new view of data through significant local patterns that occur in data. Similarity measure such as Pattern Set Kernel proposed here would prove to be very useful in utilizing such a view of data in many applications.

References

1. Achar, A., Laxman, S., Sastry, P.S.: A unified view of the apriori-based algorithms for frequent episode discovery. Knowl. Inf. Syst. 31(2), 223–250 (2012)
2. Archer, B., Shivakumar, S., Rowe, A., Rajkumar, R.: Profiling primitives of networked embedded automation. In: IEEE International Conference on Automation Science and Engineering, CASE 2009, pp. 531–536. IEEE (2009)
3. Fernando, B., Fromont, E., Tuytelaars, T.: Effective use of frequent itemset mining for image classification. In: Fitzgibbon, A., Lazebnik, S., Perona, P., Sato, Y., Schmid, C. (eds.) ECCV 2012, Part I. LNCS, vol. 7572, pp. 214–227. Springer, Heidelberg (2012)
4. Gärtner, T., Flach, P.A., Wrobel, S.: On graph kernels: hardness results and efficient alternatives. In: Schölkopf, B., Warmuth, M.K. (eds.) COLT/Kernel 2003. LNCS (LNAI), vol. 2777, pp. 129–143. Springer, Heidelberg (2003)

5. Ibrahim, A.: Effective characterization of sequence data through frequent episodes. Ph.D. thesis, (Under review), Indian Institute of Science, Bangalore (2015, submitted)

6. Ibrahim, A., Sastry, P.S., Sastry, S.: Pattern set kernel. In: Proceedings of the Second ACM IKDD Conference on Data Sciences, pp. 122–123. ACM (2015)

7. Ibrahim, A., Sastry, S., Sastry, P.S.: Discovering compressing serial episodes from event sequences. Knowl. Inf. Syst. 1–28 (2015). http://link.springer.com/article/10.1007/s10115-015-0854-3

8. Kondor, R., Jebara, T.: A kernel between sets of vectors. In: ICML, vol. 20, p. 361 (2003)

9. Lam, H.T., Mörchen, F., Fradkin, D., Calders, T.: Mining compressing sequential patterns. Stat. Anal. Data Min. 7(1), 34–52 (2014)

10. Lichman, M.: UCI machine learning repository (2013)

11. Lodhi, H., Saunders, C., Shawe-Taylor, J., Cristianini, N., Watkins, C.: Text classification using string kernels. J. Mach. Learn. Res. 2, 419–444 (2002)

12. Lyu, S.: A kernel between unordered sets of data: the gaussian mixture approach. In: Gama, J., Camacho, R., Brazdil, P.B., Jorge, A.M., Torgo, L. (eds.) ECML 2005. LNCS (LNAI), vol. 3720, pp. 255–267. Springer, Heidelberg (2005)

13. Tatti, N., Vreeken, J.: The long, the short of it: summarising event sequences with serial episodes. In: Proceedings of the 18th ACM SIGKDD International Conference on Knowledge Discovery and Data Mining, pp. 462–470. ACM (2012)

14. van Leeuwen, M., Vreeken, J., Siebes, A.: Compression picks item sets that matter. In: Fürnkranz, J., Scheffer, T., Spiliopoulou, M. (eds.) PKDD 2006. LNCS (LNAI), vol. 4213, pp. 585–592. Springer, Heidelberg (2006)

15. Vreeken, J., Van Leeuwen, M., Siebes, A.: Characterising the difference. In: Proceedings of the 13th ACM SIGKDD International Conference on Knowledge Discovery and Data Mining, pp. 765–774. ACM (2007)

16. Vreeken, J., Van Leeuwen, M., Siebes, A.: KRIMP: mining itemsets that compress. Data Min. Knowl. Disc. 23(1), 169–214 (2011)

17. Xin, D., Han, J., Yan, X., Cheng, H.: Mining compressed frequent-pattern sets. In: Proceedings of the 31st International Conference on Very Large Data Bases, pp. 709–720. VLDB Endowment (2005)

18. Yan, X., Han, J., Afshar, R.: Clospan: mining closed sequential patterns in large datasets. In: Proceedings of SIAM International Conference on Data Mining, pp. 166–177. SIAM (2003)

Significant Pattern Mining
with Confounding Variables

Aika Terada[1,2,3], David duVerle[1,3], and Koji Tsuda[1,3,4(✉)]

[1] Department of Computational Biology and Medical Sciences,
Graduate School of Frontier Sciences, The University of Tokyo,
Chiba 277-8561, Japan
terada@cbms.k.u-tokyo.ac.jp, dave@cb.k.u-tokyo.ac.jp
[2] Research Fellow of Japan Society for the Promotion of Science, Kojimachi, Japan
[3] Biotechnology Research Institute for Drug Discovery,
National Institute of Advanced Industrial Science and Technology,
Tokyo 135-0064, Japan
tsuda@k.u-tokyo.ac.jp
[4] Center for Materials Research by Information Integration,
National Institute for Materials Science, Ibaraki 305-0047, Japan

Abstract. Recent pattern mining algorithms such as LAMP allow us
to compute statistical significance of patterns with respect to an outcome variable. Their p-values are adjusted to control the family-wise
error rate, which is the probability of at least one false discovery occurring. However, they are a poor fit for medical applications, due to their
inability to handle potential confounding variables such as age or gender. We propose a novel pattern mining algorithm that evaluates statistical significance under confounding variables. Using a new testability
bound based on the exact logistic regression model, the algorithm can
exclude a large quantity of combination without testing them, limiting
the amount of correction required for multiple testing. Using synthetic
data, we showed that our method could remove the bias introduced by
confounding variables while still detecting true patterns correlated with
the class. In addition, we demonstrated application of data integration
using a confounding variable.

Keywords: Significant pattern mining · Multiple testing · Exact logistic
regression

1 Introduction

Statistical significance measures how well a claimed proposition is supported by
data. Use of p-values is ubiquitous and indispensable in scientific literature [10].
Given a set of examples (e.g. itemsets, sequences or graphs) and associated
class labels, recent methods such as Limitless Arity Multiple-testing Procedure
(LAMP) [15] and its subsequent studies [9,13] can list all patterns (e.g. subsets,
subsequences or subgraphs) that are significantly associated with the class label.

© Springer International Publishing Switzerland 2016
J. Bailey et al. (Eds.): PAKDD 2016, Part I, LNAI 9651, pp. 277–289, 2016.
DOI: 10.1007/978-3-319-31753-3_23

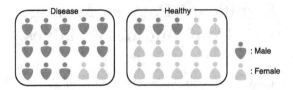

Fig. 1. Effect of confounding variable. If one tries to find patterns associated with the disease, patterns associated with gender are likely to be found instead. To avoid this issue, the statistical test has to take the confounding variable (gender) into account.

Unlike earlier methods [17], these approaches provide a guarantee that the probability of at least one false discovery occurring (i.e. family-wise error rate, FWER) is smaller than a pre-defined threshold (conventionally 0.05 or 0.01).

However, these methods fail to take into account the various biases that can be introduced by confounding variables (e.g. age or gender of patients) in observational medical data. Figure 1 provides a simple illustration of the type of issues that can arise when potential confounding effects are ignored, leading to unacceptable results. While all patients (represented by their gene expression levels) are separated into two classes along their disease status, we can see that one gender is heavily over-represented in each group. Ignoring the confounding effect of gender on this dataset would likely find significant genes that are related to the gender of the patient, rather than the disease.

To remove the bias introduced by confounding factors, statisticians typically use likelihood ratio tests based on logistic regression models [12]. However, due to the very large number of candidate combinations, the correction factor for multiple testing can grow extremely large, removing any chance of finding statistically significant patterns. Reducing the correction factor using the same technique as existing algorithms [9,13,15], requires finding a good testability bound to prune out non-significant patterns (without testing them), which is notoriously difficult when using such a method based on logistic regression models.

To achieve this goal, we turned to *exact logistic regression* [7], a model less popular than likelihood ratio testing but commonly used in statistical and biological communities via tools such as SPSS [8] or R [19]. The term "exact" essentially indicates that the test statistics are computed exactly without large-sample approximations, whereas a likelihood-based approach uses asymptotic statistics. Crucially, the computation of exact statistics does not require an iterative process to update the parameters until convergence [1].

In this article, we propose a novel pattern mining algorithm, LAMP-ELR (Limitless Arity Multiple-testing Procedure with Exact Logistic Regression), that evaluates statistical significance for all possible patterns using exact logistic regression. LAMP-ELR can be used in a number of different scenarios: controlling FWER even when the dataset is affected by confounding variables, and *Data integration*, which is a useful technique to merge similar experimental data taken from different sources so as to provide a larger set for analysis. We show that our algorithm contributes to these scenarios by applying it to both synthetic data and the Predictive Toxicology Challenge (PTC) dataset [3].

2   Related Work

Statistical tests based on logistic regression exist to deal with confounding effects in datasets, such as exact interface [7] or approximate conditioning [11]. However, these procedures only perform testing of one hypothesis at a time, requiring us to adjust for multiple testing to control for false discoveries [2].

Bonferroni correction is a widely used multiple testing procedure, but is known to be overly conservative in computing the FWER and cannot be applied for significant pattern mining, where the number of combinations can grow exponentially. More recently, multiple testing procedures for controlling the FWER when detecting significant patterns have been proposed [9,13,15]. They use Bonferroni-like multiple testing procedures with Tarone's p-value bound strategy [14] to improve the sensitivity of the correction through frequent itemset mining. However, none of the algorithms can take into account possible confounding effects introduced by a covariate, making it poorly suited for many types of real-life data.

3   Significant Pattern Mining

This section lays out the theoretical foundations of significant pattern mining. Given a set of transactions, each labelled with a positive or negative class, and a pattern X, the transaction set can be divided between those where the pattern occurs and the rest, producing a contingency table. Statistical association between pattern occurrence and class label is measured using p-values, which are computed by hypothesis testing such as Fisher's exact test and χ-squared test. If the p-value is smaller than a threshold, the pattern is regarded as statistically significantly associated with the class label.

Due to the nature of pattern mining, it can be necessary to test a huge number of such hypotheses. In itemset mining with n items, for example, the number of possible patterns ℓ can be as large as $2^n - 1$. Such analysis causes serious false discoveries, known as the multiple testing problem. We therefore need to control the FWER. If we note ℓ null hypotheses as $H_1, \ldots, H_\ell$, and if V describes the number of false discoveries, the FWER can be expressed as $P(V > 0 \mid \bigcap_{i=1}^{\ell} H_i)$.

Multiple testing procedures [2] adjust the p-value threshold so that the FWER is controlled under a pre-specified value (usually $\alpha = 0.05$ or 0.01). For example, the Bonferroni correction adjusts the threshold to $\delta = \alpha/\ell$ by calculating the FWER bound as

$$P(V > 0 \mid \bigcap_{i=1}^{\ell} H_\ell) \le \sum_{i=1}^{\ell} P(p_i \le \delta \mid H_i) \le \ell\delta = \alpha. \tag{1}$$

Thus, it is clear that Bonferoni's threshold keeps the FWER below α. Unfortunately, Bonferroni correction is inappropriate for use with pattern mining: since ℓ grows exponentially with the number of items, δ sinks to a very small value, making new discoveries extremely unlikely.

Recently, Terada et al. [15] have shown that this issue can be mitigated by employing the following trick, first proposed by Tarone [14]. Suppose one can bound the p-value p_i from below with a function f_i depending only on marginal counts, i.e. $p_i \geq f_i$. We call this bound a *min-p bound* or min-p for short. If $f_i > \delta$, the hypothesis can never be rejected because $P(p_i \leq \delta \mid H_i) = 0$. Using this property, the FWER bound of Eq. 1 can become tighter:

$$P(V > 0 \mid \bigcap_{i=1}^{\ell} H_\ell) \leq \sum_{i=1}^{\ell} P(p_i \leq \delta \mid H_i)$$
$$\leq \sum_{\{i \mid f_i \leq \delta\}} P(p_i \leq \delta \mid H_i) \leq |\{i \mid f_i \leq \delta\}|\delta.$$

LAMP uses customised pattern mining algorithms to find the maximum value of δ that keeps the FWER below α. The resulting δ is normally much larger than Bonferroni's correction factor, resulting in more significant discoveries. However, LAMP cannot handle confounding effects introduced by a covariate, making them poorly suited for many types of real-life data.

4   Exact Logistic Regression

4.1   Logistic Regression

In biological and medical domains, the logistic regression model is the method of choice for deriving p-values with confounding variables [12]. Let us consider evaluating the association between a binary outcome $y \in \{0,1\}$ and an explanatory variable $x_1 \in \{0,1\}$. A categorical covariate $x_2 \in \{0,\ldots,K\}$ is assumed to be known as a confounding variable. Using dummy coding for the categorical variable, the logistic model, with π denoting the probability of y being 1, is defined as

$$\log(\frac{\pi}{1-\pi}) = \gamma + \beta_1 x_1 + \sum_{k=1}^{K} \beta_{2k} [\![x_2 = k]\!],$$

where $[\![\mathcal{P}]\!] \in \{0,1\}$ is the boolean variable resulting from the evaluation of predicate $\mathcal{P}$. To measure the statistical significance of x_1, we consider an alternative model $\beta_1 = 0$.

Incorporating this test into pattern mining is difficult because the p-value is based on large-sample approximations and very inaccurate for biased contingency tables [4]. Deriving a min-p bound requires considering the most biased table, so the min-p can be unreliable.

4.2   Exact Inference

Here we introduce the concept of exact logistic regression [7]. Although based on the same model as regular logistic regression, its p-value is computed exactly, without relying on large-sample approximations, which makes it possible to derive an easily-computable min-p bound.

Figure 2 illustrates exact logistic regression testing. We consider a set of q examples, where the ith example is a tuple $\{y_i, x_{1i}, x_{2i}\}$. Without loss of generality, the examples are assumed to be sorted with respect to x_2 (the covariate). Let us define q_k as the number of examples whose covariate is k ($x_{2i} = k$). Let vectors $\mathbf{y}, \mathbf{x_1}, \mathbf{x_{2k}}$ respectively denote the q-dimensional outcome, explanatory and k^{th} (out of K) covariate values of the q examples. Then, the sufficient statistics for γ, β_1 and β_2 are defined as $\tau_0 = \mathbf{1}^T\mathbf{y}$, $\hat{\tau}_1 = \mathbf{x_1}^T\mathbf{y}$ and $\tau_{2k} = \mathbf{x_{2k}}^T\mathbf{y}$, respectively.

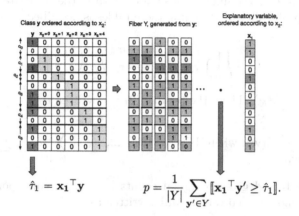

$$\hat{\tau}_1 = \mathbf{x_1}^T\mathbf{y} \qquad\qquad p = \frac{1}{|Y|}\sum_{y'\in Y}[\![\mathbf{x_1}^T\mathbf{y}' \geq \hat{\tau}_1]\!].$$

Fig. 2. Illustration of exact logistic regression testing. Assuming data (both explanatory variable x_1 and class y) ordered according to the covariate (x_2), we consider the fiber Y made of all the permutations of y that preserve the occurrence counts q_k within each covariate category k. The observed statistic $\hat{\tau}_1$ is obtained from the dot product of y and x_1 and the p-value is the ratio of y' (elements of the fiber Y) for which the dot product with x_1 is higher than $\hat{\tau}_1$, i.e. the probability of obtaining a distribution more biased than what is observed, while keeping the same marginal counts in each covariate categories.

When testing for statistical significance of the explanatory variable, we would like to find out if the observation $\mathbf{y}$ is special in that it shows particularly high correlation to $\mathbf{x_1}$. If the obtained level of explanatory correlation $\hat{\tau}_1$ is easily predictable from the existing information $\hat{\tau}_0$ and $\hat{\tau}_2$, it should not be regarded as statistically significant. In exact logistic regression, the sample space is defined as the set of all sample vectors whose positive class size and covariate correlation are constrained to the observed value, called a *fiber* [6],

$$Y = \{\mathbf{y}' \mid \mathbf{y}'^T\mathbf{1} = \hat{\tau}_0, \mathbf{y}'^T\mathbf{x_{2k}} = \hat{\tau}_{2k}, k = 1, \ldots, K\}.$$

It is equivalently represented as

$$Y = \{\mathbf{y}' \mid \mathbf{y}'^T\mathbf{x_{2k}} = \hat{\tau}_{2k}, k = 0, \ldots, K\}, \text{ where } \hat{\tau}_{20} = \hat{\tau}_0 - \sum_{k=1}^{K}\hat{\tau}_{2k}.$$

To calculate the p-value with respect to the explanatory variable, the null distribution of $\hat{\tau}_1$ is defined as uniform sampling from Y. Then, the p-value is defined as

$$p = \frac{1}{|Y|} \sum_{y' \in Y} [\![\mathbf{x_1}^\top \mathbf{y'} \geq \hat{\tau}_1]\!]. \tag{2}$$

The p-value is computed by making K contingency tables for each value of x_2. The vectors $\mathbf{y_k}$ and $\mathbf{x_{1k}}$ denote the outcome and explanatory values in the covariate category k, respectively, and let $m_k = \mathbf{y_k}^\top \mathbf{1}$, $s_k = \mathbf{x_{1k}}^\top \mathbf{1}$ and $t_k = \mathbf{y_k}^\top \mathbf{x_{1k}}$. n_k represents the number of samples with in the covariate category k.

The joint probability of obtaining these tables is described as

$$C(\mathbf{t}) = \prod_{k=1}^{K} \left\{ \binom{m_k}{t_k} \binom{n_k - m_k}{s_k - t_k} \middle/ \binom{n_k}{s_k} \right\},$$

where $\mathbf{t} = (t_1, \ldots, t_K)$. Then, the p-value (2) is rewritten as

$$p = \sum_{\mathbf{w} \in W(\mathbf{t})} C(\mathbf{w}), \text{ where } W(\mathbf{t}) = \left\{ (w_1, \ldots, w_K) \mid \sum_{k=1}^{K} w_k \geq \sum_{k=1}^{K} t_k \right\}.$$

The min-p bound given marginal counts s_k, m_k, n_k, corresponding to the p-value of the most biased table, can be written as

$$f = \prod_{k=1}^{K} f_k(s_k), \text{ where } f_k(s_k) = \begin{cases} \dbinom{m_k}{s_k} \middle/ \dbinom{n_k}{s_k} & (\text{if } s_k \leq m_k) \\ 1 \middle/ \dbinom{n_k}{m_k} & (\text{otherwise}). \end{cases} \tag{3}$$

5   Min-P Decrease Algorithm

This section presents LAMP-ELR: an algorithm that uses a min-p bound (3) to solve the multiple testing problem and find statistically significant patterns.

Let E denote a set of items. Let $f(X)$ denote the minimum p-value (i.e. min-p) for itemset $X \subseteq E$, and $\delta \in \mathbb{R}$ denote a threshold for p-values that discriminates between significant and non-significant patterns. Then, the number of all testable patterns can be described as

$$\kappa(\delta) = |\{X \subseteq E \mid f(X) \leq \delta\}|.$$

If the following bounding condition is satisfied:

$$\delta \leq \alpha/\kappa(\delta) \tag{4}$$

the FWER is bounded by α. We compute the largest δ satisfying this condition.

Algorithm 1. LAMP-ELR algorithm, which handles confounding variables.

1: Global variables: $\delta = \alpha$ and S: empty priority queue structure.
2: **function** MINPDECREASE(X)
3:     $S.insert(X, f(X))$ ▷ Insert X with priority $f(X)$
4:     **while** $\delta \geq \alpha/|S|$ **do**
5:         $\delta \leftarrow S.max\_priority()$ ▷ Retrieve highest priority
6:         $T \leftarrow |S|$
7:         **while** $\delta \geq \alpha/T$ **do**
8:             $S.pop()$ ▷ Remove element with highest priority
9:         **end while**
10:     **end while**
11:     **for** each item $e > tail(X)$ **do**
12:         **if** $f(X \cup e) < \delta$ **then**
13:             MinPDecrease($X \cup e$)
14:         **end if**
15:     **end for**
16: **end function**

5.1   Algorithm for K Contingency Tables

We propose a depth-first algorithm, called LAMP-ELR, which follows a similar strategy to a fast version of the LAMP [9], to handle the case of K contingency tables. Our algorithm finds the optimal δ by using a key point: When an item e is added to the itemset X, the min-p bound for the itemset becomes larger.

We first show that the function in Eq. 3 holds this property.

Theorem 1. *If* $s_k \geq s'_k$, $f_k(s_k) \leq f_k(s'_k)$.

Proof. If $s_k \leq m_k$, $f_k(s_k) = \{(m_k - x + 1)/(n_k - x + 1)\}f_k(s_k - 1)$. If $(m_k - x + 1)/(n_k - x + 1) < 1$, $f_k(x) < f_k(x - 1)$. If $s_k > m_k$, $f_k(s_k)$ is a constant. Therefore, Theorem 1 holds.

Theorem 2. *For itemsets* X *and* $X' = \{X \cup e\}$, $f(X) \leq f(X')$.

Proof. Let s_k and s'_k be the frequencies of the group k for X and X'. When $s_k \geq s'_k$, $f_k(s_k) \leq f_k(s'_k)$ for $k \in \{1, ..., K\}$ from Theorem 1. Therefore, Theorem 2 holds.

The function MINPDECREASE, outlined in Algorithm 1, performs a depth-first search to collect as many testable patterns as possible while conforming to the bounding condition (Eq. 4). Upon starting, the threshold of min-p is set as $\delta = \alpha$ and the priority queue holding eligible patterns is empty: $S = \emptyset$. S stores all the patterns traversed so far whose priority (min-p) is below δ. If $\delta \geq \alpha/|S|$, it is clear that $\delta \geq \alpha/\kappa(\delta)$, since $\kappa(\delta)$ refers to the number of patterns whose min-p is smaller than δ and $|S|$ refers to those among the patterns traversed so far. We therefore reduce the current δ until the bounding condition is satisfied. On lines 12–14, the current itemset, X, is extended by adding one item e and the function MINPDECREASE is recursively called. If $f(X \cup e) \geq \delta$, there is no need to explore all further patterns that contain $X \cup e$ due to the monotonicity of $f(X)$ (Theorem 2). Inserting into the priority queue S ($S.insert()$) and removing the element with highest priority ($S.pop()$) take $\mathcal{O}(\log |S|)$ time, while retrieving the maximum priority ($S.max\_priority()$) can be done in constant time.

The following theorem proves that the obtained threshold bounds the FWER.

Theorem 3. *Let δ_{end} denote the value of δ at the end of the algorithm. It satisfies the bounding condition* (4), *i.e., $\delta_{end} \leq \alpha/\kappa(\delta_{end})$.*

Proof. Due to the pruning condition on line 12 of Algorithm 1, only elements with min-p value strictly smaller than δ can be added to S, therefore the value of $\delta = S.max\_priority()$ never increases with each iteration. Let S_{end} denote S at the end of the algorithm. Since δ can only decrease, the patterns whose min-p is smaller than δ_{end} do not get pruned out and are included in S_{end}. Also, patterns whose min-p is larger than or equal to δ_{end} are eliminated at line 8 and therefore: $|S_{final}| = \kappa(\delta_{end})$. Since the update condition at line 4 ensures that $\delta < \alpha/|S|$, we have: $\delta_{end} < \alpha/\kappa(\delta_{end})$.

Algorithm 1 is designed for itemset mining, but extensions to sequence or graph mining are straightforward, as long as depth-first search is adapted.

5.2   Speed and Memory Usage Improvements

Because Algorithm 1 does not make use of the collection of itemsets X after completing the calculation of min-p, instead of keeping those patterns in S, we can simply store their total count. We modify Algorithm 1 using a special type of priority queue C, instead of S, to store the histogram counter: insertion into C only happens (with value 1) when the priority key did not previously exist, and its value is incremented otherwise. In most typical implementation of the priority queue structure, such an operation, which we note $C.insert\_or\_increment()$, can easily be implemented with the same computational complexity as a traditional insertion, in $\mathcal{O}(log|C|)$.

6   Experiment on Synthetic Dataset

We first evaluated the performance of our procedure using synthetic dataset with an application to significant itemsets detection. Our algorithm was implemented by modifying *lcmplusplus*[1], a C++ implementation of the LCM algorithm, which is currently the fastest frequent itemset mining method available [16]. All experiments were run under Mac OS 10.4.4 on a 1.7 GHz Intel Core i7.

The synthetic datasets consist of 1000 transactions, shared equally between positive and negative labels. All positive and negative transactions are assigned to Group 1 or 2 according to a bias factor r (similar to the gender bias in Fig. 1).

---

[1] https://code.google.com/p/lcmplusplus/.

Table 1. PTC datasets. All datasets use 22 node labels and 4 edge labels.

	Size	# positives	Avg $|V|$	Avg $|E|$	Max $|V|$	Max $|E|$	Min $|V|$	Min $|E|$
FM	349	143	25.25	25.62	109	108	2	1
MM	336	129	25.04	25.40	109	108	2	1
FR	351	121	26.08	26.53	109	108	2	1
MR	344	152	25.56	25.96	109	108	2	1

A ratio of r out of all positive transactions are assigned to Group 1, along with $1 - r$ of all negative transactions. All other transactions are assigned to Group 2. This dataset contains 100 items, and each item appears with a default probability of 0.1 within each transaction. We introduce *true patterns* and *shadow patterns* containing three items each. The 5 *true patterns* correlate with the class of the transactions: each pattern appears in a randomly chosen 20 % of positive transactions and is absent from a random 20 % of negative transactions. The 5 *shadow patterns* simulate a confounding effect by correlating to the group of the transactions: each shadow pattern appears in a random 20 % of Group 1 transactions and is absent from a random 20 % of Group 2 transactions. We generated 3 groups at different levels of confounder bias: an unbiased dataset ($r = 0.5$), low-bias dataset ($r = 0.7$), and high-bias dataset ($r = 0.9$).

For each dataset, we compared the significance (at level $\alpha = 0.05$) of the patterns obtained through frequent itemset mining using a traditional implementation of the LAMP (performing one-sided Fisher's exact tests) on one side, and LAMP-ELR (performing exact logistic regression tests) on the other. A summary of the results can be seen in Fig. 3. Both methods detect true patterns equally well independent of r. However, as suspected, when group bias is introduced and increased, LAMP tends to select more of the shadow patterns along with the true patterns, with a steadily decreasing ratio of true positives among the discoveries. LAMP-ELR, on the other hand, is able to identify and reject patterns whose occurrence is due to the confounder, rather than the actual class.

7   Experiment on Data Integration

To demonstrate the usefulness of our method, we tested it on the PTC dataset [3]. The dataset is made of graph structures representing chemical compounds and labelled with an indication of carcinogenicity over four groups: Female Mouse (FM), Male Mouse (MM), Female Rat (FR) and Male Rat (MR). Statistics for this dataset are summarised in Table 1.

We implemented our method in C++, using gSPAN [18] as a base. The experiment was run on a machine with two Intel Xeon E5-2680v2 CPUs at 2.8 GHz and 64 GB of RAM. The significance level α was set to 0.05.

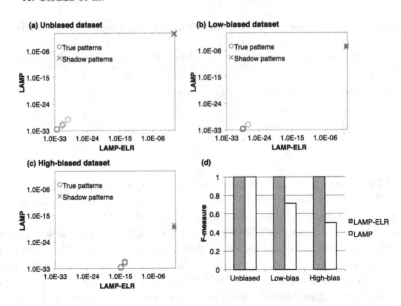

Fig. 3. Comparison between LAMP (ignoring potential confounding effects) and LAMP-ELR (taking confounder into account). (a)–(c) Comparing p-values of true and shadow patterns by the two methods. For high values of confounder bias, LAMP tends to small p-values to the false patterns. (d) F-measure (ratio of true patterns discovered over total number of patterns discovered) for both LAMP and LAMP-ELR, at different levels of confounding bias.

7.1   Significant Subgraphs

In order to outline the advantage of our method for integrated analysis, we compared the results of significant subgraph mining over each individual FM, MM, FR and MR dataset, using LAMP, with the analysis of the dataset obtained by merging all four (adding the subgroup origin as a covariate), using LAMP-ELR. Result statistics are compiled in Table 2.

The correction factor for each of the smaller individual dataset analysis is about 100 times smaller than that of the integrated dataset. However, the smaller

Table 2. Analysis of PTC Data: highlighting the importance of integration

Method	Dataset	Correction factor	# significant patterns found
LAMP	FM	16333	0
LAMP	MM	17713	0
LAMP	FR	22543	0
LAMP	MR	12192	0
LAMP-ELR	FM+MM+FR+MR (using covariate)	1141376	9250

numbers of samples lead to higher pattern p-values, producing no significant patterns despite the lower correction. When integrating the datasets, the total number of transaction approximately quadruples and we are able to identify 9250 subgraphs with statistically significant toxic effect. Meanwhile, the use of a confounder variable identifying the individual dataset from which each transaction is taken, guarantees that the patterns identified are not caused by artefacts or other subset biases.

7.2 Performance Evaluation

We assessed the efficiency of our pruning method by comparing it with a naive brute-force version which calculates the minimum p-value for any itemset occurring at least once. Figure 4 shows the time performances when considering subgraphs of increasingly large edge sizes. As could be expected, the calculation time exponentially increases in the brute-force approach. By contrast, the running time of LAMP-ELR increases linearly up to approximately 25 edges, with no noticeable change afterward. For instance, the brute force algorithm requires 11,644 s when considering subgraph of edge size no more than 15, whereas LAMP-ELR can finish in 2,885 s without edge size limit. Our pruning technique succeeds in dramatically reducing the running time, enabling us to detect high-dimensional combinatorial effects while taking into account potential effects from confounder variables.

The improvements for space and speed presented in Sect. 5.2 mainly depend on the number of patterns sharing identical values of min-p, which is used as a (unique) key in the priority queue structure C. Figure 5 shows the frequency of each unique value of min-p present in the PTC dataset analysis. A large number of itemsets share identical values of min-p. For example, 115,816 patterns have the identical min-p value $1.44E - 10$. Eliminating this redundancy from storage and insertion computation is directly related with the overall speed gain of our optimised method over a standard approach.

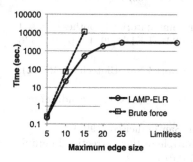

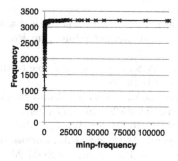

Fig. 4. Calculation time when increasing the edge size of the subgraphs investigated.

Fig. 5. Cumulative frequency of itemsets with a minimum p-value.

8   Conclusion

In this work, we developed a significant pattern mining method based on exact logistic regression statistics that can account for potential confounding effect from a covariate. This is, to our knowledge, the first such method to combine a covariate-aware model with optimised multiple testing procedures to keep significance sensitivity up while limiting the effect of confounders.

In future work, we plan to improve confounder detection for cases where they might not be known in advance. Several methods have been proposed to identify confounders using probabilistic models [5], that are not currently compatible with pattern mining problems, but show potential promises for our work.

Acknowledgments. AT is supported by JST PRESTO and JSPS Research Fellowships for Young Scientists. The research of K.T. was supported by JST CREST, JST ERATO, RIKEN PostK, NIMS MI2I, Kakenhi Nanostructure and Kakenhi 15H05711.

References

1. Diaconis, P., Sturmfels, B.: Algebraic algorithms for sampling from conditional distributions. Ann. Stat. **26**(1), 363–397 (1998)
2. Dut, S., Van Der Laan, M.J.: Multiple Testing Procedures with Applications to Genomics. Springer Science, Heidelberg (2007)
3. Helma, C., et al.: The predictive toxicology challenge 2000–2001. Bioinformatics **17**(1), 107–108 (2001)
4. Hirji, K.: Exact Analysis of Discrete Data. Taylor and Francis, London (2006)
5. Janzing, D., et al.: Identifying confounders using additive noise models. In: Proceedings of the Twenty-Fifth Conference on UAI, pp. 249–257 (2009)
6. Karwa, V., Slavkovic, A.: Conditional inference given partial information in contingency tables using Markov bases. WIREs Comput. Stat. **5**, 207–218 (2013)
7. Mehta, C.R., Patel, N.R.: Exact logistic regression: theory and examples. Stat. Med. **14**(19), 2143–2160 (1995)
8. Menard, S.: Applied Logistic Regression Analysis, vol. 106. Sage, Beverley Hills (2002)
9. Minato, S., et al.: Fast statistical assessment for combinatorial hypotheses based on frequent itemset mining. In: Proceedings of ECML/PKDD 2014, pp. 422–436 (2014)
10. Noble, W.S.: How does multiple testing correction work? Nat. Biotechnol. **27**(12), 1135–1137 (2009)
11. Pierce, D.A., Peters, D.: Improving on exact tests by approximate conditioning. Biometrika **86**(2), 265–277 (1999)
12. Sokal, R., Rohlf, F.: Biometry, 3rd edn. Freeman, San Francisco (1995)
13. Sugiyama, M., López, F.L., Borgwardt, K.M.: Multiple testing correction in graph mining. In: Proceedings of SDM 2015, pp. 37–45 (2015)
14. Tarone, R.: A modified bonferroni method for discrete data. Biometrics **46**, 515–522 (1990)
15. Terada, A., et al.: Statistical significance of combinatorial regulations. Proc. Nat. Acad. Sci. USA **110**(32), 12996–13001 (2013)

16. Uno, T., et al.: LCM: an efficient algorithm for enumerating frequent closed item sets. In: Proceedings of FIMI 2003 (2003)
17. Webb, G.I.: Discovering significant rules. In: Proceedings of KDD 2006, pp. 434–443 (2006)
18. Yan, X., Han, J.: gSpan: graph-based substructure pattern mining. In: ICDM 2002, pp. 721–724 (2002)
19. Zamar, D., McNeney, B., Graham, J.: elrm: software implementing exact-like inference for logistic regression models. J. Stat. Softw. **21**, 1–18 (2007)

Building Compact Lexicons for Cross-Domain SMT by Mining Near-Optimal Pattern Sets

Pankaj Singh[(✉)], Ashish Kulkarni, Himanshu Ojha, Vishwajeet Kumar, and Ganesh Ramakrishnan

Computer Science and Engineering, IIT Bombay, Mumbai, India
pr.pankajsingh@gmail.com, kulashish@gmail.com, himanshuojha.lko@gmail.com,
vishwajeetkumar86@gmail.com, ganesh@cse.iitb.ac.in

Abstract. Statistical machine translation models are known to benefit from the availability of a domain bilingual lexicon. Bilingual lexicons are traditionally comprised of multiword expressions, either extracted from parallel corpora or manually curated. We claim that "patterns", comprised of words and higher order categories, generalize better in capturing the syntax and semantics of the domain. In this work, we present an approach to extract such patterns from a domain corpus and curate a high quality bilingual lexicon. We discuss several features of these patterns, that, define the "consensus" between their underlying multiwords. We incorporate the bilingual lexicon in a baseline SMT model and detailed experiments show that the resulting translation model performs much better than the baseline and other similar systems.

Keywords: Submodular · Pattern extraction · Cross-domain SMT

1 Introduction

A statistical machine translation (SMT) model typically relies on the availability of a large parallel corpus, often collected from multiple sources and spanning different domains. While a domain-specific corpus might share some of its lexical characteristics with the cross-domain corpus, it often differs in its language usage and vocabulary. A cross-domain SMT model might, therefore, fail to reliably translate an in-domain text. While it is possible to train an in-domain translation model, domain-specific parallel corpus is either non-existent or scarce and expensive to generate. The problem of domain adaptation deals with augmenting a cross-domain translation model to reliably translate an in-domain text and poses an interesting research challenge [8].

Although in-domain parallel text might be difficult to obtain, in-domain bilingual lexicons are often readily available or could be manually curated. Typically, these are restricted to words or short phrases specific to the domain of interest. A medical domain bilingual lexicon, for instance, consists of technical and popular medical terminology covering the anatomy of body, certain diseases, medicines *etc.* In addition to these however, a domain corpus, due to its specific language

© Springer International Publishing Switzerland 2016
J. Bailey et al. (Eds.): PAKDD 2016, Part I, LNAI 9651, pp. 290–303, 2016.
DOI: 10.1007/978-3-319-31753-3_24

Table 1. Examples of recurring patterns, sample snippets covered by them and the number of such covered snippets (in brackets) for the EMEA corpus

PATTERN: in patients with ⟨ CAT1 ⟩ (568)	Contains ⟨ CAT2 ⟩ mg of ⟨ CAT3 ⟩ (91)
in patients with HIT type II	capsule contains 25 mg of lenalidomide
in patients with CNS metastases	tablet contains 300 mg of maraviroc
in patients with ESRD	syringe contains 100 mg of anakinra
in patients with normal and impaired renal function	tablet contains 2.3 mg of sucrose
in patients with previous history of pancreatitis	capsule contains 200 mg of pregabalin
in patients with cirrhosis of the liver	vial contains 10 mg of the active substance
	tablet contains 30 mg of aripiprazole

structure, is often replete with redundant phrases. Consider for instance, the phrase *"...be given marketing authorisation"*, appearing 218 times in the EMEA medical corpus [23]. These, if extracted and translated in a bilingual lexicon, might aid in-domain translation [21,26]. In fact, repetition in a domain corpus could be further exploited by observing that certain phrases, which might themselves be infrequent, tend to have "consensus" when generalized to higher-level patterns. Table 1 illustrates two patterns and corresponding sample phrases extracted from the EMEA medical domain corpus. These patterns are typically n-grams of tokens, domain-specific categories or higher-level phrase classes (noun phrase, verb phrase *etc.*).

Given a domain corpus, it is not obvious how to extract a set of such patterns to be manually translated. Moreover, in the absence of a parallel in-domain corpus, translation of these patterns requires manual effort, which poses other challenges. Specifically, syntactically well-formed patterns like *"the CAT5 of treatment"* might be easier for humans to translate than others like *"CAT4 condition has"*. Chen *et. al.* [3] present this and other quality criteria that every pattern must satisfy to be worth being translated in order to aid cross-domain SMT applications. We will refer to such patterns as *quality patterns*. In this work, we generalize the search space of patterns as well as the quality criteria that a pattern must meet.

More importantly, two or more quality patterns could have instances that significantly overlap in their spans in the corpus. Is translating each such *quality* pattern really necessary? We expect the human effort for translating patterns to have a budget constraint and therefore, a compact *set of patterns* is desirable. For example, it is desirable to extract a set of patterns (for bilingual lexicon), such that, the set maximally covers the corpus. We argue that some formulations of this problem are natural instances of submodular maximization. A set function $f(.)$ is said to be submodular if for any element v and sets $A \subseteq B \subseteq V \setminus \{v\}$, where V represents the ground set of elements,

$f(A \cup \{v\}) - f(A) \geq f(B \cup \{v\}) - f(B)$. This is called the diminishing returns property and states, informally, that adding an element to a smaller set increases the function value more than adding that element to a larger set. Submodular functions naturally model notions of coverage and diversity, and therefore, a number of subset selection problems can be modeled as forms of submodular optimization [7, 11].

We illustrate the relevance of the submodular coverage function to pattern-subset selection in Fig. 1. We plot the corpus coverage (in terms of number of words) with increasing number of patterns in the set, for pattern lengths varying from 3 to 9. In each case, while the coverage improves with increasing number of patterns, the gain in coverage progressively diminishes with growth in the size of the subset.

Our contribution is a framework to curate a high quality bilingual lexicon based on three key ideas. Our first two ideas generalize the approach of Chen et al. [3].

1. **Language of patterns:** A pattern could either be lexical, comprised of words alone, or it could be a combination of words and higher-level categories.
2. **Quality criteria for a pattern:** The quality (or cost) of every instance of a pattern is a function of several features including its frequency in the corpus and whether or not it is syntactically well-formed. The quality (or cost) of a pattern is then a simple (modular) aggregation of the instance costs.
3. **Quality criteria for a set of patterns:** We define the "goodness" of a set of patterns based on element-wise non-decomposable submodular costs.

We incorporate these patterns along with their translations, as entries in a bilingual lexicon and study[1] its effect on the translation accuracy for the domain adaptation of a baseline SMT model. While significantly improving over the baseline, we also show significant improvement over the modular setting of Chen et al.

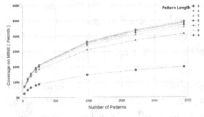

(a) Corpus EMEA: corpus coverage vs. #patterns

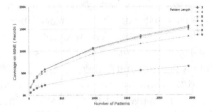

(b) Corpus KDE4: corpus coverage vs. #patterns

Fig. 1. Gain in coverage shows diminishing returns with increasing number of patterns in the set

---

[1] We release our code for optimal pattern-set identification, as well as the lexicons. https://www.cse.iitb.ac.in/~ganesh/Publications.html.

2   Related Work

Extraction of Bilingual Multi-word Expressions (BMWE): SMT systems often use word-to-word alignment approaches for inferring translation probabilities from bilingual data [17,25]. However, in some cases it might not be possible to perform word-to-word alignment between two phrases that are translations of each other [10]. This has motivated a body of work [10,18,21] on automatic extraction of multi-word expressions from bilingual corpora. Ren et al. [21] propose multiple techniques to integrate BMWE's into a phrase-based SMT system and show improvement over the baseline translation system. Recently, Liu et al. [12] proposed an approach to mine quality phrases from large text corpora. They use a phrasal segmentation-based approach for phrase mining and combine that with several phrase quality assessment metric in a scalable framework. While our approach is inspired by these works, we differ from them in that we aim to extract generalized patterns comprising words and categories. Also, we do not assume availability of a parallel corpus in the target domain.

Domain Adaptation: Typically, the application domain of a translation system might be different from the domain of the system's training data. In-domain parallel corpus might either be non-existent or scarce, but, in-domain monolingual corpus is usually available. The problem of domain adaptation[2] has therefore been in focus and there has been work [8,16,26] to build in-domain translation lexicons and combine them with out-of-domain parallel corpus to achieve in-domain translation. Koehn and Schroeder [8] use limited in-domain parallel corpus to train a language model and a translation model and present techniques to integrate them with corresponding models trained on an out-of-domain corpus. Wu et al. [26] manually create an in-domain lexicon where the lexicon entries are restricted to words. They propose an algorithm to combine an out-of-domain bilingual corpus, an in-domain bilingual lexicon, and monolingual in-domain corpora in a unified framework for in-domain translation.

Pattern Mining: The other body of work most related to our approach comes from the area of pattern mining. While most earlier work [22] dealt with identifying consecutive word sequences, Joshi et al. [6] present an efficient approach to mine significant non-consecutive word sequences, where, *significance* is captured by the support measure. Contrary to mining patterns that satisfy pre-specified criterion, there has also been work on interactive pattern mining [1,2,27] that uses human feedback to identify a set of *interesting* patterns. Chen et. al. [3] proposed an English-Chinese medical summary translation system that adapts a baseline SMT model with significant patterns (of lexical as well as medical type tokens) learned from an English medical summary corpus. The quality of a pattern is assessed based on its frequency in the corpus and its linguistic completeness. While being closest to our work, we differ from them and the other aforementioned works in two ways. Firstly, we realize that the quality criterion for a set of patterns is not always a modular function of quality of the

---

[2] http://www.statmt.org/wmt07/shared-task.html.

constituent patterns in the set. We define several quality criteria based on both element-wise decomposable (modular) costs and element-wise non-decomposable (non-modular) costs and combine them in a mathematical formalism for the task of significant pattern mining. Secondly, domain-specific classes often rely on the availability of corresponding term lexicons. Our framework also makes use of general phrase classes such as noun phrases (NP), verb phrases (VP), thereby extracting generic patterns whose instances themselves might not be frequent in a corpus (Refer to Fig. 1). Moreover, the use of phrase classes allows for the induction of new instances in a class (type) lexicon.

3   Framework

The task of lexicon curation finds applications in several NLP tasks including machine translation. We present a formulation of the problem and a solution framework that one could invoke based on underlying application requirements. The lexicon is composed of quality patterns extracted from a domain corpus and for the specific task of machine translation with low resource constraint, we then acquire translations of these patterns to create an in-domain *bilingual lexicon*.

3.1   Formal Problem Definition

We are given a domain corpus $\mathcal{C}$ and optionally a set of "types" $\mathcal{T}$. A type might represent a *domain type*, like *disease* in medical domain, a *lexical type*, like *noun phrases* or a complex type involving a combination of these. The problem of lexicon curation is to extract from $\mathcal{C}$, a set H of quality patterns, as per a quality function $Q_\mathcal{C}(h)$ for the quality of a pattern $h \in H$ in the corpus and a quality function $Q_\mathcal{C}(H)$ for the quality of the set H.

3.2   Solution Framework

We define and describe below the components of our solution framework.

1. **Context Free Grammar G:** A context free grammar (CFG) allows us to encode our types and is comprised of a set V of non-terminals, a set Σ of terminals, a start symbol $S \in V$ and a set P of productions $\alpha \to \beta$, where $\alpha \in V$ and $\beta \in (V \cup \Sigma)^*$. Our choice of CFG as a formalism to represent the types is motivated from the fact that the grammar can be directly consumed by a high-level grammar formalism like Grammatical Framework (GF) [20], which is type theoretic, multilingual, and modular and suits our downstream translation usecase. We define a grammar G, where, the set V of non-terminals corresponds to the set of types $\mathcal{T}$. Each type $T_i \in \mathcal{T}$ could be available as a lexicon list, comprising entries (undisambiguated), $T_i = \{t_{i1}, t_{i2}, \ldots, t_{ik}\}$, where, each entry $t_i \in T_i$ is a sequence of lexical tokens alone (in case of simple types) or a combination of lexical and type tokens (in case of complex types). Alternatively, a type could

(a) Corpus and Types

(b) CFG Grammar G: (V, Σ, R, S)

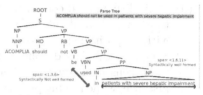

(c) Patterns

(d) Syntactically well-formed span

Fig. 2. Examples of components of our solution framework

also be available as a set of spans in the corpus (disambiguated entries), $T_i = \{< s_{i1}, u_{i1}, v_{i1} >, < s_{i2}, u_{i2}, v_{i2} >, \ldots, < s_{ik}, u_{ik}, v_{ik} >\}$, obtained as an output from an annotator (for instance, Stanford NER *etc.*) (Refer to Fig. 2). Here, a span is a 3-tuple of sentence id, start and end token index within the sentence. The set Σ of terminals then comprises:

- in the presence of type lexicons, the set of lexical tokens in the entries of each type $T_i \in \mathcal{T}$;
- in the presence of an annotator, the set of spans $< s, u, v >$, encoded as productions of the form $T_i \rightarrow < s_{i1}, u_{i1}, v_{i1} > \mid < s_{i2}, u_{i2}, v_{i2} > \mid \ldots \mid < s_{ik}, u_{ik}, v_{ik} >$.

2. **Pattern Extractor:** A pattern extractor is a program that uses the context free grammar G, to extract from $\mathcal{C}$, a set $\mathcal{H}$ of patterns, where, each pattern $h \in \mathcal{H}$ is a sequence of tokens of words or types. A pattern could be thought of as a potential higher level domain type along with a set of spans in the corpus from which it is extracted. For a pattern h, let S_h be this set of spans. We say that the spans in S_h are covered by the pattern h. Consider a span $\mu_i = < s_i, u_i, v_i > \in S_h$. We define $tokens(\mu_i) = \{< s_i, u_i, u_{i+1} >, \ldots, < s_i, v_{i-1}, v_i >\}$ as the set of all tokens covered by the span μ_i. We then say that the *token coverage* of the pattern h is the set $cover(h) = \cup_{\mu \in S_h} tokens(\mu)$ (Refer to Fig. 2c).

3. **Quality $Q_C(h)$ of a Pattern:** Quality of a pattern h is defined as a function $Q_C(h) : \mathcal{H} \rightarrow [0, 1]$. Then the set $\mathcal{H}_Q = \{h \in \mathcal{H} | Q_C(h) > r\}$, where $0 < r < 1$ is a quality threshold, is the set of all patterns in $\mathcal{H}$ that meet the quality criteria. Such a quality criterion of a pattern is typically a simple (modular) aggregation of the quality of its instances. Some examples of quality criteria are:

(a) Pattern consensus: $|S_h|$, the number of spans covered by the pattern h;

(b) Informativeness: For a set $\mathscr{C}$ of corpora, $\dfrac{|\mathscr{C}|}{|\{C \in \mathscr{C} : |S_h^C| > 0\}|}$, where, S_h^C is the set of spans covered by h in corpus C;

(c) Syntactic well-formedness: A span covered by a pattern is said to be syntactically well-formed if it forms a sub-tree in the parse tree of its corresponding sentence. A pattern is then syntactically well-formed if at least k of the spans covered by it are syntactically well-formed (Refer to Fig. 2d).

(d) Lexical rule-based consensus: Spans covered by a pattern should conform to a set of lexical rules. For instance, a pattern should not start or end with prepositions;

(e) Semantic rule-based consensus: Enforces a semantic constraint among the tokens in $tokens(\mu)$, where, μ is a span covered by the pattern. For instance, while mining patterns specific to "mergers and acquisition" from a financial services transactions corpus, we might enforce a constraint on the semantic role of agents in the patterns to be either a *buyer* or a *seller*.

(f) Model-based quality criteria: A trained classifier could be used to classify a pattern as interesting or not based on other criteria as features.

4. **Quality $Q_C(H)$ of a Patterns Set H:** Quality of a set H of patterns, given a corpus C, is defined as a function $Q_C(H)$ from $2^{\mathcal{H}_Q} \to \mathbb{Z}$. Typically, $Q_C(H)$ is either modular (*e.g.* $|H|$, $\sum_{h \in H} |cover(h)|$) or submodular (*e.g.* $|\cup_{h \in H} cover(h)|$).

5. **Pattern Selection:** The problem of lexicon curation can now be posed as the problem of selecting an optimal subset H of $\mathcal{H}_Q$. Clearly, H is optimal quality set when $H = \mathcal{H}_Q$, however, in practice, selection of an optimal H often involves an optimization of conflicting requirements on the quality and the cost of the subset. Formally,

$$H^* = \arg \max_{H \subseteq \mathcal{H}_Q} Q_C^2(H) \text{ s.t. } Q_C^1(H) < c \tag{1}$$

Or

$$H^* = \arg \min_{H \subseteq \mathcal{H}_Q} Q_C^1(H) \text{ s.t. } Q_C^2(H) > d \tag{2}$$

where, c and d are thresholds on the cost and the quality of H respectively. It is known that this optimization has an efficient solution under the assumption that the cost function $Q_C^1(H)$ be modular and the quality function $Q_C^2(H)$ be submodular [5].

3.3   Our Approach

We implemented our framework to curate high quality compact lexicons for the cross-domain SMT task. Given a source language corpus, we pose the problem of curating an optimal set of high quality patterns, solve it using a greedy algorithm and use a human-in-the-loop approach to get their translation. More precisely, we follow the steps described below:

Context Free Grammar: We use Stanford parser to create a lexicon list of type Noun phrases (NP) present in the corpus. Refer to Sect. 3.2 for details.

Pattern Extraction and Filtering: We use our grammar to index corpus and mine patterns. Our mining approach is inspired from Joshi et al. [6]. We first mine patterns for each sentence using their dynamic programming-based approach and then aggregate patterns across all sentences. Subsequently, we filter out bad patterns (we refer to this as pattern filtering), where, the quality of a pattern is judged based on two modular quality criteria—aggregated frequency of its instances and their syntactic well-formedness.

Pattern Selection: We formulate this as a subset selection problem (Refer to the formulations (1) and (2)). Although formulation (1) has a better approximation guarantee, both formulations performed equally well in our evaluation. Both formulations can be efficiently solved if the cost function $Q^1_C(H)$ is modular and the quality function $Q^2_C(H)$ is submodular [9]. In our experiments, we use as $Q^1_C(H)$ the modular cardinality constraint $|H| < c$ and as $Q^2_C(H)$ the submodular token coverage of corpus: $| \cup_{h \in H} cover(h)|$. Further, if $Q^2_C(H)$ is a submodular and monotone function, that is, $A \sqsubseteq B$ then $Q^2_C(A) \leq Q^2_C(B)$, then this problem can be solved greedily with theoretical guarantee of $1 - \frac{1}{\epsilon}$ [14]. This is the best approximation result we can achieve efficiently [15]. Further, we use an accelerated version of this algorithm [13] which at every iteration lazily evaluates the function value to get the best item to add in the output set.

Pattern Translation: After mining a high quality set of patterns, we ask humans to provide translations of these patterns and thus create a *bilingual lexicon*. We leveraged Matecat [4] and MyMemory[3] to help human translators while using our interactive system for gathering translations.

4 Evaluation

4.1 Experimental Setup

We study the effect of curating a domain-specific bilingual lexicon using our approach on domain adaptation of pre-built cross-domain statistical machine translation (SMT) models[4] trained for three language pairs on the Europarl corpus [26]. We experimented with adapting the pre-built SMT model on three domain specific datasets [23,24], each pertaining to a different domain: (i) JRC (legal), (ii) EMEA (medical) and (iii) KDE (technical). Each domain has specific language usage that differs from that of a large cross-domain corpus (*i.e.*, Europarl) typically used to train an SMT model. Moreover, the availability of parallel corpora for training SMT models is very limited in these domains. While each of these is a parallel corpus, we made use of the aligned target language corpus only for evaluation. The remaining source language data was used for mining patterns. We experimented with these datasets for three language pairs:

---

[3] https://mymemory.translated.net/.
[4] http://www.statmt.org/moses/RELEASE-3.0/models/.

Table 2. Corpus statistics showing number of sentences in the
domain-specific parallel corpora

Corpus	en-fr	en-es	en-de
JRC (legal)	814,167	805,756	537,850
EMEA (medical)	1,092,568	1,098,333	1,108,752
KDE4 (tech-nical)	210,173	218,655	224,035

Table 3. Effect of filtering on the
number of patterns extracted from
the JRC corpus.

Pattern length	3	4	5	6	7	8	9
F/Uª %	15.9	11.1	9.2	8.3	7.4	6.5	6.1

ªFiltered (F), Unfiltered (U)

English-French (en-fr), English-Spanish (en-es), and English-German (en-de). In
Table 2, we present the number of sentences in each of these parallel corpora.

For each dataset, we create a test set (TEST) for evaluation, by randomly
sampling 3000 unique sentence pairs from the corpus. A different random sample of up to 100, 000 source language sentences is used as the set for mining
the patterns (MINE). Pattern extraction is performed using the sentences in the
source language from the MINE set and for the set of quality patterns mined,
we manually obtain their corresponding translations, while being guided by the
aligned target language sentences from the MINE set. We evaluate the translation quality on the TEST set using the standard BLEU metric [19]. Our baseline
corresponds to the pre-built SMT model. For domain adaptation, we incorporate
the bilingual lexicon (curated using the MINE set) into the baseline model using
the XML markup feature[5] available in the Moses tool.

The entire process of sampling TEST and MINE sets for each corpus and
language pair is repeated thrice and the baseline and domain-adapted numbers
are reported after averaging across the three runs. In the following sections, we
present several intermediate results and ablation tests before presenting the final
BLUE score comparisons. Owing to space issues we present select plots for select
datasets and language pairs here.

4.2 Effect of Syntactic Completeness-Based Consensus on Pattern Extraction

The pattern extraction step extracts all frequent patterns (frequency
threshold = 2) of up to a certain length. This results in a large number of
patterns, not all of which are syntactically well-formed. We filter out patterns
whose instances do not conform to a phrasal structure as per the Stanford parser
(inferred via the Grammar discussed earlier), thus leaving behind between 6 %
to 9 % patterns for further processing for lexicon curation (c.f. Table 3).

---

[5] inclusive and exclusive mode http://www.statmt.org/moses/?n=Advanced.Hybrid.

4.3   Effect of Varying the Lexicon Size

Pattern selection (Eq. 1) allows to constrain the cardinality of the final set of quality patterns. The manual translation of these patterns requires human effort that is proportional to the number of patterns in this set. On the other hand, cardinality of this set might also affect the corpus coverage and thereby the translation accuracy. We study this effect by setting the cardinality of this set to various values: 25, 75, 125, 200, 250, 1000, 1750 and 2500.

Coverage on MINE versus TEST: In Fig. 3, we present the corpus coverage (in terms of number of words) on the MINE and TEST data sets with varying number of patterns and for different pattern lengths. Patterns mined using the MINE split seem to generalize well and the coverage on both MINE and TEST increases as we increase the number of patterns. This observation holds true for other datasets and language pairs as well and the coefficient of correlation between MINE and TEST coverage is consistently above 0.99.

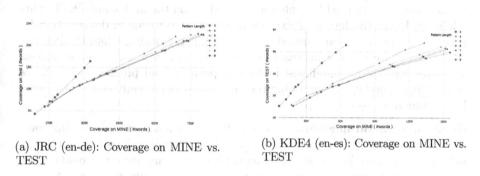

(a) JRC (en-de): Coverage on MINE vs. TEST

(b) KDE4 (en-es): Coverage on MINE vs. TEST

Fig. 3. Effect of varying the size of the set of quality patterns on corpus coverage

Coverage and Translation Accuracy on TEST: The patterns in the lexicon are translated and added to an in-domain bilingual lexicon. Figure 4 shows that as we increase the size of the lexicon, the TEST coverage improves and we see corresponding improvement in the translation accuracy.

4.4   Comparison of Different Approaches to Pattern-Set Extraction for Cross-Domain SMT

We compare different approaches to extracting a good set of patterns from a source language corpus and translating them for cross-domain SMT application. Figure 5 shows accuracy of these models for varying number of patterns in the lexicon.

Effect of Bilingual Lexicon in Domain Adaptation: The task of domain adaptation involves using a translation model trained on a large out-of-domain parallel corpus and adapting it to reliably translate an in-domain corpus.

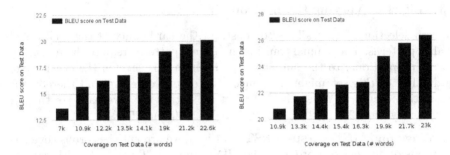

(a) JRC (en-de): Coverage vs. BLEU on TEST (b) KDE4 (en-es): Coverage vs. BLEU on TEST

Fig. 4. Effect of varying the size of the set of quality patterns on translation accuracy.

The pre-built SMT models trained on the out-of-domain Europarl corpus serve as baselines and are used to evaluate translations on the in-domain TEST splits (Refer to B0 in the figure). Next, we adapt the model for in-domain translation by incorporating our curated in-domain bilingual lexicons into the baseline model. The improvement in translation accuracy (Refer to B2) is quite evident. The lexicons capture significant in-domain patterns and provide their reliable translation, thereby, further aiding the baseline model already trained to translate common cross-domain phrases.

Effect of Submodular Optimization: Would we have got the same improvement in translation accuracy had we curated a bilingual lexicon from a random subset of patterns? We curated a bilingual lexicon using our set of quality patterns (B2) and another using a random subset of frequent patterns (B6) and compared their impact on translation accuracy. The translation model incorporating bilingual lexicon curated from random subset of frequent patterns does improve upon the baseline. However, the one with our high quality bilingual lexicon, obtained after submodular optimization, does much better in generating a high quality translation.

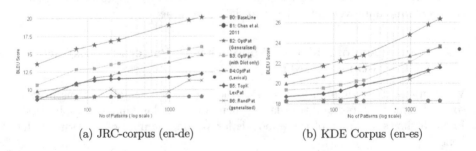

(a) JRC-corpus (en-de)   (b) KDE Corpus (en-es)

Fig. 5. Comparison of different approaches for creating a bilingual lexicon for cross-domain SMT

Effect of Pattern Generalization: Since our patterns comprise words and phrase classes, phrases in a corpus that might otherwise be infrequent, turn out to be frequent when folded into patterns. In order to ascertain that this indeed positively affects our curated lexicon and the final translation model, we compared this with a lexicon curated from frequent lexical-only phrases (B4 and B5) in the corpus. The final set of patterns was obtained, in one case, by extracting the top-k frequent phrases (B5: modular criterion) and in the other case, by using the submodular quality criterion (Refer to B4). As can be seen in Fig. 5, the modular frequency-based criterion does much better on generalized patterns than on phrasal (lexical only) patterns and together with submodular optimization results in a much better bilingual lexicon.

Comparison with Other Work. The system proposed by Chen *et al.* [3] comes closest to our work. We used publicly available domain lexicons to annotate our corpora with domain types and used their clustering-based approach to extract a set of significant patterns. The bilingual lexicon was created by sampling and manually translating one representative pattern from each cluster (Refer to B1). Next, we applied our submodular optimization-based approach on the same annotated corpora (Refer to B3). We observe that the pattern-set obtained using our quality criteria does better, even with patterns composed of domain types instead of the more general phrase classes.

5   Conclusion

We presented a novel framework for extraction of a high quality bilingual lexicon for domain specific translation. We defined several quality criteria that could be modeled as modular or submodular functions over the set of patterns mined from a domain specific corpus. The problem of pattern selection is then formulated as an optimization of these criteria and solved to produce a good set of representative in-domain patterns. Experimental results justify that a cross-domain SMT model indeed benefits from the availability of this high quality in-domain bilingual lexicon and does better in translating domain specific text.

Acknowledgments. This research was supported by the Intranet Search project from IRCC at IIT Bombay.

References

1. Bhuiyan, M., Mukhopadhyay, S., Hasan, M.A.: Interactive pattern mining on hidden data: a sampling-based solution. In: Proceedings of the 21st ACM International Conference on Information and Knowledge Management, CIKM 2012, pp. 95–104. ACM, New York, NY, USA (2012)
2. Bonchi, F., Giannotti, F., Mazzanti, A., Pedreschi, D.: ExAnte: anticipated data reduction in constrained pattern mining. In: Lavrač, N., Gamberger, D., Todorovski, L., Blockeel, H. (eds.) PKDD 2003. LNCS (LNAI), vol. 2838, pp. 59–70. Springer, Heidelberg (2003)

3. Chen, H., Huang, H., Tjiu, J., Tan, C., Chen, H.: Identification and translation of significant patterns for cross-domain SMT applications. In: Proceedings of Machine Translation Summit XIII (2011)

4. Federico, M., Bertoldi, N., Cettolo, M., Negri, M., Turchi, M., Trombetti, M., Cattelan, A., Farina, A., Lupinetti, D., Martines, A., et al.: The MateCat tool. In: Proceedings of COLING, pp. 129–132 (2014)

5. Iyer, R.K., Bilmes, J.A.: Submodular optimization with submodular cover and submodular knapsack constraints. In: Advances in Neural Information Processing Systems, pp. 2436–2444 (2013)

6. Joshi, S., Ramakrishnan, G., Balakrishnan, S., Srinivasan, A.: Information extraction using non-consecutive word sequences. In: Proceedings of TextLink 2007, The Twentieth International Joint Conference on Artificial Intelligence (2007)

7. Kempe, D., Kleinberg, J., Tardos, E.: Maximizing the spread of influence through a social network. In: SIGKDD (2003)

8. Koehn, P., Schroeder, J.: Experiments in domain adaptation for statistical machine translation. In: Proceedings of the Second Workshop on Statistical Machine Translation StatMT 2007, pp. 224–227. Association for Computational Linguistics, Stroudsburg, PA, USA (2007)

9. Krause, A., Golovin, D.: Submodular function maximization. Tractability: Pract. Approaches Hard Prob. **3**, 19 (2012)

10. Lambert, P.: Data inferred multi-word expressions for statistical machine translation. In: MT Summit X (2005)

11. Lin, H., Bilmes, J.: Multi-document summarization via budgeted maximization of submodular functions. In: NAACL (2010)

12. Liu, J., Shang, J., Wang, C., Ren, X., Han, J.: Mining quality phrases from massive text corpora. In: Proceedings of the 2015 ACM SIGMOD International Conference on Management of Data, SIGMOD 2015, pp. 1729–1744. ACM, New York, NY, USA (2015)

13. Minoux, M.: Accelerated greedy algorithms for maximizing submodular set functions. In: Stoer, J. (ed.) Optimization Techniques. LNCIS, vol. 7, pp. 234–243. Springer, Heidelberg (1978)

14. Nemhauser, G.L., Wolsey, L.A., Fisher, M.L.: An analysis of approximations for maximizing submodular set functions. Math. Program. **14**(1), 265–294 (1978)

15. Nemhauser, G.L., Wolsey, L.A.: Best algorithms for approximating the maximum of a submodular set function. Math. Oper. Res. **3**(3), 177–188 (1978)

16. Nepveu, L., Lapalme, G., Qubec, M., Foster, G.: Adaptive language and translation models for interactive machine translation. In: Proceedings of the Conference on Empirical Methods in Natural Language Processing (2004)

17. Och, F.J., Ney, H.: A systematic comparison of various statistical alignment models. Comput. Linguist. **29**(1), 19–51 (2003)

18. Pal, S., Bandyopadhyay, S.: Handling multiword expressions in phrase-based statistical machine translation. In: Machine Translation Summit XIII, pp. 215–224 (2011)

19. Papineni, K., Roukos, S., Ward, T., Zhu, W.J.: BLEU: a method for automatic evaluation of machine translation. In: Proceedings of the 40th Annual Meeting on Association for Computational Linguistics, pp. 311–318. Association for Computational Linguistics (2002)

20. Ranta, A.: Grammatical framework. J. Funct. Program. **14**(02), 145–189 (2004)

21. Ren, Z., Lü, Y., Cao, J., Liu, Q., Huang, Y.: Improving statistical machine translation using domain bilingual multiword expressions. In: Proceedings of the Workshop on Multiword Expressions: Identification, Interpretation, Disambiguation and Applications, MWE 2009, pp. 47–54. Association for Computational Linguistics, Stroudsburg, PA, USA (2009)

22. Tan, C.M., Wang, Y.F., Lee, C.D.: The use of bigrams to enhance text categorization. Inf. Process. Manag. **38**(4), 529–546 (2002)

23. Tiedemann, J.: News from OPUS - a collection of multilingual parallel corpora with tools and interfaces. Recent Advances in Natural Language Processing, pp. 237–248. John Benjamins, Amsterdam (2009)

24. Tiedemann, J.: Parallel data, tools and interfaces in OPUS. In: Proceedings of the 8th International Conference on Language Resources and Evaluation (LREC 2012), European Language Resources Association (ELRA), Istanbul, Turkey, May 2012

25. Vogel, S., Ney, H., Tillmann, C.: HMM-based word alignment in statistical translation. In: Proceedings of the 16th Conference on Computational Linguistics - COLING 1996, vol. 2, pp. 836–841. Association for Computational Linguistics, Stroudsburg, PA, USA (1996)

26. Wu, H., Wang, H., Zong, C.: Domain adaptation for statistical machine translation with domain dictionary and monolingual corpora. In: IEEE Signal Processing Magazine (2008)

27. Xin, D., Shen, X., Mei, Q., Han, J.: Discovering interesting patterns through user's interactive feedback. In: Proceedings of the 12th ACM SIGKDD International Conference on Knowledge Discovery and Data Mining, KDD 2006, pp. 773–778. ACM, New York, NY, USA (2006)

Forest CERN: A New Decision Forest Building Technique

Md. Nasim Adnan$^{(\boxtimes)}$ and Md. Zahidul Islam

School of Computing and Mathematics, Charles Sturt University,
2795 Bathurst, Australia
{madnan,zislam}@csu.edu.au

Abstract. Persistent efforts are going on to propose more accurate decision forest building techniques. In this paper, we propose a new decision forest building technique called *"Forest by Continuously Excluding Root Node (Forest CERN)"*. The key feature of the proposed technique is that it strives to exclude attributes that participated in the root nodes of previous trees by imposing penalties on them to obstruct them appear in some subsequent trees. Penalties are gradually lifted in such a manner that those attributes can reappear after a while. Other than that, our technique uses bootstrap samples to generate predefined number of trees. The target of the proposed algorithm is to maximize tree diversity without impeding individual tree accuracy. We present an elaborate experimental results involving fifteen widely used data sets from the UCI Machine Learning Repository. The experimental results indicate the effectiveness of the proposed technique in most of the cases.

Keywords: Decision tree · Decision forest · Ensemble accuracy

1 Introduction

From 2005 to 2020, the "Digital Universe" will expand by a factor of 300, from 130 Exabytes to 40,000 Exabytes, or 40 trillion Gigabytes (more than 5,200 Gigabytes for every man, woman, and child in 2020) [1]. Thus it is easily understandable that we need highly efficient automated means if we want to capitalize these data. Data mining is the method of automatically discovering useful information from large data sets [21]. Classification and clustering are two widely used data mining tasks that are applied for knowledge discovery and pattern recognition.

Classification aims to generate a function (commonly known as a classifier) that maps a set of non class attributes $m = \{A_1, A_2, ..., A_m\}$ to a predefined class attribute C from an existing data set D [21]. A data set D can be regarded as a two dimensional table with columns/attributes ($\{A_1, A_2, ..., A_m, C\}$) and rows/records ($\{R_1, R_2, ..., R_n\}$). A data set generally has two types of attributes such as numerical (e.g. Age) and categorical (e.g. Gender). Out of all categorical attributes, one is chosen to be the class attribute. All other attributes are termed

© Springer International Publishing Switzerland 2016
J. Bailey et al. (Eds.): PAKDD 2016, Part I, LNAI 9651, pp. 304–315, 2016.
DOI: 10.1007/978-3-319-31753-3_25

as non class attributes. A classifier is then built from an existing data set (i.e. training data set) where the values of the class attribute are present and then applied on unseen/test records in order to predict their class values.

The use of ensembles in classification have been actively studied in recent years [2,18,20]. Interestingly, an ensemble of classifiers is found to be effective for unstable classifiers such as decision trees [21]. Decision trees are considered to be an unstable classifier because a slight change in a training data set can induce significant differences between the decision trees generated from the original and modified data sets. A decision forest is an ensemble of decision trees where an individual decision tree acts as a base classifier. The classification is performed by taking a vote based on the predictions made by each decision tree of the decision forest [21].

In order to achieve better ensemble accuracy a decision forest needs both accurate and diverse (in terms of classification errors) individual decision trees as base classifiers [11,18]. An accurate decision tree can be generated by feeding a training data set to a decision tree algorithm such as CART [7]. Nevertheless, a single decision tree can discover only one set of logic rules and thus may wrongly predict the class value of a test record which could have been predicted correctly by a more appropriate logic rule. A different decision tree can be obtained from a differentiated data set which may include a more appropriate logic rule for the given test record. If a decision forest contains a set of decision trees which are different from each other then some of the trees may discover appropriate logic rules for a set of test records while some other trees may discover appropriate logic rules for another set of test records resulting in better generalization performance for the forest. However, to establish the scope of generating too diverse trees can be the cause of generating less accurate trees as optimization on the two conflicting objectives cannot be attained simultaneously [11]. Thus, decision forest algorithms need to draw a balance between how different/diverse as well as how accurate trees they need in order to increase the ensemble accuracy.

There are many decision forest algorithms aiming to generate more accurate and diverse decision trees. In Sect. 2, we briefly introduce some of the prominent and recent algorithms and their limitations. Apparently, there is room for further improvement in achieving higher ensemble accuracy for decision forests. In this paper, we propose a novel decision forest algorithm called "*Forest by Continuously Excluding Root Node (Forest CERN)*" that aims to build a set of highly accurate decision trees by exploiting the strength of all non class attributes available in a data set, unlike some existing techniques that use a subset of non class attributes. At the same time to promote strong diversity, *Forest CERN* emphasizes to exclude attributes that participated in the root nodes of previous trees by imposing penalties on them to deter them appear in some subsequent trees. Penalties are gradually lifted in such a manner that those attributes can reappear after a while.

The remainder of this paper is organized as follows: In Sect. 2 we discuss some of the well-known decision forest algorithms. Section 3 explains the proposed *Forest CERN* algorithm in detail. Section 4 discusses the experimental results. Finally, we offer some concluding remarks in Sect. 5.

2   Literature Review

In literature we find many forest building algorithms that differentiate training data set in different ways to generate diverse decision trees. We introduce some of the prominent forest building algorithms as follows.

(a) **Bagging** [5]: Bagging generates a new training data set D_i where the records of D_i are chosen randomly from the original training data set D. A new training data set D_i contains the same number of records as in D. Thus some records of D can be chosen multiple times and some records may not be chosen at all. This approach of generating a new training data set is known as bootstrap sampling. On an average 63.2 % of the original records are typically chosen in a bootstrap sample [10]. Bagging generates a predefined number (T) of bootstrap samples $D_1, D_2, ..., D_T$ using the above approach. A decision tree building algorithm is then applied on each bootstrap sample D_i $(i = 1, 2, ..., T)$ in order to generate T number of trees for the forest.

(b) **Random Subspace** [11]: The Random Subspace algorithm randomly draws a subset of attributes (subspace) f from the entire attribute set m in order to determine the splitting attribute for each node of a decision tree. The Random Subspace algorithm is applied on the original training data set in building every decision tree.

(c) **Random Forest** [6]: Random Forest is technically a fusion of Bagging and Random Subspace algorithms. In Random Forest, the Random Subspace algorithm is applied on bootstrap samples instead of the original training data set.

(d) **MDMT** [12]: In MDMD (Maximally Diversified Multiple Decision Tree), each decision tree tests a completely different set of attributes than the set of attributes tested in any other decision tree. MDMT builds the first decision tree using a traditional decision tree building algorithm such as CART [7]. All non class attributes that have been tested in the first tree are then removed from the data set and the decision tree building algorithm is again applied on the modified data set to build the second tree. The process continues until either the user defined number of trees is generated or all non class attributes of the data set are removed.

(e) **CS4** [15]: To build the decision forest, CS4 (Cascading and Sharing Trees) first ranks all attributes of a training data set according to their classification capacities (e.g. Gain Ratios [19]). Then in a cascading manner, the attribute with the highest Gain Ratio value is selected as the root node of the first tree; the attribute with the second highest Gain Ratio value is selected as the root node of the second tree and so on.

(f) **SysFor** [13]: SysFor (Systematically Developed Forest of Multiple Decision Trees) takes a user input to determine the number of decision trees to be generated. Then a set of "good attributes"and corresponding split points are determined based on a user defined "goodness"threshold and "separation"threshold. SysFor then starts building decision trees by placing the good attributes one by one as the root attribute (at Level 1) of a tree, and thereby build as many trees as the number of good attributes. If the number of trees built at Level 1 is less than the user defined number of trees, then more trees (until the user-defined number of trees are built or the maximum number of possible trees are built) are generated by using alternative good attributes at Level 2. The alternative good attributes are chosen from the set of good attributes for the nodes at Level 2.

We next discuss about the major problems of the algorithms stated above: Bootstrap samples are the only source of diversity in Bagging. Theoretically, a bootstrap sample contains $\approx 63.2\%$ of the original records of a training data set; the remaining $\approx 36.8\%$ records are repeated [10]. This sampling ratio that theoretically can occur in bootstrap samples is not optimal for every data set [17] and thus may not extract strong diversity specially for data sets with redundant/similar (i.e. difference is low) records.

Both Random Subspace and Random Forest randomly draw a subset of attributes (subspace) f from the entire attribute set m in order to determine the splitting attribute for each node splitting event of a decision tree. In effect, individual tree accuracy and diversity among decision trees depend on the size of f. If f is sufficiently small then the chance of having the same attribute in different subspace becomes low. Thus the trees in a forest tend to become more diverse. However, a sufficiently small f may not guaranty the presence of the adequate number of good attributes (i.e. the attributes with high classification capacity) resulting in decreased individual accuracy. Thus the value of $|f|$ plays a strong role in determining the ultimate efficiency of a decision forest and commonly known as the hyperparameter [4]. In Breiman's original Random Forest [6], $|f|$ is chosen to be $int(\log_2 |m|) + 1$.

It is worthy to mention that the value of $int(\log_2 |m|) + 1$ does not increase at the same rate to the increase of $|m|$. For example, let us assume that we have a low dimensional data set consisting of 4 attributes. Thus a splitting attribute is determined from a randomly selected subspace of 3 attributes $(int(\log_2 4)+1 = 3)$ encompassing 75% of the total attributes. As a result, the chance of appearing similar attributes in different subspaces becomes high, resulting in decreasing diversity among the trees. On the other hand, when $|m|$ is large say, 150 then $|f|$ contains 8 randomly chosen attributes $(int(\log_2 150) + 1 = 8)$ covering only 5% of the total attributes. Hence, if the number of good attributes is not high enough in m then the chance of containing adequate number of good attributes in a subspace f becomes low, which is supposed to cause low individual accuracy of the trees as described before.

Decision forest algorithms such as MDMT, CS4 and SysFor limit the number of trees based on the non class attributes available in a training data set. CS4 can generate at most $|m|$ trees for the forest. For example, from a low-dimensional data set such as Balance Scale with four non class attributes [16] CS4 can build only four trees. SysFor tries to eliminate this constraint by placing alternative good attributes in Level 2. However, for Balance Scale data set SysFor can generate only 12 trees which is still a small number in ensemble standard [20]. This problem is more severe for MDMT. It can generate only one tree from the

Balance Scale data set as no non class attributes are left to generate the second tree. Besides, both CS4 and SysFor just change the root nodes. In CS4, each non class attributes are placed in the root node once in a cascading manner according to their Gain Ratios. As a result, some attributes with vary low classification capacity can be placed in the root node resulting in not generating some trees entirely. SysFor places only good attributes in the root node but the good attributes are determined by a user defined "goodness" threshold. If this user input is not tuned correctly, the number of good attributes can be too high or too low. Furthermore, just changing the root nodes should not be sufficient in rendering strong diversity as trees may be taken over by some attributes with relatively higher classification capacity just below the root node.

3   Our Technique

In order to address the issues raised in Sect. 2, we propose a new method for decision tree ensemble construction called *Forest CERN*. The main feature of our technique is that it strives to exclude attributes that participated in the root nodes of previous trees by imposing penalties on them to deter them appear in some subsequent trees. But, what is the impact of excluding attributes that appeared in root nodes? For example, let A_i be the attribute that was placed in the root node of the first tree. As a result, all logic rules generated from the first tree starts with attribute A_i. Thus, to generate different logic rules from the next tree, it is preferable to exclude A_i from that tree. In order to facilitate the exclusion of A_i (in this case), we propose our novel penalty/weight imposing strategy that works as follows:

Let, we have a data set D with $m = \{A_1, A_2, ..., A_m\}$ attributes. Initially, the weight values for all attributes are set to 1.0 (default weights). To determine the splitting attribute at first the classification capacity such as Gain Ratio [19] or Gini Index [7] of each attribute is calculated. Then a new merit value is obtained by multiplying the classification capacity with respective weight for each attribute. After the merit values of all attributes are calculated, the attribute with the highest merit value is selected as the splitting attribute. We generate the first tree from a bootstrap sample of D using the default weights. Thus, the first tree is generated in the same way as Bagging [5]. To generate the second tree, we first isolate the attribute that appeared in the root node of the first tree. Let A_i be the attribute that appeared in the root node of the first tree. Then, we calculate the weight of A_i as follows (Eq. 1):

$$\omega_i = \frac{1}{|m|} \tag{1}$$

Here, $|m|$ is the number of non class attributes in D. For example, when $|m| = 25$, the weight of A_i is $\frac{1}{25}$. The weights of other attributes remain as 1.0. The next (second) tree is generated from another bootstrap sample of D with the updated weight values. Due to it's disadvantageous weight, A_i will have lesser chance to appear in the entire second tree (let alone in the root node) compared to other attributes. Let, A_j be the attribute that appeared in the root node of the second tree. Thus, the weight of A_j will be reduced after generating the second tree. If the weight imposing approach is continued in the same manner, all the attributes qualified to appear in the root node will acquire disadvantageous weights. As a result, trees may not be generated at all. To prevent this scenario, we increase the weights of all attributes having weight < 1.0 other than the current root attribute. For example, after generating the second tree the weight of A_j is reduced to $\frac{1}{|m|}$ and at the same time the weight of A_i is increased by adding $\frac{1}{|m|}$ to it's current weight value. In this way, when $|m| = 25$ the weight of A_i can reach to the default weight 1.0 if it misses to appear in the root nodes of all subsequent 24 trees. But, whenever A_i reappears in the root node of any tree, the weight is again set to $\frac{1}{|m|}$ and at the same time the weights of all other attributes with weight < 1.0 are increased by $\frac{1}{|m|}$. This weight imposing strategy can be further illustrated in Table 1.

Table 1. Weight Imposing Strategy

After First Tree	Attributes	A_1	...	A_i	...	A_j	...	A_k	...	$A_{|m|}$
	Weights	1	...	$\frac{1}{|m|}$	...	1	...	1	...	1
After Second Tree	Attributes	A_1	...	A_i	...	A_j	...	A_k	...	$A_{|m|}$
	Weights	1	...	$\frac{2}{|m|}$	...	$\frac{1}{|m|}$	...	1	...	1

From Table 1, we see that for third tree, both A_i and A_j have disadvantageous weights relative to other non class attributes. In the same way, a set of attributes will obtain disadvantageous weights when building the i-th tree. This phenomenon is clearly different form either of CS4 or SysFor; where each attribute comes iteratively in the root node and thus attributes with relatively higher classification capacity retain their presence just below the root nodes. In fact, the proposed technique exhibits slightly similar effect of MDMT where the proposed technique can take out a set of attributes with severely disadvantageous weights from participating in some subsequent trees. However, in the proposed technique attributes can regain weights to be able to reappear in the root node. This phenomenon helps preventing non-deserving attributes (attributes with very low classification capacity) to appear in the root node. Furthermore, even when the same attribute reappears in the root node the proposed technique strives to generate different tree through different bootstrap sample and different combination of disadvantageous attributes.

4   Experimental Results

We performed an elaborate experimentation on fifteen (15) well known data sets that are publicly available from the UCI Machine Learning Repository [16] representing a variety of areas. The data sets used in the experimentation are described in Table 2. For example, the Chess data set has 36 non class attributes, 3196 records with two (02) distinct class values. For our experimentation, we remove records with missing values (Table 2 shows the number of records with no missing values) and identifier attributes such as *Transaction_ID* from each applicable data set. We generate 100 trees for every decision forest since the number is considered to be large enough to ensure convergence of the ensemble effect [2,4,9]. We use Gini Index [7] as a measure of classification capacity in accordance with Random Forest [6]. The minimum Gini Index/merit value is set to 0.01 for any attribute to qualify for splitting a node. Each leaf node of a tree contains at least two records and no further post-pruning is applied. We apply majority voting to aggregate results for forest classification [6,18].

Table 2. Description of the data sets

Data Set Name	Non Class Attributes	Records	Distinct Class Values
Balance Scale	04	625	3
Car Evaluation	06	1728	4
Chess	36	3196	2
Credit Approval	15	653	2
Ecoli	07	336	8
Hayes-Roth	04	132	3
Ionosphere	34	351	2
Iris	04	150	3
Liver Disorders	06	345	2
Nursery	08	12960	5
Sonar	60	208	2
Statlog Vehicle	18	846	4
Thyroid-New	05	215	3
Tic-Tac-Toe	09	958	2
Wine	13	178	3

The experimentation is conducted by a machine with Intel(R) 3.4 GHz processor and 8 GB Main Memory (RAM) running under 64-bit Windows 7 Enterprise Operating System. All the results reported in this paper are obtained using 10-fold-cross-validation (10-CV) [3,14,15] for every data set. In 10-CV, a data set is divided randomly into 10 segments and from the 10 segments each time one segment is regarded as the test data set (out of bag samples) and the rest 9 segments are used for training decision trees. In this way, 10 training and 10 corresponding testing data sets are generated. In our experimentation, we generate 100 trees from each training data set (thus 1000 trees in total) for each decision forest algorithm and then evaluate their performance with the corresponding testing data sets. All the performance indicators reported in this paper are the average values obtained from the 10 testing data sets and the best results are distinguished through **bold-face**.

Ensemble Accuracy (EA) is one of the most important performance indicators for any decision forest algorithm [1]. In Table 3 we present the EA (in percent) along with Ranks (1 for the best to 4 for the worst) of Bagging, Random Subspace, Random Forest and the proposed *Forest CERN* for all data sets considered. We do not include the results generated from CS4, SysFor and MDMT as they are not able to generate 100 trees for most of the data sets used in the experimentation and as a result may not perform to the level of the considered algorithms.

Table 3. Ensemble Accuracies

Data Set Name	Bagging	Random Subspace	Random Forest	*Forest CERN*
Balance Scale	77.48 (3)	72.16 (4)	80.50 (2)	**82.25 (1)**
Car Evaluation	93.34 (2)	**93.51 (1)**	91.19 (4)	93.28 (3)
Chess	97.87 (2)	95.06 (4)	95.22 (3)	**98.00 (1)**
Credit Approval	86.37 (2)	85.92 (4)	86.07 (3)	**86.99 (1)**
Ecoli	83.10 (4)	83.15 (3)	**84.97 (1)**	84.36 (2)
Hayes-Roth	71.28 (2)	45.44 (4)	69.54 (3)	**78.87 (1)**
Ionosphere	92.59 (4)	93.45 (3)	93.73 (2)	**94.30 (1)**
Iris	95.33 (2)	94.00 (3)	**96.00 (1)**	**96.00 (1)**
Liver Disorders	68.65 (4)	69.79 (3)	**71.48 (1)**	70.97 (2)
Nursery	**97.57 (1)**	94.89 (4)	95.07 (3)	97.52 (2)
Sonar	80.71 (4)	84.57 (2)	83.07 (3)	**84.93 (1)**
Statlog Vehicle	73.56 (3)	73.45 (4)	74.14 (2)	**75.06 (1)**
Thyroid-New	93.61 (3)	93.13 (4)	94.56 (2)	**96.38 (1)**
Tic-Tac-Toe	**91.77 (1)**	80.67 (4)	84.54 (3)	87.14 (2)
Wine	96.47 (3)	97.25 (2)	97.25 (2)	**98.42 (1)**
Average	86.65 (2.67)	83.76 (3.27)	86.49 (2.33)	*88.30 (1.40)*

From Table 3, we see that Bagging delivers the best EA for 2 data sets (Avg. EA Rank: 2.67), Random Subspace for 1 data set (Avg. EA Rank: 3.27), Random Forest for 3 data sets (Avg. EA Rank: 2.33) whereas the proposed *Forest CERN* becomes the best for 10 out of 15 data sets with Avg. EA Rank of **1.40**. We already know that better ensemble accuracy of a decision forest is a consequence of a better balance between individual accuracy and diversity among the trees. Thus, to explain the reason behind this outcome, we first compute individual accuracies (in percent) of each tree of a forest to compute the Average Individual Accuracies (AIA) for the forest [2]. Kappa typically estimates the diversity between two trees T_i and T_j. Diversity among more than two trees is typically computed by first computing the Kappa (K) value of a single tree T_i with the ensemble of trees except the tree in consideration [2]. The combined prediction of the ensemble (excluding T_i) can be regarded as a single tree T_j. Then Kappa is computed between T_i and T_j as shown in Eq. 2, where $Pr(a)$ is the probability of the observed agreement between two classifiers T_i and T_j, and $Pr(e)$ is the probability of the random agreement between T_i and T_j. Once the Kappa for every single tree T_i of a forest is computed we then compute the Average Individual Kappa (AIK) for the forest [2].

$$K = \frac{Pr(a) - Pr(e)}{1 - Pr(e)} \tag{2}$$

Table 4. Average Individual Accuracies and Average Individual Kappa

Data Set Name	Bagging	Random Subspace	Random Forest	Forest CERN
	AIA/AIK	AIA/AIK	AIA/AIK	AIA/AIK
Balance Scale	64.05/0.4585	**67.41**/0.6055	64.88/0.4322	64.96/**0.4275**
Car Evaluation	**89.41**/0.8262	86.02/0.7569	75.62/0.5413	77.41/**0.5335**
Chess	**97.84**/0.9765	65.26/**0.4673**	67.93/0.4902	78.91/0.6494
Credit Approval	**85.16**/0.9246	80.91/0.7544	74.05/0.6366	70.49/**0.5389**
Ecoli	78.68/0.7938	**80.01**/0.8350	77.07/ 0.7631	75.00/**0.7288**
Hayes-Roth	**55.93**/0.3554	48.27/0.5285	55.39/0.3244	55.32/**0.3163**
Ionosphere	**88.96**/0.8335	88.94/0.8243	88.02/0.7884	86.66/**0.7434**
Iris	93.84/0.9237	**94.12**/0.9467	93.28/0.9157	92.16/**0.8736**
Liver Disorders	60.96/0.3635	**62.59**/0.4894	60.10/0.3122	59.10/**0.2441**
Nursery	**95.85**/0.9522	70.40/0.6360	70.16/**0.6333**	95.23/0.9420
Sonar	70.34/0.4909	**72.44**/0.4664	68.94/0.3958	65.85/**0.3259**
Statlog Vehicle	68.51/0.6641	**68.57**/0.6662	66.74/0.6215	63.94/**0.5749**
Thyroid-New	91.41/0.8712	**92.05**/0.8805	91.08/0.8376	90.36/**0.7905**
Tic-Tac-Toe	**79.04**/0.5505	40.53/**0.2801**	54.33/0.3137	59.60/0.3017
Wine	89.46/0.8363	**90.39**/0.8526	88.95/0.8287	85.08/**0.7691**
Average	**80.63**/0.7214	73.86/0.6660	73.10/0.5890	74.67/**0.5840**

From Table 4 we see that the proposed *Forest CERN* delivers most diverse decision trees (lower AIK value indicates higher diversity) with retaining the second highest AIA value (higher AIA value indicates better quality individual trees). Consequently, *Forest CERN* outperforms other contending algorithms in terms of EA. Now, to access the significance of the improvement we conduct a non-parametric (EA do not follow a normal distribution and thus do not satisfy the conditions for any parametric tests) one-tailed Wilcoxon Signed-Ranks Test for $n = 15$ (number of data sets used) with the significance level $\alpha = 0.05$

(thus the critical value is: 30) [22]. Wilcoxon Signed-Ranks Test is said to be more preferable to counting only significant wins and losses for comparison between two classifiers [8]. Here we see from Fig. 1 that *Forest CERN* performs significantly better (in terms of EA) than all three contending algorithms on 15 widely used data sets as the test values remain lower than the critical value for every head-to-head comparisons.

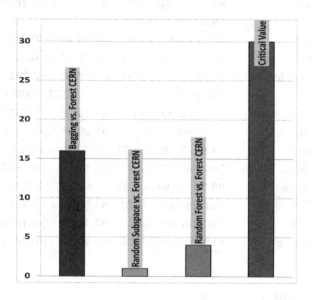

Fig. 1. Wilcoxon Signed-Ranks Test for EA

Apparently, *Forest CERN* shows some resemblance with SysFor and hence we compare them on smaller 10-tree ensemble [20] as SysFor may fall short of generating a large ensemble for every data set. The results shown in Table 5 clearly depict the distinction between SysFor and *Forest CERN*. As expected, *Forest CERN* generates more diverse trees than SysFor for every data set as a result of excluding root attributes in some subsequent trees. On the contrary, SysFor generates more accurate trees for every data sets with less diversity verifying the fact the trees generated from SysFor are more similar to each other as a result of probable presence of attributes with high classification capacity in each tree. Ultimately, *Forest CERN* outperforms SysFor in terms of EA.

Another important observation from the results presented in Tables 3, 4 and 5 is the effect of ensemble size on EA for *Forest CERN* (i.e. when *Forest CERN* is applied in the scale of larger 100-tree ensemble and smaller 10-tree ensemble). We see, the average EA is higher for 100-tree *Forest CERN* compared to 10-tree *Forest CERN* (**88.30** vs 85.45) even when the average of AIA/AIK values are very competitive (74.67/0.5840 for 100-tree *Forest CERN* vs. **76.23/0.5814** for 10-tree *Forest CERN*) as smaller *Forest CERN* may not ensure convergence of the ensemble effect for many data sets.

Table 5. Comparison between SysFor and *Forest CERN*

Data Set Name	SysFor	*Forest CERN*
	EA/AIA/AIK	EA/AIA/AIK
Balance Scale	67.50/**65.03**/0.5441	**76.97**/64.83/**0.3871**
Car Evaluation	**94.27**/**93.88**/0.9732	93.39/78.15/**0.5036**
Chess	95.97/**94.05**/0.9925	**97.13**/88.38/**0.7629**
Credit Approval	**86.37**/**80.76**/0.8819	84.53/72.92/**0.5352**
Ecoli	78.09/**76.83**/0.7761	**84.66**/75.51/**0.6353**
Hayes-Roth	59.28/**57.62**/0.6629	**66.77**/55.58/**0.3695**
Ionosphere	92.30/**91.25**/0.8787	**93.17**/86.82/**0.7128**
Iris	**96.00**/**94.89**/0.9762	95.33/92.53/**0.8885**
Liver Disorders	64.49/**63.83**/0.8378	**65.23**/59.07/**0.2757**
Nursery	96.99/**96.61**/0.9886	**97.29**/95.16/**0.9386**
Sonar	**78.86**/**72.29**/0.5253	77.36/69.66/**0.3649**
Statlog Vehicle	70.46/**67.17**/0.7278	**75.17**/66.11/**0.5645**
Thyroid-New	**93.32**/**92.68**/0.9331	92.84/90.72/**0.7308**
Tic-Tac-Toe	82.16/**80.90**/0.6073	**84.50**/61.48/**0.2669**
Wine	96.47/**92.68**/0.8842	**97.44**/86.55/**0.7842**
Average	83.50/**81.36**/0.8126	**85.45**/76.23/**0.5814**

5   Conclusion

In this paper, we propose a new decision forest building algorithm *Forest CERN* which strives to exclude attributes that participated in the root nodes of previous trees by imposing penalties (disadvantageous weights) on them to deter them appear in some subsequent trees. Penalties are eventually lifted in such a manner that those attributes can reappear in root nodes after a while. We find that trees generated from *Forest CERN* are more diverse compared to other contending algorithms which can be helpful for discovering interesting knowledge. Also, *Forest CERN* is fully independent of any parameter value unlike Random Subspace, Random Forest and SysFor to be more evenly suitable for a wide range of data sets.

However, there is an apparent limitation of *Forest CERN* specially when the number of non class attributes is high and a few of them have high classification capacity. In this case, after receiving drastic disadvantageous weights (due to high dimension) attributes with high classification capacity may disappear for a large number of intermediate trees leading to lowering AIA for the forest. One obvious solution to this problem is to reduce the attribute space. Another possible solution is probably improving *Forest CERN* to address the problem. In future, we intend to extend our work by including data sets with more number of attributes.

References

1. Adnan, M.N.: On dynamic selection of subspace for random forest. In: Luo, X., Yu, J.X., Li, Z. (eds.) ADMA 2014. LNCS, vol. 8933, pp. 370–379. Springer, Heidelberg (2014)
2. Amasyali, M.F., Ersoy, O.K.: Classifier ensembles with the extended space forest. IEEE Trans. Knowl. Data Eng. **16**, 145–153 (2014)
3. Arlot, S.: A survey of cross-validation procedures for model selection. Stat. Surv. **4**, 40–79 (2010)
4. Bernard, S., Heutte, L., Adam, S.: Forest-RK: a new random forest induction method. In: Huang, D.-S., Wunsch, D.C., Levine, D.S., Jo, K.-H. (eds.) ICIC 2008. LNCS (LNAI), vol. 5227, pp. 430–437. Springer, Heidelberg (2008)
5. Breiman, L.: Bagging predictors. Mach. Learn. **24**, 123–140 (1996)
6. Breiman, L.: Random forests. Mach. Learn. **45**, 5–32 (2001)
7. Breiman, L., Friedman, J., Olshen, R., Stone, C.: Classification and Regression Trees. Wadsworth International Group, Belmont (1985)
8. Demsar, J.: Statistical comparisons of classifiers over multiple data sets. J. Mach. Learn. Res. **7**, 1–30 (2006)
9. Geurts, P., Ernst, D., Wehenkel, L.: Extremely randomized trees. Mach. Learn. **63**, 3–42 (2006)
10. Han, J., Kamber, M.: Data Mining Concepts and Techniques. Morgan Kaufmann Publishers, San Francisco (2006)
11. Ho, T.K.: The random subspace method for constructing decision forests. IEEE Trans. Pattern Anal. Mach. Intell. **20**, 832–844 (1998)
12. Hu, H., Li, J., Wang, H., Daggard, G., Shi, M.: A maximally diversified multiple decision tree algorithm for microarray data classification. In: Proceedings of the Workshop on Intelligent Systems for Bioinformatics (WISB), vol. 73, pp. 35–38 (2006)
13. Islam, M.Z., Giggins, H.: Knowledge discovery through sysfor - a systematically developed forest of multiple decision trees. In: Proceedings of the 9th Australian Data Mining Conference (2011)
14. Kurgan, L.A., Cios, K.J.: Caim discretization algorithm. IEEE Trans. Knowl. Data Eng. **16**, 145–153 (2004)
15. Li, J., Liu, H.: Ensembles of cascading trees. In: Proceedings of the third IEEE International Conference on Data Mining, pp. 585–588 (2003)
16. Lichman, M.: UCI machine learning repository (2013). Last Accessed 10 January, 2015 http://archive.ics.uci.edu/ml/datasets.html
17. Munoz, G.M., Suarez, A.: Out-of-bag estimation of the optimal sample size in bagging. Pattern Recogn. **43**, 143–152 (2010)
18. Polikar, R.: Ensemble based systems in decision making. IEEE Circuits Syst. Mag. **6**, 21–45 (2006)
19. Quinlan, J.R.: C4.5: Programs for Machine Learning. Morgan Kaufmann Publishers, San Mateo (1993)
20. Rodriguez, J.J., Kuncheva, L.I., Alonso, C.J.: Rotation forest: a new classifier ensemble method. IEEE Trans. Pattern Anal. Mach. Intell. **28**, 1619–1630 (2006)
21. Tan, P.N., Steinbach, M., Kumar, V.: Introduction to Data Mining. Pearson Education, New York (2006)
22. Triola, M.F.: Elementary Statistics. Addison Wesley Longman Inc., Boston (2001)

Sparse Logistic Regression with Logical Features

Yuan Zou[(⊠)] and Teemu Roos

Helsinki Institute for Information Technology HIIT,
Gustaf Hällströmin katu 2b, 00014 Helsinki, Finland
{yuan.zou,teemu.roos}@cs.helsinki.fi
http://www.hiit.fi/cosco/promo

Abstract. Modeling interactions in regression models poses both computational as well as statistical challenges: the computational resources and the amount of data required to solve them increases sharply with the size of the problem. We focus on logistic regression with categorical variables and propose a method for learning dependencies that are expressed as general Boolean formulas. The computational and statistical challenges are solved by applying a technique called transformed Lasso, which involves a matrix transformation of the original covariates. We compare the method to an earlier related method, LogicReg, and show that our method scales better in terms of the number of covariates as well as the order and complexity of the interactions.

Keywords: Feature selection · Logistic regression · Lasso

1 Introduction

A basic logistic regression model includes individual effects of feature variables (a.k.a. regressors or covariates) on the probability of a response event. In addition to the individual effects, interactions between the feature variables are important in a wide range of applications. Examples include identifying important single nucleotide polymorphism (SNPs) in genome-wide association studies [6], pathway analysis of gene-expression or metabolomic data [14], regulatory motif identification [7], and association studies of gene-gene interactions [16].

Pairwise and higher order interactions between the features can be modelled by explicitly including the interactions as feature variables. In other words, a kth order interaction can be modelled by merging a subset of k variables into a new variable whose domain becomes the Cartesian product of the domans of the merged variables, and including indicator variables for each of the values in the new domain. High order interactions pose statistical and computational challenges since the number of interaction terms grows exponentially with the order, k. Different techniques for dealing with these challenges have been proposed. These include, for instance, forward and backward selection (see e.g. [5]) and more recently, Lasso [15].

We consider situations where the interactions can be expressed as Boolean formulas, each of which is composed of a subset of the original feature variables

© Springer International Publishing Switzerland 2016
J. Bailey et al. (Eds.): PAKDD 2016, Part I, LNAI 9651, pp. 316–327, 2016.
DOI: 10.1007/978-3-319-31753-3_26

connected by logical operations such as AND, OR, and XOR (exclusive or). For example, if we have a vector of m binary variables $x = \{x_1, x_2, \ldots, x_m\}$, the model may involve terms such as "x_1 and x_2, or x_3" or "x_4 and x_5 and not x_6". We define the model formally as a generalized linear model

$$g(E(y)) = \beta_0 + \sum_{i=1}^{t} \beta_i L_i(x),\tag{1}$$

where y is a binary response variable, $E(y) = \Pr[y = 1 \mid x]$ denotes the expectation of y conditional on the regressors x, g is a link function, $\beta_0, \ldots, \beta_t$ are regression coefficients, and $L_1, \ldots, L_t$ are Boolean functions of x. The right-hand side of Eq. (1) is called a *linear predictor*. Depending on how we define the link function g, this framework includes a range of different model families, such as linear regression, logistic regression, Poisson regression, etc. In this study, we focus on the logistic case where the link function is defined as $g(E(y)) = \text{logit}(E(y)) = \log(E(y)/(1 - E(y)))$.

For each of the Boolean functions L_i, $1 \le i \le t$, in the above model, we define the *order* of the function as the minimum number of variables $x_1, \ldots, x_m$ that are sufficient to determine the value of L_i. For example, a function that depends on only one of the variables is said to be first order, and so on. The order of the model is defined as the maximum order of all the functions involved. As mentioned above, the basic logistic regression model is usually defined as a first order model that includes all the m first order functions $L_i(x) = x_i$, $1 \le i \le t = m$.

1.1 Related Work

An important prior work regarding logical features was done by Ruczinski et al. in [11] who also provide an implementation of their method in the R package LogicReg.[1] It uses a greedy algorithm to search through the space of possible Boolean functions, with additional simulated annealing step to avoid local optima. It shows better performance than the tree based or rule based methods that are used by CART [2] and MARS [4]. However, as the number of covariates and the order of interactions are increased, the number of possible Boolean functions grows significantly. This makes greedy search heuristics computationally inefficient and prone to converge to local optima.

On the other hand, Shi et al. [13] proposed a Lasso based method which is suitable for identifying a large set of interactions. However, they only interactions defined using the AND operator. To allow more types of Boolean operations, the works in [1,8] incorporate extra information such as the structure of coefficients to guide the learning process. However, these methods need expert knowledge that is not always available. They are also unable to handle complex interactions. Both methods are only suitable for cases when all coefficients inside a group are either zero or non-zero. Our new Lasso based method can efficiently and

---

[1] Available from CRAN, http://cran.r-project.org.

effectively learn arbitrary logical functions and deal with situations when the covariates have multiple values.

The least absolute shrinkage and smoothing operator (Lasso) for linear regressions was proposed by Tibshirani in 1996 [15] and has since then gained a lot of popularity in subset selection problems. Lasso aims to minimize the sum of squared errors subject to a bound on the sum of absolute values of the regression coefficients, i.e., the L_1 norm of the coefficient vector $\beta = (\beta_1, \ldots, \beta_t)^T$:

$$\underset{\beta \,:\, \|\beta\|_1 < \lambda}{\mathrm{argmin}} \ \|Y - X\beta\|_2^2, \tag{2}$$

where $Y = (y^{(1)}, \ldots, y^{(n)})^T$ is a vector of n responses and X is a matrix whose rows are n observation vectors $x^{(1)}, \ldots, x^{(n)}$, and $\lambda > 0$ is a regularization parameter.

Lasso encourages sparsity in the estimated coefficient vector. As the value of λ is decreased, more and more coefficients are set to zero. This feature is especially useful when we need to identify a small set of relevant variables out of a large collection of candidates. Thus Lasso is suitable for discovering interactions in regression problems. Furthermore, there are plenty of well-developed tools for solving Lasso problems efficiently.

If we encode logical features constructed using the original variables as a new set of regressor variables, we can use Lasso to select the significant ones out of all possible interactions. However, the number of all possible logical expressions grows too rapidly to be handled efficiently. Furthermore, the redundancy caused by different combinations of logical formulas expressing the exact same model may lead to numerical and statistical instability. To restrict the number of predictors within a manageable size, previous work in [13] confines the logic expressions to include only two variables and the AND operator.

1.2 Contributions

In this paper we introduce a Lasso-based method for learning sparse logistic regression models with logical features. Technically, the method is implemented as a *transformed Lasso* [10] which involves a transformation matrix that multiplies the original design matrix X. The transformed Lasso can deal with any type of logical interactions. Here we also extend it to handle multivalued covariates. In the following sections, we first propose a transformation of the original data involving a generalized Walsh-Hadamard matrix. The transformation increases the dimension of the data but enables the learning of arbitrarily complex logical dependencies. We demonstrate the power of the proposed method by comparing its performance to that of the greedy method in LogicReg with different settings of sample sizes and model complexities. Finally, we propose several possible future extensions.

2 Model Formulation

As is well known, any Boolean formula can be decomposed as a linear combination of XOR functions of the same or lower orders as the formula itself. For example, to represent AND or OR formulas over a subset of indices $I_0 \subseteq \{1, 2, \ldots, l\}$, by using linear combinations of XOR formulas, we can write them respectively as

$$\text{AND}(x_{I_0}) = 2^{1-|I_0|} \sum_{l=1}^{|I_0|} (-1)^{l-1} \sum_{I' \subseteq I_0, |I'|=l} \text{XOR}(\{x_{I'}\}), \tag{3}$$

$$\text{OR}(x_{I_0}) = 2^{1-|I_0|} \sum_{l=1}^{|I_0|} \sum_{I' \subseteq I_0, |I'|=l} \text{XOR}(\{x_{I'}\}). \tag{4}$$

To map a binary covariate matrix to a design matrix that is composed of XOR functions of subsets of the covariates, it is convenient to use the discrete Walsh-Hadamard transform, see [10]. To construct the design matrix, we first expand the original covariate matrix into a larger matrix with columns corresponding to indicator variables for each possible covariate vector (e.g., in the case of two binary variables: 00, 10, 01, and 11), and then (pre-)multiply this matrix by the Walsh-Hadamard matrix. For a more complete explanation including a detailed example, see [10]. As we explain below, in practice the matrices need not be explicitly constructed.

The rows of a Walsh-Hadamard matrix of order 2^m correspond to all possible vectors composed of the m covariates. The columns of a such matrix are XOR functions on all subsets of the covariates given by the corresponding row. For example, the rows of a fourth order Walsh-Hadamard matrix correspond to all combinations of two binary elements x_1 and x_2, while the columns are $\text{XOR}(0)$, $\text{XOR}(x_1)$, $\text{XOR}(x_2)$, and $\text{XOR}(x_1, x_2)$. The fourth order Walsh-Hadamard matrix is then

$$W_4 = \begin{array}{c} \\ 00 \\ 10 \\ 01 \\ 11 \end{array} \begin{array}{cccc} \text{XOR}(0) & \text{XOR}(x_1) & \text{XOR}(x_2) & \text{XOR}(x_1, x_2) \\ \left(\begin{array}{cccc} 0 & 0 & 0 & 0 \\ 0 & 1 & 0 & 1 \\ 0 & 0 & 1 & 1 \\ 0 & 1 & 1 & 0 \end{array}\right) \end{array}. \tag{5}$$

When the covariates takes three or more values, we can formulate the design matrix in the similar way as forming the Walsh-Hadamard matrix. Each column corresponds to a XOR function of a subset of possible sequences while each variable equals to one of its possible values. For instance, if the covariates are ternary with values $\{0, 1, 2\}$, the columns relating to two variables x_1 and x_2 are $\text{XOR}(0)$, $\text{XOR}(x_1 = 0)$, sc $\text{xor}(x_1 = 1)$, $\text{XOR}(x_2 = 0)$, $\text{XOR}(x_2 = 1)$, $\text{XOR}(x_1 = 0, x_2 = 0)$, $\text{XOR}(x_1 = 0, x_2 = 1)$, $\text{XOR}(x_1 = 1, x_2 = 0)$, and $\text{XOR}(x_1 = 1, x_2 = 1)$. We ignore the XOR functions when x_1 or x_2 equals 2, because we can represent them as linear combinations of XOR functions when x_1 or x_2 equals 0 or 1. Such linear dependencies would significantly complicate the parameter estimation stage. The design matrix W_9' based on the Walsh-Hadamard matrix for two ternary variables is

$$
W_9' = \begin{array}{c} 00 \\ 10 \\ 01 \\ 11 \\ 02 \\ 20 \\ 22 \\ 12 \\ 21 \end{array}
\begin{pmatrix}
0 & 1 & 0 & 1 & 0 & 0 & 1 & 1 & 0 \\
0 & 0 & 1 & 1 & 0 & 1 & 0 & 0 & 1 \\
0 & 1 & 0 & 0 & 1 & 1 & 0 & 0 & 1 \\
0 & 0 & 1 & 0 & 1 & 0 & 1 & 1 & 0 \\
0 & 1 & 0 & 0 & 0 & 1 & 1 & 0 & 0 \\
0 & 0 & 0 & 1 & 0 & 1 & 0 & 1 & 0 \\
0 & 0 & 0 & 0 & 0 & 0 & 0 & 0 & 0 \\
0 & 0 & 1 & 0 & 0 & 0 & 0 & 1 & 1 \\
0 & 0 & 0 & 0 & 1 & 0 & 1 & 0 & 1
\end{pmatrix}. \tag{6}
$$

In practice, when the number of covariates is large, it becomes more likely that we only observe a small subset of all possible combinations of variables. In this case we do not need to build the full Walsh-Hadamard matrix. For each observed combination, we can map it to a vector of the corresponding XOR functions directly.

For example, assume that we have five binary variables $\{x_1, x_2, \ldots, x_5\}$, and the linear predictor only depends on the first three variables: $g(E(y)) = \beta_0 + \beta_1 x_1 + \beta_2 \text{AND}(x_2, x_3)$. If we restrict the maximum order of the interactions to be three, the design matrix has $\binom{5}{0} + \binom{5}{1} + \binom{5}{2} + \binom{5}{3} = 1 + 5 + 10 + 10 = 26$ columns corresponding to the XOR functions with at most third order interactions. If the sample size is sufficiently large, by using Lasso, only the coefficients for the five predictors: $\text{XOR}(0)$, $\text{XOR}(x_1)$, $\text{XOR}(x_2)$, $\text{XOR}(x_3)$, and $\text{XOR}(x_2, x_3)$ will be non-zero. Then we can apply the XOR functions with non-zero coefficients in new data sets for prediction. In the following experiments on simulated data sets, we show that even when the number of XOR functions is relatively large, the learned models quickly converge toward the generating model as the sample size is increased.

3 Experiments

We compare the greedy search method in the R package LogicReg with transformed Lasso for different models and sizes of training data in three experiments. For Lasso regression, we use the popular R package glmnet. For each model and size of training data, we generated 100 different training data sets based on the true model. Later we compare the log-losses of learned models by LogicReg and transformed Lasso by evaluating them on another 100 new data sets with sample sizes 1024. We ran all the experiments on computers with 32 GB RAM and 2.53 GHz CPUs using only a single core at a time.

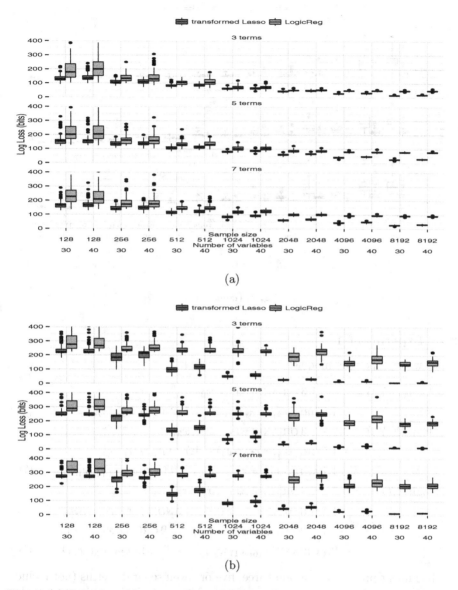

Fig. 1. (a) Box plot of log-losses for independent test data by LogicReg and transformed Lasso when data is simulated from Eq. 7(a) (3 terms)–(c) (7 terms), respectively. Sample sizes increase along the x-axis, and for each sample size, the total number of regressor variables is either 30 or 40 as indicated in the x-axis label. Upper and lower whiskers show the maximum and minimum values of log-losses respectively. To emphasize the differences between the log-losses of the two methods, the outliers (mainly the results by LogicReg) that lie above the upper limit are not shown. (b) Data simulated from Eq. 8(a)–(c) that include XOR operators. (c) Box plot of log-losses of inferred logic functions by LogicReg and transformed Lasso for models in Eq. 9(a)–(c).

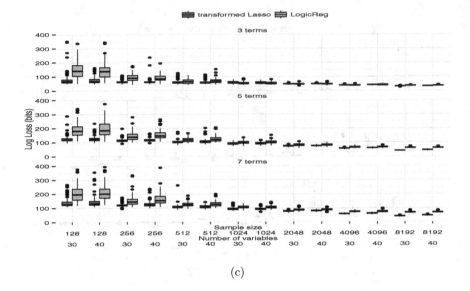

(c)

Fig. 1. (*continued*)

3.1   Experiment 1

First, we use the true models:

$$\text{logit}(E(Y)) = 0.5 - 1.3\,\text{OR}[\text{AND}(\neg x_1, \neg x_2), x_3]$$
$$+ 1.5\,\text{OR}(\neg x_4, \neg x_5, \neg x_6) - 1.7\,\text{AND}(x_7, x_8, x_9), \qquad (7a)$$
$$\text{logit}(E(Y)) = 0.5 - 1.3\,\text{OR}[\text{AND}(\neg x_1, \neg x_2), x_3]$$
$$+ 1.5\,\text{OR}(\neg x_4, \neg x_5, \neg x_6) - 1.7\,\text{AND}(x_7, x_8, x_9)$$
$$+ 1.1\,\text{AND}(\neg x_{10}, \neg x_{11}, \neg x_{12}) - 0.9\,\text{OR}(x_{13}, x_{14}, x_{15}), \qquad (7b)$$
$$\text{logit}(E(Y)) = 0.5 - 1.3\,\text{OR}[\text{AND}(\neg x_1, \neg x_2), x_3]$$
$$+ 1.5\,\text{OR}(\neg x_4, \neg x_5, \neg x_6) - 1.7\,\text{AND}(x_7, x_8, x_9)$$
$$+ 1.1\,\text{AND}(\neg x_{10}, \neg x_{11}, \neg x_{12}) - 0.9\,\text{OR}(x_{13}, x_{14}, x_{15})$$
$$+ 0.7\,\text{OR}[\text{AND}(x_{16}, x_{17}), x_{18}] - 0.5\,\text{AND}(x_{19}, x_{20}, x_{21}). \qquad (7c)$$

The linear predictors contain three, five or seven separate terms (not including intercepts) that have no XOR operators. The terms included in the simpler linear predictors (three and five terms) are subsets of the terms in the more complex linear predictors. Each term in the linear predictor involves three covariates. The covariates are independent from each other and generated with equal probabilities for 1s and 0s.

For the LogicReg method, we restrict the search space by no more than three variables in a term and no more than five separate terms.[2] Model selection

---

[2] We found that the implementation of LogicReg provided in the package cannot handle more than five terms in the same formula.

is performed by 10-fold cross validation. For the method based on the Lasso framework, we also consider only up to third order interactions. The best tuning parameter λ in Lasso is also determined by 10-fold cross validation. The sample size ranges from 64 to 8192. The total number of covariates is 30 or 40 of which all but the ones appearing in Eq. 7(a)–(c) have no effect on the response. For example, the response of model in Eq. (7a) only depends on nine variables, $x_1, \ldots, x_9$, while covariates $x_{10}, \ldots, x_m$ ($m = 30$ or 40) are irrelevant. We repeat the experiment on 100 different training sets for each combination of sample size and number of covariates. The log-likelihoods achieved by both methods are compared to the log-likelihoods under the true model on 100 independently generated samples of size 1024.

We show how the log-loss changes under different conditions in Fig. 1a. We can see that when the sample size reaches 8192, both methods have almost converged to the same estimated models with small log-loss. The Lasso's ability to shrink most of the unimportant coefficients plays a key role in achieving a similar level of performance already from the small samples sizes, unlike the LogicReg method. For instance, with 40 covariates, selecting the required number of covariates (under 100) out of the 10 701 candidates seems to be very hard for LogicReg up until sample size 1024.

When the sample size grows, both methods have increasingly better results in the sense of both smaller average log-losses and smaller variance between the repetitions. But when the sample size is relatively small, the greedy search method is very unstable. This is because it needs to pick the right Boolean terms out of a much larger number of possible terms — recall that the number of features considered in the LogicReg method is significantly greater than in transformed Lasso because LogicReg includes all possible Boolean operators while the latter method only includes XOR operators. Moreover, the Lasso method has no local optima unlike the greedy search applied in LogicReg.

The performance of the two methods also depends on the number of irrelevant covariates. This is because the number of terms to be considered in both methods grows with the number of covariates. For example, the number of XOR terms included in the design matrix in transformed Lasso for 30 variables is 4526 and for 40 variables 10 701. On the other hand, the (negative) effect of increasing the number of covariates is not very significant compared to the (positive) effect of increasing the sample size except in the sense that as the number of variables is increased, the computational cost of both methods increases sharply.

3.2 Experiment 2

In the second experiment, we replace the AND and OR operators in Eq. (7) by the XOR operator while keeping the coefficients unchanged. The new data generating models are

$$
\begin{aligned}
\mathrm{logit}(E(Y)) = {} & 0.5 - 1.3\,\mathrm{XOR}(\neg x_1, \neg x_2, x_3) \\
& + 1.5\,\mathrm{XOR}(\neg x_4, \neg x_5, \neg x_6) - 1.7\,\mathrm{XOR}(x_7, x_8, x_9),
\end{aligned} \tag{8a}
$$

$$\text{logit}(E(Y)) = 0.5 - 1.3 \, \text{XOR}(\neg x_1, \neg x_2, x_3)$$
$$+ 1.5 \, \text{XOR}(\neg x_4, \neg x_5, \neg x_6) - 1.7 \, \text{XOR}(x_7, x_8, x_9)$$
$$+ 1.1 \, \text{XOR}(\neg x_{10}, \neg x_{11}, \neg x_{12}) - 0.9 \, \text{XOR}(x_{13}, x_{14}, x_{15}), \qquad (8b)$$
$$\text{logit}(E(Y)) = 0.5 - 1.3 \, \text{XOR}(\neg x_1, \neg x_2, x_3)$$
$$+ 1.5 \, \text{XOR}(\neg x_4, \neg x_5, \neg x_6) - 1.7 \, \text{XOR}(x_7, x_8, x_9)$$
$$+ 1.1 \, \text{XOR}(\neg x_{10}, \neg x_{11}, \neg x_{12}) - 0.9 \, \text{XOR}(x_{13}, x_{14}, x_{15})$$
$$+ 0.7 \, \text{XOR}(x_{16}, x_{17}, x_{18}) - 0.5 \, \text{XOR}(x_{19}, x_{20}, x_{21}). \qquad (8c)$$

All other experiment settings are the same as in the first experiment. Comparing the results of the second experiment as illustrated in Fig. 1b with the first experiment, we find that for small sample sizes, the transformed Lasso method has difficulty in fitting the true models if they contain many XOR operators. However, it performs better for sample sizes larger than 1024. On the other hand, the greedy search method fails to find acceptable models in the second experiment even when the sample size is 8196.

For the models in the second experiment, we need less non-zero coefficients in the transformed Lasso method, because it contains only XOR functions. For instance, if we have three input terms, we only need three predictors with non-zero coefficients to represent the whole linear predictor. But for the linear predictor of three terms in the first experiment, we need to decompose it into fifteen XOR functions. There are less correct XOR terms in the second experiment, thus each XOR term is comparatively more important. For small sample sizes, when there is not enough information, it becomes much harder to identify the correct XOR functions. For the linear predictors in the second experiment, any incorrect choice of XOR term decreases the accuracy much more significantly than in the first experiment. Therefore, transformed Lasso performs worse when the sample size is small in the second experiment than in the first experiment. On the other hand, a smaller set of non-zero coefficients result in stronger effects of the relevant coefficients on responses. When we have enough training data, it becomes easier for transformed Lasso to identify the right terms in the second experiment.

On the other hand, the greedy search method in LogicReg represents Boolean functions only by AND, OR or negation operators. Thus, it needs to construct much more complex expressions in the second experiment. For example, the term $\text{XOR}(x_1, x_2, x_3)$ needs to be expressed by interactions between ten variables with the repeating use of x_1, x_2 and x_3. It requires exploring an extremely large model space. Moreover, the greedy search method starts by picking up the most significant variables. However, a single variable in a XOR function has a much weaker effect in responses, which makes it very hard for the greedy search method to find a good starting point. Furthermore, in the following process, it can only modify a current model by adjusting one variable or one operator at each step. Although the simulated annealing method are incorporated to avoid local optima, an updated model should be reachable by a single move from the previous one. Therefore, a proper starting point is crucial for the greedy search method, which is difficult to find in the second experiment. Even when the sample size is as large as 8192, the learned models by the greedy search method are far from close to the true ones.

3.3   Experiment 3

For the last experiment, we show how the two methods work when variables have multiple values. We use the similar data generating models as in the first experiment, but allow the variables to take one of the three values: $\{0, 1, 2\}$. The true models are:

$$\text{logit}(E(Y)) = 0.5 - 1.3 \, \text{OR}[\text{AND}(x_1 \neq 0, x_2 \neq 1), x_3 = 2]$$
$$+ 1.5 \, \text{OR}(x_4 \neq 0, x_5 \neq 1, x_6 \neq 2) - 1.7 \, \text{AND}(x_7 = 0, x_8 = 1, x_9 = 2), \tag{9a}$$
$$\text{logit}(E(Y)) = 0.5 - 1.3 \, \text{OR}[\text{AND}(x_1 \neq 0, x_2 \neq 1), x_3 = 2]$$
$$+ 1.5 \, \text{OR}(x_4 \neq 0, x_5 \neq 1, x_6 \neq 2) - 1.7 \, \text{AND}(x_7 = 0, x_8 = 1, x_9 = 2)$$
$$+ 1.1 \, \text{AND}(x_{10} \neq 0, x_{11} \neq 1, x_{12} \neq 2) - 0.9 \, \text{OR}(x_{13} = 0, x_{14} = 1, x_{15} = 2), \tag{9b}$$
$$\text{logit}(E(Y)) = 0.5 - 1.3 \, \text{OR}[\text{AND}(x_1 \neq 0, x_2 \neq 1), x_3 = 2]$$
$$+ 1.5 \, \text{OR}(x_4 \neq 0, x_5 \neq 1, x_6 \neq 2) - 1.7 \, \text{AND}(x_7 = 0, x_8 = 1, x_9 = 2)$$
$$+ 1.1 \, \text{AND}(x_{10} \neq 0, x_{11} \neq 1, x_{12} \neq 2) - 0.9 \, \text{OR}(x_{13} = 0, x_{14} = 1, x_{15} = 2)$$
$$+ 0.7 \, \text{OR}[\text{AND}(x_{16} = 0, x_{17} = 1), x_{18} = 2] - 0.5 \, \text{AND}(x_{19} = 0, x_{20} = 1, x_{21} = 2). \tag{9c}$$

To build the design matrix for the transformed Lasso method, we code interactions between ternary covariates as described in Sect. 2. On the other hand, because the method used by LogicReg requires binary input, we add dummy variables to indicate when a covariate takes one of the three values.

Figure 1c shows that for ternary covariates, transformed Lasso works better than the greedy search method under all sample sizes and numbers of covariates. The ternary case is more difficult for both methods than the binary case because it has much larger search spaces for both methods. However, both methods still show their power to learn good model if there are enough data. When the sample size reaches 4096, both methods converge to true models for ternary covariates as well as for binary covariates.

Based on the previous experiments, we can see that transformed Lasso performs better than the greedy search method in all the cases with models of different complexities and training sample sizes. The greedy search method achieves reasonable results only when there is a decent number of samples and input functions are simple. However, it has very large log-losses as well as large variances when the sample size is less than 512 in all different settings in the three experiments. But in real life studies, a relative small number of training samples is very common. Moreover, the greedy search method fails when the interaction includes complex operators like XOR, which makes the responses less affected by any single covariate involved in the interaction.

Another factor that we need to consider is computational cost. For the simple case with three terms in the linear predictor, a total of 30 covariates and sample size 128 in the first experiment, we need on average 373 s to perform model learning by the greedy search method in LogicReg, but only 6.0 s for the transformed Lasso method. For the more complex model with seven terms including XOR operators, a total of 40 covariates and sample size 8196 in the second experiment, LogicReg uses on average 15 920 s, whereas the Lasso based method needs only 1 580 s.

4   Discussion

In this study, we propose on approach to learning sparse logistic models with logical features of multivalued inputs. The same approach can also be applied in other types of regression models such as Poisson regression and Cox proportional hazards models. In our experiments with simulated data, our Lasso based method was able to estimate different models based on AND, OR, and XOR features and their combinations more effectively than the earlier LogicReg method. More extensive experiments, including a comparison to other types of state-of-the-art classification techniques will provide more detailed information about the performance of the proposed method.

In future work, the proposed approach can be extended in several directions. Firstly, for handling a large number of variables, we can use the LIBLINEAR [3] library that scales better for large sparse data. Even then, a very large number of covariates will necessarily pose problems to methods that include high order interactions. For example, most genome wide studies may include hundreds of thousands genomic covariates. Many existing approaches include a screening stage to narrow down the set of candidate covariates to a manageable number. This can be done either by exploiting expert knowledge or by other statistical methods, see, e.g. [6,12]. Exploring suitable screening methods for the transformed Lasso with very high dimensional data is an interesting research direction that is necessary for many genomics applications.

Moreover, it can be helpful to integrate additional assumptions concerning the model structure to guide the learning process in the spirit of [1,8]. This can be achieved by modifying the Lasso penalization in various ways. For instance, it may be reasonable to assume that if a given high order coefficient takes a non-zero value, the lower order interactions among the variables that are included in the higher order interactions are more likely to be non-zero as well. Different group Lasso techniques are available for achieving this [9]. Furthermore, Lasso tends to select most significant variables out of a group of correlated variables. Integrating the structure of predictors can also be beneficial in the case when the covariates are highly correlated.

Acknowledgments. We thank Mr. Jussi Määttä for checking optimization results and the anonymous reviewers for useful comments. This work was supported in part by the Academy of Finland (Centre-of-Excellence COIN) and by the DoCS graduate school of the Department of Computer Science at the University of Helsinki.

References

1. Bien, J., Taylor, J., Tibshirani, R.: A lasso for hierarchical interactions. Ann. Appl. Stat. **41**(3), 1111–1141 (2013)
2. Breiman, L.: Bagging predictors. Mach. Learn. **24**(2), 123–140 (1996)
3. Fan, R.E., Chang, K.W., Hsieh, C.J., Wang, X.R., Lin, C.J.: LIBLINEAR: A library for large linear classification. J Mach. Learn. Res. **9**, 1871–1874 (2008)
4. Friedman, J.H.: Multivariate adaptive regression splines. Ann. Stat. **19**(1), 1–67 (1991)
5. Hastie, T., Tibshirani, R., Friedman, J.: The Elements of Statistical Learning: Data Mining, Inference and Prediction. Springer, New York (2009)
6. Holger, S., Ickstadt, K.: Identification of SNP interactions using logic regression. Biostat. **9**(1), 187–198 (2008)
7. Keleş, S., van der Laan, M.J., Vulpe, C.: Regulatory motif finding by logic regression. Bioinform. **20**(16), 2799–2811 (2004)
8. Kim, S., Xing, E.P.: Tree-guided group lasso for multi-response regression with structured sparsity, with an application to eQTL mapping. Ann. Appl. Stat. **6**(3), 1095–1117 (2012)
9. Meier, L., van de Geer, S., Bühlmann, P.: The group lasso for logistic regression. J. Roy. Stat. Soc B. **70**(1), 53–71 (2008)
10. Roos, T., Yu, B.: Estimating sparse models from multivariate discrete data via transformed Lasso. In: Proceedings of Information Theory and Applications Workshop, pp. 290–294. IEEE Press (2009)
11. Ruczinski, I., Kooperberg, C., LeBlanc, M.: Logic regression. J. Comp. Graph Stat. **12**(3), 475–511 (2003)
12. Saeys, Y., Inza, I., Larrañaga, P.: A review of feature selection techniques in bioinformatics. Bioinform. **23**(19), 2507–2517 (2007)
13. Shi, W., Wahba, G., Wright, S., Lee, K., Klein, R., Klein, B.: LASSO-Patternsearch algorithm with application to ophthalmology and genomic data. Stat Interface. **1**(1), 137–153 (2008)
14. Suehiro, Y., Wong, C.W., Chirieac, L.R., Kondo, Y., Shen, L., Webb, C.R., et al.: Epigenetic-genetic interactions in the APC/WNT, RAS/RAF, and P53 pathways in colorectal carcinoma. Clin. Cancer Res. **14**(9), 2560–2569 (2008)
15. Tibshirani, R.: Regression shrinkage and selection via the lasso. J. Roy. Stat. Soc B. **58**(1), 267–288 (1996)
16. Zhao, J., Li, J., Xiong, M.: Test for interaction between two unlinked loci. Am. J. Hum. Gen. **79**(5), 831–845 (2006)

A Nonlinear Label Compression and Transformation Method for Multi-label Classification Using Autoencoders

Jörg Wicker$^{(\boxtimes)}$, Andrey Tyukin, and Stefan Kramer

Institut of Computer Science, Johannes Gutenberg University Mainz,
Staudingerweg 9, 55128 Mainz, Germany
wicker@uni-mainz.de, tyukinandrey@gmail.com, kramer@informatik.uni-mainz.de

Abstract. Multi-label classification targets the prediction of multiple interdependent and non-exclusive binary target variables. Transformation-based algorithms transform the data set such that regular single-label algorithms can be applied to the problem. A special type of transformation-based classifiers are label compression methods, which compress the labels and then mostly use single label classifiers to predict the compressed labels. So far, there are no *compression-based algorithms* that follow a *problem transformation approach* and address non-linear dependencies in the labels. In this paper, we propose a new algorithm, called MANIAC (Multi-lAbel classificatioN usIng AutoenCoders), which extracts the non-linear dependencies by compressing the labels using autoencoders. We adapt the training process of autoencoders in a way to make them more suitable for a parameter optimization in the context of this algorithm. The method is evaluated on eight standard multi-label data sets. Experiments show that despite not producing a good ranking, MANIAC generates a particularly good bipartition of the labels into positives and negatives. This is caused by rather strong predictions with either really high or low probability. Additionally, the algorithm seems to perform better given more labels and a higher label cardinality in the data set.

1 Introduction and Related Work

Multi-label classification, the classification of objects into many, possibly interdependent, but non-disjoint binary classes, has received a lot of attention in recent years. Binary relevance (BR), the most basic method, predicts each label independently and thus does not make use of label correlations. Further methods can be divided into the families of (i) method adaptation schemes and (ii) problem transformation schemes. Method adaptation schemes develop multi-label versions of standard machine learning schemes, like decision trees [2], k-Nearest Neighbors [25], or neural networks [12]. The nerual network based approach equips a previously known method with a different loss function. Using neural networks as multi-label classifiers is an obvious and simple extension to neural

© Springer International Publishing Switzerland 2016
J. Bailey et al. (Eds.): PAKDD 2016, Part I, LNAI 9651, pp. 328–340, 2016.
DOI: 10.1007/978-3-319-31753-3_27

networks. Problem transformation schemes, by contrast, transform the multi-label problems into multiple single-label problems and then apply standard single-label methods to the transformed problems – the single-label learners and classifiers act as plug-ins. The results of the single-label classifiers are then transformed back to calculate the multi-label classifications. Examples for this approach are Ensembles of Classifier Chains (ECCs) [15][1] or methods using matrix factorizations to obtain a smaller set of pseudo-labels or latent labels as targets for standard single-label classifiers [17,24].[2] Problem transformation algorithms have some advantages over method adaptation methods: First, any type of single-label classifier can be used as plug-in. (a) If a classifier has a suitable bias for a data set at hand, it can readily be used. (b) It can be chosen depending on the practical requirements of a project (e.g., a preference for lower errors over shorter running times, or vice versa). (c) New single-label learning schemes can be immediately used and tested, if needed, without the need for a potentially non-trivial adaptation of the method and time to achieve this. Further, transformation methods do not depend on any specific type of data, for instance, real-valued or graph data, as long as the base (single-label) classifier can handle them as input. This affects mainly highly specialized algorithms, like for instance method adaptation schemes based on specific neural networks architectures, that are mostly restricted to real-valued data [8].

Orthogonal to the above distinction (method adaptation vs. problem transformation), so-called *label compression* algorithms were suggested [17,19,24]. These algorithms compress the labels into a typically smaller label space and make the predictions on the compressed labels. The prediction is done by decompressing the predicted labels again. MLC-BMaD [24], for instance, uses Boolean matrix decomposition to generate a compression of the labels, which is used as input for a BR learner. The decompression is achieved by multiplying with a matrix that contains carefully chosen basis vectors. Hence, the dependencies among the labels are encoded in this basis matrix, which is part of the learned model. Another approach similar to MLC-BMaD is multi-label classification using the principal label space transformation [17]. This method uses singular value decomposition instead of Boolean matrix factorization in the compression step. Label compression methods can also belong to the family of method adaptation schemes, for instance, in neural network variants [8] (see below).

Almost all of the existing approaches aim to use the dependencies among the labels to improve the prediction. Nevertheless, so far, only few of the classifiers were able to cope with non-linear dependencies among the labels. One of the approaches targeting non-linear dependencies among the labels was proposed by Li and Guo [8]. Nevertheless, the method aims for a compression of both the input (the features) and the output (the labels) together. This differs from the one presented here, as we will solely focus on finding a compressed repre-

---

[1] Notice that ECCs should still be considered as a strong baseline method as confirmed by a quite recent extensive experimental comparison [10].

[2] A yet different family of methods transforms larger multi-label problems into smaller multi-label methods, like for instance HOMER [20] or RAKEL [22].

sentation of the labels that encodes their dependencies. This allows us to make use of base learners that can cope with features that are not readily embedded into $\mathbb{R}^n$ (e.g. text, nominal). Li *et al.* [9] presented an algorithm that uses conditional Restricted Boltzmann Machines (RBMs) to train models on data sets with partially missing labels. All these methods belong to the method adaptation schemes, with the above mentioned disadvantages. Read *et al.* [14] showed that ensembles of classifiers can capture non-linear dependencies among the labels by functioning as a layered network. This is achieved by using RBMs as a base classifier in ECCs.

This paper makes three contributions to the field: (i) We introduce the first method from the intersection of problem transformation approaches and label compression approaches that is able to capture nonlinear label dependencies. The algorithm, called MANIAC[3] (multi-label classification using autoencoders), compresses the labels using autoencoders [6], and then learns a multi-label model on the compressed label set. After the prediction, the same structure (autoencoder) is used to decompress the labels and obtain the final predictions. (ii) We introduce autoencoders to the set of available multi-label compressors and decompressors in problem transformation approaches. So far, autoencoders have surprisingly only been used on the whole data set, despite using them on the labels only seems to be an obvious approach. (iii) We state exactly where the algorithm performs well, namely on the task of "genuine" multi-label classification problems with many labels and dependencies among the labels and when the prediction of the correct bipartition of the labels is important.

The remainder of the paper is organized as follows: In the next section, we describe in detail MANIAC, the newly proposed algorithm. Subsequently, we explain the evaluation approach. Finally, we discuss the experimental results and give a conclusion.

2   MANIAC – Multi-Label Classification Using Autoencoders

The algorithm works similarly to other label compression based algorithms like MLC-BMaD [24] or the method proposed by Tai and Lin [17] – Principal Label Space Transformation (PLST). In the first step, the labels are compressed using a compression algorithm. Then, base learners are trained on the compressed labels (see Fig. 1). In the case of MANIAC, we use autoencoders for compression, in the case of MLC-BMaD Boolean matrix decomposition is used, Tai and Lin use singular value decomposition. The difference in using the autoencoders in this step is that unlike other approaches, this captures non-linear dependencies among the labels. Boolean matrix decomposition and SVD cannot capture them, hence the performance the prediction performance can be improved on data sets with non-linear dependencies.

---

[3] The implementation is available at https://github.com/kramerlab/maniac, as well as directly integrated in Meka http://meka.sourceforge.net/.

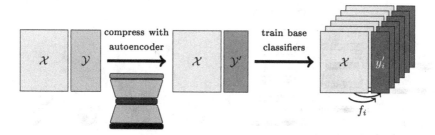

Fig. 1. Illustration of the training phase of MANIAC. After the autoencoders are trained (not visualized), $\mathcal{Y}$ is compressed using the trained autoencoders. Next, a BR model is trained on the compressed labels.

Autoencoders [6] are neural networks that consist of a compressor and a decompressor part, connected by a small central layer. (see Fig. 2). In the following, we will explain details of the autoencoder training.

Autoencoders can be trained in an unsupervised manner[4]. For this, training data is clamped to the equally sized input and output layers of the autoencoder. The training data is compressed using the compressor part of the network, and then reconstructed again using the decompressor part. The result is compared to the original data, the reconstruction errors are used to tune the parameters of the compressor and the decompressor. More specifically, we used a Conjugate Gradient (CG) algorithm, and computed the necessary gradients from reconstruction errors using classical backpropagation. If the training of an autoencoder is successful, the output of the network is similar to the input data. Since the output is reconstructed by the decompressor from the activations of neurons in the innermost layer, the small innermost layer of the autoencoder can be thought of as an informational bottleneck. The activations of the neurons in this bottleneck layer constitute an efficient low-dimensional representation of the input data. In particular, this representation captures non-linear dependencies between the activations of the neurons in the input layer.

Hinton and Salakhutdinov [6] proposed to treat pairs of layers (see Fig. 3) of the autoencoder as *Restricted Boltzmann Machines* (RBMs), and train them using the *Contrastive Divergence* (CD) algorithm. Connection weights and neuron activation biases produced by contrastive divergence were then used as input to the backpropagation-based optimization algorithm.

We use a variation of these ideas to train whole *streams* of increasingly deep autoencoders. The training process of autoencoders is as follows. We start with a trivial autoencoder that has a single neuron layer and no connections. Suppose that we already have an autoencoder with $2n+1$ layers. We apply the compressor part to the data, and train a new RBM on the compressed data. The size of

---

[4] It should be noted that, as we train the autoencoders on the labels, this could be understood as supervised learning. Nevertheless, for the training of the autoencoders, no additional target variable is used, and the labels are not treated as target variables for this step, hence this is still unsupervised training.

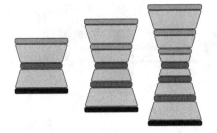

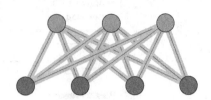

Fig. 2. Stream of increasingly deep autoencoders. The deeper autoencoders are created from the shallower ones by unfolding the innermost layer and then tuning with backpropagation.

Fig. 3. Topology of a Restricted Boltzmann Machine (RBM). Nodes in the lower layer (blue) are called *visible units*. Nodes in the upper layer (green) are called *hidden units* (Color figure online).

the new layer of hidden units (see Fig. 3) is determined by the *compression factor* hyperparameter (see also appendix). Then we unfold the RBM and merge it into the center of the original autoencoder, obtaining an autoencoder with $2n + 3$ layers (see Fig. 2). Hinton and Salakhutdinov originally proposed to keep unfolding all RBMs until the desired depth is reached, and fine-tune the final autoencoder with the backpropagation algorithm in the very end. However, we have found it beneficial to treat the unfolded RBMs as small autoencoders of depth 1 and also tune them with backpropagation. After gluing the new small autoencoder into the center of the previously obtained $(2n+1)$-layer autoencoder, we also tune the resulting $(2n + 3)$-layer autoencoder with backpropagation.[5]

We evaluated multiple training strategies for the autoencoders, and the results showed that in contrast to the strategy originally suggested by Hinton and Salakhutdinov [6], when using a stream of autoencoders, better results can be achieved. Additionally, there seems to be no big difference from using the parameters suggested by Hinton compared to using an optimization of these parameters.

The next step is then to train a base classifier (a multi-target model) using the compressed labels as new target variables. In the previous step, we extracted the dependencies from the labels into the autoencoders, hence, in the best case, there are no dependencies left among the latent labels. Therefore, it should not be beneficial to use a sophisticated multi-target learner over a simple BR model. Nevertheless, if the autoencoder does not manage to extract all dependencies, it might be beneficial to use a more advanced multi-target learner. The training phase of the algorithm is composed of these two steps, and the model consists of the autoencoders and the multi-target (BR) model.

With an autoencoder, a threshold for binarization, and the trained base learners, one can easily predict labels for new instances. First, the multi-target model is applied and the latent labels are predicted. Next, the autoencoders are used

---

[5] Due to space limitations, we moved a more detailed version of this section, which is more technically involved, to https://github.com/kramerlab/maniac/blob/master/docs/supplementary.pdf.

to decompress the latent labels. The final labels are obtained by thresholding, so that the output is binary.

The predicted bipartition is calculated from the confidences given from the autoencoder similar to other multi-label classifiers using a threshold. It should be noted that the calculated confidences are mostly close to 0 and 1, and rarely somewhere in between.

3 Evaluation

We optimized two parameters of the autoencoders, the compression factor and the number of layers using an internal holdout evaluation. We have also attempted to optimize the hyperparameters of Contrastive Divergence (used for RBM training), but the optimization turned out to be expensive, while not having much impact on the quality of the final model. In general, using RBMs for pre-training seemed to speed up the computation, but the effect of fine-tuning the small unfolded autoencoders and all the intermediate autoencoders seemed to outweigh the effects of Contrastive Divergence on the quality of the model. Therefore, all the parameters of Contrastive Divergence are essentially the same as proposed originally [6]. The optimization of the parameter for the final thresholding is very cheap, it can be accomplished by simple grid-search that tests thresholds in the interval [0, 1] using 1000 steps for each column individually, and only models need to be applied, not trained.

We compared our method to BR, ECC, and MLC-BMaD[6]. BR is a good choice for a baseline as it uses no dependencies among labels, hence, a good multi-label classifier should perform better than BR. ECC is a fast and well performing method and currently considered as the benchmark method for multi-label classification [10]. We evaluated the algorithm using a repeated holdout evaluation with one third of the data set as test set, two thirds of the data set as training set and ran all evaluations 15 times, as suggested by Nadeau and Bengio [11], and calculated the corrected re-sampled t-test statistics implemented in WEKA [11]. For ECC, we set the number of chains to 10, BR did not require any parameters to be set, the parameters of MLC-BMaD we optimized using a greedy search, as suggested by Wicker et al. [24]. As Wicker et al. showed that their approach outperforms PLST [17], we did not compare to PLST seperately.

The choice of a multi-target base learners depends on the question whether the autoencoder is capable of extracting all dependencies and returning a set of latent labels that are completely independent. If this is the case, the learners can be trained independently, and the multi-target version of BR can be used. If there are dependencies left in the data, a more sophisticated multi-target learning method should be chosen. In our experiments, we used an adaptation of ECC for multi-target problems, as it has been proven to be a fast and well-performing learner. Both base learners were used and compared to the other models.

---

[6] As the authors of [8] did not share their code for experimental comparisons, and we were not able to reproduce the published results, we did not compare to this algorithm.

3.1 Implementation

We implemented the algorithm in Mulan [21] using our own implementation of autoencoders[7]. The implementation provides a way to train a stream of autoencoders, adding one layer at a time, given a compression factor. This is used to speed up the optimization process of choosing the right number of layers. Using an internal training set, autoencoders with one layer are trained. A model is trained on this autoencoder and the performance is evaluated on a test set. The autoencoders are extended to have more layers, and are evaluated again, and so on. Hence, we do not need to repeat the process of training an autoencoder with n layers, we can reuse the previous trained autoencoder and extend it. As base learners, we used an adaptation of ECC and BR for multi-target problems. These learners used random forests as a base learners due to their speed and typically good performance. We used no parameter optimization except an internal holdout validation to optimize the number of layers and compression factor of the autoencoders.

4 Experimental Results

We used nine standard multi-label data sets from the data set repository of Mulan (see http://mulan.sourceforge.net/datasets.html). The results of the experiments can be split into two parts: First, the performance of MANIAC regarding the split into positive and negative labels (bipartition-based measures are given in Table 1), and second, the predicted ranking (confidence-based measures are given in Table 2)[8]. The predicted bipartitions give an overall good performance. Yet the ranking or estimated confidences are rather bad. The latter is caused by the autoencoder decompression algorithm, which assigns each label with high confidence in any case, yet the assignment of the confidence is not reliable. Therefore, if a label is incorrectly assigned to be positive or negative, it is not incorrectly assigned with a confidence a bit below the threshold, the confidence is set to a value close to 1 or 0. In the overall ranking then, it can be in a position in the ranking completely wrong and impact the quality of the ranking strongly. Thus, a low number of false positives or negatives has a huge impact on the ranking, much stronger than in the case of other classifiers. Hence all ranking based measures as *area under ROC curve*, *one error*, or *coverage* are rather bad for MANIAC. Nevertheless, MANIAC performs well regarding other measures like *accuracy* or *FMeasure*, which simply take into account the bipartition. In some cases they strongly improve the performance compared to ECC or BR, even in the range of 20 % (e.g. the example-based accuracy of the medical data set). Hence, in the following, we will focus on discussing the bipartition-based measures.

---

[7] The autoencoder implementation is available at https://github.com/kramerlab/autoencoder.

[8] Due to space limitations, we give only representative results, the full results are given at https://github.com/kramerlab/maniac/blob/master/docs/supplementary.pdf.

Table 1. Evaluation using bipartition-based measures. Results for MANIAC are given both with BR and ECC as base classifier. The significant improvement • or degradation ○ of MANIAC using BR as base classifier compared to the associated classifier is given. Note that the data sets are sorted according to the number of labels in descending order.

	Data set	MANIAC (BR)	BR	ECC	MANIAC (ECC)	MLC-BMaD	MANIAC (single layer)
Example-Based Accuracy	CAL500	0.25	0.21 •	0.20 •	0.25	0.01 •	0.25
	enron	0.44	0.41 •	0.42 •	0.41 •	0.32 •	0.39 •
	medical	0.54	0.46 •	0.44 •	0.47 •	0.41 •	0.17 •
	genbase	0.74	0.93 ○	0.95 ○	0.65	0.90 ○	0.52 •
	birds	0.57	0.59	0.55	0.60	0.57	0.53
	yeast	0.51	0.50 •	0.51	0.52	0.50 •	0.52
	flags	0.59	0.61 ○	0.62 ○	0.60	0.55 •	0.61
	scene	0.58	0.60	0.58	0.56	0.47 •	0.60
	emotions	0.52	0.53	0.55	0.53	0.48 •	0.54 ○
	[○/ /•]		[2/3/4]	[2/4/3]	[0/7/2]	[1/1/7]	[1/5/3]
Example-Based FMeasure	CAL500	0.39	0.34 •	0.33 •	0.39	0.01 •	0.39
	enron	0.56	0.52 •	0.53 •	0.53 •	0.39 •	0.50 •
	medical	0.58	0.48 •	0.46 •	0.50 •	0.43 •	0.18 •
	genbase	0.76	0.94 ○	0.95 ○	0.66	0.91 ○	0.54 •
	birds	0.60	0.61	0.57	0.62	0.59	0.55 •
	yeast	0.63	0.61 •	0.62 •	0.64	0.61 •	0.63
	flags	0.72	0.73	0.73	0.72	0.66 •	0.73
	scene	0.61	0.60	0.59	0.59	0.48 •	0.61
	emotions	0.61	0.61	0.62	0.62	0.54 •	0.64 ○
	[○/ /•]		[1/4/4]	[1/4/4]	[0/7/2]	[1/1/7]	[1/4/4]

Regarding the size of the data sets, the results show that the larger the number of labels, the better the algorithm performs. In all three cases of *CAL500* [23], *enron* [7], and *medical* [13], MANIAC outperforms ECC without exception, with in all cases a strong improvement in the range of up to 25 %. The higher the number of labels, the easier it is for the autoencoders to generate sensible latent labels, and capture the dependencies among the labels. On the other hand, data sets with only few labels, like *flags* [5], *scene* [1], and *emotions* [18], only having 6 or 7 labels, are hard to compress any further. Hence the dependencies cannot be extracted by the autoencoders. When using data sets of that size, a compression based algorithm is rather useless. Hence the performance of MANIAC on these data sets is in almost all cases worse than BR or ECC. In the medium range of number of labels, the trend is not that clear, in the case of the *yeast* data set [4], small improvements were possible, on the other hand, on *genbase* [3], MANIAC does not seem to work. Here, other aspects like label cardinality seem to be more important.

Table 2. Evaluation using confidence-based measures. Results for MANIAC are given both with BR and ECC as base classifier. The significant improvement • or degradation ○ of MANIAC using BR as base classifier compared to the associated classifier is given. Note that the data sets are sorted according to the number of labels in descending order.

	Data set	Maniac (BR)	BR	ECC	Maniac (ECC)	MLC-BMaD	Maniac (single layer)
Micro-Averaged AUC	CAL500	0.76	0.81 ○	0.82 ○	0.76	0.79 ○	0.80 ○
	enron	0.87	0.91 ○	0.92 ○	0.91 ○	0.88	0.90 ○
	medical	0.94	0.98 ○	0.98 ○	0.95	0.97 ○	0.96 ○
	genbase	0.98	1.00 ○	1.00 ○	0.97	0.99 ○	0.97
	birds	0.83	0.92 ○	0.92 ○	0.85	0.91 ○	0.89 ○
	yeast	0.82	0.85 ○	0.85 ○	0.83	0.85 ○	0.83
	flags	0.80	0.82 ○	0.83 ○	0.80	0.82 ○	0.81
	scene	0.91	0.96 ○	0.96 ○	0.90	0.84 •	0.93 ○
	emotions	0.83	0.87 ○	0.87 ○	0.84	0.86 ○	0.85
	[○/ /•]		[9/0/0]	[9/0/0]	[1/8/0]	[7/1/1]	[5/4/0]

Especially in the mid range of number of labels, it becomes evident that cardinality and density are important measures for MANIAC. Despite *yeast* being similar in the number of labels to *genbase* and *birds*, it is easier for MANIAC to predict with a higher performance. The biggest difference between these data sets is the cardinality and density. Cardinality in *yeast* is more than four times higher than the cardinality in *birds*. While the number of instances is also higher, this does not seem to have an effect on the other data sets. For instance *CAL500* is the smallest of the top three data sets in terms of instances, yet there is no difference in performance. Other measures like the number of numeric or nominal features do not have an influence on the performance of MANIAC. They are simply an input to the random forests and do not change anything for the autoencoders.

If we compare the performance of MANIAC using the multi-target versions of BR with MANIAC using the multi-target version of ECC as base classifier, we can see that in most cases there is no difference. If there is a difference, BR outperforms ECC. This is a good indication that it is possible to extract all dependencies from the labels using autoencoders. The resulting compression consists of no dependencies, otherwise ECC would outperform MANIAC using BR. It was shown that ECC does not work well using regression models, due to the stronger error propagation in the chains [16]. Nevertheless, in this setting, this does not seem a problem: The performance is more or less the same, independent of using BR or ECC as base learners. Hence the dependencies must be extracted almost completely by the autoencoders, and the random forest most likely only uses the features for the prediction of latent labels, ignoring other latent labels in the case of ECC. Hence it is recommended to use BR as a base learner for MANIAC, as this is beneficial in terms of runtime and resources (the data set for each base learner is smaller in terms of features for base learners in BR compared to ECC).

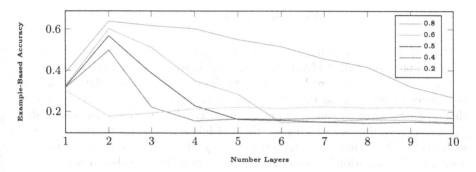

Fig. 4. Comparison of the number of layers depending on the compression factor on the *medical* data set. The x-axis gives the number of layers used, the y-axis the example-based accuracy.

We additionally compared MANIAC to itself using only a single layer in the autoencoders. This can give an indication to what extent non-linear dependencies are exploited by MANIAC. If the autoencoders only have a single layer, they are a simple mapping function, not capable of handling non-linear dependencies. The results show that in this case MANIAC does not perform too well. Hence, MANIAC seems to use the non-linear dependencies to improve its classification.

Figure 4 shows the performance depending on the compression factor and the number of layers. We evaluated several settings on the medical data set. The behavior was similar on all other data sets, with exception of the less than small data sets with 10 labels, where the compression was difficult due to the anyway small size of the data set. Clearly, the smaller the compressed data sets due to the compression factor, the faster the number of layers become too high to come up with a meaningful representation, and the overall accuracy tends towards a value of approximately 0.15. Nevertheless, with a compression factor of 0.8, the accuracy is much more stable and a higher number of layers seems to become beneficial. On the other extreme, with a compression factor of 0.2, after the first layer, the number of layers is reduced to 20 % of the original size: One layer appears to be the optimum. This is simply because after one layer the number of latent labels would become too small, if another layer would be added.

Although the complexity of the algorithm seems to be high and the runtime is certainly higher than that of other algorithms, as it adds a rather expensive compression step, we were able to train autoencoders with one layer on a desktop[9] for the labels of the biggest data set (*CAL500*) in 58 s, two layers could be trained in only 133 s, using a compression factor of 0.85, which seems to be a typical setting the optimization process ends with. In our experiments, most optimal autoencoders used at maximum 4 layers in total. The most time consuming step is the optimization of the parameters, which is reduced by using streams of autoencoders.

---

[9] We used only a single core of an Intel ® Core[TM]i7-4770 K CPU – 3.50 GHz Processor and 4 GB of RAM.

5   Conclusion and Future Work

In this paper, we presented a new approach to multi-label classification based on label compression, called MANIAC. Unlike previously presented transformation-based methods, MANIAC exploits non-linear dependencies among the labels. This is achieved by compressing the label space with autoencoders. The results showed that MANIAC strongly outperforms standard baseline methods for multi-label classification in the case of a high number of labels and seems to benefit from a high cardinality. While MANIAC produces a good bipartition, the confidence is only a rough indication if a label is positive or negative and should not be used for ranking.

While the method works already well for standard multi-label classification, in particular for bipartition-based measures and genuine multi-label data sets, it also has a high potential for online multi-label classification. Using the autoencoder compression in an online scenario is trivial, and hence, MANIAC can be easily adapted for online learning. The base models can be trained using an online learner, combined using BR, and the compression could be updated with every instance. Nevertheless, this would lead to a large runtime in each step which might not be practical for online learning. Hence, training the autoencoders in batches could be more convenient. On the other hand, the online learning could be completely left to the autoencoders and the base models could be updated only from time to time or vice-versa. Another extension of this method would be to transfer it to multi-target learning, which has been recently done for many multi-label methods (e.g. by Spyromitros-Xioufis et al. [16]). This step would be quite straightforward, as autoencoders by default can compress numerical values.

References

1. Boutell, M.R., Luo, J., Shen, X., Brown, C.M.: Learning multi-label scene classification. Pattern Recogn. **37**(9), 1757–1771 (2004)
2. Clare, A.J., King, R.D.: Knowledge discovery in multi-label phenotype data. In: Siebes, A., De Raedt, L. (eds.) PKDD 2001. LNCS (LNAI), vol. 2168, p. 42. Springer, Heidelberg (2001)
3. Diplaris, S., Tsoumakas, G., Mitkas, P.A., Vlahavas, I.P.: Protein classification with multiple algorithms. In: Bozanis, P., Houstis, E.N. (eds.) PCI 2005. LNCS, vol. 3746, pp. 448–456. Springer, Heidelberg (2005)
4. Elisseeff, A., Weston, J.: A kernel method for multi-labelled classification. In: Advances in Neural Information Processing Systems, pp. 681–687 (2001)
5. Gonçalves, E.C., Plastino, A., Freitas, A.A.: A genetic algorithm for optimizing the label ordering in multi-label classifier chains. In: 2013 IEEE 25th International Conference on Tools with Artificial Intelligence (ICTAI), pp. 469–476. IEEE (2013)
6. Hinton, G.E., Salakhutdinov, R.R.: Reducing the dimensionality of data with neural networks. Science **313**(5786), 504–507 (2006)
7. Klimt, B., Yang, Y.: The enron corpus: a new dataset for email classification research. In: Boulicaut, J.-F., Esposito, F., Giannotti, F., Pedreschi, D. (eds.) ECML 2004. LNCS (LNAI), vol. 3201, pp. 217–226. Springer, Heidelberg (2004)

8. Li, X., Guo, Y.: Bi-directional representation learning for multi-label classification. In: Calders, T., Esposito, F., Hüllermeier, E., Meo, R. (eds.) ECML PKDD 2014, Part II. LNCS, vol. 8725, pp. 209–224. Springer, Heidelberg (2014)

9. Li, X., Zhao, F., Guo, Y.: Conditional restricted Boltzmann machines for multi-label learning with incomplete labels. In: Proceedings of the Eighteenth International Conference on Artificial Intelligence and Statistics, pp. 635–643 (2015)

10. Madjarov, G., Kocev, D., Gjorgjevikj, D., Džeroski, S.: An extensive experimental comparison of methods for multi-label learning. Pattern Recogn. **45**(9), 3084–3104 (2012)

11. Nadeau, C., Bengio, Y.: Inference for the generalization error. Mach. Learn. **52**(3), 239–281 (2003)

12. Nam, J., Kim, J., Loza Mencía, E., Gurevych, I., Fürnkranz, J.: Large-scale multi-label text classification — revisiting neural networks. In: Calders, T., Esposito, F., Hüllermeier, E., Meo, R. (eds.) ECML PKDD 2014, Part II. LNCS, vol. 8725, pp. 437–452. Springer, Heidelberg (2014)

13. Pestian, J.P., Brew, C., Matykiewicz, P., Hovermale, D., Johnson, N., Cohen, K.B., Duch, W.: A shared task involving multi-label classification of clinical free text. In: Proceedings of the Workshop on BioNLP 2007: Biological, Translational, and Clinical Language Processing, pp. 97–104. Association for Computational Linguistics (2007)

14. Read, J., Hollmén, J.: A deep interpretation of classifier chains. In: Blockeel, H., van Leeuwen, M., Vinciotti, V. (eds.) IDA 2014. LNCS, vol. 8819, pp. 251–262. Springer, Heidelberg (2014)

15. Read, J., Pfahringer, B., Holmes, G., Frank, E.: Classifier chains for multi-label classification. Mach. Learn. **85**(3), 333–359 (2011)

16. Spyromitros-Xioufis, E., Tsoumakas, G., Groves, W., Vlahavas, I.: Multi-label classification methods for multi-target regression (2012). arXiv preprint arxiv:1211.6581

17. Tai, F., Lin, H.T.: Multilabel classification with principal label space transformation. Neural Comput. **24**(9), 2508–2542 (2012)

18. Trohidis, K., Tsoumakas, G., Kalliris, G., Vlahavas, I.P.: Multi-label classification of music into emotions. In: Proceedings of the Ninth International Conference on Music Information Retrieval, vol. 8, pp. 325–330 (2008)

19. Tsoumakas, G., Dimou, A., Spyromitros, E., Mezaris, V., Kompatsiaris, I., Vlahavas, I.: Correlation-based pruning of stacked binary relevance models for multi-label learning. In: Proceedings of the 1st International Workshop on Learning from Multi-Label Data, pp. 101–116 (2009)

20. Tsoumakas, G., Katakis, I., Vlahavas, I.: Effective and efficient multilabel classification in domains with large number of labels. In: Proceedings of the ECML/PKDD 2008 Workshop on Mining Multidimensional Data (MMD 2008), pp. 30–44 (2008)

21. Tsoumakas, G., Katakis, I., Vlahavas, I.: Mining multi-label data. In: Maimon, O., Rokach, L. (eds.) Data mining and knowledge discovery handbook, pp. 667–685. Springer, Heidelberg (2010)

22. Tsoumakas, G., Vlahavas, I.P.: Random k-labelsets: an ensemble method for multilabel classification. In: Kok, J.N., Koronacki, J., Lopez de Mantaras, R., Matwin, S., Mladenič, D., Skowron, A. (eds.) ECML 2007. LNCS (LNAI), vol. 4701, pp. 406–417. Springer, Heidelberg (2007)

23. Turnbull, D., Barrington, L., Torres, D., Lanckriet, G.: Semantic annotation and retrieval of music and sound effects. IEEE Trans. Audio Speech Lang. Process. **16**(2), 467–476 (2008)

24. Wicker, J., Pfahringer, B., Kramer, S.: Multi-label classification using Boolean matrix decomposition. In: Proceedings of the 27th Annual ACM Symposium on Applied Computing, pp. 179–186. ACM (2012)
25. Zhang, M.L., Zhou, Z.H.: ML-KNN: a lazy learning approach to multi-label learning. Pattern Recogn. **40**(7), 2038–2048 (2007)

Preconditioning an Artificial Neural Network Using Naive Bayes

Nayyar A. Zaidi$^{(\boxtimes)}$, François Petitjean, and Geoffrey I. Webb

Faculty of Information Technology,
Monash University, Melbourne, VIC 3800, Australia
{nayyar.zaidi,francois.petitjean,geoff.webb}@monash.edu

Abstract. Logistic Regression (LR) is a workhorse of the statistics community and a state-of-the-art machine learning classifier. It learns a linear model from inputs to outputs trained by optimizing the Conditional Log-Likelihood (CLL) of the data. Recently, it has been shown that preconditioning LR using a Naive Bayes (NB) model speeds up LR learning many-fold. One can, however, train a linear model by optimizing the mean-square-error (MSE) instead of CLL. This leads to an Artificial Neural Network (ANN) with no hidden layer. In this work, we study the effect of NB preconditioning on such an ANN classifier. Optimizing MSE instead of CLL may lead to a lower bias classifier and hence result in better performance on big datasets. We show that this NB preconditioning can speed-up convergence significantly. We also show that optimizing a linear model with MSE leads to a lower bias classifier than optimizing with CLL. We also compare the performance to state-of-the-art classifier Random Forest.

Keywords: Logistic regression · Preconditioning · Conditional log-likelihood · Mean-square-error · WANBIA-C · Artificial neural networks

1 Introduction

Logistic Regression (LR) is a state-of-the-art machine learning classifier and is widely used by statisticians [1,2]. It has been shown recently that LR training converges more rapidly when each axis is scaled by the log of the naive Bayes estimates of the conditional probabilities [3,4]. Such rescaling leads to an alternative parameterization with both naive Bayes parameters (learned generatively) and LR parameters (learned discriminatively). The resulting parameterization of LR is known as WANBIA$^C_{CLL}$ and has been shown to be effective for both online and batch gradient based optimization for logistic regression[1]. LR optimizes the conditional log-likelihood (CLL) of the data given the model. We conjecture that optimizing the mean square error (MSE) should lead to more accurate (low-biased) models, especially for bigger datasets because, it is mainly

---

[1] Note, we add CLL as subscript to WANBIA-C to show explicitly the objective function that it optimizes.

© Springer International Publishing Switzerland 2016
J. Bailey et al. (Eds.): PAKDD 2016, Part I, LNAI 9651, pp. 341–353, 2016.
DOI: 10.1007/978-3-319-31753-3_28

the bias that contributes to the error on the bigger datasets [5,6]. Note, that a linear model optimizing MSE is an Artificial Neural Network (ANN) with no hidden layer (the structure constitutes only an input layer with multiple nodes and an output layer with multiple nodes).

This paper investigates the performance of linear classification models that optimize MSE relative to those that optimize CLL and whether NB regularization is as effective with the MSE objective function as it is with CLL. One can view WANBIA$_{\text{CLL}}^{\text{C}}$ from two perspectives.

1. From the NB perspective, the parameters learned with discriminative training are only alleviating NB's independence assumption. It is irrelevant whether the weights are optimized by the CLL or by the MSE objective function.
2. From the LR perspective, WANBIA$_{\text{CLL}}^{\text{C}}$ introduces NB weights that precondition the search space. For CLL, which is a convex objective function, this leads to faster convergence. A natural question is: will the same trend hold for other objective functions which are not convex, such as MSE?

The contributions of this paper are two-fold:

1. We show that NB preconditioning is applicable and equally useful for learning a linear classification model optimizing the MSE objective function.
2. Optimizing MSE leads to a lower bias classifier than LR optimizing CLL. This leads to lower 0–1 loss and RMSE on big datasets.

The rest of this paper is organized as follows. We discuss LR and WANBIA$_{\text{CLL}}^{\text{C}}$ in Sect. 2. We will derive NB preconditioning of a linear classification model optimizing MSE in Sect. 3. Empirical analysis is given in Sect. 4. We conclude in Sect. 5 with some pointers to future work.

2   WANBIA$_{\text{CLL}}^{\text{C}}$

Let us start by explaining WANBIA$_{\text{CLL}}^{\text{C}}$. Typically, an LR optimizes the following objective function:

$$\text{CLL}(\beta) = \sum_{i=1}^{N} \log P_{LR}(y^{(i)}|\mathbf{x}^{(i)}), \tag{1}$$

where N is the number of data points. Note, we are constraining ourselves to categorical attributes and multi-class problems only. We write P_{LR} for categorical features and multiple classes as:

$$P_{LR}(y|\mathbf{x}) = \frac{\exp(\beta_y + \sum_{i=1}^{a} \beta_{y,i,x_i})}{\sum_{c \in \Omega_Y} \exp\left(\beta_c + \sum_{j=1}^{a} \beta_{c,j,x_j}\right)},$$

$$= \exp\left(\beta_y + \sum_{i=1}^{a} \beta_{y,i,x_i} - \log \sum_{c \in \Omega_Y} \exp\left(\beta_c + \sum_{j=1}^{a} \beta_{c,j,x_j}\right)\right), \tag{2}$$

where a is the number of attributes and β_{y,i,x_i} denotes the parameter associated with class y, and attribute i taking value x_i. On the other hand, naive Bayes is defined as:

$$P_{NB}(y\,|\,\mathbf{x}) = \frac{P(y)\prod_{i=1}^{a}P(x_i|y)}{\sum_{c\in\Omega_Y}P(c)\prod_{j=1}^{a}P(x_j|c)}.$$

One can add weights to NB to alleviate the attribute independence assumption, resulting in the WANBIA$_{CLL}^{C}$ formulation, that can be written as:

$$P_W(y\,|\,\mathbf{x}) = \frac{P(y)^{w_y}\prod_{i=1}^{a}P(x_i|y)^{w_{y,i,x_i}}}{\sum_{c\in\Omega_Y}P(c)^{w_c}\prod_{j=1}^{a}P(x_j|c)^{w_{c,j,x_j}}}$$

$$= \exp\!\Big(w_y\log P(y) + \sum_{i=1}^{a}w_{y,i,x_i}\log P(x_i|y)$$

$$-\log\sum_{c\in\Omega_Y}\exp\!\Big(w_c\log P(c) + \sum_{j=1}^{a}w_{c,j,x_j}\log P(x_j|c)\Big)\Big). \qquad (3)$$

When conditional log likelihood (CLL) is maximized for LR and weighted NB using Eqs. 2 and 3 respectively, we get an equivalence such that $\beta_c \propto w_c\log P(c)$ and $\beta_{c,i,x_i} \propto w_{c,i,x_i}\log P(x_i|c)$. Thus, WANBIA$_{CLL}^{C}$ and LR generate equivalent models. While it might seem less efficient to use WANBIA$_{CLL}^{C}$ which has twice the number of parameters of LR, the probability estimates are learned very efficiently using maximum likelihood estimation, and provide useful information about the classification task that in practice serve to effectively precondition the search for the parameterization of weights to maximize conditional log likelihood.

3   Method

In this section, we will derive a variant of WANBIA$_{CLL}^{C}$ that is optimized to minimize MSE. But before doing that, we will first derive a variant of LR using the MSE objective function — an ANN with no hidden layer.

ANN. Instead of optimizing the objective function in Eq. 1, one can optimize the following MSE objective function: MSE$(\beta) = \frac{1}{N}\sum_{i=1}^{N}\frac{1}{C}\sum_{c=1}^{C}(P(y|\mathbf{x}^{(i)}) - \hat{P}(c|\mathbf{x}^{(i)}))^2$, where y is the true label and $C = |\Omega_Y|$. Let us simplify the above equation slightly:

$$\text{MSE}(\beta) = \frac{1}{2}\sum_{i=1}^{N}\sum_{c=1}^{C}\Big(\delta(y=c) - P(c|\mathbf{x}^{(i)})\Big)^2, \qquad (4)$$

where $\delta(.)$ is an indicator function which is 1 if its input parameter condition holds and 0 otherwise. Note that unlike the CLL objective function in Eq. 1, the above objective function (Eq. 4) is not convex. It is likely that one will be stuck in local minimum and, therefore, local minimum avoidance techniques may be required. We will show in Sect. 4 that in practice one can obtain good results with simple gradient descent based (such as quasi-Newton) optimization algorithms without requiring specific mechanisms to avoid local minima.

In the following, we will drop the superscript (j) for simplicity. Optimizing Eq. 4 requires us to compute its derivative with respect the parameters β. We have the following:

$$\frac{\partial \text{MSE}(\beta)}{\partial \beta_{k,i,x_i}} = -\sum_{i=1}^{N} \sum_{c}^{C} (\delta(y = c) - \text{P}(c|\mathbf{x})) \frac{\partial \text{P}(c|\mathbf{x})}{\partial \beta_{k,i,x_i}}, \tag{5}$$

where,

$$\begin{aligned}
\frac{\partial \text{P}(c|\mathbf{x})}{\partial \beta_{k,i,x_i}} &= \frac{\partial}{\partial \beta_{k,i,x_i}} \left(\frac{\exp(\beta_c + \sum \beta_{c,i,x_i})}{\sum_{c'} \exp(\beta_{c'} + \sum \beta_{c',i,x_i})} \right), \\
&= \frac{\partial}{\partial \beta_{k,i,x_i}} \exp\left((\beta_c + \sum \beta_{c,i,x_i}) - \log(\sum_{c'} \exp(\beta_{c'} + \sum \beta_{c',i,x_i})) \right), \\
&= \text{P}(c|\mathbf{x}) \left(\delta(c = k)\delta(x_i) - \left(\frac{\beta_k + \sum \beta_{k,i,x_i}}{\sum_{c'} \exp(\beta_{c'} + \sum \beta_{c',i,x_i})} \right) \delta(x_i) \right), \\
&= \text{P}(c|\mathbf{x}) \left(\delta(c = k)\delta(x_i) - \text{P}(k|\mathbf{x})\delta(x_i) \right), \\
&= \text{P}(c|\mathbf{x})(\delta(c = k) - \text{P}(k|\mathbf{x}))\delta(x_i),
\end{aligned} \tag{6}$$

where, $\delta(x_i)$ is an indicator function if value of x_i is same to the value with which we are differentiating. Plugging in Eq. 5, we get:

$$\frac{\partial \text{MSE}(\mathbf{w})}{\partial \beta_{k,i,x_i}} = -\sum_{i=1}^{N} \sum_{c}^{C} (\delta(y = c) - \text{P}(c|\mathbf{x}))\text{P}(c|\mathbf{x})(\delta(c = k) - \text{P}(k|\mathbf{x}))\delta(x_i). \tag{7}$$

Note, the gradients with respect to the class parameters can be calculated similarly. The gradients in Eq. 7 is the same as optimized by ANN with back-propagation training algorithm. In the following, we will formulate WANBIA$_{\text{CLL}}^{\text{C}}$ with MSE objective function.

WANBIA$_{\text{MSE}}^{\text{C}}$. Given Eq. 3, assuming a Dirichlet prior, a MAP estimate of $\text{P}(y)$ is π_y which equals: $\frac{\#_y + m/C}{N + m}$, where $\#_y$ is the number of instances in the dataset with class y and N is the total number of instances, and m is the smoothing parameter. We will set $m = 1$ in this work. Similarly, a MAP estimate of $\text{P}(x_i|y)$ is $\theta_{x_i|c}$ which equals: $\frac{\#_{x_i,y} + m/|x_i|}{\#_y + m}$, where $\#_{x_i,y}$ is the number of instances in the dataset with class y and attribute values x_i. Now, we have:

$$\text{P}(y|\mathbf{x}) = \frac{\pi_y^{w_y} \prod_{i=1}^{a} \theta_{x_i|y}^{w_{y,i,x_i}}}{\sum_{c \in \Omega_Y} \pi_c^{w_c} \prod_{j=1}^{a} \theta_{x_i|y}^{w_{c,j,x_j}}}.$$

Using the above equation, let us optimize the MSE objective function by taking gradients with respect to the parameters w. We write:

$$\frac{\partial \text{MSE}(\mathbf{w})}{\partial w_{k,i,x_i}} = -\sum_{i=1}^{N}\sum_{c}^{C}(\delta(y=c) - \text{P}(c|\mathbf{x}))\frac{\partial \text{P}(c|\mathbf{x})}{\partial w_{k,i,x_i}}, \qquad (8)$$

where w_{k,i,x_i} denotes parameter associated with attribute i taking value x_i and class attribute k. Let us expand $\frac{\partial \text{P}(c|\mathbf{x})}{\partial w_{k,i,x_i}}$ in the following way:

$$\frac{\partial \text{P}(c|\mathbf{x})}{\partial w_{k,i,x_i}} = \frac{\partial}{\partial w_{k,i,x_i}}\exp\Big(w_c \log \pi_c + \sum_{i=1}^{a} w_{c,i,x_i}\log \theta_{x_i|c}$$

$$- \log \sum_{c' \in \Omega_Y} \exp\Big(w_{c'}\log \pi_{c'} + \sum_{j=1}^{a} w_{c',j,x_j}\log \theta_{x_j|c'}\Big)\Big),$$

$$= \text{P}(c|\mathbf{x})\Big(\delta(c=k)\delta(x_i)\log \theta_{x_i|k}$$

$$-\frac{\exp(w_k \log \pi_k + \sum_{j=1}^{a} w_{k,j,x_j}\log \theta_{x_j|k})}{\log \sum_{c' \in \Omega_Y}\exp\Big(w_{c'}\log \pi_{c'} + \sum_{j=1}^{a}w_{c',j,x_j}\log \theta_{x_j|c'}\Big)}\delta(x_i)\log \theta_{x_i|k}\Big),$$

$$= \text{P}(c|\mathbf{x})(\delta(c=k) - \text{P}(k|\mathbf{x}))\delta(x_i)\log \theta_{x_i|k},$$

and plug it in Eq. 8:

$$\frac{\partial \text{MSE}(\mathbf{w})}{\partial w_{k,i,x_i}} = -\sum_{i=1}^{N}\sum_{c}^{C}\Big(\delta(y=c) - \hat{\text{P}}(c|\mathbf{x})\Big)\text{P}(y|\mathbf{x})(\delta(y=k) - \text{P}(k|\mathbf{x}))\log \theta_{x_i|k}\delta(x_i). \quad (9)$$

The gradients for weights associated with class y (w_y) can be computed similarly. Comparing Eqs. 7 and 9, the following holds:

$$\frac{\partial \text{MSE}(\mathbf{w})}{\partial w_{k,i,x_i}} = \frac{\partial \text{MSE}(\beta)}{\partial \beta_{k,i,x_i}}\log \theta_{x_i|k}, \quad \text{and} \quad \frac{\partial \text{MSE}(\mathbf{w})}{\partial w_k} = \frac{\partial \text{MSE}(\beta)}{\partial \beta_k}\log \pi_k.$$

This shows that when optimizing MSE, just like CLL, naive Bayes preconditioning has the effect of scaling the gradients of a linear classification model by the log of the NB probability estimates. Such scaling leads to faster convergence, as is shown in the next section.

4   Experimental Results

In this section, we compare the performance of a linear model optimized with the MSE objective function with and without NB preconditioning in terms of 0–1 loss, RMSE, bias, variance, training time and the number of iterations it takes each algorithm to converge on 73 natural domains from the UCI repository (Table 1). We will also compare performance with LR and WANBIA$_{\text{CLL}}^{\text{C}}$ optimized with the CLL objective function.

In this work, we use the bias and variance definitions of [7] together with the repeated cross-validation bias-variance estimation method proposed by [8].

The reason for performing bias/variance estimation is that it provides insights into how the learning algorithm will perform with varying amount of data. We expect low variance algorithms to have relatively low error for small data and low bias algorithms to have relatively low error for large data [9].

The experiments are conducted on the datasets described in Table 1. There are a total of 73 datasets, 40 datasets with less than 1000 instances, 21 datasets with instances between 1000 and 10000, and 12 datasets with more than 10000 instances. The datasets with more than 10000 are shown in bold font in Table 1.

Table 1. Details of datasets (UCI Domains)

Domain	Case	Att	Class	Domain	Case	Att	Class
Poker-hand	1175067	11	10	Annealing	898	39	6
Covertype	581012	55	7	Vehicle	846	19	4
Census-Income (KDD)	299285	40	2	PimaIndiansDiabetes	768	9	2
Localization	164860	7	3	BreastCancer (Wisconsin)	699	10	2
Connect-4Opening	67557	43	3	CreditScreening	690	16	2
Statlog (Shuttle)	58000	10	7	BalanceScale	625	5	3
Adult	48842	15	2	Syncon	600	61	6
LetterRecognition	20000	17	26	Chess	551	40	2
MAGICGammaTelescope	19020	11	2	Cylinder	540	40	2
Nursery	12960	9	5	Musk1	476	167	2
Sign	12546	9	3	HouseVotes84	435	17	2
PenDigits	10992	17	10	HorseColic	368	22	2
Thyroid	9169	30	20	Dermatology	366	35	6
Pioneer	9150	37	57	Ionosphere	351	35	2
Mushrooms	8124	23	2	LiverDisorders (Bupa)	345	7	2
Musk2	6598	167	2	PrimaryTumor	339	18	22
Satellite	6435	37	6	Haberman'sSurvival	306	4	2
OpticalDigits	5620	49	10	HeartDisease (Cleveland)	303	14	2
PageBlocksClassification	5473	11	5	Hungarian	294	14	2
Wall-following	5456	25	4	Audiology	226	70	24
Nettalk (Phoneme)	5438	8	52	New-Thyroid	215	6	3
Waveform-5000	5000	41	3	GlassIdentification	214	10	3
Spambase	4601	58	2	SonarClassification	208	61	2
Abalone	4177	9	3	AutoImports	205	26	7
Hypothyroid (Garavan)	3772	30	4	WineRecognition	178	14	3
Sick-euthyroid	3772	30	2	Hepatitis	155	20	2
King-rook-vs-king-pawn	3196	37	2	TeachingAssistantEvaluation	151	6	3
Splice-junctionGeneSequences	3190	62	3	IrisClassification	150	5	3
Segment	2310	20	7	Lymphography	148	19	4
CarEvaluation	1728	8	4	Echocardiogram	131	7	2
Volcanoes	1520	4	4	PromoterGeneSequences	106	58	2
Yeast	1484	9	10	Zoo	101	17	7
ContraceptiveMethodChoice	1473	10	3	PostoperativePatient	90	9	3
German	1000	21	2	LaborNegotiations	57	17	2
LED	1000	8	10	LungCancer	32	57	3
Vowel	990	14	11	Contact-lenses	24	5	3
Tic-Tac-ToeEndgame	958	10	2				

Each algorithm is tested on each dataset using 5 rounds of 2-fold cross validation. We report Win-Draw-Loss (W-D-L) results when comparing the 0–1 loss, RMSE, bias and variance of two models. A two-tail binomial sign test is used to determine the significance of the results. Results are considered significant if $p \leq 0.05$.

The datasets in Table 1 are divided into two categories. The first category constitutes all the datasets. The category is denoted by *All* in the results. The second category constitutes only datasets with more than 10000 instances. This is denoted by *Big* in the results When comparing average results across *All* and *Big* datasets, we normalize the results with respect to one of the comparative technique and present the geometric mean.

Numeric attributes are discretized by using the Minimum Description Length (MDL) discretization method [10]. A missing value is treated as a separate attribute value and taken into account exactly like other values.

We employed L-BFGS quasi-Newton methods [11] for solving the optimization[2].

We used a Random Forest that is an ensemble of 100 decision trees [13].

We will denote a linear model optimized with the MSE objective function with or without NB preconditioning as $WANBIA^C_{MSE}$ and ANN respectively.

4.1   MSE Vs. CLL

A win-draw-loss (W-D-L) comparison of bias, variance, 0–1 loss and RMSE of $WANBIA^C_{CLL}$ and LR versus $WANBIA^C_{MSE}$ and ANN is given Table 2. It can be seen that $WANBIA^C_{MSE}$ achieves significantly lower bias than $WANBIA^C_{CLL}$, whereas ANN has lower bias than LR but this difference does not achieve statistical significance. Both $WANBIA^C_{MSE}$ and ANN exhibit higher variance, but this is statistically significant in the case of ANN vs. LR only. This suggests that both $WANBIA^C_{MSE}$ and ANN are well suited for bigger datasets for which lower bias is preferable [14]. This is also evident from Table 2 where $WANBIA^C_{MSE}$ has significantly lower 0–1 loss than $WANBIA^C_{CLL}$ on *Big* datasets. Similarly, the ANN results (with 9 wins, 1 draw and 2 losses), though not significantly different, are better than LR.

In Fig. 1, we show the geometric average of the results. It can be seen that $WANBIA^C_{MSE}$ and ANN are lower-bias and higher-variance models as compared to $WANBIA^C_{CLL}$ and LR. The superior performance of $WANBIA^C_{MSE}$, however, comes at an extra cost. A comparison of the training and classification time of $WANBIA^C_{CLL}$ and $WANBIA^C_{MSE}$ is shown in Fig. 1(e) and (f) respectively. It can be seen that optimizing the MSE objective function, though low biased, is a magnitude of order slower than optimizing the CLL objective function.

---

[2] The original L-BFGS implementation of [12] from http://users.eecs.northwestern. edu/~nocedal/lbfgsb.html is used.

Table 2. Win-Draw-Loss: WANBIA$^C_{MSE}$ vs. WANBIA$^C_{CLL}$ and ANN vs. LR. p is two-tail binomial sign test. Results are significant if $p \leq 0.05$.

	WANBIA$^C_{MSE}$ vs. WANBIA$^C_{CLL}$		ANN vs. LR	
	W-D-L	p	W-D-L	p
All datasets				
Bias	45/7/20	**0.002**	38/5/28	0.276
Variance	19/6/47	**<0.001**	21/4/47	**0.002**
0–1 Loss	34/6/32	0.902	31/5/36	0.625
RMSE	29/4/39	0.275	31/3/38	0.470
Big datasets				
0–1 Loss	10/1/1	**0.011**	9/1/2	0.065
RMSE	8/0/4	0.387	8/0/4	0.387

4.2 WANBIA$^C_{MSE}$ Vs. ANN

Now that we have established that optimizing the MSE for LR leads to a lower bias model than that by CLL, in this section, we will compare WANBIA$^C_{MSE}$ and ANN to see the effects of scaling and whether NB preconditioning is as effective with the MSE as with the CLL objective function. We compare the scatter of 0–1 loss and RMSE values in Figs. 2 and 3 respectively. It can be seen that both parameterizations lead to a similar scatter of 0–1 loss and RMSE. This suggests the equivalence of two models (same model, different parameterizations).

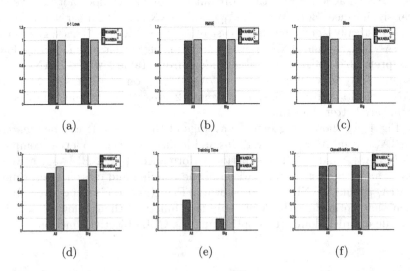

Fig. 1. An (geometric) average comparison of the 0–1 loss, RMSE, Bias and Variance of WANBIA$^C_{MSE}$ and WANBIA-C on *All* and *Big* datasets.

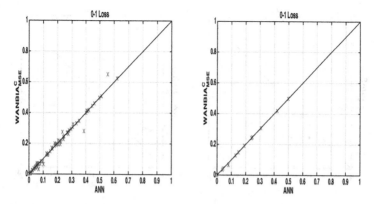

Fig. 2. Comparative scatter of 0–1 Loss of ANN and WANBIA$^C_{MSE}$ on *All* (Left) and *Big* (Right) datasets.

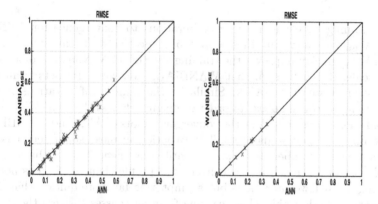

Fig. 3. Comparative scatter of RMSE of ANN and WANBIA$^C_{MSE}$ on *All* (Left) and *Big* (Right) datasets.

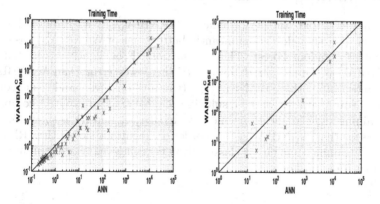

Fig. 4. Comparative scatter of training time of ANN and WANBIA$^C_{MSE}$ on *All* (Left) and *Big* (Right) datasets.

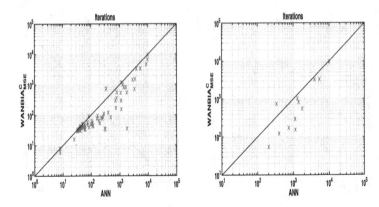

Fig. 5. Comparative scatter of number of iterations to convergence of ANN and WANBIA$_{\text{MSE}}^{\text{C}}$ on *All* (Left) and *Big* (Right) datasets.

The training time and number of iterations to convergence for ANN and WANBIA$_{\text{MSE}}^{\text{C}}$ is shown in Figs. 4 and 5 respectively. It can be seen that WANBIA$_{\text{MSE}}^{\text{C}}$ greatly improves the training time of ANN. Note, the plots are on the log scale. It can be seen that WANBIA$_{\text{MSE}}^{\text{C}}$ on some datasets is an order of magnitude faster than ANN. Similarly, the number of iterations it takes WANBIA$_{\text{MSE}}^{\text{C}}$ to converge are an order of magnitude less than for ANN.

Finally, let us have a look at the convergence plots of ANN and WANBIA$_{\text{MSE}}^{\text{C}}$ in Fig. 6 on some sample datasets. The variation in mean-square-error is plotted with varying number of iterations until convergence. It can be seen that WANBIA$_{\text{MSE}}^{\text{C}}$ has a much better convergence profile than ANN. It is not only converging in far fewer iterations but asymptoting far more quickly than ANN. This is extremely desirable when learning from few passes through the data.

4.3 WANBIA$_{\text{MSE}}^{\text{C}}$ Vs. Random Forest

In Table 3, we compare the performance of WANBIA$_{\text{MSE}}^{\text{C}}$ with Random Forest. It can be seen that though not significantly better, bias of WANBIA$_{\text{MSE}}^{\text{C}}$ is smaller than that of Random Forest. The variance of RF is slightly lower than that of WANBIA$_{\text{MSE}}^{\text{C}}$. On bigger datasets, RF has lower error than WANBIA$_{\text{MSE}}^{\text{C}}$ slightly more often than WANBIA$_{\text{MSE}}^{\text{C}}$ (winning on seven and losing on five datasets). Note that none of the results in the table are significant. A comparison of the training and classification time of WANBIA$_{\text{MSE}}^{\text{C}}$ and RF is shown in Fig. 7. It can be seen that WANBIA$_{\text{MSE}}^{\text{C}}$ is an order of magnitude slower than RF on *Big* datasets at training time but at classification time, it is many order of magnitude faster than Random Forest.

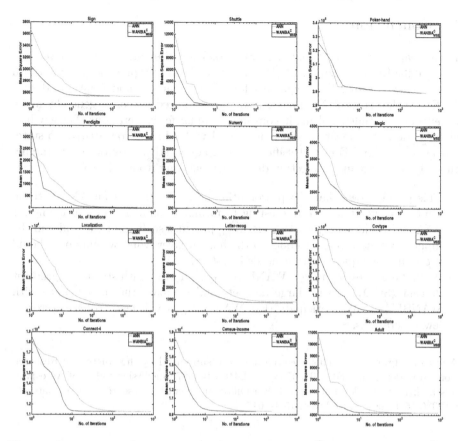

Fig. 6. Comparison of rate of convergence of WANBIA$^C_{MSE}$ and ANN on several datasets. The X-axis (No. of iterations) is on log scale.

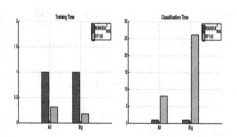

Fig. 7. Comparison of the (geometric) average of the training and classification time of RF and WANBIA$^C_{MSE}$ on *All* and *Big* datasets.

Table 3. Win-Draw-Loss: WANBIA$^C_{MSE}$ vs. Random Forest. p is two-tail binomial sign test. Results are significant if $p \leq 0.05$.

	WANBIA$^C_{MSE}$ vs. RF100	
	W-D-L	p
All Datasets		
Bias	41/5/26	0.086
Variance	32/2/38	0.550
0–1 Loss	30/2/40	0.282
RMSE	27/0/45	0.044
Big Datasets		
0-1 Loss	5/0/7	0.774
RMSE	5/0/7	0.774

5   Conclusion

In this paper, we showed that a linear classifier optimizing MSE has lower bias than vanilla LR optimizing CLL. We also showed that NB preconditioning, which is very effective for LR, is equally effective for a linear model optimized with MSE. We showed that NB preconditioning can speed-up convergence by many orders of magnitude resulting in convergence in far fewer iterations. The low-bias classification of a linear classifier optimized with MSE is competitive to state-of-the-art Random Forest classifier with an added advantage of faster training time. There are many interesting directions following from this work:

- This paper shows that NB preconditioning is effective for an ANN with no hidden layers. It will be interesting to formulate similar preconditioning for ANNs with hidden layers. WANBIA$^C_{CLL}$ provides scaling for the nodes in the input layer, however, for nodes in the hidden layer, what weights one should use is an open question that needs investigation.
- It will be interesting to run WANBIA$^C_{MSE}$ with MSE with stochastic gradient descent (SGD) on very large datasets and compare the performance with WANBIA$^C_{CLL}$. We anticipate that WANBIA$^C_{MSE}$ will lead to lower error in fewer iterations.

Acknowledgements. This research has been supported by the Australian Research Council under grants DP120100553 and DP140100087, and Asian Office of Aerospace Research and Development, Air Force Office of Scientific Research under contracts FA2386-15-1-4007 and FA2386-15-1-4017.

References

1. Duda, R., Hart, P., Stork, D.: Pattern Classification. John Wiley and Sons, New York (2006)
2. Minka, T.P.: A comparison of numerical optimizers for logistic regression (2003)
3. Zaidi, N.A., Cerquides, J., Carman, M.J., Webb, G.I.: Alleviating naive Bayes attribute independence assumption by attribute weighting. J. Mach. Learn. Res. **14**, 1947–1988 (2013)
4. Zaidi, N.A., Carman, M.J., Cerquides, J., Webb, G.I.: Naive-bayes inspired effective pre-conditioners for speeding-up logistic regression. In: IEEE International Conference on Data Mining (2014)
5. Martinez, A., Chen, S., Webb, G.I., Zaidi, N.A.: Scalable learning of Bayesian network classifiers. J. Mach. Learn. Res. (2015) (in press)
6. Zaidi, N.A., Webb, G.I., Carman, M.J., Petitjean, F.: Deep broad learning - Big models for Big data (2015). arxiv:1509.01346
7. Kohavi, R., Wolpert, D.: Bias plus variance decomposition for zero-one loss functions. In: ICML, pp. 275–283 (1996)
8. Webb, G.I.: Multiboosting: A technique for combining boosting and wagging. Mach. Learn. **40**(2), 159–196 (2000)
9. Brain, D., Webb, G.I.: The need for low bias algorithms in classification learning from small data sets. In: PKDD, pp. 62–73 (2002)

10. Fayyad, U.M., Irani, K.B.: On the handling of continuous-valued attributes in decision tree generation. Mach. Learn. **8**(1), 87–102 (1992)
11. Zhu, C., Byrd, R.H., Nocedal, J.: LBFGSB, fortran routines for large scale bound constrained optimization. ACM Trans. Math. Softw. **23**(4), 550–560 (1997)
12. Byrd, R., Lu, P., Nocedal, J.: A limited memory algorithm for bound constrained optimization. SIAM J. Sci. Stat. Comput. **16**(5), 1190–1208 (1995)
13. Breiman, L.: Random forests. Mach. Learn. **45**, 5–32 (2001)
14. Brain, D., Webb, G.: On the effect of data set size on bias and variance in classification learning. In: Proceedings of the Fourth Australian Knowledge Acquisition Workshop, pp. 117–128. University of New South Wales (1999)

OCEAN: Fast Discovery of High Utility Occupancy Itemsets

Bilong Shen[1], Zhaoduo Wen[2], Ying Zhao[1(✉)], Dongliang Zhou[3],
and Weimin Zheng[1]

[1] Department of Computer Science and Technology, Tsinghua Uinversity,
Beijing, China
shenbilong@gmail.com, {yingz,zwm-dcs}@tsinghua.edu.cn
[2] School of Information and Telecommunication Engineering,
Beijing University of Posts and Telecommunications, Beijing, China
wenzhaoduo@gmail.com
[3] Department of Thermal Engineering, Tsinghua Uinversity, Beijing, China
zhoudongliang08@gmail.com

Abstract. Frequent pattern mining has been widely studied in the past decades and has been applied to many domains. In particular, numerical transaction databases, where not only the items but also the utility associated with them are available in user transactions, are useful for real applications. For example, customer mobile App traffic data collected by mobile service providers contains such information. In this paper, we aim to find frequent patterns that occupy a large portion of total utility of the supporting transactions, to answer questions like "On which mobile Apps do the customers spend most of their data traffic?" Towards this goal, we define a measure called *utility occupancy* to measure the contribution of a pattern within a transaction. The challenge of high utility occupancy itemset discovering is the lack of monotone or anti-monotone property. So we derive an upper bound for *utility occupancy* and design an efficient mining algorithm called OCEAN based on a fast implementation of utility list. Evaluations on real world mobile App traffic data and other three datasets show that OCEAN is efficient and effective in finding frequent patterns with large utility occupancy.

Keywords: Frequent pattern mining · High utility mining · Utility occupancy · Upper bound

1 Introduction

Frequent pattern mining has been extensively studied over decades and has been applied to marketing, recommendation, and other important domains. In addition to the pattern frequency, other properties of patterns, such as utility [4] and occupancy [14], have been studied to meet the needs of various applications and different types of transaction databases. Among them, transaction databases with utility information, which reflects users' interests to items, are

© Springer International Publishing Switzerland 2016
J. Bailey et al. (Eds.): PAKDD 2016, Part I, LNAI 9651, pp. 354–365, 2016.
DOI: 10.1007/978-3-319-31753-3_29

important to study. For example, mobile service providers may collect customer mobile App usage profiles upon approval and use them to improve their service. In this case, each customer mobile App traffic profile (daily or monthly) is a transaction, with a mobile App as an item, and its traffic usage as its utility. Another example can be found in browsing logs, where each website with its browsing time can be viewed as an item and its utility.

In these databases, in order to better capture customer behaviors, *e.g*, answering questions like "On which mobile Apps do the customers spend most of their data traffic?", it is useful to find frequent patterns that are supported by a large portion of transactions and can explain individual customer behaviors to a large extent. In particular, we want to find patterns that occupy a large portion of total utility of the supporting transactions. Towards this goal, we define a novel measure called *utility occupancy* to measure the utility contribution of a pattern within a transaction. Different from existing utility measures, *utility occupancy* uses a relative utility ratio in each transaction to emphasize the importance of a pattern to each customer.

In this paper, we propose a novel frequent pattern mining algorithm that can find frequent patterns with high utility occupancy, and make the following contributions:

1. A novel measure, called *utility occupancy* is proposed to describe the interestingness of a frequent pattern for transactions with utility information.
2. An efficient mining algorithm, called **OCEAN**, is developed. **OCEAN** derive the properties of *utility occupancy* upper bound to pruning the searching space for high efficient pattern mining.
3. Extensive experiments were performed to evaluate the effectiveness and efficiencies of **OCEAN** on various databases, including a real world mobile App traffic usage dataset and three real world datasets that are often used to evaluate high utility mining algorithms.

The rest of this paper is organized as follows. Section 2 discusses some of the related work. Section 3 defines *utility occupancy* and presents the problem formulation. Section 4 derives an upper bound and presents the details of OCEAN. Section 5 provides the detailed experimental evaluation of OCEAN. Finally, Sect. 6 provides some concluding remarks.

2 Related Work

In addition to the pattern frequency, other properties and interestingness measures of frequent patterns and association rules have been studied in the literature [13], to fulfill the need for patterns with good quality, *e.g*, closed frequent patterns [7], maximal frequent patterns [3,15], patterns with constraints [6], and recently, patterns with high occupancy [14], which means the pattern that occupies a large portion of the transactions it appears in. The occupancy of an itemset in a transaction was defined as the ratio between the number of items in the itemset and the number of items in the transaction, and the occupancy of an itemset in a transaction database was defined as the harmonic average of its occupancy values in all supporting transactions in [14]. According to [14], patterns with high occupancy were more effective in webpage print region recommendation than maximal frequent patterns. Our work is an extension of itemset occupancy in case of transaction databases with utility information. Since we are dealing with a different type of data, the formulation and derivation of the upper bound for *utility occupancy* is tight to the numerical nature of data closely.

Another related problem to our work, which is also an emerging and important topic in marketing, is high utility itemset mining [2,4,5,10,11]. In this problem, each transaction contains the quantity of a set of items, and an item profit (utility) table is also available. The utility of an item in a transaction is defined as the multiplication of its profit and its quantity in the transaction, and the utility of an itemset is defined as the summation of the utilities of the items in the itemset. High utility itemset mining is to find itemsets with high utility or other utility related objectives defined by users. Recent works have been focusing on discussions on different utility functions [8,9], and efficient mining algorithms [2,5,10,11]. For example, HUI-miner [11] and FHM [5] can discover high utility itemsets without candidate generation. Unlike other existing absolute utility measures, the proposed measure *utility occupancy* is a relative measure, which emphasizes whether a certain set of items is important for individual users.

3   Problem Formulation

Suppose a transaction database contains a set of transactions Γ. A *transaction t* containing k items in Γ is in the form of $\{(item_1, util_1), (item_2, util_2), \cdots, (item_k, util_k)\}$, where $item_i$ is the identification of the ith item in t and $util_i$ is a numerical value to represent its utility in t. When a utility table is available, such as shown in Table 1(a), $util_i$ is usually calculated as follows.

Definition 1. *The utility of item i in transaction t, denoted as $U_t(i)$, is the product of the external utility of item i given in a utility table and the count of item i in t. For example, $U_{T2}(c) = 4 \times 3 = 12$ in Table 1.*

Table 1. Database

(a) Utility table   (b) Transaction table (item count)

Item	Utility
a	1
b	1
c	4
d	5
e	2
f	3
g	3

Tid	T1	T2	T3	T4	T5	T6	T7	T8
a	1	9		1		1		
b	1				1	8	2	
c	1	3			2	1		1
d		2	1		2			
e	1	1		1			3	
f		1			1		3	3
g		2	1		1		1	
U(t)	8	42	8	4	32	7	18	13

We say an itemset W is supported by a transaction t, if and only if, for each $item_i$ in W, $item_i$ is also in t. We denote the set of *supporting* transactions of W as Γ_W. The *support ratio* of W, denoted as $\theta(W)$, is calculated as $\frac{|\Gamma_W|}{|\Gamma|}$.

The total utility of a transaction t is defined as the summation of the utilities of all items in t, i.e, $U(t) = \sum_{item_i \in t} U_t(item_i)$. For example, the utilities of all transactions in Table 1(b) are shown in the last row. The utility of an itemset W in a transaction t, denoted as $U_t(W)$, is defined as the summation of the utilities of the items present in both t and W, i.e, $U_t(W) = \sum_{item_i \in t \cap W} U_t(item_i)$. For example, let $W = \{f, g\}$, $U_{T2}(W) = 9$. The transaction-y utilities of an itemset W in a database Γ, denoted as $TWU(W)$, is defined as the summation of the utilities of W's supporting transactions, i.e, $TWU(W) = \sum_{t \in \Gamma_W} U(t)$. For example, let $W = \{f, g\}$, $TWU(W) = U(T2) + U(T5) + U(T7) = 42 + 32 + 18 = 92$.

Now we are ready to define **utility occupancy** for an itemset W.

Definition 2. *The utility occupancy of an itemset W in its supporting transaction t is defined as*

$$\phi_t(W) = \frac{U_t(W)}{U(t)} = \frac{\sum_{item_i \in W, item_i \in t} util_i}{\sum_{item_i \in t} util_i}.$$

The utility occupancy of W in a database is defined as the arithmetic average of the utility occupancy values of W in all its supporting transactions,

$$\phi(W) = \frac{\sum_{t \in \Gamma_W} \frac{U_t(W)}{U(t)}}{|\Gamma_W|}$$

For example, let $W = \{f, g\}$, $\phi_{T2}(W) = 9/42$ and $\phi(W) = (\phi_{T2}(W) + \phi_{T5}(W) + \phi_{T7}(W))/3 = (9/42 + 6/32 + 12/18)/3 = 0.355$.

Definition 3. *Given a minimum support threshold α and a minimum utility occupancy threshold β, the problem of* **high utility occupancy itemset mining** *is to find all itemsets whose support ratio is no less than α and utility occupancy is no less than β.*

Note that the minimum support threshold must be specified, otherwise any transaction itself is a high utility occupancy itemset with a value of 1. Clearly, the utility occupancy measure does not hold downward closure property. In the example of Table 1, $\phi(\{f, g\}) = 0.355$, $\phi(\{d, g\}) = (16/42 + 13/32 + 8/8)/3 = 0.596$, and $\phi(\{d, f, g\}) = (19/42 + 16/32)/2 = 0.476$, which means when an itemset is extended, its utility occupancy can either increase or decrease.

Now we compare high utility occupancy itemsets with high utility itemsets [4] and high occupancy itemsets [14] to have a better understanding of the nature of high utility occupancy itemsets. If we view each transaction as a usage pattern of a user, high utility occupancy itemset mining selects the itemsets that can explain the usage of a group of users to a large extent, whereas high utility itemsets simply represent the ones with highest overall utilities across the database.

For example, compare itemset $\{a, b\}$ with itemset $\{f, g\}$ in our example. Both of them appear in three supporting transactions (T1, T4, and T6 for $\{a, b\}$, T2, T5, and T7 for $\{f, g\}$). $\{f, g\}$ has much larger utility $(U(\{f, g\}) = 9 + 6 + 12 = 27)$ than $\{a, b\}$ $(U(\{a, b\}) = 2 + 2 + 3 = 7)$, however, the utility occupancy of $\{a, b\}$ $(\phi(\{a, b\}) = (2/8 + 2/4 + 3/7)/3 = 0.392)$ is higher than that of $\{f, g\}$ $(\phi(\{f, g\}) = 0.355)$. Our proposed high utility occupancy itemset mining is an extension of the high occupancy itemset mining problem proposed in [14], which defines occupancy using the cardinality of an itemset instead of utility. For example, in transaction T6 of our example, the occupancy of $\{a, b\}$ is higher than that of $\{c\}$ in terms of cardinality, however, $\phi_{T6}(\{c\}) = 4/7 > \phi_{T6}(\{a, b\}) = 3/7$.

4   High Utility Occupancy Itemset Mining

In this section, we first derive an upper bound for utility occupancy, and then describe an efficient algorithm **OCEAN** to find qualified patterns with the help of the derived upper bound.

4.1   Upper Bound of Utility Occupancy

The process of itemset mining can be viewed as an exploration of a set enumeration tree [12], whose root node is an empty set, and each node represents an itemset and its children are extended itemsets from it by a certain order. The order can be lexicographic order, or transaction-weighted utility ascending order, which is a common choice in many high utility itemset mining algorithms. In our example, according to the values of transaction-weighted utility, we have $b \prec a \prec e \prec d \prec g \prec c \prec f$, and we use this order for the rest of the paper. Given a subtree rooted at itemset X, the set of all possible items that can be added to form itemsets in this subtree is called the *extension set E*. For example, let us consider the subtree rooted at g for our example in Fig. 1, and $E = \{c, f\}$ for the node g.

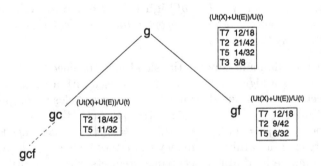

Fig. 1. The subtree rooted at g

Given a subtree rooted at X, we need to derive an upper bound $\hat{\phi}()$ of utility occupancy for all possible qualified patterns residing in the subtree without actually generating these patterns. If $\hat{\phi}(X)$ is less than the minimum utility occupancy threshold, we can safely discard this subtree from further consideration.

Lemma 1. *Given a subtree rooted at X, its supporting transaction set Γ_X and its extension set E, for any possible qualified itemset W in the subtree, we have*

$$\phi(W) \leq \frac{\sum_{t \in \Gamma_W} \frac{U_t(X) + U_t(E)}{U(t)}}{|\Gamma_W|}.$$

Proof. Since $W - X \subset E$, we have $U_t(W - X) \leq U_t(E)$. Therefore,

$$\phi(W) = \frac{\sum_{t \in \Gamma_W} \frac{U_t(W)}{U(t)}}{|\Gamma_W|} = \frac{\sum_{t \in \Gamma_W} \frac{U_t(X) + U_t(W - X)}{U(t)}}{|\Gamma_W|} \leq \frac{\sum_{t \in \Gamma_W} \frac{U_t(X) + U_t(E)}{U(t)}}{|\Gamma_W|}.$$

$\square$

Theorem 1. *Given the minimum support ratio α, a subtree rooted at X, its supporting transaction set Γ_X and extension set E, for any possible qualified itemset W in the subtree, $\hat{\phi}(W) = \frac{\sum_{top\alpha|\Gamma|, t \in \Gamma_X} (\frac{U_t(X) + U_t(E)}{U(t)})^{\downarrow}}{\alpha|\Gamma|} \geq \phi(W).$*

Proof. Since Γ_W is unknown, we calcultate $\frac{U_t(X) + U_t(E)}{U(t)}$ for all transactions in Γ_X, and sort this vector in descending order, denoted as $(\frac{U_t(X) + U_t(E)}{U(t)})^{\downarrow}$. The average of top $|\Gamma_W|$ values of this vector is an upper bound of $\frac{\sum_{t \in \Gamma_W} \frac{U_t(X) + U_t(E)}{U(t)}}{|\Gamma_W|}$. Furthermore, since a qualified pattern should be supported by at least $\alpha|\Gamma|$ transactions, we have $\alpha|\Gamma| \leq |\Gamma_W|$. Put all together, we have

$$\phi(W) \leq \frac{\sum_{t \in \Gamma_W} \frac{U_t(X) + U_t(E)}{U(t)}}{|\Gamma_W|} \leq \frac{\sum_{top|\Gamma_W|, t \in \Gamma_X} (\frac{U_t(X) + U_t(E)}{U(t)})^{\downarrow}}{|\Gamma_W|}$$

$$\leq \frac{\sum_{top\alpha|\Gamma|, t \in \Gamma_X} (\frac{U_t(X) + U_t(E)}{U(t)})^{\downarrow}}{\alpha|\Gamma|}.$$

$\square$

For example, the sorted $\frac{U_t(X) + U_t(E)}{U(t)}$ vectors are shown for each node in Fig. 1. Suppose the minimum support ratio is 0.25. We can average the top two values from the sorted vectors to calculate $\hat{\phi}()$ for each node. To be specific, $\hat{\phi}(\{g\}) = (12/18 + 21/42)/2 = 0.584$. $\hat{\phi}(\{g, c\}) = (18/42 + 11/32)/2 = 0.387$. If the minimum utility occupancy ratio is 0.5, we can safely prune the subtree rooted at $\{g, c\}$.

4.2   Design and Implementation of OCEAN

We design and implement our proposed high utility occupancy mining algorithm based on an important data structure *utility list*, which was used in the state-of-the-art high utility itemset mining algorithms: HUI-miner [11] and FHM [5].

Algorithm 1. UpperBound

Input: P_X : the root itemset

   $minSupp$: the minimum support threshold

Output: the upper bound of utility occupancy for the subtree rooted at P_X

1: Derive $\frac{U_t(X)+U_t(E)}{U(t)}$ from $P_X.UL$ as V;

2: $V_{top} \leftarrow$ the largest $minSupp \times |\Gamma|$ values in V;

3: $sum \leftarrow 0$;

4: **for** $u \leftarrow 1$ **to** $minSupp \times |\Gamma|$ **do**

5:     $sum \leftarrow sum + V_{top}(u)$

6: **end for**

7: $occu \leftarrow sum/(minSupp \times |\Gamma|)$;

8: **return** $occu$

For any given itemset X and its extension set E, its utility list is a list of utility records in its supporting transactions. For each supporting transaction t, the utility list contains a three-tuple record, i.e, $(transaction\_id, U_t(X), U_t(E))$. Therefore, the itemset data structure in our algorithms has two fields, *itemset* for the set of items, and UL for its utility list. In both HUI-miner [11] and FHM [5], the direct summation of $U_t(X)$ and $U_t(E)$ for each itemset was used for utiltiy prunning, whereas in OCEAN the upper bound of occupancy is more complicated to calculate, which is given in Algorithm 1. Note that we do not have to sort the entire $\frac{U_t(X)+U_t(E)}{U(t)}$ vector to get the top $minSupp \times |\Gamma|$ values.

The utility records are sorted according to *transaction_id* in all utility lists, so that the merge of two utility lists can be done efficiently. Given two itemsets P_X and P_Y, suppose $P_X.itemset = X$, $P_Y.itemset = Y$, and $X \cap Y = \varnothing$, the itemset $X \cup Y$ and its utility list can be easily constructed by scanning $P_X.UL$ and $P_Y.UL$ simultaneously and adding a utility record for each common transaction t into $P_{XY}.UL$. We name this procedure as $merge(P_X.UL, P_Y.UL)$ and omitted its detailed implementation. In addition, $Size(P_X.UL)$ returns the number of transaction records in $P_X.UL$.

The main idea of our proposed algorithm OCEAN is to perform a depth first search on the set enumeration tree of a database. In Algorithm 2, for each item i, we create an itemset I_i such that $I_i.itemset = \{i\}$ and $I_i.UL$ contains its utility list. I is a set of such itemsets for all items and is sorted in ascending order according to the transaction-weighted utility of each item. Such order is also used in the depth first search, which is implemented in Algorithm 3. In Algorithm 3, qualified itemsets are reported (in line 2 and 3) and node explorations are coordinated by determining whether the node should be pruned (in line 5 and 6) and preparing the set of extension itemsets of the node (in line 8 to 12) for its exploration (in line 13). To be specific, given an itemset X and its extension set E, for each item $y \in E$, $X \cup y$ forms an extension itemset of X.

Algorithm 2. OCEAN

Input:

  D: a transaction database

  $minOccu$: the minimum utility occupancy threshold

  $minSupp$: the minimum support threshold

Output:

  The set of itemsets with utility occupancy $\geq minOccu$ and support $\geq minSupp$.

  1: Scan D to build the utility list of each item i to form I and calculate transaction utilities and item utilities.

  2: **sort** I in ascending order according to the transaction-weighted utility of each item.

  3: **Prune**($\varnothing$, I, $minOccu$, $minSupp$)

Algorithm 3. Prune

Input: P : the itemset

  ExP : the set of extension itemsets of P

  $minOccu$: the minimum utility occupancy threshold

  $minSupp$: the minimum support threshold

Output:

  The set of itemsets with utility occupancy $\geq minOccu$ and support $\geq minSupp$.

  1: **for** each itemset $P_X \in ExP$ **do**

  2:   **if** $Occu(P_X.UL) \geq minOccu \wedge Size(P_X.UL) \geq minSupp \times |\Gamma|$ **then**

  3:     Output P_x

  4:   **end if**

  5:   $maxOccu \leftarrow UpperBound(P_X)$

  6:   **if** $maxOccu \geq minOccu \wedge Size(P_X.UL) \geq minSupp \times |\Gamma|$ **then**

  7:     $ExP_X \leftarrow null$;

  8:     **for** each item y such that $y \succ X$ **do**

  9:       $P_{Xy}.itemset \leftarrow P_X.itemset \cup \{y\}$

  10:       $P_{Xy}.UL \leftarrow merge(P_X.UL, I_y.UL)$

  11:       $ExP_X \leftarrow ExP_X \cup P_{Xy}$

  12:     **end for**

  13:     $Prune(P_X, ExP_X, minOccu, minSupp)$

  14:   **end if**

  15: **end for**

---

5   Experiment

We evaluated the effectiveness and efficiency of the proposed algorithm OCEAN using four real world datasets. In addition to OCEAN, we also implemented a variation: OCEAN without minOccu pruning as our baseline. All algorithms were implemented in JAVA. Experiments were performed on a PC with a 3.4 Hz Intel i7 processor and 8 GB memory running Windows7 OS.

The four datasets used in our experiments represent various applications, itemset distributions, and other characteristics, as shown in Table 2 including the number of transactions, the number of distinct items, the average length of transactions, and the number of items in the longest transaction(s).

Table 2. Characteristics of datasets

Database	Trans	Items	Avelen	Maxlen	Type
MobileApp	109774	14	3.82	13	dense
Mushroom	8124	119	23	23	dense
Kosarak	990002	41270	8.1	2498	sparse
BMS	59602	497	2.5	267	sparse

MobileApp was obtained from a mobile service provider, with each transaction recording the traffic data that a customer spends on various mobile Apps. **Mushroom**, **BMS**, and **Kosarak** are benchmark datasets for high utility itemset mining (downloaded from FIMI Repository [1]). Mushroom includes descriptions of hypothetical samples corresponding to 23 species of gilled mushrooms, each transaction contains 23 items. BMS is a real-life sparse click-stream data from a webstore. Kosarak is a real-life sparse click-stream data of a hungarian on-line news portal. Unlike MobileApp, the other three datasets provide transactions without utility information. So we generated the external utilities for items between 1 and 10 by using a log-normal distribution and randomly generated item counts between 1 and 5 as in [5, 11].

5.1 High Utility Occupancy Itemsets Vs. High Utility Itemsets

We used MobileApp to illustrate the effectiveness of high utility occupancy itemsets for identifying the set of mobile Apps that users are interested in. Towards this goal, we used OCEAN and FHM to generate 10 itemsets with highest utility occupancy values and highest utilities, respectively. The results are shown in Table 3. The itemsets found by OCEAN are more favorable than those found by FHM for two reasons. First, high utility occupancy itemsets indicate high interests of users in these itemsets. For example, itemset {7,9,14}, as shown in Table 3, has a relatively low utility of 18.1 %, but it has a high utility occupancy of 77.4 %, which means customers who use mobile Apps {7,9,14} spend 77.4 % of their traffic quota on these applications on average. Second, the size of itemsets found by OCEAN is larger than that by FHM in general, because larger itemsets lead to a decrease in support ratio which usually harms utility values as well. The diversity of found mobile Apps groups with high occupancy is more favorable for the task of mobile App recommendation or promotion.

5.2 Efficiency of OCEAN

We evaluated the efficiency of OCEAN on BMS, Kosarak, and Mushroom datasets. We compared the running time and the number of searched itemsets for OCEAN and OCEAN without minOccu pruning, shorted as OCEAN_no, to see how effective our derived upperbound is. The results are shown in Fig. 2. We fixed minOccu and varied minSupp in Fig. 2(a), (b) and (c), and we fixed

Table 3. Top 10 results of MobileApp

	(a) OCEAN		
Pattern	Occu(%)	Util(%)	Supp(%)
{7,9,14}	77.4	18.1	21.8
{2,9,10}	74.5	16.4	14.1
{7,10,14}	73.7	13.3	16.1
{7,9,10}	73.6	21.4	19.0
{7,14}	72.7	35.2	65.7
{9,14}	71.6	19.3	23.1
{9,10}	71.4	25.3	22.6
{10,14}	66.8	14.0	17.6
{8,9}	65.0	14.2	12.8
{14}	64.4	35.7	68.9

	(b) FHM		
Pattern	Occu(%)	Util(%)	Supp(%)
{14}	64.4	35.7	68.9
{7,14}	72.7	35.2	65.7
{9}	42.4	25.4	44.3
{9,10}	71.4	25.3	22.6
{7,9}	45.6	21.8	38.1
{7,9,10}	73.6	21.4	19.0
{9,14}	71.6	19.3	23.1
{10}	43.8	19.1	38.1
{7}	77.4	18.8	41.4
{7,9,14}	77.4	18.1	21.8

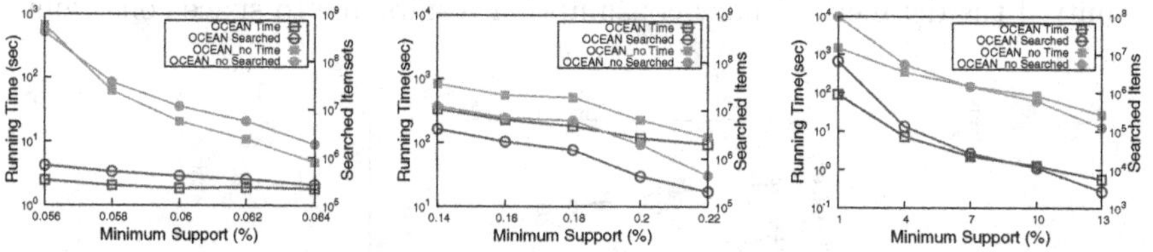

(a) BMS minOccu=50%   (b) Kosarak minOccu=50%   (c) Mushroom minOccu=70%

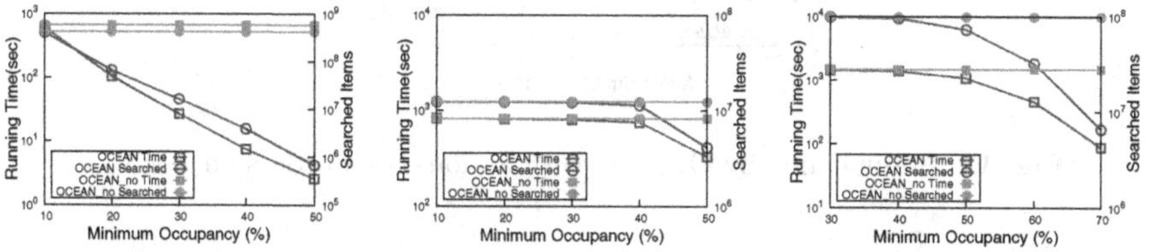

(d) BMS minSupp=0.056%   (e) Kosarak minSupp=0.14%   (f) Mushroom minSupp=1%

Fig. 2. Running time and number of searched itemsets by OCEAN and OCEAN_no

minSupp and varied minOccu in Fig. 2(d), (e) and (f). Clearly, with increasing minOccu, OCEAN pruned more itemsets in addition and run in less time. In general, the running time and the number of searched itemsets of OCEAN decreased as minOccu and minSupp increased. For sparse datasets, small minSupp and minOccu can be effective too. For example, there are 41270 distinct items in Kosarak. With minSupp and minOccu being 0.14 %, 10 % respectively, OCEAN actually searched 12290355 itemsets, whereas there are 2^{41270} possible itemsets in total. We can see the trend more clearly in Fig. 3, in which we measured the running time of OCEAN with varying minSupp and minOccu. The higher the minSupp or minOccu is, the fewer the number of searched itemsets is and thus the less running time is.

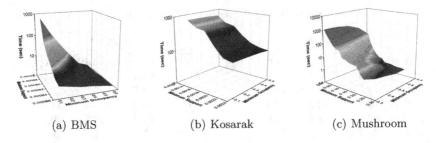

(a) BMS   (b) Kosarak   (c) Mushroom

Fig. 3. Running time by OCEAN

We also studied the memory consumption of OCEAN. We found that for all databases we used, the memory consumption is stable. For example, for Mushroom, the consumption is almost 161 MB for all minSupp and minOccu values. We omitted the detailed memory consumption results due to space constraint.

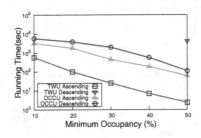

Fig. 4. Running time by OCEAN with various item orders on BMS

The efficiency of OCEAN is manifestly influenced by the processing order of items. We recorded the running time of OCEAN using TWU (transaction-weighted utility) ascending order, TWU descending order, item occupancy ascending order, item occupancy descending order, and lexicographic order on BMS. Figure 4 shows the experimental results. According to Fig. 4, the TWU ascending order made a best performance as it can reduce the number of generated utility lists [11], thus OCEAN processes items by following this order. Note that the curve of lexicographic order is not shown in Fig. 4 since no results were generated under 1×10^5 s.

6   Conclusion

In this paper, we propose *utility occupancy* as a measure for interestingness of itemsets with utility information. The proposed measure can lead to useful itemsets that contribute a large portion of total utility for each individual transaction representing user interests or user habit. We derive an upper bound for utility occupancy from an itemset for all possible extension itemsets from it, and design an efficient mining algorithm using this upper bound based on a utility list data structure and a depth first search procedure. Experimental evaluations on real world data show that the proposed mining algorithm can find such patterns effectively and efficiently.

References

1. Frequent itemset mining implementations repository. http://fimi.ua.ac.be/data/
2. Ahmed, C.F., Tanbeer, S.K., Jeong, B.S., Lee, Y.K.: Efficient tree structures for high utility pattern mining in incremental databases. IEEE Trans. Knowl. Data Eng. **21**(12), 1708–1721 (2009)
3. Burdick, D., Calimlim, M., Gehrke, J.: Mafia: A maximal frequent itemset algorithm for transactional databases. In: Proceedings of the 17th International Conference on Data Engineering (ICDE), pp. 443–452. IEEE (2001)
4. Chan, R.C., Yang, Q., Shen, Y.D.: Mining high utility itemsets. In: Proceedings of the 3rd IEEE International Conference on Data Mining (ICDM), pp. 19–26. IEEE (2003)
5. Fournier-Viger, P., Wu, C.-W., Zida, S., Tseng, V.S.: FHM: Faster high-utility itemset mining using estimated utility co-occurrence pruning. In: Andreasen, T., Christiansen, H., Cubero, J.-C., Raś, Z.W. (eds.) ISMIS 2014. LNCS, vol. 8502, pp. 83–92. Springer, Heidelberg (2014)
6. Han, J., Cheng, H., Xin, D., Yan, X.: Frequent pattern mining: Current status and future directions. Data Min. Knowl. Disc. **15**(1), 55–86 (2007)
7. Han, J., Wang, J., Lu, Y., Tzvetkov, P.: Mining top-k frequent closed patterns without minimum support. In: Proceedings of the 2rd IEEE International Conference on Data Mining (ICDM), pp. 211–218. IEEE (2002)
8. Hong, T.P., Lee, C.H., Wang, S.L.: Effective utility mining with the measure of average utility. Expert Syst. Appl. **38**(7), 8259–8265 (2011)
9. Lan, G.C., Hong, T.P., Tseng, V.S.: A projection-based approach for discovering high average-utility itemsets. J. Inf. Sci. Eng. **28**(1), 193–209 (2012)
10. Liu, J., Wang, K., Fung, B.: Direct discovery of high utility itemsets without candidate generation. In: Proceedings of the 12th IEEE International Conference on Data Mining (ICDM), pp. 984–989. IEEE (2012)
11. Liu, M., Qu, J.: Mining high utility itemsets without candidate generation. In: Proceedings of the 21st ACM International Conference on InformatIon and Knowledge Management (CIKM), pp. 55–64. ACM (2012)
12. Rymon, R.: Search through systematic set enumeration. Technical Reports (CIS), pp. 539–550 (1992)
13. Tan, P.N., Kumar, V., Srivastava, J.: Selecting the right objective measure for association analysis. Inf. Syst. **29**(4), 293–313 (2004)
14. Tang, L., Zhang, L., Luo, P., Wang, M.: Incorporating occupancy into frequent pattern mining for high quality pattern recommendation. In: Proceedings of the 21st ACM International Conference on Information and Knowledge Management (CIKM), pp. 75–84. ACM (2012)
15. Yang, G.: The complexity of mining maximal frequent itemsets and maximal frequent patterns. In: Proceedings of the Tenth ACM SIGKDD International Conference on Knowledge Discovery and Data Mining, pp. 344–353. ACM (2004)

Graph and Network Data

Leveraging Emotional Consistency
for Semi-supervised Sentiment Classification

Minh Luan Nguyen[(⊠)]

Institute for Infocomm Research, Singapore, Singapore
mlnguyen@i2r.a-star.edu.sg

Abstract. In this work, we exploit the emotional consistency between label information obtained by label propagation and distant supervision to leverage tweet-level sentiment analysis. Existing methods are either relied heavily on sufficient labeled data or sentiment lexicon resources, which are domain-specific in social media. We propose a three-phase approach to build a semi-supervised sentiment classifier for social media data. Our framework leverages on both labeled, unlabeled tweets and social relation graph data. First, we use label propagation to learn propagated labels for unlabeled tweets and partition all tweets into two clusters. Our label propagation is inspired by social science about emotional behaviors of connected users, who tend to hold similar opinions. Second, using sentiment lexicon resources, we use an unsupervised method to obtain noisy labels, which is utilized to train a distant supervision classifier. Next, we determine the relevance of each classifier to the unlabeled tweets, using the label consistency between the clustering given by the propagated tweet labels and the clustering given by these trained sentiment classifiers. Third, we trade-off between using relevance-weighted trained classifiers and the labeled tweet data. Our method outperforms numerous baselines and a social networked sentiment classification method on two real-world Twitter datasets.

Keywords: Sentiment analysis · Opinion mining · Twitter · Social networks · Semi-supervised classification · Distant supervision

1 Introduction

With the explosion of Web 2.0 services, popular social media mediums like Twitter enable user to easily express and share their own opinion on various kinds of topics [16]. These opinions often serve as helpful advice and influence following users' decision making. For instance, when a customer wants to buy a laptop on Amazon he/she will usually looks for reviews and comments written by previous customers [13]. Sentiment analysis can also provide organizations or company with the ability to listen customer's voice on social media forums in timely manner and act instantly. As such, advertisers, companies and politicians are seeking ways to make sense user's sentiments through social media on their product, service and policy quality.

© Springer International Publishing Switzerland 2016
J. Bailey et al. (Eds.): PAKDD 2016, Part I, LNAI 9651, pp. 369–381, 2016.
DOI: 10.1007/978-3-319-31753-3_30

Sentiment analysis for social media platforms like Twitter and MySpace poses several challenges for researchers [2,25]. In the past, sentiment classification has been extensively investigated for user reviews [12,13] with good performance. However, the social text data in blog posts is different substantially from the traditional product review data. The social text data often contains short and noisy messages [1]. It also contains lots of emotional abbreviations, emoticons and has no syntactic structure. Because of highly discourse variations, sentiment data in social media often lacks of sufficient labeled data. Thus, it is not easy to automatically identify the sentiment meanings of these messages, though it can be understood conveniently in human communications.

Besides textual content, social media platforms often provide additional information about user-user relationships in online social networks. In particular, relations between messages can be represented via a matrix of between users and messages and a matrix of between user interactions. Blog messages is potentially networked through user connections. This is a distinct feature because it may contain useful structural or community information that are not able to explore from purely text-based methods for sentiment analysis. Social networks is useful for two reasons. Firstly, social relation graph information is now more easily to obtain in social media platforms' API. Secondly, according to the principle of homophily [11,17], if two users hold a personal relationship, they may be tend to hold the same opinions about some affairs.

However, these popular sentiment analysis methods require sufficient texts labeled with polarity, thereby, not suitable for social media data. Currently, there are two ways to overcome this issue. The first way is supervised method [14] to model message similarity by exploiting user relationship. The second way is reducing the dependence on labeled texts by using either distant supervision on noisy labels [9] or semi-supervised label propagation on unlabeled tweets upon social relations [24]. To the best of our awareness, there has been no effort on bringing together several of the above ideas for social media sentiment analysis.

This paper proposes a semi-supervised approach for sentiment analysis in social media by leverage emotional consistency on label information obtained by label propagation and distant supervision. In particular, it takes the advantage of label propagation based on both textual and social relation information and the emotional consistency of propagated labels and distant supervision labels in tackling the insufficient labels issue and noisy nature of the tweets. We propose a unified three-phase framework for semi-supervised sentiment classification. Our framework leverages on both labeled, unlabeled tweets, social graph and sentiment lexicons. First, we use label propagation to learn propagated labels for unlabeled tweets and partition all tweets into different clusters. Second, we train a distant supervision classifier based on sentiment lexicon resources. We also train a linear classification model based on training dataset. Next, we determine the relevance of each classifier to the unlabeled tweet, using the label consistency between the clustering given by the propagated tweet labels and the clustering given by these trained sentiment classifiers. Third, we train the final classifier by trading-off between using relevance-weighted trained classifiers and the labeled tweet data. Finally, we conduct empirical experiments to evaluate the effectiveness of the proposed model. We compare the proposed three-phase framework with state-of-the-art classification baselines for the sentiment analysis task, and we find that our method outperforms the baselines.

2   Related Work

Sentiment analysis has been studied extensively on various kinds of text data such as movie and product reviews [12,21]. The basic method is to build a hand-crafted sentiment features, which can effectively express the sentiment of the texts, and apply machine learning. With the growing popularity of social media recently, sentiment analysis for user generated contents has attracted lots of attention from researchers [2,4].

Existing methods cannot effectively exploits noisy, short texts in microblogging and make use of the social relation information of micro-blogging [4]. Similar to traditional methods, there are some ongoing efforts on sentiment analysis for the micro-blogging data. Alec et al. [9] used distant supervision with noisy labels obtained from emoticons for Twitter sentiment analysis. Barbosa and Feng [2] analyze the linguistically features of tweets as well as the meta-data information of words for Twitter sentiment. However, the above two methods have not exploited the social relation information.

The current approaches for sentiment analysis on social media often rely on pre-defined sentiment lexicons or vocabularies [15,18], which are highly domain-dependent. Standard supervised classification methods improve the situation somewhat [20,21], by training a text similarity model purely based on the content. However, these require sufficient text labeled with polarity. To overcome this issue, some efforts have been made to explore other external information such as emoticons [9] and especially social relation information [10,27], on sentiment analysis. Distant supervision [9] based on noisy labels can reduce dependence on labeled texts with promising performance. In addition, label propagation [24] is develop to prorogate sentiment of labeled tweets to unlabeled tweets upon social relations. In [14], a supervised method based on l_1-norm least squares and Laplacian graph regularization have also been proposed to model message similarity by exploiting user relationship. The basis idea of above two algorithms is determining the sentiment of a tweet posted by a user based on both of sentiments of its tweets and its neighbor's tweets collectively upon social relation graph.

Our approach unifies the advantages of the above ideas. It differs from [24] by incorporating both of textual and social graph. It leverages the unlabeled data as compared with [14]. It goes beyond than merely use noisy labels [9] by considering the relevancy.

3   Problem Statement

Given a corpus $D = \{t_1, t_2, ..., t_n\}$ of n tweets. Let $\mathcal{L} = \{L_1, L_2, ..., L_k\}$ be sentiment lexicon resources, where each of them is a list of sentiment words corresponding to their sentiment polarity, e.g., *positive* or *negative*. We introduce a feature extraction χ that maps a tweet t to its feature vector x. For each message in the corpus $t_i \in D$, $t_i = (x_i, y_i) \in \mathbb{R}^{m+c}$ are tweet features and corresponding sentiment label, where $x_i \in \mathbb{R}^m$ is the tweet feature vector and $y_i \in \mathbb{R}^c$ is the sentiment label vector. The sentiment dataset has a some labeled data $D_l = \{(x_i, y_i)\}_{i=1}^{n_l}$ and plenty of unlabeled data $D_u = \{(x_i)\}_{n_l+1}^{n_l+n_u}$, where n_l is the number of labeled instances, n_u is the number of unlabeled instances, x_i is the feature vector, y_i is the corresponding label (if available). Let $n = n_l + n_u$. $D_l = [X, Y]$, where $X \in \mathbb{R}^{n \times m}$ is the feature matrix, $Y \in \mathbb{R}^{n \times c}$ is the label matrix, m is the number of tweet features, n is the number of tweets and c is number of sentiment labels. Let $u = \{u_1, u_2, ..., u_d\}$ be the user set. Let $\mathbf{U} \in \mathbb{R}^{d \times n}$ be a user-tweet matrix, where $U_{ij} = 1$ denotes that tweet t_j is created by user u_i. Let $F \in \mathbb{R}^{d \times d}$ be the user-user matrix, where $F_{ij} = 1$ indicates that user u_i is connected by user u_j.

Learning a sentiment classifier can then be abstracted to finding a function p such that $p(\chi(t)) = p(x) = y$. We define the problem *sentiment analysis in social media* as : *Given a corpus of tweets $D = D_l \cap D_u$ with feature matrix X and labels Y as well as the user-tweet relation U and user-user relation F, it aims to learn a sentiment classifier $p(\chi(t)) = p(x) = y$ in order to predict sentiment polarity for new tweets t.*

4 Proposed Approach

In this section, we describe our three-phase approach, which comprises of a label propagation, emotional clustering consistency and combined classifier learning phases.

Our key idea is to exploit the concept of *emotional consistency* between label information obtained by label propagation and distant supervision. The propagated labels are obtained by spectral-based label propagation. The distant supervision labels are obtained by a sentiment classifier trained based on lexicon-based unsupervised sentiment predictions, also known as *noisy labels*. There are two main challenges in our approach. Firstly, it is not easy to effectively propagate labels to unlabeled data based on both of textual content and social relation graph. Secondly, how to effectively leverage the sentiment prediction performance based on the distant supervision labels, a.k.a *noisy labels*. Although the ground-truth labels and noisy labels come from the same dataset, they probably have different marginal distributions on the sentiment classess. If we merely combine ground-truth labels and noisy labels to train a sentiment classifier, this can produce a negative transfer phenomenon [22], where using knowledge from noisy labels degrades the performance on the ground-truth labels.

To tackle these challenges, we propose a three-phase Robust Semi-supervised Sentiment Propagation (RSSP) framework. Our method is inspired by [8] for multi-class case. In the first phase, based on the social graph and textual content, we use label propagation to learn noisy labels for unlabeled tweets and partition all tweets into different regions. In the second phase, we train two different sentiment classifiers: one linear classifier based on ground-truth labels and one distant supervision classifier based on noisy labels. Next, we determine the relevance of each classifier to a region, using the label consistency between the clustering given by the propagated labels and the clustering given by these trained classifiers. By using unlabeled data, we alleviate the lack of labeled samples for rarer classes due to imbalanced distributions in labels.

The third phase uses the relevances determined the second phase to produce a reference predictor by weighing the pre-trained classifiers for each unlabeled sample separately. The intention is to alleviate the effect of mismatched distributions. The final classifier in the unlabeled data is trained on the labeled data while taking reference from the reference predictions on the unlabeled data. This ensures reasonable predictive performance even when all the noisy labels are irrelevant and augments the rarer classes with examples in the unlabeled data.

4.1   Phase 1: Label Propagation

To apply this idea to sentiment classification, we need to (i) partition the entire data input space into clusters/regions and (ii) assign preliminary labels for all the examples. We approximate the our data input space with all the samples from D_l and D_u. With data from both the labeled and unlabeled data sets, we apply transductive inference or semi-supervised learning [28] to achieve both (i) and (ii). By augmenting with unlabeled data D_u, we aim to alleviate the effect of imbalanced relation distribution, which causes a lack of labeled samples for rarer classes in a small set of labeled data. Briefly, the known labels in D_l are propagated to the entire target input space by encouraging label smoothness in neighborhoods. The next three paragraphs give more details.

At present, we assume a similarity matrix W, where W_{ij} is the similarity between the ith and the jth input samples in $D_l \cup D_u$. Matrix W then determines the neighborhoods and is defined as follows:

$$W = W^{Sim} + 0.5W^{Sam} + 0.5W^{Soc}$$

where $W_{ij}^{Sim} = Sim(t_i, t_j)$ denotes the pairwise kernel similarity based on textual contents of t_i and t_j. Here, the similarity matrix W^{Sim} uses the Gaussian kernel $K(x, x') = \exp(\|x - x'\|^2/2\sigma^2)$, which is based on textual features. $W^{Sam} = \mathbf{U}^T\mathbf{U}$, $W_{ij}^{Sam} = 1$ indicates whether t_i and t_j as posted by the same user. $W^{Soc} = \mathbf{U}^T\mathbf{F}\mathbf{U}$, $W_{ij}^{Soc} = 1$ indicates whether the authors of t_i and t_j are connected friends in the social graph. Let Λ be a diagonal matrix where the (i, i)th entry is the sum of the ith row of W. Let us also encode the labeled data D_l in an n-by-c matrix H, such that $H_{ij} = 1$ if instance i is labeled with sentiment class j in D_l, and $H_{ij} = 0$ otherwise. Our objective is the c-dimensional sentiment-class indicator vector F_i for the ith instance, for every sample. This is achieved via a regularization framework [28]:

$$\min_{\{F_i\}_{i=1}^n} \left(\sum_{i,j=1}^n W_{ij} \left\| \frac{F_i}{\sqrt{\Lambda_{ii}}} - \frac{F_j}{\sqrt{\Lambda_{jj}}} \right\|^2 + \mu \sum_{i=1}^n \|F_i - H_i\|^2 \right) \quad (1)$$

This trades off two criteria: the first term encourages nearby samples (under distance metric W) to have the same labels, while the second encourages samples to take their labels from the labeled data. The closed-form solution is

$$F^* = (I - (1 + \mu)^{-1}L)^{-1}H, \qquad (2)$$

where $L = \Lambda^{-1/2}W\Lambda^{-1/2}$; and the n-by-c matrix F^* is the concatenation of the F_is.

Using vector F_i^*, we now assign preliminary labels to the samples. For a sample i, we transform F_i^* into probabilities $p_i^1, p_i^2, \ldots, p_i^c$ using softmax. Our propagated label ℓ_i for sample i is then

$$\ell_i = \arg\max_j p_i^j \qquad (3)$$

Next, we partition the data in $D_l \cup D_u$ into c regions, $R_1, R_2, \ldots, R_c$, corresponding to the c sentiment class labels. The intuition is to use the true label in D_l when available, or otherwise resort to using the propagated label. That is,

$$x_i \in \begin{cases} R_{y_i} & \text{if } x_i \in D_l, \\ R_{\ell_i} & \text{if } x_i \in D_u. \end{cases}$$

4.2   Phase 2: Emotional Clustering Consistency

In this section, we first use the lexicon-ratio method for sentiment analysis [19, 26] to build three different sentiment predictors base three different lexicons resources. The three sentiment lexicon resources include: (1) MPQA Opinion Corpus[1] for daily sentiment words; (2) Twitter sentiment words[2]; and (3) popular Twitter emoticons[3]. Due to this unsupervised setting, logistic regression classifier will be used for tweets, which does not have any sentiment words. We then combine these three sentiment predictors to a final unsupervised predictor based on majority voting, which outputs noisy labels for unlabeled tweets. Using both ground-truth and noisy labels, we train a distant supervision classifier [9]. We also train a logistic regression classifier based on label data.

We use the concept of clustering consistency to determine the relevance of a trained classifier to particular regions in the unlabeled input space. Figure 1 illustrates this. There, both enclosing circles in the left and right figures denote the same input space of the target domain. There are four disjoint regions within the input space, located at the left, right, top and bottom of the space. There are two classes of labels: asterisk ($*$) for positive sentiment and circle ($\circ$) for negative sentiment.

The labels in the left figure are given by a *preliminary predictor* in the original data using label propagation, while the labels in the right figure are given by a

---

[1] http://mpqa.cs.pitt.edu/corpora/mpqa_corpus/.

[2] http://saifmohammad.com/WebPages/lexicons.html.

[3] http://www.datagenetics.com/blog/october52012/index.html.

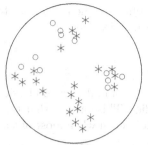 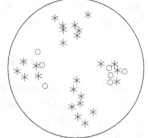

Original input space with transductive   Original input space with labels from the a
learning using labeled and unlabeled data.   predictor trained on the noisy labels.

Fig. 1. Clustering consistency is used to determine the relevance of a trained classifier
to a region in the original input space. The bottom and right regions are more relevant
than the top and left regions. See text for explanation.

predictor trained on the noisy labeled data. Comparing the figures, we see the
preliminary predictor and noisy labeled predictor are consistent for the bottom
and right regions, but are inconsistent for the top and left regions. This suggests
that the predictor trained on noisy labels is very relevant for the bottom and
right regions of the target input space, but less so for the top and left regions.

We now quantify the *clustering consistency* between a trained predictor and
a region in the original input space. Intuitively, this is the agreement between
the trained predictor and the preliminary predictor within the original input
space. We use supervised weighting in the following manner. Let $D^1 = D_l =$
$\{(x_i, y_i)\}_{i=1}^{n_l}$ and $D^2 = D = \{(x_i, l_i)\}_1^{n_l+n_u}$. For each subset of samples, say s,
we first train a sentiment predictor p_s based on its training data D^s. Then, for
every region R_j, we compute the relevance score as:

$$w_{s,j} = \sum_{x_i \in R_j} [\![p_s(x_i) = \ell_i]\!]/|R_j| \tag{4}$$

where $[\![\cdot]\!]$ is the Iverson bracket.

	R_1	R_2
w_{D_1}	0.75	0.49
w_{D_2}	0.84	0.34

Fig. 2. Heat map of the relevance scores $w_{s,j}$ between each trained classifier with a
region in the original input space on OMD data set. A lighter shade means a higher
score, or more relevant.

Figure 2 shows the heat map of the relevance scores $w_{s,j}$ between the each
train classifier with the regions in original input space of OMD corpus. We
observe, for example, that the classifier trained on noisy labels (D^2) data is
more relevant in the R_1 region than R_2 region of the original input space.

These relevance scores will be used in the next phase of the framework to weigh the contributions of each predictor to the eventual classifier.

4.3 Phase 3: Target Classifier Learning

This phase uses both of the previous predictions from all trained classifiers and the labeled data D_l to learn a sentiment classifier. This ensures that the performance of proposed method will not degrade badly even when most of the source instances are irrelevant,

The previous phase has computed the relevance $w_{s,j}$ for trained classifier p_s in region R_j. We translate this to the relevance weight $u_{s,i}$ for an example x_i: if $x_i \in R_j$, then $u_{s,i} = w_{s,j}$. From the previous phase, we also have 2 sentiment predictors p_s that have been trained on D^s. We combine and weight the predictions from multiple classifiers to obtain the *reference prediction* $\hat{r}_{ji} = \sum_{s=1}^{2} u_{s,i}(2[\![p_s(x_i) = j]\!] - 1)$ for example x_i belonging to sentiment class j, using the ± 1 encoding.

The sentiment classifier consists of c functions $f_1, \ldots, f_c$ using the one-versus-rest decoding for multi-class classification.[4] Based on the Domain Adaptive Machine [6], we incorporate the reference predictions and the labeled data of the target domain to learn the final classifier:

$$\min_{\{f_j\}_{j=1}^{c}} \sum_{j=1}^{c} \left\{ \frac{1}{n_l} \sum_{i=1}^{n_l} (f_j(x_i) - r_{ji})^2 + \gamma \|f_j\|_{\mathcal{H}}^2 + \frac{\beta}{2} \sum_{i=n_l+1}^{n} \|f_j(x_i) - \hat{r}_{ji}\|^2 \right\}, \quad (5)$$

where $r_{ji} = 2[\![y_i = j]\!] - 1$ is the ± 1 binary encoding for the i labeled sample belonging to relation j. Here, we have three objectives: the first term specifies the training error; the second governs the complexity of the functions f_js in the Reproducing Kernel Hilbert Space (RKHS) $\mathcal{H}$; and the third favors the predicted labels of the unlabeled data D_l to be close to the reference predictions. The third term provides additional pseudo-training samples for the rarer sentiment classes, if these are available in D_u. Parameters β and γ govern the trade-offs between these objectives.

Let $K(\cdot, \cdot)$ be the reproducing kernel for $\mathcal{H}$. By the Representer Theorem [23], the solution for Eq. 5 is linear in $K(x_i, \cdot)$: $f_j(x) = \sum_{i=1}^{n} \alpha_{ji} K(x_i, x)$. Putting this into Eq. 5, parameter vectors α_j are [3]:

$$\alpha_j^* = (JK + \gamma(n_l + \beta n_u)I)^{-1} J R_j. \quad (6)$$

Here, R_j is an $(n_l + n_u)$-vector, where $R_{ji} = r_{ji}$ if instance i is in the labeled set, and $R_{ji} = \hat{r}_{ij}$ if it is in the unlabeled set; and J is an $(n_l + n_u)$-by-$(n_l + n_u)$ diagonal matrix where the first n_l diagonal entries are ones and the rest are βs.

---

[4] For two-classes, though, only one function is needed.

5   Performance Evaluation

5.1   Experimental Settings

We evaluate our algorithm using two corpora: the Stanford Twitter Sentiment (STS) and the Obama-McCain Debate (OMD). Table 1 provides some statistics on them.

Table 1. Statistics on STS and OMD

Properties	STS	OMD
# Tweets	22,262	1,827
# Users	8,467	735
# Max Degree of the Users	897	138
# Avg. Degree of the Users	36	21
# Avg. Tweets per User	2.63	2.49

The Stanford Twitter Sentiment (STS) was collected by Go et al. [9]. Due to original purpose, it does not have social relation graph information among users. To overcome this, we obtain the social graph by using the Twitter complete graph crawled by Kwak et al. [16]. After that, we exclude tweets that its author has no friend or has less than two tweets. Finally, we got a dataset consists of 22,262 tweets with labels.

The original Obama-McCain Debate (OMD) consists of 3,269 tweets collected during the presidential debate [24]. Similar as in STS dataset, we also obtain the social relation graph for this dataset by exploiting the Twitter complete follower graph [16]. Next, we also filter out tweets as with the STS datasets. Finally, we got a dataset of 1,827 tweets with labels.

We benchmark our approach with several other methods, including machine learning and common sentiment analysis methods. These are described below.

Support Vector Machine (SVM) [5] is a typical text classification method.
Logistic Regression with l_1-norm regularization (LG) [7] is also a traditional text classification.
Distant Supervision (DS) is a semi-supervised learning algorithm that makes uses a noisy labels based on lexicon resources as mentioned in Sect. 4.2.
Label Propagation (LPROP) [24] is a semi-supervised graph regularization based on exploiting relation between tweets.
Linear Regression with graph regularization (SANT) [14] is a supervised method, which exploits user social relation graph.
RSSP-LP is a simple version of our approach, which combines ground-truth and noisy labels to train the final classifier without doing Phase 2 and 3.

We use available libraries for SVM and LG. We also re-implement other baselines according to published articles. In our experiments, we set $\mu = 0.8$ in Eq. 2; $\theta = 0.18$ in Eq. 3; and $\gamma = 0.1$ and $\beta = 0.3$ in Eq. 5. We use five-fold cross validation.

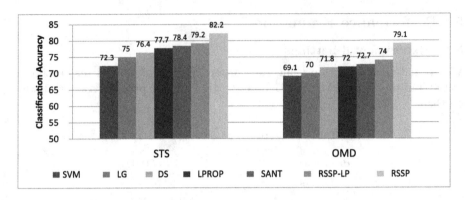

Fig. 3. Sentiment classification accuracy comparison on STS and OMD datasets.

5.2   Experimental Results

Figure 3 presents the sentiment classification accuracy results on STS and OMD datasets. We see that the proposed RSSP method has consistently outperformed the other methods on accuracy. We also observe that although the noisy labels learnt from RSSP are helpful on sentiment task, using only noisy labels to train a classifier (DS) marginally increases the accuracy as compared with supervised ground-truth methods like SVM and LG. This suggests that there would be negative effects from the noisy labels. More importantly, we also see that RSSP-LP, which exploits both text and social graph information, can also only gain a marginal accuracy improvement as compared with LPROP and SANT. This emphasizes the effectiveness of Phase 2 and 3 of RSSP, which considers the relevance of each trained classifier to each unlabeled sample before training the final classifier. As a result, RSSP outperforms the best SANT method by 5.5 % on average.

Weakly-Supervised Setting: We conduct experiments to verify the sensitivity of RSSP according to different training data sizes. We cross validate our model with five-folds. in the Table 2(a) and (b), $D_{percentage}$ represents the percentage of data used for training out of the entire training dataset. Due to five-fold cross validation, 80 % of the entire dataset is used for training in each round of the experiment. For instance, $OMD_{50\%}$ indicates that 50 % of 80 % thus 40 % of the whole dataset as training data. The test set always occupies 20 % of the entire dataset.

From Table 2(a) and (b), we find that our method has consistently outperformed all the other methods on accuracy peformance. We first notice that DS generally perform quite well, and it performs better than SVM and LG especially when the number of labeled instances is small. However, the training size increases, the performance gap becomes smaller. The reason is that DS aims to obtain a consensus on the entire dataset, and this will give a worse label than SVM and LG when there are enough irrelevant and noisy labeled samples to influence the classification decision wrongly. There is another side effect

Table 2. The sentiment classification accuracy on STS and OMD datasets in semi-supervised settings with different training samples. The best performance for each setting is in bold.

(a) STS

	$STS_{10\%}$	$STS_{30\%}$	$STS_{50\%}$	$STS_{80\%}$
SVM	0.650	0.684	0.701	0.713
LG	0.672	0.717	0.730	0.747
DS	0.681	0.721	0.738	0.755
LPROP	0.705	0.735	0.754	0.764
SANT	0.695	0.727	0.752	0.769
RSSP-LP	*0.688*	*0.709*	0.760	0.773
RSSP	**0.748**	**0.775**	**0.795**	**0.816**

(b) OMD

	$OMD_{10\%}$	$OMD_{30\%}$	$OMD_{50\%}$	$OMD_{80\%}$
SVM	0.593	0.614	0.649	0.668
LG	0.607	0.635	0.668	0.693
DS	0.618	0.649	0.674	0.702
LPROP	0.626	0.661	0.689	0.715
SANT	0.619	0.639	0.682	0.720
RSSP-LP	*0.640*	*0.675*	0.703	0.734
RSSP	**0.686**	**0.715**	**0.756**	**0.784**

of noisy propagated labels based on both textual and social graph information under small training set situation. In fact, one can roughly deduce that a RSSP classifier has few relevant noisy propagated labels by simply comparing rows LPROP with rows RSSP-LP in the tables: a decrease in accuracy from LPROP to RSSP-LP suggests that the noisy labels are somewhat irrelevant. For example, for datasets $STS_{10\%}$, $STS_{30\%}$, $OMD_{10\%}$ and $OMD_{30\%}$, we find that its accuracy decreases from LPROP and RSSP-LP in Table 2(a), (b) when there are only few labeled samples, which suggests that noisy labels from $STS_{10\%}$, $STS_{30\%}$, $OMD_{10\%}$ and $OMD_{30\%}$ are generally irrelevant to train a classifier. We investigate this further by examining the relevance scores $w_{s,j}$s, and we find that the decreases in accuracy from LPROP and RSSP-LP happen when there are more regions in the original input space to which noisy labels learnt by RSSP-LP are irrelevant.

We find that LPROP, SANT and RSSP-LP are better than SVM and LG. We also observe that label propagation methods such as LPROP and RSSP-LP are generally better than distant supervision method DS. However, we find that the performances of RSSP-LP is not quite stable: for example, on $STS_{10\%}$, $STS_{30\%}$, $OMD_{10\%}$ and $OMD_{30\%}$ datasets. In contrast, we find the performance of LPROP and SANT to be more stable. The reason is their reduced vulnerability to negative effect from irrelevant noisy labels by relying on similarity of feature vectors based on labeled, unlabeled data and social graph information. Further improvements can still be made, as shown by the better performance of RSSP over LPROP and SANT. This is achieved by further adjusting the relevances between a trained classifier and a sample according to its region in the original input space.

6   Conclusion

We have developed a robust semi-supervised sentiment propagation (RSSP) approach for the social media sentiment analysis problem where labeled data is scarce and information about social relations is available. Existing sentiment analysis approaches suffer from lacking of labeled data and under imbalanced distributions with noisy labels. To overcome these, we have proposed a three-phase approach to leverage only relevant information from the noisy labels, and thus leverage accuracy performance on the unlabeled data. Experimental results on OMD and STS have shown that the our semi-supervised method outperforms the other methods on accuracy performance when only few labeled instances are used. Because of the practical importance of sentiment analysis on social media data due to lack of labeled data in these domains, we hope our research will open up several investigations in the future.

References

1. Baldwin, T., Cook, P., Lui, M., MacKinlay, A., Wang, L.: How noisy social media text, how diffrnt social media sources. In: Proceedings of the Sixth International Joint Conference on Natural Language Processing, pp. 356–364 (2013)
2. Barbosa, L., Feng, J.: Robust sentiment detection on twitter from biased and noisy data. In: Proceedings of the 23rd International Conference on Computational Linguistics: Posters, pp. 36–44, Association for Computational Linguistics (2010)
3. Belkin, M., Niyogi, P., Sindhwani, V.: Manifold regularization: a geometric frame-work for learning from labeled and unlabeled examples. J. Mach. Learn. Res. **7**, 2399–2434 (2006)
4. Bermingham, A., Smeaton, A.F.: Classifying sentiment in microblogs: is brevity an advantage? In: Proceedings of the 19th ACM International Conference on Information and Knowledge Management, pp. 1833–1836. ACM (2010)
5. Chang, C.C., Lin, C.J.: LIBSVM: a library for support vector machines. ACM Trans. Intell. Syst. Technol. (TIST) **2**(3), 27 (2011)
6. Duan, L., Tsang, I.W., Xu, D., Chua, T.S.: Domain adaptation from multiple sources via auxiliary classifiers. In: Proceedings of the 26th Annual ICML, pp. 289–296. ACM (2009)
7. Fan, R.E., Chang, K.W., Hsieh, C.J., Wang, X.R., Lin, C.J.: LIBLINEAR: a library for large linear classification. J. Mach. Learn. Res. **9**, 1871–1874 (2008)
8. Ge, L., Gao, J., Ngo, H., Li, K., Zhang, A.: On handling negative transfer and imbalanced distributions in multiple source transfer learning. Stat. Anal. Data Min. ASA Data Sci. J. **7**(4), 254–271 (2014)
9. Go, A., Bhayani, R., Huang, L.: Twitter sentiment classification using distant supervision. CS224N Project Report, Stanford, pp. 1–12 (2009)
10. Gryc, W., Moilanen, K.: Leveraging textual sentiment analysis with social network modelling. From Text to Political Positions: Text analysis across disciplines 55, 47 (2014)
11. Hatfield, E., Cacioppo, J.T., Rapson, R.L.: Emotional Contagion. Cambridge University Press, Cambridge (1994)
12. Hu, M., Liu, B.: Mining and summarizing customer reviews. In: Proceedings of the Tenth ACM SIGKDD International Conference on Knowledge Discovery and Data Mining, pp. 168–177. ACM (2004)
13. Hu, M., Liu, B.: Mining opinion features in customer reviews. AAAI 4, 755–760 (2004)

14. Hu, X., Tang, L., Tang, J., Liu, H.: Exploiting social relations for sentiment analysis in microblogging. In: Proceedings of the Sixth ACM International Conference on Web Search and Data Mining, pp. 537–546. ACM (2013)
15. Kiritchenko, S., Zhu, X., Mohammad, S.M.: Sentiment analysis of short informal texts. J. Artif. Intell. Res. **50**, 723–762 (2014)
16. Kwak, H., Lee, C., Park, H., Moon, S.: What is twitter, a social network or a news media? In: Proceedings of the 19th International Conference on World Wide Web, pp. 591–600. ACM (2010)
17. McPherson, M., Smith-Lovin, L., Cook, J.M.: Birds of a feather: homophily in social networks. Annu. Rev. Sociol. **27**, 415–444 (2001)
18. Mohammad, S.M., Kiritchenko, S., Zhu, X.: NRC-Canada: building the state-of-the-art in sentiment analysis of tweets. In: Second Joint Conference on Lexical and Computational Semantics (* SEM), vol. 2, pp. 321–327 (2013)
19. O'Connor, B., Balasubramanyan, R., Routledge, B.R., Smith, N.A.: From tweets to polls: linking text sentiment to public opinion time series. ICWSM **11**(122–129), 1–2 (2010)
20. Pang, B., Lee, L.: A sentimental education: sentiment analysis using subjectivity summarization based on minimum cuts. In: Proceedings of the 42nd Annual Meeting on Association for Computational Linguistics, p. 271. Association for Computational Linguistics (2004)
21. Pang, B., Lee, L., Vaithyanathan, S.: Thumbs up?: sentiment classification using machine learning techniques. In: Proceedings of the ACL-02 Conference on Empirical Methods in Natural Language Processing, vol. 10, pp. 79–86. Association for Computational Linguistics (2002)
22. Rosenstein, M.T., Marx, Z., Kaelbling, L.P., Dietterich, T.G.: To transfer or not to transfer. In: NIPS 2005 Workshop on Transfer Learning, vol. 898 (2005)
23. Schölkopf, B., Smola, A.J.: Learning with Kernels: Support Vector Machines, Regularization, Optimization, and Beyond. MIT press (2002). ISBN: 0262194759
24. Speriosu, M., Sudan, N., Upadhyay, S., Baldridge, J.: Twitter polarity classification with label propagation over lexical links and the follower graph. In: Proceedings of the First Workshop on Unsupervised Learning in NLP, pp. 53–63. Association for Computational Linguistics (2011)
25. Tang, J., Wang, X., Gao, H., Hu, X., Liu, H.: Enriching short text representation in microblog for clustering. Front. Comput. Sci. **6**(1), 88–101 (2012)
26. Wilson, T., Wiebe, J., Hoffmann, P.: Recognizing contextual polarity in phrase-level sentiment analysis. In: Proceedings of the Conference on Human Language Technology and Empirical Methods in Natural Language Processing, pp. 347–354. Association for Computational Linguistics (2005)
27. Zafarani, R., Cole, W.D., Liu, H.: Sentiment propagation in social networks: a case study in livejournal. In: Chai, S.-K., Salerno, J.J., Mabry, P.L. (eds.) SBP 2010. LNCS, vol. 6007, pp. 413–420. Springer, Heidelberg (2010)
28. Zhou, D., Bousquet, O., Lal, T.N., Weston, J., Schölkopf, B.: Learning with local and global consistency. In: NIPS (2003)

The Effect on Accuracy of Tweet Sample Size for Hashtag Segmentation Dictionary Construction

Laurence A.F. Park[(✉)] and Glenn Stone

School of Computing, Engineering and Mathematics,
Western Sydney University, South Penrith, Australia
{l.park,g.stone}@westernsydney.edu.au
http://www.scem.uws.edu.au/~lapark

Abstract. Automatic hashtag segmentation is used when analysing twitter data, to associate hashtag terms to those used in common language. The most common form of hashtag segmentation uses a dictionary with a probability distribution over the dictionary terms, constructed from sample texts specific to the given hashtag domain. The language used in Twitter is different to the common language found in published literature, most likely due to the tweet character limit, therefore dictionaries constructed to perform hashtag segmentation should be derived from a random sample of tweets. We ask the question "How large should our sample of tweets be to obtain a given level of segmentation accuracy?" We found that the Jaccard similarity between the correct segmentation and the predicted segmentation using a unigram model, follows a Zero-One inflated Beta distribution with four parameters. We also found that each of these four parameters are functions of the sample size (tweet count) for dictionary construction, implying that we can compute the Jaccard similarity distribution once the tweet count of the dictionary is known. Having this model allows us to compute the number of tweets required for a given level of hashtag segmentation accuracy, and also allows us to compare other segmentation models to this known distribution.

1 Introduction

Twitter is a dynamic environment, accumulating approximately 500 million tweets per day from millions of users worldwide.[1] Hashtags have become the de facto standard for labelling the topic or intent of a tweet within Twitter. Therefore, if we understand the hashtag, we gain a deeper insight into the intent and sentiment of the tweet. Hashtag analysis usually involves breaking down the hashtag or *segmenting* it into the words that are used in its formation, allowing us to associate the hashtag to other words within a sample of tweets. Hashtag segmentation has been used to assist search engines in providing more accurate search results [2,6], to increase the effectiveness of topic identification [12], to increase the effectiveness of sentiment analysis of tweets [11,13], and it has also been used in the meta analysis of predicting hashtag trends [5].

---

[1] According to https://about.twitter.com/company.

J. Bailey et al. (Eds.): PAKDD 2016, Part I, LNAI 9651, pp. 382–394, 2016.
DOI: 10.1007/978-3-319-31753-3_31

Automatic segmentation is highly dependent on the segmentation dictionary, which should be chosen so that it matches the domain of the text to be segmented. Therefore, dictionaries for hashtag segmentation are best constructed from a random sample of tweets. However, it is unclear how large this sample should be.

In this article, we ask "How large should our sample of tweets be to obtain a given level of segmentation accuracy?" To the best of our knowledge, this is the first analysis of the relationship between the number of tweets used to construct the segmentation dictionary and the accuracy of the hashtag segmentation.

The contributions of this article are:

- Identification of the hashtag segmentation Jaccard similarity distribution as a Zero-One Beta distribution (Sect. 3),
- The presentation of closely fitting models for each distribution parameter, as functions of the number of tweets used during dictionary creation (Sect. 4),
- A method of predicting the mean and standard deviation of the Jaccard similarity for a given sample dictionary (Sect. 5).

The article will proceed as follows: Sect. 2 introduces the method of hashtag segmentation, Sect. 3 provides an analysis of the distribution of the segmentation accuracy, Sect. 4 examines the change in distribution parameters with respect to the dictionary sample size, and Sect. 5 provides a model of predicting the expected segmentation accuracy.

2   Hashtag Segmentation Using Dynamic Programming

The problem of hashtag segmentation is identical to the problem of string segmentation described in [7], which is also applicable to tasks such as novelty location [10] and popularity induction [8,9]. To segment a string of n characters, we imagine $n - 1$ potential breaks in between each of the n characters which we can turn on or off. If the break is turned on, we segment the string at that point; if the break is off, we don't segment the string at that point. Having $n - 1$ potential breaks implies that we have 2^{n-1} possible segmentations of the given string. To compute the most likely segmentation, we compute the likelihood of each of the possible 2^{n-1} segmentations and retain the segmentation with maximum likelihood. We can see that as n grows this approach becomes infeasible due to the exponentially increasing computation required. Rather than observing all 2^{n-1} possible segmentations, we can use the more sophisticated approach of dynamic programming (also described in [7]), reducing the complexity to the order of n^2.

To obtain the segmentation, we require a method of computing the probability that a given segmentation of a hashtag would be written by the author of the hashtag. A simple method of computing this probability is to assume each term is independent, therefore the probability of a given segmentation is simply the product of the probability of each word in the segmentation. In doing so, we now only need the probability of each word in the segmentation.

An estimate of the probability of an author writing a word can easily be computed from a sample of the author's writing, as the proportion of the frequency of the word relative to the number of words in the sample. This set of words and their associated probability is referred to as a dictionary. Tradition text segmentation methods compute their dictionary from a large sample of text with the same writing style as the author (e.g. to segment a string from a news article, the dictionary would be constructed from a sample of news articles). In fact, [3] found that segmentation is highly dependent on the dictionary used.

Note that segmentation methods such as [1] use additional features of the hashtag to compute the probability of the segmented word sequence (such as the case of the letters). For this analysis, we are concerned with the effect of the number of tweets used to construct the dictionary, therefore we treat the hashtag as a string and use no additional information to remove the variability that would be introduced otherwise.

The language used in Twitter is unlike the language found in published media [4]. It contains many short snippets of information that is regularly updated by users worldwide, therefore the language used is constantly evolving. To effectively segment hashtags, we must construct the dictionary based on the language used in Twitter, which means using a random sample of tweets. Unfortunately, it is unclear how large a sample we should take to obtain an acceptable level of segmentation accuracy.

3   Segmentation Accuracy Distribution

In the previous section, we established that a random sample of tweets is required to construct a hashtag segmentation dictionary, but we were unsure of how large a sample to obtain. In this section, we will examine the distribution of the segmentation accuracy to take a step towards identifying the required sample size.

To begin the analysis, we obtained a random sample of 251171 tweets to use as our tweet pool. All further random samples of tweets were resampled from this pool.

To compute the segmentation accuracy, we must obtain a random sample of hashtags with known segmentations, which we then compare to the predicted segmentation. To observe the distribution, we also require the random sample to be large. Since hashtags are word sequences that have had the space between words removed, we generated a set of hashtags by sampling one tweet at a time from the pool, sampling a random sequence length from a shifted Poisson distribution (with minimum of 1 and mean 3.5, found through analysis of existing hashtag segmentations), then sampling a sequence of that length from the tweet and combining it to form the hashtag. Doing so allowed us to obtain the hashtag and the true segmentation. To ensure that the segmentation dictionary did not contain the tweet in which the hashtag was generated from, we first obtained a random sample of 100 tweets in which we generated one hashtag from each tweet, we then constructed the dictionary using a sample from the remaining tweets in the pool.

The segmentation dictionary was constructed by removing URLs, user handles and hashtags from the tweet sample. Numbers and punctuation were also removed and each remaining term was case folded. Stemming and stop word removal were not used to ensure that the dictionary contained the distribution over all words and that good estimates of probabilities were obtained. Initial experiments also showed that using stop word removal and stemming when creating a dictionary, harms the accuracy of the hashtag segmenter. This is likely due to the unique spelling and acronyms used to write informative tweets within the 140 character limit. By not removing stop words and not performing stemming, we also remove the variability that would be introduced by the different stop word lists and different stemming algorithms that can be used.

The true segmentation and the predicted segmentation are both sets of variable length; if both sets are the same, the hashtag segmentation is correct, but as the number of differing words between sets increases, the accuracy is reduced. To evaluate the similarity between the true and predicted segmentation, we used Jaccard similarity, since it is a measure of set similarity.

This hashtag generation and evaluation process was replicated 50 times for each dictionary size, to obtain 5000 Jaccard similarity scores for each dictionary size. The distribution of the 5000 Jaccard similarity scores is shown in Fig. 1 for dictionaries constructed from tweet samples of size 100, 200, 500, 1000, 2000, 5000, 10000, 20000 and 50000. The histograms show a decreasing distribution with a spike at 1 for all dictionary sizes. This implies that there are many Jaccard similarity scores that have values of 0 and 1 and there are a subset that range between 0 and 1. To examine this further, we plotted the histograms again, with all scores of 0 and 1 removed and found the histogram shape to be similar to a Beta density. By examining the Q-Q plot (shown in Fig. 2), comparing the quantiles of the scores to the quantiles of the Beta distribution, we found that this mid-band set of scores is very closely aligned to the Beta distribution.

After examining the plots we arrived at the Zero-One inflated Beta distribution model for the probability distribution J of a Jaccard score for a given dictionary size, where the probability of J being 0 is the proportion p_0, being 1 is the proportion p_1, and the probability of J being within 0 and 1 follows a Beta distribution with parameters α and β, giving us four parameters for the model.

The model is a mixture of three density functions where we can write the mixture as the density $f_J(x)$:

$$f_J(x) = p_0\delta(x) + p_b f_B(x; \alpha, \beta) + p_1\delta(1 - x) \tag{1}$$

where $\delta(x)$ is the Dirac delta function, $p_b = 1 - p_1 - p_0$, and $f_B(x; \alpha, \beta)$ is the density function of the Beta distribution. It is possible to easily split $f_J(x)$ into its three components, as long as only one of the terms is non-zero for each x. This is true for our model, as long as both α and β are greater than 1 (causing the Beta density to be pinned to zero when $x = 0$ or 1). This property is apparent in our model, since we have allocated the scores of 0 and 1 to the Dirac delta functions, leaving a density of 0 for the Beta density function, so we will assume that both $\alpha > 1$ and $\beta > 1$.

Jaccard similarity distribution

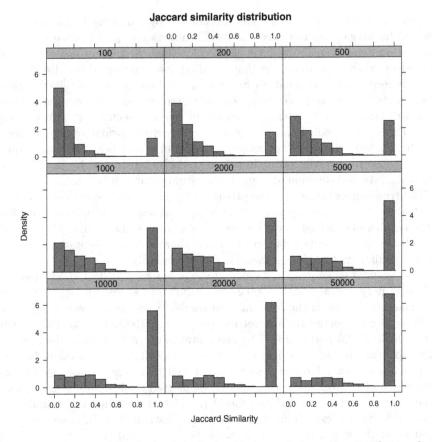

Fig. 1. Distribution of the Jaccard similarity. Each panel shows the distribution for a given dictionary construction size, where the number of tweets used to construct each dictionary ranges from 100 to 50000 tweets.

This model separation allows us to easily estimate the model parameters from the data:

- $\hat{p}_0$ is the proportion of 0 scores (estimate of p_0).
- $\hat{p}_1$ is the proportion of 1 scores (estimate of p_1).
- $\hat{p}_b = 1 - \hat{p}_0 - \hat{p}_1$ is the proportion of scores between 0 and 1 (estimate of p_b).
- $\hat{\alpha} = \bar{x}_b \left(\bar{x}_b(1 - \bar{x}_b)/s_b^2 - 1 \right)$ (estimate of α).
- $\hat{\beta} = (1 - \bar{x}_b) \left(\bar{x}_b(1 - \bar{x}_b)/s_b^2 - 1 \right)$ (estimate of β).

where $\bar{x}_b$ and s_b are the sample mean and standard deviation of the mid-band (excluding scores of 0 and 1) Jaccard scores. The Beta density parameters α and β are estimated using the method of moments, which is a good approximation to the true α and β when $\bar{x}_b(1 - \bar{x}_b) > s_b^2$ (ensuring both α and β are positive). Fitting the model to the set of scores for each dictionary construction size provides us with the parameter estimates and standard errors of $\hat{p}_1$, $\hat{p}_0$, $\hat{p}_b$, $\hat{\alpha}$ and

Jaccard similarity mid–band Q–Q plot

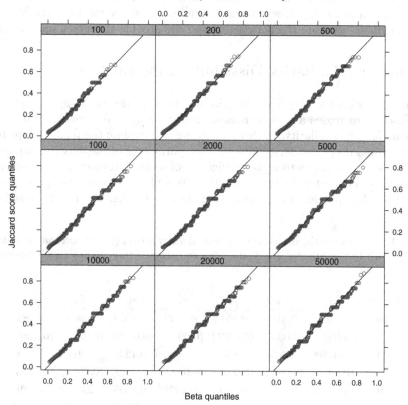

Fig. 2. Q-Q plots of the Jaccard similarity mid-band (excludes all scores of value 0 and 1) and fitted Beta distribution. Each panel shows the Q-Q plot for a given dictionary construction size, where the number of tweets used to construct each dictionary ranges from 100 to 50000 tweets. The closeness of the points to the line show that both distributions are very similar.

$\hat{\beta}$ in Table 1, where the standard errors were computed from a bootstrap sample of size 1000.

Table 1 contains the column "Dict Size", showing the number of randomly sampled tweets used to construct the dictionary. As Dict Size increases, we find that $\hat{p}_1$ increases and $\hat{p}_0$ decreases, displaying that as more tweets are used to construct the dictionary, the proportion of correct hashtag segmentations increases, and the proportion of incorrect segmentations (containing no correct words) decreases. Examining $\hat{p}_b$, we also find that the proportion of partially correct segmentations decreases. The statistics $\bar{x}_b$ and s_b show the mean and standard deviation of the mid-band Jaccard scores (the set of scores with 0 and 1 removed). We find that the mean increases, while the standard deviation increases, then tapers off to stay around 0.15 as Dict Size increases.

We also find that $\hat{\alpha}$ and $\hat{\beta}$ are greater than 1 for all dictionary sizes and don't seem to be approaching 1, which was a required property for using the Method of Moments to estimate the parameters and the provide the model partitioning.

4   Jaccard Similarity Distribution Parameters

Table 1 shows that as the dictionary size increases, $\hat{p}_0$ decreases, $\hat{p}_1$ increases, $\hat{p}_b$ decreases, $\hat{\alpha}$ increases and $\hat{\beta}$ decreases, all as we expect to provide an increase in mean Jaccard similarity. In this section, we will explore the relationships further to gain a deeper understanding of each parameter of the Jaccard similarity distribution, allowing us to make predictions of what dictionary size is needed to obtain a given expected Jaccard similarity. We will first examine the proportions p_0 and p_1, then proceed to examine the Beta distribution parameters α and β.

Table 1. Statistics of the fitted Jaccard score density model (Eq. 1) and their standard error (shown in parentheses) for dictionaries constructed from 100 to 50000 randomly sampled tweets.

Dict size	$\hat{p}_1$	$\hat{p}_0$	$\hat{p}_b$	$\bar{x}_b$	s_b	$\hat{\alpha}$	$\hat{\beta}$
100	0.1296	0.3160	0.5544	0.1713	0.1068	1.960	9.483
	(0.0047)	(0.0069)	(0.0071)	(0.0020)	(0.0019)	(0.0578)	(0.3260)
200	0.1706	0.2428	0.5866	0.2030	0.1245	1.917	7.523
	(0.0053)	(0.0060)	(0.0070)	(0.0023)	(0.0019)	(0.0502)	(0.2273)
500	0.2522	0.1886	0.5592	0.2421	0.1395	2.038	6.379
	(0.0061)	(0.0055)	(0.0071)	(0.0027)	(0.0020)	(0.0532)	(0.1849)
1000	0.3188	0.1476	0.5336	0.2693	0.1459	2.221	6.024
	(0.0065)	(0.0049)	(0.0070)	(0.0029)	(0.0018)	(0.0537)	(0.1606)
2000	0.3860	0.1274	0.4866	0.2945	0.1509	2.390	5.725
	(0.0068)	(0.0048)	(0.0070)	(0.0031)	(0.0019)	(0.0622)	(0.1563)
5000	0.5084	0.0834	0.4082	0.3206	0.1526	2.678	5.673
	(0.0072)	(0.0040)	(0.0072)	(0.0034)	(0.0020)	(0.0767)	(0.1653)
10000	0.5564	0.0784	0.3652	0.3349	0.1532	2.841	5.642
	(0.0070)	(0.0039)	(0.0068)	(0.0037)	(0.0023)	(0.0921)	(0.1902)
20000	0.6120	0.0686	0.3194	0.3569	0.1533	3.126	5.633
	(0.0069)	(0.0034)	(0.0067)	(0.0038)	(0.0023)	(0.1094)	(0.1879)
50000	0.6700	0.0616	0.2684	0.3718	0.1547	3.257	5.502
	(0.0066)	(0.0033)	(0.0063)	(0.0043)	(0.0026)	(0.1215)	(0.2063)
100000	0.6996	0.0558	0.2446	0.3811	0.1582	3.206	5.207
	(0.0061)	(0.0032)	(0.0057)	(0.0045)	(0.0026)	(0.1191)	(0.1931)
200000	0.7212	0.0470	0.2318	0.3869	0.1548	3.440	5.450
	(0.0062)	(0.0030)	(0.0058)	(0.0045)	(0.0027)	(0.1365)	(0.2179)

Both p_1 and p_0 are bound between 0 and 1, where we would expect p_1 to approach 0 and p_0 to approach 1 and Dict Size d decreases. We would also expect p_1 to approach an upper limit that may be less than 1 (meaning that the segmenter never achieves perfection for any dictionary size), and p_0 to approach a lower limit greater than 0 (meaning that there will always be hashtags that cannot be segmented for any dictionary size) as d increases. Therefore, we would expect p_1 and p_0 to be well approximated by a type of sigmoid function (to limit p_1 and p_0 to the $[0, 1]$ domain) of the log scale of d (to extend d to the real domain). The form of the functions are:

$$p_1 = \frac{\theta_1}{1 + \exp(-\theta_2 \log(d) + \theta_3)} = \frac{\theta_1}{1 + e^{\theta_3} d^{-\theta_2}} \tag{2}$$

$$p_0 = 1 - \frac{\theta_4}{1 + \exp(-\theta_5 \log(d) + \theta_6)} = 1 - \frac{\theta_4}{1 + e^{\theta_6} d^{-\theta_5}} \tag{3}$$

The plot of the change in p_1 and p_0 with respect to d is shown in Fig. 3 with the fitted models, weighted by standard error, providing parameters $\theta_1 = 0.771$, $\theta_2 = 0.569$, $\theta_3 = 4.267$, $\theta_4 = 0.958$, $\theta_5 = 0.498$ and $\theta_6 = 1.359$. The error bars show the 95 % confidence interval of the parameter. We can see that the models for both p_1 and p_0 provide an excellent fit to the simulation.

When examining the plot of α vs. d, it was difficult to determine a direct relationship, therefore we will examine the relationships between μ_b (mean of the mid-band) and β, since α can be calculated from these. The mean μ_b has the same conditions as the proportion p_1, therefore we use the same model form. The parameter β seems to be decreasing, but plateaus near 5 as d increases. After examining the data, we arrived at the functions:

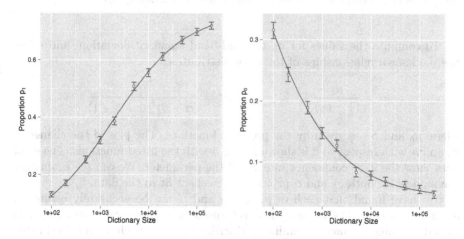

Fig. 3. The point estimate (shown as circles) and 95 % confidence interval (given by error bars) of the proportions p_1 (left plot) and p_0 (right plot), along with the fitted sigmoid functions given in Eq. 2 as the curve, for segmentation dictionaries constructed from various tweet sample sizes (given as Dictionary Size).

$$\mu_b = \frac{\theta_7}{1 + e^{\theta_9 d - \theta_8}} \qquad\qquad \beta = \frac{\theta_{10}}{d} + \theta_{11} \qquad (4)$$

The plot of the change in μ_b and β with respect to d is shown in Fig. 4 with the fitted models, weighted by their standard error, providing parameters $\theta_7 = 0.412$, $\theta_8 = 0.406$, $\theta_9 = 2.181$, $\theta_{10} = 399.21$ and $\theta_{11} = 5.512$. The error bars show the 95 % confidence interval of the parameter. We can see that the models for both μ_b and β provide an excellent fit to the data.

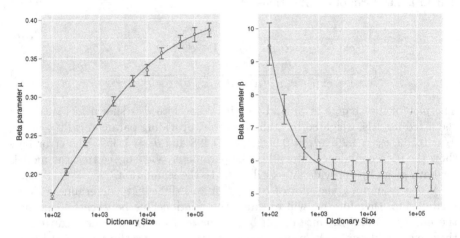

Fig. 4. The point estimate (shown as circles) and 95 % confidence interval (given by error bars) of the mean μ_b (left plot) and parameter β (right plot) of the mid-band data, along with the fitted functions given in Eq. 4 as the curve, for segmentation dictionaries constructed from various tweet sample sizes (given as Dictionary Size).

To compute the values for σ_b the mid-band standard deviation, and α, we use the known relationships of the Beta distribution:

$$\alpha = \frac{\mu_b}{(1 - \mu_b)}\beta \qquad\qquad \sigma^2 = \frac{\alpha\beta}{(\alpha + \beta)^2(\alpha + \beta + 1)} \qquad (5)$$

where μ_b and β are given by the previous functions. The plot of the change in σ_b and α with respect to d is shown in Fig. 5 with the fitted functions. The error bars show the 95 % confidence interval of the parameter. We can see again that the models for both σ_b and α provide an excellent fit to the data.

We have found that each of p_1, p_0, α and β can be accurately estimated as functions of d (the number of tweets used to generate the dictionary), in turn describing the Jaccard similarity distribution, once we have obtained fitted values for the 11 parameters.

Note that a limitation of our analysis comes from taking each tweet sample from the pool of 251171 tweets. This sample size is sufficient for generating hastags, and for dictionaries using a small number of tweets. But when randomly

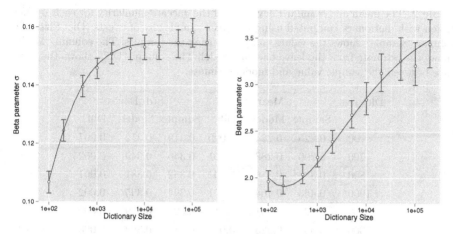

Fig. 5. The point estimate (shown as circles) and 95 % confidence interval (given by error bars) of the standard deviation σ_b (left plot) and parameter α (right plot) of the mid-band data, along with the computed functions given in Eq. 5 as the curve, for segmentation dictionaries constructed from various tweet sample sizes (given as Dictionary Size).

sampling 200000 tweets from the pool to compute a dictionary, the sample size is of the same order as the population being sampled from, therefore the variance introduced by randomly sampling for large dictionaries is reduced when compared to a pure random sample from Twitter. We believe that this will only effect the values of θ_1, θ_4 and θ_7, but further analysis is required to test this effect.

5   Accuracy of the Model

The previous section showed that we can model the Jaccard accuracy as a function of d (the number of tweets used to construct the segmentation dictionary). In this section, we will examine how well our model can estimate the four distribution parameters when not included in the parameter fitting process.

Our experiment consisted of a leave one out process, where we fitted the eleven parameters using all data, excluding the data associated to the desired d. We then used our model to compute the 11 distribution parameters and then predict the mean and standard deviation of the Zero-One inflated Beta distribution for the desired d (using the equations in Appendix A). This process was repeated for all d. Table 2 contains the mean and standard deviation of the Jaccard scores predicted by the model and computed from the sample data. We find that the predictions from the model are very accurate, showing how closely the sample data follows the model. The results show that this model can be used to compute the expected Jaccard similarity, and hence compute the number of tweets required to construct a dictionary in order to obtain a given expected Jaccard similarity. Using knowledge of the distribution, other statistics can also

Table 2. The mean and standard deviation of the Jaccard similarity for each segmentation with dictionary computed using the given sample size of tweets (Dict Size). The column "Sample" shows the values computed using the samples. The column "Model" shows the estimate from the leave one out model. The column "Diff" shows the difference between the sample value and model estimate.

Dict size	Mean			Std. Dev.		
	Sample	Model	Diff	Sample	Model	Diff
100	0.225	0.224	0.000	0.319	0.312	0.007
200	0.290	0.288	0.001	0.346	0.349	−0.002
500	0.388	0.386	0.001	0.382	0.381	0.001
1000	0.463	0.464	−0.001	0.394	0.395	−0.002
2000	0.529	0.541	−0.012	0.399	0.399	−0.000
5000	0.639	0.623	0.016	0.389	0.392	−0.003
10000	0.679	0.684	−0.005	0.381	0.379	0.002
20000	0.726	0.727	−0.001	0.365	0.365	−0.000
50000	0.770	0.771	−0.001	0.348	0.346	0.002
100000	0.793	0.794	−0.001	0.336	0.334	0.002
200000	0.811	0.810	0.001	0.322	0.325	−0.003

be computed (such as quantiles for confidence intervals). To assist in computing the required statistics, or for model validation, the source code is available at the authors' Web site[2].

6   Conclusion

Hashtag segmentation allows us to obtain a more in depth analysis of Twitter data. Automatic hashtag segmentation requires a domain specific dictionary and therefore should be computed from a random sample of tweets.

In this article we examined the effect of the tweet sample size, used to create the segmentation dictionary, on the accuracy of automatic hashtag segmentation. We found not only that the accuracy distribution is a Zero-One inflated Beta distribution containing four parameters, but we also found that each of the four parameters can be modelled on the number of tweets used to construct the dictionary.

This model can be used to predict the distribution and statistics of the accuracy distribution for automatic hashtag segmentation, when using a given number of tweets to construct the dictionary. The model can also be used to gain deeper understanding into the relationships between the model parameters.

---

[2] http://www.scem.uws.edu.au/~lapark/segmentHash.

A   Derivation of Model Mean and Variance

The mean and variance of our Jaccard similarity (Zero-One inflated Beta) density function $f_J(x)$ from Eq. 1 is derived from the expected value of J and J^2.

$$\mathbb{E}[J] = \int x[p_0\delta(x) + p_b f_B(x; \alpha, \beta) + p_1\delta(1-x)]dx$$

$$= p_0 \int x\delta(x)dx + p_b \int x f_B(x; \alpha, \beta)dx + p_1 \int x\delta(1-x)dx$$

$$= p_b\mu_b + p_1$$

$$= \frac{p_b\alpha}{\alpha + \beta} + p_1$$

$$\mathbb{E}[J^2] = \int x^2[p_0\delta(x) + p_b f_B(x; \alpha, \beta) + p_1\delta(1-x)]dx$$

$$= p_0 \int x^2\delta(x)dx + p_b \int x^2 f_B(x; \alpha, \beta)dx + p_1 \int x^2\delta(1-x)]dx$$

$$= \frac{p_b\alpha\beta}{(\alpha+\beta)^2(\alpha+\beta+1)} + p_b\mu_b^2 + p_1$$

$$\text{Var}(J) = \mathbb{E}[J^2] - \mathbb{E}[J]^2$$

$$= \frac{p_b\alpha\beta}{(\alpha+\beta)^2(\alpha+\beta+1)} + p_b\mu_b^2 + p_1 - (p_b\mu_b + p_1)^2$$

$$= \frac{p_b\alpha\beta}{(\alpha+\beta)^2(\alpha+\beta+1)} + p_b\mu_b^2 + p_1 - p_b^2\mu_b^2 - p_1^2 - 2p_b p_1\mu_b$$

$$= \frac{p_b\alpha\beta}{(\alpha+\beta)^2(\alpha+\beta+1)} + (p_b - p_b^2)\mu_b^2 + p_1 - p_1^2 - 2p_b p_1\mu_b$$

$$= \frac{p_b\alpha\beta}{(\alpha+\beta)^2(\alpha+\beta+1)} + \frac{(p_b - p_b^2)\alpha^2}{(\alpha+\beta)^2} - \frac{2p_b p_1\alpha}{\alpha+\beta} + p_1 - p_1^2.$$

References

1. Bansal, P., Bansal, R., Varma, V.: Towards deep semantic analysis of hashtags. In: Hanbury, A., Kazai, G., Rauber, A., Fuhr, N. (eds.) ECIR 2015. LNCS, vol. 9022, pp. 453–464. Springer, Heidelberg (2015)
2. Berardi, G., Esuli, A., Marcheggiani, D., Sebastiani, F.: Isti@ trec microblog track 2011: exploring the use of hashtag segmentation and text quality ranking. In: TREC (2011)
3. Devine, B.J.: A method for segmenting topical Twitter hashtags. Ph.D. thesis, San Diego State University (2014)
4. Gouws, S., Metzler, D., Cai, C., Hovy, E.: Contextual bearing on linguistic variation in social media. In: Proceedings of the Workshop on Languages in Social Media, pp. 20–29. Association for Computational Linguistics (2011)

5. Ma, Z., Sun, A., Cong, G.: On predicting the popularity of newly emerging hashtags in twitter. J. Am. Soc. Inf. Sci. Technol. **64**(7), 1399–1410 (2013)

6. Milajevs, D., Bouma, G.: Real time discussion retrieval from twitter. In: Proceedings of the 22nd International Conference on World Wide Web Companion, pp. 795–800. International World Wide Web Conferences Steering Committee (2013)

7. Norvig, P.: Natural language corpus data. In: Segaran, T., Hammerbacher, J. (eds.) Beautiful Data: The Stories Behind Elegant Data Solutions, Chap. 14. O'Reilly Media Inc., Sebastopol (2009)

8. Park, L.A.F., Simoff, S.: Power walk: revisiting the random surfer. In: Proceedings of the 18th Australasian Document Computing Symposium, ADCS 2013, pp. 50–57. ACM, New York (2013)

9. Park, L.A.F., Stone, G.: The effect of assessor coverage and assessor accuracy on rank aggregation precision. In: Proceedings of the 20th Australasian Document Computing Symposium, ADCS 2015, pp. 6:1–6:4. ACM, New York (2015)

10. Park, L.A., Simoff, S.: Second order probabilistic models for within-document novelty detection in academic articles. In: Proceedings of the 37th International ACM SIGIR Conference on Research & Development in Information Retrieval, SIGIR 2014, pp. 1103–1106. ACM, New York (2014)

11. Qadir, A., Riloff, E.: Learning emotion indicators from tweets: Hashtags, hashtag patterns, and phrases. In: Proceedings of the Conference on Empirical Methods in Natural Language Processing (EMNLP), pp. 1203–1209. Association for Computational Linguistics (2014)

12. Tsur, O., Rappoport, A.: What's in a hashtag?: content based prediction of the spread of ideas in microblogging communities. In: Proceedings of the Fifth ACM International Conference on Web Search and Data Mining, pp. 643–652. ACM (2012)

13. Van Hee, C., Van de Kauter, M., De Clercq, O., Lefever, E., Hoste, V.: Lt3: Sentiment classification in user-generated content using a rich feature set. In: SemEval 2014, pp. 406–410 (2014)

Social Identity Link Across Incomplete Social Information Sources Using Anchor Link Expansion

Yuxiang Zhang[1(✉)], Lulu Wang[2], Xiaoli Li[3], and Chunjing Xiao[1]

[1] Civil Aviation University of China, Tianjin, China
{yxzhang, cjxiao}@cauc.edu.cn
[2] Beijing Jiaotong University, Beijing, China
llwang_14@bjtu.edu.cn
[3] Institute for Infocomm Research, A*STAR, Singapore, Singapore
xlli@i2r.a-star.edu.sg

Abstract. Social link identification SIL, that is to identify accounts across different online social networks that belong to the same user, is an important task in social network applications. Most existing methods to solve this problem directly applied machine-learning classifiers on features extracted from user's rich information. In practice, however, only some limited user information can be obtained because of privacy concerns. In addition, we observe the existing methods cannot handle huge amount of potential account pairs from different OSNs. In this paper, we propose an effective SIL method to address the above two challenges by expanding known anchor links (seed account pairs belonging to the same person). In particular, we leverage potentially useful information possessed by the existing anchor link, and then develop a local expansion model to identify new social links, which are taken as a generated anchor link to be used for iteratively identifying additional new social link. We evaluate our method on two most popular Chinese social networks. Experimental results show our proposed method achieves much better performance in terms of both the number of correct account pairs and efficiency.

Keywords: Social networks · Social Identity Link · Hometown inference

1 Introduction

Online social networks (OSNs), such as Twitter, Facebook, Sina Weibo, Renren and Foursquare, have become more and more popular in recent years. Each social network can be represented as an individual graph and focuses on a specific application. Oftentimes, people are getting involved in numeric social networks concurrently. For example, we can access the latest news from Twitter and Sina Weibo, post our photos using Facebook and Renren, and share interesting places (or locations) with our friends using Foursquare. Thus, it comes as no surprise that many users often have multiple separate accounts in different OSNs, although there are no direct correspondences or connections among these multiple accounts belonging to the same users from different networks.

© Springer International Publishing Switzerland 2016
J. Bailey et al. (Eds.): PAKDD 2016, Part I, LNAI 9651, pp. 395–408, 2016.
DOI: 10.1007/978-3-319-31753-3_32

Discovering the correspondences between accounts of the same user, i.e. social identity link (SIL) problem, by integrating information from multiple OSNs is a crucial prerequisite for many practical Web based applications, such as detecting more accurate community structures [1], finding rising stars in social networks [2], and providing better customer support and personalized services matching the user preferences. For example, if we know a user's Twitter account, then its social connections and location data in Twitter can be used to better recommend the taste to this user in the Foursquare. However, existing research (such as [3–8]) have showed that it is very challenging to identify user accounts of the same natural person across different social media platforms. The main reason is that users and social platform operators take extremely strict measures to avoid divulging user personal information.

Previous studies (such as [7, 9, 10]) assume that they can collect rich user information/attributes about *user profiles, user generated content, behaviors and friend networks*. After collecting all the rich attributes for each user from different social networks, existing methods mainly employ supervised learning techniques [3, 7, 8, 11–13] (with an exception which uses unsupervised learning [14]) to build binary classification models for SIL prediction. However, it is very difficult, if not impossible, to obtain user's private information in many real-world applications. As such, existing research will suffer when only incomplete information is available.

The second facing challenge is that current classification methods are not feasible to handle huge amount of potential account pairs from different OSNs. Particularly, the computational cost for identifying pair-wise accounts is $N_1 * N_2$ (N_1 and N_2 are the number of accounts in source and target networks, respectively). We can imagine how many account pairs could be generated given each OSN could have more than 1 billion users (e.g. Facebook). Clearly, it will be extremely time consuming, if not impossible, to perform the intensive classification task.

In this paper, we employ open APIs, provided by the social platform operators, to only collect the *publically available* attributes, including 6 user profile attributes, such as nickname, gender, birthday, university name, university entry year and location, and friend network attribute. Thus, we are handling the SIL problem in a difficult but *practical* scenario with *incomplete* information sources. In addition, we also observe that many profile attributes have missing or false values, making this research even more challenging. Additionally, to tackle the second challenge, contrast to existing standard classification methods, we leverage anchor link information and propose a local search strategy to iteratively identify the new social links. Our proposed approach largely reduces the search space and is thus more feasible than existing methods for handling those real-world large scale OSNs.

2 Related Works

SIL problem across different social platforms has been studied in recent few years. User link was formalized as connecting identity problem across communities in [3–6] in the early stage. Subsequently, various methods were proposed.

The performance of existing methods largely relies on the extracted features, from user profiles, user generated content, behaviors and friend networks. Some research papers [15–19] heavily focus on username parsing to link multiple online identities of a user, based on the assumption that same users will have the similar names from different social platforms. Paper [20] studies three features extracted from the content created by a user, i.e. timestamp of posts, geo-location attached to post and writing styles. It finds that the geo-location of posts is the most powerful features to identify social links. Another research explores the social meta path concept (which is a means to capture connection information in the social networks) to generate useful compound features from friendship, location, timestamp of post and post content [21]. Some works have shown that other information about users, like their group memberships [22] and tagging behavior [23], can also be used to uniquely identify users. More recent papers [7, 10, 14, 24] have integrated as many features as possible to identify social links across different social networks, since researchers believe that less features are not sufficient enough to achieve good performance.

Unfortunately, in practice we can only obtain the limited information, leading to limited or incomplete features and thus much worse results. In addition, the existing methods are also inefficient and the computational costs are prohibitively high, as they need to classify large amount of all the possible account pairs from different networks. In this paper, we leverage those potentially useful information possessed by the anchor link to overcome the above two weaknesses from the existing methods.

3   Overall Algorithm

Denote P as the set of all natural *persons* in real life. For a social network G, represent $V(G)$ as the set of all *accounts,* each belonging to a distinct user. An injective function $\phi_G: V(G) \to P$ maps each account in $V(G)$ to a natural person in P.

Social Identity Link, SIL. Given an account I_i^S from a source network G^S (i.e. $I_i^S \in V$ (G^S)), social identity link problem is to find a corresponding account I_j^T from a target network G^T(i.e. $I_j^T \in V(G^T)$), such that $\phi_S(I_i^S) = \phi_T(I_j^T)$. This definition is very strict. In fact, formula should be associated with a certain probability or confidence score.

Firstly, we need to collect a seed anchor link set *ALS*, consisting of the account pairs where one account from source anchor set AR^S in G^S and the other account is from target anchor set AR^T in G^T: $ALS = \{(ar_i^S, ar_i^T) | (ar_i^S, ar_i^T)$ is an anchor link provided, $ar_i^S \in AR^S$, $AR^S \subseteq V(G^S)$, $ar_i^T \in AR^T$, $AR^T \subseteq V(G^T)\}$. *ALS* can be obtained by either questionnaires or rule-based filtering methods.

Secondly, starting from an anchor link from *ALS*, our proposed the anchor link local expansion algorithm *iteratively* searches the new putative social identity links until they cannot be found. Figure 1 shows the key idea of our proposed method. Given an anchor link (ar_i^S, ar_i^T) (ar_i^S and ar_i^T are the anchor nodes from source or target network respectively), we first visit 3^S that is any one of neighbors of the account ar_i^S, and then we try to find a best matching account from G^T. If $1'^T$ is found, the new social link, called *generated* anchor link, $(3^S, 1'^T)$, can be leveraged to further identify other social links. Thus, a set of social links is identified in the order of the following sequence $(ar_i^S, ar_i^T) \to (3^S, 1'^T) \to (5^S, 4'^T) \to (4^S, 5'^T)$.

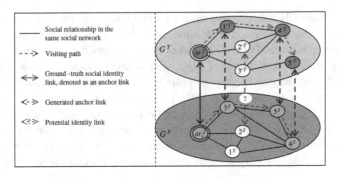

Fig. 1. Key idea overview.

This detailed algorithm is shown in Algorithm 1. In step 1, we initialized a queue Q as an empty set, which will be used to store the given anchor nodes in AR^S or newly generated anchor nodes from source network. We also initialize an output set O. Steps 2 and 3 mark all the nodes in the source network as "unvisited". From steps 4 to 12, we will go through all the anchor nodes ar_i in AR^S and generated anchor nodes. Particularly, for each unvisited anchor node in the source network (step 8), we first find all its neighbors (step 9). Then, for each of the neighbors, function $FindSIL()$ in step 10 is used to find the best match account in the target network – the detailed process will be introduced in next section. Step 11 will add the newly generated anchor nodes from the source network into the queue Q. Finally, the results are returned in the step 13.

Algorithm 1. Overall Algorithm

Input: *source network G^S; target network G^T; anchor link set ALS*

Output: *matched account pairs in O*

1:    *Initialize a queue $Q=\varnothing$; $O=\varnothing$ //Initialize a queue Q and set O;*
2:    **For** *(int j=0; j<|V(G^S)|; j++)   //For all the accounts in G^S*
3:        *Mark[v_j]=unvisited;*
4:    **For** *(int i=0;i<|AR^S|;i++)   //For all the anchors in AR^S*
5:        *Q.enqueue(ar_i); // $ar_i \in AR^S$*
6:        **While** *(Q.empty())*
7:            *u=Q.pop();*
8:            **If** *(Mark[u]==unvisited)*
9:                **While** *(k<|N(u)|)   // $N_k(u)$ is k^{th} neighbor of u.*
10:                    **If** *(l=**FindSIL**($N_k(u)$))   //FindSIL() is used to find the best match account of $N_k(u)$ in target network G^T.*
11:                        *O = O$\cup${($N_k(u)$, l)}*
12:                        *Q.enqueue($N_k(u)$)*
13:    Return O as the set of matched account pairs

The worst-case time complexity of Algorithm 1 is $O(|V(G^S)||E(G^S)|)$ time, where $|V(G^S)|$ and $|E(G^S)|$ are the number of accounts and the number of edges between users in source network G^S respectively.

The intra-network connections associated with the anchor link are very useful. In particular, given an anchor link (ar_i^S, ar_i^T) and ar_i^S's neighbor I_i^S, we believe the best matching account I_j^T is likely to locate in a small range of the anchor account ar_i^T. This is because, the friend connections in one network have higher chance to be re-occurred in another network either directly (re-occurrence connections) or indirectly (friends' friends). As such, our proposed technique can largely reduce the search space based on this novel idea.

4 Optimal Search Range

Our goal is to solve SIL problem with minimum expected computational cost. We define search range and use it to control computational cost in the target network G^T.

We first define the shortest distance between two nodes u and v, i.e. $d(u, v)$, as the number of edges in the shortest paths. Let $P(u, v)$ be the set of all paths that start from u and end at v. Note $d(u, v)$ is ∞ if v is not reachable from u:

$$d(u, v) = \begin{cases} \arg\ \min_{p \in P(u,v)} |p| & \text{if}\quad P(u, v) \neq \varnothing \\ \infty\ \text{otherwise.} \end{cases} \tag{1}$$

Search Range. Given an anchor link (ar_i^S, ar_i^T) where ar_i^S and ar_i^T are from network G^S and G^T respectively, the search range $R_d(ar_i^T)$ around ar_i^T in G^T is defined by $R_{d \leq n}(ar_i^T) = \{I^T \in G^T \mid d(ar_i^T, I^T) \leq n\}$.

Here, n is a non-zero natural number. In the best case, d is equal to 1; that is to say, $R_{d=1}(ar_i^T)$ represents a set of *direct* neighbors of ar_i^T. At worst case scenario, d is less than or equal to infinity; that is to say, $R_{d \leq inf}(ar_i^T)$ represents a set of all accounts in G^T. Our strategy for selecting the search range is to gradually grow the value of the d from 1 to infinity according to specific requests. This strategy can largely reduce the search space. The effect of the parameter d on the system performance of user identification is discussed in the Subsect. 6.2 in detail.

5 Identity Matching

We introduce how to select a best match account from the candidate set. Particularly, we first define some distinguishing features in nickname, hometown and friend network. Learning models are subsequently used to find the best match accounts.

5.1 Features Definition

(1) **Nickname Similarity:** Features derived from the nickname have been widely used to identify the social links across different social platforms. There are even a few studies, such as [8, 11], which only use nickname features for identification.

However, in many real datasets, there are too few consistent names (namesakes) across different social platforms. In our dataset, 98 % of ground-truth linked account pairs (which is created manually for the same users from different social platforms.) possess different nicknames. Nonetheless, we find that many pairs of different nicknames belonging to same users are somewhat related. We have performed comparison of these ground-truth social links and summarized the most frequent relationships between two nickname pairs as follows: (1) there exists a common substring, such as (张金鹏, zjp金鹏042); (2) a common substring occurs many times in one nickname, such as (辛倩文, 小辛辛辛辛辛辛); (3) there are no differences if Chinese characters are converted into alphabets, such as (范一真, 范熠禛) (both are Fan Yizheng). In order to tackle the case (3), we convert Chinese characters into their corresponding alphabets when there are mismatches between two nicknames written in Chinese characters.

Before calculating nickname features, we introduce some basic notations and definitions. We denote the nickname of an account by $Ni(.)$ for two accounts I_i^S and I_j^T, and $p = |Ni(I_i^S) \cap Ni(I_j^T)|$ is thus the size of *common/overlapping characters* between $Ni(I_i^S)$ and $Ni(I_j^T)$. A function $lcs(.)$ is to compute *the longest continuous common substring* between two nicknames, which is implemented by the generalized suffix tree [25]. We define $q = |lcs(Ni(I_i^S), Ni(I_j^T))|$ as the length of the longest continuous common substring between $Ni(I_i^S)$ and $Ni(I_j^T)$. We use r and s to represent the *frequency* which $lcs(Ni(I_i^S), Ni(I_j^T))$ occurs in $Ni(I_i^S)$ and $Ni(I_j^T)$, and in *all* nicknames respectively. Finally, function $len(.)$ and $max(.)$ are used to compute the *length* of a nickname, and the *maximum* nickname length.

Finally, the nickname similarity $NiS(I_i^S, I_j^T)$ between $Ni(I_i^S)$ and $Ni(I_j^T)$ is defined as follows: $NiS(I_i^S, I_j^T) = (CoC(Ni(I_i^S), Ni(I_j^T)) + LoS(Ni(I_i^S), Ni(I_j^T)) + ReS(Ni(I_i^S), Ni(I_j^T)) + SpS(Ni(I_i^S), Ni(I_j^T)))/4$, where $CoC(Ni(I_i^S), Ni(I_j^T)) = p/max(len(Ni(I_i^S)), len(Ni(I_j^T)))$ is used to reflect the contribution from the *common characters*. $LoS(Ni(I_i^S), Ni(I_j^T)) = q/max(len(Ni(I_i^S)), len(Ni(I_j^T)))$ is used to reflect the contribution of the *longest common substring*. $ReS(Ni(I_i^S), Ni(I_j^T)) = r/(len(Ni(I_i^S)) + len(Ni(I_j^T)))$, on the other hand, is used to reflect the contribution of the *repetitions of the longest common substring*. $SpS(Ni(I_i^S), Ni(I_j^T)) = 1/\sqrt{s}$, is used to reflect the contribution of *rarity of the longest common substring* in all nicknames. Typically, those account pairs with less rare longest common substring will get higher similarity than those frequent ones, as they are more helpful for identification purpose.

(2) **Hometown Similarity:** We observe that different social networks could have different types of location information. Sina Weibo and Twitter only possess the current location, while Renren and Facebook possess many different types of locations, such as hometown, current city and workplace. Because hometown in Renren may be different from current location in Sina Weibo for same users (we could move to other places for education or to make a living), we are facing a very challenging task, i.e. how to compute hometown/location similarity based on different types of location information.

Our two interesting observations help us to tackle this challenging problem. One is that the hometown is still the same as the current location for some account pairs. For example, in our ground-truth linked account pairs, there are 46 % of account pairs, of which location in Sina Weibo is the same as hometown in Renren (they are kind of permanent dwellers in their hometown). The other is that, for some accounts, some of their friends have been re-occurred in multiple networks, like a mirror. According to these two phenomena, we propose the following solution to this task.

Given an account I_i^S in G^S and an account I_j^T in G^T, we denote hometown of I_i^S and I_j^T by $Hi(I_i^S)$ and $Ho(I_j^T)$ respectively. The probability score $HoS(I_i^S, I_j^T)$ that Hi (I_i^S) is equal to $Ho(I_j^T)$ can be computed as follows: $HoS(I_i^S, I_j^T) = P(Hi(I_i^S) = Ho$ $(I_j^T))$. As mentioned above, Sina Weibo as the source network only has the current location. Although Renren has the hometown, some accounts do not fill up the hometown and some accounts may fill up the false hometown, to make our problem even more difficult. In other words, we are not sure whether the value of a hometown is true or not. As such, we cannot match the value of $Hi(I_i^S)$ and value of $Ho(I_j^T)$ directly. In this paper, we propose a novel hometown inference model by leveraging the location information of neighbors.

For an account I_i^S in G^S, we denote the set of neighbors of I_i^S by $N(I_i^S)$, and denote the set of current location of accounts in $N(I_i^S)$ by $CL(N(I_i^S))$. Likewise, we use $N(I_j^T)$ to represent the set of the neighbors of I_j^T in G^T. The set of hometown of accounts in $N(I_j^T)$ is denoted by $HT(N(I_j^T))$. The intersection of sets $CL(N(I_i^S))$ and $HT(N(I_j^T))$ is denoted by $CH(I_i^S, I_j^T) = CL(N(I_j^T)) \cap HT(N(I_i^S))$, and let the size of CH (I_i^S, I_j^T) be K.

In addition, the value of hometown of I_j^T is denoted as l_h, which may be either empty or filled up by user. If l_h has been filled up, we should then take into account the contribution of l_h to the hometown similarity even though we can not make sure whether l_h is true or not. For this reason, we define a new hometown set $HT_I(N(I_j^T)) = \{HT(N(I_j^T)), l_h\}$. Let $CH_I(I_i^S, I_j^T)$ be the intersection of $CL(N(I_i^S))$ and $HT_I(N(I_j^T))$.

We consider six different cases for the location and hometown mapping, illustrated in Fig. 2. These cases represent different relationships among $CL(N$ $(I_i^S))$, $HT(N(I_j^T))$ and l_h. The probability score $HoS(I_i^S, I_j^T)$ will be computed according to each specific case. The weight of each edge is the frequency of the hometown/current.

Case (1) shown in Fig. 2(a). The intersection $CH_I(I_i^S, I_j^T)$ is empty, i.e., CH_I $(I_i^S, I_j^T) = \emptyset$. We can derive $HoS(I_i^S, I_j^T) = 0$.

Case (2) shown in Fig. 2(b). The value of hometown of I_j^T is empty and the $CH(I_i^S, I_j^T)$ is not empty, i.e., $l_h = \emptyset \wedge CH(I_i^S, I_j^T) \neq \emptyset$. The $HoS(I_i^S, I_j^T)$ is computed by the formula: $HoS(I_i^S, I_j^T) = \sum_{k=1}^{K} P(Hi(I_i^S) = l_k)P(Ho(I_j^T) = l_k)$, where $l_k \in CH$ (I_i^S, I_j^T). For example, we can compute $HoS(I_i^S, I_j^T) = 0.1 \times 0.3 + 0.6 \times 0.4 = 0.27$.

Case (3)–(4) shown in Fig. 2(c)–(d). The value of hometown of I_j^T is not empty, the $CH_I(I_i^S, I_j^T)$ is not empty, and l_h does not appear in $CH_I(I_i^S, I_j^T)$, i.e., $l_h \notin \emptyset \wedge CH_I$ $(I_i^S, I_j^T) \neq \emptyset \wedge l_h \notin CH_I(I_i^S, I_j^T)$. The $HoS(I_i^S, I_j^T)$ is computed by the following formula: $HoS(I_i^S, I_j^T) = \sum_{k=1}^{K} P(Hi(I_i^S) = l_k)P(Ho(I_j^T) = l_k) + aP(Ho(I_j^T) = l_h)$, where $l_k \in CH_I(I_i^S, I_j^T)$, and a is the weight of the additional account-attribute relationship from I_i^S to l_h, and assigned to the minimum of all weights related to I_i^S. The weight a is used to reflect the contribution of l_h to the hometown similarity. For example, a is equal to 0.1 in Fig. 2(c). So $HoS(I_i^S, I_j^T)$ is equal to $0.3 \times 0.3 + 0.6 \times 0.6 + (0.1 \times 0.1) = 0.46$.

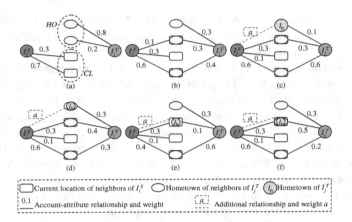

Fig. 2. Different cases tackled in hometown inference.

Case (5)–(6) shown in Fig. 2(e)–(f). The value of hometown of I_j^T is not empty, the $CH_1(I_i^S, I_j^T)$ is not empty, and l_h appears in $CH_1(I_i^S, I_j^T)$, i.e., $l_h \not\subseteq \emptyset \wedge CH_1(I_i^S, I_j^T) \neq \emptyset \wedge l_h \in CH_1(I_i^S, I_j^T)$. The $HoS(I_i^S, I_j^T)$ is computed by the formula (5), in which a is used to reflect the contribution of l_h to the hometown similarity, and is equal to $P(hi(I_i^S) = l_h)$. For example, a is equal to 0.1 in Fig. 2(e). $HoS(I_i^S, I_j^T)$ is equal to 0.3.

(3) **Friendliness**: We suppose that accounts ar_i^S and ar_i^T belong to the same user, and I_i^S is a neighbor of ar_i^S. If the degree of friendliness between I_j^T and ar_i^T is high, we believe that I_i^S and I_j^T likely belong to the same user. The triadic closure principle [26] can be used to indirectly explain this underlying inference.

Because the search range $R_d(ar_i^T)$ constrains the friendliness score $FrS(ar_i^T, I_j^T)$, which is related to the parameter d. Then, $FrS(ar_i^T, I_j^T)$ can be evaluated by the following metrics: (1) $FrS_1(ar_i^T, I_j^T) = 1$, $d \geq 1$; (2) $FrS_2(ar_i^T, I_j^T) = |CN(ar_i^T, I_j^T)|$, $d \geq 2$; (3) $FrS_3(ar_i^T, I_j^T) = |CN(N(ar_i^T), I_j^T)|$, $d \geq 3$. Here $CN(.)$ represents the common neighbors between two accounts, $|CN(.)|$ is the size of $CN(.)$.

5.2   Decision Model on Pairwise Similarity

(1) **Machine Learning Models:** Existing studies on the social identity link identification mainly rely on the supervised classification model. There are four classification models, namely multilayer perceptron (MLP) in [4], support vector machine (SVM) in [7, 14], logistic regression (LR) in [8], and Naive Bayes (NB) in [15], which have been demonstrated to perform well for this problem. As such, we also build these four classification models using our labeled dataset so that we can apply them to select the best match account from the candidate set in the target network. The experiments are described in detail in Subsect. 6.2. The results reported in Fig. 3 show that LR and MLP models are more accurate. Thus, we select LR and MLP models for our experiments on the whole dataset.

(2) **Algorithm for Finding the Best Match Account:** After all problems mentioned above have been solved, we integrate all solutions into the algorithm *FindSIL()*, which is used for finding the best match account I_j^T of I_i^S. Note the detailed *FindSIL()* algorithm is shown in Algorithm 2, which is called in our overall Algorithm 1.

Algorithm 2. FindSIL()

Input: (1) account I_i^S, which is from G^S and waiting for identification; (2) anchor links (ar_i^S, ar_i^T); (3) parameter d, which is used to control the search range $R_d(ar_i^T)$.
Output: best match account I_j^T
1: Define the search range $R_d(ar_i^T)$ according to ar_i^T and d.
2: Find the best match account from the candidates through the decision model.
3: Return identified account I_j^T.

---

6 Experiments

6.1 Experimental Setup

(1) **Data Preparation:** As there is no publicly available benchmark datasets for social identity link, we have to create our own data sets for performance evaluation purpose. We leverage two publicly available large-scale social network data sets from China for our experiments. One is Sina Weibo dataset, and the other one is Renren dataset.

Before crawling user profile datasets from the two social networks for account linking, we make sure that the profiles of the linked users have overlaps, at least partially. In this paper, we request that the crawled accounts must satisfy a constraint, i.e. their university profile from two social platforms is equal to a specific value. We crawled 40,618 Renren accounts and 20,448 Sina Weibo accounts. The number of average friends per account in Renren and Sina Weibo is 339.9 and 27.5, respectively.

(2) **Evaluation Metrics:** We conduct our experiment on both the small set of labeled data and the large set of unlabeled data, i.e. those nodes in the target network to be identified. The objective of the former is to determine the best classification models, while the objective of the latter is to identify as many social links as possible.

For the first experiments on the small set of labeled data, we evaluate the effectiveness of various methods, using *precision*, *recall* and *F-score*, which are standard metrics in machine learning and information retrieval, and have widely been used for user identification across different social networks [7, 19, 22].

For the second experiments on the large set of unlabeled data, we need to manually check each of predicted linked account pairs, which can be classified into three categories: *correct* account pairs (*tp*), *uncertain* account pairs (*up*) and *wrong* account pairs (*fp*). Let the total number of predicted account pairs be *total* = *tp* + *up* + *fp*. The correct ratio or precision (*Pr*), uncertain ratio (*Ur*) and wrong ratio (*Wr*) are computed as follows: $Pr = tp/total$, $Ur = up/total$, $Wr = fp/total$. In addition, we evaluate the proportion of predicted account pairs in whole data set. We define the *coverage ratio* (*Cr*) as $Cr = total/min(N_1, N_2)$, where N_1 and N_2 are the number of accounts in the source network and target network, respectively.

(3) **Comparative Methods:** In this subsection, we compare our proposed methods with the following state-of-the-art methods.

  (1) *Nickname Similarity Method (NSM)*: Many features extracted from nicknames have been used to predict the social links. Especially, a few studies [8, 11] *only* use the nickname similarity features. We also implement a *NSM* method (only use the nickname features in Subsect. 5.1 (1)) to predict social link.

  (2) *Rule-based Filtering Method (RFM)*: The rule-based filtering method uses hand-picked similarity features and rules designed to predict identity link. This method achieves good performance [19]. We build a prototype *RFM* system based on this paper, which has won the second prize in the third China Software Developing Contest in 2014 (www.cnsoftbei.com).

  (3) *Our Method based on Logistic Regression (OM-LR)*: we use LR model to select the best match account from the search range.

  (4) *Our Method based on Multilayer Perceptron (OM-MLP)*: we use MLP model to select the best match account from search range.

  Note that the performance of our proposed *OM-LR* and *OM-MLP* methods is related to the parameter *d* in the search range $R_d(ar_i^T)$.

6.2   Experimental Results

We first aim to find the best classification models through performing experiments on the small set of labeled data. Here, the labeled dataset consists of 1,304 positive and some negative instances where the number of negative instances is determined by an imbalance ratio and the number of positive instances.

We partition the dataset into two groups using 10-fold cross validation (CV). Note this is different from standard CV as we use less training data and more test data, to

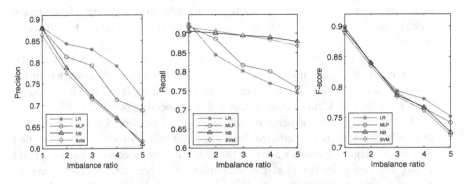

Fig. 3. Performance comparison of different classification models under imbalance ratios.

better reflect the real scenario. We report the average results of 10-fold CV. In each iteration of the cross validations, we sample the negative account pairs according to different imbalance ratios. Figure 3 shows the performance comparison of 4 different classification models under 4 different imbalance ratios (1:5). From Fig. 3, LR and MLP get better performance than the other two methods, and their performance is very close. So we choose both LR and MLP models for selecting the best match account from candidates.

Secondly, we compare our methods, OM-LR and OM-MLP, with existing *NSM* and *RFM* methods, based on the large set of unlabeled ground-truth data. As the objective of this experiment is to identify as many social links as possible, the search range includes all accounts from the target network ($d = inf$). Our methods use 186 randomly selected anchor links only and Table 1 shows experimental results for different methods. The first four columns show the performance in terms of various evaluation metrics. The fifth column #*coap* denotes the number of correct account pairs and the last column *Total* is the total number of predicted account pairs.

From Table 1, the performance of the *NSM* method is worst among all the methods, as it predicted only 316 correct account pairs and with lowest coverage ratio 1.8 %. We observe that rule based method *RFM*, albeit accurate (with highest precision), its coverage ratio $Cr = 5.4$ %, is much lower than 13.7 % and 13.1 % of our proposed OM-LR and OM-MLP respectively. In addition, the number of correct account pairs identified by *RFM* is much less than that by our OM-LR and OM-MLP. In summary, our methods, especially OM-LR, can identify much more correct social links than existing methods, which cannot be identified by both *NSM* and *RFM* methods.

Table 1. Performance comparison of different methods on the unlabeled data ($d = inf$).

	Pr	Ur	Wr	Cr	#coap	Total
NSM	84.7 %	3.2 %	12.1 %	1.8 %	316	373
RFM	92.6 %	1.7 %	5.7 %	5.4 %	1017	1098
OM-LR	59.6 %	3.5 %	36.9 %	13.7 %	1667	2798
OM-MLP	58.8 %	3.1 %	38.1 %	13.1 %	1576	2673

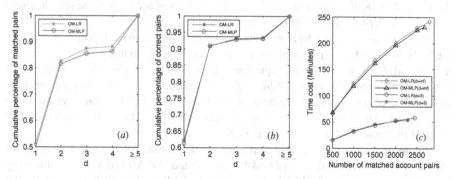

Fig. 4. The cumulative percentage of the matched account pairs (*a*) and of the correctly matched account pairs (*b*) with regard to the different d; (*c*) Time cost.

Finally, we evaluate the effectiveness of OM-LR and OM-MLP under different values of parameter d. Fig. 4(a) shows the cumulative percentage of the matched account pairs with regard to the different d. About 80 % of account pairs are matched with $d \leq 2$ by the OM-LR or OM-MLP. In particular, there are about 50 % of matched account pairs, in which each account from the target network is found in $R_{d=1}(ar_i^T)$; Fig. 4(b) shows the cumulative percentage of the correctly account pairs with regard to the different d. About 90 % of account pairs are correctly matched with $d \leq 2$ by the OM-LR or OM-MLP. Only about 7 % of account pairs are correctly matched with $d \geq 5$. These statistics reveal that most of correct social links can be found in small range ($d \leq 2$) around their anchor links and generated anchor links. As such, our methods can solve the social link identification efficiently using a local search strategy.

6.3   Efficiency Evaluation

In existing methods, the computational cost of identifying pair-wise accounts is $N_1{*}N_2$ (N_1 and N_2 are the number of accounts in G^S and G^T respectively). For our method, the computational cost is estimated as follows.

Given an account I_i^S in G^S and an anchor link (ar_i^S, ar_i^T), the number of accounts in the search range $R_d(ar_i^T)$ of I_i^S is denoted by $|R_d(ar_i^T)|$. Assuming $d \leq m$, $|R_{d \leq m}(ar_i^T)| = |R_{d=1}(ar_i^T)| + ... + |R_{d=m}(ar_i^T)|$ ($\cap_{j=1}^m R_{d=j}(ar_i^T) = \varnothing$). Denote the corresponding search tree by $Tr(G^T)$, and let the average number of friends per account in $Tr(G^T)$ is k_2', then we can compute the number of accounts of $Tr(G^T)$ by $N_2' = k_2'{*}((k_2')^m - 1)/(k_2'-1) \approx (k_2')^m$. Obviously, N_2' is much smaller than N_2 when the depth parameter m is small.

Let us consider real social networks described in the Subsect. 6.1. The average shortest path length of RenRen is about 5 (that was also confirmed by the work [27]). Assume $m = 4$, which is less than the actual value. Then the average number of friends per account is $k_2' = 14.2$ in the search tree $Tr(G^T)$. The actual average number of friends per account in G^T is about 340 computed by $k_2 = 2E_2/N_2$, where E_2 is the number of friends. Obviously, there are 325.8 (340–14.2) redundant accounts. Knowing that about 93 % of account pairs are correctly matched with $d \leq 3$ in our methods, the number of accounts in the search range can be estimated by $|R_{d \leq 3}(ar_i^T)| = |R_{d=1}(ar_i^T)| + ... + |R_{d=2}(ar_i^T)| = k_2'+(k_2')^2 + (k_2')^3 \leq k_2 + k_2 k_2'+(k_2')^3 = 8,031$, which is much less (<20 %) than $N_2 = 40,618$. If we assume $m = 5$, then the number of accounts in the search range $|R_{d \leq 3}(ar_i^T)| = 3,766$, which is smaller than that $m = 4$.

Next, we also use the total execution time to evaluate the efficiency of different methods. We conduct experiments on our methods with different parameter values, i.e. $d = inf$ and $d \leq 3$, where $d = inf$ represents the computational cost $N_1{*}N_2$, while $d \leq 3$ denotes the computational cost $N_1|R_{d \leq 3}()|$. The parameter $d \leq 3$ is reasonable because about 93 % of account pairs are correctly matched with $d \leq 3$ by our methods. The experiments and latency observations are conducted on a PC, with Intel® Core™ i5-4460 processor and 8 GB main memory.

Figure 4(c) shows the relationship between the number of the matched accounts and the time cost. The time cost of OM-LR ($d \leq 3$) and OM-MLP($d \leq 3$) is significantly less than the time cost of OM-LR($d = inf$) and OM-MLP ($d = inf$) for identifying the same number of account pairs. Of course, the high efficiency of OM-LR($d \leq 3$) and OM-MLP($d \leq 3$) is at the expense of slightly lower *coverage ratio*. Nevertheless, as we handle large-scale networks, it is thus acceptable. In addition, the time cost of OM-LR and OM-MLP with the same d value is very close.

7 Conclusion

In this paper, we address the problem of linking user accounts of the same natural person across different social networks. Our proposed method is based on our unique theoretical assumption inspired by the triadic closure principle. In particular, given two user accounts of the same natural person across different social media platforms, their friends/neighbors in different social platforms should still be directly or indirectly connected to itself. Based on the theoretical assumption, we propose a novel method, which *is to link accounts across different social platforms using the local expansion strategy*. Experimental results demonstrate that our proposed method outperforms existing methods significantly. Note our proposed method is generic and thus it can be applied to link up user accounts across other Chinese or English social networks (e.g. Twitter and Facebook), as long as we can collect their large-scale network data.

Acknowledgments. This work was partially supported by grants from the National Natural Science Foundation of China (Grant No. U1533104, U1333109, 61301245, 61305107).

References

1. Li, X.-L., Foo, C.S., Tew, K.L., Ng, S.-K.: Searching for rising stars in bibliography networks. In: Zhou, X., Yokota, H., Deng, K., Liu, Q. (eds.) DASFAA 2009. LNCS, vol. 5463, pp. 288–292. Springer, Heidelberg (2009)
2. Li, X.-L., Tan, A., Yu, P.S., Ng, S.-K.: ECODE: event-based community detection from social networks. In: Yu, J.X., Kim, M.H., Unland, R. (eds.) DASFAA 2011, Part I. LNCS, vol. 6587, pp. 22–37. Springer, Heidelberg (2011)
3. Carmagnola, F., Cena, F.: User identification for cross-system personalization. Inf. Sci. **179**(1–2), 16–32 (2009)
4. Vosecky, J., Hong, D., Shen, V.Y.: User identification across multiple social networks. In: Proceedings of NDT 2009 (2009)
5. Zafarani, R., Liu, H.: Connecting corresponding identities across communities. In: Proceedings of ICWSM 2009 (2009)
6. Narayanan, A., Shmatikov, V.: De-anonymizing social networks. In: Proceedings of S&P (2009)
7. Liu, S., Wang, S., Zhu, F., Zhang, J., Krishnan, R.: HYDRA: large-scale social identity linkage via heterogeneous behavior modeling. In: Proceedings of SIGMOD 2014 (2014)
8. Zafarani, R., Liu, H.: Connecting users across social media sites: a behavioral-modeling approach. In: Proceedings of KDD 2013 (2013)
9. Jain, P., Kumaraguru, P.: Finding nemo: Searching and Resolving Identities of Users across Online Social Networks (2012). arXiv preprint arXiv:1212.6147

10. Jain, P., Kumaraguru, P., Joshi, A.: @i seek 'fb.me': identifying users across multiple online social networks. In: Proceedings of WWW 2013 (2013)

11. Malhotra, A., Totti, L., Meira, W., Kumaraguru, P., Almeida, V.: Studying user footprints in different online social networks. In: Proceedings of Advances in Social Networks Analysis and Mining, 2012 (2012)

12. Nunes, A., Calado, P., Martins, B.: Resolving user identities over social networks through supervised learning and rich similarity features. In: Proceedings of SAC 2012 (2012)

13. Zhang, H., Kan, M.-Y., Liu, Y., Ma, S.: Online social network profile linkage. In: Jaafar, A., Mohamad Ali, N., Mohd Noah, S.A., Smeaton, A.F., Bruza, P., Bakar, Z.A., Jamil, N., Sembok, T.M.T. (eds.) AIRS 2014. LNCS, vol. 8870, pp. 197–208. Springer, Heidelberg (2014)

14. Liu, J., Zhang, F., Song, X., Song, Y.I., Lin, C.Y., Hon, H.W.: What's in a Name?: an unsupervised approach to link users across communities. In: Proceedings of WSDM 2013 (2013)

15. Goga, O.: Matching User Accounts across Online Social Networks: Methods and Applications. Ph.D. thesis, University Pierre and Marie CURIE (2014)

16. Iofciu, T., Fankhauser, P., Abel, F., Bischoff, K.: Identifying users across social tagging systems. In: Proceedings of ICWSM 2011 (2011)

17. Kong, X., Zhang, J., Yu, P.S.: Inferring anchor links across multiple heterogeneous social networks. In: Proceedings of CIKM 2013 (2013)

18. Anwar, T., Abulaish, M.: An MCL-based text mining approach for namesake disambiguation on the web. In: Proceedings of ICWI 2012 (2012)

19. Carmagnola, F., Osborne, F., Torre, I.: User data discovery and aggregation: the CS-UDD algorithm. Inf. Sci. **270**(20), 41–72 (2014)

20. Goga, O., Lei, H., Krishnan, S., Friedland, G., Sommer, R., Teixeira, R.: Exploiting innocuous activity for correlating users across sites. In: Proceedings of WWW 2013 (2013)

21. Zhang, J.W., Yu, P.S., Zhou, Z.H.: Meta-path based multi-network collective link prediction. In: Proceedings of KDD 2014 (2014)

22. Li, J., Wang, G.A., Chen, H.: Identity matching using personal and social identity features. Inf. Syst. Front. **13**(1), 101–113 (2011)

23. Iofciu, T., Fankhauser, P., Abel,. F., Bischoff, K.: Identifying users across social tagging systems. In: Proceedings of AAAI Conference on Weblogs and Social Media, 2011 (2011)

24. Chen, Y., Zhuang, C., Cao, Q., Hui, P.: Understanding cross-site linking in online social networks. In: Proceedings of SNA-KDD 2014 (2014)

25. Gusfield, D.: Algorithms on Strings, Trees and Sequences: Computer Science and Computational Biology. Cambridge University Press, New York (1999)

26. Rapoport, A.: Spread of information through a population with socio-structural bias i: assumption of transitivity. Bull. Math. Biophys. **15**(4), 523–533 (1953)

27. Zhao, X., Sala, A., Zheng, H., Zhao, B.: Efficient shortest paths on massive social graphs. In: Proceedings of CollaborateCom 2011 (2011)

Discovering the Network Backbone
from Traffic Activity Data

Sanjay Chawla[1,2], Kiran Garimella[3(✉)], Aristides Gionis[3], and Dominic Tsang[2]

[1] Qatar Computing Research Institute, HBKU, Doha, Qatar
schawla@qf.org.qa
[2] University of Sydney, Sydney, Australia
dwktsang@yahoo.com
[3] Aalto University, Espoo, Finland
{kiran.garimella,aristides.gionis}@aalto.fi

Abstract. We introduce a new computational problem, the BACKBONE-DISCOVERY problem, which encapsulates both *functional* and *structural* aspects of network analysis. While the topology of a typical road network has been available for a long time (e.g., through maps), it is only recently that fine-granularity functional (activity and usage) information about the network (like source-destination traffic information) is being collected and is readily available. The combination of functional and structural information provides an efficient way to explore and understand usage patterns of networks and aid in design and decision making. We propose efficient algorithms for the BACKBONEDISCOVERY problem including a novel use of edge centrality. We observe that for many real world networks, our algorithm produces a backbone with a small subset of the edges that support a large percentage of the network activity.

1 Introduction

In this paper we propose a novel formulation for discovering the *backbone* of traffic networks. We are given the topology of a network (its structure) $G = (V, E)$ and a traffic log (functional activity) $\mathcal{L} = \{(s_i, t_i, w_i)\}$, recording the amount of traffic w_i that incurs between source s_i and destination t_i. We are also given a budget B that accounts for a total edge cost. The goal is to discover a sparse subnetwork R of G, of cost at most B, which *summarizes* as well as possible the recorded traffic $\mathcal{L}$.

The problem we study has applications for both *exploratory data analysis* and *network design*. An example application of our algorithm is shown in Fig. 1. Here, we consider a traffic log (Fig. 1, left), which consists of the most popular routes used on the London tube. The backbone produced by our algorithm takes into account this demand (based on the traffic log) and summarizes the underlying

---

S. Chawla—On Leave from Sydney University.

A. Gionis—This work is supported by the European Communitys H2020 Program under the scheme 'INFRAIA-1-2014-2015: Research Infrastructures', grant agreement #654024 'SoBigData: Social Mining & Big Data Ecosystem'.

J. Bailey et al. (Eds.): PAKDD 2016, Part I, LNAI 9651, pp. 409–422, 2016.
DOI: 10.1007/978-3-319-31753-3_33

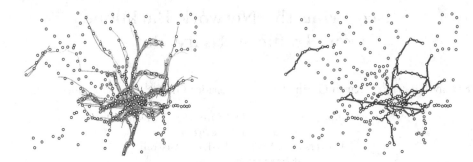

Fig. 1. London tube network, with nodes representing the stations. The figure on the left shows a subset of the trips made, and the figure on the right shows the corresponding backbone, as discovered by our algorithm. The input data contains only source–destination (indicating start and end points of a trip) pairs and for visualization purposes, a B-spline was interpolated along the shortest path between each such pair. The backbone presented on the right covers only 24 % of the edges in the original network and has a stretch factor of 1.58. This means that even with pruning 76 % of the edges in the network, we are able to maintain shortest paths which are at most 1.58 times the shortest path length original graph.

network, thus presenting us with insights about usage pattern of the London tube (Fig. 1, right). This representation of the "backbone" of the network could be very useful to identify the important edges to upgrade or to keep better maintained in order to minimize the total traffic disruptions.

We only consider source-destination pairs in the traffic log, and not full trajectories, as source-destination information captures *true mobility demand* in a network. For example, data about the daily commute from home (source) to office (destination) is more resilient than trajectory information, which is often determined by local and transient constraints, like traffic conditions on the road, time of day, etc. Furthermore, in communication networks, only the source-ip and destination-ip information is encoded in TCP-IP packets. Similarly, in a city metro, check-in and check-out information is captured while the intervening movement is not logged.

Example. To understand the key aspects of BACKBONEDISCOVERY problem consider the example shown in Fig. 2. In this example, there are four groups of nodes: (i) group A consists of n nodes, $a_1, \ldots, a_n$, (ii) group B consists of n nodes, $b_1, \ldots, b_n$, (iii) group C consists of 2 nodes, c_1 and c_2, and (iv) group D consists of m nodes, $d_1, \ldots, d_m$. Assume that m is smaller then n, and thus much smaller than n^2. All edges shown in the figure have cost 1, except the edges between c_1 and c_2, which has cost 2. Further assume that there is one unit of traffic between each a_i and each b_j, for $i, j = 1, \ldots, n$, resulting in n^2 source-destination pairs (the majority of the traffic), and one unit of traffic between d_i and d_{i+1}, for $i = 1, \ldots, m - 1$, resulting in $m - 1$ source-destination pairs (some additional marginal traffic).

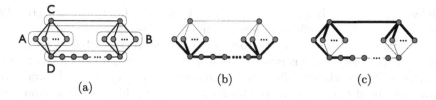

Fig. 2. The BACKBONEDISCOVERY problem solution results in a better network than the one obtained from the Steiner forest solution. (a) A traffic network. We consider a unit of traffic from each node in A to each node in B, and from each node in D to its right neighbor. (b) Shown with thick edges is an optimal Steiner forest for certain cost C. (c) Shown with thick edges is a backbone of cost at most C that captures the traffic in the network better than the optimal Steiner forest.

The example abstracts a common layout found in many cities: a few busy centers (commercial, residential, entertainment, etc.) with some heavily-used links connecting them (group C), and some peripheral ways around, that serve additional traffic (group D).

This example highlights advantages of the backbone discovery problem:

- We do not need to guarantee short paths for all pairs of nodes, but only for those in our traffic log which makes our approach more general. In particular, based on the budget requirements a backbone could be designed for the most voluminous paths.
- Due to the budget constraint, it may not be possible to guarantee connectivity for all pairs in the traffic log. In fact, it is possible that the optimal backbone may even contain cycles while leaving pairs disconnected.
- Certain high cost edges may be an essential part of the backbone that other problem formulations may leave out. For example, while the edge that connects the nodes in C is a very important edge for the overall traffic (as it provides a short route between A and B), the optimal Steiner-forest solution (see Related Work), shown in Fig. 2b, prefers the long path along the nodes in D. Our algorithm includes the component C (as seen in Fig. 2c) because it is an edge that is part of many shortes-paths between nodes (high edge-betweenness).

The rest of the paper is organized as follows. In Sect. 2, we rigorously define the BACKBONEDISCOVERY problem. In Sect. 3 we survey related work and distinguish our problem from other relevant approaches. Section 4 introduces our algorithm based on the greedy approach, while Sect. 5 provides details of our experimental evaluation, results and discussion. Section 6 is a short conclusion.

2   Problem Definition

Let $G = (V, E)$ be a network, with $|V| = n$ and $|E| = m$. For each edge $e \in E$ there is a cost $c(e)$. Additionally, we consider a traffic log $\mathcal{L}$, specified as a set of triples (s_i, t_i, w_i), with $s_i, t_i \in V$, $i = 1, \ldots, k$. A triple (s_i, t_i, w_i) indicates the fact that w_i units of traffic have been recorded between nodes s_i and t_i.

We aim at discovering the *backbone* of traffic networks. A backbone R is a subset of the edges of the network G, that is, $R \subseteq E$ that provides a good summarization for the whole traffic in $\mathcal{L}$. In particular, we require that if the available traffic had used only edges in the backbone R, it should have been almost as efficient as using all the edges in the network. We formalize this intuition below.

Given two nodes $s, t \in V$ and a subset of edges $A \subseteq E$, we consider the shortest path $d_A(s, t)$ from s to t that uses only edges in the set A. In this shortest-path definition, edges are counted according to their cost c. If there is no path from s to t using only edges in A, we define $d_A(s, t) = \infty$. Consequently, $d_E(s, t)$ is the shortest path from s to t using all the edges in the network, and $d_R(s, t)$ is the shortest path from s to t using only edges in the backbone R.

To measure the quality of a backbone R, with respect to some traffic log $\mathcal{L} = \{(s_i, t_i, w_i)\}$ we use the concept of *stretch factor*. Intuitively, we want to consider shortest paths from s_i to t_i, and evaluate how much longer are those paths on the backbone R, than on the original network. The idea of using stretch factor for evaluating the quality of a subgraph has been used extensively in the past in the context of spanner graphs [8].

In order to aggregate shortest-path information for all source–destination pairs in our log in a meaningful way, we need to address two issues. The first issue is that not all source–destination pairs have the same volume in the traffic log. This can be easily addressed by weighting the contribution of each pair (s_i, t_i) by its corresponding volume w_i.

The second issue is that since we aim at discovering very sparse backbones, many source–destination pairs in the log could be disconnected in the backbone. To address this problem we aggregate shortest-path distances using the *harmonic mean*. This idea, proposed by Marchiori and Latora [5] and recently used by Boldi and Vigna [1] in measuring centrality in networks, provides a very clean way to deal with infinite distances: if a source–destination pair is not connected, their distance is infinity, so the harmonic mean accounts for this by just adding a zero term in the summation. Using the arithmetic mean is problematic, as we would need to add an infinite term with other finite numbers.

Overall, given a set of edges $A \subseteq E$, we measure the connectivity of the traffic log $\mathcal{L} = \{(s_i, t_i, w_i)\}$, $|\mathcal{L}| = k$ by

$$H_{\mathcal{L}}(A) = \left(\sum_{i=1}^{k} w_i \right) \left(\sum_{i=1}^{k} \frac{w_i}{d_A(s_i, t_i)} \right)^{-1}.$$

The *stretch factor* of a backbone R is then defined as

$$\lambda_{\mathcal{L}}(R) = \frac{H_{\mathcal{L}}(R)}{H_{\mathcal{L}}(E)}.$$

The stretch factor is always greater or equal than 1. The closer it is to 1, the better the connectivity that it offers to the traffic log $\mathcal{L}$. This definition of stretch factor provides a principled objective to optimize connectivity while allowing to leave disconnected pairs, when there is insufficient budget.

We are now ready to formally define the problem of backbone discovery.

Problem 1 (BACKBONEDISCOVERY). Consider a network $G = (V, E)$ and a traffic log $\mathcal{L} = \{(s_i, t_i, w_i)\}$. Consider also a cost budget B. The goal is to find a backbone network $R \subseteq E$ of total cost B that minimizes the stretch factor $\lambda_{\mathcal{L}}(R)$ or report that no such solution exists.

BACKBONEDISCOVERY is an **NP**-hard problem. The proof, omitted to due lack of space, uses a reduction from the set cover problem.

3 Related Work

The BACKBONEDISCOVERY is related to the k-spanner and the Steiner-forest problem [8,13]. In the k-spanner problem the goal is to find a minimum-cost subnetwork R of G, such that for *each pair* of nodes u and v, the shortest path between u and v on R is at most k times longer than the shortest path between u and v on G. In our problem, we are not necessarily interested in preserving the k-factor distance between all nodes but for only a subset of them.

In the Steiner-forest problem we are given a set of pairs of terminals $\{(s_i, t_i)\}$ and the goal is to find a minimum-cost forest on which each source s_i is connected to the corresponding destination t_i. Our problem is different from the Steiner-forest problem because we do not need all $\{(s_i, t_i)\}$ to be connected, and try to optimize a stretch factor so that the structural aspect of the network are also taken into account. The Prize collecting Steiner-forest problem (PCSF) [4] is a version of the Steiner-forest problem that allows for disconnected source–destination pairs, by imposing a penalty for disconnected pairs.

The BACKBONEDISCOVERY problem is also related to finding graph sparsifiers, simplifying graphs and subgraph extraction [2,6,7,11,14]. However these approaches do not consider budget constraints in the context of structural and functional information.

There has been some work in social network research to extract a subgraph from larger subgraphs subject to constraints [3,10]. The main focus of most of these approaches is on the trade-off between the level of network reduction and the amount of relevant information to be preserved either for visualization or community detection.

4 Algorithm

The algorithm we propose for the BACKBONEDISCOVERY problem is a *greedy* heuristic that connects one-by-one the source–destination pairs of the traffic log $\mathcal{L}$. A distinguishing feature of our algorithm is that it utilizes a notion of *edge benefit*. In particular, we assume that for each edge $e \in E$ we have available a benefit measure $b(e)$. The higher is the measure $b(e)$ the more beneficial it is to include the edge e in the final solution. The benefit measure is computed using the traffic log $\mathcal{L}$ and it takes into account the global structure of the network G.

The more central an edge is with respect to a traffic log, the more beneficial it is to include it in the solution, as it can be used to serve many source–destination pairs. In this paper we use *edge-betweenness* as a centrality measure, adapted to take into account the traffic log. We also experimented with *commute-time centrality*, but edge-betweenness was found to be more effective.

Our algorithm relies on the notion of *effective distance* $\widehat{\ell}(e)$, defined as $\widehat{\ell}(e) = c(e)/b(e)$, where $c(e)$ is the cost of an edge $e \in E$, and $b(e)$ is the edge-betweenness of e. The intuition is that by dividing the cost of each edge by its benefit, we are biasing the algorithm towards selecting edges with high benefit. We now present our algorithm in more detail.

4.1   The Greedy Algorithm

As discussed above, our algorithm operates with effective distances $\widehat{\ell}(e) = c(e)/b(e)$, where $b(e)$ is a benefit score for each edge e. The objective is to obtain a cost/benefit trade-off: edges with small cost and large benefit are favored to be included in the backbone. In the description of the greedy algorithm that follows, we assume that the effective distance $\widehat{\ell}(e)$ of each edge is given as input.

The algorithm works in an iterative fashion, maintaining and growing the backbone, starting from the empty set. In the i-th iteration the algorithm picks a source–destination pair (s_i, t_i) from the traffic log $\mathcal{L}$, and "serves" it. Serving a pair (s_i, t_i) means computing a shortest path p_i from s_i to t_i, and adding its edges in the current R, if they are not already there. For the shortest-path computation the algorithm uses the effective distances $\widehat{\ell}(e)$. When an edge is newly added to the backbone its cost is subtracted from the available budget. Here, the actual cost of the edge $c(e)$ (instead of the $\widehat{\ell}(e)$) is used. Also its effective distance is reset to zero, since it can be used for free in subsequent iterations of the algorithm. The source–destination pair that is chosen to be served in each iteration is the one that reduces the stretch factor the most at that iteration; and hence the greedy nature of the algorithm. The algorithm proceeds until it exhausts all its budget or until the stretch factor becomes equal to 1 (which means that all pairs in the log are served via a shortest path). The pseudo-code for the greedy algorithm is shown as Algorithm 1.

We experiment with two variants of this greedy scheme, depending on the benefit score we use. (i) Greedy: we use uniform benefit scores, $b(e) = 1$; (ii) GreedyEB: the benefit score of an edge is set equal to its weighted *edge-betweenness centrality*, weighted by the traffic log $\mathcal{L}$.

4.2   Speeding up the Greedy Algorithm

As we show in the experimental section the greedy algorithm provides solutions of good quality, in particularly the variant with the edge-betweenness weighting scheme. As the greedy algorithm is expensive, in this section we discuss a number of optimizations. We start by analyzing the running time of the greedy.

Algorithm 1. The greedy algorithm

Input: Network $G = (V, E)$, edge costs $c(e)$, benefit costs $b(e)$, cost budget B, traffic
  log $\mathcal{L} = \{(s_i, t_i, w_i)\}$
Output: A subset of edges $R \subseteq E$ of total cost $c(R) \leq B$ and small stretch factor
  $\lambda(R)$
1: **for** all $e \in E$ **do**
2:   $\widehat{\ell}(e) \leftarrow c(e)/b(e)$
3: $R \leftarrow \emptyset$
4: $\lambda \leftarrow \infty$
5: **while** $(B > 0)$ and $(\lambda > 1)$ **do**
6:   **for** each $(s_i, t_i, w_i) \in \mathcal{L}$ **do**
7:     $p_i \leftarrow$ SHORTESTPATH$(s_i, t_i, G, \widehat{\ell})$
8:     $\lambda_i \leftarrow$ STRETCHFACTOR$(R \cup p_i, G, \mathcal{L})$
9:   $p^* \leftarrow \min_i\{\lambda_i\}$ // the path with min. stretch factor in the above iteration
10:   $R' \leftarrow p^* \setminus R$ // edges to be newly added
11:   **if** $c(R') > B$ **then**
12:     Return R // budget exhausted
13:   $R \leftarrow R \cup R'$ // add new edges in the backbone
14:   $\widehat{\ell}(R') \leftarrow 0$ // reset cost of newly added edges
15:   $B \leftarrow B - c(R')$ // decrease budget
16:   $\lambda \leftarrow$ STRETCHFACTOR$(R, G, \mathcal{L})$ // update λ
17: Return R

Running Time. Assume that the benefit scores $b(e)$ are given for all edges
$e \in E$, and that the algorithm performs I iterations until it exhausts its budget.
In each iteration we need to perform $\mathcal{O}(k^2)$ shortest-path computations, where
k is the size of the traffic log $\mathcal{L}$. A shortest path computation is $\mathcal{O}(m + n \log n)$,
and thus the overall complexity of the algorithm is $\mathcal{O}(Ik^2(m + n \log n))$. The
number of iterations I depends on the available budget and in the worst case
it can be as large as k. However, since we aim at finding sparse backbones, the
number of iterations is typically smaller.

To improve the performance of greedy, we use a number of different optimization techniques: (i) We maintain the connected components built by greedy. We
use this information to avoid re-computing shortest paths for all (s_i, t_i) pairs for
which s_i and t_i belong to different connected components. (ii) When computing
the decrease in the stretch factor due to a candidate shortest path to be added
in the backbone, for pairs for which we have to recompute a shortest-path distance, we first compute an optimistic lower bound, based on the shortest path on
the whole network (which we compute once in a preprocessing step). If this optimistic lower bound is not better than the current best stretch factor then we can
skip the computation of the shortest path on the backbone. In practice, these two
optimizations lead to 20–35% improvement in performance (details in Sect. 5).
(iii) We use landmarks [9] to approximate the computation of shortest paths.
This reduces the complexity of our greedy algorithm to $\mathcal{O}(I\ell(k + m + n \log n))$,
where ℓ is the number of landmarks. As we show in Fig. 4, this optimization
provides an improvement of up to 4 times in terms of runtime.

5   Experimental Evaluation

The aim of the experimental section is to evaluate the performance of the proposed algorithm, the optimizations, and the edge-betweenness measure. We also compare our algorithm with other state-of-the-art methods which attempt to solve a similar problem.

Datasets. We experiment with six real datasets: four transportation networks, one web network and one internet-traffic network. For five of the datasets we also obtain real traffic, while for one we use synthetically-generated traffic. The characteristics and description of our datasets are provided in Table 1.

Table 1. Dataset statistics

Dataset	Type	# Nodes	# Edges	Real network	Real traffic	Network description	Traffic log description
LondonTube	Transportation	316	724	✓	✓	London subway network	Subway usage log
USFlights	Transportation	1 268	51 098	✓	✓	US airport network	Flight usage data
UKRoad	Transportation	8 341	13 926	✓	-	UK road network	Artificial
NYCTaxi	Transportation	50 736	158 898	✓	✓	NYC road network	Taxi usage data
Wikispeedia	Web	4 604	213 294	✓	✓	[12]	[12]

Traffic Log for UKRoad. Since we were not able to obtain real-world traffic data for the UKRoad network, we generate synthetic traffic logs $\mathcal{L}$ simulating different scenarios. In particular we generate traffic logs according to four different distributions: (i) power-law traffic volume, power-law s-t pairs; (ii) power-law traffic volume, uniformly random s-t pairs; (iii) uniformly random traffic volume, power-law s-t pairs; and (iv) uniformly random traffic volume, uniformly random s-t pairs. The goal is to understand the behavior of the algorithm with respect to the characteristics of the traffic log $\mathcal{L}$.

Baseline. To obtain better intuition for the performance of our methods we define a simple baseline, where a backbone is created by adding edges in increasing order of their effective distances $\widehat{\ell}(e) = c(e)/b(e)$, where $b(e)$ is edge-betweenness; this was the best-performing baseline among other baselines we tried, such as adding source–destination pairs one by one (i) randomly, (ii) in decreasing order of volume (w_i), (iii) in increasing order of effective distance defined using closeness centrality, etc.

5.1   Quantitative Results

We focus our evaluation on three main criteria: (i) Comparison of the performance with and without the edge-betweenness measure; (ii) effect of the optimizations, in terms of quality and speedup; and (iii) effect of allocating more budget on the stretch factor.

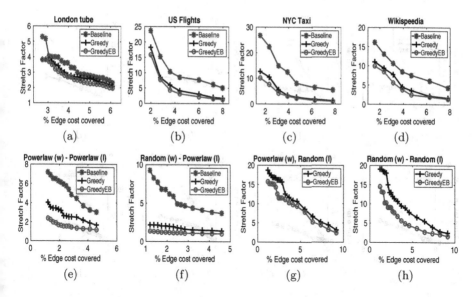

Fig. 3. Effect of edge-betweenness on the performance of the Greedy algorithm, for various datasets (a) LondonTube, (b) USFlights, (c) NYCTaxi, (d) Wikispeedia, (e–h) UKRoad. Baseline is missing in figure (g) and (h) because the stretch factor was very large or infinity. We see a consistent trend that using edge-betweenness improves the performance. In Figure (e–h), (w) indicates traffic volume, and (l) indicates the log.

Effect of Edge-Betweenness. We study the effect of using edge-betweenness in the Greedy algorithm. The results are presented in Fig. 3.

Effect of Landmarks. Landmarks provide faster computation with a trade off for quality. Figure 4 shows the speedup achieved when using landmarks. In the figures, BasicGreedyEB indicates the greedy algorithm that doesn't use any optimizations. GreedyEBCC makes use of the optimizations proposed in Sect. 4.2 which do not use approximation. GreedyEBLandmarks* makes use of the landmarks optimizatation and the * indicates the number of landmarks we tried. Figure 5 shows the performance of GreedyEB algorithm with and without using landmarks.

Budget vs. Stretch Factor. We examine the trade-off between budget and stretch factor for our algorithm and its variants. A lower stretch factor for the same budget indicates that the algorithm is able to pick better edges for the backbone. Figure 3 shows the trade-off between budget and stretch factor for all our datasets. In all figures the budget used by the algorithms, shown in the x-axis, is expressed as a percentage of the total edge cost.

Key Findings. Our key findings are the following.

- The greedy algorithm and its variants performs much better than the baseline (See Fig. 3). Note that the baseline is not included in Fig. 3(g,h) because the edges in the baseline are added one-by-one and for a large interval of the cost, the stretch factor was very large or even infinity.

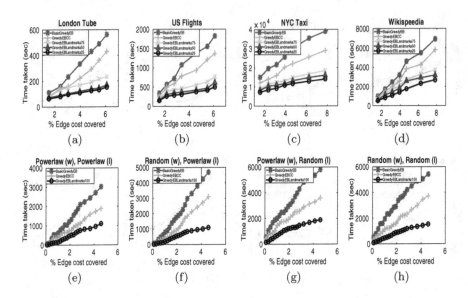

Fig. 4. Comparison of the time taken by the algorithm using different optimizations mentioned in Sect. 4.2, for (a) LondonTube, (b) USFlights, (c) NYCTaxi, (d) Wikispeedia, (e–h) UKRoad. BasicGreedyEB doesnt use any optimizations, GreedyEBCC is the version using connected components, GreedyEBLandmarks* uses * landmarks. We can clearly see a great improvement (up to 4x) in speed by using landmarks. As we increase the number of landmarks, we trade-off speed with accuracy. In Figure (e–h), (w) indicates traffic volume, and (l) indicates the log.

– The backbones discovered by our algorithms are sparse and summarize well the given traffic (Figs. 3, 5). In all cases, with about 15 % of the edge cost in the network it is possible to summarize the traffic with stretch factor close to 1. In some cases, even smaller budget (than 15 %) is sufficient to reach a lower stretch-factor value.

– Incorporating edge-betweenness as an edge-weighting scheme in the algorithm improves the performance, in certain cases there is an improvement of at least 50 % (see Fig. 3; in most cases, even though there is a significant improvement, the plot is overshadowed by a worse performing baseline). This is because, using edges of high centrality will make sure that these edges are included in many shortest paths, leading to re-using many edges.

– The optimizations we propose in Sect. 4.2 help in reducing the running time of our algorithm (Fig. 4). For the optimizations not using landmarks, we see around 30 % improvement in running time. Using landmarks substantially decreases the time taken by the algorithms (3–4 times). While there is a compromise in the quality of the solution, we can observe from Fig. 5 that the performance drop is small in most cases and can be controlled by choosing the number of landmarks accordingly. Our algorithms, using the optimizations we propose, scale to large, real-world networks with tens of thousands of nodes, which is the typical size of a road/traffic network.

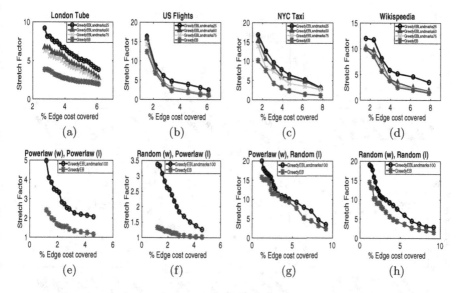

Fig. 5. Performance in terms of stretch factor of our greedy algorithm with and with out using landmarks, for (a) LondonTube, (b) USFlights, (c) NYCTaxi and (d) Wikispeedia (e–h) UKRoad. For all the datasets, as expected, we see a slight decrease in performance using landmarks. In Figure (e–h), (w) indicates traffic volume, and (l) indicates the log.

5.2 Comparison to Existing Approaches

In this section, we compare the performance of BACKBONEDISCOVERY with other related work in literature. The comparison is done based on two factors (i) stretch factor and (ii) percentage of edges covered by the solution. Intuitively, a good backbone should try to minimize both, i.e., produce a sparse backbone, which also preserves the shortest paths between vertices as well as possible.

Comparison with Prize Collecting Steiner-Forest (PCSF). Prize Collecting Steiner-forest [4] is a variant of the classic Steiner Forest problem, which allows for disconnected source–destination pairs, by paying a penalty. The goal is to minimize the total cost of the solution by 'buying' a set of edges (to connect the s–t pairs) and paying the penalty for those pairs which are not connected. We compare the performance of GreedyEB with PCSF, based on two factors (i) stretch factor (Fig. 6a), and (ii) percentage of edges covered by the solution (Fig. 6b). We use the same (s,t) pairs that we use in GreedyEB and set the traffic volume w_i as the penalty score in PCSF. We first run PCSF on our datasets and compute the budget of the solution produced. Using the budget as input to GreedyEB, we compute our backbone.

We can see from Fig. 6a that GreedyEB produces a backbone with a much better stretch factor than PCSF. In most datasets, our algorithm produces a backbone which is at least 2 times better in terms of stretch factor.

Figure 6b compares the fraction of edges covered by GreedyEB and PCSF. We observe that the fraction of edges covered by our algorithm is lower than that

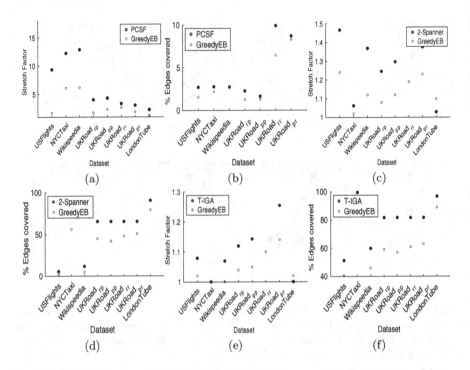

Fig. 6. (a,b) Comparison of GreedyEB with PCSF, in terms of (a) stretch factor (b) Percentage of edges covered. (c,d) Comparison of GreedyEB with 2-spanner in terms of (c) stretch factor (d) Percentage of edges covered. (e,f) Comparison of GreedyEB with T-IGA, in terms of (e) stretch factor (f) Percentage of edges covered. The 4 variants of UKRoad for the different traffic log are indicated by $UKRoad_{ab}$ where a indicates traffic volume, b indicates (s,t) pairs (r: random, p: powerlaw). (In (b) LondonTube is not plotted because of a mismatch in scale).

of PCSF. This could be because GreedyEB re-uses edges belonging to multiple paths. Figure 6(a,b) show that even though our solution is much better in terms of stretch factor, we produce sparse backbones (in terms of the percentage of edges covered).

Comparison with k-spanner. As described in Sect. 3, our problem is similar to k-spanner [8] in the sense that we try to minimize the stretch factor. A k-spanner of a graph is a subgraph in which any two vertices are at most k times far apart than on the original graph. One of the main advantages of GreedyEB compared to spanners is that spanners can not handle disconnected vertices. We also propose and optimize a modified version of stretch factor in order to handle disconnected vertices. Similar to PCSF, we first compute a 2-spanner using a 2 approximation greedy algorithm and compute the budget used. We then run GreedyEB for the same budget. Figure 6(c,d) show the performance of GreedyEB in terms of stretch factor and percentage of edges covered. Our objective here is to compare the cost GreedyEB pays in terms of stretch factor for allowing disconnected vertices.

We can clearly observe that even though we allow for disconnected pairs, Greedy-EB performs slightly better in terms of stretch factor and also produces a significantly sparser backbone.

Comparison with the Algorithm of Toivonen et al. (T-IGA). Next, we compare GreedyEB with the Iterative Global Algorithm proposed in Toivonen, et al. [11] (T-IGA), a framework for path-oriented graph simplification, in which edges are pruned while keeping the original quality of the paths between all pairs of nodes. The objective here is to check how well we perform in terms of graph sparsification. Figure 6(e,f) shows the comparison in terms of stretch factor and percentage of edges covered. Similar to the above approaches, we use the same budget as that used by T-IGA. We observe that for most of the datasets, their algorithm works poorly in terms of sparsification, pruning less than 20 % of the edges (Fig. 6(f)). Our algorithm performs better both in terms of the stretch of the final solution as well as sparseness of the backbone.

The above results, comparing our work with the existing approaches showcase the power of our algoritm in finding a concise representation of the graph, at the same time maintaining a low stretch factor. In all the three cases, GreedyEB performs considerably better than the related work.

Fairness. Though we claim that our approach performs better, we need to keep in mind that there might be differences between these algorithms. PCSF does not optimize for stretch factor. Spanners and T-IGA do not have a traffic log $((s,t)$ pairs). They also do not try to optimize stretch factor. For this section, we were just interested in contrasting the performance of our approach with existing state of the art methods and show how our approach is different and better at what we do.

6   Conclusions

We introduced a new problem, BACKBONEDISCOVERY, to address a modern phenomenon: these days not only is the *structural* information of a network available but increasingly, highly granular *functional (activity)* information related to network usage is accessible. For example, the aggregate traffic usage of the London Subway between all stations is available from a public website. The BACKBONE-DISCOVERY problem allowed us to efficiently combine structural and functional information to obtain a highly sophisticated understanding of how the Tube is used (See Fig. 1) making it an important tool for network and traffic planning.

References

1. Boldi, P., Vigna, S.: Axioms for centrality (2013). CoRR abs/1308.2140
2. Bonchi, F., De Francisci Morales, G., Gionis, A., Ukkonen, A.: Activity preserving graph simplification. DMKD **27**(3), 321–343 (2013)

3. Du, N., Wu, B., Wang, B.: Backbone discovery in social networks. In: Web Intelligence (2007)
4. Hajiaghayi, M.T., Khandekar, R., Kortsarz, G., Nutov, Z.: Prize-collecting steiner network problems. In: Eisenbrand, F., Shepherd, F.B. (eds.) IPCO 2010. LNCS, vol. 6080, pp. 71–84. Springer, Heidelberg (2010)
5. Marchiori, M., Latora, V.: Harmony in the small world. Physica A **285**, 539 (2000)
6. Mathioudakis, M., Bonchi, F., Castillo, C., Gionis, A., Ukkonen, A.: Sparsification of influence networks. In: KDD (2011)
7. Misiolek, E., Chen, D.Z.: Two flow network simplification algorithms. IPL **97**, 197–202 (2006)
8. Narasimhan, G., Smid, M.: Geometric Spanner Networks. Cambridge University Press, Cambridge (2007)
9. Potamias, M., Bonchi, F., Castillo, C., Gionis, A.: Fast shortest path distance estimation in large networks. In: CIKM (2009)
10. Ruan, N., Jin, R., Wang, G., Huang, K.: Network backbone discovery using edge clustering (2012). arxiv:1202.1842
11. Toivonen, H., Mahler, S., Zhou, F.: A framework for path-oriented network simplification. In: Cohen, P.R., Adams, N.M., Berthold, M.R. (eds.) IDA 2010. LNCS, vol. 6065, pp. 220–231. Springer, Heidelberg (2010)
12. West, R., Pineau, J., Precup, D.: Wikispeedia: an online game for inferring semantic distances between concepts. In: IJCAI, pp. 1598–1603 (2009)
13. Williamson, D., Shmoys, D.: The design of approximation algorithms. In: CUP (2011)
14. Zhou, F., Mahler, S., Toivonen, H.: Network simplification with minimal loss of connectivity. In: IDA (2010)

A Fast and Complete Enumeration
of Pseudo-Cliques for Large Graphs

Hongjie Zhai[1], Makoto Haraguchi[1(✉)], Yoshiaki Okubo[1], and Etsuji Tomita[2]

[1] Graduate School of Information Scienece and Technology,
Hokkaido University, N-14 W-9, Sapporo 060-0814, Japan
mh@kb.ist.hokudai.ac.jp
[2] The Advanced Algorithms Research Laboratory,
The University of Electro-Communications,
Chofugaoka 1-5-1, Chofu, Tokyo 182-8585, Japan

Abstract. This paper discusses a complete and efficient algorithm for enumerating densely-connected k-Plexes. A k-Plex is a kind of pseudo-clique which imposes a *Disconnection Upper Bound (DUB)* by the parameter k for each constituent vertex. However, since the parameter is usually fixed not depending on sizes of our targeted pseudo-cliques, we often have k-Plexes not densely-connected. In order to overcome this drawback, we introduce another constraint using a parameter j designating *Connection Lower Bound (CLB)*. Based on CLB, we can additionally enjoy a monotonic j-core operation and design an efficient depth-first algorithm which can exclude hopeless vertex sets which cannot be extended to their supersets satisfying both DUB and CLB. Our experimental results show it can work well as a useful tool for detecting densely-connected pseudo cliques in large networks including one with over $800,000$ vertices.

1 Introduction

Detecting communities in a network has been an important task in Social Network Analysis [10]. Cliques are typical vertex sets understood as potential communities [11]. Moreover, the class of cliques has an anti-monotonicity property helpful in designing an efficient enumerator for them. In a real world network, however, the clique model is too restrictive to capture various communities because it is rare for actual communities to appear as cliques. This would motivates us to study clique relaxation models and various pseudo-clique models have been proposed [11].

As another standard approach to community detection, graph clustering or partitioning methods are well known to be useful (e.g. [8,9]). However, when we aim at obtaining smaller communities, we shall use (pseudo-)clique detectors instead of those methods because they usually suppose small numbers of clusters whose sizes are consequently non-small and hence our targets are invisibly merged and absorbed into those larger clusters.

In a density-based model of pseudo-cliques (e.g. [7]), some indices for measuring the density of vertex sets are presented. Unlike the clique model, since

© Springer International Publishing Switzerland 2016
J. Bailey et al. (Eds.): PAKDD 2016, Part I, LNAI 9651, pp. 423–435, 2016.
DOI: 10.1007/978-3-319-31753-3_34

this class of pseudo-cliques do not satisfy the anti-monotonicity property, efficient but heuristic detectors have often been proposed for searching them [7]. We therefore might loose some vertex sets possibly valuable for us. Moreover, even if we have a complete detector, it cannot process large scale networks because of the hugeness of the number of pseudo-cliques. In spite of these difficulties, it is still important to develop an efficient complete enumerator that can handle large networks because such an enumerator is useful not only for community discovery but also for analyzing the nature of large networks in terms of statistics about (pseudo-)cliques [18].

As another clique relaxation models, a distance-based model, k-clique [2], and diameter-based models, k-club and k-clan [3,4], have been proposed. The parameter k controls admissible distances among vertices. As discussed in [5], when we allow a longer distance, large dense subgraphs appear which are almost cliques even when its subgraphs w.r.t. the original edge connection are not dense.

On the other hand, a k-Plex model [6] discusses the density in terms of original connection by setting an upper bound for the number of missing edges among vertices. The class of k-Plexes has an anti-monotonicity property which helps us to design a simple bottom-up enumerator [1,15]. For this reason, we discuss k-Plexes in this paper, introducing new constraints that can cover its weakness we discuss just below.

A vertex set is called a k-Plex if, for each vertex x in it, the number of vertices not adjacent to x is at most k. *Disconnection Upper Bound (DUB)* is thus specified by the parameter k. Note that a k-Plex could be non-connected. However, since it seems not interesting as a community, we exclude those ones from our consideration.

For a very small k, connected k-Plex is in fact dense if its size is relatively larger than k. On the other hand, as the size of densely connected vertex set increases, the number of disconnected vertices, k per each vertex, shall be non-small depending on density requirement. In other words, we think some constraint taking the sizes of targeted pseudo-cliques into account is important. When we suppose a k-Plex of size n, then each vertex has at least $n - k$ adjacent vertices, where $n - k$ must be a certain number provided we admit the vertex set as a densely connected one. We thus introduce another constraint using a parameter j designating *Connection Lower Bound (CLB)*. Then we try to enumerate all maximal connected k-Plexes meeting the CLB constraint.

A naive strategy, computing maximal k-Plexes and then checking the CLB constraint, does not work well when we consider non-small k because every vertex set with size no larger than k is trivially a k-Plex so we have to examine exponential number of such sets. The key to solve this problem lies in another fact that, if a connected k-Plex X can be extendable to a maximal connected k-Plex under the CLB constraint, X is involved in a "core" of X together with candidates that are potential vertices to be added to X, where the term "core" means the largest subset of vertices with at least j adjacents in the subset [19].

In this paper, j is fixed beforehand depending on the targeted vertex set size. The monotonicity of core operation is suitable to standard k-Plex enumerator enjoying the anti-monotonicity of k-Plexness. Based on the monotonicities, we can design an efficient complete depth-first algorithm which can exclude numerous hopeless k-Plexes that cannot be extended to maximal ones meeting our requirement. This realizes much improvement of performance of k-Plex enumerator. In our experimentation, we compare our algorithm with a state-of-the-art maximal k-Plex enumerator proposed very recently [1] from the viewpoints of computational performance and quality of solution k-Plexes as pseudo-cliques. For synthetic and real world large networks including a Web graph with over 800,000 vertices, it is verified that the CLB constraint as well as our pruning mechanisms are quite effective and as the results, the proposed algorithm can work very well as a practical tool for detecting dense pseudo-cliques.

2   Preliminary and Notation

An undirected graph is denoted by (V, Γ), where $V = \{v_1, ..., v_{|V|}\}$ is a set of vertices and $\Gamma(v_n) = \{v_m \in V \mid v_n, v_m$ are adjacent$\}$. $\Gamma(v_n)$ is assumed to not include v_n itself. An ordering $\prec$ over V is defined by $v_i \prec v_j$ and v_i is said younger than v_j iff $i < j$. If the identifiers, i of v_i, need not to be specified, we prefer the notation for vertices as v, x, u and so on. For a vertex set $X \subseteq V$, $G[X]$ is a subgraph of G induced by X. For a vertex $x \in X$, $\Gamma_X(x)$ denotes $\Gamma(x) \cap X$. $|\Gamma_X(x)|$ is often referred to as $deg_X(x)$.

A vertex set X is called a k-Plex if $|X - \Gamma_X(x)| \leq k$ for any $x \in X$. It is easy to see that for a k-Plex Y, any subset X of Y is also a k-Plex, Thus, the class of k-Plexes has an anti-monotonicity. A vertex $y \notin X$ is called a k-Plex candidate if Xy is still a k-Plex, where Xy is an abbreviation of $X \cup \{y\}$. $Cand(X)$ denotes the set of all k-Plex candidates of X. We especially discuss in this paper connected k-Plexes (c-k-Plexes, for short).

For a vertex set X and a vertex y, the distance between X and y, $dist(X, y)$, is given by the minimum length of paths in G from X to y, where $dist(X, x) = 0$ whenever $x \in X$. $dist(X, y) = \infty$ only when y is not reachable from X. $D_n(X)$ is defined as $\{y \in V \mid dist(X, y) = n\}$. $K_1(X) = D_1(X) \cap Cand(X)$, the set of k-Plex candidates directly connected to X, plays a key role in discussing c-k-Plexes and is called a K_1-candidate set at X.

3   Maximal Connected k-Plex

A maximal c-k-Plex, abbreviated as k-MPC, is a c-k-Plex that is maximal among c-k-Plexes. It is clear a c-k-Plex X is extendable to its super c-k-Plex iff $K_1(X) \neq \emptyset$. Thus, the extension is made by adding K_1-candidates at X to X.

A formation X^f of c-k-Plex X w.r.t. an indexing f is a sequence of vertices $v_{f(i)}$, $(v_{f(1)}, v_{f(2)}, ..., v_{f(|X|)})$, where $f(i)$ is the identifier of i-th vertex added to form $X = \{v_{f(1)}, ..., v_{f(|X|)}\}$ as sets, and all the intermediate $X^f_i = \{v_{f(1)}, ..., v_{f(i)}\}$ must be a c-k-Plex. That is,

$$v_{f(i+1)} \in K_1(X^f_i). \tag{1}$$

The sequence $(X_1^f, ..., X_{|X|}^f)$ is also called a formation of X. For any k-MPC Z, its formation ends at $Z = Z_{|Z|}^f$ with $K_1(Z) = \phi$.

We have multiple formations Z^f for a k-MPC Z, depending on the order of vertex addition. In fact, for an arbitrary initial vertex $v_{n_1} \in Z$, repeat the addition of $v_{n_i} \in Z$ such that v_{n_i} is adjacent to some vertex added prior to v_{n_i}. Then any intermediate $X_i(\subseteq Z)$ is a c-k-Plex by the monotonicity of k-Plexes and Z^f with the index f such that $f(i) = n_i$ gives a formation of Z. In Sect. 5, we introduce some control rules to disregard useless formations under the restriction that formations must be those for k-MPCs whose density is guaranteed by another parameter j.

4 j-Cored k-MPC

In this section, we define a class of k-MPCs, as our targets, which are dense in the sense that each vertex has at least j adjacents in them. We then discuss how to enumerate those targets completely and efficiently.

Given a graph $G = (V, \Gamma)$, a vertex set X is said to be j-cored if $deg_X(x) \geq j$ for any $x \in X$. The largest j-cored set, denoted by $core_j(V)$, is called j-core of G [19][1]. For a vertex set X, $core_j(X)$ is the j-core of the induced subgraph $G[X]$. We can observe a monotonicity of j-core operation.

Fact (Monotonicity of j-Cores): For vertex sets X_1, X_2 of V, $core_j(X_1) \subseteq core_j(X_2)$ holds whenever $X_1 \subseteq X_2$.

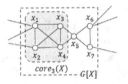

Fig. 1. Example of $core_j(X)$ for $j = 3$ case

The construction of $core_j(X)$ for a vertex set X is simple. Roughly speaking, it can be obtained by iteratively removing vertices with degree less than j from X. Since removing some vertices in general decreases degree of other vertices, we can iterate such a removal until no vertex can newly be removed.

A behavior of j-core operation is illustrated in Fig. 1. Any connection towards outside of X is ignored as we consider the subgraph $G[X]$. For this X, for $i = 6, 7$, v_i are firstly removed as $deg_X(x_i) = 2 < j = 3$. By this removal, $auxdeg(x_5)$ becomes 2. So x_5 is furthermore removed. Then, no more removal is made by

---

[1] The notion of j-core has originally been defined in [12]. We use in this paper the definition and construction method for j-core, according to [19].

degree lower bound constraint in this case, since all remaining vertices x_1, x_2, x_3 and x_4 still have at least j adjacents even after the removals so far.

Our target is now defined as a k-MPCs which is j-cored, called a j-cored k-MPCs ((j, k)-MPCs, for short). The following discussion is devoted to an efficient algorithm for enumerating every (j, k)-MPCs.

Unlike k-Plexes, the class of j-cored vertex sets does not have anti-monotonicity property. In spite of this fact, we can give a sufficient condition for c-k-Plexes to grow to j-cored ones. Then, we can use the condition conversely to reject hopeless c-k-Plexes appearing in formations.

Fact (Hopeful c-k-Plex): Let X be a c-k-Plex extendable to some (j, k)-MPCby adding vertices to X. Then

$$X \subseteq U(X), \quad \text{where} \tag{2}$$
$$U(X) = core_j(X \cup Cand(X)). \tag{3}$$

X satisfying (2) is said to be hopeful. Conversely, hopeless ones are those satisfying the negation of (2).

$$\textbf{(Hopelessness)} \quad X - U(X) \neq \phi \tag{4}$$

The definition (3) of $U(X)$ allowing k-Plex candidates $Cand(X)$ before taking j-core is weak particularly for small X with its size less than k. In that case, a large number of vertices appear as k-Plex candidates independently of the length of paths connecting X and candidates. As X is extended to include vertex y more distant from X, the shortest path connecting X and y in the extension becomes longer. Consequently, such an extended Y will not be k-Plex, because there exist many vertices not adjacent to y on the path. As long as we target connected k-Plexes including X, the distance limit becomes a key to define $U_1(X) = U(X) \cap D_1(X)$ as a set of potential candidates to obtain (j, k)-MPCs.

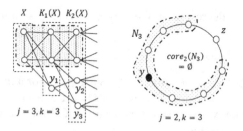

Fig. 2. Examples showing how j-core is useful.

Figure 2 shows that j-core under such a distance limit is useful, and also suggests that the construction of $U(X)$ depends on the sizes of X. In the right figure, for a singleton Y of the black circle y, $Y \cup Cand(Y)$ is the whole vertex set that is j-cored, where $j = 2$. Nothing is therefore removed by taking j-core for $Y \cup Cand(Y)$. Nevertheless, note that the vertex z cannot be consistent with Y to make a c-k-Plex, as $dist(Y, z) = 4 > k = 3$. More precisely, on the paths connecting y and z, there are 4 vertices not adjacent to z including z itself. Hence, there is no c-3-Plex involving Yz. For this reason we removed z before we test if Y is extendable to j-cored c-k-Plexes. N_3 is the set of remaining candidates together with Y. Then $core_j(N_3) = \phi$ asserting that there exists no j-cored c-k-Plex containing Y.

On the other hand, the left figure shows a case of X with $|X| = 2$. Similarly, when $k = 3$, vertices with $dist(X, z) > 3$ cannot be members of c-3-Plex including X. After excluding vertices violating the distance limit, we take 3-core. y_2, y_3 are firstly removed, and then y_1 is. The remaining part, surrounded by dotted line and filled by pattern, is 3-cored and is a maximal 3-Plex in this case.

4.1   Small C-k-Plex

A c-k-Plex X is said to be small if $|X| < k$. Note here that any connected X with small size is a c-k-Plex. In this case, therefore, $Cand(X)$ is given as $V - X$ which is generally large. To disregard useless candidates, we make a question whether there exists some c-k-Plex Z such that $Xy \subseteq Z$ for a small c-k-Plex X and $y \notin X$. When $dist(X, y) = 1$, $Z = Xy$ is a trivial positive answer, so we analyze the case of $\ell = dist(X, y) \geq 2$. Consider the shortest path $p = (y_0, y_1, ..., y_{\ell'} = y)$ from X to y in $G[Z]$, where $y_0 \in X$ and $y_1, ..., y_{\ell'} \in Z$. Then, $\ell \leq \ell'$, and y is not adjacent to X, $y_1, ..., y_{\ell'-2}$ and $y_{\ell'} = y$ itself. Z is a k-Plex, so it must be $|X| + (\ell' - 2) + 1 \leq k$. Hence we have $\ell \leq \ell' \leq k - |X| + 1$. In other words, y with $dist(X, y) > k - |X| + 1$ can never be a member of any c-k-Plex including X. So, we have more accurate definition for $U(X)$ for small case as follows.

$$U(X) = core_j(X \cup K(X)),$$
$$K(X) = \bigcup_{i=1}^{k-|X|+1} D_i(X), \text{where } k - |X| + 1 \text{ is the distance limit.}$$

The distance limit decreases as X enlarges, so $K(X)$ is monotonic decreasing.

4.2   Medium c-k-Plex

We say that a c-k-Plex X is medium if $k \leq |X| < j+k$. For a medium c-k-Plex X, any vertex whose distance from X is greater than 1 is never a k-Plex candidate, as it has no connection with at least k vertices in X. Hence, $Cand(X) = K_1(X)$, and $U(X)$ is exactly given as

$$U(X) = core_j(X \cup K_1(X)).$$

4.3   Large c-k-Plex

We say that c-k-Plex X is large if $|X| \geq j + k$. Any large X is j-cored, as $j \leq |\Gamma_X(x)|$ is derived from $j + k \leq |X|$ and $|X - \Gamma_X(x)| = |X| - |\Gamma_X(x)| \leq k$.

We need not take j-core, that is, $U(X) = X \cap Cand(X)$. So we have the rule for updating U as follows.

$$U(Xu) = U(X) \cap Cand(Xu), \text{ where } u \in U_1(X).$$

The rule for large case is the same as one for k-Plexes.

4.4   Formations Revised for (j, k)-MPCs

Using $U(X)$ thus defined depending on $|X|$, a formation $Z^f = (v_{f(1)}, ..., v_{f(|V|)})$ of (j, k)-MPC Z satisfies

$$U(Z^f_{i+1} = Z^f_i v_{f(i+1)}) \subseteq U(Z^f_i) \text{ and} \tag{5}$$

$$v_{f(i+1)} \in U_1(Z^f_i), \text{ where } U_1 \text{ depends on } |Z^f_i|, \tag{6}$$

$$U_1(Z = Z^f_{|Z|}) = \phi. \tag{7}$$

(6) is stronger than (1) since $U_1(Z^f_i) \subseteq K_1(Z^f_i)$. The condition (7) holds because $U_1(Z) \subseteq K_1(Z)$ and $K_1(Z) = \phi$ by the property of (j, k)-MPC as a k-MPC.

5   Search Control Rules, Right and Left Ones

As a c-k-Plex Z is more densely connected, the number of possible formations of Z increases. For our efficient computation, we can enjoy two search control rules, right and left rules, to avoid useless and duplicated formations.

With the right candidate control (abbreviated as RCC), we can exclude many useless formations. A formation $Z^f = (v_{f(1)}, ..., v_{f(|Z|)})$ can be obtained by extending an intermediate Z^f_i with a vertex in $U_1(Z^f_i)$. In other words, for complete enumeration of formations, we have to examine every vertex in $U_1(Z^f_i)$ to extend Z^f_i. Some of them, however, result in non-maximal k-Plexes. Extending an idea discussed in [13,14], we can identify the candidate set, $R(Z^f_i)$, with the property that adding member vertices in $R(Z^f_i)$ only to the present Z^f_i never achieves a maximal k-Plex. Thus, the set of vertices actually used to extend Z^f_i is given as $NR(Z^f_i) = U_1(Z^f_i) - R(Z^f_i)$, called non-right candidates. Our RCC here can be regarded as an extended version of RCCs in [15] in the sense that it is much more applicable for non-small Z^f_i than that in [15]. Although we skip the details due to space limitation, interested readers can refer to [16].

In addition to RCC, we can enjoy a left candidate control (abbreviated as LCC) with which just one formation for each (j, k)-MPC can be composed. LCC is in some sense a standard technique for set enumeration [17] and is almost similar to what is stated in [13]. Roughly speaking, when we extend an intermediate Z^f_i, we do not need to care any vertex y such that $y \prec v_{f(i)}$, called a left candidate, because any formation obtained by extending Z^f_i with such a y can be composed by extending another intermediate formation with some left candidate $\ell(\neq y)$. Thus, the set of vertices we actually use to extend Z^f_i is given by

$NR(Z_i^f) - L(Z_i^f)$, where $L(Z_i^f)$ is the set of left candidates. Although we do not go into the details, effects of the control rules in the form of search tree are illustrated in the next section.

6   Algorithm for (j, k)-MPCs

We present an algorithm in Fig. 4 for making formations for (j, k)-MPCs. The algorithm is written using recursive calls of procedure $Expand$. This realizes a depth-first search of (j, k)-MPCs.

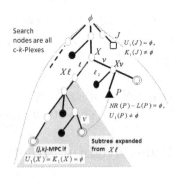

Fig. 3. Search tree

We depict in Fig. 3 the invocation process of $Expand$ in the form of search tree with c-k-Plexes as its nodes. In the search tree, a path from the root, an empty set $\emptyset$, to a leaf just corresponds to a formation of c-k-Plex. Dark circle shows hopeless c-k-Plexes from which no branch is expanded. Double circle is a (j, k)-MPC. Single circle has chance of extension by choosing some of $NR(X) - L(X) \neq \phi$. Dark triangle is a c-k-Plex P such that every non-right candidate is left. This means that (j, k)-MPC Z obtained by extending X is generated by another branch with some left candidates ℓ along the path. In other words, with some initial segment X of Z and its non-left $\ell \in NR(X)$, Z appears in the subtree rooted by $X\ell$. Finally, white square J is a j-cored c-k-Plex which is not k-MPC. It is maximal in the sense that there exists no j-cored c-k-Plexes including J. Although it is straightforward to exclude such a J with the condition $K_1(X) \neq \phi$, we allow to output J in addition to all possible (j, k)-MPCs.

Another point we have to note here is that every (j, k)-MPC is a subset of global j-core $core_j(V)$. In addition, since $core_j(V)$ consists of several connected components, and since our targets must be connected, we make the algorithm run for each connected component C of $core_j(V)$.

For setting parameters k and j, we assume a preferable size range $[n_1, n_2]$ and a density parameter τ. Then the connection lower bound will be $j = n_1\tau$, and the disconnection upper bound $k = n_2 * (1 - \tau)$.

procedure MAIN(G^{input}):
   [Input] $G^{input} = (V^{input}, \Gamma^{input})$: an input graph.
   [Output] All maximal j-cored c-k-Plexes including (j,k)-MPCs
   begin
      for each connected component C of $core_j(V^{input})$
         Let $G = (C, \Gamma = \Gamma^{input}_C)$;
         $\tilde{u} = \arg \max_{u \in C} deg(u)$;
         $NR = C - \Gamma(\tilde{u})$; $L = \phi$;
         $X = \phi$; $U = NR$;
         $Expand(X, NR, L, U)$;
      endfor
   end

procedure EXPAND(X, NR, L, U):
    // X: c-k-Plex, $NR = NR(X), L = L(X); U = U(X)$ (depending on $|X|$)
   begin
      $U_1 = U \cap D_1(X)$; //for non-small case, $U_1 = U$
      if $(U_1 = \phi)$ **then**
         print X; return; // maximal j-cored c-k-Plex including (j,k)-MPC
      endif
      for each $v \in NR - L$ // the order accords to $\prec$
         $Xnew = Xv$;
         $Unew = U(Xnew)$;
         if $Xnew - Unew \neq \phi$ **then** continue; // $Xnew$ is hopeless
         $NRnew = NR(Xnew)$; $Lnew = (L(X) \cup \{v\}) \cap Unew$;
         $Expand(Xnew, NRnew, Lnew, Unew)$;
      endfor
   end

Fig. 4. Enumeration algorithm for (j,k)-MPCs

7 Experiments

We present in this section our experimental results. The proposed system, referred to as JKMPC, has been coded in Java and executed on a PC with Intel® Core™-i7 (1.7 GHz) processor and 8 GB memory. For several datasets, we observe computation times and quality of solutions as pseudo-cliques.

In order to verify practical efficiency of JKMPC, we compare it with MaxKplexEnum, a state-of-the-art maximal connected k-Plex enumerator [1][2].

As a synthetic dataset, we have created a small-world network, referred to as WS, based on *Watt-Strogatzs Model* [20]. As real datasets, we have prepared benchmark networks, DBLP and Google [21]. DBLP is a collaboration network constructed from the DBLP Computer Science Bibliography. Authors as nodes are connected if they have published a paper together. Google is a Web graph consisting of web pages and their hyperlinks. Scale and degree distributions of those networks are presented in Fig. 5.

As far as we know, there exists no algorithm devoted to enumerate all (j,k)-MPCs. In this sense, it is difficult to farely compare JKMPC with others. However, our target k-MPCs can be obtained by any maximal k-Plex enumerator with j-coreness check. Therefore, we here observe computational performance of a state-of-the-art enumerator, MaxKplexEnum [1], and our JKMPC in order to verify practical efficiency of our system.

---

[2] Its source codes in Python have been kindly provided by the authors of [1].

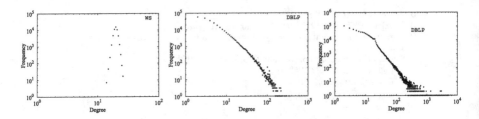

Fig. 5. Scale and degree distributions of networks

Given a network, `MaxKplexEnum` can enumerate a designated number N of maximal k-Plexes. That is, in order to completely obtain our target (j, k)-MPCs with `MaxKplexEnum`, we have to give an adequate value of N which can provide a set of candidates completely including all of our solutions. It is, however, impossible to identify such a suitable N in advance. Therefore, in our comparison here, we first run `JKMPC` and identify the number of our solutions, say $\tilde{N}$. Then, we try to detect $\tilde{N}$ maximal k-Plexes by `MaxKplexEnum`. It is noted here that those k-Plexes extracted by `MaxKplexEnum` do not always contain all of our solution (j, k)-MPCs. That is, the value of $\tilde{N}$ can provide `MaxKplexEnum` the best-case scenario.

7.1   Computational Performance

Figure 6 shows computation times by both systems, where solid lines are for `JKMPC` and dotted lines for `MaxKplexEnum`. Data points for each k-value are distinguished by point types (e.g. ●, □). Moreover, missing points mean we have failed to extract all solutions within 12-hours.

The larger a j-value is, the smaller the number of (j, k)-MPCs becomes. That is, the task of `MaxKplexEnum` would be easier at a higher range of j because it is required to enumerate a smaller number of maximal k-Plexes. In fact, `MaxKplexEnum` can run much faster than ours at higher ranges of j. However, its performance suddenly gets worse, as j becomes slightly smaller. In most of the cases, it has failed to complete enumerations within the time limit. For example, in case of `WS` with $k = 3$ and $j = 9$, the number of solutions `MaxKplexEnum` has

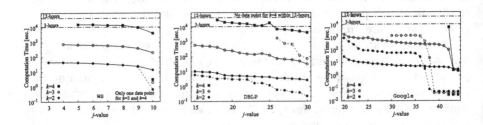

Fig. 6. Computation times

to extract is about 20 times larger than that in case with $j = 10$. The difference has caused a rapid increase of computation time.

On the other hand, the performance of JKMPC is almost stable. Even in cases quite hard for MaxKplexEnum, our system can enumerate all solutions. In other words, JKMPC has an ability for extracting (j, k)-MPCs even with relatively small size. This is a remarkable advantage of our system because it is quite difficult for standard methods of graph clustering and partitioning to detect such small dense subgraphs. Thus, it is verified that JKMPC is an efficient and practical system for enumerating (j, k)-MPCs.

7.2   Quality of Solutions as Pseudo-Cliques

Since the notion of k-Plex has originally been proposed as a relaxation model of clique, any maximal k-Plexes with lower densities would be undesirable. In order to examine quality of solution k-Plexes from the view point of pseudo-cliques, we observe density distributions of solutions obtained by MaxKplexEnum and JKMPC.

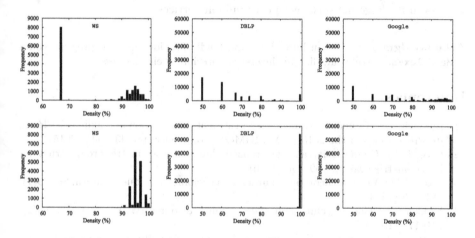

Fig. 7. Density distributions of solution k-Plexes

Figure 7 shows density distributions of solutions for each network. Upper figures are distributions by MaxKplexEnum and lowers by JKMPC. For WS, solutions by MaxKplexEnum have density of 0.80 on average with the standard deviation of 0.14. For DBLP and Google, the average densities are 0.65 and 0.75 on average, where their standard deviations are 0.15 and 0.17, respectively. On the other hand, for WS, the average density by JKMPC is 0.95 and the standard deviation is less than 0.018. For both DBLP and Google, moreover, their average densities are about 0.99 with the standard deviations less than 0.0018. It is obvious that solutions by JKMPC have sufficient densities. All of them would be regarded as pseudo-cliques without any doubt. However, solutions by MaxKplexEnum include many maximal k-Plexes with relatively low densities. In other words, quality of solutions by `MaxKplexEnum` is unstable. Thus, `JKMPC` can practically work as an effective pseudo-clique detector.

8   Conclusion

In this paper, we designed an efficient complete algorithm for enumerating (j, k)-MPCs. For efficient computation, we discussed several search mechanisms which can effectively prune many useless search nodes. Our experimental results showed the algorithm can work well as a practical tool for extracting densely-connected pseudo cliques in large networks.

Although our j-coreness constraint can drastically reduce the number of solutions to be enumerated, we would still suffer its hugeness when we are concerned with much larger scale networks. It is, therefore, worth introducing some additional constraint which more tightly targets our solutions. The authors are currently developing a reasonable constraint, taking distance among communities or separateness between clusters into account. Based on the consideration, we expect efficiency of our algorithm can further be improved and it can practically work even for huge networks with over million vertices.

Acknowledgments. The authors of [1] have kindly provided us their program codes of `MaxKplexEnum`. We would like to sincerely appreciate their kindness.

References

1. Berlowitz, D., Cohen, S., Kimelfeld, B.: Efficient enumeration of maximal k-Plexes. In: Proceedings of the 2015 ACM SIGMOD Conference, pp. 431–444 (2015)
2. Luce, D.R.: Connectivity and generalized cliques in sociometric group structure. Psychometrika **15**(2), 169–190 (1950)
3. Alba, R.D.: A graph-theoretic definition of a sociometric clique. J. Math. Soc. **3**(1), 113–126 (1973)
4. Mokken, R.: Cliques, clubs and clans, quality & quantity. Int. J. Meth. **13**(2), 161–173 (1979)
5. Watts, D.J., Strogatz, S.H.: Collective dynamics of small-world networks. Nature **393**(6684), 440–442 (1998)
6. Seidman, S.B., Foster, B.L.: A graph-theoretic generalization of the clique concept. J. Math. Soc. **6**(1), 139–154 (1978)
7. Abello, J., Resende, M.G.C., Sudarsky, S.: Massive quasi-clique detection. In: Rajsbaum, S. (ed.) LATIN 2002. LNCS, vol. 2286, pp. 598–612. Springer, Heidelberg (2002)
8. Luxburg, U.: A tutorial on spectral clustering. Stat. Comput. **17**(4), 395–416 (2007)
9. Newman, M.E.J.: Finding community structure in networks using the eigenvectors of matrices. Phys. Rev. E **74**(3), 036104 (2006)
10. Scott, J.P., Carrington, P.J. (eds.): The SAGE Handbook of Social Network Analysis. Sage Publications, London (2011)
11. Pattillo, J., Youssef, N., Butenko, S.: Clique relaxation models in social network analysis. In: Thai, M.T., Pardalos, P.M. (eds.) Handbook of Optimization in Complex Networks: Communication and Social Networks. Springer Optimization and Its Applications, vol. 58, pp. 143–162. Springer, New York (2012)

12. Seidman, S.B.: Network structure and minimum degree. Soc. Netw. **5**(3), 269–287 (1983)
13. Tomita, E., Tanaka, A., Takahashi, H.: The worst-case time complexity for generating all maximal cliques and computational experiments. Theor. Comput. Sci. **363**(1), 28–42 (2006)
14. Eppstein, D., Strash, D.: Listing all maximal cliques in large sparse real-world graphs. In: Pardalos, P.M., Rebennack, S. (eds.) SEA 2011. LNCS, vol. 6630, pp. 364–375. Springer, Heidelberg (2011)
15. Wu, B., Pei, X.: A parallel algorithm for enumerating all the maximal k-Plexes. In: Washio, T., Zhou, Z.-H., Huang, J.Z., Hu, X., Li, J., Xie, C., He, J., Zou, D., Li, K.-C., Freire, M.M. (eds.) PAKDD 2007. LNCS (LNAI), vol. 4819, pp. 476–483. Springer, Heidelberg (2007)
16. Okubo, Y., Haraguchi, M., Tomita, E.: Structural change pattern mining based on constrained maximal k-Plex search. In: Ganascia, J.-G., Lenca, P., Petit, J.-M. (eds.) DS 2012. LNCS, vol. 7569, pp. 284–298. Springer, Heidelberg (2012)
17. Rymon, R.: Search through systematic set enumeration. In: Proceedings of International Conference on Principles of Knowledge Representation Reasoning - KR 1992, pp. 539–550 (1992)
18. Slater, N., Itzchack, R., Louzoun, Y.: Mid size cliques are more common in real world networks than triangles. Netw. Sci. **2**(3), 387–402 (2014)
19. Batagelj, V. and Zaversnik, M.: An $O(m)$ algorithm for cores decomposition of networks. In: CoRR 2003, cs.DS/0310049 OpenURL
20. Watts, D.J., Strogatz, S.H.: Collective dynamics of small-world networks. Nature **393**, 440–442 (1998)
21. Leskovec, J., Krevl, A.: SNAP datasets: Stanford large network dataset collection (2014). http://snap.stanford.edu/data

Incorporating Heterogeneous Information for Mashup Discovery with Consistent Regularization

Yao Wan[1](✉), Liang Chen[2], Qi Yu[3], Tingting Liang[1], and Jian Wu[1]

[1] College of Computer Science and Technology, Zhejiang University, Hangzhou, China
{wanyao,liangtt,wujian2000}@zju.edu.cn
[2] School of Computer Science and Information Technology, RMIT, Melbourne, Australia
liang.chen@rmit.edu.au
[3] College of Computing and Information Sciences, Rochester Institute of Technology, Rochester, USA
qi.yu@rit.edu

Abstract. With the development of service oriented computing, web mashups which provide composite services are increasing rapidly in recent years, posing a challenge for the searching of appropriate mashups for a given query. To the best of our knowledge, most approaches on service discovery are mainly based on the semantic information of services, and the services are ranked by their QoS values. However, these methods can't be applied to mashup discovery seamlessly, since they merely rely on the description of mashups, but neglecting the information of service components. Besides, those semantic based techniques do not consider the compositive structure of mashups and their components. In this paper, we propose an efficient consistent regularization framework to enhance mashup discovery by leveraging heterogeneous information network between mashups and their components. Our model also integrates mashup discovery and ranking properly. Comprehensive experiments have been conducted on a real-world ProgrammableWeb.com (http://www.programmableweb.com) dataset with mashups and APIs (In ProgrammableWeb.com, APIs are the service components of mashups. Our model verified on the ProgrammableWeb.com dataset could also be applied to other compositive service discovery scenarios.). Experimental results show that our model achieves a better performance compared with ProgrammableWeb.com search engine and a state-of-the-art semantic based model.

Keywords: Mashup discovery · Ranking · Heterogeneous · Regularization

1 Introduction

With the development of service oriented computing and the increasing demand of service consumers, a single web service is far from enough to satisfy the

J. Bailey et al. (Eds.): PAKDD 2016, Part I, LNAI 9651, pp. 436–448, 2016.
DOI: 10.1007/978-3-319-31753-3_35

complex demand of users, contributing to the boom of composite services and mashups. More and more developers are using mashup technologies to build their own services, and publishing them for other users to invoke. Statistics from ProgrammableWeb.com, a popular mashups and APIs management platform, show that the number of mashups has reached up to 6,094, and that the number of APIs has reached up to 10,634 by November 2014. In addition, the mashups are increasing on a daily basis. The rapid increase of mashups demands a systematic approach to discover mashups with great accuracy and efficiency.

There have been many research works focusing on service discovery and mashup discovery, which can be roughly divided into three categories: the traditional web service discovery, mashup discovery and mashup components discovery. Firstly, some literatures such as [9,13] propose reasoning-based similarity algorithm to retrieve satisfied web services from the formalized description languages such as OWL-S and WSMO. But these methods cannot be applied to mashup discovery seamlessly for the reason that there is no formalized description for mashups. Secondly, in [10,12], some semantic-based approaches are used for mashup discovery, while only the semantic information of mashups is considered, neglecting the relationship between mashups and their components. Thirdly, many research works [3,11,14] focus on discovery and recommendation for mashup components. Furthermore, ProgrammableWeb.com has its own mashup search engine, but how the search algorithm works is unknown to the public. According to our study, the mashup search system of ProgrammableWeb.com only supports weak semantics searching and gives low recall, losing many relevant results with similar semantics. For example, when searching "film", many mashups about "movie" will be lost, and when searching "Cellular phone", no results about "mobile" will be discovered.

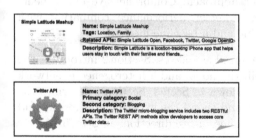

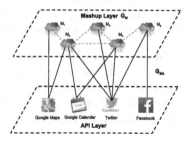

Fig. 1. An example of mashup and API   **Fig. 2.** A heterogeneous graph

In addition to the semantic information of mashups, the semantic information of their related components and the relationship between them should also be taken into consideration. Taking ProgrammableWeb.com as an example, it not only records the information of each mashup and API, but also the composition relationship between them. Figure 1 shows the detailed information of an mashup and API. Specifically, we can find that the tags, description of *Simple Latitude*

Mashup, as well as APIs of which the mashup is composed from the website. Based on the information of Fig. 1, a composition network between mashups and their related APIs can be constructed. Links exist between mashups and APIs through the relation of "include" and "included by". Besides, in our paper we will also explore the relationship between mashups. If two mashups consist of a common API, there should be a link between them. These two networks containing the two described kinds of links are called heterogeneous network in our paper. A sample heterogeneous network of ProgrammableWeb.com is shown in Fig. 2. In this figure, G_{MA} represents the network between mashups and APIs, G_M represents the network on the mashups layer.

The aim of this paper is to improve the discovery of mashups by introducing the semantic information of mashups as well their components, and leveraging the heterogeneous information network between mashups and their components. We first construct a heterogeneous network on mashups and APIs. A probabilistic model is proposed to calculate the relevance score of each mashup, integrating the semantic information of mashups and APIs, as well as the composition network between mashups and APIs. Furthermore, a regularized framework is proposed to ensure the consistency between mashups.

The contribution of this paper is summarized as follows:

- A probabilistic model is proposed to leverage the semantic information of mashups as well as their components. Besides, this model also integrate discovery and ranking process properly.
- A heterogeneous information network between mashups and their components is constructed and a regularized framework with consistency hypothesis is proposed to ensure the similarity consistency between mashups.
- We crawl 4,699 mashups and 937 related APIs from ProgrammableWeb.com to evaluate the performance of our approach. Comprehensive experiments show that our approach achieves a better performance comparing with ProgrammableWeb.com search engine and a baseline method.

The remainder of this paper is organized as follows. Section 2 introduces the related work of this paper. A probabilistic model and a heterogeneous network based mashup discovery approach are shown in Sect. 3. Section 4 describes the datasets we will use in our experiment and shows the experimental results and analysis. Finally, we conclude and give some future directions in Sect. 5.

2   Related Work

In service oriented computing, discovery of appropriate web services has always been a hot research topic. A number of approaches have been proposed for service discovery. Many semantic-based approaches develop reasoning-based similarity algorithms to retrieve relevant web services described using semantic web languages such as OWL-S and WSMO [9,13].

With the increase of mashup services, those traditional methods of service discovery cannot be applied seamlessly to mashup discovery, since most of them only consider the information extracted from WSDL documents. Recently, more and more research works are focusing on mashup searching or mashup discovery. In [10], Li et al. provided a semantics extended framework to improve the precision and recall of mashup discovery as well as to improve the performance of the mashup discovery processing time. Elmeleegy et al. [8] presented a recommendation tool named *MashupAdvisor*, which used a semantic matching algorithm and a metric planner to modify the mashup to produce the suggested output. Bianchini et al. [2] proposed a recommendation system to design mashup applications based on the semantic description of mashup components, according to their similarities with designer's requirements and their mutual coupling.

Recently, some approaches using social networks to discover services are proposed. In [15], the authors combine current discovery techniques with social information as a mechanism to trade off exploration and exploitation. In [16], Zhou et al. provided an approach that learns a semantic Bayesian network with a semi-supervised learning method to build a web mashup network. Cao et al. [3] proposed a recommendation approach for mashup service that utilizes both users' interests and the social network based on relationships among mashups, APIs and tags. The inspiration of our paper mainly stems from some research on expertise finding. In [7], Deng et al. proposed a joint regularized framework to improve expertise retrieval by modeling heterogeneous networks as regularization constraints. In [18], an incremental method based on multiple graphs was proposed for document recommendation in a digital library. Besides, [17] provides a strong theoretical support for learning with local and global consistency. Inspired by those works, our paper builds a heterogeneous social network between mashups and their components, and a regularized framework is proposed to ensure the consistency.

3   Heterogeneous Network Based Mashup Discovery

3.1   Baseline Model for Mashup Discovery

The probabilistic model proposed in this subsection mainly follows the basic idea of a document-centric probatilistic model, which is proposed to estimate the expertise of a candidate by summing the relevance of its associated documents [7]. In the contex of mashup discovery, we denote the relevance score of candidate mashup m_i related to a given query q as $p(m_i|q)$, and according to the document-centric model,the relevance score of a mashup related to a given query can be formulated as:

$$p(m_i|q) = \lambda \sum_{a \in \mathscr{A}_{m_i}} p(m_i|a)p(a|q) + (1-\lambda)p(m_i|q)$$
$$\propto \lambda \sum_{a \in \mathscr{A}_{m_i}} p(m_i|a)p(q|a)p(a) + (1-\lambda)p(q|m_i)p(m_i) \tag{1}$$

where $\mathscr{A}_{m_i}$ denotes the APIs set of which mashup m_i is composed of; $p(m_i|a)$ denotes the probability of mashup m_i being relevant to a given API a; $p(q|a)$ and

$p(q|m_i)$ denote the semantic similarities of API and mashup, respectively, for a given query; $p(a)$ and $p(m_i)$ can be seen as the quality of API a and mashup m_i, respectively; $\lambda(0 \leq \lambda \leq 1)$ is a tuning parameter used to determine how much the relevance score of a candidate mashup relies on the candidate mashup itself and its API components. In this equation, the Bayes' theorem is applied. Intuitively, we argue that when we calculate the relevance score of a mashup candidate, we should not only consider the information of the mashup, but also the information of its components.

The right hand of Eq. (1) can be divided into two parts. The first term represents the relevance score contributed by APIs by aggregating the relevance scores of APIs directly associated with a mashup. The second term denotes the relevance score from the mashup itself. Those two terms are combined by a tuning parameter λ. When $\lambda = 0$, only the information of mashup is considered, and when $\lambda = 1$, only the information of mashup's components is considered, otherwise, the semantic information of mashup as well as its components are integrated to improve the performance of mashup discovery.

Specifically, in this model, $p(m_i|a)$ represents the association between the candidate and its components. Suppose that mashup m_i consists of n_{m_i} APIs, then $p(m_i|a) = 1/n_{m_i}$ if a is a component of m_i, and zero otherwise. $p(q|a)$ measures the relevance between q and API a, while $p(q|m_i)$ measures the relevance between q mashup m_i. These two probabilities can be determined by using the language model. In this paper, we use Latent semantic indexing (LSI) mothod [6] to calculate the semantic similarities between a query and APIs or mashups. Before applying LSI, some standard process of natural language processing such as *case folding, tokenization, pruning, stemming* and *spell correcting* will be conducted on all APIs and mashups.

In addition, the prior probability $p(a)$ and $p(m)$ can be viewed as the quality of an API and mashup respectively, which generally follow the uniform distribution. Indeed, the quality of an API or mashup can also be set to be how popular the API or mashup is. In the context of our problem, since the popularity of mashups are difficult to measure, the qualities of mashups are set to be uniform. However, in the information network of APIs and mashups, we define the popularity of a particular API as the number of times that it is used in the formation of mashups. For example, in Fig. 2, the *"Twitter API"* is used by four mashups, then the popularity of *"Twitter API"* is 4. From our analysis, we can find the distribution of API popularity is long tailed [1]. So we estimate $p(a)$ by the logrithm of the popularity of API.

For simplicity, let $\mathbf{x}$ be the relevance vector between API a_i and query q with $x_i = p(q|a_i)$, $\mathbf{y}$ be the relevance vector between mashup m_i and query q with $y_i = p(q|m_i)$, $\mathbf{Q}_A$ and $\mathbf{Q}_M$ be the diagonal matrix which represent the quality of APIs and mashups respectively, and $\mathbf{P}_{MA}$ be the composition matrix between mashups and APIs. The primary model as shown in Eq. (1) can be rewritten as:

$$\mathbf{z} = \lambda \mathbf{P}_{MA}\mathbf{Q}_A\mathbf{x} + (1 - \lambda)\mathbf{Q}_M\mathbf{y} \tag{2}$$

where $\mathbf{z}$ represents the relevance score vector of all candidate mashups, and λ is the tuning parameter which controls the weight between the mashups and their components.

3.2   Heterogeneous Information Incorporation

In the previous subsection, we utilize the description of mashups and their components to measure the relevance of candidate mashups for a given query. In addition to the textual document information, some information of the heterogeneous network should also be considered. In this subsection, we will describe a mashup consistency hypothesis and enforce the hypotheses by defining regularization constraints.

Mashup Consistency Hypothesis: *If two mashups share many common services with respect to a given query, then their relevance score in the queried field should be similar in some sense.* As shown in Fig. 2, mashup m_2 is composed of *"Google Maps API"*, *"Google Calendar API"* and *"Twitter API"*, while mashup m_4 is composed of *"Twitter API"* and *"Google Calendar API"*, these two mashups share two common APIs, so we can consider that their functionality and quality are similar in some sense, and their relevance for a given query should also be similar.

According to [17], we enforce the above hypothesis by defining the regularization constraints. Suppose we are given a mashup graph $G_M = (V_M, E_M)$, which is a weighted undirected graph. Suppose that the pairwise similarities among the mashups are described by matrix $S_M \in \mathbb{R}^{|M| \times |M|}$ measured based on G_M. Thus, we formulate to minimize a regularization loss function as follows:

$$\Omega(\mathbf{z}) = \mathbf{z}^T (\mathbf{I} - \mathbf{S}_M)\mathbf{z} + \mu \left\| \mathbf{z} - \mathbf{z}^0 \right\|^2 \tag{3}$$

$$s.t. \quad \mathbf{z}^0 = \lambda \mathbf{P}_{MA} \mathbf{Q}_A \mathbf{x} + (1 - \lambda) \mathbf{Q}_M \mathbf{y} \tag{4}$$

where $\mu > 0$ is the regularization parameter. The first term of the loss function defines the *mashup consistency*, which prefers small difference in relevance scores between nearby mashups; the second term is the *fitting constraint* that measures the difference between final scores $\mathbf{z}$ and the initial relevance scores $\mathbf{z}^0$. The initial relevance score vector $\mathbf{z}^0$ can be calculated according to Eq. (2) in the probabilistic model. Setting $\partial\Omega(\mathbf{z})/\partial\mathbf{z} = 0$, we can see that the solution $\mathbf{z}^*$ is essentially the solution to the linear equation:

$$(\mathbf{I} - \alpha \mathbf{S}_M)\mathbf{z}^* = (1 - \alpha)\mathbf{z}^0 \tag{5}$$

where $\alpha = 1/(1+\mu)$. Since the matrix $\mathbf{S}_M$ is usually very sparse, calculating the inversion of $\mathbf{S}_M$ is of high time complexity. One solution to the above equation is using a powerful iterative method [17]:

$$\mathbf{z}^{t+1} \leftarrow \beta \mathbf{S}_M \mathbf{z}^t + (1 - \beta) [\lambda \mathbf{P}_{MA} \mathbf{Q}_A \mathbf{x} + (1 - \lambda) \mathbf{Q}_M \mathbf{y}] \tag{6}$$

where $\beta = 1/(1+\mu)$, $\mathbf{z}^* = \mathbf{z}^\infty$ is the solution.

Now the interesting question is how to calculate $\mathbf{S}_M$ among set M. For graph data, a number of works [4] have been given on obtaining the similarity measures. For undirected graph, $\mathbf{S}_M$ is simply the normalized adjacency matrix $\mathbf{W}$:

$$\mathbf{S}_M = \mathbf{\Pi}^{-1/2}\mathbf{W}\mathbf{\Pi}^{1/2} \tag{7}$$

where $\mathbf{W}$ is the adjacency matrix of mashups in G_M, $\mathbf{W}_{ij} = 1$ if node i is linked to node j, otherwise, $\mathbf{W}_{ij} = 0$, and $\mathbf{\Pi}$ is a diagonal matrix with $\mathbf{\Pi}_{ii} = \sum_j \mathbf{W}_{ij}$.

3.3   Implementation

Figure 3 shows the framework of the proposed heterogeneous information network based mashup discovery, which integrates the semantic information of mashups and their components, as well as the similarity consistency between mashups. The proposed framework is comprised of three components: *heterogeneous network construction*, *data processing*, and *mashup ranking*. The construction of heterogeneous network and data processing can be performed offline, while the mashup ranking part should be conducted online according to the query specified by a user.

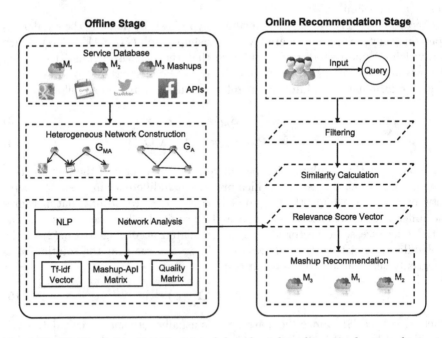

Fig. 3. The heterogeneous network based mashup discovery framework

4   Experiments

4.1   Experimental Setup

Dataset. ProgrammableWeb.com is one of the most popular platforms that has collected lots of APIs and mashups used in Web and mobile applications. To evaluate our proposed approach, we crawl all the mashups and their related APIs from ProgrammableWeb.com as we can.

From the statistics, we totally get 4,699 mashups and 937 APIs in our data collection. After the construction of the heterogeneous graph, we observe that there are many edges on the mashup graph, while relatively few edges in the graph of mashups and APIs. As for G_M the density of matrix is nearly 17.0%, while for G_A the density of matrix is only 0.18%.

Evaluation Metrics. For the evaluation, several categories of Web search evaluation metrics are used to measure the performance of our proposed model from different aspects, including some relevance based metrics, ranking based metrics and diversity based metrics. To measure the relevance of our search results, we use the precision at rank k ($P@K$) which is widely used and is defined as: $P@K = \frac{\#\ relevant\ in\ top\ K\ results}{K}$. $P@K$ measures the fraction of the top-K retrieved results that are relevant for the given query. From the ranking aspect, we use Mean Reciprocal Rank (MRR) to evaluate the ranking of our search results. A larger MRR value means a better result. The MRR is defined as: $MRR = \frac{1}{|Q|} \sum_{i=1}^{|Q|} \frac{1}{rank_i}$, where $|Q|$ is the size of query set. We expect that our search results should not only have a high precision and reasonable ranking, but also have a high diversity. Following [5], we use the α-DCG metric to measure the novelty and diversity of our retrieved results. The α-DCG is defined as: α-$DCG_K = \sum_{i=1}^{K} \frac{G_i}{log_2(i+1)}$, where $G_i = \sum_{j=1}^{n} J(m_i, j)(1-\alpha)^{\sum_{k=1}^{i-1} J(m_k,j)}$; n is the total number of topics the searching results contains; $J(d_i, j) = 0$, if result m_i contains topic j, otherwise, $J(m_i, j) = 1$. $\sum_{k=1}^{i-1} J(m_k, j)$ represents the degree of diversity and novelty of the searching result; α is a parameter. In our model, α is set to be 0.5.

4.2   Comparison

In this subsection, comparisons between our method and the following approaches have been made to show the effectiveness of our proposed approach.

- **PW Search Engine**: The ProgrammableWeb.com has its own search engine for mashups and APIs discovery. From our observation, we find that the search results is mostly based on the name, description and tags/categories of mashups/APIs. In addition, the search results are all ranked by their updated date.
- **MD-Sim**: This method which has been used in many state-of-the-art works just utilizes the semantic information of mashups. The semantic similarities between mashups and query are calculated by the LSI method.

- **MD-Sim+**: Compared with the above two methods, our probabilistic model (MD-Sim+) employs the semantic information of mashups, as well as the semantic information of the related APIs which are the components of mashups. The qualities of related APIs are also introduced in this approach.
- **MD-HIN**: This method is the extended version of the probabilistic model (MD-Sim+). It proposes a consistency hypothesis on mashups firstly, and a regularization constraints is employed.

Before comparing the performance of the above four methods, several points should be made clear first. Since there is no published ground truth for comparison, we select twenty queries of different topics to evaluate the performance on the above metrics. According to the search results, if the categories of mashup are related to a query, then the mashup will be considered to be relevant with the query. We judge the degree of relevance by the number of followers of mashups. Since most mashups have more than one tags, we will use the tags to evaluate the diversity of results. The parameter settings of our approaches are $\lambda = 0.4$, $\beta = 0.5$, $\#iteration = 100$, and topic number of LSI is set as 20. The experimental results are shown in Table 1, and the detailed investigations of parameter settings will be provided in Sects. 4.3 and 4.4.

Table 1. Experimental results of our proposed method and other methods. The percentages of relative improvements(%) are also shown in this table.

	P@10	P@20	P@50	MRR	α-DCG
PW	**0.595**	0.493	0.431	-	-
MD-Sim	0.555	0.488	0.553	0.121	2.920
MD-Sim+	0.575	0.495	0.534	0.137	2.916
(vs **MD-Sim**)	+3.60 %	+1.43 %	-3.44 %	+13.22 %	-0.01 %
MD-HIN	0.555	**0.53**	**0.537**	**0.160**	**3.027**
(vs **PW**)	-6.72 %	+7.51 %	+24.59 %	-	-
(vs **MD-Sim**)	0.00 %	+8.61 %	-2.89 %	+32.23 %	+3.66 %

Based on the results in Table 1, we have the following observations:

- From the perspective of $P@K$, when the value of K is small, the performance of ProgrammableWeb.com search engine is a litter better than our approach (MD-HIN). While when the value of K is large, our approach has a great advantage over ProgrammableWeb search engine. This is because as K increases, the ProgrammableWeb.com search engine is unable to discover so many mashups, which just depends on the utilization of the mashup information, while our approach can find more related mashups by incorporating the mashups and APIs' information and leveraging the heterogeneous social network between them.

- From the perspective of ranking of results, our extented approach (MD-HIN) achieves better performance than the probabilistic approach (MD-SIM) since introducing the quality of APIs along with the similarity consistency on mashup social network, making sure that the mashups with similar quality will have similar ranking scores.
- Among all the discovery methods, our proposed method (MD-HIN) generally achieves better performance on both $P@20$, MRR and α-DCG, indicating that integrating the semantic information of mashups with APIs, and considering the similarity consistency on mashup social network will facilitate and improve the discovery of mashups. These experimental results demonstrate that our model leveraging the heterogeneous social network is practical and effective.

4.3 Impact of λ

In our model, the parameter λ controls how much our method relies on the semantic information of mashups and their related APIs. To study the impact of λ on $P@20$, MRR and α-DCG, we vary λ from 0 to 1 with a step value 0.1. The experimental results are shown in Fig. 4. Figure 4(a) shows that optimal λ value settings can achieve better performance of mashup discovery, which demonstrates that fusing the information of mashups and their related APIs with our proposed approach will improve the discovery accuracy. As λ increases, the $P@20$ value increases at first, but when λ surpasses a certain threshold, the $P@20$ value decreases with further increase of the value of λ. This phenomenon confirms the intuition that purely using the semantic information of mashups or purely employing the semantic information of their related APIs cannot generate better performance than fusing these two factors together. From Fig. 4(b) and (c), we can find that the value of λ also has an impact on the MRR and α-DCG although the impact is small. This demonstrates that when we introduce the semantic information and quality of APIs, the ranking and diversity of search results will be improved.

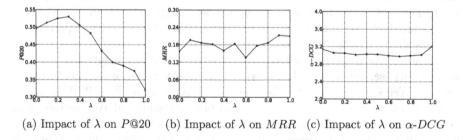

(a) Impact of λ on $P@20$ (b) Impact of λ on MRR (c) Impact of λ on α-DCG

Fig. 4. Impact of λ

4.4   Impact of β

In our model, $\beta = 1/(\mu+1)$ where μ is a regularization parameter which controls the difference between final score $\mathbf{z}$ and the initial score $\mathbf{z}^0$. To study the impact of β on the metrics of our approach, we change β from 0 to 1 with a step value 0.1. We set $\lambda = 0.4$, Top-K=20, and $\#iteration = 100$ in this experiment. From Fig. 5(a), we can find that value β has a significant impact on the precision of the discovery results. When β increases to 1, the $P@20$ will decrease rapidly. This is because that when β is near to 1, μ is near to 0, in this condition, we only consider the similarity constraints of mashups, neglecting the constraint that the final score value should be fitting to the initial value. Figure 5(b) and (c) show that the value of β still has an effect on MRR and α-DCG, although the effect is not obvious.

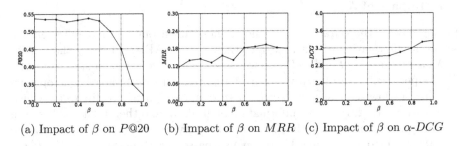

(a) Impact of β on $P@20$ (b) Impact of β on MRR (c) Impact of β on α-DCG

Fig. 5. Impact of β

5   Conclusion and Future Work

Based on some traditional semantic-based service discovery methods, we propose an approach to improve mashup discovery by integrating the semantic information of mashups and their related APIs. Besides, a similarity consistency between mashups is proposed and a regularization framework is employed to achieve better performance. Comprehensive experiments on a real-world ProgrammableWeb.com dataset are conducted, and the extensive experimental analysis shows the effectiveness of our approach.

Although the data crawled from ProgrammableWeb.com is sufficient for evaluation purpose, we believe that there is a necessity for an experiment on traditional Web services described by a standard language such as WSDL. In our future work, we plan to build a social network on traditional Web services and verify our model. In addition, we will extend our ground truth with more queries, and more information retrieval evaluation metrics will be introduced to enhance the quality of our work. Furthermore, we are going to conduct more research on the social network, a more complex social network including service users will be built.

Acknowledgments. This research was partially supported by the Natural Science Foundation of China under grant of No. 61379119, Science and Technology Program of Zhejiang Province under grant of No. 2013C01073, the Open Project of Qihoo360 under grant of No. 15-124002-002.

References

1. Anderson, C.: The long tail: How endless choice is creating unlimited demand. Random House, New York (2007)
2. Bianchini, D., De Antonellis, V., Melchiori, M.: A recommendation system for semantic mashup design. In: 2010 Workshop on Database and Expert Systems Applications (DEXA), pp. 159–163. IEEE (2010)
3. Cao, B., Liu, J., Tang, M., Zheng, Z., Wang, G.: Mashup service recommendation based on user interest and social network. In: 2013 IEEE 20th International Conference on Web Services (ICWS), pp. 99–106. IEEE (2013)
4. Chung, F.R.: Spectral graph theory, vol. 92. American Mathematical Soc. (1997)
5. Clarke, C.L., Kolla, M., Cormack, G.V., Vechtomova, O., Ashkan, A., Büttcher, S., MacKinnon, I.: Novelty and diversity in information retrieval evaluation. In: Proceedings of the 31st Annual International ACM SIGIR Conference on Research and Development in Information Retrieval, pp. 659–666. ACM (2008)
6. Deerwester, S.C., Dumais, S.T., Landauer, T.K., Furnas, G.W., Harshman, R.A.: Indexing by latent semantic analysis. JAsIs **41**(6), 391–407 (1990)
7. Deng, H., Han, J., Lyu, M.R., King, I.: Modeling and exploiting heterogeneous bibliographic networks for expertise ranking. In: Proceedings of the 12th ACM/IEEE-CS Joint Conference on Digital Libraries, pp. 71–80. ACM (2012)
8. Elmeleegy, H., Ivan, A., Akkiraju, R., Goodwin, R.: Mashup advisor: A recommendation tool for mashup development. In: IEEE International Conference on Web Services, 2008. ICWS 2008, pp. 337–344. IEEE (2008)
9. Klusch, M., Fries, B., Sycara, K.: Automated semantic web service discovery with owls-mx. In: Proceedings of the Fifth International Joint Conference on Autonomous Agents and Multiagent Systems, pp. 915–922. ACM (2006)
10. Li, C., Cheng, B., Chen, J., Li, C., Wang, G., Gu, P., Li, D.: A semantics extended indexes framework for mashup discovery. J. Comput. Inf. Syst. **7**(5), 1446–1454 (2011)
11. Ni, Y., Fan, Y., Huang, K., Bi, J., Tan, W.: Negative-connection-aware tag-based association mining and service recommendation. In: Ghose, A.K., Lewis, G.A., Bhiri, S., Franch, X. (eds.) ICSOC 2014. LNCS, vol. 8831, pp. 419–428. Springer, Heidelberg (2014)
12. Riabov, A.V., Boillet, E., Feblowitz, M.D., Liu, Z., Ranganathan, A.: Wishful search: Interactive composition of data mashups. In: Proceedings of the 17th International Conference on World Wide Web, pp. 775–784. ACM (2008)
13. Rohallah, B., Ramdane, M., Zaidi, S.: Agents and owl-s based semantic web service discovery with user preference support. arXiv preprint arXiv:1306.1478 (2013)
14. Tapia, B., Torres, R., Astudillo, H.: Simplifying mashup component selection with a combined similarity-and social-based technique. In: Proceedings of the 5th International Workshop on Web APIs and Service Mashups, p. 8. ACM (2011)
15. Torres, R., Tapia, B., Astudillo, H.: Improving web api discovery by leveraging social information. In: 2011 IEEE International Conference on Web Services (ICWS), pp. 744–745. IEEE (2011)

16. Zhou, C., Chen, H., Peng, Z., Ni, Y., Xie, G.: A semantic bayesian network for web mashup network construction. In: Green Computing and Communications (GreenCom), 2010 IEEE/ACM Int'l Conference on and Int'l Conference on Cyber, Physical and Social Computing (CPSCom), pp. 645–652. IEEE (2010)

17. Zhou, D., Bousquet, O., Lal, T.N., Weston, J., Schölkopf, B.: Learning with local and global consistency. Ad. Neural Inf. Proces. Syst. **16**(16), 321–328 (2004)

18. Zhou, D., Zhu, S., Yu, K., Song, X., Tseng, B.L., Zha, H., Giles, C.L.: Learning multiple graphs for document recommendations. In: Proceedings of the 17th International Conference on World Wide Web, pp. 141–150. ACM (2008)

Link Prediction in Schema-Rich Heterogeneous Information Network

Xiaohuan Cao[1], Yuyan Zheng[1], Chuan Shi[1(✉)], Jingzhi Li[2], and Bin Wu[1]

[1] Beijing Key Lab of Intelligent Telecommunications Software and Multimedia,
Beijing University of Posts and Telecommunications, Beijing 100876, China
devil_baba@126.com, zyy0716_source@163.com, {shichuan,wubin}@bupt.edu.cn
[2] Department of Mathematics, Southern University of Science and Technology,
Shenzhen 518055, China
lijz@sustc.edu.cn

Abstract. Recent years have witnessed the boom of heterogeneous information network (HIN), which contains different types of nodes and relations. Many data mining tasks have been explored in this kind of network. Among them, link prediction is an important task to predict the potential links among nodes, which are required in many applications. The contemporary link prediction usually are based on simple HIN whose schema are bipartite or star-schema. In these HINs, the meta paths are predefined or can be enumerated. However, in many real networked data, it is hard to describe their network structure with simple schema. For example, the knowledge base with RDF format include tens of thousands types of objects and links. On this kind of schema-rich HIN, it is impossible to enumerate meta paths. In this paper, we study the link prediction in schema-rich HIN and propose a novel *Li*nk *P*rediction with *a*utomatic meta *P*aths method (LiPaP). The LiPaP designs an algorithm called Automatic Meta Path Generation (AMPG) to automatically extract meta paths from schema-rich HIN and a supervised method with likelihood function to learn weights of the extracted meta paths. Experiments on real knowledge database, Yago, validate that LiPaP is an effective, steady and efficient method.

Keywords: Heterogeneous Information Network · Link prediction · Similarity measure · Meta path

1 Introduction

Nowadays, the study of Heterogeneous Information Network (HIN) become more and more popular in data mining area [5], where the network includes different types of nodes and relations. Many data mining tasks have been exploited on this kind of network, such as clustering [14], and classification [7]. Among those researches in HIN, link prediction is a fundamental problem that attempts to estimate the likelihood of the existence of a link between two nodes, based on observed links and the attributes of nodes. Link prediction is the base of many data mining tasks, such as data clearness and recommendation.

© Springer International Publishing Switzerland 2016
J. Bailey et al. (Eds.): PAKDD 2016, Part I, LNAI 9651, pp. 449–460, 2016.
DOI: 10.1007/978-3-319-31753-3_36

Some works have been done to predict link existence in HIN. Because of the unique semantic characteristic of HIN, meta path [14], a sequence of relations connecting two nodes, is widely used for link prediction. Utilizing the meta path, these works usually employ a two-step process to solve link prediction problem in HIN. The first step is to extract meta path-based feature vectors, and the second step is to train a regression or classification model to compute the existence probability of a link [3,12,13,15]. For example, Sun et al. [12] propose PathPredict to solve the problem of co-author relationship prediction, Cao et al. [3] propose an iterative framework to predict multiple types of links collectively in HIN, and Sun et al. [13] model the distribution of relationship building time to predict when a certain relationship will be formed. These works usually have a basic assumption: the meta paths can be predefined or enumerated in a simple HIN. When the HIN is simple, we can easily and manually enumerate some meaningful and short meta paths [14]. For example, a bibligraphic network with star schema is used in [12,13,15] and only several meta paths are enumerated.

However, in many real networked data, the network structures are more complex, and meta paths cannot be enumerated. Knowledge graph is the base of the contemporary search engine [10], where its resource description framework (RDF) [1] $< object, relation, object >$ naturally constructs a HIN. In such a HIN, the types of nodes and relations are huge. For example, DBpedia [2], a kind of knowledge graph, has recorded more than 38 million entities and 3 billion facts. In this kind of network, it is hard to describe them with simple schema, so we call them schema-rich HIN. Figure 1 shows a snapshot of the RDF structure extracted from DBpedia. You can find that there are many types of objects and links in such a small network, e.g., Person, City, Country. Moreover, there are many meta paths to connect two object types. For example, for Person and Country types, there are two meta paths: $Person \xrightarrow{bornin} City \xrightarrow{locatedIn} Country$ and $Person \xrightarrow{Diedin} City \xrightarrow{hasCapital^{-1}} Country$. Note that Fig. 1 is one extreme little part of the whole DBpedia network, and there will be huge number of

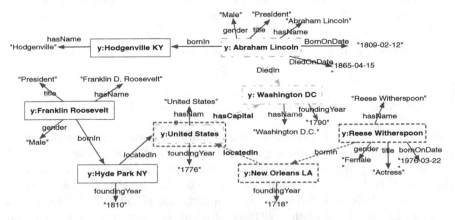

Fig. 1. A snapshot of the RDF structure extracted from DBpedia.

meta paths can connect Person and Country in a real network. So that the meta paths in this kind of schema-rich HIN are too many to enumerate and it's hard to analyze them.

To be specific, the challenges of link prediction in schema-rich HIN are mainly from two aspects. (1) The meta path cannot be enumerated. As mentioned above, there are tens of thousands of nodes and links in such schema-rich HIN and the meta paths in the network have the same order of magnitude. It's impossible to enumerate meta paths between two node types. (2) It is also not easy to effectively integrate these meta paths. Even though masses of meta paths can be found between target nodes, most of them are meaningless or less important for link prediction. So that we need to learn weight for each meta path, where the weight represents the importance of paths for link prediction.

In this paper, we study the link prediction in schema-rich HIN and propose the *Li*nk *P*rediction with *a*utomatic meta *P*aths method (LiPaP). The LiPaP designs a novel algorithm, called Automatic Meta Path Generation (AMPG), to automatically extract meta paths from schema-rich HIN. And then we design an supervised method with likelihood function to learn the weights of meta paths. On a real knowledge base Yago, we do extensive experiments to validate the performances of LiPaP. Experiments show that LiPaP can effectively solve link prediction in schema-rich HIN through automatically extracting important meta paths and learning the weights of paths.

2   Preliminary and Problem Definition

In this section, we introduce some basic concepts used in this paper and give the problem definition.

The **Heterogeneous Information Network (HIN)** [5] is a kind of information network defined as a directed network graph $G = (V, E)$, which consists of either different types of nodes V or different types of edges E. Specifically, a information network can be abstracted to a **network schema** $M = (R, L)$ where R is the set of the node types and L is the set of the edge types, and there is a node type mapping function $\theta : V \rightarrow R$, and an edge type mapping function $\varphi : E \rightarrow L$. When the number of node types $|R| > 1$ or the number of edge types $|L| > 1$, the network is a **heterogeneous information network**. For example, in bibliographic database, like DBLP [4], papers are connected together via authors, venues and terms, they can be organized as a star-schema HIN. Another example is the users and items in e-commerce website which constitutes a bipartite HIN [6].

In a HIN, there can be different paths connecting two entity nodes and these paths are called as **meta path** [14]. A meta path $\prod$ that is defined as $\prod^{R_1, \cdots, R_{l+1}} = R_1 \xrightarrow{L_1} R_2 \xrightarrow{L_2} \cdots \xrightarrow{L_l} R_{l+1}$, which describes a path between two node types R_1 and R_{l+1}, is going through a series of node types $R_1, \cdots, R_{l+1}$ and a series of link types $L_1, \cdots, L_l$. Taking the knowledge base in Fig. 1 as an example, we can consider the knowledge base as an HIN, which includes many different node types (e.g., person, city, country) and link types (e.g., bornIn and locatedIn). Two node types can be connected by multiple meta paths. For example, there are two meta paths connnecting Person and Country: $Person \xrightarrow{bornin}$ $City \xrightarrow{locatedIn} Country$ and $Person \xrightarrow{Diedin} City \xrightarrow{hasCapital^{-1}} Country$.

Traditional HIN usually has a simple network schema, such as bipartite [16] and star schema [9]. However, in some complex HINs, there are so many node types or link types that it is hard to describe their network schema. We call the HIN with many types of nodes and links as **schema-rich HIN**. In simple HIN, the meta paths can be easily enumerated, but it is difficult to do the same in the schema-rich HIN. Data mining in schema-rich HIN will face new challenges. Specifically, we define a new task as follows:

Link Prediction in Schema-Rich HIN. Given a schema-rich HIN G and a training set of entity node pairs $\phi = \{(s_i, t_i)|1 \leq i \leq k\}$, search a set of meta paths $\Upsilon = \{\prod_i |1 \leq i \leq e\}$ which can exactly describe the pairs. With these meta paths, we design a model $\eta(s,t|\Upsilon)$ to do link prediction on the test set $\psi = \{(u_i, v_i)|1 \leq i \leq r\}$.

3 The Method Description

In order to solve the link prediction problem defined above, we propose a novel link prediction method named *Link Prediction* with *automatic meta Paths* method (LiPaP). This method includes two steps: Firstly, we design an algorithm called **Automatic Meta Path Generation (AMPG)** to discover useful meta paths with training pairs automatically. Secondly, we use a supervised method to integrate meta paths to form a model for further prediction.

3.1 Automatic Meta Path Generation

In order to extract the appropriate and relevant meta paths as model features for link prediction, we would like to show the AMPG algorithm, which can generate useful meta paths smartly in schema-rich HIN. We would illustrate AMPG through a toy example in Fig. 2, where the training pairs are {(1,8), (2,8), (3,9), (4,9)}.

The main goal of AMPG is, given the training set of entity pairs, to find all the useful and relevant meta paths connecting them. These paths to be found would not only connect more training pairs, but also show much closer relationship to

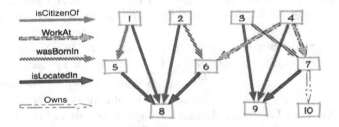

Fig. 2. Subgraph example of schema-rich HIN.

present implicit features of the training set. For example, $\xrightarrow{isCitizenOf}$ is the meta path initially found by our method in Fig. 3 and it is not only the shortest relation but also the one connecting most training pairs. Besides, the meta paths to be found are still most relevant in the candidate paths. Basically, we start to search from the source nodes step by step to find out the useful meta paths greedily. At each step, we select the meta path that is most relevant and maybe reaching more target nodes. Then we check whether the path connects the training pairs or not. If so, we pick out the meta path, otherwise make a move forward until the unchecked meta paths are irrelevant enough. It guarantees that the generated meta paths all well describe the relationship between each training pairs and the selected paths are not too many to add noise paths.

The AMPG method is a greedy algorithm that heuristically chooses the optimal paths at each step. For judging the priority of meta paths for selection, AMPG utilizes a similarity score S as a selection criterion based on a similarity measurement Path-Constrained Random Walk (PCRW) [8], which is to calculate the relevance between the given entity pairs in the meta paths. The higher similarity score S is, more likely the meta path is to be chosen.

Specifically, in AMPG, we use a data structure to record the situation of each step. The structure records a meta path passed by, a set of entity pairs reached and their PCRW values and the similarity score S of the current structure, as Fig. 3 shown. Besides, we create a candidate set to record the structure to be handled.

The similarity score S of the structure mentioned above is for judging the priority of the structure. S measures the similarity of the whole arrival pairs in the structure. The highest S means the most relevant relationship and the most promising meta paths, so we get the structure with the highest S at every step. The definition of similarity score S is as follows:

$$S = \sum_s \frac{1}{T} \sum_t [\sigma(s, t | \textstyle\prod) \bullet r(s)], \tag{1}$$

where s and t are source and reaching entity node respectively on meta path $\prod$, T is the number of reaching entity nodes and $\sigma(s, t | \prod)$ is the PCRW value. $r(s) = 1 - \alpha \bullet N$ is the contribution of s to the current structure for training pairs selection balance, where α is the decreasing coefficient of the contribution as 0.1 because of the good performance on it, and N is the number of the target nodes that s has reached through other selected paths. It means if one of source nodes in $\sum_s$ has more target nodes matched before, N will be larger and S will be reduced due to the smaller $r(s)$. So that the structure with other source nodes which have fewer matches will get high priority to be traversed greedily.

In order to get rid of the unimportant or the low pair-matched meta paths, we set a threshold value l to judge the structures whether being put to the candidate set or not.

$$l = \epsilon \bullet |A|, \tag{2}$$

where ϵ is a limited coefficient, $|A|$ is the number of entity pairs in the structure. If S is no less than l, add this structure into the candidate set, otherwise delete it.

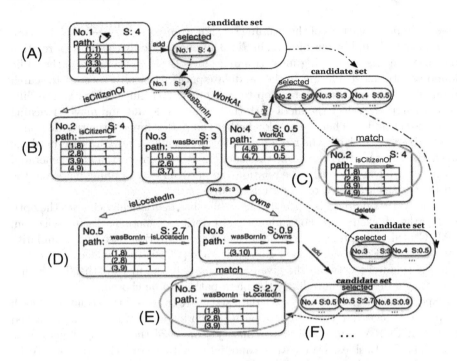

Fig. 3. An example of meta-path automatic generation.

Furthermore, we explain AMPG with a case study shown in Fig. 3. The training pairs are (1, 8), (2, 8), (3, 9), (4, 9) and sources nodes are 1, 2, 3, 4. The case starts with creating an initial structure No.1 and inserts it into the candidate set as Fig. 3(A) shown. The entity pair is composed of the source node and itself and no meta path is generated at this step. Our algorithm will read candidate set iteratively and choose the structure with highest S at each step. For each selected structure, it will be checked if any training pairs are matched. If not, we move one step in HIN, as Fig. 3(B) shown. We can pass by three edge types $\xrightarrow{isCitizenOf}$, $\xrightarrow{wasBornIn}$ and $\xrightarrow{WorkAt}$. For each passed edge type, we create new structures like No.2 and No.4. Then, we check the new structures whether fit the conditions of expanding further and insert them into the candidate set. Remove the used structure No.1 and read next structure. Otherwise, as Fig. 3(C) shown, four pairs are matched, so a new relevant meta path $\xrightarrow{isCitizenOf}$ is generated and its similarity value vector is recorded. Remove the used structure No.2 and continue to read next. The algorithm terminates when the candidate set is empty.

The detail process of AMPG is described in Algorithm 1. Step 1–2 is the variable initialization step. Step 3–26 shows the main process of searching meta paths by greedy S in a loop. In every searching movement, we pop the structure with the largest S to handle until the candidate set is empty. Finally, the algorithm will generate a set of meta paths with the related similarity matrix of training pairs.

Algorithm 1. AMPG(G, ϕ)

Input: G: shema-rich HIN; ϕ: set of entity training pairs;
Output: $\varUpsilon$: set of selected meta paths; M: similarity matrix of ϕ corresponding to $\varUpsilon$.
1 $N \Leftarrow \{0,0,\ldots,0\}$; //length: $|\phi|$; element is times of each training pair matched to calculate S
2 Create the starting structure and insert to candidate set T
3 **while** T *is not empty* **do**
4      $m \Leftarrow \{0,0,\ldots,0\}$; //length: $|\phi|$; record if meta path has pairs matched in this expanding
5      $W \Leftarrow$ popping the structure with the largest score S from T.
6   **for** *each pair* $(q,p) \in W$ **do**
7   **if** $(q,p) \in \phi$ **then**
8              $m(q,p) \Leftarrow \sigma(q,p|\prod)$;
9              $N(q,p) \Leftarrow N(q,p) + 1$;

10   **if** m *has nonzero element* **then**
11   add the meta-path $\prod$ of W into $\varUpsilon$;
12          $M \Leftarrow M \bigcup m$;
13   break;

14   **else**
15   create a empty temp Map E inserted with (next passed link, related structure);
16   **for** *each pair* $(q,p) \in W$ **do**
17   **for** *each neighbor* s *without passed in HIN* G **do**
18                  $u^d \Leftarrow$ edge type u with direct d from p to s //forward: d=1; reverse:d=-1
19   **if** E *does not have the key* u^d *or the related structure* **then**
20   create a new structure N from W adding into E.
21                  $\prod \Leftarrow$ the meta path of N
22   insert the tuple$((q,s), \sigma(q,s|\prod))$ to N

23   **for** *each structure* $K \in E$ **do**
24              $K.S \Leftarrow$ cauculated by Equation (1)
25   **if** $K.S >$ *threshold value* l **then**
26   add K into T

27 return $\varUpsilon$, M

3.2 Integration of Meta Path

Each meta path found by AMPG is important but has different importances for further link prediction. It's necessary to find a solution of measuring the importance for each meta path and integrating them into a link prediction model.

The link prediction can be considered as a classification problem. So we use the positive and negative samples to train a model to predict whether the link exists between the given pairs or not. Positive samples are the training pairs, while negative samples are generated by replacing the target nodes of the training pairs with the same-typed nodes without the same relations. Thus positive value is the similarity value vector of each positive pair on all selected meta paths, while negative value is the vector of negative pair.

For training model, we assume that the weight of each meta path $\prod_i$ is $\varpi_i (i = 1, \cdots, N)$, $\varpi_i \geq 0$, and $\sum_{i=1}^{N} \varpi_i = 1$. In order to train the appropriate path weights, we use the log-likelihood function. The specific formula is as follows:

$$\max h = \sum_{x^+ \in q^+} \frac{ln(t(\varpi, x^+))}{|q^+|} + \sum_{x^- \in q^-} \frac{ln(1 - t(\varpi, x^-))}{|q^-|} - \frac{||\varpi||^2}{2}, \quad (3)$$

where $t(\varpi, x)$ is the Sigmoid function (i.e., $t(\varpi, x) = \dfrac{e^{\varpi^T x}}{e^{\varpi^T x} + 1}$). x is similarity value vector of sample pair in all selected paths, x^+ positive sample and x^- negative. q^+ is similarity matrix of positive pairs made of x^+. And q^- is similarity matrix of negative pairs made of x^-. $\dfrac{||\varpi||^2}{2}$ is the regularizer to avoid overfitting.

After learning weights of relevant meta paths Υ, we use a logistic regression model to integrate meta paths for link prediction.

$$\eta(s,t|\Upsilon) = (1 + e^{-(\sum_{x\in\Upsilon} \varpi_x \bullet \sigma(s,t|\Pi_x)+\varpi_0)})^{-1}, \tag{4}$$

where (s,t) is the pair we should do link prediction, and x is each selected meta path feature, while ϖ_x is the weight of x we learn above. And Υ is the set of selected meta paths. If $\eta(s,t|\Upsilon)$ is larger than a specific value, we judge they would be connected by the link predicted.

4   Experiment

In order to verify the superiority of our designed method of link prediction in schema-rich HIN, we conduct a series of relevant experiments and validate the effectiveness of LiPaP from four aspects.

4.1   Dataset

In our experiments, we use Yago to conduct relevant experiments and it is a large-scale knowledge graph, which derives from Wikipedia, WordNet and GeoNames [11]. The dataset includes more than ten million entities and 120 million facts made from these entities. We only adopt "COREFact" of this dataset, which contains 4484914 facts, 35 relationships and 1369931 entities of 3455 types. A fact is a triple: $< entity, relationship, entity >$, e.g., $< NewYork, locatedin, UnitedStates >$.

4.2   Criteria

We use receiver operating characteristic curve known as ROC curve to evaluate the performance of different methods. It is defined as a plot of true positive rate (TPR) as the y coordinate versus false positive rate (FPR) as the x coordinate. TPR is the ratio of the number of true positive decisions and actually positive cases while FPR is the ratio of the number of false positive decisions and actually negative cases. The area under the curve is referred to as the AUC. The larger the area is, the larger the accuracy in prediction is.

4.3   Effectiveness Experiments

This section will validate the effectiveness of our prediction method LiPaP on accurately predicting links existing in entity pairs. Since there are no existing solutions for this problem, as a baseline (called PCRW [8]), we enumerate all meta paths, and the same weight learning method with LiPaP is employed. Because meta paths with length more than 4 are most irrelevant, the PCRW enumerates the meta paths with the length no more than 1, 2, 3, and 4, and the corresponding methods are called PCRW-1, PCRW-2, PCRW-3, and PCRW-4, respectively. Based on Yago dataset, we randomly and respectively select 200 entity pairs from two relations $\xrightarrow{isLocatedIn}$ and $\xrightarrow{isCitizenOf}$. Note that, we assume that these two types of links are not available in the prediction task. In this experiment, 100 entities pairs of them are used as the training set, the other are used as the test set. In LiPaP, we set ϵ in Eq. (2) as 0.005 and the max path length is also limited to 4.

The results of two link prediction tasks are shown in Fig. 4. It is clear that LiPaP has better performances than all PCRW methods, which implies that LiPaP can effectively generate useful meta paths. Moreover, the PCRW generally has better performance when the path length is longer, since it can exploit more useful meta paths. However, it will take more cost to search more meta paths, most of which are irrelevant. For example, PCRW-3 generates more than 80 paths and PCRW-4 finds more than 600 paths with lots of irrelevant paths. On the contrary, LiPaP only generates 30 meta paths for the $\xrightarrow{isCitizenOf}$ task.

In order to intuitively observe the effectiveness of meta paths found, Table 1 shows the top 4 generated meta paths and the corresponding training weights for the $\xrightarrow{isCitizenOf}$ task. It is obvious that 4 meta paths are all relevant to the link $\xrightarrow{isCitizenOf}$. The most relevant one is the first meta path which shows the fact that a person is born in a city and the city is located in a country. It describes the citizen relationship in fact. The last one with length 4 seems not

Table 1. Most relevant 4 meta paths for $isCitizenOf$

Meta path	Weight
Person $\xrightarrow{wasBornIn}$ City $\xrightarrow{islocatedIn}$ County	0.1425
Person $\xrightarrow{livesIn}$ County	0.0819
Person $\xrightarrow{livesIn}$ City $\xrightarrow{islocatedIn}$ County	0.0744
Person $\xrightarrow{wasBornIn}$ City $\xleftarrow{isLeaderOf}$ Person $\xrightarrow{graduatedFrom}$ university $\xrightarrow{islocatedIn}$ County	0.0609

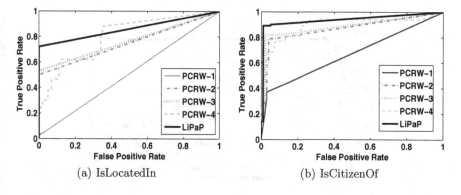

(a) IsLocatedIn   (b) IsCitizenOf

Fig. 4. Prediction accuracy of different methods on two link prediction tasks.

to be close, but actually has certain logistic relation with the link $\xrightarrow{isCitizenOf}$. However, these long and important meta paths can be missed if the maximum length of meta path was limited too short, as PCRW does. While our method can automatically find these paths and assign them a high importance.

4.4   Influence of the Size of Training Set

In this section, we evaluate the influence of the size of training set on the prediction performances. The size of training set are set with $\{2, 6, 10, 20, 40, 60, 80, 100\}$. Besides our LiPaP, we choose PCRW-2 as baseline, since it can generate most of useful meta paths and achieve good performances compared to other PCRW methods. As illustrated in Fig. 5, when the number of training pairs is smaller than 10, the performances of both methods improve rapidly with the size of pairs growing. However, when the size is more than 10, the size of training set has little effect on the performances of both methods. We think the reason lies in that too small training set cannot discover all useful meta paths, while large training set may introduce much noise. When the size of training set is from 10 to 20 in this dataset, it is good enough to discover all useful meta paths and avoid much noise. Furthermore, it can save space and time to learn model and make the performance of our method better.

4.5   Impact of Weight Learning

To illustrate the benefit of weight learning, we redone the experiments on the $\xrightarrow{isCitizenOf}$ task mentioned in Sect. 4.3. We run LiPaP with the weight learning or random weights, and with average weights. Figure 6 shows the performances of these methods. It is obvious that the weight learning can improve prediction performances. The model with random weight performs worst, owing to giving the more relevant paths low weights. The model with weight just has a little better performance than the model with average weight, because the meta path

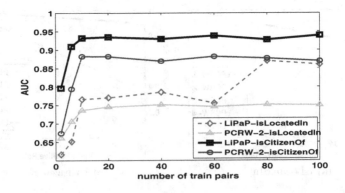

Fig. 5. Influence of different sizes of training set.

features generated by AMPG are all relevant and important, the most important feature also has not get a very low weight in the model with average weight. So the performance of the model with average weight is also not poor in spite of being inferior to the model with weight. Therefore, the weight learning can adjust the importance of different meta paths so as to integrate them well and make the model better.

4.6 Efficiency

In this section, we choose 5 different sizes of training set, i.e., $\{20, 40, 60, 80, 100\}$, to validate the efficiency of finding meta paths of different methods. Figure 7 demonstrates the running time on different models for the $\xrightarrow{isLocatedIn}$ task. It is obvious that the running time of these models approximate linearly increase with the increase of the size of training set. In spite of the small running time, the short meta paths found by PCRW-1 and PCRW-2 restrict their prediction performances. Our LiPaP has smaller running time than PCRW-3 and PCRW-4, since it only finds a small number of important meta paths. In this way, LiPaP has a better balance on effectiveness and efficiency.

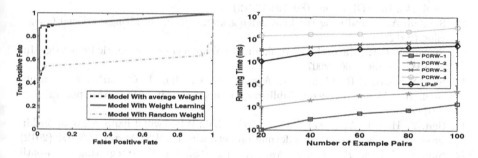

Fig. 6. Effectiveness of weight learning.   **Fig. 7.** Running times of different methods.

5 Conclusions

In this paper, we introduce a novel link prediction method in schema-rich HIN named *Link* *P*rediction with *a*utomatic meta *P*aths (LiPaP), which proposes an algorithm called AMPG to automatically extract meta paths based on given training pairs and designs an supervised method to learn weights of the extracted meta paths to form a link prediction model. Experiments on real knowledge database, Yago, validate the effectiveness, efficiency, and feasibility of LiPaP.

Acknowledgment. This work is supported in part by National Key Basic Research and Department (973) Program of China (No. 2013CB329606), and the National Natural Science Foundation of China (No. 71231002, 61375058,11571161), and the CCF-Tencent Open Fund, the Co-construction Project of Beijing Municipal Commission of Education, and Shenzhen Sci.-Tech Fund No. JCYJ20140509143748226.

References

1. Rdf current status. http://www.w3.org/standards/techs/rdf#w3c_all
2. Bizer, C., Lehmann, J., Kobilarov, G., Auer, S., Becker, C., Cyganiak, R., Hellmann, S.: Dbpedia-a crystallization point for the web of data. Web Semant.: Sci. Serv. Agents World Wide Web **7**(3), 154–165 (2009)
3. Cao, B., Kong, X., Yu, P.S.: Collective prediction of multiple types of links in heterogeneous information networks. In: ICDM, pp. 50–59 (2014)
4. Deng, H., Lyu, M.R., King, I.: A generalized co-hits algorithm and its application to bipartite graphs. In: KDD, pp. 239–248 (2009)
5. Jaiwei, H.: Mining heterogeneous information networks: the next frontier. In: SIGKDD, pp. 2–3 (2012)
6. Jamali, M., Lakshmanan, L.: HeteroMF: recommendation in heterogeneous information networks using context dependent factor models. In: WWW, pp. 643–654 (2013)
7. Kong, X., Yu, P.S., Ding, Y., Wild, D.J.: Meta path-based collective classification in heterogeneous information networks. In: CIKM, pp. 1567–1571 (2012)
8. Lao, N., Cohen, W.W.: Relational retrieval using a combination of path-constrained random walks. Mach. Learn. **81**(1), 53–67 (2010)
9. Shi, C., Kong, X., Yu, P.S., Xie, S., Wu, B.: Relevance search in heterogeneous networks. In: EDBT, pp. 180–191 (2012)
10. Singhal, A.: Introducing the knowledge graph: things, not strings. Official Google Blog (2012)
11. Suchanek, F.M., Kasneci, G., Weikum, G.: Yago: a core of semantic knowledge. In: WWW, pp. 697–706 (2007)
12. Sun, Y., Barber, R., Gupta, M., Aggarwal, C.C., Han, J.: Co-author relationship prediction in heterogeneous bibliographic networks. In: ASONAM, pp. 121–128 (2011)
13. Sun, Y., Han, J., Aggarwal, C.C., Chawla, N.V.: When will it happen?: relationship prediction in heterogeneous information networks. In: WSDM, pp. 663–672 (2012)
14. Sun, Y., Norick, B., Han, J., Yan, X., Yu, P.S., Yu, X.: Integrating meta-path selection with user-guided object clustering in heterogeneous information networks. In: KDD, pp. 1348–1356 (2012)
15. Yu, X., Gu, Q., Zhou, M., Han, J.: Citation prediction in heterogeneous bibliographic networks. In: SDM, pp. 1119–1130 (2012)
16. Zha, H., He, X., Ding, C.H.Q., Gu, M., Simon, H.D.: Bipartite graph partitioning and data clustering (2001). CoRR cs.IR/0108018

FastStep: Scalable Boolean Matrix Decomposition

Miguel Araujo[1,2]([✉]), Pedro Ribeiro[1], and Christos Faloutsos[2]

[1] Cracs/INESC-TEC, University of Porto, Porto, Portugal
pribeiro@dcc.fc.up.pt
[2] Computer Science Department, Carnegie Mellon University, Pittsburgh, USA
{maraujo,christos}@cs.cmu.edu

Abstract. Matrix Decomposition methods are applied to a wide range of tasks, such as data denoising, dimensionality reduction, co-clustering and community detection. However, in the presence of boolean inputs, common methods either do not scale or do not provide a boolean reconstruction, which results in high reconstruction error and low interpretability of the decomposition. We propose a novel step decomposition of boolean matrices in non-negative factors with boolean reconstruction. By formulating the problem using threshold operators and through suitable relaxation of this problem, we provide a scalable algorithm that can be applied to boolean matrices with millions of non-zero entries. We show that our method achieves significantly lower reconstruction error when compared to standard state of the art algorithms. We also show that the decomposition keeps its interpretability by analyzing communities in a flights dataset (where the matrix is interpreted as a graph in which nodes are airports) and in a movie-ratings dataset with *10 million* non-zeros.

1 Introduction

Given a boolean who-watched-what matrix, with rows representing users and columns representing movies, how can we find an interpretation of the data with low error? How can we find its underlying structure, helpful for compression, prediction and denoising? Boolean matrices appear naturally in many domains (e.g. user-reviews [3] or user-item purchases [16], graphs, word-document co-occurrences [4] or gene-expression datasets [17]) and describing the underlying structure of these datasets is the fundamental problem of community detection [6] and co-clustering [15] techniques.

We address the problem of finding a low-rank representation of a given $n \times m$ boolean matrix $\mathbf{M}$, with small reconstruction error while easily describing $\mathbf{M}$'s latent structure. We propose FastStep, a method for finding a non-negative factorization that, unlike commonly used decomposition methods, yields the best interpretability by combining a **boolean reconstruction** with **non-negative factors**. This combination allows FastStep to find structures that go beyond blocks, providing more realistic representations. Figure 1a showcases three communities (representing 3 venues) in the DBLP dataset that illustrate the important hyperbolic structures found in real data; compare them to the community

© Springer International Publishing Switzerland 2016
J. Bailey et al. (Eds.): PAKDD 2016, Part I, LNAI 9651, pp. 461–473, 2016.
DOI: 10.1007/978-3-319-31753-3_37

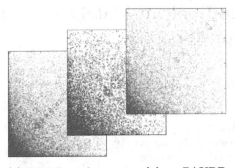 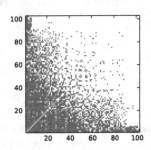

(a) **DBLP real communities** - PAKDD, KDD and VLDB.

(b) **FastStep community** - American airports.

Fig. 1. Realistic hyperbolic structure - Adjacency Matrices of real communities in DBLP and a community found by FastStep.

found by FastStep in Fig. 1b representing the American community in the Airports dataset.

Using our scalable method, we analyze two datasets of movie ratings and airports flights and show FastStep's interpretability power with intuitively clear and surprising decompositions. As an additional example, Fig. 2 illustrates an application of FastStep to the task of community detection. Using route information alone, the world airports are decomposed in 10 factors that clearly illustrate geographical proximity. As we explain in more detail in Sect. 4.3, the communities we find have an arbitrary marginal and do not need to follow a block shape.

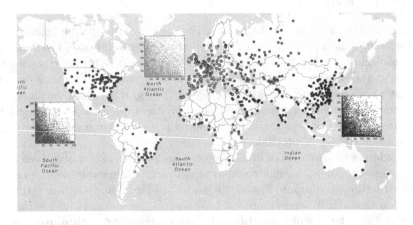

Fig. 2. Intuitive non-block communities - Communities automatically found in the Airports dataset from flight records (best viewed in color) (Color figure online).

2   Background and Related Work

Real and Non-negative Matrix Decompositions. In the Singular Value Decomposition (SVD) [7], a real matrix $\mathbf{M}$ is decomposed into $\mathbf{U\Sigma V^T}$ where $\mathbf{U}$ and $\mathbf{V}$ are real orthogonal matrices and $\mathbf{\Sigma}$ is a $k \times k$ non-negative diagonal matrix. While the Eckart-Young theorem [5] proves this to be the best approximation using regular matrix multiplication and real entries, negative values in the factor matrices make it hard to interpret. What does it mean for an element to have a negative score in a component? For non-negative $\mathbf{M}$, Non-Negative Matrix Factorization (NNMF) [9] methods were developed to overcome this problem.

Neither of these methods have clear extensions to the boolean case as the reconstructed matrix is not boolean. One simple idea is rounding or thresholding the reconstructed matrix, but no guarantee can be given on the reconstruction error. Another possibility is thresholding the factor matrices and using boolean algebra in order to obtain a boolean reconstruction, but selecting the appropriate threshold is a difficult problem as a clear cut-off might not exist.

Decomposition of Boolean Matrices. Tao Li [11] proposed an extension of the K-means algorithm to the two-sided case where $\mathbf{M}$ is decomposed into $\mathbf{AXB^T}$ with $\mathbf{A}$ and $\mathbf{B}$ binary, and an alternating least squares method when $\mathbf{X}$ is the identity matrix. Pauli Miettinen showed that Boolean Matrix Factorizations (BMF) methods could achieve lower reconstruction error than SVD in boolean data and proposed an algorithm using association rules (ASSO) which exploits the correlations between columns, but unfortunately it's time complexity is $O(nm^2)$. Zhang et al. [17] proposed two approaches for BMF, one using a penalty in the objective function (BMF-PENALTY) which achieved good results for dense datasets, and an alternative thresholding method (BMF-THRESH) which by thresholding factor matrices is better suited for sparse datasets. None of these methods is scalable and they have the problem of forcing a tiling of the data matrix, as each factor is effectively treated as a block. In particular, the notion of "importance" inside a cluster, which previously existed in NNMF, is now lost and the analysis of the resulting factors is limited. In Logistic PCA (L-PCA), Schein et al. [12] replace PCA's Gaussian assumption with a Bernoulli distribution and fit their new model using an alternating least squares algorithm that maximizes the log-likelihood. Their alternating algorithm has running time $O(nm)$ when applied to a $n \times m$ matrix and therefore does not scale. It is also hard to interpret due to the possibility of negative values in the factors.

Related Techniques. There is a strong relationship between boolean matrices and graph data, where matrix decompositions are linked to community detection and graph partitioning, but we would like to refer the reader to a review on spectral algorithms in this area for further details [6]. However, recent work such as the Hyperbolic Community Model [2] has shown the non-uniform nature of real-world communities and has highlighted the need for boolean decomposition methods which do not discard node importance.

An important aspect of fast decomposition methods is their ability to evaluate the reconstruction error $||M - R||_F^2$ in less than quadratic time.

Leskovec et al. [10] approximated the log-likelihood of the fit by exploiting the Kronecker nature of their generator. In the Compact Matrix Decomposition [14], Jimeng Sun et al. approximated the sum-square-error (SSE) by sampling a set of rows and columns and scaling the error in the submatrix accordingly.

Table 1 provides a quick comparison of some of the methods discussed in this section. We characterize as *Beyond blocks* methods who do not force a rectangular tiling of the data. *Arbitrary Marginals* refers to a method's ability to represent any marginal in the data (e.g. rectangles, but also triangles or hyperbolic structures). We define *interpretability* as the ability to easily select a subset of elements representing a factor. Given our focus on efficient decompositions, we will limit our comparison in Sect. 4 to scalable methods.

Table 1. Comparison of decomposition methods - **FastStep combines interpretability and beyond block structures for large datasets.**

	FASTSTEP	SVD	NNMF	ASSO	THRESH	HYCOM	L-PCA
Scalability	✓	✓	✓			✓	
Overlapping	✓	✓	✓	✓	✓	✓	✓
Beyond blocks	✓	✓	✓			✓	✓
Boolean reconstruction	✓			✓	✓	✓	✓
Arbitrary marginals	✓	✓	✓				✓
Interpretability	✓			✓	✓	✓	

3   Proposed Method

As hinted in the previous section, there are two aspects for a strong interpretability of a boolean matrix decomposition: boolean reconstruction allows clear predictions and explanations of the non-zeros, while the existence non-negative factors establishes the importance of elements and enable the representation of beyond-block structures. In this section, we introduce a new formulation using a step operator that achieves both goals.

3.1   Formal Objective

Let $\mathbf{M}$ be a $n \times m$ boolean matrix. Our goal is to find a $n \times r$ non-negative matrix A and a $m \times r$ non-negative matrix B, so that the product $\mathbf{AB}^T$ is a good approximation of $\mathbf{M}$ after thresholding:

$$\min_{\mathbf{A},\mathbf{B}} ||\mathbf{M} - u_\tau(\mathbf{AB}^T)||_F^2 = \sum_{i,j} \left(\mathbf{M}_{ij} - u_\tau(\mathbf{AB}^T)_{ij} \right)^2 \tag{1}$$

where $|| \cdot ||_F$ is the Frobenius norm and $u_\tau(\mathbf{X})$ simply applies the standard step function to each element X_{ij}:

$$[u_\tau(X)]_{ij} = \begin{cases} 1 & \text{if } X_{ij} \geq \tau \\ 0 & \text{otherwise} \end{cases} \tag{2}$$

where τ is a threshold parameter. Note that the choice of τ does not affect the decomposition, as matrices $\mathbf{A}$ and $\mathbf{B}$ can always be scaled accordingly.

3.2 Step Matrix Decomposition

The thresholding operator renders the objective function non-differentiable and akin to a binary programming problem. In order to solve it, we will approximate the objective function of Eq. 1 by a function with similar objective:

$$\min_{\mathbf{A},\mathbf{B}} \sum_{i,j} \log \left(1 + e^{-M_{ij}* \left(\sum_{k=1}^{r} A_{ik}B_{jk} - \tau \right)} \right) \tag{3}$$

where M was transformed so that it has values in $\{-1, 1\}$ by replacing all zeros with -1.

Note that $log(1 + e^{-x})$ will tend to zero when x is positive and it will increase when x is negative; the intuition is that this error function will be approximately zero when $M_{i,j}$ and $(\sum_{k=1}^{r} A_{i,k}B_{j,k} - \tau)$ have the same sign and a linear penalty is in place whenever their signs differ.

Given the above formulation, there are several methods for finding $\mathbf{A}$ and $\mathbf{B}$ and one possibility is using gradient descent. The gradient is given by

Lemma 1. *Let* $S_{ij} = \sum_{k=1}^{r} A_{ik}B_{jk}$, *then the gradient of the objective function in 3 is given by:*

$$\frac{\partial F}{\partial A_{ik}} = \sum_{j \notin \mathbf{M_i}} \frac{B_{jk}}{1 + e^{\tau - S_{ij}}} - \sum_{j \in \mathbf{M_i}} \frac{B_{jk}}{1 + e^{S_{ij} - \tau}} = \sum_{j=1}^{m} \frac{B_{jk}}{1 + e^{\tau - S_{ij}}} - \sum_{j \in \mathbf{M_i}} B_{jk}. \tag{4}$$

Proof. Omitted for brevity.

The update rules for $\mathbf{B}$ are similar and are also omitted for brevity.

Due to the non-negativity requirement, matrices $\mathbf{A}$ and $\mathbf{B}$ are projected after each iteration - this projection is made to a small value ϵ instead of to 0, as $\mathbf{A} = \mathbf{B} = 0$ is a stationary point of the objective function and the algorithm wouldn't improve.

Different gradient descent algorithms and small variations can now be tried. Our experiments indicate that stochastic gradient descent with batches corresponding to factors provides the quickest convergence, as factors quickly converge to different submatrices. Our results also indicate that initializing $\mathbf{A}$ and $\mathbf{B}$ to small random numbers provides the best results. Comparing alternative gradient descent methods is out of the scope of this paper.

It should also be noted that τ now impacts the gradient, as the relative error $\left(\dfrac{\log(1 + e^{\tau})}{\log(1 + e^{0})} \right)$ of misrepresenting an element increases. However, it is clear that it should be chosen to be the highest possible value in order to improve convergence and to get a sharper decomposition, as long as numerical stability is not compromised. Our implementation uses $\tau = 20$.

Complexity. A straightforward implementation would take $O(TNMR^2)$ time where T is the number of iterations, N and M are the dimensions of the matrix and R is the rank of the decomposition. However, by using additional $O(NM)$ memory, caching and updating $\mathbf{S}$ in each iteration, it can be reduced to $O(TNMR)$.

3.3   FastStep Matrix Decomposition

Unfortunately, the previous algorithm is not adequate for many datasets given its quadratic nature; it grows linearly in $O(NM)$. In many scenarios such as community detection and recommender systems, $\mathbf{M}$ is extremely sparse and algorithms must be linear (or quasilinear) in the number of non-zeros (E). In the following, we describe how to quickly approximate $F(\mathbf{A}, \mathbf{B})$ and the respective gradients of the sparse matrix.

Fast Gradient Calculation. As shown in Eq. 4, calculating the gradient exactly requires $O(NM)$ operations per factor because each A_{ik} requires a summation over all elements B_{jk}. Furthermore, there is a $A_{ik}B_{jk}$ term in S_{ij}, which means that this loop cannot be easily unrolled or reused between elements of $\mathbf{A}$. The goal of this subsection is to approximate the gradient of the factor using a number of operations in the order of $O(E)$, the number of non-zeros in the matrix.

Careful analysis of the structure of this summation in the gradient allows us to quickly approximate it. The impact of position (i, j) in factor k is a sigmoid function, scaled by B_{jk} and with parameter S_{ij}. This means that only positions with simultaneously high S_{ij} and B_{jk} have a significant impact on the gradient, which implies that we should first consider pairs (i, j) with high $A_{ik}B_{jk}$, as that correlates well with both metrics.

In other words, Eq. 4 can be approximated as

$$\frac{\partial F}{\partial A_{ik}} \simeq \sum_{(i,j)\in P(t)} \frac{B_{jk}}{1 + e^{\tau - S_{ij}}} - \sum_{j\in \mathbf{M_i}} B_{jk} \tag{5}$$

where $P(t)$ is the set of elements of $\mathbf{M}$ that the decomposition "believes" should be reconstructed, i.e. with high $A_{ik}B_{jk}$ for some k. We define r sets of elements $P_k(t)$ that each factor k would like to reconstruct and $P(t) = \bigcup P_k(t)$. The intuition is that, initially, only non-zeros contribute to the gradient so we can quickly calculate it with no error using the second summand of Eq. 5. As we iterate, the error will gradually move from the non-zeros of $\mathbf{M}$ to some of the zeros. However, given $\mathbf{M}$'s sparsity and the symmetry of the error function – the error of misrepresenting a one is the same as misrepresenting a zero – $|P(t)|$ can be kept small and in the order of $O(rE)$; Fig. 3 shows the error as the size of $P(t)$ increases.

In order to quickly find the top-t pairs (i, j) with highest $A_{ik}B_{jk}$, let $\mathbf{a_k}$ and $\mathbf{b_k}$ be columns k of matrices $\mathbf{A}$ and $\mathbf{B}$, respectively. After sorting $\mathbf{a_k}$ and $\mathbf{b_k}$, the biggest $A_{ik}B_{jk}$ not currently in P_k can be selected from a very small set of elements along one sort of "diagonal" in the matrix. In particular, it can be shown that element (x, y) should not be added to P_k before both $(x - 1, y)$ and $(x, y - 1)$ are added, as they would necessarily be at least as big. There-fore, one can keep a priority queue

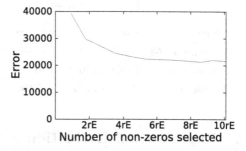

Fig. 3. A small number of non-zeros approximates the gradient – quick con-vergence in the Airports dataset.

with $O(min(n, m))$ elements and it is possible to select a set of t non-zeros and approximate the gradient of all elements in factor k in $O(t + n \log n + m \log m)$ operations.

Fast Function Evaluation. Given the method currently used to quickly calcu-late the gradient, one possibility would be to only calculate the error at positions $E + P(t)$. Although fast, some positions of the matrix would never be considered and the algorithm would over-fit, thus it cannot be used to detect convergence.

Therefore, in order to detect convergence and after each iteration of the gra-dient descent (i.e. after all the batches are completed), we calculate an estimate of the error $\tilde{F}(\mathbf{A}, \mathbf{B})$ by considering all the non-zeros and a uniform sample of the zeros of the matrix and then scaling the error accordingly. Additionally, in order to decrease the probability of underestimating $F(\mathbf{A}, \mathbf{B})$ and compromising future iterations of the gradient descent, we take the median of 9 simulations.

Complexity. Using the same notation as before, the time complexity is now bounded by the number of non-zeros and $P = |P(t)|$, which as we showed can be $O(rE)$, and the number of samples S to check for convergence. The complexity is now $O(TR(E + P \log(\min(N, M)) + N \log N + M \log M + S))$.

Obtaining Clusters from A and B. When a binary answer on whether a given element "belongs" to a factor is desired (e.g. community detection), a clear interpretation exists solely based on the principles of the decomposition:

Definition 1 *Part of a Factor. A row element i belongs to a factor k if there is non-zero in the reconstructed matrix in row i and if this factor contributed with a weight above $\frac{\tau}{r}$, i.e.:*

$$A_{ik} \geq \frac{\tau}{r \max(\mathbf{b_k})} \text{ and } S_{i, \arg\max(\mathbf{b_k})} \geq \tau.$$

We show that this method generates empirically correct clusters in the next section.

Table 2. Datasets used to evaluate FASTSTEP.

Name	Size	Non-zeros	Description
MovieLens100k	945×1684	100000	User-movie ratings
MovieLens10m	71568×10681	10000055	User-movie ratings
Airports	7733×7733	34660	Airport to airport flight information

4   Experimental Evaluation

FASTSTEP was tested on 2 fairly different real-world datasets, see Table 2 for details. MovieLens100k and MovieLens10m are user-movie ratings datasets made available by MovieLens and the Airports dataset is a graph made available by OpenFlights. Unless otherwise specified, FASTSTEP was run using the default parameters defined in Sect. 3 and 1000000 samples.

We answer the following questions:

Q1. How **scalable** is the fast version of FASTSTEP?

Q2. How does the **reconstruction error** compare to other methods?

Q3. How **effective** and **interpretable** is the FASTSTEP decomposition?

4.1   Scalability

The fast approximation of the gradient has a runtime proportional to the number of non-zeros of the matrix. For the runtime to be reproducible, we took different subsets of the MovieLens10m dataset by removing all the ratings of movies produced after a given decade. Please note that the matrix was not resized, resulting in columns (and possibly rows) full of zeros.

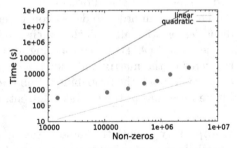

Fig. 4. Scalability: the FASTSTEP decomposition has linear running time on the number of non-zeros.

Figure 4 shows the execution time of the decomposition for these different matrices. Notice the sub-quadratic running time.

4.2   Low Reconstruction Error

When considering the same number of factors, a lower reconstruction error implies better compression and potentially enables lower-rank representations of the data. Given the boolean nature of $\mathbf{M}$, the error function is intuitively easy to represent. Let $\mathbf{M}$ represent the original dataset and $\mathbf{R}$ represent the reconstructed matrix, then the error E is given by $E = ||M - R||_F^2$.

We compared FASTSTEP to other methods that were quasilinear in the number of non-zeros. Table 3 compares the squared error of FASTSTEP, SVD, NNMF and HyCoM in the `MovieLens100k` and `Airports` datasets when using 10 factors. For SVD and NNMF, as arbitrary values such as 0.5 do not guarantee the lowest error, we tried all thresholds and considered the optimal. For FASTSTEP, we selected the lowest error from Fig. 3 and its equivalent in the `MovieLens100k` data (which converged after considering only $2rE$ non-zeros). For HyCoM, we considered as error the sum of the edges not represented and the mistakes made inside each community. Among the state of the art methods, we did not compare with non scalable algorithms (L-PCA, ASSO, BMF-Threshold).

Table 3. The **FastStep** Decomposition achieves **lower squared error** than popular scalable methods.

Dataset	FASTSTEP	SVD	NNMF	HyCoM
`Airports`	**21206**	26061	27235	29117
`MovieLens100k`	**68863**	70627	74040	86964

However, while comparing the reconstruction error of these methods might be appropriate given the same number of parameters, their expressiveness is not the same given their different characteristics. In this regard, by allowing negative numbers, SVD is at an advantage when compared to the rest of the methods. Please note that common techniques such as the Bayesian Information Criterion (BIC) [13] or the Akaike Information Criterion (AIC) [1] would not provide a fairer comparison because, as the number of parameters is the same, all methods would keep the same relative *rank*. Techniques such as Minimum Description Length (MDL) [8] measure the number of bits required to encode both the error and the model, but it is not clear which method should be used to represent real numbers, especially given that the importance of the bits is not the same - as a result, methods such as HyCoM that uses integer values would greatly benefit.

We can see that FASTSTEP is able to simultaneously achieve a lower reconstruction error while maintaining higher interpretability.

4.3 Discoveries

MovieLens. The `MovieLens100k` user-movie dataset was decomposed using a rank-10 decomposition and the factors were clustered as described. Table 4 illustrates the top-5 movies (ranked by score) in three of the factors and shows a grouping by movie theme.

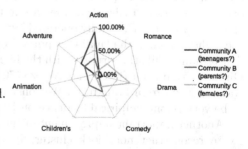

Fig. 5. MovieLens genre separation

Figure 5 shows 3 clusters and the percentage of movies in each cluster that correspond to a given genre (movies might have more than one tag, so genres

Table 4. FastStep is able to automatically group similar movies in the MovieLens dataset. Groups manually labeled according to their highest scoring movies.

"Action"	"Romance"	"Drama"
Raiders of the lost ark	Picture Perfect	Titanic
The empire strikes back	Addicted to Love	Wag the Dog
Terminator 2: Judgment day	Bed of Roses	L.A. Confidential
The terminator	My Best Friend's Wedding	Jackie Brown
Star Trek 3: The search for spock	Fly Away Home	Replacement Killers

do not sum to 1). We labeled group A as *teenagers* due to the clear prevalence of Action and Adventure movies. In group B, most of the movies rated were in categories of Comedy, Children's, Animation and Adventure; we hypothesize that users rating these movies are parents and labeled the group accordingly. Finally, we labeled group C as *females* due to the Drama, Comedy and Romance movie genres.

Airports. The Airports dataset is a symmetric matrix representing an undirected unipartite graph, which implies that $\mathbf{B} = \mathbf{A}$ as we are looking for communities. The minimization problem is similar $\left(\min_{\mathbf{A}} ||\mathbf{M} - u(\mathbf{A}\mathbf{A}^T)||_F^2 \right)$ and the gradient is omitted for brevity.

Figure 2 shows a geographical plot of the airports in the different communities; some big hubs, such as Frankfurt and Heathrow, appear in multiple communities and were coded with a single color to simplify visualization. Even though no geographical information was used to perform this task, there is a very clear distinction between north American airports, Brazilian airports, European airports, previous French colonies in Africa, Russian airports, Middle-Eastern airports and south-east Asia airports. Additionally, in order to illustrate one of the surprising findings, Fig. 6 highlights the two European communities (in blue and yellow) along with the overlapping airports (in green). While it would initially seem that all these airports should be considered the same community, a quick overview makes us realize that they are in fact divided by "major airports" and "secondary airports", usually operated by low-cost companies. The airports with the highest score in the "major airports" community are Barcelona, Munich and Amsterdam, while the airports with the highest score in the "low-cost" group are Girona (85 km from Barcelona), Weeze (70 km from Dusseldorf) and Frankfurt-Hahn (120 km from Frankfurt). We consider these and other surprising findings to be very strong empirical evidence on FASTSTEP's usefulness for these tasks.

Another important improvement of the FASTSTEP decomposition is its ability to reconstruct non-block clusters in the data. Figure 1b shows the adjacency matrix of the American community found in the previous decomposition. As we have non-negative factors, lets explore the additional information available in matrix $\mathbf{A}$. The airports with the highest score correspond to central

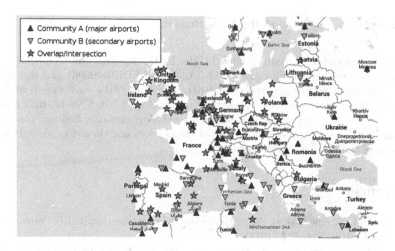

Fig. 6. Intuitive split of European airports - FASTSTEP identifies 2 European communities, one with the major international airports (in blue) and the other with secondary airports (in yellow). Overlapping airports appear in green (Color figure online).

airports in continental United States with hubs from big airlines: Minneapolis, Denver, Chicago, Dallas, Detroit, Houston, etc. Therefore, using this decomposition alone, measures of centrality can be directly obtained.

Finally, the scores of the elements in the communities, when sorted in descending order, closely follow a power-law. This characteristic has been previously observed in ground-truth communities using significantly different ground-truth definitions [2]. Given that no bias was introduced in FASTSTEP, we consider this a strong indicator of its ability to detect realistic structures in graph data.

5   Conclusion

FASTSTEP carefully combines a non-negative decomposition and a boolean reconstruction for the best interpretability of the data. We have shown that it achieves lower reconstruction error than similar methods and have provided strong empirical evidence of its ability to find structural patterns in the data. The main contributions of this work are the following:

1. **New formulation and tractable approximation:** We introduce a novel FASTSTEP Decomposition which exploits thresholding of the reconstructed data in order to minimize the reconstruction error.
2. **Scalable:** A very efficient approximation enables a runtime linear in the number of non-zeros.
3. **Low reconstruction error** when compared to standard methods.
4. **Realistic representation** which relates to nodes in clusters or degree inside communities.
5. **Meaningful and interesting discoveries** in real-world datasets.

Reproducibility. Available at http://cs.cmu.edu/~maraujo/faststep/.

Acknowledgments. Partially funded by the ERDF through the COMPETE 2020 Program and by FCT within project POCI-01-0145-FEDER-006961 and through the CMU—Portugal Program under Grant SFRH/BD/52362/2013. Based upon work supported by the National Science Foundation under Grants No. CNS-1314632 and IIS-1408924, and by a Google Focused Research Award. Any opinions, findings, conclusions or recommendations expressed are those of the authors and do not necessarily reflect the views of the funding parties.

References

1. Akaike, H.: A new look at the statistical model identification. IEEE Trans. Autom. Control **19**(6), 716–723 (1974)
2. Araujo, M., Günnemann, S., Mateos, G., Faloutsos, C.: Beyond blocks: hyperbolic community detection. In: Calders, T., Esposito, F., Hüllermeier, E., Meo, R. (eds.) ECML PKDD 2014, Part I. LNCS, vol. 8724, pp. 50–65. Springer, Heidelberg (2014)
3. Bell, R.M., Koren, Y.: Lessons from the netflix prize challenge. ACM SIGKDD Explor. Newslett. **9**(2), 75–79 (2007)
4. Dhillon, I.S., Mallela, S., Modha, D.S.: Information-theoretic co-clustering. In: Proceedings of the Ninth ACM SIGKDD International Conference on Knowledge Discovery and Data Mining, pp. 89–98. ACM (2003)
5. Eckart, C., Young, G.: The approximation of one matrix by another of lower rank. Psychometrika **1**(3), 211–218 (1936). http://dx.org/10.1007/BF02288367
6. Fortunato, S.: Community detection in graphs. Phys. Rep. **486**(3–5), 75–174 (2010)
7. Golub, G., Kahan, W.: Calculating the singular values and pseudo-inverse of a matrix. J. Soc. Ind. Appl. Math. Ser. B Numer. Anal. **2**(2), 205–224 (1965)
8. Grünwald, P.D.: The Minimum Description Length Principle. The MIT Press, Cambridge (2007)
9. Lee, D.D., Seung, H.S.: Learning the parts of objects by non-negative matrix factorization. Nature **401**(6755), 788–791 (1999)
10. Leskovec, J., Chakrabarti, D., Kleinberg, J., Faloutsos, C., Ghahramani, Z.: Kronecker graphs: an approach to modeling networks. J. Mach. Learn. Res. **11**, 985–1042 (2010)
11. Li, T.: A general model for clustering binary data. In: Proceedings of the Eleventh ACM SIGKDD International Conference on Knowledge Discovery in Data Mining, pp. 188–197. ACM (2005)
12. Schein, A.I., Saul, L.K., Ungar, L.H.: A generalized linear model for principal component analysis of binary data. In: Proceedings of the 9th International Workshop on Artificial Intelligence and Statistics, pp. 14–21 (2003)
13. Schwarz, G., et al.: Estimating the dimension of a model. Ann. Stat. **6**(2), 461–464 (1978)
14. Sun, J., Xie, Y., Zhang, H., Faloutsos, C.: Less is more: Compact matrix decomposition for large sparse graphs. In: Proceedings of the Seventh SIAM International Conference on Data Mining, vol. 127, p. 366. SIAM (2007)
15. Tanay, A., Sharan, R., Shamir, R.: Biclustering algorithms: A survey. Handb. Comput. Mol. Biol. **9**(1–20), 122–124 (2005)

16. Vlachos, M., Fusco, F., Mavroforakis, C., Kyrillidis, A., Vassiliadis, V.G.: Improving co-cluster quality with application to product recommendations. In: 23rd ACM Conference on Information and Knowledge Management, pp. 679–688 (2014)
17. Zhang, Z.Y., Li, T., Ding, C., Ren, X.W., Zhang, X.S.: Binary matrix factorization for analyzing gene expression data. Data Min. Knowl. Disc. **20**(1), 28–52 (2010)

Applications

An Expert-in-the-loop Paradigm for Learning Medical Image Grouping

Xuan Guo[1]([⊠]), Qi Yu[2], Rui Li[2], Cecilia Ovesdotter Alm[2], Cara Calvelli[2], Pengcheng Shi[2], and Anne Haake[2]

[1] B. Thomas Golisano College of Computing and Information Sciences,
20 Lomb Memorial Drive, Rochester, NY 14623, USA
`xxg3358@rit.edu`
[2] Rochester Institute of Technology, Rochester, USA
`{qyuvks,rxl5604,coagla,cfcscl,spcast,arhics}@rit.edu`
`http://hccl.gccis.rit.edu`

Abstract. Image grouping in knowledge-rich domains is challenging, since domain knowledge and expertise are key to transform image pixels into meaningful content. Manually marking and annotating images is not only labor-intensive but also ineffective. Furthermore, most traditional machine learning approaches cannot bridge this gap for the absence of experts' input. We thus present an interactive machine learning paradigm that allows experts to become an integral part of the learning process. This paradigm is designed for automatically computing and quantifying interpretable grouping of dermatological images. In this way, the computational evolution of an image grouping model, its visualization, and expert interactions form a loop to improve image grouping. In our paradigm, dermatologists encode their domain knowledge about the medical images by grouping a small subset of images via a carefully designed interface. Our learning algorithm automatically incorporates these manually specified connections as constraints for re-organizing the whole image dataset. Performance evaluation shows that this paradigm effectively improves image grouping based on expert knowledge.

Keywords: Dermatological images · Multimodal data · Image grouping · Visual analytics · Interactive machine learning

1 Introduction

In visually-oriented specialized medical domains such as dermatology and radiology, physicians explore interesting image cases from medical image repositories for comparative case studies to aid clinical diagnoses, educate medical trainees, and support medical research. This image browsing and lookup could benefit from a grouping of medical images that is consistent with experts' understanding of the image content. However, it is challenging, because medical image interpretation usually requires domain knowledge that tends to be tacit. Therefore, to make expertise more explicit we propose an interactive machine learning

© Springer International Publishing Switzerland 2016
J. Bailey et al. (Eds.): PAKDD 2016, Part I, LNAI 9651, pp. 477–488, 2016.
DOI: 10.1007/978-3-319-31753-3_38

paradigm that has experts in the loop to improve image grouping. Particularly, dermatologists encode their domain knowledge about the medical images by grouping a small subset of images via an interface. Our learning algorithm automatically incorporates these manually specified connections as constraints for re-organizing the whole image dataset. In this way, the computational evolution of an image grouping model, its visualization, and expert interactions form a loop to improve image grouping. A user evaluation study shows that this paradigm improves image grouping based on expert knowledge.

In order to minimize human efforts and provide experts with a good starting point to group images, we create an initial image grouping using a multimodal expert dataset described in Sect. 2 [18]. This initial image grouping is learned through a multimodal data fusion algorithm flexible to incorporate new images [11]. From here, the loop to improve image grouping begins (see Fig. 1). An expert can inspect the image grouping and choose to improve it through an interface. The interface design and the supported expert image manipulations are presented in Sect. 3. The interface then parses expert manipulations as implicit constraints by the rules described in Sect. 5 and incrementally learns the model, and visualizes the new image grouping using the techniques in Sect. 4. An expert-in-the-loop evaluation study is described in Sect. 6. Related studies, including visual (text) analytics and interactive machine learning systems, are compared with our learning paradigm in Sect. 7.

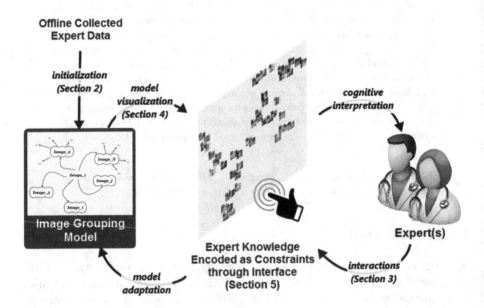

Fig. 1. Overview of the flow chart of our expert-in-the-loop paradigm. An expert encodes domain knowledge as special constraints through rounds of interactions.

2   Paradigm Initialization

The initial image grouping was learned from an offline collected expert dataset. To elicit expert data, 16 physicians were asked to inspect 48 medical images and describe the image content aloud towards a diagnosis, as if teaching a student who was seated nearby [17]. Their eye movements were recorded, as eye movement features highlight perceptually important image regions, which is especially useful in knowledge-rich domains [9]. In this paper, we use experts' eye fixation map to filter image features (SIFT features [16]). See Fig. 2 for an example. A bag of visual words is created from the remaining image features, and each image is described by a histogram of the visual words. Physicians' verbal image descriptions were also recorded concurrently, as they provide insights into experts' diagnostic image understanding. Figure 3 shows a sample transcription. The medical concepts were extracted from the transcriptions using MetaMap, a medical language processing resource [1,10]. These concepts formed a high dimensional feature space, in which each image is described by the occurrences of these medical concepts.

We choose a Laplacian sparse coding approach [8,20] over latent semantic analysis (LSA) or latent Dirichlet allocation (LDA). LSA does not perform as well as Laplacian sparse coding to cluster images by object [3]. LDA is affected not only by an initial specification but also by the samples randomly generated at each iteration [4]. It does not support users to make incremental changes, due to the inconsistent results obtained from multiple runs.

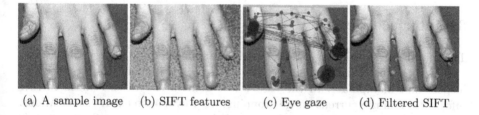

(a) A sample image   (b) SIFT features   (c) Eye gaze   (d) Filtered SIFT

Fig. 2. Image features filtered by an expert's eye gaze. (Image courtesy of Logical Images, Inc.)

(SIL) okay and so this is a classic in dermatology periungual warts uh (SIL) there are plaque like areas of uh thickened hyperkeratotic skin (SIL) uh they uh they show areas of hemorrhage kind of like the psoriasis does (SIL) but uh the uh distribution the thickness of the keratin (SIL) uh and this tendency to to (SIL) fissure centrally and the (SIL) thicker areas of the of the wart (SIL) uh is is kind of c- like characteristic (SIL) it's interesting that when you see this you almost always see (SIL) uh a person who has hangnails as well (SIL) and uh and that's because the virus gains entry through the defect in the hangnail (SIL) so periungual warts

Fig. 3. A sample diagnostic narrative expressed by a dermatologist inspecting the medical image shown in Fig. 2a. *(SIL)* represents silent pause.

To initialize an image grouping based on the features extracted from multiple modalities, we adopt a data fusion framework based on Laplacian sparse coding [11]. The objective function is presented in Eq. (1). Matrices $E \in \mathbb{R}^{n_e \times m}$ and $V \in \mathbb{R}^{n_v \times m}$ are eye gaze-filtered image features and verbal features, respectively (n_e being the number of visual words, n_v being the number of verbal features, and m being the number of images). This model provides flexibility to allow extra data modalities by adding terms like the first two in Eq. (1). The coefficient matrix $C \in \mathbb{R}^{k \times m}$ (k being the number of latent topics) stores the new image representations, each of which is a distribution of latent topics learned and stored in the basis matrices $P \in \mathbb{R}^{n_e \times k}$ and $Q \in \mathbb{R}^{n_v \times k}$. The matrices P and Q reveal the transformation from the original feature spaces to latent topics.

$$\min_{P,Q,C \geq 0} \|E - PC\|_F^2 + \|V - QC\|_F^2 + \alpha \mathcal{G}(W,C) + \beta \mathcal{S}(C) \tag{1}$$

where $\mathcal{S}(\cdot)$ represents a sparsity constraint (l_1-norm), and $\mathcal{G}(\cdot, \cdot)$ represents a graph-regularizer. These constraints form the Laplacian sparse coding that helps capture underlying semantics behind observations in both modalities [14,20]. W is a neighboring matrix that indicates similarities between pairs of data instances. A multimodal variation of the *feature-sign search algorithm* is developed to selectively update some elements of each data instance to tackle the non-derivativeness of the l_1-norm [14]. Since the sparse codes learned through general-purpose machine learning algorithms usually do not reflect ideal expert image understanding [12], we extend this framework with extra constraints from expert knowledge to improve semantic image representations.

3   Interface Design

The initial image grouping purely based on offline collected expert data is first visualized in the *Older Image Organization* in Fig. 4 (panel 1-a) for experts to inspect and manipulate. In the case where domain expert users need further information on the current image grouping, we provide two extra visualizations. First, experts can see an image cluster and the top features contributing to this cluster (see Fig. 6). Second, experts can click the buttons in Fig. 4 (panel 3) to compare the image grouping obtained when using different subsets of features, such as only primary morphology terms[1] (Fig. 5a), with that using the whole feature set; see Fig. 4 (panel 1-a).

Experts have two options to improve the image grouping in each round. First, they can directly drag images toward or apart from each other in Fig. 4 (panel 1-a). The system parses such expert inputs and incorporates them for updating the neighboring graph-based regularizer (see Sect. 5.1). Second, experts can select a topic from the listbox in Fig. 4 (panel 5), and indicate the least relevant

---

[1] Primary morphology terms (PRI) and 8 other categories of terms were identified by 2 highly trained dermatologists as *thought units* in an annotation study to label the stages in diagnostic reasoning [17]. Used here in this interface, these thought units can disclose the influence of each category of terms on the medical image grouping.

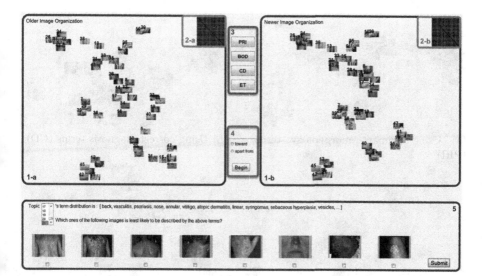

Fig. 4. Image grouping interface (details of algorithms behind this interface are described in Sects. 4 and 5): Panel (1-a) visualizes the image grouping before each round of expert image manipulation, and panel (1-b) visualizes the resulting image grouping afterwards. Experts are allowed to select multiple images in (1-a) for manipulation. Panels (2-a) and (2-b) are matrix views corresponding to (1-a) and (1-b), respectively, to show global pairwise image similarities. A button set (3) pops up new windows (shown in Fig. 5) to visualize image grouping initialized using various subsets of features, such as primary morphology terms (PRI). BOD stands for body parts, CD for correct diagnoses, and ET for eye gaze-filtered image features. Panel (4) allows experts to specify the direction to manipulate the selected images. Panel (5) lists the top key terms in each topic and allows experts to disconnect images from a topic.

image(s) according to the vocabulary distribution of the selected topic. Based on such expert inputs, the system updates the image-topic distribution matrix (see Sect. 5.2). After experts interact with the interface using either option, the image grouping in the previous round is copied to Fig. 4 (panel 1-a), and the improved one is shown in Fig. 4 (panel 1-b). In each round, both image groupings are visualized following the approaches discussed in Sect. 4.

4   Visualizing Image Groups

To comprehensively visualize the image grouping, our interface presents both a *graph view* shown in Fig. 4 (panel 1) and a *matrix view* shown in Fig. 4 (panel 2). Both views are automatically updated during expert interactions.

In the graph view, we adopt t-distributed stochastic neighborhood embedding (t-SNE) algorithm [19]. It better visualizes the high dimensional structure of image grouping in 2D graph view than other dimensionality reduction techniques, such as principal component analysis (PCA) [4]. We use a *distance*

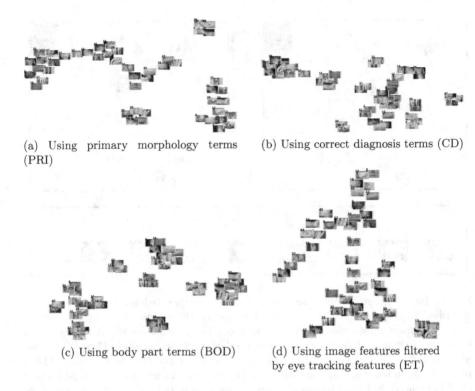

(a) Using primary morphology terms (PRI)

(b) Using correct diagnosis terms (CD)

(c) Using body part terms (BOD)

(d) Using image features filtered by eye tracking features (ET)

Fig. 5. Image groupings generated using subsets of features.

metaphor to imply to experts that more similar images are spatially closer. However, this metaphor does not proportionally reflect all pairwise image similarities[2] in high-dimensional space, because of the difficulty to retain the whole data structure for any dimensionality reduction algorithms. To tackle this issue, our interface allows experts to see an image and its high dimensional close neighbors in 2D visualization. The popup window visualizing these neighbors are illustrated in Fig. 6. The interface also presents a matrix view that serves to give an overview of the pairwise image similarities, because it is impractical that experts choose to see the close neighbors of all images in a 2D graph view. See Fig. 7 for a magnified matrix view. The matrix view provides a global indexing of pairwise image similarities in the learned representation.

5   Expert Knowledge Constraints

There are mainly two approaches in prior studies allowing user interactions to help improve learning a model: document-level interactions [4,15], or

---

[2] We do not define *image similarity* for domain experts to not restrict them by layperson definitions. We use t-SNE only as a feature projection technique for low dimensional visualization.

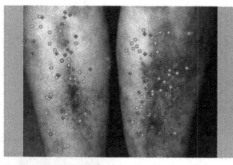

Closest neighboring image

Shared topics' top terms are:
vasculitis, papule, linear,
vitiligo, hyper, allergen, etc.

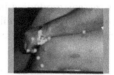

Second closest neighboring image

Shared topics' top terms are:
linear, vitiligo, hyper,
allergen, four, aspect, etc.

Third closest neighboring image

Shared topics' top terms are:
vasculitis, linear, vitiligo,
hyper, allergen, four, etc.

Fig. 6. An example visualization of an image and its high dimensional close neighbors. The target image is shown in the upper left quarter, and its top 3 close neighbors in the learned topic space are visualized in other quarters. The shared verbal features are ranked by term frequency, and the top ones are listed below each corresponding neighbor. The shared perceptually important image features are also ranked, and the top ones are marked in both the target image and its neighbors. The colors of the markers differentiate the image pairs. (Images courtesy of Logical Images, Inc.)

topic/cluster-level interactions [5,6]. In our scenario, to improve medical image grouping, the documents are images. To develop this interface, we prefer document (image)-level interactions for two reasons. On the one hand, the medical conditions are more intuitive in the form of images than texts to physicians. On the other hand, the topics we learned offline based on a multimodal expert dataset are not easily visualizeable nor interpretable by physicians. Below are two functions in the interface for receiving expert inputs and updating the model, both to support image-level interactions.

5.1   Constraint on Neighboring Matrix, W

Let the images in the original feature space be denoted as $x_1, ..., x_m$. A nearest neighbor graph G with m vertices can be constructed. A heat kernel can be used to compute the element W_{ij} in the neighboring matrix W of the graph G [2].

$$W_{ij} = e^{-\frac{\|x_i - x_j\|}{\sigma}} \tag{2}$$

If x_i and x_j are identical, then W_{ij} equals 1; and if they are extremely different, then W_{ij} asymptotically approaches 0.

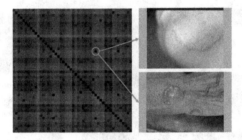

Fig. 7. An example of the matrix view. The intensity of each block represents the similarity between corresponding images. The darker the block is, the more similar the images are. For example, the similarity between the images on the right is indicated by the dark block circled in the matrix view on the left. (Image courtesy of Logical Images, Inc.)

The interface can encode expert image manipulations as a transformation of the neighboring matrix W. This transformation is determined by multiple factors, including previous image grouping and experts' interpretation of it. The transformation of W can be simplified as $\mathcal{F}(\cdot, \cdot)$ in Eq. (3) and be considered as a constraint set by experts to guide the learning process.

$$\min_{P,Q,C \geq 0} \|E - PC\|_F^2 + \|V - QC\|_F^2 + \alpha\mathcal{G}(\mathcal{F}(W, K), C) + \beta\mathcal{S}(C) \qquad (3)$$

where K denotes the set of images selected by an expert in Fig. 4 (panel 1-a). In this paper, we use *hard constraints*, i.e., by moving one image toward or away from another, experts can connect or disconnect them in the model. Such expert constraint essentially sets a boundary regarding pairwise image similarities. Once an expert begins to connect these images, the system sets all W_{ij}'s $(i, j \in K$, $i \neq j)$ to be 1. Likewise, W_{ij}'s $(i, j \in K, i \neq j)$ are all set to be 0, if they should be grouped differently. This rule is designed to update the neighboring matrix W in Eq. (2). Once all W_{ij}'s specified by the expert are updated, the algorithm will trigger the further learning process for the image representation C and the visual and verbal topics P and Q with respect to the objective function in Eq. (3).

5.2   Constraint on Topic-Coefficient Matrix, C

Experts can also improve the image grouping through the task illustrated in Fig. 4 (panel 5). For each topic selected by experts in the listbox, its top terms in the topic-term distribution are listed. The list of top terms explains the gist of the topic to experts. The images that are considered highly relevant to the selected topic by the algorithm are then displayed at the bottom. The task for experts is to submit the least relevant image(s) to the topic to disconnect its/their link(s) to the topic. After experts have indicated the least relevant image(s), the system updates the coefficient matrix C according to the constraint in Eq. (4).

$$\min_{P,Q,C \geq 0} \|E - PC\|_F^2 + \|V - QC\|_F^2 + \alpha \mathcal{G}(W, C) + \beta \mathcal{S}(C)$$

$$\text{s.t. } C_{ij} = 0, \, i \in T, \text{ and } j \in L(i) \tag{4}$$

where T is the collection of selected topics, and $L(i)$ represents the least relevant images for topic i. In this paper, the element C_{ij} will be set to 0, if image j is selected to be least relevant to topic i. Once all C_{ij}'s are updated, the algorithm begins to learn P, Q and C further with respect to Eq. (4).

For both the update of neighboring matrix W and that of the topic-coefficient matrix C, the model is learned incrementally, and it is consistent between successive interactions. In order for experts to work on consistent image groupings, we also keep the visualization consistent between successive interactions. This is achieved by storing the 2D coordinates of images and using them as the starting point in the graph view (Fig. 4 (panel 1)) for the next interaction [19].

6 Evaluation and Discussions

To evaluate the effectiveness of the paradigm per expert's objectives, a domain expert (co-author) was asked to provide a *reference image grouping* that best matches her overall understanding of the relationships between medical images in the database. In particular, for each image she listed its most similar images in terms of their differential diagnoses. We designed an experiment to compare the image grouping performances between the results of fully automated machine learning and our expert-in-the-loop paradigm. For fully automated learning (case 1), the resulting image grouping was estimated by our model without expert inputs. In our paradigm (case 2), the physician interacted with the model in the loop towards a better image grouping result. She manipulated the images based on her medical knowledge and the clinical information presented in these images. To quantitatively evaluate the image grouping performances, we retrieved the image neighbors and compared them to the corresponding reference image grouping for both cases.

Table 1 summarizes the performances of both cases given various modalities, based on which the machine learns from experts. The image groupings with expert interactive constraints consistently outperform the traditional learning case. In particular, our paradigm performs much better than fully automated learning with verbal feature of correct diagnosis (CD). This suggests that diagnoses are the primary factor considered by the expert to group medical images. For both cases, eye tracking filters boost the performance of image features (e.g., 12.5 % $\rightarrow$ 17.86 %). Furthermore, learning from multimodal features achieves the best performance for both cases. We also elicited the expert's qualitative evaluation through an interview. The expert noticed the improvement of each iteration. Centroid-based clustering algorithms (e.g., K-means) and connectivity-based clustering algorithms (e.g., hierarchical clustering [7]) are used for comparison purposes. Since these algorithms are not easily applied to the multiple modalities, their multimodal performances are omitted. Their performances fall behind that of Laplacian sparse coding, and this suggests that Laplacian sparse coding is a good learning framework. Density-based and distribution-based algorithms do not work because of the small number of data instances.

Table 1. Image grouping performances of fully automated learning and our paradigm. The measurement is the percentage of images in the reference list to appear within the top 5 retrieved neighbors. Different combinations of modalities include primary morphology terms only (PRI), body location terms only (BOD), correct diagnoses terms only (CD), SIFT features only, SIFT features filtered by gaze features (SIFT+Gaze), and multimodal data (overall).

		Verbal			SIFT	SIFT+Gaze	Multimodal
		PRI	BOD	CD			
Case 1 (fully	K-means	29.46	11.61	14.29	8.93	14.29	–
automated)	Hierarchical clustering	25.00	11.61	12.50	11.61	13.39	–
learning	Laplacian sparse coding	33.04	14.29	36.61	10.71	14.29	**52.68**
Case 2 (our paradigm)		34.82	16.96	42.86	12.50	17.86	**59.82**

During the paradigm evaluation, we also recorded the expert's verbal labeling of the image groups. The labeling of image groups is useful to disclose her diagnostic reasoning while grouping images. This can be incorporated in future work to optimize the semantic feature space. Another important part of our future work involves implementing our paradigm on a larger dermatological image database with more experts in the loop to test our paradigm's robustness. An image hierarchy can be learned and visualized. For the ease of expert interactions, a few representative images can be selected from each group. In the case where new images do not even have offline annotations, they can still be positioned in an existing image grouping for further improvements, since single-modal features can be easily projected into the unified topic space [11].

The presentation of image groupings could also be based on experts' trade-off between various factors, such as the primary lesion morphology and the causes of the diseases. Our current visualization may not be feasible for a larger database. It is necessary to design a more effective visualization strategy to allow experts to explore both global structure and local details of image grouping.

By replacing the hard constraints in Eq. 3 with soft ones, the parameters in neighboring graph can also be learned. In order to balance the influences between the offline collected expert data and online expert inputs, soft constraints could be applied by encoding expert interactions in a new penalty term.

7   Related Work

Existing systems that allow interactive user visual analysis usually adopt topic modeling techniques [4–6,13,15]. Original features are reduced to a lower-dimensional topic space, in which documents are grouped. One type of such system, including UTOPIAN [4] and iVisClustering [15], visualizes the topics, so that users can adjust the topic-term distribution at the term granularity. In contrast, our paradigm focuses experts on natural high-level image grouping tasks and encodes expert image manipulations as constraints to improve the overall image grouping. Besides, in our domain the objects for experts to interact with are medical images rather than latent topics, which may be confusing to the experts. Another type of system, including LSAView [6] and iVisClassifier [5], involves document-level interactions. These systems require users to change the parameters of the algorithms. In contrast, our system updates the underlying topic model based on experts' natural manipulations of the images.

8 Conclusions

This paper presents an interactive machine learning paradigm with experts in the loop for improving image grouping. We demonstrate that image grouping can be significantly improved by expert constraints through incremental updates of the underlying computational model. In each iteration, our paradigm allows to accommodate our model to experts' input. Performance evaluation shows that expert constraints are an effective way to infuse expert knowledge into the learning process and improve overall image grouping.

Acknowledgments. This work was partially supported by NIH grant 1R21 LM010039-01A1 and NSF grant IIS-0941452. We would like to thank the participating physicians, the reviewers; and Logical Images, Inc. for images. Any opinions, findings, and conclusions or recommendations expressed in this paper are those of the authors and do not necessarily reflect the official views of the NIH or the NSF.

References

1. Aronson, A.R.: Effective mapping of biomedical text to the UMLS Metathesaurus: the MetaMap program. In: American Medical Informatics Association Annual Symposium Proceedings, pp. 17–21. AMIA (2001)
2. Belkin, M., Niyogi, P.: Laplacian eigenmaps and spectral techniques for embedding and clustering. NIPS **14**, 585–591 (2001)
3. Cai, D., Bao, H., He, X.: Sparse concept coding for visual analysis. In: IEEE Conference on Computer Vision and Pattern Recognition (CVPR), pp. 2905–2910. IEEE (2011)
4. Choo, J., Lee, C., Reddy, C.K., Park, H.: UTOPIAN: user-driven topic modeling based on interactive nonnegative matrix factorization. IEEE Trans. Vis. Comput. Graph. **19**(12), 1992–2001 (2013)
5. Choo, J., Lee, H., Kihm, J., Park, H.: iVisClassifier: an interactive visual analytics system for classification based on supervised dimension reduction. In: IEEE Symposium on Visual Analytics Science and Technology (VAST), pp. 27–34. IEEE (2010)
6. Crossno, P.J., Dunlavy, D.M., Shead, T.M.: LSAView: A tool for visual exploration of latent semantic modeling. In: IEEE Symposium on Visual Analytics Science and Technology (VAST), pp. 83–90. IEEE (2009)

7. El-Hamdouchi, A., Willett, P.: Comparison of hierarchic agglomerative clustering methods for document retrieval. Comput. J. **32**(3), 220–227 (1989)
8. Gao, S., Tsang, I.W., Chia, L.-T., Zhao, P.: Local features are not lonely-Laplacian sparse coding for image classification. In: IEEE Conference on Computer Vision and Pattern Recognition (CVPR), pp. 3555–3561. IEEE (2010)
9. Guo, X., Li, R., Alm, C., Yu, Q., Pelz, J., Shi, P., Haake, A.: Infusing perceptual expertise, domain knowledge into a human-centered image retrieval system: a prototype application. In: Proceedings of the Symposium on Eye Tracking Research and Applications, pp. 275–278. ACM (2014)
10. Guo, X., Yu, Q., Alm, C.O., Calvelli, C., Pelz, J.B., Shi, P., Haake, A.R.: From spoken narratives to domain knowledge: mining linguistic data for medical image understanding. Artif. Intell. Med. **62**(2), 79–90 (2014)
11. Guo, X., Yu, Q., Li, R., Alm, C.O., Haake, A.R.: Fusing multimodal human expert data to uncover hidden semantics. In: Proceedings of the 7th Workshop on Eye Gaze in Intelligent Human Machine Interaction: Eye-Gaze and Multimodality, pp. 21–26. ACM (2014)
12. Holzinger, A.: Human-Computer Interaction and Knowledge Discovery (HCI-KDD): what is the benefit of bringing those two fields to work together? In: Cuzzocrea, A., Kittl, C., Simos, D.E., Weippl, E., Xu, L. (eds.) CD-ARES 2013. LNCS, vol. 8127, pp. 319–328. Springer, Heidelberg (2013)
13. Hu, Y., Boyd-Graber, J., Satinoff, B., Smith, A.: Interactive topic modeling. Mach. Learn. **95**(3), 423–469 (2014)
14. Lee, H., Battle, A., Raina, R., Ng, A.Y.: Efficient sparse coding algorithms. In: Advances in Neural Information Processing Systems, pp. 801–808 (2006)
15. Lee, H., Kihm, J., Choo, J., Stasko, J., Park, H.: iVisClustering: an interactive visual document clustering via topic modeling. In: Computer Graphics Forum, vol. 31, pp. 1155–1164. Wiley Online Library (2012)
16. Lowe, D.G.: Distinctive image features from scale-invariant keypoints. Int. J. Comput. Vis. **60**(2), 91–110 (2004)
17. McCoy, W., Alm, C.O., Calvelli, C., Li, R., Pelz, J.B., Shi, P., Haake, A.: Annotation schemes to encode domain knowledge in medical narratives. In: Ide, N., Xia, F. (eds.) Proceedings of the 6th Linguistic Annotation Workshop, pp. 95–103. Association for Computational Linguistics, Stroudsburg (2012)
18. Vaidyanathan, P., Pelz, J., Li, R., Mulpuru, S., Wang, D., Shi, P., Calvelli, C., Haake, A.: Using human experts' gaze data to evaluate image processing algorithms. In: Hemami, S., Pappas, T.N. (eds.) 10th IVMSP Workshop: Perception and Visual Signal Analysis, pp. 129–134. IEEE (2011)
19. Van der Maaten, L., Hinton, G.: Visualizing data using t-SNE. J. Mach. Learn. Res. **9**(2579–2605), 85 (2008)
20. Zheng, M., Bu, J., Chen, C., Wang, C., Zhang, L., Qiu, G., Cai, D.: Graph regularized sparse coding for image representation. IEEE Trans. Image Process. **20**(5), 1327–1336 (2011)

Predicting Post-operative Visual Acuity for LASIK Surgeries

Manish Gupta[1(✉)], Prashant Gupta[1], Pravin K. Vaddavalli[2], and Asra Fatima[2]

[1] Microsoft, Hyderabad, India
{gmanish,prgup}@microsoft.com
[2] L. V. Prasad Eye Institute (LVPEI), Hyderabad, India
{pravin,asrafatima}@lvpei.org

Abstract. LASIK (Laser-Assisted in SItu Keratomileusis) surgeries have been quite popular for treatment of myopia (nearsightedness), hyperopia (farsightedness) and astigmatism over the past two decades. In the past decade, over 10 million LASIK procedures had been performed in the United States alone with an average cost of approximately $2000 USD per surgery. While 99 % of such surgeries are successful, the commonest side effect is a residual refractive error and poor uncorrected visual acuity (UCVA). In this work, we aim at predicting the UCVA post LASIK surgery. We model the task as a regression problem and use the patient demography and pre-operative examination details as features. To the best of our knowledge, this is the first work to systematically explore this critical problem using machine learning methods. Further, LASIK surgery settings are often determined by practitioners using manually designed rules. We explore the possibility of determining such settings automatically to optimize for the best post-operative UCVA by including such settings as features in our regression model. Our experiments on a dataset of 791 surgeries provides an RMSE (root mean square error) of 0.102, 0.094 and 0.074 for the predicted post-operative UCVA after one day, one week and one month of the surgery respectively.

Keywords: LASIK surgeries · UCVA · Uncorrected visual acuity · Regression

1 Introduction

Refractive surgeries for eye are performed to correct (normalize) the refractive state of the eye, to decrease or eliminate dependency on glasses or contact lenses. This can include various methods of surgical remodeling of the cornea or cataract surgery. LASIK is a refractive eye surgery that uses a laser to correct nearsightedness, farsightedness, and/or astigmatism. In LASIK, a thin flap in the cornea is created using either a micro-keratome blade or a femto-second laser. The surgeon folds back the flap, then removes some corneal tissue underneath using a laser. The flap is then laid back in place, covering the area where the corneal

J. Bailey et al. (Eds.): PAKDD 2016, Part I, LNAI 9651, pp. 489–501, 2016.
DOI: 10.1007/978-3-319-31753-3_39

tissue was removed. With nearsighted people, the goal of LASIK is to flatten the steep cornea; with farsighted people, a steeper cornea is desired. LASIK can also correct astigmatism by smoothing an irregular cornea into a more normal shape. LASIK surgeries are highly popular; over 10 million LASIK procedures have been performed in the United States alone in the past decade[1].

Motivation. While overall patient satisfaction rates after primary LASIK surgery have been around 95 %, it may not be recommended for everybody for two reasons: (1) high cost with potentially no significant improvement for certain types of patients, and (2) possible eye complications after the surgery. LASIK surgeries cost approximately $2000 USD per surgery. An ability to predict post-operative UCVA can help patients make an informed decision about investing their money in undergoing a LASIK surgery or not. It can also help surgeons recommend the most promising type of laser surgery to the patients. How can we perform this prediction? Further, while performing such surgeries, surgeons need to set multiple parameters like suction time, flap and hinge details, etc. These are often set using manually designed rules. Can we design a data driven automated method to suggest the best settings for a patient undergoing a laser surgery of a certain type?

Problem Definition. In this paper, we address the following problem.

Given: Pre-operative examination results and demography information about a patient.

Predict: Post-operative UCVA after one day, one week and one month of the surgery.

Challenges. The problem is challenging because (1) large amount of data about such surgeries is not easily available; (2) there are a lot of pre-operative measurements that can be used as signals; and (3) data is sparse, i.e., there are a lot of missing values.

Brief Overview of the Proposed Approach. We model the task as a regression problem. We use domain knowledge to pre-process data by transforming a few categorical features into binary features. We also use average values to impute missing values for numeric features. For categorical features, we impute missing values using the most frequent value for the feature. We evaluate multiple regression approaches. Our experiments on a dataset of 791 surgeries provides an RMSE of 0.102, 0.094 and 0.074 for the predicted post-operative UCVA after one day, one week and one month of the surgery respectively.

Main Contributions. In summary, we make the following contributions in this paper.

- We propose a critical problem of predicting post-operative UCVA for patients undergoing LASIK surgeries.

---

[1] http://www.statista.com/statistics/271478/number-of-lasik-surgeries-in-the-us/.

- We model the task as a regression problem. We explore the effectiveness of demographic, pre-operative features and surgery settings for the prediction task. To the best of our knowledge, this is the first work to systematically explore this critical problem using machine learning methods.
- Using a dataset of 791 LASIK surgeries performed on 404 patients from 2013 and 2014, we show the effectiveness of the proposed methods. The dataset is made publicly available[2].

Paper Organization. The paper is organized as follows. We start with a basic introduction to laser surgery procedure in Sect. 2. In Sect. 3, we discuss various features that can be used for post-operative UCVA prediction. Further, in Sect. 4, we discuss various kinds of regression methods that can be used for the task. In Sect. 5, we present dataset details, and also insights from analysis of results. We discuss related work in Sect. 6 and conclude with a summary in Sect. 7.

2 Introduction to Laser Surgeries

In this section, we discuss main steps in a typical laser surgery. This will help us understand the importance of features discussed in Sect. 3. Further, we also discuss various types of laser surgeries depending on the laser ablation profile.

A laser surgery involves three main steps during the operation as follows.

- Flap creation: A soft corneal suction ring is applied to the eye, holding the eye in place. Once the eye is immobilized, a flap is created by cutting through the corneal epithelium and Bowman's layer. This process is achieved with a mechanical micro-keratome using a metal blade, or a femto-second laser that creates a series of tiny closely arranged bubbles within the cornea. A hinge is left at one end of this flap. The flap is folded back, revealing the stroma, the middle section of the cornea.
- Laser remodelling: The second step of the procedure uses a laser to remodel the corneal stroma. The laser vaporizes the tissue in a finely controlled manner without damaging the adjacent stroma. The layers of tissue removed are tens of microns thick.
- Repositioning of the flap: After the laser has reshaped the stromal layer, the LASIK flap is carefully repositioned over the treatment area by the surgeon and checked for the presence of air bubbles, debris, and proper fit on the eye. The flap remains in position by natural adhesion until healing is completed.

There are four types of laser surgeries depending on laser ablation profiles as follows.

- Plano-scan-LASIK: During the plano-scan LASIK procedure the corneal tissue is evenly ablated by the laser beam.
- Aspheric-LASIK: Using the aspheric profile means to ablate the corneal tissue in an "egg-shaped" way using a "flying spot" laser beam, similar to an American football.

---

[2] https://www.dropbox.com/s/xdm835jg1w5qvlu/lasik.txt?dl=0.

- Tissue-saving-LASIK: The tissue-saving profile is one where the aim is to save as much corneal tissue as possible during the laser ablation time in case of borderline cases with regard to the initial corneal thickness when the standard LASIK procedure is no longer possible.
- Wavefront-guided-LASIK: In case of the application of the wavefront guided LASIK procedure, which is often also called "individualized" or "personalized" treatment LASIK, the eyes are measured pre-operatively using a wavefront pattern scanner. By using the wavefront guided LASIK procedure aberrations can be eliminated and therefore result in an optimum of visual acuity as well in daylight as in night-vision.

For more details about the fundamentals, surgical techniques and complications in LASIK surgeries, the reader is redirected to [2].

3   Features for Post-operative UCVA Prediction

In this section, we discuss various features that we use to learn the regression model for predicting post-operative UCVA.

3.1   Demography Features

Intuitively, post-operative UCVA must depend on features of the patient. Hence, we consider two important demography features: age and gender.

3.2   Pre-operative Examination Features

Before the surgery, the patient's corneas are examined with a pachymeter to determine their thickness, and with a topographer, or corneal topography machine, to measure their surface contour. Using a beam of light, a topographer creates a topographic map of the cornea. Using this information, the surgeon calculates the amount and the location of corneal tissue to be removed. We use the following features obtained using such pre-operative examination.

- Left/right eye: This is a binary feature to indicate the eye which is being operated upon: OD (right) or OS (left).
- Uncorrected Visual Acuity (UCVA): Visual acuity score without the aid of glasses or contact lenses.
- Uncorrected Near vision: Visual acuity measured using a small chart held near the patient.
- Corrected Near vision: Visual acuity measured using a small chart held near the patient with glasses.
- BCVA with glasses: The best acuity score one can achieve with glasses.
- Sphere: This indicates the amount of lens power, measured in diopters (D), prescribed to correct nearsightedness (−) or farsightedness (+). The term "sphere" means that the correction for nearsightedness or farsightedness is "spherical," or equal in all meridians of the eye. This is measured using retinoscopy as well as using an auto-refractor leading to two separate features.

- Cylinder: This indicates the amount of lens power for astigmatism. The term "cylinder" means that this lens power added to correct astigmatism is not spherical, but instead is shaped so one meridian has no added curvature, and the meridian perpendicular to this "no added power" meridian contains the maximum power and lens curvature to correct astigmatism. This is measured using retinoscopy as well as using an auto-refractor leading to two separate features.
- Axis: This describes the lens meridian that contains no cylinder power to correct astigmatism. The axis is defined with a number from 1 to 180. The number 90 corresponds to the vertical meridian of the eye, and the number 180 corresponds to the horizontal meridian. This is measured using retinoscopy as well as using an auto-refractor leading to two separate features.
- Spherical equivalent: This indicates the spherical power whose focal point coincides with the circle of least confusion of a sphero-cylindrical lens. Hence, the spherical equivalent is equal to the algebraic sum of the value of the sphere and half the cylindrical value.
- Slit lamp Examination: Slit lamp is an apparatus for projecting a narrow flat beam of intense light into the eye. It helps in the microscopic study of various structures of the eye like eyelid(s), lashes, conjunctiva, cornea, anterior chamber, pupil, iris, vitreous, and retina. Typical values for this column could be "normal", "corneal scar", "sub epithelial scar", "fuchs heterochromic iridocyclitis", etc. We convert it to a binary column "normal" versus "abnormal".
- IOP (intraocular pressure): The pressure of the intraocular fluid, usually measured in millimeters of mercury.
- Retina examination: This is a binary feature and can take the following values: "normal" or "abnormal". Abnormal cases include various forms of retinal issues like "Chorioretinal Atrophy", "Familial Exudative Vitreo-retinopathy (FEVR)", "Barrage laser done", "Retinal pigment epithelium (RPE) atrophy", "Tilted disc with temporal pallor", etc.
- Steep-K, Flat-K and Axis@Flat-K: For a given corneal topography reading, the lower diopter number represents the less steep meridian of the cornea, or the "flat-K". The higher diopter number represents the steepest meridian of the cornea, or the "steep-K". Usually these are numbers between 40 and 50. The difference between the horizontal (higher) and vertical (lower) diopter readings gives you the approximate amount of corneal astigmatism, or cylinder correction. Axis@Flat-K is a number from 1 to 180.
- Thinnest Preop Corneal Thickness: The minimum thickness of the cornea. This usually varies from 450 to 650 microns.
- Topography machine: This indicates the type of topography machine used. In our dataset, three kinds of machines were used: Orbscan, Galilei and Oculyzer.

3.3   Surgery Settings

This set of features include various settings used when performing the surgery. The following is the list of features used.

– Surgery type: This depends on the laser ablation profiles and can be of 4 types: Plano-scan-LASIK, Aspheric-LASIK, tissue-saving-LASIK, or wavefront-guided-LASIK.
– Flap thickness: Most surgeons choose flap thicknesses between $100\,\mu$ and $120\,\mu$.
– Suction time: Suction time should be as short as possible to minimize optic nerve head and retinal ischemia (i.e., reduced vision) during LASIK. This usually varies from half a minute to a minute.
– Optic zone: This is the size of the treatment area. Common optic zone diameter is between 6–7 mm. It has been shown that a larger surgical optical zone diameter significantly decreases higher order aberrations after LASIK.
– Flap diameter: Diameter of the flap.
– Flap side cut angle: Previously, all side cuts were made at 90°. But now surgeons believe that certain cut angles could lead to stronger adhesion, less dry eyes, or better cosmetic looks[3].
– Hinge details: This includes hinge position, hinge angle, and the hinge width. Hinge position is usually kept at 90 for most of the surgeries. Hinge angle can be varied from 40 to 60°. Hinge width varies from 3 to 4.5 mm.

4   Approaches for Post-operative UCVA Prediction

We model the post-operative UCVA prediction task as a regression problem using the features described in Sect. 3. We explore four approaches for regression as follows.

– GDBT Regression: We used an efficient implementation of the Multiple Additive Regression Trees (MART) gradient boosting algorithm. MART learns an ensemble of regression trees, which is a decision tree with scalar values in its leaves. The ensemble of trees is produced by computing, in each step, a regression tree that approximates the gradient of the loss function, and adding it to the previous tree with coefficients that minimize the loss of the new tree. The output of the ensemble produced by MART on a given instance is the sum of the tree outputs.
– Online Gradient Descent based Regression: Linear regression is used for modeling the relationship between a scalar dependent variable and one or more explanatory variables (or independent variables). Linear regression models are often fitted using the least squares approach. Stochastic gradient descent is a gradient descent optimization method for minimizing an objective function that is written as a sum of differentiable functions. Online Gradient Descent based Regression is a form of linear regression which uses Stochastic gradient descent for optimization.
– Neural network based regression: A neural network model is defined by the structure of its graph (namely, the number of hidden layers and the number of neurons in each hidden layer), the choice of activation function, and the

---

[3] http://www.reviewofophthalmology.com/content/i/1777/c/32309/.

weights on the graph edges. The neural network algorithm tries to learn the optimal weights on the edges based on the training data. We use a neural network with one hidden layer and a single output neuron.

- Poisson Regression: Poisson Regression assumes that the unknown function, denoted Y has a Poisson distribution, i.e., given the instance $x = (x_0, x_1, ..., x_{D-1})$, for every $k = 0, 1, ...$, the probability that its value is k is given by Eq. 1.

$$p(k|x; \theta) = \frac{[E(Y|x)]^k e^{-E(Y|x)}}{k!} \tag{1}$$

where $E(Y|x) = e^{\sum \theta_i x_i}$. Given the set of training examples, the algorithm tries to find the optimal values for $\theta_0, ..., \theta_{D-1}$ by trying to maximize the log likelihood of the parameters given the input. The likelihood of the parameters $\theta_0, ..., \theta_{D-1}$ is the probability that the training data was sampled from a distribution with these parameters.

5 Experiments

In this section, we describe our dataset, metrics and experiments to analyze relative accuracy of various methods proposed in Sect. 4 for the post-operative UCVA prediction problem.

5.1 Dataset

The dataset contains information for 404 patients in the age range of 18 to 47 years. 215 of these patients are females, and the rest are males. The 791 LASIK surgeries were done in 2013 and 2014. 397 of the surgeries were performed on the left eye and remaining ones on the right eye. Most of the surgeries are either of the Wavefront-guided-LASIK type or of the Plano-scan-LASIK type. Orbscan is the most popular topography machine used; Oculyzer being the second most popular one. Pre-operative UCVA values vary between 0.15 and 2. Post-operative UCVA values vary between -0.2 and 1 for day 1, -0.3 and 1 for week 1 and -0.2 and 0.95 for month 1 after the operation. Although usually large datasets improve accuracy of the learned machine learning models, it is difficult to obtain large datasets in this domain.

Data Pre-processing. The dataset contains features like "Slit lamp examination" and "Retina examination". Although these columns contain a few cases of various abnormalities, specific types of abnormalities are not very useful due to low occurrence frequency of such abnormal cases in the dataset. Hence, we group all abnormal cases into a single attribute value called "abnormal" and convert the two features to binary-valued features with two values: "normal" and "abnormal".

The dataset contains a lot of missing values. On average around 83 instances have missing values across all attributes. The most number of missing values

(272) were for the IOP attribute. Categorical features were converted to numeric features by mapping them to consecutive integers. Missing values were replaced by the average value for the column for numeric features, and by the most frequent value for the column for categorical features.

5.2   Metrics

Measuring Visual Acuity. Visual acuity is measured by a psycho-physical procedure and as such relates the physical characteristics of a stimulus to a subject's percept and his/her resulting responses. Measurement can be done by using an eye chart, by optical instruments, or by computerized tests like the FrACT[4]. Visual acuity is often measured according to the size of letters viewed on a Snellen chart or the size of other symbols, such as Landolt Cs[5] or the Tumbling E[6].

In some countries, acuity is expressed as a vulgar fraction, and in some as a decimal number. Using the meter as a unit of measurement, (fractional) visual acuity is expressed relative to 6/6 (normal vision). Otherwise, using the foot, visual acuity is expressed relative to 20/20. For all practical purposes, 20/20 vision is equivalent to 6/6. In the decimal system, acuity is defined as the reciprocal value of the size of the gap (measured in arc minutes) of the smallest Landolt C, the orientation of which can be reliably identified. A value of 1.0 is equal to 6/6.

LogMAR is another commonly used scale, expressed as the (decadic) logarithm of the minimum angle of resolution. The LogMAR scale[7] converts the geometric sequence of a traditional chart to a linear scale. It measures visual acuity loss: positive values indicate vision loss, while negative values denote normal or better visual acuity. In this paper, we use the LogMAR scale for visual acuity. In practice, LogMAR values can vary from -0.3 (equivalent to 20/10) to 2 (equivalent to 20/2000).

Evaluating Regression Results. In this work, we use regression to predict post-operative UCVA in LogMAR. To evaluate regression results, we use three metrics: L1, L2 and Root Mean Squared Error (RMSE). Let N be the number of instances. Consider 2 vectors of UCVA across various instances: true (T) and predicted (P). The three metrics are then defined as follows.

- Avg L1 $= \sum_i \frac{|T_i - P_i|}{N}$
- Avg L2 $= \sum_i \frac{|T_i - P_i|^2}{N}$
- RMSE (Root mean squared error) is the square-root of the L2 error.

Lower values of L1, L2 and RMSE are better.

---

[4] http://michaelbach.de/fract/.
[5] https://en.wikipedia.org/wiki/Landolt_C.
[6] https://en.wikipedia.org/wiki/E_chart.
[7] https://en.wikipedia.org/wiki/LogMAR_chart.

5.3   Results

We use four different regression mechanisms for the prediction task. 10-fold cross validation is used to report accuracy values. Tables 1, 2 and 3 show the accuracy numbers for day 1, week 1 and month 1 after the operation without using the surgery settings features. Values in the brackets indicate standard deviation. We can see that in all the cases, GDBT Regression performs the best. This is in line with various other studies which claim that ensemble based methods perform well.

Table 1. 10-Fold cross validation accuracy of day 1 UCVA predictions using various models without surgery settings features (Numbers in brackets indicate standard deviation)

Model	L1(avg)	L2(avg)	RMS(avg)
Online gradient descent	0.0771 (0.0084)	0.0136 (0.0065)	0.1155 (0.026)
Poisson regression	0.0744 (0.008)	0.0128 (0.0049)	0.1108 (0.0217)
GDBT regression	0.0695 (0.0123)	0.0108 (0.0051)	0.1024 (0.0234)
Regression neural network	0.082 (0.0077)	0.0142 (0.0068)	0.1179 (0.027)

Table 2. 10-Fold cross validation accuracy of week 1 UCVA predictions using various models without surgery settings features (Numbers in brackets indicate standard deviation)

Model	L1(avg)	L2(avg)	RMS(avg)
Online gradient descent	0.0638 (0.0068)	0.0132 (0.0057)	0.112 (0.0244)
Poisson regression	0.061 (0.0118)	0.0118 (0.0067)	0.1064 (0.0298)
GDBT regression	0.0577 (0.007)	0.0094 (0.0032)	0.094 (0.0161)
Regression neural network	0.0643 (0.011)	0.0149 (0.0073)	0.1184 (0.03)

Figure 1 shows the variation of true versus predicted post-operative UCVA for the one-day after the surgery prediction. The figure illustrates the accuracy of the prediction model.

We also experimented by adding the surgery settings features. The metrics improve a little compared to using only demography and pre-operative examination features. However, the results are not significant. This could be because the current surgery settings are already set to optimum values, or because of the small amount of available training data.

Finally, we performed feature selection to identify the most important features using the CfsSubsetEval attribute selector [13] and the Best First search method. CfsSubetEval evaluates the worth of a subset of attributes by considering the individual predictive ability of each feature along with the degree of

Table 3. 10-Fold cross validation accuracy of month 1 UCVA predictions using various models without surgery settings features (Numbers in brackets indicate standard deviation)

Model	L1(avg)	L2(avg)	RMS(avg)
Online gradient descent	0.0514 (0.0115)	0.0099 (0.0068)	0.0933 (0.0344)
Poisson regression	0.051 (0.006)	0.0097(0.0051)	0.0955 (0.0247)
GDBT regression	0.0458 (0.0097)	0.0057 (0.0029)	0.0736 (0.0184)
Regression neural network	0.0539 (0.0099)	0.0102 (0.0058)	0.0967 (0.0283)

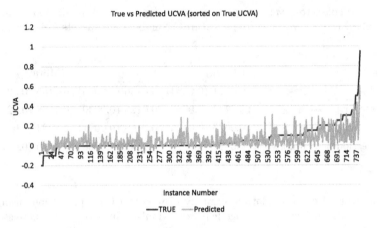

Fig. 1. True vs predicted UCVA for day 1

Table 4. Most important features

Day 1	Week 1	Month 1
BCVA with glasses	BCVA with glasses	Age
Spherical equivalent	Spherical equivalent	Uncorrected near vision
Slit lamp examination	Corrected near vision	BCVA with glasses
IOP	Thinnest Preop Corneal Thickness	Axis
Axis@Flat K	Axis@Flat K	Spherical equivalent

redundancy between them. The most important features across the datasets are shown in Table 4.

Surprisingly, pre-operative UCVA does not turn up in the top important features possibly due to high correlation with the above features. We show the variation of each of these important features with respect to the post-operative UCVA (day 1) in Fig. 2.

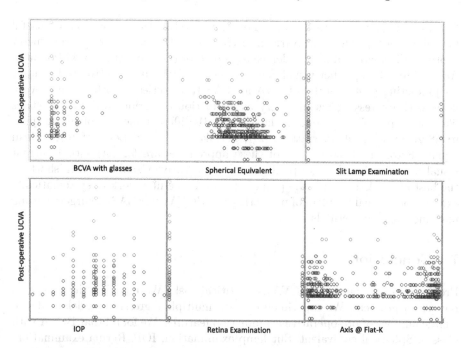

Fig. 2. Variation of important features with respect to the Post-Operative UCVA (day 1)

6   Related Work

Data Mining in Healthcare. Recently there has been a lot of work in the data mining community across various aspects of health care. Main areas of focus include personalized medicine [9], phenotyping [7,14,15], analysis of electronic medical records [5,18], mortality prediction [11,16], patient re-admission risks [4, 6]. However, there has not been much work on applying machine learning and data mining techniques to problems in ophthalmology.

Visual Acuity Prediction. Our work is most related to previous work on visual acuity prediction. There has been some previous work on predicting visual acuity in the ophthalmology community but it differs from our work in multiple aspects like the type of surgeries, the type of features used and the type of methods used for prediction as detailed in the following. Baron et al. [3] use pupil size, ablation size, refractive error, and photoreceptor directional sensitivity as features and a point-spread function as the method to predict visual acuity. We explore a much larger set of features using regression. Also, their aim is to predict correlations only rather than the post-operative UCVA values. Besides this they perform analysis for PRK (photorefractive keratectomy) surgeries while we focus on LASIK surgeries. Olsen et al. [17] use coloboma size, optic nerve color, foveal development, and subfoveal retinal pigment epithelial changes as features and linear regression as the method to predict visual acuity for children. Unlike our work, this is not related to LASIK surgeries, and is useful for children only. There are various instruments like white-light interferometer (Lotmar Visometer) and a Snellen chart projector (Guyton-Minkowski Potential Acuity Meter) to predict visual acuity after cataract surgeries [10,19] but none for predicting post-operative UCVA for LASIK surgeries. Another line of work deals with expressing visual acuity as a function of various other eye metrics like "pupil plane" and "image plane" [12], 31 different metrics of image quality [8,20], wavefront aberrations [21], Zernike Mode and Level of Root Mean Square Error [1]. However, all of these approaches are for estimating current visual acuity rather than for predicting post-operative visual acuity. In short, to the best of our knowledge, the proposed work is the first work to systematically explore the critical problem of predicting the UCVA after LASIK surgeries using machine learning methods.

7   Conclusion

Predicting post-operative UCVA is a critical task. We modeled the task as a regression problem. We experimented with multiple regression models and also a large number of pre-operative examination features. We found that BCVA with glasses, Spherical equivalent, Slit lamp examination, IOP, Retina examination, Axis@Flat K are very important features. Our models can provide predictions for UCVA after 1 day, 1 week and 1 month with an RMSE of 0.102, 0.094 and 0.074 respectively. The surgery settings seemed to reduce the RMSE but the reduction was not statistically significant. In the future, we plan to examine the impact of surgery settings on the UCVA prediction task with larger amount of data. We also plan to extend the set of features to include other features like ethnicity and profession of patients, and also features related to surgeons like their expertise, experience, etc.

References

1. Applegate, R.A., Ballentine, C., Gross, H., Sarver, E.J., Sarver, C.A.: Visual acuity as a function of zernike mode and level of root mean square error. Optom. Vis. Sci. **80**(2), 97–105 (2003)
2. Azar, D.T., Koch, D.: LASIK (Laser in Situ Keratomileusis): Fundamentals, Surgical Techniques, and Complications. CRC Press, Boca Raton (2002)
3. Baron, W.S., Munnerlyn, C.: Predicting visual performance following excimer photorefractive keratectomy. Refract. Corneal Surg. **8**(5), 355–362 (1991)
4. Roy, S.B., Teredesai, A., Zolfaghar, K., Liu, R., Hazel, D., Newman, S., Marinez, A.: Dynamic hierarchical classification for patient risk-of-readmission. In: Proceedings of the 21th ACM SIGKDD International Conference on Knowledge Discovery and Data Mining, pp. 1691–1700. ACM (2015)
5. Caballero Barajas, K.L., Akella, R.: Dynamically modeling patient's health state from electronic medical records: a time series approach. In: Proceedings of the 21th ACM SIGKDD International Conference on Knowledge Discovery and Data Mining, pp. 69–78. ACM (2015)

6. Caruana, R., Lou, Y., Gehrke, J., Koch, P., Sturm, M., Elhadad, N.: Intelligible models for healthcare: predicting pneumonia risk and hospital 30-day readmission. In: Proceedings of the 21th ACM SIGKDD International Conference on Knowledge Discovery and Data Mining, pp. 1721–1730. ACM (2015)

7. Che, Z., Kale, D., Li, W., Bahadori, M.T., Liu, Y.: Deep computational phenotyping. In: Proceedings of the 21th ACM SIGKDD International Conference on Knowledge Discovery and Data Mining, pp. 507–516. ACM (2015)

8. Cheng, X., Bradley, A., Thibos, L.N.: Predicting subjective judgment of best focus with objective image quality metrics. J. Vis. 4(4), 7 (2004)

9. Fan, K., Eisenberg, M., Walsh, A., Aiello, A., Heller, K.: Hierarchical graph-coupled HMMs for heterogeneous personalized health data. In: Proceedings of the 21th ACM SIGKDD International Conference on Knowledge Discovery and Data Mining, pp. 239–248. ACM (2015)

10. Faulkner, W.: Laser interferometric prediction of postoperative visual acuity in patients with cataracts. Am. J. Ophthalmol. 95(5), 626–636 (1983)

11. Ghassemi, M., Naumann, T., Doshi-Velez, F., Brimmer, N., Joshi, R., Rumshisky, A., Szolovits, P.: Unfolding physiological state: mortality modelling in intensive care units. In: Proceedings of the 20th ACM SIGKDD International Conference on Knowledge Discovery and Data Mining, pp. 75–84. ACM (2014)

12. Guirao, A., Williams, D.R., et al.: A method to predict refractive errors from wave aberration data. Optom. Vis. Sci. 80(1), 36–42 (2003)

13. Hall, M.A.: Correlation-based Feature Subset Selection for Machine Learning. Ph.D. thesis, University of Waikato, Hamilton, New Zealand (1998)

14. Ho, J.C., Ghosh, J., Sun, J.: Marble: high-throughput phenotyping from electronic health records via sparse nonnegative tensor factorization. In: Proceedings of the 20th ACM SIGKDD International Conference on Knowledge Discovery and Data Mining, pp. 115–124. ACM (2014)

15. Liu, C., Wang, F., Jianying, H., Xiong, H.: Temporal phenotyping from longitudinal electronic health records: a graph based framework. In: Proceedings of the 21th ACM SIGKDD International Conference on Knowledge Discovery and Data Mining, pp. 705–714. ACM (2015)

16. Nori, N., Kashima, H., Yamashita, K., Ikai, H., Imanaka, Y.: Simultaneous modeling of multiple diseases for mortality prediction in acute hospital care. In: Proceedings of the 21th ACM SIGKDD International Conference on Knowledge Discovery and Data Mining, pp. 855–864. ACM (2015)

17. Olsen, T.W., Summers, C.G., Knobloch, W.H.: Predicting visual acuity in children with colobomas involving the optic nerve. J. Pediatr. Ophthalmol. Strabismus 33(1), 47–51 (1995)

18. Somanchi, S., Adhikari, S., Lin, A., Eneva, E., Ghani, R.: Early prediction of cardiac arrest (code blue) using electronic medical records. In: pp. 1691–1700 (2015)

19. Spurny, R.C., Zaldivar, R., Davis Belcher, C., Simmons, R.J.: Instruments for predicting visual acuity: a clinical comparison. Arch. Ophthalmol. 104(2), 196–200 (1986)

20. Thibos, L.N., Hong, X., Bradley, A., Applegate, R.A.: Accuracy and precision of objective refraction from wavefront aberrations. J. Vis. 4(4), 9 (2004)

21. Watson, A.B., Ahumada, A.J.: Predicting visual acuity from wavefront aberrations. J. Vis. 8(4), 17 (2008)

LBMF: Log-Bilinear Matrix Factorization for Recommender Systems

Yunhui Guo[✉], Xin Wang, and Congfu Xu

Institute of Artificial Intelligence, Zhejiang University, Hangzhou, China
{gyhui,cswangxinm}@zju.edu.cn, xucongfu@cs.zju.edu.cn

Abstract. Collaborative filtering techniques have been successfully applied in recommender systems recently. In order to improve recommendation accuracy for better user experience, the review texts should be exploited due to its rich information about users' explicit preferences and items' features, which cannot be fully revealed only by rating scores. In this paper, we propose an effective algorithm called LBMF to explore review texts and rating scores simultaneously. We directly correlate user and item latent dimensions with each word in review texts and ratings in our model, so semantic word vectors can be easily learned and effectively clustered based on rating values. On the other hand, the learned semantic word vectors can justify the rating values, which can promote better learning of user and item latent vectors for rating prediction. The learned latent dimensions by our model can reasonably explain why users rated items the way they did. This revelation can promote better modeling of user profiles and item information, and enable further analysis of user behaviors. Experimental results on several real-world datasets demonstrate the efficiency and effectiveness of LBMF comparing to the state-of-the-art models.

Keywords: Recommender systems · Collaborative filtering

1 Introduction

Recent years, more and more shopping websites have employed recommender systems to help improve shopping experience for customers. A functional recommendation algorithm should tell sellers whether or not a user prefers an item, and it is more valuable if the algorithm can discover the reason for users' preferences. Traditional recommendation techniques usually only based on rating scores. For example, a recommender system may recommend a new movie to a certain user considering the ratings the user given to other movies previously, but the review texts the user wrote are typically ignored. Although this approach have gained success to some extend, due to the ignorance of review information, it cannot give reasonable explanations for users' behavior. Especially, it cannot tell us the interaction between user's preferences and item's features.

Customers always express mixed emotions at items due to their attitude towards different items' features. For example, maybe a user likes the size of a

© Springer International Publishing Switzerland 2016
J. Bailey et al. (Eds.): PAKDD 2016, Part I, LNAI 9651, pp. 502–513, 2016.
DOI: 10.1007/978-3-319-31753-3_40

telephone's screen, and prefers the brand but has no interest in the color. This diverse meaning cannot be digged out by rating information simply. A rating score can only reflect a customer's overall impression on a product, but the subtle emotion beneath the rating information cannot be revealed, so this is a big drawback of recommendation algorithms that only based on rating scores.

Up to present, only a few works have taken notice of the importance and benefits of leveraging review texts to promote recommendation performance. However, due to the complexity of natural language and the rich information the review texts may contain, it is still a challenging task to extract effective information from review texts for recommendation task. Different from other text processing fields, review comments usually contain much emotional information, and customers always express their opinions on products by writing comments. Therefore, understanding the subjectivity of the review texts is the main point for such approach. Most existing works try to combine latent topic factors found by topic models such as Latent Dirichlet Allocation (LDA) [3], but the improvements are modest.

In this paper, we propose a novel recommendation algorithm based on the log-bilinear document model [8]. Specifically, we state the following two points: First, we assume the review texts and ratings are consistent and both of them are determined by users' or items' latent space. Then the review texts and rating information can be directly linked together through users' and items' dimensions. Second, we adopt a modified log-bilinear document model to exploit the coherence of topics in each comments. This is a probabilistic model with log-bilinear energy function to model the bag of words distribution of a document. Previous experiments [8] proved that the log-bilinear document model is more powerful than LDA in capturing word semantic and sentiment features.

Meanwhile, we apply a softmax transformation to project the users' and items' latent space into the semantic space. By this correlation, the latent factors can have a correspondence with the topics revealed by review texts. Different from other methods, we assume each user and item has a sentiment space respectively, and we map the latent factor space to the semantic word space directly and causally, making the model more effective and explainable. The experimental results on 9 real-world datasets show that our algorithm gains real improvement on prediction performance comparing to the state-of-the-art models.

2   Related Work

Recommender techniques can be roughly classified into two categories: content-based filtering techniques [1] and collaborative filtering techniques [11]. Content-based filtering systems try to use content information such as users' profiles, review texts and item tags. The methods based on collaborative filtering can be classified into neighborhood based algorithms and latent factor models [4]. Neighborhood based methods usually recommend an item to a user based on the behavior of similar users, and it is common to use rating information to find people who have the same interest. The latent factor based models denote users' and items' latent factors as vectors, and the product of certain element of the user's vector and item's vector reflects the user's preference for the corresponding item's feature. The most popular methods to learn the latent vectors are matrix factorization [10] and non-negative matrix factorization methods [6].

In order to provide more flexible models, more and more works try to combine contend-based methods and collaborative filtering techniques for better recommendation performance. Among them, the combination of rating information and review texts receives more and more attentions. In the early work, [12] proposes a probabilistic graphical model called CTR, which first defines a document as the set of all reviews of an item and then uses the LDA model to learn the topic similarity between each document.

Inspired by CTR, two related models (i.e., HFT and RMR) try to combine the topic factors with the latent user factors or latent item factors in different ways. Different from CTR, HFT [9] supposes that the latent factors are directly mapped to the topic vectors in LDA rather than by a sampling process. RMR [7] assumes users' latent space is correspondent to a mixture of Gaussian distributions and the ratings are sampled from a Gaussian mixture model.

To give each user or item a semantic space, researchers recently treat each review text as a document. SUIT [13] supposes user, item and topic latent factors co-determine each rating and then uses tensor factorization to learn the model. However, since all three factors are at the same level, it ignores the causal relationships among them and the model is hard to explain. In another work [2], the authors propose a model called topicMF. Like HFT and our work, it maps users' and items' latent space to the semantic space. The difference is that it uses nonnegative matrix factorization to learn topic factors, a primary drawback of this approach is the time complexity, hence it does not scale well and is not suitable for real-world large datasets.

3   Preliminaries

In this section, we first give the notations, then briefly address latent factor models.

3.1   Notations

We assume that there are N users and M items in the datasets. $U \in \mathbb{R}^{N \times K}$, $V \in \mathbb{R}^{M \times K}$ and $\theta \in \mathbb{R}^{K \times L}$ are user, item and topic factor matrix respectively, where K denotes the number of latent dimensions of each user, item and review text. K is a free parameter and can be set manually, a larger K means the user and item latent factors can carry richer information, and meanwhile each review text can be represented by a more detailed topic weightings.

We denote each user and item as u and i, meanwhile the corresponding latent vector is denoted by U_u and V_i respectively (i.e., the row vector of U and V). The rating score that user u gives to item i is R_{ui}. The review text given by user u to item i is represented as d_{ui} and the review texts collection is denoted as D. The topic distribution vector of a particular review text is θ_{ui}, which is a corresponding vector in θ.

We also use w to denote a single word in d_{ui}. The number of words a review text contains is represented as N_{ui}. We use k to denote words' index in the corresponding review text. And $\beta \in \mathbb{R}^{K \times |W|}$ is a word representation matrix denotes the words' association strength with respect to each latent topic dimension, where $|W|$ is the vocabulary size. The notations are shown in Table 1.

Table 1. Notations used in this work

Notations	Description
D	review collection
L	number of review texts
N	number of users
M	number of items
K	number of latent dimensions or topics
U_u	K-dimensional latent factors for user u
V_i	K-dimensional latent factors for item i
b_u	bias parameter of user u
b_i	bias parameter of item i
μ	global bias parameter
U	$R^{N \times K}$ matrix denotes all user factors
V	$R^{M \times K}$ matrix denotes all item factors
R_{ui}	rating value given by user u for item i
d_{ui}	review text for item i by user u
θ_{ui}	K-dimensional topic weighting vector
N_{ui}	number of words in d_{ui}
w	a single word in d_{ui}
$|W|$	vocabulary size
β	$R^{K \times |W|}$ word representation matrix
ϕ_{w_k}	K-dimensional word representation vector
b_w	bias parameter of word w

3.2 Latent Factor Models

Latent factor models use the low dimensional latent factors of users and items to approximate the rating matrix and then use the latent factors to predict ratings that are missed in the original rating matrix. It assumes that the preference of a user u for an item i can be denoted as a product of their latent factors: $U_u V_i^T$. Then the rating matrix can be estimated by the product of the two low-rank latent matrices: $R = UV^T$. Many latent factor models further add user bias term b_u and item bias term b_i to offset local effects. The optimization function is

$$\arg\min_{U,V} \sum_{u,i} (R_{ui} - U_u V_i^T - \mu - b_u - b_i)^2 + \Theta \tag{1}$$

where μ is the global mean rating value and the regularization term Θ is $\lambda_u \sum_u ||U_u||^2 + \lambda_i \sum_i ||V_i||^2 + \lambda_{b_u} \sum_u b_u^2 + \lambda_{b_i} \sum_i b_i^2$. Here λs are free regularization parameters for the purpose of avoiding over-fitting. Various optimization algorithms have been developed to find optimal solutions of U and V [5].

4   Log-Bilinear Matrix Factorization

In this section we introduce a new model called Log-Bilinear Matrix Factorization (LBMF). It is based on the log-bilinear document model, however these two methods have some intrinsic differences in modeling review texts. LBMF integrates rating scores and review texts simultaneously and naturally, thus this approach enables the latent dimensions have clearer interpretations, meanwhile this implication can have some practical meanings which we will analyze in detail.

4.1   Log-Bilinear Document Model

Log-bilinear document model [8] is a probabilistic model for learning semantic word vectors. It maps θ_d to each word w in the corresponding document and assumes that words in a document are conditionally independent given the mixture variable θ_d. Here θ_d is the semantic space and can be regarded as a weighting over topics. Different from LDA, θ_d is not sampled from Dirichlet distribution. Log-bilinear document model [8] uses a Gaussian prior on θ_d. Given a review text d_{ui}, it assumes each word $w_k \in d_{ui}$ is conditionally independent of other words in d_{ui} given θ_{ui} and w_k depends on θ_{ui} directly. The probability of the review text is given by:

$$\begin{aligned} p(d_{ui}) &= \int p(d_{ui}, \theta_{ui}) d\theta_{ui} \\ &= \int p(\theta_{ui}) \prod_{k=1}^{N_{ui}} p(w_k | \theta_{ui}) d\theta_{ui} \end{aligned} \tag{2}$$

where d_{ui} denotes the review text of user u given to item i. N_{ui} is the number of words in d_{ui} and w_k is the k-th word in d_{ui}. Here $p(w_k | \theta_{ui})$ is defined as a log-linear model. It has two parameters: $\beta \in \mathbb{R}^{K \times |W|}$ is a semantic-word matrix like LDA, and $b_w \in \mathbb{R}$ is bias for the word w in the vocabulary W, which captures local features of each word.

The energy assigned to a word w given these model parameters is:

$$E(w; \theta_{ui}, \phi_w, b_w) = -\theta_{ui}^T \phi_w - b_w \tag{3}$$

where $\phi_w = \beta w$ is a K dimensional vector related to the corresponding word's column in β. w is represented as a one-on vector, where for the k-th word $w_k = 1$. Then it uses *softmax* to represent $p(w|\theta_{ui})$ as:

$$p(w|\theta; \beta, b) = \frac{\exp(-E(w; \theta_{ui}, \phi_w, b_w))}{\sum_{w' \in W} \exp(-E(w'; \theta_{ui}, \phi_{w'}, b_{w'}))}$$
$$= \frac{\exp(\theta_{ui}^T \phi_w + b_w)}{\sum_{w' \in W} \exp(\theta_{ui}^T \phi_{w'} + b_{w'})} \qquad (4)$$

The intuition is that if a word w's representation vector ϕ_w matches the direction of θ_{ui} better, then its occurrence probability is higher in the corresponding review text. The log-bilinear document model aims to capture word representations and further to discover semantic and topic information. It can be employed in the field of sentiment analysis and subjectivity detection.

4.2 LBMF

In LBMF, we try to mine features and preferences information buried in review texts and utilize the information to enforce better learning of user latent dimensions and item latent dimensions. Different from previous works, we use direct mappings from user and item latent dimensions to the topic dimensions, and meanwhile the topic dimensions of each review text have a direct correlation with each word in the corresponding review text as the log-bilinear document model, hence the user and item latent dimensions can be more interpretable. The review texts information serves as a *regulariser* for the rating values, each word given by user u for item i can *regularize* the latent factor models, thus the latent user and item dimensions can embody textual information.

We do not restrict θ in the Dirichlet simplex space, but project it into the user or item latent space. It is based on the fact that the users' preferences or items' features determine both ratings and review texts, which is more reasonable than CTR and HFT. The mapping function needs to be monotonic, since a large U_{uk} or V_{ik} should corresponds to a larger θ_{kui}, which means the corresponding feature or preference is talked more frequently in the review text. So we define it as a softmax function:

$$\theta_{kui} = \frac{\exp(k_1|U_{uk}| + k_2|V_{ik}|)}{\sum_{k'}^K (\exp(k_1|U_{uk'}| + k_2|V_{ik'}|))} \qquad (5)$$

Specifically, $|U_{uk}|$ describes the user's level of interest at the dominant properties of the item and $|V_{ik}|$ denotes that whether or not the item possesses the corresponding property. To consider the negative values is critical, a large negative value may indicates that the user talks certain features of the item a lot in the review text but in a negative attitude. By this transformation, θ_{kui} is the weighting of topic k. And k_1 and k_2 are introduced to moderate the proportions of users' and items' affect.

There are other choices for the mapping relation, and we also consider only mapping topic dimensions into latent user space or latent item space separately.

We propose two new algorithms called LBMFu and LBMFi. The mapping relations are defined as:

$$\theta_{kui} = \frac{\exp(k|U_{uk}|)}{\sum_{k'}^{K}(\exp(k|U_{uk'}|))} \tag{6}$$

$$\theta_{kui} = \frac{\exp(k|V_{ik}|)}{\sum_{k'}^{K}(\exp(k|V_{ik'}|))} \tag{7}$$

We can regard LBMFu and LBMFi as special cases of LBMF.

Different from the log-bilinear document model, by Eqs. 4 and 5, we correlate each word distribution to user latent dimensions $\{U_u\}_1^N$ and item latent dimensions $\{V_i\}_1^M$ directly, hence the θ matrix is more coherent. Furthermore, unlike the log-bilinear document model, we confine θ in the unit simplex, hence θ can be regarded as a distribution over topics, thus endow θ with a clearer interpretation. Our model aims at learning semantic and topic information in review texts rather than sentiment information directly as the log-bilinear document model does. To capture the rating coherence, we use probabilistic matrix factorization proposed in [10]. Given the rating set $\{R_{ui}\}$ and review texts collection D, we assume each document and rating is i.i.d sampled, then we wish to learn the parameters $\{U_u\}_1^N$, $\{V_i\}_1^M$, $\{b_u\}$, $\{b_i\}$, $\{\phi_w\}$ and $\{b_w\}$ of LBMF.

In LBMF, $\{U_u\}_1^N$ and $\{V_i\}_1^M$ determine rating values and word distributions simultaneously, thus the latent user or item vectors correspond to features or preferences information. Since word distributions in the corresponding review text reflect this inexplicit features or preferences information exactly, it is sensible to use word distributions information to *regularize* the rating prediction models. The word distributions embody the textual information, which endow the latent user or item dimensions with clearer implications. This intuition motivates LBMF to substitute the log-likelihood of the review collection for the Θ in the latent factor models to promote better recommendation results. The objective optimization function is defined as:

$$\mathcal{L} = \sum_{u,i \in D} (U_u V_i^T + b_u + b_i - R_{ui})^2 - \gamma l(D) \tag{8}$$

Where $l(D) = \log \prod_{u,i \in D} \int p(\theta_{ui}) \prod_{k=1}^{N_{ui}} p(w_k|\theta_{ui}) d\theta_{ui}$ is the log-likelihood function of the review collection.

Since θ_{ui} is directly determined by U_u and V_i, we can substitute $\hat{\theta}_{ui}$ for the integral. Then Eq. 8 can be approximated as

$$\mathcal{L} = \sum_{u,i \in D} (U_u V_i^T + b_u + b_i - R_{ui})^2 - \gamma \log \prod_{u,i \in D} p(\hat{\theta}_{ui}) \prod_{k=1}^{N_{ui}} p(w_k|\hat{\theta}_{ui}) \tag{9}$$

where γ is introduced to balance the affect of the two components. A larger γ means the word distributions *regularize* the latent factor models more heavily.

In the following derivations, we set γ to 1 for clarity. Since θ_{ui} is directly deter-mine by U_u and V_i, we substitute Eq. 5 into Eq. 9. Hence the optimization para-meters here are U, V, β, k_1, k_2 and bias parameters. We optimize LBMF based on gradient descent (GD) by computing the corresponding gradients given Eq. 9:

$$\frac{\partial \mathcal{L}}{\partial U_u} = \sum_i 2\delta_{ui}(U_u V_i^T + b_u + b_i - R_{ui})V_i$$

$$- \sum_i \sum_{N_{ui}} [\phi_w - \Phi_w]\, \theta_{ui}(1 - \theta_{ui})\frac{U_u}{|U_u|} \tag{10}$$

$$\frac{\partial \mathcal{L}}{\partial V_i} = \sum_u 2\delta_{ui}(U_u V_i^T + b_u + b_i - R_{ui})U_u$$

$$- \sum_u \sum_{N_{ui}} [\phi_w - \Phi_w]\, \theta_{ui}(1 - \theta_{ui})\frac{V_i}{|V_i|} \tag{11}$$

where $\delta_{ui} = 1$ if u gives rating to i and $\Phi_w = \frac{\sum_{w' \in N_{ui}}(\phi_{w'} \exp(\theta_{ui}^T \phi_{w'} + b_{w'}))}{\sum_{w' \in N_{ui}} \exp(\theta_{ui}^T \phi_{w'} + b_{w'})}$.

$$\frac{\partial \mathcal{L}}{\partial \phi_{w_k}} = - \sum_{w_k \in d_{ui}} \left[\theta_{ui} - \frac{\theta_{ui} \exp(\theta_{ui}^T \phi_{w_k} + b_{w_k})}{\sum_{w' \in N_{ui}} \exp(\theta_{ui}^T \phi_{w'} + b_{w'})} \right] + \lambda_\beta \phi_{w_k} \tag{12}$$

$$\frac{\partial \mathcal{L}}{\partial b_{w_k}} = - \sum_{w_k \in d_{ui}} \left[1 - \frac{\exp(\theta_{ui}^T \phi_{w_k} + b_{w_k})}{\sum_{w' \in N_{ui}} \exp(\theta_{ui}^T \phi_{w'} + b_{w'})} \right] \tag{13}$$

Where w_k is a one-hot vector indicating the selected word representation vec-tor and we add a regularization term for β. The optimization procedures for LBFMu and LBMFi are similar to LBMF so we omit them, and the optimiza-tion equations of user and item bias parameters are also omitted due to space limitation.

From Eq. 9, we can find the differences between LBMF with other models. LBMF regards each review text as a document which is more sensible than CTR and HFT, in which review texts are combined for each users or items. Since users have inherent bias towards different features of items, it is not reasonable to com-bine all the review texts of a item to model a given user's preferences. This can account for our model's effectiveness over CTR and HFT partly. Furthermore, the mapping relation we adopted link θ_{ui} and latent user or item dimension is more reasonable than CTR and HFT.

LBMF assumes each word is directly correlated with θ_{ui}, and random variable θ_{ui} defines a distribution over a variety of topics. Meanwhile, θ_{ui} is determined by latent user and item factors, hence this transformation can bridge the gap between review texts and rating scores. On the one hand, word distributions indicates topics diversity and the variety of individual preferences information, and on the other hand each word in review texts can shape the latent user and item factors, in other words, the user and item latent dimensions correlate with the word distributions via the features and preferences information, thus the learned latent user and item factors are robust and interpretable. The utilization of the log-bilinear document model rather than LDA is critical. Since most of the review texts are short and sparse, we doubt about the effectiveness of LDA for such learning task. Compared with SUIT, we presume that topics and preferences information are determined by the latent user dimensions and item dimensions rather than a tensor outer product approach.

5   Experiments

In this section, we investigate the performance of our model for recommendation task. We train our model on several large real-world datasets collected from Amazon.com to tune the parameters. The description of the datasets can be found in [9] and we omit it due to the space limitation.

5.1   Compared Algorithms

We implement several baseline models for comparison. We list the models below and give simple summaries.

PMF: This is the traditional matrix factorization model which only models ratings.

HFT: This is the state-of-the-art method that models ratings and reviews simultaneously. HFT combines matrix factorization model with LDA to explore the rich information in review texts that can enhance the performance of the rating prediction.

CTR: This model focuses on recommending scientific articles to potential readers. CTR utilizes rating scores and review texts simultaneously. The key property of CTR lies in how the item vector is generated. By adjusting the precision parameter c, CTR can solve the one-class collaborative filtering problem.

SUIT: SUIT is a new supervised user-item based topic model, which utilizes the textual topics and latent user-item factors simultaneously. The model uses tensor outer product of text topic proportion vector, user latent factor and item latent factor to model the sentiment label generalization.

TopicMF-AT: This model is proposed in [2]. It incorporates the nonnegative matrix factorization with the standard matrix factorization model. It adopts the NMF to uncover hidden topics information in review texts. Both user latent vectors and item latent vectors are correlated with the topic factors via an exponential transform function analogous to the one adopted by HFT.

RMR: This method is recently proposed in [7]. It employs a model similar to LDA to explore interpretable topics to improve performance of rating prediction. It uses a mixture of Gaussian distributions rather than matrix factorization based methods. The item is modeled as a distribution of topics, together with the user-topic specific Gaussian distributions, determine how a user would rate an item.

5.2 Evaluation

We first evaluate rating prediction performance of various models. For each dataset we randomly select 80 % as training set and the remaining parts are evenly split into validation set and testing set. We adopt mean squared error (MSE) to evaluate our model and baseline models, a lower MSE on the test set indicates better rating prediction performance. We report the MSE of the test set which has the lowest MSE on the validation set. The initial parameters are randomly assigned. For all the models, across various datasets, the latent dimension K is set to 5. To assess the parameter sensitivity, we assign the latent dimension K to a variety of values later. We set all the regularization parameters to 0.01 and the learning rate to 0.001. The training of compared baseline models follow the same parameter settings. For our results, the balance parameter is set to 1. Other choices are also considered for the purpose of parameter sensitivity analysis.

5.3 Rating Prediction

Rating prediction task is to predict the rating values that users have not rated yet on the test set. Experimental results in terms of MSE are shown in Table 2, where the best performance result is in bold font. The main points and observations from the performance comparison include:

(1) The last eight models (CTR, HFT, TopicMF, SUIT, RMR, LBMF, LBMFu, LBMFi) combine ratings with review texts, which show great improvements in rating prediction performance comparing to PMF. Because review texts can embody users' preference for items, the learned latent vectors are more reasonable. The improvement in rating prediction is result of this combination.

(2) LBMF achieves the most accurate rating prediction on almost all the datasets. Careful examinations show that LBMF gains prediction improvements over HFT and topicMF up to 6.03 % and 1.41 % across all the datasets on average. This is in conformity with our previous analysis. Further, we observe that in several datasets, such as 'arts' and 'jewelry', our model performs exceedingly better than other approaches. The statistics of these categories reveal two important facts. First, there are fewer words per review in these datasets. Second, products in these categories are only reviewed by a small number of people. Despite the short and sparse review texts, our method has proved to perform exceedingly well in contrast with the LDA-based models. As for denser texts, our method is also competitive. Lengthy review texts may contain extra noises, this fact can explain why LBMF has better performance on sparser texts. Further text processing technique can be conducted on the review texts to exclude noisy information so as to achieve more accurate results.

Table 2. Performance of compared models (K = 5).

Method	Arts	Automotive	Baby	Beauty	Cell & accessories	Gourmet foods	Industrial & scientific	Jewelry	Musical instruments
PMF	1.5437	1.4814	1.5932	1.4010	2.1907	1.5823	0.3873	1.3902	1.5488
CTR	1.4221	1.4718	1.5729	1.3601	2.1813	1.4692	0.3832	1.2111	1.4239
HFT	1.4916	1.4725	1.5706	1.3923	2.1309	1.5547	**0.3527**	1.3116	1.4968
topicMF	1.3948	1.4165	1.5017	1.3348	2.1580	1.4128	0.3625	1.2056	1.3786
SUIT	1.4128	1.4562	1.5821	1.3622	2.2107	1.4528	0.3729	1.2121	1.4089
RMR	1.4138	1.4668	1.6768	1.3707	2.1625	1.4836	0.3821	1.2290	1.4188
LBMF	**1.3408**	**1.4112**	1.4540	**1.3036**	**2.1307**	**1.3928**	0.3576	**1.1709**	**1.3659**
LBMFu	1.3550	1.4226	1.4573	1.3160	2.1320	1.4125	0.3633	1.1792	1.3800
LBMFi	1.3428	1.4189	**1.4429**	1.3093	2.1378	1.4102	0.3557	1.1760	1.3754

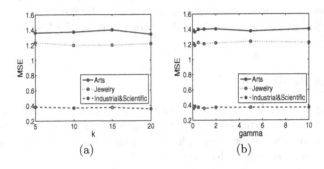

(a)   (b)

Fig. 1. (a) Performance by varying latent dimension k; and (b) Performance by varying parameter γ.

5.4 Parameter Sensitivity

We also examine the parameter sensitivity of our model on three datasets: *Arts*, *Jewelry*, *Industrial & Scientific*. The two important parameters of our model are the latent dimension k and the balance parameter γ. First we keep λ fixed, and vary k to be $\{5, 10, 15, 20\}$. As the Fig. 1(a) shows, our model is stable for different ks, which indicates our method is insensitive to different dimensions. This is different from conventional latent factor models, which tend to perform better as dimension increases. Then we fix k and vary γ to be $\{0.1, 0.2, 0.5, 1, 2, 5, 10\}$. The results of Fig. 1(b) show that although we set γ to different values, the model performance is consistent on different datasets, indicating that our algorithm is not sensitive to γ.

6   Conclusion

In this paper, we have presented a novel recommendation algorithm called Log-Bilinear Matrix Factorization (LBMF). LBMF can predict ratings that users have not yet conducted on relative products. LBMF also suits other tasks such as word representation learning, alleviating the cold-start problem and finding helpful reviews. Extensive experiments show the effectiveness and efficiency of our approach comparing to the state-of-the-art methods. Future works are needed in capturing personalization information and exploring other application scenarios.

Acknowledgment. This research is supported by the National Natural Science Foundation of China (NSFC) No. 61272303 and Natural Science Foundation of Guangdong Province No. 2014A030310268.

References

1. Balabanović, M., Shoham, Y.: Fab: content-based, collaborative recommendation. Commun. ACM **40**(3), 66–72 (1997)
2. Bao, Y., Zhang, H.F.J.: Topicmf: simultaneously exploiting ratings and reviews for recommendation. In: AAAI (2014)
3. Blei, D.M., Ng, A.Y., Jordan, M.I.: Latent dirichlet allocation. J. Mach. Learn. Res. **3**, 993–1022 (2003)
4. Koren, Y.: Factorization meets the neighborhood: a multifaceted collaborative filtering model. In: Proceedings of the 14th ACM SIGKDD International Conference on Knowledge Discovery and Data Mining, pp. 426–434. ACM (2008)
5. Koren, Y., Bell, R., Volinsky, C.: Matrix factorization techniques for recommender systems. Computer **42**(8), 30–37 (2009)
6. Lee, D.D., Sebastian Seung, H.: Algorithms for non-negative matrix factorization. In: Advances in Neural Information Processing Systems, pp. 556–562 (2001)
7. Ling, G., Lyu, M.R., King, I.: Ratings meet reviews, a combined approach to recommend. In: Proceedings of the 8th ACM Conference on Recommender Systems, RecSys 2014, pp. 105–112. ACM (2014)
8. Maas, A.L., Ng, A.Y.: A probabilistic model for semantic word vectors. In: NIPS Workshop on Deep Learning and Unsupervised Feature Learning (2010)
9. McAuley, J., Leskovec, J.: Hidden factors and hidden topics: understanding rating dimensions with review text. In: Proceedings of the 7th ACM Conference on Recommender Systems, pp. 165–172. ACM (2013)
10. Salakhutdinov, R., Mnih, A.: Probabilistic matrix factorization. In: Advances in Neural Information Processing Systems, vol. 20 (2008)
11. Sarwar, B., Karypis, G., Konstan, J., Riedl, J.: Item-based collaborative filtering recommendation algorithms. In: Proceedings of the 10th International Conference on World Wide Web, pp. 285–295. ACM (2001)
12. Wang, C., Blei, D.M.: Collaborative topic modeling for recommending scientific articles. In: Proceedings of the 17th ACM SIGKDD International Conference on Knowledge Discovery and Data Mining, pp. 448–456. ACM (2011)
13. Wang, S., Li, F., Zhang, M.: Supervised topic model with consideration of user and item. In: AAAI (2013)

An Empirical Study on Hybrid Recommender System with Implicit Feedback

Sunhwan Lee[✉], Anca Chandra, and Divyesh Jadav

IBM Almaden Research Center, 650 Harry Road,
San Jose, CA 95120, USA
{shlee,anca,divyesh}@us.ibm.com

Abstract. The amount of data generated by systems is growing quickly because of the appearance of mobile devices, wearable devices, and The Internet of Things (IoT), to name a few. Because of that, the importance of personalized recommendations by recommender systems becomes more important for consumers inundated with vast amount of choices. Many different types of data are generated implicitly (for example, purchase history, browsing activity, and booking history), and less intrusive recommendation systems can be built upon implicit feedback. There are previous efforts to build a recommender system with implicit feedback by estimating the latent factors or learning the personalized ranking but these approaches do not fully take advantage of various types of information that can be created from implicit feedback such as implicit profiles or a popularity of items. In this paper, we propose a hybrid recommender system which exploits implicit feedback and demonstrate better performance of the proposed recommender system based on the expected percentile ranking and a precision-recall curve against two state-of-the-art recommender systems, Bayesian Personalized Ranking (BPR) and Implicit Matrix Factorization methods, using hotel reservation data.

Keywords: Hotel recommendation · Hybrid recommender system · Collaborative filtering · Matrix factorization · Bayesian Personalized Ranking

1   Introduction

As e-commerce gains its popularity, it is challenging for customers to sort through an enormous number of products to find ones that align with their interest. The main task of recommender systems is to provide a customer with the most appropriate products matching the customer's preference, and thus to enhance user satisfaction and loyalty. The product provider analyzes vast amounts of data including (but not limited to) product profile, user profile, purchase history, and user ratings generated from marketplace implicitly or explicitly to create an efficient, applicable, and personalized recommender system. Recommender systems emerged as an independent and active research area in the mid-90's,

© Springer International Publishing Switzerland 2016
J. Bailey et al. (Eds.): PAKDD 2016, Part I, LNAI 9651, pp. 514–526, 2016.
DOI: 10.1007/978-3-319-31753-3_41

when explicit user rating was the main resource to build the system [8,15,17]. The goal of recommender systems using explicit user feedback is to estimate the rating of unrated products so that the system can use estimated ratings in order to recommend to the users the new items. Adomavicius and Tuzhilin [1] provided a mathematical formulation of the recommendation problem as follows. Let $\mathcal{V}$ and $\mathcal{I}$ be the set of all users and the set of all possible items that can be recommended respectively. Also u is a utility function that maps the pair of item i and user v to a real number R which is a metric of user's utility, i.e., $u : \mathcal{V} \times \mathcal{I} \rightarrow R$. Then, for each user $v \in \mathcal{V}$, we want to choose an item $i'_v \in \mathcal{I}$ that maximizes the user's utility. More formally, $\forall v \in \mathcal{V}$, $i'_v = \arg\max_{i \in \mathcal{I}} u(v, i)$.

Depending on how the utility function is computed and how users and items are modeled, there are two main approaches to building recommender systems: collaborative filtering [10] and content-based approach [1]. Recommender systems can also be differentiated based on the type of source data on which the system is built - explicit or implicit user feedback. Though explicit feedback is a clear indication of the user's preference over the item, it is not always available and it is intrusive to users, because it requires a significant level of user engagement. On the other hand, implicit user feedback such as purchase history can be automatically generated and the properties of implicit feedback (such as no negative feedback, inherent noise, indication of confidence, need of appropriate measure) are listed in [9]. In addition, the amount of implicit data today is significantly larger than the amount of explicit data, thanks to increased number of devices generating data (mobile phone, wearable devices, to name a few). Different approaches have their own limitations. We propose a *hybrid recommender system* which not only uses observations from implicit feedback but also utilizes implicit profiles of users and items. The experiment indicates that the proposed recommender system outperforms two state-of-the-art methods for implicit feedback datasets on its accuracy.

The rest of this paper is organized as follows. Previous research on recommender systems is reviewed in Sect. 2 and We propose hybrid recommender system in the context of hotel recommendations in Sect. 3. The specification of source data and the comparison of performance is demonstrated in the following section and conclusions and future works are made in Sect. 5.

2   Previous Work

2.1   Content-Based and Collaborative Filtering Methods

Content-based methods use a set of attributes that characterize an item and generate content-defined user profile by analyzing items previously rated or seen by each user. With the item and user profiles given, the utility function computes the similarity between user and item to create the recommendation. Suppose the item i and user v can be represented by l dimensional vector, $\boldsymbol{g}_i$ and $\boldsymbol{h}_v$. Then utility function u takes $\boldsymbol{g}_i$ and $\boldsymbol{h}_v$ as inputs and computes the similarity between item i and user v. One of the most widely used similarity measures is cosine similarity, $u(v, i) = \cos(\boldsymbol{h}_v, \boldsymbol{g}_i) = \frac{\boldsymbol{h}_v \cdot \boldsymbol{g}_i}{\|\boldsymbol{h}_v\|_2 \times \|\boldsymbol{g}_i\|_2}$.

Collaborative filtering methods compute the utility for a pair of user v and item i, $u(v, i)$, either by grouping similar users or items using explicit or implicit feedback. An observation matrix $U \in \mathbb{R}^{m \times n}$ is defined such that its element, u_{vi}, represents the observation that associates user v and item i. Two representative methods of collaborative filtering are neighborhood model and latent factor model.

Neighborhood Model: The model has been used in a variety of applications in early stage of the recommender system [7]. The underlying idea of neighborhood model is to estimate unknown ratings by using similar users or items. Two similarity measures are frequently used, Pearson correlation coefficient and cosine similarity, which can be computed from an observation matrix U. Utility of item i for user v can be estimated by identifying k most similar items among those rated by user v. If we define $\mathcal{I}^k(i; v)$ as such k items, and s_{ij} as a similarity between item i and j, then the utility is given as weighted sum of similarities of k similar items.

$$u(v, i) = \frac{\sum_{j \in \mathcal{I}^k(i;v)} s_{ij} u_{vj}}{\sum_{j \in \mathcal{I}^k(i;v)} s_{ij}}. \tag{1}$$

Latent Factor Model: Latent factor model explains the observed ratings with finite number of latent factors and this approach was one of the winning algorithms of Netflix prize announced by Netflix in 2006 [2]. Among various attempts, Singular Value Decomposition (SVD) model [12, 16] became popular because of its scalability and accuracy. The idea is to represent the users and items using f dimensional vector, $x_v \in \mathbb{R}^f$ and $y_i \in \mathbb{R}^f$, and to estimate the unknown ratings or utility by a dot product of two vectors, $u(v, i) = x_v^T y_i$.

An observation matrix can be decomposed by two lower dimensional matrices $X \in \mathbb{R}^{m \times f}$ and $Y \in \mathbb{R}^{f \times n}$ through the optimization with a regularized model to avoid overfitting.

$$\min_{x_*, y_*} \sum_{u_{i,j} \text{ is known}} (u_{ij} - x_i^T y_j)^2 + \lambda(\|x_i\|^2 + \|y_j\|^2) \tag{2}$$

where λ is a parameter for a regularization.

2.2   Methods with Implicit Feedback

We introduce two state-of-the-art methods to build a recommender system with implicit feedback. Two methods can be considered as an adaption of collaborative filtering methods and learn embeddings of users and items in a f-dimensional factor space. But the main difference comes from the optimization criterion.

Weighted Regularized Matrix Factorization (WR-MF): [9,11] proposed a recommender system with implicit feedback based on a matrix factorization. They both extended an objective function with a regularization term to avoid an overfitting and introduced an additional parameter to assign more weights on positive feedback. The optimization criterion is given as

$$\sum_{v \in V} \sum_{i \in \mathcal{I}} c_{vi}(\langle x_v, y_i \rangle - 1)^2 + \lambda(\|X\|^2 + \|Y\|^2). \tag{3}$$

A parameter, c_{vi}, is given a priori and [9] sets up this parameter proportional to the number of observations between user v and item i while [11] suggests to have $c_{vi} = 1$ for positive feedback and smaller values for unobserved feedback.

Bayesian Personalized Ranking (BPR): BPR [14] computes the parameter of an arbitrary model class to differentiate the user's preference on two items. The optimization maximizes the likelihood of the parameter by using Bayesian approach in order to find the correct personalized ranking.

$$\sum_{(v,i,j) \in D_s} \ln(\sigma(\hat{u}_{vij})) + \lambda\|\Theta\|^2 \tag{4}$$

where σ is the logistic sigmoid function, $\hat{u}_{vij}$ is a real-valued function parameterized with Θ evaluating the difference between items with positive feedback and no feedback. D_s is the set of user, observed, and unobserved item tuple.

2.3 Hybrid Model

Main motivation of the hybrid recommender system is to deal with a cold-start or sparsity problem of collaborative filtering or content-based approach and to further improve the performance of the recommender system. The strategies of hybrid recommender systems can be divided into seven categories [5] depending on how they combine multiple or same techniques: weighted, switching, mixed, feature combination, feature augmentation, cascade, and meta-level. As we can see from the complete survey of hybrid recommender systems in [4], most of previous works implemented and tested hybrid recommender system for explicit user feedback such as MovieLens, EachMovie dataset and almost no systems have been implemented on implicit feedback.

3   Our Model

In this section we describe a hybrid model for implicit feedback by explaining different components of the system and the method to combine them.

3.1   Collaborative Component

We conducted an empirical study on collaborative filtering methods with different types of observation matrices and similarity measures. Three different observations are investigated as an element of observation matrix - binary variable, term frequency (TF), and term frequency/inverse document frequency (TF-IDF). The idea is to put more weights on more observations to reflect higher confidence on items. To derive observations, we first define r_{vi} which indicates the observed action of user v on item i, for example, the number of consumptions of item i from user v. Binary variable p_{vi} indicates whether user v has ever consumed item i:

$$p_{vi} = \begin{cases} 1 & \text{if } r_{vi} > 0 \\ 0 & \text{if } r_{vi} = 0 \end{cases}. \tag{5}$$

Unobserved value, $r_{vi} = 0$, for implicit feedback cannot be ignored as no action on item i from user v is also meaningful feedback and should not be handled as missing values as opposed to explicit feedback. For example, there are cases where item i was not available when user v consumed other items. Furthermore, a specific item was not chosen because of demographic reason in the case of hotel reservation system. Thus we treat zero values of r_{vi} as a low confidence from user v on item i instead of a negative feedback on item i.

TF and TF-IDF are mainly used in specifying keywords weights for text-based documents. Suppose f_{ij} is the number of keyword j appearing in document i. Then normalized TF is defined as $TF_{ij} = \frac{f_{ij}}{\max_z f_{iz}}$ where f_{ij} is normalized by the maximum number of appearance of keywords in the same document. IDF is often used to diminish the weights of words that occur frequently in the documents and to increase the weights of keywords specific to the document, $IDF_j = \log \frac{n}{n_j}$, where n is the total number of documents and n_j is the number of documents in which keyword j appears.

We borrow the concept of TF-IDF and re-define it in the setting of user-item pair for implicit feedback datasets as follows. We replace f_{ij} by r_{vi} and n, n_v are the total number of items and the number of items consumed by user v.

$$b_{vi} = TF_{vi} = \frac{r_{vi}}{\max_j r_{vj}}, \quad w_{vi} = TF_{vi} \times IDF_v = \frac{r_{vi}}{\max_j r_{vj}} \log \frac{n}{n_v}. \tag{6}$$

Using the above definition, b_{vi} represents the number of consumption of item i by user v normalized by maximum number of consumption on any items from user v and IDF_v is the weights associated with user's previous action on items. The interpretation of w_{vi} in user-item pair is that an item can be better represented by users having consumed on specific items, than by users who consumed on broad range of items. Note that three different types of $m \times n$ observation matrices U can be created by arranging three different observations, p_{vi}, b_{vi}, and w_{vi} into the matrix. We used cosine similarity measure for neighborhood model and actual similarity depends on the type of observation used.

$$s_{ij}^p = \cos(U_{\cdot i}^p, U_{\cdot j}^p), \quad s_{ij}^b = \cos(U_{\cdot i}^b, U_{\cdot j}^b), \quad s_{ij}^w = \cos(U_{\cdot i}^w, U_{\cdot j}^w) \tag{7}$$

where the superscript indicates which observation is used.

Next step is to compute the utility for the pair of item i and user v as an aggregate of similarity of k similar items. In general, the utility is computed as a sum of similarity weighted by its own observation as seen in Eq. (1). As stated earlier, however, a positive observation of implicit feedback can be considered as higher confidence level of the preference. Consequently, we introduce a method to aggregate similarities, which is computed from a binary variable, of k similar items weighted by our confidence in observing p_{vi}.

$$u_{\text{cf}}(v, i) = \frac{\sum_{j \in \mathcal{I}^k(i;v)} s_{ij}^p b_{vj}}{\sum_{j \in \mathcal{I}^k(i;v)} s_{ij}^p}. \tag{8}$$

In our empirical experiment, a similarity measure based on binary variable, s_{ij}^p, weighted by b_{vj} is chosen after comparing different combinations of similarity measures in Eq. (7) but the results of comparison are not provided due to the limited space. We name this approach as Term-Frequency k-Nearest Neighbor (TFkNN) to simplify the notation.

3.2 Content-Based Component

A profile vector of an item, g_i, is built from attributes of an item and we used summation as a method to aggregate content-based user profile, h_v. Cosine similarity measure is used to compute the similarity between user and item, which is the utility for content-based method, $u_{\text{cb}}(v, i) = \cos(h_v, g_i)$.

3.3 Hybrid System

As the first step of a hybrid recommender system, we use a linear weighting scheme for collaborative filtering and content-based components as follows.

$$u_{\text{cf\_cb}}(v, i) = \beta u_{\text{cb}}(v, i) + (1 - \beta) u_{\text{cf}}(v, i). \tag{9}$$

Another utility to be added to hybrid recommender system is a popularity measure, which can readily computed from implicit feedback datasets. Popularity measure of item i is defined by a rank of the item i and a total number of items, n. Let $rank_i$ be a rank of item i within the reversely ordered list of all items based on the number of users consumed that item. Then, the item that the most users consumed has 1 as popularity measure and it decreases by $\frac{1}{n}$ for the rest of items.

$$pop_i = u_{\text{pop}}(v, i) = \frac{rank_i}{n}. \tag{10}$$

A popular item is known to be one of the appropriate items to be presented to users without many observations, and to be an efficient solution for extracting the most value about users [13]. But at the same time, we want to design a hybrid recommender system that relies more on personal data as more user feedback is accumulated. We therefore propose an adaptive approach to create a hybrid recommender system that uses a popularity measure combined with a user dependent weighting scheme. The utility of the proposed hybrid recommender system by taking into account implicit feedback, implicit profile of users, and popularity measure is computed as:

$$u_{\text{hybrid}}(v, i) = \alpha \log n_v \underbrace{(\beta u_{\text{cb}}(v, i) + (1 - \beta)u_{\text{cf}}(v, i))}_{\text{Collaborative \&Content-based component}}$$

$$+ \underbrace{u_{\text{pop}}(v, i)}_{\text{Popularity component}} \qquad (11)$$

$$= \alpha \log n_v u_{\text{cf\_cb}}(v, i) + u_{\text{pop}}(v, i).$$

Note that the user adaptive weighting scheme is intuitive in a sense that a weight for personalized recommendation increases as the number of items that a user has consumed increases. In an extreme example where the user consumes only 1 item, i.e. $n_v = 1$, the hybrid recommender system becomes a system based purely on a popularity measure. After an exhaustive search, we found that $\alpha = 2.0$, $\beta = 0.01$ to work the best for datasets used in an experimental study.

4   Experimental Study

4.1   Data Description

In order to test and evaluate the hybrid recommender system proposed in Sect. 3, we conducted an experimental study with a real dataset consisting of hotel reservation records. We collected 30 months of hotel reservation records from a reservation system for a large enterprise US corporation. Personally identifiable information or personally sensitive information (e.g.: credit card number) is not extracted or is anonymized for the purpose of this research. Data includes reservation records from an online reservation system or a direct phone call to the designated travel agency.

Among available fields in the hotel reservation data include time information such as reservation date, arrival date, departure date, and hotel-specific information such as room type, daily rate, hotel chain and brand, hotel name, and address. We found 711,111 booking records from 92,323 users to 19,788 hotels in 5,052 different cities globally (density is 0.0389 %). Density of representative public datasets [3] with explicit feedback is about 4.61 % for MovieLens 1 M dataset, 1.308 % for MovieLens 10 M dataset, 1.17 % for Netflix dataset, 55.84 % for Jester dataset, and 2.36 % for EachMovie dataset. Only the Book-crossing dataset is sparser than our implicit dataset with density of 0.0014 %.

Out of 30 months of data, training data was constructed with 24 months of data before the most recent month (which was held for test data) in the entire dataset. If user v has ever booked hotel i, then $p_{vi} = 1$ and r_{vi} is the number of reservations. b_{vi} is the number of reservations normalized by the maximum number of bookings from user v across all hotels. Lastly, a weight dependent on the number of hotels for user v are applied to compute w_{vi}. Test data is similarly constructed as training data for the most recent one month of the entire dataset. Another input to the recommender system, in addition to the identity of the traveler, is a destination city to pre-filter the list of hotels before the system sorts them by utility value - it does not make sense to recommend hotels not in the destination city. The number of hotels in one city ranges between 1 and

190 in our dataset and we excluded data with city having only one hotel from the test data because there are no multiple choices against which the recommender system needs to prioritize. Furthermore, we did not remove the possibility to recommend the same hotel to a user, as opposed to recommendations for many other application domains such as movies and book recommendations.

There are 38,418 user-hotel pairs in the test data and we used best performing TFkNN method as a collaborative component of the hybrid recommender system. The content-based recommender system was created based on the implicit user profiles and the hierarchy of hotel chains and brands. With 324 different hotel chains encoded in the dataset, content vector for hotel g_i and user h_v is defined as $l(= 324)$-dimensional vector. Finally, a popularity measure is integrated with the hybrid recommender system which computes the utility of user-hotel pair as in Eq. (11).

4.2 Evaluation Methods

The recommender system was evaluated by two measures, the expected percentile ranking [9] and a precision-recall measure for recommending top-N items [6]. A recall-based expected percentile ranking is a measure which is appropriate to implicit user feedbacks where type I error (false positive) cannot be detected. Suppose $rank_{vi}$ is the percentile rank of hotel i within a destination city to which user v travels. Then, the expected percentile rank is given as follows.

$$\overline{rank} = \frac{\sum_{v,i} r_{vi}^t rank_{vi}}{\sum_{v,i} r_{vi}^t} \tag{12}$$

where superscript t is a notation for test data.

A recall-precision measure is a widely used measure to evaluate the performance of recommender system whose output is top-N recommended items. By changing the number of items recommended to the user, N, recall and precision can be examined. The range of N in our case, however, does vary depending on the destination city of individual test data point. To overcome this discrepancy, we define N as a percentile instead of a raw rank of recommended hotels as explained in percentile rank measure. Now, N ranges between 0 and 100 and recall having 1 with lower value of N is desired for the recommender system. The precision and recall for a given recommender system with T test data points are defined as below.

$$recall(N) = \frac{\#\text{test data with } rank_{vi} < N}{T}, \tag{13}$$

$$precision(N) = \frac{\#\text{test data with } rank_{vi} < N}{NT} = \frac{recall(N)}{N}. \tag{14}$$

4.3 Results and Discussion

Our experiments tested TFkNN, a hybrid recommender system based on TFkNN, WR-MF, and two BPR approaches, BPR-MF and BPR-kNN respectively. For WR-MF and BPR-MF, the number of factors ranges from 8 to 128.

Overall results of comparison are shown in Figs. 1, 2 and we have the following observations.

Comparison Among TFkNN, WR-MF, and BPR Methods: The expected percentile ranking of TFkNN methods is lower than two BPR based methods and WR-MF method. This shows the importance of augmenting the similarity between hotels by number of bookings from each user when the recommender system is built for implicit feedback to accommodate the different level of confidence on a positive feedback. Also, the expected percentile ranking of WR-MF method is increasing as the number of factor increases. It is known that a matrix factorization model learned by singular vector decomposition (SVD) suffers from an overfitting and we can see the same problem from our experiment.

Improvement Through Hybrid Approach: Though not shown in Fig. 1, the expected percentile ranking was reduced a little bit to $\overline{rank} = 7.2353\,\%$ from $\overline{rank} = 7.3577\,\%$ when a content-based component was added. Note that we used a limited number of hotel profiles, the hierarchy of chains and brands, due to the

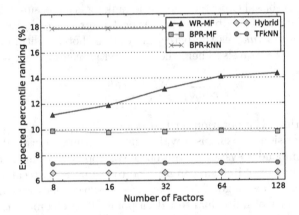

Fig. 1. Comparison of the expected percentile ranking for five different methods - WR-MF, BPR-MF, BPR-kNN, TFkNN, a proposed hybrid recommender system based on TFkNN.

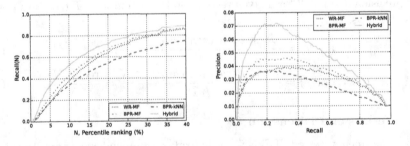

Fig. 2. Comparison of hybrid recommender system with WR-MF and BPR methods in terms of recall vs. percentile ranking (left) and recall vs. precision (right).

Table 1. Percentage of test users in terms of number of previous hotel reservations in training time period

# Reservation	# Total users	% of users
1	1,155	5.37 %
2	1,210	5.63 %
3	1,153	5.36 %
4	1,041	4.84 %
5	1,021	4.75 %
> 5	15,927	74.05 %
Total	21,507	

limitation of data availability. But there are many other profiles that can improve the performance such as a distance from a meeting location, the order of display in the online reservation system and so on. We reserve further improvement of the hybrid recommender system through content-based components as future works. Addition of a popularity measure to a hybrid recommender system achieved the lowest expected percentile rank, $\overline{rank} = 6.6183\%$.

Comparison of Precision-Recall: We picked the number of factor for WR-MF and BPR-MF based on their lowest expected percentile ranking shown in Fig. 1 and did not include TFkNN method to narrow down the comparison. Recall in terms of percentile ranking N, and precision-recall relation are plotted in Fig. 2. It is observed from the left figure that the probability that WR-MF, BPR-kNN, BPR-MF, and hybrid model can correctly recommend the actually booked hotel in top 10th percentile ranking is 38.02 %, 33.92 %, 42.90 %, and 52.52 % respectively. Precision recall metric from the right figure again confirms that the hybrid recommender system outperforms the other approaches based on the area under the curve. Each line shows precision measure when recall is provided and from the figure, when recall is 0.4, then precision of WR-MF, BRP-kNN, BPR-MF, and hybrid model is 0.038, 0.032, 0.044, and 0.063. In other words, the percentile ranking of our proposed hybrid model, with recall of 0.4, is lower than other models by about 30 % to 50 %.

Impact of the Number of Previous Observations: We also analyzed the results to ensure our assumption that including a popularity measure results in a performance improvement particularly for users with less number of bookings in their history. The percentage of users in terms of number of bookings in training period is shown in Table 1 and it says that about 25 % of users have reserved less than 6 different hotels in train data. For these users, coming up with personalized hotel recommendations with favorable accuracy is difficult due to the lack of data to learn a personal preference from. Collaborative filtering, either neighborhood or latent factor models, or content-based methods suffer from the same problem. We think a popularity measure especially in the context of a business travel is a good recommendation for users with less number of bookings

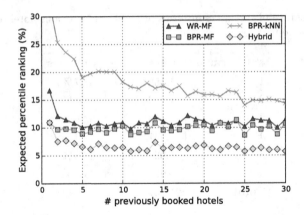

Fig. 3. Expected percentile ranking in terms of the number of hotels previously booked.

because a popularity of hotels stems from an enterprise compliance policy or frequently traveled places. Figure 3 shows the expected percentile rank for 4 different methods used in our experiment according to the number of previously booked hotels. First, it is clear that $\overline{rank}$ is gradually decreasing for users with more booking histories though the effect is not very noticeable for BPR-MF method. Secondly, we noticed that hybrid recommender system benefits from popularity measure mostly for users with less number of previous booked hotels because the decrease of $\overline{rank}$ happens largely for smaller value on x axis.

5 Conclusions and Future Works

In this work we studied a hybrid recommender system for implicit user feedback datasets with the combination of TFkNN, content-based method, and a popularity measure. Possible extensions include introducing the recency of the observation, modifying the confidence level by analyzing the order of display in the reservation system for inferring user preference on unobserved user-item pair interactions. Content-based method using a hierarchy of hotel chains also contributes to the improved accuracy of the recommender system. We think that the limited amount of hotel profiles used in this empirical study explains the small amount of improvement on the expected percentile rank by adding content-based approach. The last component of the hybrid recommender system was a popularity measure which was adaptively added to the recommender system depending on the number of each user's reserved hotels in training period.

The proposed hybrid recommender system was implemented and tested with hotel reservation data. We evaluated the performance of the hybrid recommender system using two measures with the consideration of the property of implicit datasets. Recall based measures were examined and we were able to observe the improved performance of the hybrid recommender system over two state-of-the-art methods. Our proposed method strives to point users to hotels that they like to reserve by taking a balance between personalized and popular flavor of hotels. We are motivated to do further research on statistical analysis of the results, finding a dependency of property of items on the choice of each components of the hybrid recommender system, and how components can be integrated.

References

1. Adomavicius, G., Tuzhilin, A.: Toward the next generation of recommender systems: a survey of the state-of-the-art and possible extensions. IEEE Trans. Knowl. Data Eng. **17**(6), 734–749 (2005)
2. Bennett, J., Lanning, S.: The netflix prize. In: KDD Cup and Workshop in Conjunction with KDD (2007)
3. Bobadilla, J., Ortega, F., Hernando, A., GutiéRrez, A.: Recommender systems survey. Know.-Based Syst. **46**, 109–132 (2013)
4. Burke, R.: Hybrid recommender systems: survey and experiments. User Modell. User-Adap. Interact. **12**(4), 331–370 (2002)
5. Burke, R.: Hybrid web recommender systems. In: Brusilovsky, P., Kobsa, A., Nejdl, W. (eds.) Adaptive Web 2007. LNCS, vol. 4321, pp. 377–408. Springer, Heidelberg (2007)
6. Cremonesi, P., Koren, Y., Turrin, R.: Performance of recommender algorithms on top-N recommendation tasks. In: Proceedings of the Fourth ACM Conference on Recommender Systems, RecSys 2010, pp. 39–46, New York (2010)
7. Herlocker, J.L., Konstan, J.A., Borchers, A., Riedl, J.: An algorithmic framework for performing collaborative filtering. In: Proceedings of the 22nd Annual International ACM SIGIR Conference on Research and Development in Information Retrieval, SIGIR 1999, pp. 230–237. ACM, New York (1999)
8. Hill, W., Stead, L., Rosenstein, M., Furnas, G.: Recommending and evaluating choices in a virtual community of use. In: Proceedings of the SIGCHI Conference on Human Factors in Computing Systems, CHI 1995, pp. 194–201. ACM Press/Addison-Wesley Publishing Co., New York (1995)
9. Hu, Y., Koren, Y., Volinsky, C.: Collaborative filtering for implicit feedback datasets. In: Proceedings of the 2008 Eighth IEEE International Conference on Data Mining, ICDM 2008, pp. 263–272, Washington, DC, USA (2008)
10. Koren, Y., Bell, R., Volinsky, C.: Matrix factorization techniques for recommender systems. Computer **42**(8), 30–37 (2009)
11. Pan, R., Zhou, Y., Cao, B., Liu, N., Lukose, R., Scholz, M., Yang, Q.: One-class collaborative filtering. In: Eighth IEEE International Conference on Data Mining, ICDM 2008, pp. 502–511, December 2008
12. Paterek, A.: Improving regularized singular value decomposition for collaborative filtering. In: Proceedings of KDD Cup and Workshopp (2007)
13. Rashid, A.M., Albert, I., Cosley, D., Lam, S.K., McNee, S.M., Konstan, J.A., Riedl, J.: Getting to know you: learning new user preferences in recommender systems. In: Proceedings of the 7th International Conference on Intelligent User Interfaces, IUI 2002, pp. 127–134. ACM, New York (2002)
14. Rendle, S., Freudenthaler, C., Gantner, Z., Schmidt-Thieme, L.: BPR: bayesian personalized ranking from implicit feedback. In: Proceedings of the Twenty-Fifth Conference on Uncertainty in Artificial Intelligence, UAI 2009, pp. 452–461. AUAI Press, Arlington (2009)

15. Resnick, P., Iacovou, N., Suchak, M., Bergstrom, P., Riedl, J.: Grouplens: an open architecture for collaborative filtering of netnews. In: Proceedings of the ACM Conference on Computer Supported Cooperative Work, CSCW 1994, pp. 175–186. ACM, New York (1994)
16. Salakhutdinov, R., Mnih, A.: Probabilistic matrix factorization. In: Advances in Neural Information Processing Systems, vol. 20 (2008)
17. Shardanand, U., Maes, P.: Social information filtering: algorithms for automating "word of mouth". In: Proceedings of the SIGCHI Conference on Human Factors in Computing Systems, pp. 210–217, New York (1995)

Who Will Be Affected by Supermarket Health Programs? Tracking Customer Behavior Changes via Preference Modeling

Ling Luo[1,2(✉)], Bin Li[2], Shlomo Berkovsky[3], Irena Koprinska[1],
and Fang Chen[2]

[1] School of Information Technologies, University of Sydney, Sydney,
NSW 2006, Australia
{ling.luo,irena.koprinska}@sydney.edu.au
[2] Machine Learning Research Group, NICTA, Eveleigh, NSW 2015, Australia
{bin.li,fang.chen}@nicta.com.au
[3] CSIRO Digital Productivity Flagship, North Ryde, NSW 2113, Australia
shlomo.berkovsky@csiro.au

Abstract. As obesity has become a worldwide problem, a number of health programs have been designed to encourage participants to maintain a healthier lifestyle. The stakeholders often desire to know how effective the programs are and how to target the right participants. Motivated by a real-life health program conducted by an Australian supermarket chain, we propose a novel method to track customer behavior changes induced by the program and investigate the program's effect on different segments of customers, split according to demographic factors like age and gender. The method: (1) derives customer preferences from the transaction data, (2) captures the customer behavior changes via a temporal model, (3) analyzes the program effectiveness on different customer segments, and (4) evaluates the program influence using a one-year data set obtained from a major Australian supermarket. Our results indicate that while overall the program had positive effect in encouraging customers to buy healthy food, its impact varied for the different customer segments. These results can inform the design of personalized health programs that target specific customers in the future and benefit more people. Our method can also be applied to other programs that use transaction data and customer profiles.

Keywords: Customer behaviors · Temporal preference modeling · Health programs · Shopping data analysis

1 Introduction

The World Health Organization (WHO) reports that in 2014 more than 1.9 billion adults were overweight, and over 600 million were obese [1]. Being overweight or obese increases the risk of cardiovascular problems, diabetes, and musculoskeletal disorders. To address the obesity problem, numerous behavior change

© Springer International Publishing Switzerland 2016
J. Bailey et al. (Eds.): PAKDD 2016, Part I, LNAI 9651, pp. 527–539, 2016.
DOI: 10.1007/978-3-319-31753-3_42

programs have been designed, aiming at encouraging participants to maintain a healthier lifestyle, e.g. change their diet and perform physical activity [2,3]. The program stakeholders often desire to know to what extent their program influences the participants, and more importantly, how to improve the program to benefit a broader population in the future. Therefore, mining the vast amount of participant behavior data collected by the program and understanding the behavior changes have become critical and timely research tasks.

The existing studies of health programs mainly report customer demographic statistics and customer survey results, and use statistical tests to examine the overall effect of the programs [2–4]. Although these studies can evaluate the general impact of a health program, they often overlook how the behavior of different types of participants is influenced by the program, which can potentially facilitate effective personalized programs. Thus, our aim is to explore in more depth how the participant behavior changes over time and investigate how participants from different segments are affected by the program.

We propose a systematic approach for tracking the customer behavior changes induced by the health program and evaluate how customers from different demographic segments (e.g. age and gender) are influenced by the program. The method comprises four specialized modules: we extract customer preferences from their transaction data, construct the temporal preference models, then analyze behavior change since joining the program, and most importantly, quantify the program effectiveness on different types of customers.

We evaluate our approach using a large-scale real-life health program delivered by an Australian supermarket chain. The program offered 10 % discount on fresh produce to participants, in order to encourage them to eat healthier. The duration of the discount was 24 weeks, but the purchase data of the participants for the entire year was collected, which allows us to compare their behavior before, on, and after the program. We analyze the data collected by the program using the proposed method and study how the purchase behavior of different types of customers is affected. We examine four customer segmentation criteria and show that female customers, younger customers, customers who live with their family, and obese customers are more likely to be encouraged by the program. Hence, the contributions of our work are as follows:

- We construct a temporal preference model that tracks and visualizes preference changes of the program participants over time.
- We quantify the customer preference changes as well as the program effectiveness on different types of customers.
- We evaluate the effectiveness of an Australian health program. Our method and results can be used to inform future personalized health programs.

2   Related Work

Behavior analytics has been recognized as an indispensable part of business intelligence [5,6]. Understanding customer behavior changes allows various stakeholders to monitor dynamic business environment and evaluate their policies and marketing campaigns [7,8]. A health program is a specific type of campaign, which promotes a healthier lifestyle and behavior [2].

The effectiveness of a program is usually evaluated by investigating customer behavior changes, which can be categorized into two groups: (1) *incremental* approach, that continuously adjusts the model with new transactions [9], and (2) *direct* approach, that models the behavior data at different time periods to identify differences [7,10]. Rule-based methods like association rules [7] and decision trees [10], are frequently used due to their easy interpretation. The incremental approach is sensitive to noise, while the direct approach can be too coarse-grained to reflect the temporal dynamics. Our goal is to design a model that facilitates direct comparison across multiple time periods – instead of just before and after the program – and track changes in customer preferences.

Temporal collaborative filtering (CF) techniques are powerful tools for analyzing patterns of customer preference over time. TimeSVD++ [11] introduces a time-dependent factor into each user-feature for modeling customer preference changes on the items. In [12], preference changes are analyzed via Bayesian tensor factorization, where the tensor is a three-dimensional array of user-item-time tuples. In contrast, [13] considers temporal dependence in Bayesian matrix factorization to model the frequency of preference changes for different types of customers. As the base underpinning for our temporal customer preference model, RMGM-OT [14] takes the advantage of probabilistic topic models to explicitly model the customer preference distributions over item groups; in addition, the method can illustrate how the customer preferences drift over time.

3   Methodology

This section introduces the proposed method for measuring the effectiveness of a health program. As shown in the flow chart in Fig. 1, our method consists of four modules: (1) extracting customer preferences from the transaction data, (2) constructing temporal model for customer preferences, (3) analyzing preference changes over time, and (4) evaluating program influence on different types of customers. The method can visualize customer preferences and provide program analytics as the output.

3.1   Extracting Customer Preferences

The first module extracts the customer preferences from the transaction data. Our program data consists of two parts: (1) transaction data, i.e. purchase records, of 931 participants captured through loyalty cards between 1st January and 31st December, and (2) self-reported survey data addressing their demographic and health information. The original transaction data set covers over 35,600 items from 200 categories. As we are interested in food and drinks relevant to the health program, the data set was reduced to 3,394 items from 24 categories, ranging from vegetables and fruits to snacks and soft drinks. Hence, we used 884 out of the 931 customers, who had a sufficient range of categories

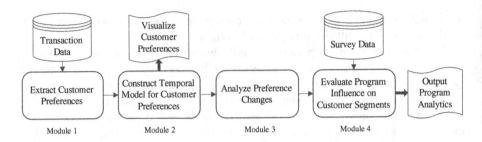

Fig. 1. Flow chart of the proposed method.

and number of purchased items in their transaction data. Specifically, each purchase record shows the customer ID, item ID, item metadata (name, code, and category), purchased quantity and the time-stamp.

Given the transaction data set comprising N customers and M items, each transaction record is a tuple $(u_i, v_j, quantity, time)$, which indicates that customer u_i purchased $quantity$ of item v_j at $time$. We represent all the transaction tuples as a tensor, $\mathcal{X} \in \{1, \ldots, R\}^{N \times M \times T}$, where each element $\mathcal{X}_{ijt}$ denotes u_i's preference in R levels on v_j at $t \in \{1, \ldots, T\}$. The whole time span of the transaction data is evenly partitioned into T periods, and the time index t is determined by the period into which $time$ falls. For example, if each time period corresponds to one month, $t = 1$ indicates January transactions. Then, we denote $\mathcal{X}$ as T preference matrices $\{\mathbf{X}^{(1)}, \ldots, \mathbf{X}^{(T)}\}$ corresponding to the T periods; and each $\mathbf{X}^{(t)}$ contains the preference information of N customers for M items in time period t. The series of preference matrices $\{\mathbf{X}^{(1)}, \ldots, \mathbf{X}^{(T)}\}$ can inform the temporal analysis of customer preference changes.

In our study, the transaction data is partitioned monthly, so the dimension of $\mathcal{X}$ is 844 customers $\times$ 3394 items $\times$ 12 months. Each element $\mathcal{X}_{ijt}$ is a nominal preference value based on the aggregated amount of item v_j bought in month t. In more detail, the sum of amounts bought by customer u_i in period t is computed and discretized into 5 levels $\{1, 2, 3, 4, 5\}$ item-wisely. For a certain item, we sort all $N \times T$ monthly sum values in ascending order; the value smaller than the first 5-quantile (i.e. ranked within the first 20%) becomes 1 – the lowest preference level, and the value greater than the first 5-quantile but smaller than the second 5-quantile (i.e. ranked within 20%–40%) becomes 2, and so on. If u_i did not purchase v_j in month t, then $\mathcal{X}_{ijt}$ is a missing value.

3.2   Constructing Temporal Model for Customer Preferences

The key component of investigating how the health program influences customer behavior is to build a temporal model for customer preferences on all item categories. However, the item-level preference matrix $\mathbf{X}^{(t)}$ can be very sparse in real-life cases, and the missing values do not necessarily mean the lowest preference level in that period (it is common for customers not to buy certain items at certain periods). Therefore, temporal CF techniques can be exploited to estimate and smooth customer preferences across the time periods. In our study, we are particularly interested in customer preference changes at the category level, rather than at individual item level. Considering the above requirements, we adapt the temporal CF method RMGM-OT [14] to our problem setting to capture the temporal dynamics of the customer preference for item categories.

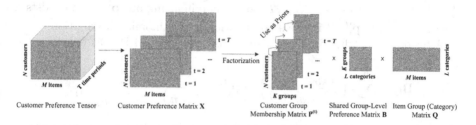

Fig. 2. Illustration of factorizing customer preference tensor $\mathcal{X}$

The preference matrix $\mathbf{X}$ can be factorized by $\hat{\mathbf{X}} = \mathbf{PBQ}^{\top}$, as schematically shown in Fig. 2. In the setting of customer preference analysis, the above factorization results in K latent customer groups and L latent item groups. $\mathbf{B}$ is a $K \times L$ group-level preference matrix, which represents the preferences of K customer groups for L item groups. $\mathbf{P} \in [0,1]^{N \times K}$, where each row $\mathbf{p}_i$ can be interpreted as u_i's membership distribution over K customer groups, and $\sum_k \mathbf{p}_{ik} = 1$ (soft-membership). $\mathbf{Q} \in \{0,1\}^{M \times L}$ represents the membership information of M items over L item groups. It is worth noting that, in our study, we adopt predefined categories of supermarket products, e.g. vegetables, fruits, and soft drinks, to define the item groups. Thus, v_j belongs to only one item category: $\mathbf{q}_{jl} = 1$ if l is the predefined category; otherwise, $\mathbf{q}_{jl} = 0$.

By taking the temporal domain into consideration, the varying customer preference over time can be modeled based on $\{\mathbf{X}^{(1)}, \ldots, \mathbf{X}^{(T)}\}$. We can obtain customer-group membership matrix $\mathbf{P}^{(t)}$ for each time period, while the group-level preference matrix $\mathbf{B}$ is shared across all the time periods and the item-group membership matrix $\mathbf{Q}$ is predefined. We further assume that the customer-group membership satisfies the Markov property, i.e. the state at t depends on the previous state at $t-1$. Considering the empirical Bayes approach, we can simply use the preceding customer-group membership $\mathbf{P}^{(t-1)}$ as the prior distribution of the current customer-group membership $\mathbf{P}^{(t)}$ in the model (shown by the dotted arrows in Fig. 2).

We adapt the collapsed Gibbs sampler used in [14] to our problem setting, where item categories $\mathbf{Q}$ are given. In other words, item latent variables z_{ij}^v are known in advance and we only need to infer customer latent variables z_{ij}^u. The conditional distribution of z_{ij}^u for Gibbs sampling is

$$P(z_{ij}^u = k | z^{\neg(ij)}, l, \mathbf{X}^{(t)}) \propto \left(\frac{n_{klr}^{\neg(ij)} + \beta/R}{\sum_r n_{klr}^{\neg(ij)} + \beta} \right) \left(n_{ikt}^{\neg(ij)} + \lambda \mathbf{p}_{ik}^{(t-1)} \right) \quad (1)$$

where α and β are hyper-parameters, and λ is the weight of the prior knowledge. The conditional distribution is proportional to the product of two Dirichlet-multinomial distributions: the first is the proportion of preference r in customer-item joint group (k, l) and the second is the proportion of preference records falling in customer group k (for customer u_i in time period t). It is worth noting that the prior distribution of $\mathbf{p}_i^{(t)}$ is $Dirichlet(\alpha)$ for $t = 1$ and $Dirichlet(\lambda \mathbf{p}_i^{(t-1)})$ for $t = 2, \ldots, T$; so by marginalizing out $\mathbf{p}_i^{(t)}$, there exists a pseudo counting $\lambda \mathbf{p}_i^{(t-1)}$ in the second Dirichlet-multinomial distribution.

After obtaining the sample z_{ij}^u, we can estimate $\mathbf{B}$ and $\mathbf{P}^{(t)}$ as follows

$$
\mathbf{B}_{kl} = \sum_{r=1}^{R} r \left(\frac{n_{klr} + \beta/R}{\sum_r n_{klr} + \beta} \right), \quad \mathbf{p}_{ik}^{(t)} = \frac{n_{ikt} + \lambda \mathbf{p}_{ik}^{(t-1)}}{\sum_k n_{ikt} + \lambda \sum_k \mathbf{p}_{ik}^{(t-1)}} \tag{2}
$$

where n_{klr} denotes the number of preference r in customer-item joint group (k, l), n_{ikt} denotes the number of preference records of u_i in customer group k in time period t, and both n_{klr} and n_{ikt} are counted based on z_{ij}^u. Intuitively, $\mathbf{p}_i^{(t)} \mathbf{B} \in [0, R]^{1 \times L}$ reflects the preferences of customer u_i for L item categories in period t. Therefore, the temporal preferences of N customers can be modeled using $\{\mathbf{P}^{(1)}, \ldots, \mathbf{P}^{(T)}\}$ and $\mathbf{B}$.

3.3   Analyzing Customer Preference Changes

Thus far, the preference change of customer u_i has been modeled by $\mathbf{p}_i^{(t)} \mathbf{B}$ for $t \in \{1, \ldots, T\}$. To further understand if the change is in the direction targeted by the health program, we label each item category l as either *healthy*, *neutral* or *unhealthy*. Among the 24 food categories in our transaction data, 5 categories are labeled *healthy* (vegetables, mushrooms, fruit snacks, fruit desserts, packaged salads), 5 are labeled *unhealthy* (biscuits &cookies, chilled desserts, snacks, soft drinks, confectionery), and the remaining 14 categories are labeled as *neutral*[1]. Our aim is to investigate whether the customer behavior changes due to their participation in the health program, and more specifically, whether their preference for the healthy categories increases.

Based on the category labels, we let $\mathbf{h} \in \{0, 0.5, 1\}^L$ be the healthy indicator vector, and for each category $l \in \{1, \ldots L\}$

$$
\mathbf{h}_l = \begin{cases} 1 & \text{if } l \text{ is } healthy, \\ 0.5 & \text{if } l \text{ is } neutral, \\ 0 & \text{if } l \text{ is } unhealthy. \end{cases} \tag{3}
$$

Similarly, the indicator vector for unhealthy categories is defined as $\mathbf{1} - \mathbf{h}$, which means the value is 0 for all healthy categories, 1 for all unhealthy categories, and 0.5 for neutral categories.

---

[1] Although this manual labeling may be simplistic and coarse-grained, we posit that it generally reflects the accepted health perception of food categories.

For customer u_i, we compute the correlation coefficient between the customer preference $\mathbf{p}_i^{(t)}\mathbf{B}$ for the L categories and the healthy indicator $\mathbf{h}$, and also the correlation coefficient between $\mathbf{p}_i^{(t)}\mathbf{B}$ and the unhealthy indicator $1 - \mathbf{h}$. The difference between these two correlation coefficients defines the health score for customer u_i in time period t as follows

$$health\_score_i^{(t)} = corr((\mathbf{p}_i^{(t)}\mathbf{B})^\top, \mathbf{h}) - corr((\mathbf{p}_i^{(t)}\mathbf{B})^\top, 1 - \mathbf{h}) \qquad (4)$$

where $corr(\mathbf{x}, \mathbf{y})$ denotes Pearson's correlation coefficient.

Given the program starting time t_p, we split the T time periods into two phases: the first includes the periods before the program $t \in \{1, \ldots, t_p - 1\}$ and the second includes the remaining periods $t \in \{t_p, \ldots, T\}$. The change in customer preferences towards the healthy food categories δ_i is quantified by the difference between the average $health\_score$ in the second phase and the average $health\_score$ in the first phase. More formally,

$$\delta_i = \frac{1}{T - t_p + 1} \sum_{t=t_p}^{T} health\_score_i^{(t)} - \frac{1}{t_p - 1} \sum_{t=1}^{t_p-1} health\_score_i^{(t)} \qquad (5)$$

The positive value of δ_i indicates an increase in customer u_i's preference towards healthy categories after joining the program, while a higher value of δ_i implies a greater change in the right direction. Therefore, δ_i is the key measure for evaluating the effect of the program on customer u_i.

3.4   Evaluating Program Influence on Customer Segments

The values of δ_i for all the customers can provide a general understanding of the health program effect. However, we are also interested in determining the types of customers that are more responsive to the program. This insightful information can inform the design of future personalized programs targeting specific customers.

Generally, the entire customer base can be segmented using a number of features, e.g. geographic, demographic or behavioral [15]. In our study, the self-reported demographic and health information collected by completing a survey is used for customer segmentation. We consider four features: *gender, age, who customers live with* (alone, with partner, with family), and *Body Mass Index* (BMI). All the customers are partitioned into S_A segments with respect to the value of an attribute A, such that the customers in segment $s \in \{1, \ldots, S_A\}$ have the same value of A. If an attribute is numeric, e.g. age, its values are discretized into S_A levels.

As transaction data of customers not participating in the program is unavailable, we split the customers into *experimental group* and *control group* according to the duration of their participation in the program. Specifically, the experimental group completed two surveys – 1) at the start of the program and 2) 12 weeks after the start date, and they participated in the entire program; whereas the control group completed only the first survey, so that they participated only in half of the program or less.

Overall, the observed changes δ_i in the behavior of individual customers are minor and hard to pick in the short period of one year. Thus, to quantify the effect of the program in customer segment s, we sort all the customers in s according to their δ_i, and measure the portion of the experimental group customers in the set of top-n customers with the highest δ_i. Intuitively, this reflect whether the experimental group customers have greater preference changes towards healthy categories than the control group customers. We define the *effectiveness* of the program for a segment s as:

$$eff_s = \frac{\sum_{n=1}^{N_s} counter_n}{(1 + N_s)N_s} \tag{6}$$

where N_s is the number of customers in s and $counter_n$ is the number of experimental group customers in top-n customers. The baseline for eff_s is 0.5 and greater eff_s values imply that the experimental group customers are ranked higher than the control group customers, indicating that the program is more effective in segment s. Importantly, eff_s is computed for each customer segment and it is used as the main metric for evaluating the effect of the health program.

4   Results for Our Case Study

This section presents the results for our case study on how the health program influences the behavior of the program participants. The evaluation involves 884 participants and 3,394 items from 24 categories, as described in Sect. 3.1. The experimental group comprises 190 customers who participated in the complete program, whereas the remaining 694 customers are in the control group. The program started in May (i.e. $t_p = 5$), so $t \in \{1, \ldots, 4\}$ is the first phase, and $t \in \{5, \ldots, 12\}$ is the second phase.

Following the four-step approach described in Sect. 3, we convert the transaction data into customer preference tensor, and construct the temporal preference model by factorizing $\hat{\mathbf{X}}^{(t)} = \mathbf{P}^{(t)} \mathbf{B} \mathbf{Q}^\top$ in each month. The parameters are configured as follows: the number of customer groups $K = 20$, the item groups L are the 24 predefined categories, the number of preference levels $R = 5$, λ is set to 10, and the hyper-parameters α and β are set to 1 as in [14].

4.1   Visualization of Customer Preference Changes

We visualize customer group membership $\mathbf{p}_i^{(t)}$ and customer preference $\mathbf{p}_i^{(t)}\mathbf{B}$ over the 12 months in Fig. 3. We select three customers who clearly demonstrate different degrees of variability in group membership and category preferences over time.

The subplots for customer group memberships are shown in the upper row of Fig. 3. Each column in a subplot indicates the mixed membership of 20 customer

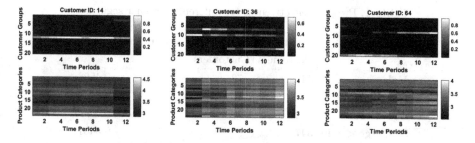

Fig. 3. Visualization of customer group membership distribution (upper row) and corresponding customer preferences for all item categories over the 12 months (bottom row). The three columns, from left to right, are for customers 14, 36, and 64.

groups in one month (sum of each column is 1). Customer 14 had a stable membership over all the time periods, consistently belonging to group 12. On the contrary, customer 36 had a mixed membership in groups 7, 10, and 17. It is interesting to note that the membership in group 17 was identified in June, after joining the program and it was observed till the end of the year. Customer 64 gradually switched from group 20 to group 8 in the middle of the year.

The subplots in the bottom row of Fig. 3 demonstrate how customer preferences for all the 24 categories change over the 12 months. Each row in a subplot reflects the fluctuations of the customer preference for a certain item category over the 12 months, and each column indicates the customer preferences for all the 24 item categories in one month. Customer 14 preferred category 23 (packaged salads) and did not like categories 13 (cheese), 16 (beef), and 17 (lamb), which implies that this customer purchased more vegetables. As for customer 36, the preference for category 20 (vegetables) was consistently high, while the preference for category 22 (fruit desserts) increased gradually, especially after June. The visualizations, such as those shown in Fig. 3, can provide an intuitive understanding of temporal preference changes of individual customers.

4.2   Program Effects for Different Types of Customers

This section quantifies customer preference changes and the program effectiveness. For the control group, the mean preference change is $\delta_i = 0.0204$, while for the experimental group the mean change is $\delta_i = 0.055$, which is more than twice as much as that of the control group. This observation generally shows that the experimental group had a greater preference change towards healthy categories than the control group.

The overall effectiveness of the program, without segmenting customers, is $eff = 0.527$. To get an insight of the fine-grained program effects, the customers are partitioned into segments according to four different criteria: *gender*, *age*, *who customers live with*, and *BMI*. For each segment, we report in Table 1 the size and the program effectiveness eff_s. The customers who had not provided their demographic and health information were excluded from this analysis.

Table 1. Program effectiveness for different customer segments.

Segments	Number of Customers	Number of Experimental Customers	eff$_s$
Male	349	72	0.4898
Female	517	118	**0.5533**
Age <=30	184	30	**0.5844**
30 < Age <= 40	347	83	0.5431
40 < Age <= 50	232	54	0.5067
Age > 50	100	23	0.4574
Live Alone	91	21	0.4929
With Partner	344	68	0.5038
With Family	385	91	**0.5555**
Normal	329	73	0.5592
Overweight	303	77	0.4787
Obese	220	34	**0.5705**

Segmentation by Gender. As shown in the first section of Table 1, male customers account for about 40 % of all the program customers and 37.8 % of the experimental group. The effectiveness of the program is 0.5533 for the female customers, which is higher than 0.4898 observed for the male customers. This implies that the female customers in the experimental group were effectively motivated to purchase healthier food, while the male customers from the experimental group were less responsive to the program. Therefore, *the program was found to be more effective for female customers.*

Segmentation by Age. The reported age of the participants varied from 19 to 67. The participants are partitioned into four equal-width segments as shown in the second section of Table 1). The program effectiveness drops significantly from 0.5844 for customers younger than 30 to 0.4574 for customers older than 50. The difference between these two effectiveness scores supports that different customer segments have different responsivenesses to the program. We notice that the program effectiveness decreases gradually as the age increases. One possible reason is that it might be easier for younger customers to change their dietary habits. However, this does not imply that customers older than 50 purchase unhealthy food. On the contrary, they might purchase healthier food consistently before and after joining the program, resulting in smaller preference changes. As our effectiveness measure focuses on the "behavior changes", the results show that *the program did not influence older customers as much as younger customers.*

Segmentation by Who Customers Live With. Based on the survey question "who you live with", the customers are partitioned into three segments: live alone, live with partner, and live with family. The segment of "living with their family" is the largest, comprising 46.95 % of all participants. The program influence on this segment is 0.5555, which is higher than the other two segments. For the customers who live alone or with partner, the effectiveness scores are relatively low, being 0.4929 and 0.5038, respectively. Thus, from the perspective of who customers live with, the results show that *customers living with family achieved a greater preference change towards healthy food than the other customers.*

Segmentation by BMI. The BMI is derived from the height and weight of a person, and its value is classified as underweight, normal, overweight, and obese [1]. As the proportion of underweight customers is less than 1.5 %, we excluded them from the analysis. It is worth noting that the overweight and obese segments together take up 61.4 % of all customers, which shows the severity of the overweight problem and the necessity of encouraging people to eat healthily. The obese segment achieves the highest effectiveness 0.5705, closely followed by the normal segment with 0.5592, and leaving the overweight segment behind with 0.4787. The effectiveness scores of the obese and normal segments are close, but there is a substantial gap between these segments and the overweight segment. The results suggest that *the program was more effective for obese and normal weight customers than for overweight customers.*

Discussion. Our results clearly demonstrate that different types of customers were influenced by the program to a different extent. For the four segmentation criteria, the program was found to be more effective for female customers, younger customers, customers who live with their family, and obese or normal weight customers. However, the program effectiveness across all the segments was relatively low. There are four main possible reasons: (1) the offered 10 % discount on fruits and vegetables may not be attractive enough for customers to notably change their behaviors; (2) the lack of real control group may have undermined the significance of the results, since using the customers, who participated in part of the program as the control group, may not truly represent customers not participating in the program; (3) the actual family size is unknown, so that the results may be inaccurate when comparing customers who purchased food, for example, for a family of 2 vs. for a family of 5; (4) the one-year duration of the program may not be sufficiently long to identify stable behavior changes of the customers. Despite these shortcomings of the program data, the evaluation results are encouraging and allow the health program stakeholders to get a fine-grained insight into the impact of the program. This allows tailoring or personalizing future programs, to motivate customers who are not very responsive, such as male customers, senior people and overweight customers in our case.

5   Conclusion

In this paper, we proposed an approach for tracking the customer preferences over time and evaluating the effectiveness of a health program for different types of customer segments defined by demographic and health attributes such as age, gender, living arrangements and BMI. We used data from a large-scale one-year program conducted by an Australian supermarket, which was designed to encourage customers to build healthy dietary habits. We analyzed how customers from different segments change their preferences over time for various food categories. Overall, the results showed that the program successfully motivated customers to purchase healthier food. The segment-wise effectiveness results demonstrated that different types of customers were influenced to a different extent. We found that female customers, younger customers, customers who live with their family and obese customers were more responsive to the program than their counterparts. Our results can be used to provide guidelines to enhance future health programs, in order to target and motivate the customers who were less responsive in this program and benefit the wider society. Although our method has been designed for a health program, it is a generic method that can be applied to other programs involving transaction records and customer profiles.

References

1. World Health Organization: World health organization factsheet: obesity and overweight (2015). http://www.who.int/mediacentre/factsheets/fs311/en/
2. Ball, K., McNaughton, S.A., Le, H.N., Gold, L., Mhurchu, C.N., Abbott, G., Pollard, C., Crawford, D.: Influence of price discounts and skill-building strategies on purchase and consumption of healthy food and beverages: outcomes of the supermarket healthy eating for life randomized controlled trial. Am. J. Clin. Nutr. **101**(5), 1055–1064 (2015)
3. Berkovsky, S., Hendrie, G., Freyne, J., Noakes, M., Usic, K.: The healthieru portal for supporting behaviour change and diet programs. In: Proceedings of the 23rd Australian National Health Informatics Conference, vol. 214. IOS Press (2015)
4. Brindal, E., Hendrie, G., Freyne, J., Coombe, M., Berkovsky, S., Noakes, M.: Design and pilot results of a mobile phone weight-loss application for women starting a meal replacement programme. J. Telemedicine Telecare **19**(3), 166–174 (2013)
5. Sheth, J.N., Mittal, B., Newman, B.I.: Customer Behavior: Consumer Behavior and Beyond. Dryden press fort worth, Texas (1999)
6. Cao, L.: Behavior informatics and analytics: let behavior talk. In: Proceedings of the IEEE International Conference on Data Mining Workshops, pp. 87–96. IEEE (2008)
7. Chen, M.C., Chiu, A.L., Chang, H.H.: Mining changes in customer behavior in retail marketing. Expert Syst. Appl. **28**(4), 773–781 (2005)
8. Huang, C.K., Chang, T.Y., Narayanan, B.G.: Mining the change of customer behavior in dynamic markets. Inf. Technol. Manage. **16**(2), 117–138 (2015)
9. Masseglia, F., Poncelet, P., Teisseire, M.: Incremental mining of sequential patterns in large databases. Data Knowl. Eng. **46**(1), 97–121 (2003)
10. Liu, B., Hsu, W., Han, H.-S., Xia, Y.: Mining changes for real-life applications. In: Kambayashi, Y., Mohania, M., Tjoa, A.M. (eds.) DaWaK 2000. LNCS, vol. 1874, pp. 337–346. Springer, Heidelberg (2000)
11. Koren, Y.: Collaborative filtering with temporal dynamics. Commun. ACM **53**(4), 89–97 (2010)
12. Xiong, L., Chen, X., Huang, T.K., Schneider, J.G., Carbonell, J.G.: Temporal collaborative filtering with bayesian probabilistic tensor factorization. In: Proceedings of the 2010 SIAM International Conference on Data Mining, vol. 10, pp. 211–222. SIAM (2010)

13. Li, R., Li, B., Jin, C., Xue, X., Zhu, X.: Tracking user-preference varying speed in collaborative filtering. In: Proceedings of the 25th AAAI Conference on Artificial Intelligence (2011)
14. Li, B., Zhu, X., Li, R., Zhang, C., Xue, X., Wu, X.: Cross-domain collaborative filtering over time. In: Proceedings of the 22nd International Joint Conference on Artificial Intelligence, pp. 2293–2298. AAAI Press (2011)
15. Baker, G.A., Burnham, T.A.: Consumer response to genetically modified foods: market segment analysis and implications for producers and policy makers. J. Agric. Resource Econ. **26**, 387–403 (2001)

TrafficWatch: Real-Time Traffic Incident Detection and Monitoring Using Social Media

Hoang Nguyen[1]([⊠]), Wei Liu[2], Paul Rivera[1], and Fang Chen[1]

[1] National ICT Australia, Eveleigh, NSW 2015, Australia
{hoang.nguyen,paul.rivera,fang.chen}@nicta.com.au
[2] Advanced Analytics Institute, University of Technology Sydney, Sydney, Australia
wei.liu@uts.edu.au

Abstract. Social media has become a valuable source of real-time information. Transport Management Centre (TMC) in Australian state government of New South Wales has been collaborating with us to develop TrafficWatch, a system that leverages Twitter as a channel for transport network monitoring, incident and event managements. This system utilises advanced web technologies and state-of-the-art machine learning algorithms. The crawled tweets are first filtered to show incidents in Australia, and then divided into different groups by online clustering and classification algorithms. Findings from the use of TrafficWatch at TMC demonstrated that it has strong potential to report incidents earlier than other data sources, as well as identifying unreported incidents. TrafficWatch also shows its advantages in improving TMC's network monitoring capabilities to assess network impacts of incidents and events.

Keywords: Social media · Incident detection · Classification

1 Introduction

Social media including micro blogging service such as Twitter has received great attention recently in order to capture real-time events. This joint research project between us and NSW Transport Management Centre (TMC)[1] aims to leverage social media as an additional channel for transport monitoring and incident management, with the ultimate goal of having more comprehensive views on traffic situations.

Processing Twitter feeds is very challenging. With over 280 million monthly active users and 500 million tweets sent per day,[2] there is a large volume of tweets from all over the world that satisfy a given query. However, very few tweets are from NSW, Australia and many of them contains words such as 'crash' or 'accident' but they are actually non-relevant to traffic tweets.

---

[1] http://www.transport.nsw.gov.au/tmc.

[2] https://about.twitter.com/company.

© Springer International Publishing Switzerland 2016
J. Bailey et al. (Eds.): PAKDD 2016, Part I, LNAI 9651, pp. 540–551, 2016.
DOI: 10.1007/978-3-319-31753-3_43

Our Twitter based incident monitoring system, named TrafficWatch, has been developed to assess its usefulness in assisting traffic and incident management on a day-to-day basis as well as during special events. The detected events or incidents were classified by advanced machine learning (ML) algorithms before visualised on an online map and presented to TMC operators in real time. We also introduce the annotation schema and computational model for extracting the traffic related entities within the tweets (e.g. street, incident type, vehicle, lanes, directions...). These entities are essential for automatic construction of the official TMC logs as well as supporting the classification and significant evaluation of the incidents.

The main contribution in this application paper is the novel employment of Conditional Random Fields (CRFs) in social media analysis specifically for traffic incident detection. We will show that the use of CRFs boosts up the performances of several popular classification algorithms. The CRFs model also generalises the content of the tweets into named entity features and then combines them with regular keyword-based features.

2   Related Work

There is great research interest in event and incident detection from social media within the last few years. For detecting large scale incidents using Twitter, various ML approaches have been proposed [7,12]. For small scale incident detection, Twitcident is a mashup for filtering, searching, and analyzing social media information about small scale incidents [1]. Li et al. (2012) introduce a system for searching and visualization of tweets related to small scale incidents, based on keyword, spatial, and temporal filtering [9].

There are also existing research for using Twitter to support live traffic monitoring and reporting. He et al. (2013) examined whether social media could improve the long-term (beyond 1 h) traffic prediction by analysing the correlation between traffic volume and tweets counts with various granularities [5]. In some other research, traffic information was extracted from Twitter using syntactic analysis with simple keywords matching and rule-based method [6,14]. In [14], tweets were classified into two categories: point (e.g. car crash) and link (e.g. traffic jam) with accuracies of 76.85 % and 93.23 % respectively. Twitter is also employed by the TMC in some cities (Sydney, Jakarta) to spread the news of traffic. NLP techniques were used by [3] to get the data of traffic from Jakarta TMC official account, so that the traffic information can be presented in map view as a mobile application. While most systems analysed social media content by simple rule-based and string matching methods, the only research that employed ML algorithms combining text classification and semantic enrichment of microblogs was presented in [13]. They introduced a system for real-time detection of small scale incidents with an accuracy of 89 %. In their study, the open source NLP tools (Google Spellchecking, Stanford CoreNLP [10]) and ML library (Weka [4]) were utilised to build the incident classifier and decision maker. However, none of these previous studies were able to provide comprehensive analyses of the tweets in terms of detecting incident related information to improve the classification process as demonstrated in our study.

3   Methods

Figure 1 illustrates our system architecture for crawling and processing public Twitter posts. The architecture is considered as two different phases: machine learning (ML) training process and real-time incident detection.

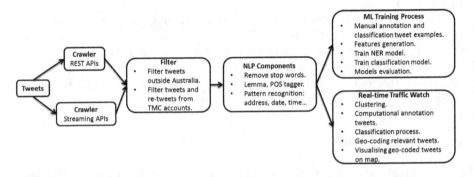

Fig. 1. System architecture.

3.1   Filters

To estimate the location distribution, 10,000 tweets were initially crawled using a keyword-based query. By default, Twitter APIs return tweets that satisfied the query from all over the world while the NSW TMC is only interested in traffic incidents within NSW which is less than 1 % of total tweets. A query in Twitter API is allowed to specify the geo-tagged tweets within certain parameters. However, only 3 % of tweets contain geo-tags as the users do not usually want others to see their current locations. Our country filter is based on the combination of the following fields ordered by their verification priority: geo-location, time zone, location and country from user's profile. This filter can remove most oversea posts and return over 90 % relevant tweets from Australia given most users set the correct time zones and have their profile locations filled.

3.2   NLP Components

The NLP cutting edge techniques are employed to pre-processing the crawled tweets. First, very frequent words like stop words and special characters are removed as they are not valuable as features for ML algorithms. To prepare for the feature generation step, the words are normalised to their root forms using the Stanford lemmatizer.[3] Furthermore, the Stanford POS tagger[4] is applied to

---

[3] http://nlp.stanford.edu/software/corenlp.shtml.
[4] http://nlp.stanford.edu/software/tagger.shtml.

identify the word form such as noun, verb, adjective... Finally, the advanced pattern recognizer based on regular expressions and finite state automata is utilised to capture the date, time and incident's location (e.g. street name, suburb) [11].

When there is no geo-tag, it is very difficult to identify the exact location of incident from the text due to lack of detailed address (e.g. house number). The common useful description of incident location contains two roads, main road and secondary road at the nearest intersection along with suburb. One example of standard tweet from TMC account is "SYD traffic ACCIDENT 3 vehicles - CASULA Hume Hwy at De Meyrick Ave" which contains suburb (CASULA), main road (Hume Highway) and secondary road (De Meyrick Avenue). TrafficWatch's pattern recogniser is trainable and flexible on text structure by simply adding new training examples. It is capable of identifying all popular road address representations in Australia such as Street, St, Road, Avenue, Highway...

3.3 Machine Learning Processes

Incident Related Tweet Classification. The public APIs provided by Twitter is limited by keyword search. Because of this, there are many tweets that contain words such as 'accident', 'crash', 'delay' or 'traffic' but they are not related to any recent traffic incidents. Example of non-relevant tweet is 'Hear loud airplane sound. Fully expects it to crash into a building'. Importantly, the TMC would like to capture as many traffic incidents as possible but at the same time they do not want to spend much effort looking at many non-relevant tweets. Based on this requirement, the classifier performance was targeted to reach a reasonably balanced precision and recall.

Table 1. Example of recorded incident from TMC logs.

Date	Time	Type	Sub-type	Location
03/10/13	11:42:03 AM	Accident	Car	WILLIAM ST YURONG ST DARLINGHURST 2010 SYDNEY (LGA) NSW
03/10/13	8:57:21 AM	Breakdown	Breakdown	SYD EINFELD DR OXFORD ST W BND BONDI JUNCTION 2022 WAVERLEY (LGA) NSW
03/10/13	2:50:58 PM	Queue	Moving	SMS: PITT ST - PARK ST; SYDNEY (235) W/B

Besides bag of words, additional features were added during the feature selection process. Finally, the following features were extracted to train incident related tweets classifier:

- Bag of words: the weight score of each word feature within a tweet is calculated by an accumulated tf-idf score over all positive tweets.
- Lemma, part-of-speech and chunk features: the Stanford Twitter tagger is specifically tuned for English tweets and outputs the lemma, part-of-speech tags and chunk tags [2].
- Pattern recogniser: this feature distinguishes the plain words with non-word tokens such as date, time, number and special characters.
- Bag of tags: this feature indicates the traffic related entities tagged by the computational annotation model (presented in next section). Entity is the generalisation of the frequent key words used to report incidents.

The experiments were executed using Weka [4]. Several ML algorithms which cover major types of classifiers have been investigated including:

- Instance based: k Nearest Neighbour (kNN).
- Generative: Bayesian Network (BN).
- Discriminative: Support Vector Machines (SVMs) and decision tree (C4.5).

Traffic Incidents Entities Annotation. During the process of obtaining the incident information, TMC operators usually ask for many details (e.g. incident type, vehicles involved, lanes blocked, direction) to put in the record. Table 1 shows example of major fields in TMC incident logs. To automatically populate structural logs; this project involves the annotation of tweets to identify the components needed to complete an incident report. In this process, free-text tweets are annotated for examples of the information to be extracted and then algorithms are developed that use the examples to compute a more general model of the desired content. The model is evaluated and the algorithm revised in a feedback process to produce a more accurate result. This is continued over a series of experiments until an optimal model is identified.

The annotation model is trained based on CRFs which is an advanced method for sequential labelling [8]. The same first three feature sets from classification model was applied to train the CRFs entities tagger.

Instead of using a general NER to identify traffic entities in the text as in [13], we have designed a special tag set which is more relevant to the incident information extraction task. This tag set is better controlled and does not contain redundant information which can add noise to mislead the classification process. An online collaborative environment based on Brat annotation tool[5] was set up to support rapid labelling of the tweets. Figure 2 illustrates the annotation schema and example annotated tweets.

Besides the tags that describe non-relevant and duplicated tweets, the annotation schema is divided into four main groups:

- Location tags include State, Suburb, Street, Point of Interests (e.g. Harbour Bridge, Opera House, Darling Harbour...), place (school, church, park...).
- Entity tags include the objects involved in the accident such as people (man, children), vehicles (car, truck) and stationary objects (e.g. tree, traffic light).

---

[5] http://brat.nlplab.org/.

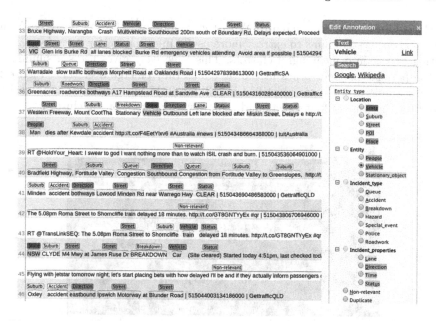

Fig. 2. Examples of annotated tweets and the traffic entities annotation schema.

- Incident_type tags include the main categories of the incidents (e.g. queue, accident, breakdown, police activities, road work...).
- Incident_properties tags includes the general descriptions about the incident on the road, they are lanes (e.g. 2 lanes closed or affected), directions (e.g. north bound, both ways), time and current status (clear or delayed).

The tagged entities are essential input for the incident's significant level ranking system. For example, the incident with type of 'accident', involved 'multi-vehicle', several 'people' and blocked 'both ways' on 'motorway' should be ranked higher than just a normal 'delay' on one 'direction' of a small road. The ranking criteria based on the detected entities are fully configurable as the TMC may have different interest over time or special events. The important tweets will be highlighted and presented to the operator to process as a priority.

In terms of ML, traffic related entities are implemented as key features to train the classification model. Because the same incident details can be described by different key words (e.g. high way/hwy, break down/stall/stationary), generalising them into categories and entities will help the model to learn these variations more efficiently.

For management purposes, the tweets are indexed by entities rather than normal key words, which will allow rapid retrieval of the incidents that satisfied certain criteria. For instance, the system supports the query 'retrieve all the breakdown incidents that caused delays on single direction of the freeway'. Furthermore, the automatic annotation models are extendable to analyse and index the official TMC's logs based on the same schema to serve the same management purpose.

Online Clustering Algorithm. With the volume of rich data returned from Twitter, our system aims to extract more useful information for users by aggregating tweets into meaningful clusters. This effectively gives users a summary of the popular incident types in the tweets as they emerge over time. Figure 3 is an example of a cluster identified on 13 Jan 2015 where the user can easily interpret the time of first tweets and the growing pattern of the cluster.

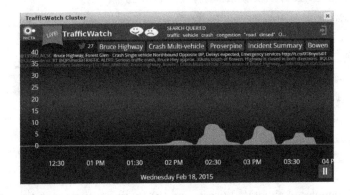

Fig. 3. Example of an incident cluster detected by TrafficWatch on 13 Jan, 2015.

We have implemented an online algorithm to incrementally cluster the tweets from live stream data using cosine similarity and Hamming distance evaluation. This algorithm is designed to gradually improve its centroids when more data is available. Besides the predefined incident types, the unsupervised clustering algorithm provides a more general view of the tweet clusters on the common key words used.

The key difference between our proposed method and previous online clustering algorithms is the cluster time frame threshold. Because similar incidents could be seen on different days, this parameter is used to remove old and outdated clusters to avoid redundant and non-relevant information. Depending on specific applications, it can be configured to span over several hours (e.g. traffic incidents) or several days (e.g. special events).

4   Experiments and Results

4.1   Data Set

The data set comprises of 5000 filtered tweets (mostly from NSW, Australia) crawled using Twitter REST API in September, 2014. These tweets were then labelled by TMC as relevant/non-relevant and used to train the ML models. 10-fold cross-validation was applied to evaluate models' performance based on averaged Precision, Recall and F1-score.

Table 2. Performance of traffic entities annotation model over 10-fold cross-validation.

Tag	TP	FP	FN	Precision (%)	Recall (%)	F1 (%)	Total
Accident	1275	64	31	95.22	97.63	96.41	1306
Breakdown	358	16	21	95.72	94.46	95.09	379
Queue	131	5	6	96.32	95.62	95.97	137
Hazard	20	1	15	95.24	57.14	71.43	35
Police	32	0	2	100	94.12	96.97	34
Roadwork	7	0	3	100	70	82.35	10
Special_event	2	0	1	100	66.67	80	3
Direction	678	40	17	94.43	97.55	95.97	695
Lane	171	4	3	97.71	98.28	97.99	174
Status	589	21	37	96.56	94.09	95.31	626
Vehicle	631	41	50	93.9	92.66	93.27	681
People	118	8	28	93.65	80.82	86.76	146
Object	12	1	3	92.31	80	85.71	15
Street	1879	32	34	98.33	98.22	98.27	1913
POI	38	1	3	97.44	92.68	95	41
Place	6	0	6	100	50	66.67	12
OVERALL	**5947**	**234**	**260**	**96.21**	**95.81**	**96.01**	**6207**

4.2 Named-Entity Recognition

Table 2 shows the performance of the computational annotation model with overall F1-score of 96 %. The State and Suburb entities were excluded because they can be identified by string matching from a pre-defined list. From this table, the most popular reported incident type is 'Accident' followed by 'Breakdown' and 'Queue'. All of these high frequency incident types were extracted with both Precision and Recall of approximately 95 % or higher. In addition, the model correctly identified most of the street information from the text which is crucial for geo-locating the incidents on the map.

The accuracy of individual tags varies greatly due to the fact that some tags were extremely rare with high variety level (e.g. hazard, place and special events), and thus it was difficult to learn when these tags were relevant.

4.3 Classification

The incident related entities recognised from the computational model were then become the important features for the classifier. Table 3 illustrates the classification performance using 4 different algorithms. The baseline features with bag of words archived the best F-score performance of more than 90 % using BN. Adding POS tagger and pattern recogniser features slightly improved the performance by approximately 1 %. With the support of Bag of tags features,

Table 3. Classification performance using 4 algorithms.

Features	Methods	Precision (%)	Recall (%)	F1-score (%)
Bag of words	BN	90.2	91.4	90.8
	SVMs	87.4	89.5	88.4
	kNN	83.1	92.4	87.5
	C4.5	86.9	90.7	88.7
Bag of words	BN	91.5	92.3	91.9
POS tagger	SVMs	88.2	92.5	90.3
Pattern recognizer	kNN	83.7	92.9	88.1
	C4.5	88.6	91.2	89.9
Bag of words	BN	**94.2**	**96.6**	**95.4**
POS tagger	SVMs	92.8	95.7	94.2
Pattern recognizer	kNN	85.5	97.8	91.2
Bag of Tags	C4.5	90.1	95.7	92.8

the performances of all four algorithms further increased significantly by 2 % to 3 %. Finally, the best F1-score of 95.4 % was recorded from BN method with precision of 94.2 % and recall of 96.6 %.

4.4   Incident Detection for Special Event

TrafficWatch has monitored traffic conditions during the International Fleet Review (IFR) special event in Sydney, Australia for the time period from 6:00 am October 3 to 10:44 am October 10, 2013. In total, 45753 IFR-related tweet messages were extracted, 1056 of which were traffic related.

With regard to locating incidents, 2.87 % (1222 tweets) were originally associated with device geo-locations, while TrafficWatch further identified 19 % (8065 tweets) with geo-locations by using text analysis. Hence in total about 21 % of the entire tweets had been visualised by TrafficWatch on maps.

Detecting Incidents Earlier than TMC Log Time

Case study 1: a heavy traffic condition at around military road in Mosman (Fig. 4). TrafficWatch detected this heavy traffic condition about 3 h earlier than the time this incident was logged in TMC.

Case study 2: a tweet-based incident saying an over-height truck blocked access to the Harbour Tunnel at North Sydney, suggesting all traffic to use the Harbour Bridge (Fig. 5). TrafficWatch detected this tunnel blockage about 3 h and 47 min earlier than the time this incident was logged in TMC.

Discovering Incidents that are not Reported to TMC

Case study 3 and 4: The tweet in the left table of Fig. 6 shows a queue of 7 kms at Brooks road, possibly caused by a truck accident. The tweet in right table is about a car accident on the corner of Elizabeth and George Street. Both incidents cannot be found in TMC logs.

Date - time	Thu Oct 3 10:13:10 EST 2013
Geolocation	-33.8253138312,151.2402590187
Twitter user	snarltraffic
Tweet texts	Mosman - Military Road - and Spit Road Lanes: Southbound traffic affected. - Heavy traffic
TMC Log ID	234 03-OCT-13
TMC Log Time	13:21:34
Detected earlier	3 hours 8 minutes

Fig. 4. Case study 1: incident detected sooner than TMC log.

Date - time	Thu Oct 3 11:17:24 EST 2013
Geolocation	-33.848191733,151.213341717
Twitter user	paullatter
Tweet texts	AGAIN!!! An overheight truck has blocked access to the Harbour Tunnel at Nth Sydney. All traffic to use the Bridge. Unbelievable...
TMC Log ID	308 03-OCT-13
TMC Log Time	11:31:15
Detected earlier	13 minutes

Fig. 5. Case study 2: incident detected sooner than TMC log.

Date - time	Fri Oct 4 13:09:24 EST 2013	Date - time	Tue Oct 8 10:39:02 EST 2013
Geolocation	-33.9894290500,150.8500003151	Geolocation	-33.9538289699,151.104966351
Twitter user	DeborahClay	Twitter user	NewsTalk4BC
Tweet texts	Queues of 7 kms on the Hume in Denham Court at Brooks Rd. A truck accident causing the southbound delays. #sydneytraffic	Tweet texts	Reports of a traffic accident at Kippering. Two car accident on the corner of Elizabeth & George Street causing delays. Avoid if

Fig. 6. Case study 3 and 4: incident was not reported to TMC.

Identifying Twitter Users that Are of Great Interest to NSW Traffic. Based on the analysed tweets, TrafficWatch has identified a list of Twitter accounts/users that have great interest in NSW traffic. TrafficWatch has ranked these accounts by their frequency of twittering activity, i.e., the accounts that are ranked higher are the ones who post more traffic-related tweets. From this list, the top ranked accounts in NSW are Snarltraffic, 2DayFM, Cbemergency.

5   Live Traffic Monitoring System

To provide TMC with a real-time overview of the traffic picture in NSW, the TrafficWatch interface was developed based on Cesium Bing map which is a geospatial visualisation software for running inside web browsers without plugins.

After pre-processed and classified, only the relevant and geo-located tweets are loaded onto the live map within a few seconds from time of posting. Figure 7 illustrates the TrafficWatch snapshot for NSW around 16:43 19 Nov 2014.

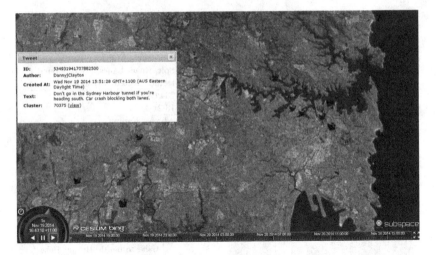

Fig. 7. Real-time NSW Traffic Watch interface with record and playback timeline on 3D Bing map. Tweets were classified and displayed by different colours.

When the user clicks on any tweet, a pop-up window will be opened to show its content, e.g. 'Don't go in the Sydney Harbour tunnel if you're heading south. Car crash blocking both lanes'. Furthermore, the timeline feature of Subspace enables review of any past incidents and events. This function is very useful to learn the tweet patterns from the historical data including first mention about the incident, the total number of re-tweets, similar and related tweets.

6 Conclusions

The TrafficWatch research project carried out detailed analysis and visualisations of the Twitter stream to monitor the daily traffic of the NSW state of Australia. It is capable of providing useful information to TMC in a real-time manner which helps TMC to be informed about issues and incidents that could potentially affect operations of public transport services, and be able to make better management decision with respect to the operations of the transport and event planning.

Besides the capacity of detecting the road incidents sooner or incidents not reported to TMC, TrafficWatch also suggests frequent traffic Twitter accounts that can be of good value to the TMC, since they are very active in releasing traffic information that can help to understand the traffic conditions.

In this study, we also introduced the general annotation schema and high accuracy computational model for detecting traffic related entities. This NER model is applicable to detect similar incident information from other social media sources as well as from official TMC's incident logs. Furthermore, the novel employment of CRFs to support the classification of Twitter feeds demonstrated significantly higher results in all investigated ML algorithms.

References

1. Abel, F., Hauff, C., Houben, G.J., Stronkman, R., Tao, K.: Twitcident: fighting fire with information from social web streams. In: Proceedings of the 21st International Conference Companion on World Wide Web, pp. 305–308. ACM (2012)
2. Derczynski, L., Ritter, A., Clark, S., Bontcheva, K.: Twitter part-of-speech tagging for all: overcoming sparse and noisy data. In: RANLP, pp. 198–206 (2013)
3. Endarnoto, S.K., Pradipta, S., Nugroho, AS., Purnama, J.: Traffic condition information extraction & visualization from social media twitter for android mobile application. In: 2011 International Conference on Electrical Engineering and Informatics (ICEEI), pp. 1–4. IEEE (2011)
4. Hall, M., Frank, E., Holmes, G., Pfahringer, B., Reutemann, P., Witten, I.H.: The weka data mining software: an update. ACM SIGKDD Explor. Newsl. **11**(1), 10–18 (2009)
5. He, J., Shen, W., Divakaruni, P., Wynter, L., Lawrence, R.: Improving traffic prediction with tweet semantics. In: Proceedings of the Twenty-Third International Joint Conference on Artificial Intelligence, pp. 1387–1393. AAAI Press (2013)
6. Kosala, R., Adi, E., et al.: Harvesting real time traffic information from twitter. Procedia Eng. **50**, 1–11 (2012)
7. Krstajic, M., Rohrdantz, C., Hund, M., Weiler, A.: Getting there first: real-time detection of real-world incidents on twitter. In: Proceedings of the 2nd IEEE Workshop on Interactive Visual Text Analytics VisWeek (2012)
8. Lafferty, J.D., McCallum, A., Pereira, F.C.N.: Conditional random fields: probabilistic models for segmenting and labeling sequence data. In: Proceedings of the Eighteenth International Conference on Machine Learning. ICML 2001, pp. 282–289. Morgan Kaufmann Publishers Inc., San Francisco (2001). ISBN: 1-55860-778-1
9. Li, R., Lei, K.H., Khadiwala, R., Chang, K.C.: Tedas: a twitter-based event detection and analysis system. In: 2012 IEEE 28th International Conference on Data Engineering (ICDE), pp. 1273–1276. IEEE (2012)
10. Manning, C.D., Surdeanu, M., Bauer, J., Finkel, J., Bethard, S.J., McClosky, D.: The stanford corenlp natural language processing toolkit. In: Proceedings of 52nd Annual Meeting of the Association for Computational Linguistics, pp. 55–60 (2014)
11. Patrick, J., Sabbagh, M.: An active learning process for extraction and standardisation of medical measurements by a trainable FSA. In: Gelbukh, A. (ed.) CICLing 2011, Part II. LNCS, vol. 6609, pp. 151–162. Springer, Heidelberg (2011)
12. Sakaki, T., Okazaki, M., Matsuo, Y.: Earthquake shakes twitter users: real-time event detection by social sensors. In: Proceedings of the 19th International Conference on World Wide Web, pp. 851–860. ACM (2010)
13. Schulz, A., Ristoski, P., Paulheim, H.: I see a car crash: real-time detection of small scale incidents in microblogs. In: Cimiano, P., Fernández, M., Lopez, V., Schlobach, S., Völker, J. (eds.) ESWC 2013. LNCS, vol. 7955, pp. 22–33. Springer, Heidelberg (2013)
14. Wanichayapong, N., Pruthipunyaskul, W., Pattara-Atikom, W., Chaovalit, P.: Social-based traffic information extraction and classification. In: 11th International Conference on ITS Telecommunications, pp. 107–112. IEEE (2011)

Automated Setting of Bus Schedule Coverage Using Unsupervised Machine Learning

Jihed Khiari[1], Luis Moreira-Matias[1(✉)], Vitor Cerqueira[1], and Oded Cats[2]

[1] NEC Laboratories Europe, 69115 Heidelberg, Germany
{jihed.khiari,luis.matias,vitor.cerqueira}@neclab.eu
[2] Department of Transport and Planning, TU Delft, 2600 Delft, Netherlands
o.cats@tudelft.nl

Abstract. The efficiency of Public Transportation (PT) Networks is a major goal of any urban area authority. Advances on both location and communication devices drastically increased the availability of the data generated by their operations. Adequate Machine Learning methods can thus be applied to identify patterns useful to improve the Schedule Plan. In this paper, the authors propose a fully automated learning framework to determine the best Schedule Coverage to be assigned to a given PT network based on Automatic Vehicle location (AVL) and Automatic Passenger Counting (APC) data. We formulate this problem as a clustering one, where the best number of clusters is selected through an *ad-hoc* metric. This metric takes into account multiple domain constraints, computed using Sequence Mining and Probabilistic Reasoning. A case study from a large operator in Sweden was selected to validate our methodology. Experimental results suggest necessary changes on the Schedule coverage. Moreover, an impact study was conducted through a large-scale simulation over the affected time period. Its results uncovered potential improvements of the schedule reliability on a large scale.

Keywords: Unsupervised learning · Public transportation · Big data · Schedule plan · Schedule coverage · Sequence mining · Probabilistic reasoning

1 Introduction

Public Transport (PT) reliability is a major issue in modern cities. A good operational planning is necessary to deliver such service quality requirements while maintaining a balanced relationship between resource usage and obtained revenues. Nowadays, major PT operators have their fleets equipped with Global Positioning System (GPS) antennas, communicational devices (e.g. 3G) and Radio-frequency Identification readers able communicate the vehicle's positioning (i.e., Automatic Vehicle Location (AVL)) and its ridership (i.e., Automatic Passenger Counting (APC)) to a central server [1].

To mine this novel source of data is a massive challenge. It contains information about the patterns of human behavior while traveling (as drivers or passengers) on an urban environment. Such patterns can provide useful insights to

© Springer International Publishing Switzerland 2016
J. Bailey et al. (Eds.): PAKDD 2016, Part I, LNAI 9651, pp. 552–564, 2016.
DOI: 10.1007/978-3-319-31753-3_44

improve the operational planning of mass transit agencies - namely, its **Schedule Plan (SP)**. Such improvement may bring multiple benefits by providing ways of reducing costs (e.g. fleet (re)sizing or fuel saving due to a decrease of the necessary number of trips) and/or improving the passenger experience.

A Schedule Planning (SP) process for a given route relies on two main steps [2]: (1) the first step is to define the number k of schedules and their individual coverage, S_i. Consequently, this first step defines different schedules for days that are characterized by different traffic and demand patterns due to seasonal variations, for instance. Secondly, (2) the timetables are assigned for each route schedule containing the time the buses pass at each schedule time point (per trip). This process is done for all routes. While the timetables are defined *route-wise* (e.g. high/low frequency routes), the number of schedules (i.e. k) and their coverage $(S_i, \forall i \in \{1,..,k\})$ must be defined *networkwise*. Such definition is key to ease PT operations (e.g. maintenance tasks) and, most of all, to facilitate the SP memorization by the passengers.

Automated data driven frameworks that aim to improve the SP are commonly focused on timetabling tasks, thus *skipping* the coverage definition. Some of the most well-known approaches include finding the optimal slack time and round-trip time to put into the schedule using Genetic/Ant Colony Algorithms [3,4], mining distribution rules able to discover feature subspaces (i.e. scenarios) for an increased travel time uncertainty [5], or clustering trips based on APC data regarding their frequency setting, i.e. high/low [6]. However, the coverage definition can easily constrain the timetable construction (e.g. two days with distinct demand peak periods should have different timetables). At the best of our knowledge, only Mendes-Moreira *et al.* [2] covers the improvement of Schedule Coverage: a Consensual Clustering framework groups days with similar behavior (using AVL data standalone) given a predefined number of schedules k.

This paper is a comprehensive extension of the work in [2]. It aims to generalize this framework's usage for every scenario that fully exploits the information available on the data repository while still minimizing the required human input to reach a decision. The contributions are threefold:

1. a novel *ad-hoc* domain-oriented metric to select the most adequate number of schedules to put in place based on Sequential Itemset Mining [7] and Probabilistic Reasoning. It settles on a trade-off between the entropy within the clusters and the operational adequacy of the resulting coverage.
2. a hybrid computation of the daily profiles using APC/AVL data simultaneously by decomposing the round trip times into a sum of link travel times (the run times between two consecutive stops) and dwell times[1]. Their computation may highlight demand peaks which would be smoothed otherwise.
3. the application of a Gaussian Mixture Model (GMM) [8] to perform the necessary clustering for the individual routes, thus replacing the originally proposed k-Means (see Sect. 5 in [2]). By doing so, we obtain a soft assignment of the samples, reducing the overfitting chances.

---

[1] Reports stoppage time at stops. Includes a fixed delay due to door opening and closing time, and a variable delay caused by passengers boarding/alighting activities.

The proposed framework was evaluated using data acquired from a large bus operator in Sweden throughout a period of six months. Numerical experiments suggested a change to the agency's original coverage. The impact of such change was measured by assigning a theoretical timetable to the affected period. A *before-and-after* schedule reliability study was conducted. The results are promising.

The remainder of the paper is structured as follows: methodology is described in Sect. 2, by doing an analysis of the previous work and a formal explanation of our contributions. The case study is presented in Sect. 3, along with some summary statistics of the used datasets. The results are presented in the Sect. 4, followed by a brief discussion. Finally, conclusions are drawn.

2   Methodology

A stepwise methodology is hereby proposed to automatically set both the number of schedules and their daily coverage. This description follows closely the one proposed in Sect. 4 of [2]. It elaborates on the principle that days where the route trips have a similar behavior (e.g. round-trip times) throughout the day should be assigned to the same schedule. Let $\mathbb{L} = \{r_1, ..., r_n\}$ denote a set of routes of interest. Firstly, for each $r \in \mathbb{L}$, the running times and the boardings/alightings at each stop (if existing) are extracted from its original AVL/APC dataset. Secondly, the daily profiles are generated. If there is no APC data available for a specific route, the procedure originally suggested in [9] is used. Otherwise, a biased dwell time model is generated based on APC data to account demand peaks/valleys. Its output is added to the link travel times computed through the AVL data - as described in Sect. 2.1.

The next two steps generate a distance matrix between the days (using their daily profiles) and cluster them. The first task is conducted using a Euclidean-flavoured Dynamic Time Warping, while the latter is addressed using a GMM. Conversely to previous works, the clustering is made for a user-defined set of admissible number of schedules $\mathbb{K} \subset \mathbb{N}$, i.e. $\forall k \in \mathbb{K}$ instead of a single predefined k value. The above mentioned steps are repeated for all routes.

Step 5 selects the best possible $k \in \mathbb{K}$ to define the best number of schedules to put in place. This is made using a two-stage process, where an *ad-hoc* metric is devised to evaluate the clustering result for each pair $(r, k), \forall r \in \mathbb{L}, k \in \mathbb{K}$. Then, a consensual k, i.e. K is found through a domain-oriented weighted mean of the previously computed metrics - as described in Sect. 2.3. Finally, a Consensual Clustering procedure is devised using the clustering pieces obtained for $k = K$ to compute the suggested Schedule Coverage, following the original procedure proposed in [2]. An illustration of our methodology is presented in Fig. 1. The remainder of this Section describes our contributions.

2.1   Modeling the Daily Profiles

Let $L = \{L_1, L_2, ..., L_n\}$ be a set of the available AVL datasets for n considered routes, and $C = \{C_1, C_2, .., C_n\}$ a set of the corresponding APC datasets.

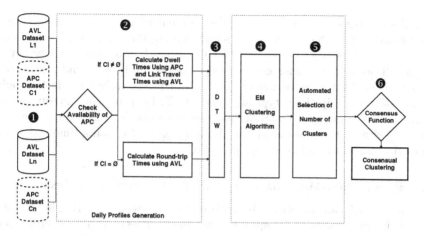

Fig. 1. A generic representation of the proposed methodology. The contributions of this paper are highlighted by the dashed blue rectangles.

If $C_i \neq \emptyset$, the round-trip time for every trip is obtained by adding the dwell times at stops and the link travel times as they are described in the AVL data.

By using trip-level APC data, we expect to *express* the demand peaks/valleys as slight increases/decreases of the computed round-trip time. Let r be a route of interest with the associated datasets (L_i, C_i) where t is the number of trips and s is its number of stops. This procedure starts by modeling the dwell time at stop through a decomposition in multiple factors. It can be computed as follows:

$$\delta_{o,j} = \max(\alpha \times a_{o,j}, \beta \times b_{o,j}) + doc \tag{1}$$

where α and β are constants that denote the alighting and boarding time per passenger, respectively, and doc denotes the time allocated for operations that take place on every stop, e.g. the opening and closing of doors. On the other hand, $a_{o,j}$ and $b_{o,j}$ are the number of passengers that alight/board on a stop j during a trip o, respectively, where $o \in \{1, 2, ..., t\}$ and $j \in \{1, 2, ..., s\}$.

Using the available values for dwell times (AVL) $\delta_{o,j}$ and the values of $a_{o,j}$ and $b_{o,j}$ (APC), we perform a linear regression procedure to estimate the values of α, β and doc. It consists of three steps: firstly, we isolate the samples (i.e. boardings/alightings and dwell times for every pair of [trips/stops] available) where $a_{o,j} = 0$ and $b_{o,j} = 0$ into two different partitions. This allows to transform Eq. 1 into a linear one. Secondly, we estimate values for α, β and two possible values for doc, i.e. doc_a, doc_b. Finally, the doc value is computed as $doc = \frac{doc_a + doc_b}{2}$. Then, we use the resulting constants to compose a novel function for the dwell time (i.e. $\hat{\delta}_{o,j}$). This function is used with the original APC data to compute novel dwell time estimations, which are summed up to the link travel times observed in the original AVL data.

The induction model used to do the abovementioned linear regression procedure is a modified version of the well-known least squares, where we replace its typical loss-function (a sum of the squared residuals) for the mean absolute deviation (MAD) (i.e. which results in a simple sum of the residuals). This change increases the framework's tolerance to large errors (i.e. demand peak/valleys), which will result in an under/overestimation of the dwell times under such conditions. This effect aims to model the demand peaks/valleys inside the daily profiles of round trip times typically used by [2]. By producing a daily profile based on heterogeneous sources of data, we aim to adequately express the differences between the route behavior - both in terms of cruising time and in its demand - on the schedule coverage definition.

2.2   Expectation-Maximization (EM) for Clustering Analysis

[2] proposed k-Means algorithm to perform the routewise clustering in the context of this application. This approach assumes a deterministic clustering step where the model is only given by the Euclidean Distance to the incrementally computed centroids (i.e. spherical clusters, parametric). Such characteristics may easily lead to an undesired overfitting, where the samples are erroneously initially assigned to a non-homogeneous cluster, potentially increasing the variance within. To overcome this limitation, we propose a GMM (a general version of k-Means), which (briefly) operates as follows: firstly, it (a) softly assigns a sample to a cluster, i.e., computing the probability of any point belonging to every centroid; then, it (b) estimates the parameters of the probability distribution, taking the sample-based covariances into account.

2.3   Automated Selection of Number of Schedules

The selection of the best number of clusters is a complex problem in data analysis. One of the most well-known metrics to do it so is the Bayesian Information Criterion (BIC) [10], which computes an entropy-based probabilistic score that, when maximized over a set of values, i.e. $\mathbb{K}$, aims to return the optimal k by minimizing the entropy between samples of the same cluster and maximizing the one between samples of different ones. However, such optimization problem may not lead to a good solution for a real-world context, given the constraints that each application domain encloses. Consequently, ad-hoc metrics are often devised to address such issues (e.g. market segmentation in [11]).

In this context, we depart from BIC to set up an ad-hoc metric, i.e. m for this problem as a linear combination of multiple factors. These factors were considered in light of two main constraints: (1) the cost of increasing the number of defined schedules (which reduces the schedule's interpretability as well as its easy memorization, the operators' ability to easily put it in place, and consequently, the route's riderships) must be necessarily balanced by a *gain* on the punctuality of the offered service, by reducing significantly the entropy on the produced clusters; (2) the cluster's output must model a *frequent pattern* (e.g. the Saturdays should be grouped with the Sundays throughout five months of an year). Such factors can be expressed as follows:

$$m(k,r) = \big(nbic(k,r) - f(k,r)^2\big) + \big(q(k,r) - \hat{\sigma}(k,r)\big), k \in \mathbb{K}, r \in \mathbb{L} \quad (2)$$

where $nbic(k,r)$ is the normalized[2] value of BIC. (1) The first term of Eq. 2 addresses the number of clusters. High values of $nbic$ will bring a gain on the punctuality of a suitable timetable defined for such partitioning. On the other hand, the increase of the number of schedules to maximize such punctuality must be done if and only if such *gain* is **significant**. Consequently, we need to model a *trade-off* between an eventual gain given by increasing the number of schedules and the associated cost of decreasing its interpretability. We do it so by introducing a penalty term $f(k,r)^2$ that favors lower values of k, where $f(k,r) = k/max(\mathbb{K}), \forall r \in \mathbb{L}$.

The second term of Eq. 2 addresses the cohesion and consistency of the partitioning for a number of schedules k. Empirically, we know that a SP in PT should cover a static set of *daytypes* (e.g. Mondays) throughout a relatively long set of weeks. Consequently, a suitable cluster would be one that provides such *frequent pattern*. The suitability of each cluster is given by an *ad-hoc quality* metric, i.e. $q(k,r)$. It is computed in two stages: (2a) frequent itemset mining and (2b) compatible pattern merging. This procedure is detailed as follows.

Cluster Quality Computation. A *frequent pattern* in this problem can be modeled through a sequence mining problem to find *frequent itemsets* of daytypes among the weeks (i.e. *transactions*) covered by the input data (e.g. Mondays to Fridays). Let $\gamma, \phi \in [0,1]$ denote two user-defined parameters for the minimum *support* to consider a given itemset as frequent (i.e. the minimum amount of weeks to define a schedule) and for the minimum cluster's mass ratio to be covered by it, respectively. The PrefixSpan algorithm [7] is hereby adopted to find such frequent itemsets, i.e. FI_i among the daytype's transactions obtained from each partition S_i. Let N denote the number of weeks in the input data. The *frequent pattern* of each cluster, i.e. FP is then selected as follows:

$$FP_i = \arg\max_{FI_i \subseteq S_i} \left(\frac{\Gamma(FI_i) \cdot |FI_i|}{N} \right) \text{ subject to: } \Gamma(FI_i) \geq \gamma, FP_i \geq \phi \quad (3)$$

where $\Gamma(FI_i)$ is the support of the frequent itemset FI_i on the partition S_i.

After such procedure, each cluster possesses a FP_i (which may be $\emptyset$). The quality of each cluster is then computed as $q(k,r) = \sum_{i=1}^{k} \frac{\Gamma(FP_i)}{k}$. However, in this domain, it is very common to find **complementary** schedules (e.g. workdays for all year and workdays during summer vacations, with a support of 0.9 and 0.1, respectively). **Together**, these complementary clusters would present a very meaningful *frequent pattern* which is penalized by the $q(k,r)$ computation formula introduced above. Consequently, we introduced a merging step which aims to find such clusters and to merge them in order to obtain the overall quality of the coverage proposed by a given value of k. This merging step aims to find clusters which have frequent itemsets complementary to a given FP_i by relaxing, at most, one of the two constraints imposed in Eq. 3. The algorithm to do it so is introduced by Fig. 2. Note that two clusters are considered as complementary if they overlap, at most, 10 % of the weeks of the input data.

---

[2] All the normalizations done throughout this section used the Euclidean distance.

1: **function** MERGING-COMP(k, γ, ϕ, S)
2:     $k' \leftarrow k;$
3:     **for** $(i$ in $\{1, ..., k'\})$ **do**
4:         **if** $(FP_i \neq \emptyset \wedge \Gamma(FP_i) < 1)$ **then**
5:             **for** $(j$ in $\{1, ..., k'\})$ **do**
6:                 **if** $\left(j \neq i \wedge (\exists\, cFI = FI_j \subseteq FP_i : FI_j(\gamma = 0) \vee FI_j(\phi = 0)) \right)$ **then**
7:                     **if** AreCoveringComplementaryPeriods?(S_i, S_j) **then**
8:                         $S_i \leftarrow S_i \cup S_j; S_j \leftarrow \emptyset; k' \leftarrow k' - 1;$
9:                         **return** Merging-Comp$(k', \gamma, \phi, S);$
10:                    **end if**
11:                **end for**
12:        **end for**
13:        **return** $\{k', S\};$
14: **end function**

Fig. 2. Merging Procedure for Complementary Clusters/Coverages.

Given the resulting clusters after the merging procedure (with a number of k' clusters), we can compute the final cluster's quality as

$$q(k, r) = \begin{cases} \sum_{i=1}^{k'} \frac{\Gamma(FP_i)}{k'} \text{ if} & k' = k \\ \left(\sum_{i=1}^{k'} \frac{\Gamma(FP_i)}{k'} \right)^{\left(1 - \frac{\chi}{2}\right)}, \; \chi = \max(FPM_i) \text{ otherwise.} \end{cases} \tag{4}$$

where FPM_i denotes the support of the frequent itemset of a *merged* cluster. Obviously, the resulting clusters may also contain other samples regarding daytypes not included in the frequent itemset (e.g. a cluster modeling the weekends which have two Mondays within). These samples are referred to as *noise* in this context. Such *noise* naturally decreases the adequacy of the *frequent pattern* modeled by each cluster. This effect is introduced by term $\hat{\sigma}(k, r)$ in Eq. 2. $\hat{\sigma}(k, r)$ is calculated based on the standard deviation between the relative frequencies of every day within a particular cluster. It can be computed as:

$$\hat{\sigma}(k, r) = \frac{1}{2} \times \sqrt{\sum_{i=1}^{k} \frac{\sigma(fr_{k, S_i, r})}{k}} \tag{5}$$

where $fr_{k, S_i, r}$ is the vector of relative frequencies of the days within the cluster i, where a relative frequency of a daytype d within a cluster S_i is given by the number of days of daytype d divided by the cluster's mass.

Given such metric computation for all pairs (r, k), we can now compute a consensual number of clusters K. Let $\eta(r)$ denote the normalized (see Footnote 2) number of trips for the route r. The consensual number of clusters K is defined by a weighted average of $k \in \mathbb{K}$. We can express $K \in \mathbb{N}$ as follows:

$$\left[\sum_{r \in \mathbb{L}} \sum_{k \in \mathbb{K}} \frac{m(k, r)^2 \times k \times \eta(r)}{\Psi} \middle/ \sum_{r \in \mathbb{L}} \eta(r), \Psi = \sum_{k \in \mathbb{K}} m(k, r)^2 \right] \tag{6}$$

3   Case Study

Our case study was a large urban bus operator in Sweden. We used data from four high-frequency (maximum planned headway of 10 min between 7:00–19:00) routes A1/A2/B1/B2, i.e. two bus lines A/B. Line A links residential areas to a PT hub as well as major shopping areas. B connects the southern parts of the city to the city center, traversing by a PT hub, major hospitals as well as a logistic center. This study covers six months between August 2011 and January 2012. The coverages in place are relative to two time periods: Summer, from 19 June till 14 December and Winter: from 15 December till 18 June. Two schedules are defined for each period: workdays and weekends/holidays.

As preprocessing, a trip pruning was performed by removing trips with more than 80 % of missing link travel times. Reversely, we performed data imputation on the remaining samples by following the interpolation procedure suggested in [2]. The dwell times were also pruned by using the 99 % percentile to remove erroneous measurements. APC data was used as is.

Table 1 presents an overview of the resulting dataset, detailed per route. It contains the (i) total number of trips (NT), (ii) its number of stops, (iii) the Daily Trips (DT), (iv) the Round Trip Times (RTT) and (v) the loads (i.e. total number of boarding passengers). Both have a similar NT, while line A has a larger RTT than B.

4   Experiments

The experiments were conducted using the R language [12]. The model-based clustering was performed using the GMM implementation of mclust package [13]. To compute the frequent itemsets used in the cluster's quality computation, a C++ implementation of *PrefixSpan* [14] was employed. This framework has three parameters: $\mathbb{K}, \gamma$ and ϕ. Their values were set to $2 \leq k \leq 7, \forall k \in \mathbb{K}$, 0.25 and 0.4, respectively. The first used the range suggested by the original experimental setup in [2]. The value of γ was empirically set such that a schedule can only be set for a period of, at least, four weeks; on the other hand, ϕ was selected out of three possible values 0.4, 0.5, 0.6 through an iterative parameter tuning setting conducted on a small subset of the training data.

The application of the proposed methodology to the available dataset suggested a novel SP - as detailed further in this Section. Its impact on the agency's operations in terms of schedule reliability was assessed through a simulation procedure, described in the next section.

4.1   Impact Evaluation Through a Data-Driven Simulation

Any change of the schedule coverage will result in one of two scenarios: (i) a group of days B changes from one coverage to another among the ones that were already in place or (ii) it will take a completely novel timetable. The procedure that we describe hereby is focused on the type-i Scenarios. Let A and Z be two

Table 1. Statistics per Route. The values are as mean ± s.d.. Times in seconds.

	Nr. Trips	Stops	DT	RTT	Loads
A1	17953	33	134 ± 27	3017 ± 425	101 ± 50
A2	16353	33	133 ± 30	2755 ± 480	98 ± 51
B1	16280	25	127 ± 23	2607 ± 465	70 ± 37
B2	16353	25	124 ± 22	2746 ± 448	60 ± 29

groups of days with different coverages and, consequently, distinct timetables assigned where $B \subseteq A$. Our goal is to test whether the time period B would benefit from having the same timetable of Z instead of its original one (i.e. from B). This procedure is done in three steps: firstly, we need to assign a timetable to B - which will change from the one in place in A to the one used in Z^3. Then, we need to simulate which would be the (a) link travel times and (b) the dwell times generated by such timetable given the available AVL/APC data.

The (a) link travel times are generated through a k-Nearest Neighbors regression [16] ($k = 1$), where the departure time of each stop is used as an independent variable. The demand on each stop is generated by using the headways computed through (a). These headways correspond to the idle time on a given bus stop bs_i, τ_i. The passenger arrivals at stops are modeled by iteratively sampling passenger arrival times pav^i from an exponential distribution, i.e. $pav^i \sim \text{Exp}(\lambda_i)$. Then, the number of boardings on each stop is computed as follows:

$$bo_i = \arg\max_x \sum_{j=1}^x pav_j^i \text{ , subject to: } \sum_{j=1}^x pav_j^i \leq \tau_i \wedge pav_j^i \sim \text{Exp}(\lambda_i) \quad (7)$$

where λ_i is computed as time-dependent Poisson process for every specific pair (r, bs) by considering averages of boardings on one hour periods of the days with similar daytypes (e.g. the number of passengers boarded on a given route between 8am and 9am of every Monday) - which are linearly normalized according to the amount of idle time available to compute each $bo_i \simeq x$. The alightings are then computed based on an assumption that the passengers traverse up to 25 % of the route. The resulting dwell times are computed using the Eq. 1 and the constant values obtained through the procedure described in Sect. 2.1.

The impact evaluation study is conducted on a *before-and-after* fashion, where schedule reliability metrics are firstly computed for the current case study (using the original AVL/APC data, as well as the SP in place). Then, the same metrics are also computed for the simulated data obtained through the abovementioned procedure. Four schedule reliability metrics were employed: On-Time Performance, Run-Time Variation, Headway Variation and Excess Waiting Time. Details about these metrics can be found in Sect. 4 of the Survey in [1].

---

3 Note that this *naive* timetabling procedure is done only for this specific purpose. Once the coverage is changed, the entire timetable of the affected periods need to be recomputed. The reader can consult the work in [15] to know more about this topic.

4.2   Results

This framework typically runs in linear time, where a single-core CPU processed the 16 k trips of our case study in ∼ 600 s. Figure 3 illustrates the computed values for the *ad-hoc* metric hereby devised to assess the quality of the partitioning provided by each value of k. These values resulted in a consensual $K = 3$. Figure 4 shows an example of the clustering results obtained for a particular route using its best value of k, i.e. $k = 5$. The consensual clustering results are exhibited in Fig. 5. Finally, Fig. 6 presents the schedule reliability evaluation metrics of the *before-and-after* study performed through the simulation described in the above Section.

4.3   Discussion

Figure 3 clearly exhibits the penalty effects of the term $f(k, r)^2$ as there is a clear trend of reducing the computed score with the increase of k. Yet, the weighted voting schema proposed in Eq. 6 ends up by finding a *consensus* around $K = 3$ - and not 2 as the charts may empirically suggest. As it is detailed by Fig. 4, this happens mainly due to a particular merge between the S_2 and S_4. Figure 5 illustrates the obtained coverage. It differs largely from the one in place by suggesting that the winter schedule should be in place four weeks earlier than it is (i.e. a change from mid-December to mid-November). The affected period was used as case study to conduct the simulation-based impact study described along Sect. 4.1. The obtained results (exhibited by Fig. 6) clearly outline high

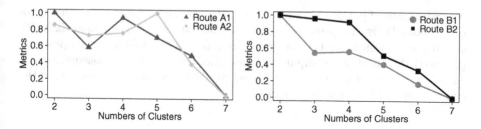

Fig. 3. Cluster quality metrics computed for every route and $k \in \mathbb{K} = \{2, .., 7\}$.

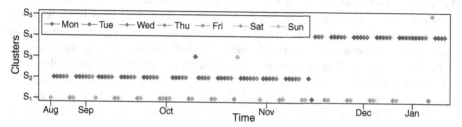

Fig. 4. Clustering results for route A2, $k = 5$.

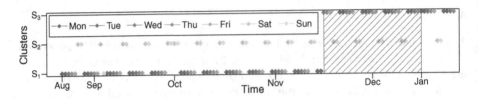

Fig. 5. Consensual clustering results, $K = 3$. Note the coverage's change on the workdays from Summer to Winter period suggested by the highlighted area.

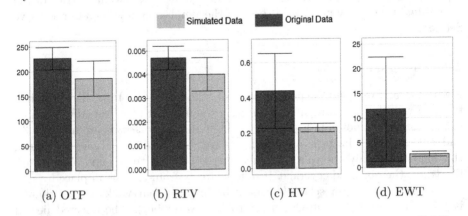

Fig. 6. *Before-and-after* impact evaluation of the novel schedule coverage reliability assessed using a data-driven simulation procedure.

potential gains of performing such change. However, such gains are mainly theoretical boundaries. They may be biased by the multiple constraints of daily PT operations, as well as by the oversimplification of the dwell time's computation (i.e. used the constants computed as described in Sect. 2.1). Consequently, an on-field deployment of this new coverage would be necessary to determine the exact impact of the suggested changes.

5   Final Remarks

This paper introduces a novel procedure to improve schedule coverage on PT networks. It is based solely on AVL/APC data. The final goal is to improve PT reliability and, consequently, their ridership and cost efficiency. Our main contribution is an *ad-hoc* metric to select the best number of schedules to put in place giving four decision factors - punctuality, adequacy, interpretability and reliability - modeled throughout sequence mining and probabilistic reasoning. To the best of our knowledge, this is first data driven framework to automatically select the number of schedules to be put in place using real-world data from a PT operator. Experimental results uncovered the potential gains introduced by this framework. As future work, the authors intend to evaluate it on a real-world testbed. Moreover, we also expect to create adequate exceptions on the concept of frequent itemset to relevant *outliers* on this domain (e.g. a schedule for the Christmas week) and identify when changes in round-trip times require introducing a novel schedule. This is still an open research question.

Acknowledgements. This work was also supported by the European Commission under TEAM, a large scale integrated project part of the Seventh Framework Programme for research, technological development and demonstration [Grant Agreement No. 318621]. The authors would like to thank all partners within TEAM for their cooperation and valuable contribution.

References

1. Moreira-Matias, L., Mendes-Moreira, J., Freire de Sousa, J., Gama, J.: Improving mass transit operations by using avl-based systems: a survey. IEEE Trans. Intell. Transp. Syst. **16**(4), 1636–1653 (2015)
2. Mendes-Moreira, J., Moreira-Matias, L., Gama, J., Freire de Sousa, J.: Validating the coverage of bus schedules: a machine learning approach. Inf. Sci. **293**, 299–313 (2015)
3. Mazloumi, E., Mesbah, M., Ceder, A., Moridpour, S., Currie, G.: Efficient transit schedule design of timing points: A comparison of ant colony and genetic algorithms. Transp. Res. Part B: Methodol. **46**(1), 217–234 (2012)
4. Cats, O., Mach Rufi, F., Koutsopoulos, H.: Optimizing the number and location of time point stops. Public Transp. **6**(3), 215–235 (2014)
5. Jorge, A.M., Mendes-Moreira, J., de Sousa, J.F., Soares, C., Azevedo, P.J.: Finding interesting contexts for explaining deviations in bus trip duration using distribution rules. In: Hollmén, J., Klawonn, F., Tucker, A. (eds.) IDA 2012. LNCS, vol. 7619, pp. 139–149. Springer, Heidelberg (2012)
6. Patnaik, J., Chien, S., Bladikas, A.: Using data mining techniques on apc data to develop effective bus scheduling. J. Syst. Cybern. Inf. **4**(1), 86–90 (2006)
7. Pei, J., Han, J., Mortazavi-Asl, N., Pinto, H., Chen, Q., Dayal, U., Hsu, M.: Prefixspan: mining sequential patterns efficiently by prefix-projected pattern growth. In: ICCCN, p. 0215. IEEE (2001)
8. Fraley, C., Raftery, A.: Model-based clustering, discriminant analysis, and density estimation. J. Am. Stat. Assoc. **97**(458), 611–631 (2002)
9. Matias, L., Gama, J., Mendes-Moreira, J., Freire de Sousa, J.: Validation of both number and coverage of bus schedules using avl data. In: 13th IEEE Conference on Intelligent Transportation Systems (ITSC), pp. 131–136 (2010)
10. Schwarz, G., et al.: Estimating the dimension of a model. Ann. Stat. **6**(2), 461–464 (1978)
11. Wagner, R., Scholz, S., Decker, R.: The number of clusters in market segmentation. In: Baier, D., Decker, R., Schmidt-Thieme, L. (eds.) Data Analysis and Decision Support. Studies in Classification, Data Analysis, and Knowledge Organization, pp. 157–176. Springer, Heidelberg (2005)
12. R Core Team: R: A Language and Environment for Statistical Computing. R Foundation for Statistical Computing, Vienna, Austria (2012). ISBN 3-900051-07-0
13. Fraley, C., Raftery, A., Scrucca, L.: Normal mixture modeling for model-based clustering, classification, and density estimation. Department of Statistics, University of Washington **23**, 2012 (2012)

14. Tabei, Y.: An imprementation of prefixspan (prefix-projected sequential pattern mining), August 2015. https://code.google.com/p/prefixspan/people/list. last access at August 2015
15. Ceder, A.: Urban transit scheduling: framework, review and examples. J. Urban Plann. Dev. **128**(4), 225–244 (2002)
16. Cover, T., Hart, P.: Nearest neighbor pattern classification. IEEE Trans. Inf. Theory **13**(1), 21–27 (1967)

Effective Local Metric Learning
for Water Pipe Assessment

Mojgan Ghanavati[1], Raymond K. Wong[1(✉)], Fang Chen[2], Yang Wang[2],
and Simon Fong[3]

[1] School of Computer Science and Engineering, University of New South Wales,
Sydney, Australia
{mojgang,wong}@cse.unsw.edu.au
[2] National ICT Australia (NICTA), Sydney, Australia
{fang.chen,yang.wang}@nicta.com.au
[3] University of Macau, Macau, China
ccfong@umac.mo

Abstract. Australia's critical water pipes break on average 7,000 times
per year. Being able to accurately identify which pipes are at risk of failure will potentially save Australia's water utilities and the community
up to $700 million a year in reactive repairs and maintenance. However,
ranking these water pipes according to their calculated risk has mixed
results due to their different types of attributes, data incompleteness
and data imbalance. This paper describes our experience in improving
the performance of classifying and ranking these data via local metric
learning. Distance metric learning is a powerful tool that can improve the
performance of similarity based classifications. In general, global metric
learning techniques do not consider local data distributions, and hence do
not perform well on complex / heterogeneous data. Local metric learning
methods address this problem but are usually expensive in runtime and
memory. This paper proposes a fuzzy-based local metric learning approach that out-performs recently proposed local metric methods, while
still being faster than popular global metric learning methods in most
cases. Extensive experiments on Australia water pipe datasets demonstrate the effectiveness and performance of our proposed approach.

1 Introduction

Based on a report by Australasian Corrosion Association (ACA), the Australian
water industry faces many challenges, particularly in the areas of asset management of aging infrastructure and the training required to support the prevention
of corrosion[1]. Failures of critical water mains typically bring severe consequences.
Closing highways, massive street flooding and damages, traffic and giant sink-
holes are just some of the impacts of water main failures. Based on official reports,
these costs have been estimated at $91 million per annum to the Australian urban
water industry. Moreover the total cost of corrosion in Australia is $982 million

---

[1] http://www.pacetoday.com.au/news/cost-of-urban-water-infrastructure-failure.

© Springer International Publishing Switzerland 2016
J. Bailey et al. (Eds.): PAKDD 2016, Part I, LNAI 9651, pp. 565–577, 2016.
DOI: 10.1007/978-3-319-31753-3_45

per year, which equates to an approximate annual cost of $60 for every adult in the country.

The maintenance process is a very critical process in water main industry. The maintenance process starts by prioritizing high risk pipes. Then proceeding to physically inspect these assets to confirm their actual condition, before finally deciding on whether to renew them or not. Hence, forecasting and ranking the pipelines according to their risk (i.e., likelihood to fail) are critical to reducing maintenance and repair costs, to save the repair and/or rehabilitation time and for safety.

Therefore, it is very useful to be able to rank accurately the water pipelines to find the most risky pipelines in each year. However, ranking water pipeline data has shown mixed results due to the characteristics of water data. On one hand, the data is sparse so that most pipes do not fail or fail just once during the observation period. On the other hand, the data is heterogeneous which means that their attributes have different types and/or scales and they are coming from different sources. In this paper, we consider the ranking problem in a pointwise manner. It is viewed as a regression or classification problem of predicting the specific relevance score for each category. In such cases, where the data is large, imbalanced and heterogeneous, learning a distance as a crucial step in classification is a very challenging issue.

Motivated by this problem, a lot of research has been done to identify similarities/distances between instances by proposing efficient metric learning techniques [11,12,15,22]. Earlier metric learning methods learn a global distance metric which determines the importance of different features and their correlations. However, when the problem is more complex, like in the water pipeline case, a global metric may not be able to fit the distance over the data manifold very well [1]. Local metric learning methods address this problem by learning a metric on each neighbourhood, or learning different metrics across the space [2,18–20].

Although local metric methods outperform global ones in many complex cases, they are prone to overfitting when they learn one metric per instance or when they learn completely independent metrics for different areas in data space. This overfitting problem has been addressed recently, by considering the relationships between the different metrics learnt [13,16]. However, these proposals still learn different metrics for different chosen areas, and their performance still depends on the way that the most appropriate areas cover the different data distributions are chosen. Secondly, same as the most local metric learning methods, they come with high CPU and memory costs [1].

Based on these observations, we exploit the properties of fuzzy c-means (FCM) clustering technique to learn one metric per cluster and parameterize the metrics by considering the membership degree of each instance for different clusters. This process will overcome the overfitting problem by parameterizing the metric matrix of each instance as a linear combination of basis metrics of cluster centres (as mentioned in [16]). To further speed up the process, ball trees are used to search for target neighbours and imposters that will be described later. As shown by experiments, the proposed fuzzy based local metric learning method is faster than the state-of-the-art local.

However, FCM may fail to find the optimal clusters when the clusters are not spherical or/and when the similarities are not linear [5]. It is also practically desirable for a metric learning method to support heterogeneous types of data like water pipeline data. Most metric learning methods do not address these issues [4]. Therefore, in this paper, a new non-linear framework is proposed to enhance FCM with kernel density functions. In this framework, all the features are mapped into a non-linear feature space based on their density quantities, so that it can handle heterogeneous datasets with numerical, categorical and mixed features. We call our approach: kernel density fuzzy-based metric learning (KDFuzzyML).

2   Related Works

Wang et al. [17] presented a robust statistical analysis by developing five multiple regression models and applied them on 15-year water pipe failure data of a Canadian water distribution network. To develop regression models, this research focused on individual water pipes rather than similar groups of pipelines and considered a wide variety of factors that may affect the break rate. Tebesh et al. [14] presented two failure prediction models based on artificial neural network (ANN) and neuro-fuzzy systems to improve the accuracy of water pipeline failure rate prediction. The proposed models are applied to real data of one of the water distribution networks in Iran. The results showed that failure prediction using ANN model is more accurate and realistic than using neuro-fuzzy and multivariate regression models.

Jafar et al. [7] investigated the application of ANN in water pipe condition prediction to estimate the rate of failure and find the best time to replace the pipelines. They tested the performance of six ANN models constructed using a cross-validation approach on an urban water distribution system in France. This study approved the efficiency of using the ANN approach to make a decision on the maintenance of water mains to reduce water pipe breaks.

Kleiner and Rajani [8] have done one of the most recent researches in this area. Four different models including a heuristic model, a naïve bayesian classification model, a logistic regression based model and a probabilistic model based on non-homogeneous poisson process (NHPP) have been developed. The proposed models have been used to rank the failure probability of some of the pipes which are expected to fail in the future. The models were applied to six various datasets from three different water distribution networks. The results of comparing these four models showed that NHPP-based model outperforms other models in terms of the prediction of the average number of failures for each pipe each year. However, it is not possible to select a model that is superior in terms of its ranking ability. In fact, different models had different but reasonable results on different scenarios.

Xu et al. [21] have done research on the water distribution network of Beijing and proposed a model to help decision makers to have a cost-effective plan for water pipe maintenance. They first developed a pipe break prediction model based on the failure data of the Beijing water distribution network between 2008 and 2011 using genetic programming and then set up an economically efficient pipe replacement model.

Recently Li et al. [9] proposed a non-parametric model using hierarchical beta process (HBP) to enhance the performance of pipe condition assessment. They addressed the limitations of parametric models since they usually have a fixed model structure based on a set of assumptions on the data behaviour.

Prioritization of risky pipelines can bring significant savings to water utilities. Hence in this paper, a local metric learning based model is presented to further increase the accuracy of this process to avoid critical main failures and also to avoid replacing any pipe that is still in working condition. To the best of our knowledge, there is no work in literature that has used metric learning based method to assess water pipeline conditions.

3   The Proposed Method

3.1   Data Collection

We use the data from two different Australian regions. Region A is a highly urbanised commercial area, and Region B is a lower density suburban area. We have access to two datasets for each region. The information of pipelines in each area is accessible through the first dataset. Attributes include: identification number, laid date, length, material, diameter size, location, protective coating and surrounding soil type. The second dataset contains the failure records between 1998 and 2012. The failure information includes failure date and time, failure type, failure location, etc.

In these regions, the oldest pipes were laid from 1800, with an average age of 60 years for Region A and 38 years for Region B. The fourteen years observation period was relatively short compared to the life cycle of water pipes such that most (about 99 %) pipes have not failed or failed just once during the observation period Hence the data is sparse or imbalanced in the other words. The soil types were similar for all the water mains belonging to the same region. Region A and B have 14, 765 and 17, 877 water pipes and 1, 696 and 6, 672 failure records respectively.

3.2   Fuzzy-Based Local Metric Learning

In this paper, we focus on supervised learning. We assume that we have a labelled dataset $S = (x_1, y_1), (x_2, y_2), ..., (x_n, y_n)$, where x_i's are D-dimensional instances and y_i's are class labels from a set of $c = 1, 2, ..., C$. First, we assume that all the features are numeric. Mahalonobis distance between two instances is given by:

$$d_M(x_i, x_j) = \sqrt{(x_i - x_j)^T \cdot M \cdot (x_i - x_j)} \qquad (1)$$

A linear metric learning method learns the positive semidefinite metric matrix M, based on some constraints. The most popular one is the large margin nearest neighbour method that is based on triplet constraints. Considering a triplet constraint (x_i, x_j, x_l), in a projected space based on M, the distance between x_i and x_j will be smaller than the distance of x_i and x_l. In a more complex problem, like when the data is large and heterogeneous, learning a single metric will not be able to model complexity of the problem. To address this problem, local metrics are learned for each learning instance or set of learning instances.

Learning a metric per instance makes the method slow and complex [1]. Therefore, in this paper, we will learn a set of metrics for a set of instances defined by FCM clustering method. A local metric function can be approximated by a linear combination of weighted metrics $\{M_1, M_2, ..., M_k\}$. Hence the local metric of each instance x_i is parameterized by $M_i = \sum_k w_{ik} M_k; \ w_{ik} \geq 0, \ \sum_k w_{ik} = 1$.

We use FCM clustering to find the anchor points and the weight matrix w. w is a $n * k$ matrix where its w_{ik} entry is the membership degree of instance x_i to the cluster k. The w_{ik} is calculated as:

$$w_{ik} = \frac{1}{\sum_j \left(\frac{d(z_k, x_i)}{d(z_j, x_i)}\right)^{\frac{2}{(\alpha-1)}}} \quad \text{where} \quad z_k = \frac{\sum_{x_i} w_{ik}^\alpha x_i}{\sum_{x_i} w_{ik}^\alpha} \tag{2}$$

where (z_k) is centre of the kth cluster.

The metric M_k is parameterized for instance x_i with the instance membership degree w_{ik}. Although this metric is used to calculate the distance of x_i from all other instances, the metric parameterization avoids overfitting. We use a similar objective function as in [16] to learn the basis metrics $\{M_1, M_2, ..., M_k\}$:

$$\min_{M_1, M_2, ... M_k, \xi} \ \alpha_1 \sum_k \|M_k\|_F^2 - \sum_{ijl} \xi_{ijl} + \alpha_2 \sum_{ij} \sum_k w_{ik} d_{M_k}^2(x_i, x_j)$$
$$s.t. \sum_k w_{ik}(d_{M_k}^2(x_i, x_l) - d_{M_k}^2(x_i, x_j)) \geq 1 - \xi_{ijl} \ \forall i, j, l; \ \xi_{ijl} \geq 0 \ \forall i, j, l; \ and \ M_k \geq 0 \ \forall k. \tag{3}$$

Speed up Using Ball Trees. Storing data samples in a hierarchical structure can make the nearest neighbour search faster. Ball trees are one of the hierarchical structures that have been used by Liu, et al. [10] for fast k-nearest neighbours (kNN) search. Ball tree partitions data points into a set of D-dimensional hyperspheres called balls. Each ball contains a subset of data points. The distance of each data sample to each ball is easily computed. The distance of a test point x_t to any trained point x_i in a ball tree S, has a bound as follows:

$$\forall x_i \in S \ \|x_t - x_i\| \geq max(\|x_t - c_s\| - r_s, 0) \tag{4}$$

where c_s is the centre and r_s is the radius of the ball s.

To further speed up the proposed method, we create one ball tree for the training samples of each class in each cluster. The created ball trees are used to search for target neighbours and the imposters. Searching for target neighbours and imposters are the most time consuming processes in the proposed method.

Note that for each data sample x_i, the imposters are data samples with different class labels that are closer to the x_i than its target neighbours. The target neighbours are the k nearest neighbours of x_i with the same class labels. If x_j is the kth closest point of x_t and its distance to the x_t is less than the bound from Eq. 4, all the training samples inside the ball can be left unexplored.

3.3   The Proposed Kernel Density-Based Fuzzy Metric Learning

This subsection is our main contribution in this paper where we propose a new fuzzy clustering method based on the kernel density and show how to learn the local metrics for the data in the new non-linear feature space. First, we need to know how to calculate the distance between instances in the new space. Considering two inputs x_i and x_j, Euclidean distance after mapping the inputs to the density probability space $\phi(x)$ is:

$$d^2(x_i, x_j) = (\phi(x_i) - \phi(x_j))^T(\phi(x_i) - \phi(x_j)) = \sum_{d=1}^{D}\sum_{c=1}^{C}[P_d(c, x_{i,d}) - P_d(c, x_{j,d})] \tag{5}$$

And Mahalonobis distance is formulated as $d^2(x_i, x_j) = (\phi(x_i) - \phi(x_j))^T M(\phi(x_i) - \phi(x_j))$, where M is a positive semidefinite matrix that we are trying to learn in different areas of the space.

Based on FuzzyML framework, we should first cluster the data using fuzzy clustering technique. Hence, we need to know how to calculate the cluster centre and the membership degree matrix. Let the cluster centre be $Z_k = [z_{k,1}, ..., z_{k,d}]$, $1 \leq d \leq D$, for clusters $1 \leq k \leq K$, where each attribute of Z_k is a vector of probabilities $z_{k,d} = [z_k(1, d); ...; z_k(c, d)]$, $1 \leq c \leq C$; $z_k(c, d)$ is calculated as:

$$z_k(c, d) = \frac{\sum_{x_i} w_{ki}^{\alpha} x_i(c, d)}{\sum_{x_i} w_{ki}^{\alpha}}. \tag{6}$$

More formally, each feature value of a cluster centre is a vector of conditional probability densities. Each item of the vector will be mean of the feature conditional probability density of all the points, weighted by degree of belonging of the points to the cluster. With fixed centres, weight matrix w is updated as Eq. 2 where:

$$d(z_k, x_i) = \sum_{d=1}^{D}\sum_{c=1}^{C}[z_k(c, d) - P_d(c, x_{i,d})] \tag{7}$$

We call this clustering approach as kernel density fuzzy c-means (KDFCM).

After clustering, all the basis metrics are learned. Considering U_k and γ_{ijl} as dual forms of M_k and ξ_{ijl} in Eq. 3, the dual formulation of objective function is obtained by substituting the following expression into the Lagrangian form of the objective function:

$$M_k^* = \frac{U_k^* + \sum_{ijl} \gamma_{ijl}^* w_{ik} G_{ijl} - \alpha_2 \sum_{ij} w_{ik} A_{ij}}{2\alpha_2} \tag{8}$$

where this expression is obtained by setting the first deviation of Lagrangian of the objective function over M_k and ξ_{ijl} to zero. Hence the dual formulation is defined as:

$$\max_{U_1,...U_k,\gamma} \sum_{ijl} \gamma_{ijl} - \sum_k \frac{1}{4\alpha_1} \cdot \|U_k + \sum_{ijl} \gamma_{ijl} w_{ik} G_{ijl} - \alpha_2 \sum_{ij} w_{ik} A_{ij}\|_F^2 \qquad (9)$$

$$s.t. \quad 0 \leq \gamma_{ijl} \leq 1; \quad \forall_{ijl} \ U_k \geq 0; \quad \forall k$$

where A_{ij} is equal to the outer product of each instance pair, and G_{ijl} is $x_{il}^T x_{il} - x_{ij}^T x_{ij}$. The outer product of each pair x_i, x_j in the new feature space should be calculated as:

$$A_{ij} = x_{ij}^T x_{ij} = \sum_{c=1}^{C} \sum_{d=1}^{D} x_d(c,ij) x_d(c,ij) \qquad (10)$$

where $x(c,ij)$ is a vector of x_{ij} over all the D dimensions. $x_{ij} = x_i - x_j$ is a vector of concatenation of all the D $(x_{i,d} - x_{j,d})$ where each $(x_{i,d} - x_{j,d})$ is a vector of concatenation of all the C $(P_d(c, x_{i,d}) - P_d(c, x_{j,d}))$. Similarly, G_{ijl} is calculated as:

$$G_{ijl} = \sum_{c=1}^{C} \sum_{d=1}^{D} x_d(c,il) x_d(c,il) - \sum_{c=1}^{C} \sum_{d=1}^{D} x_d(c,ij) x_d(c,ij) \qquad (11)$$

To make the defined optimization problem easier to solve, given a fixed γ_{ijl}, Eq. 9 can be simplified to:

$$\min_{U_k} \frac{1}{4\alpha_1} \|U_k + \sum_{ijl} \gamma_{ijl} w_{ik} G_{ijl} - \alpha_2 \sum_{ij} w_{ik} A_{ij}\|_F^2 \quad s.t. U_k \leq 0 \ \forall l \qquad (12)$$

Defining a symbol $A = \alpha_2 \sum_{ij} w_{il} A_{ij} - \sum_{ijl} \gamma_{ijl} w_{il} G_{ijl}$, U_k has a closed form solution which is the positive part of A. So we will have: $U_k^* = (A)_+$

Therefore the basis metric (Eq. 8) is simplified into $M_k^* = \frac{1}{2\alpha_1}(A)_+ - A$.

4   Experiments and Results

We conduct extensive experiments to evaluate the proposed method and compare our results to the competing methods for predicting water pipe failures.

First we consider the problem as a binary classification problem, i.e., failed and non-failed pipelines. Table 1 depicts the results of applying different methods for classifying these water pipes. We observe that KDFuzzyML has the lowest error rate (7.94 %). The water pipeline dataset is very imbalanced, so FScore and GMean values are better indictors than error rate. The methods with higher Precision, Recall, FScore and AUC have better performance. As shown in Table 1, kNN has the worst performance on the water pipe data. After that, BoostMetric is the second worst. Overall, our proposed method has the least error rate and is also at least slightly better than other methods based on all measurements. Since identifying every water pipe failure (in advance) saves lots of money, according

Table 1. Performance of different methods on the water utility data set

	kNN	LMNN	BoostMetric	PLML	FuzzyML	KDFuzzyML
Error Rate	10.37	8.40	8.49	8.10	8.12	**7.49**
Precision	0.31	0.48	0.44	0.49	0.50	**0.56**
Recall	0.25	0.54	0.30	0.57	0.57	**0.61**
FScore	0.27	0.51	0.35	0.53	0.57	**0.58**
AUC	0.60	0.75	0.63	0.76	0.76	**0.78**

to this experiment, our method is the best choice for pipe classification in the water pipeline condition assessment application.

KDFuzzyML outperforms other methods in water pipe classification as shown in Table 1. Next, we need to rank the problematic water pipelines according to their likelihood to fail, so inspectors can inspect the more problematic pipelines according to their urgency and available resources. The pipelines that are in danger of bursting (or failing) are said to be susceptible. We rank the susceptible pipelines using KDFuzzyML and two other popular survival analysis methods, Weibull [6] and HBP [9], using some real water pipeline data from two Australian regions. We choose Weibull, as a baseline for comparison, and HBP as a non-parametric method that has been very recently proposed for water pipeline condition assessment. The results are shown in Fig. 1. Y-axis indicates the percentage of susceptible pipelines actually found, and x-axis indicates the amount of inspection carried out in the water pipeline network in the region. It shows that KDFuzzyML allows more accurate, targeted inspection on susceptible pipelines than HBP and Weibull. For example, by following KDFuzzyML ranking of the susceptible pipelines in Region A, inspectors can identify more than 60 % susceptible pipelines by inspecting less than 20 % of the water pipelines in the region. When using the other two methods, they will need to inspect more than 40 % and 70 % of the pipelines, respectively, to achieve the same result. This is practically very significant, as a huge amount of resources (esp. money) can be saved for water pipeline maintenance.

Similar as Fig. 1, Table 2 uses different measurements to compare the performance of KDFuzzyML, Weibull and HBP use the real water pipeline data. As shown, for example, using AUC, KDFuzzyML is at least 20 % better than Weibull and HBP.

To investigate the significance of the results shown in Table 2, Friedman and Nemenyi tests [3] have been chosen to further analyse the results. Friedman test is used to compare different classification results according to their similarity. Friedman test, as a non-parametric test, is more suitable than parametric tests when more than two classifiers are involved [3]. Friedman test compares the average ranks of algorithms under a *null*-hypothesis. The *null*-hypothesis assumes that all the algorithms perform the same and hence their output ranks should be the same. Table 3(a) presents the results of Friedman test and hence we reject the hypothesis, i.e., the algorithms do not perform the same. Friedman test has been performed

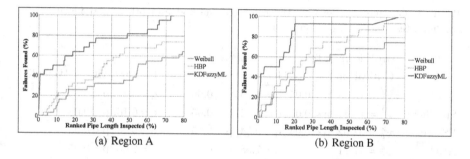

Fig. 1. Results of pipe failure prediction for regions A and B using different methods

Table 2. The performance diagnostics of different models

Metrics	Region A			Region B		
	KDFuzzyML	HBP	Weibull	KDFuzzyML	HBP	Weibull
Sensitivity	0.63	0.58	0.54	0.74	0.65	0.59
Specificity	0.62	0.59	0.54	0.73	0.66	0.6
FScore	0.57	0.48	0.44	0.71	0.48	0.43
AUC	0.74	0.57	0.5	0.82	0.61	0.58

considering the results of 10 times performing all three methods on water data in both regions. The χ_F^2 values namely Chi-Square more than 0 with $p < 0.05$ for all the performance diagnostics suggest that there are significant differences between performance measurements values of these three methods (KDFuzzyML, HBP and Weibull). In other words, these results illustrate that even if we perform these methods several times on different data samples, the performance will not be similar and their performances can be ranked.

When the *null*-hypothesis is rejected, we can proceed with a post-hoc test. Nemenyi tests are used here to show the difference between each pair of the algorithms and rank the methods based on their performance. In other words, Nemenyi test is a post-hoc test for pairwise comparisons (Table 3(b)) that are used after Friedman test. Nemenyi test is to test the same hypothesis as Friedman's but just between two methods at a time. As shown in the comparison between KDFuzzyML and HBP in Table 3(b), critical value (q) more than 0 with $p < 0.05$ rejects the *null*-hypothesis. This means that KDFuzzyML performs better than HBP on all measurements. Similarly, as shown in the comparison between KDFuzzyML and Weibull in Table 3(b), q more than 0 with $p < 0.05$ implies that KDFuzzyML outperforms Weibull. The higher the critical value is, the bigger difference between their performances is. For instance, for AUC, the critical value from Nemenyi on KDFuzzyML and Weidbull is 8.65, which is larger than the critical value from Nemenyi on KDFuzzyML and HBP (i.e., 7.98). This implies that for AUC, KDFuzzyML performs the best, followed by HBP and then Weidbull. This is consistent with the results shown in Table 2.

Table 3. Comparison between the proposed method and other methods using (*a*) Friedman test and then (*b*) Nemenyi test

Metrics	(a) Friedman test	(b) Nemenyi tests	
	Critical values	KDFuzzyML vs. HBP	KDFuzzyML vs. Weibull
Sensitivity	$\chi_F^2 = 13.85$, $p < 0.05$	$q = 4.95$, $p < 0.05$	$q = 5.37$, $p < 0.05$
Specificity	$\chi_F^2 = 14.41$, $p < 0.05$	$q = 5.34$, $p < 0.05$	$q = 6.10$, $p < 0.05$
FScore	$\chi_F^2 = 32.02$, $p < 0.05$	$q = 9.54$, $p < 0.05$	$q = 9.78$, $p < 0.05$
AUC	$\chi_F^2 = 17.21$,	$q = 7.98$, $p < 0.05$	$q = 8.65$, $p < 0.05$

Table 4. Speed of the algorithms against data size on water pipeline dataset

# Samples	CPU Time (s)	Train Error (%)	Test Error (%)
KDFuzzyML			
5000	7	8.10	10.16
15000	16	8.58	12.49
76000	384	0.17	0.52
350000	755	0.16	0.54
PLML			
5000	86	8.11	10.40
15000	237	8.51	12.03
76000	5455	0.17	0.45
350000	26283	0.16	0.50
LMNN			
5000	636	7.87	9.23
15000	3180	8.53	10.35
76000	4640	0.29	0.79
350000	>1 day	N/A	N/A

In addition, the scalability of KDFuzzyML is shown by experiments on water pipeline datasets with different sample sizes. Table 4 shows the speed (in CPU time) of different algorithms on water pipeline datasets with different sizes. For every experiment, 10-fold cross validation is performed and average over 10 runs is obtained. Table 4 is divided into 3 sections. Each section shows the performance of one algorithm on various sizes of the water pipe dataset. Based on the results, the proposed KDFuzzyML works much faster than LMNN on the water pipeline dataset with any sample size. For example, it takes just 755s for the proposed KDFuzzyML on the dataset with 350000 samples, whereas it takes more than a day for LMNN to finish the computation. Even when the sample size is less, like 15000 samples, KDFuzzyML is still about 200 times faster than LMNN with the same error rate. Hence, it can be concluded that the proposed

Table 5. Performance of various methods on data sets with different data types

Data set	Index	LMNN	BoostMetric	PLML	FuzzyML	KDFuzzyML
Contraceptive	**Error rate**	57.64	57.71	66.37	75.62	**53.29**
	Precision	0.43	0.43	0.29	0.08	0.51
	Recall	0.43	0.42	0.36	0.33	0.49
Statlog Heart	**Error rate**	22.22	20.74	22.22	21.85	**18.15**
	Precision	0.79	0.76	0.79	0.80	0.83
	Recall	0.76	0.69	0.78	0.77	0.82
Hayes-Roth	**Error rate**	24.21	28.82	40.92	37.8	**22.28**
	Precision	0.67	0.73	0.53	0.55	0.74
	Recall	0.63	0.71	0.55	0.54	0.73

KDFuzzyML is scalable to the dataset with hundred thousands of data samples. This is practically very useful for the task of water pipe condition assessment.

Finally, to show the generalization ability of the proposed method on other tasks, different datasets are employed for comparisons. All the selected data sets have been collected from UCI Machine Learning Repository[2]. All the experiments have been done 10 times using 10-fold cross validation on each data set. The average of the results over 10 runs has been presented in Table 5. KDFuzzyML is able to handle the categorical features by its nature. For all other techniques, we mapped all the categorical values to binary numbers. For example, for a categorical feature with n different values, all the values are transformed to a n bit binary. For each particular value, 1 digit will be equal to 1 and all the other digits will be equal to 0. Table 5 shows that the proposed KDFuzzyML method has a comparable performance against other popular metric learning methods on various heterogeneous data sets. We observe that KDFuzzyML has the lowest error rate and higher precision and recall on all the data sets.

5 Conclusion

We have proposed an efficient local metric learning approach to predict and rank water pipelines according to their likelihood to fail. Compared to traditional statistical modelling approaches, metric learning based methods are more flexible and able to adapt to the data complexity. When the data is large (heterogeneous and possibly imbalanced), global metric learning methods do not perform well in most cases. While local metric learning techniques perform better in those cases, they are expensive in terms of time and space. In this paper, we have proposed a local metric learning called KDFuzzyML that uses fuzzy clustering to make it fast and yet with at least similar performance as other metric learning

---

[2] https://archive.ics.uci.edu/ml/datasets.htm.

methods. Experiments have shown that our proposed approach outperforms previous parametric and nonparametric approaches for water pipe condition assessment. In practice, this represents major financial savings through more targeted inspections.

References

1. Bellet, A., Habrard, A., Sebban, M.: A survey on metric learning for feature vectors and structured data. arXiv preprint arXiv:1306.6709 (2013)
2. Bohné, J., Ying, Y., Gentric, S., Pontil, M.: Large margin local metric learning. In: Fleet, D., Pajdla, T., Schiele, B., Tuytelaars, T. (eds.) Computer Vision – ECCV 2014, Part II. LNCS, vol. 8690, pp. 679–694. Springer, Heidelberg (2014)
3. Demšar, J.: Statistical comparisons of classifiers over multiple data sets. J. Mach. Learn. Res. **7**, 1–30 (2006)
4. He, Y., Chen, W., Chen, Y., Mao, Y.: Kernel density metric learning. In: IEEE 13th International Conference on Data Mining (ICDM), 2013. pp. 271–280. IEEE (2013)
5. Huang, H.C., Chuang, Y.Y., Chen, C.S.: Multiple kernel fuzzy clustering. IEEE Trans. Fuzzy Syst. **20**(1), 120–134 (2012)
6. Ibrahim, J.G., Chen, M.H., Sinha, D.: Bayesian survival analysis. Wiley Online Library, New York (2005)
7. Jafar, R., Shahrour, I., Juran, I.: Application of artificial neural networks (ANN) to model the failure of urban water mains. Math. Comput. Model. **51**(9), 1170–1180 (2010)
8. Kleiner, Y., Rajani, B.: Comparison of four models to rank failure likelihood of individual pipes. J. Hydroinformatics **14**(3), 659–681 (2012)
9. Li, Z., Zhang, B., Wang, Y., Chen, F., Taib, R., Whiffin, V., Wang, Y.: Water pipe condition assessment: a hierarchical beta process approach for sparse incident data. Mach. Learn. **95**(1), 11–26 (2014)
10. Liu, T., Moore, A.W., Yang, K., Gray, A.G.: An investigation of practical approximate nearest neighbor algorithms. In: Advances in neural information processing systems. pp. 825–832 (2004)
11. Liu, W., Tsang, I.W.: Large margin metric learning for multi-label prediction. In: Twenty-Ninth AAAI Conference on Artificial Intelligence (2015)
12. Megano, T., Fukui, K.i., Numao, M., Ono, S.: Evolutionary multi-objective distance metric learning for multi-label clustering. In: 2015 IEEE Congress on Evolutionary Computation (CEC). pp. 2945–2952. IEEE (2015)
13. Noh, Y.K., Zhang, B.T., Lee, D.D.: Generative local metric learning for nearest neighbor classification. In: NIPS. pp. 1822–1830 (2010)
14. Tabesh, M., Soltani, J., Farmani, R., Savic, D.: Assessing pipe failure rate and mechanical reliability of water distribution networks using data-driven modeling. J. Hydroinformatics **11**(1), 1–17 (2009)
15. Wan, S., Aggarwal, J.: Spontaneous facial expression recognition: a robust metric learning approach. Pattern Recogn. **47**(5), 1859–1868 (2014)
16. Wang, J., Kalousis, A., Woznica, A.: Parametric local metric learning for nearest neighbor classification. In: NIPS. pp. 1610–1618 (2012)
17. Wang, Y., Zayed, T., Moselhi, O.: Prediction models for annual break rates of water mains. J. Perform. Constructed Facil. **23**(1), 47–54 (2009)

18. Weinberger, K.Q., Saul, L.K.: Distance metric learning for large margin nearest neighbor classification. J. Mach. Learn. Res. **10**, 207–244 (2009)
19. Wu, L., Jin, R., Hoi, S.C., Zhu, J., Yu, N.: Learning Bregman distance functions and its application for semi-supervised clustering. In: NIPS. pp. 2089–2097 (2009)
20. Xiong, C., Johnson, D., Xu, R., Corso, J.J.: Random forests for metric learning with implicit pairwise position dependence. In: Proceedings of the 18th ACM SIGKDD International Conference on Knowledge Discovery and Data Mining. pp. 958–966. ACM (2012)
21. Xu, Q., Chen, Q., Ma, J., Blanckaert, K.: Optimal pipe replacement strategy based on break rate prediction through genetic programming for water distribution network. J. Hydro-Environ. Res. **7**(2), 134–140 (2013)
22. Yu, J., Tao, D., Li, J., Cheng, J.: Semantic preserving distance metric learning and applications. Inf. Sci. **281**, 674–686 (2014)

Classification with Quantification for Air Quality Monitoring

Sanad Al-Maskari[1(✉)], Eve Bélisle[2], Xue Li[1], Sébastien Le Digabel[3], Amin Nawahda[4], and Jiang Zhong[5]

[1] School of Information Technology and Electrical Engineering, University of Queensland, Brisbane, Australia
{s.almaskari,xueli}@uq.edu.au
[2] CRCT, École Polytechnique de Montréal, Montréal, Canada
e.belisle@uq.edu.au
[3] GERAD, École Polytechnique de Montréal, Montréal, Canada
sebastien.le-digabel@polymtl.ca
[4] Environmental Research Centre, Sohar University, Sohar, Oman
ANawahda@soharuni.edu.om
[5] College of Computer Science, Chongqing University, Chongqing, China
zhongjang@cqu.edu.cn

Abstract. In this paper, a fuzzy classification with quantification algorithm is proposed for solving the air quality monitoring problem using e-noses. When e-noses are used in dynamic outdoor environment, the performance suffers from noise, signal drift and fast-changing natural environment. The question is, how to develop a prediction model capable of detecting as well as quantifying gases effectively and efficiently? The current research work has focused either on detection or quantification of sensor response without taking into account of dynamic factors. In this paper, we propose a new model, namely, Fuzzy Classification with Quantification Model (FCQM) to cope with the above mentioned challenges. To evaluate our model, we conducted extensive experiments on a publicly available datasets generated over a three-year period, and the results demonstrate its superiority over other baseline methods. To our knowledge, gas type detection together with quantification is an unsolved challenge. Our paper provides the first solution for this kind.

Keywords: E-nose · Gaussian process regression · MOX · Noise · Drift · Classification · Clustering · e-nose

1 Introduction

Big cities around the world are facing serious air quality and air pollution problems, which prompted governments to introduce more tightened emission and safety regulations. The study of gas identification and measurement using chemosensors known as electronic noses, is important in emission control. E-noses can provide an economical solution for air quality and air pollution monitoring.

© Springer International Publishing Switzerland 2016
J. Bailey et al. (Eds.): PAKDD 2016, Part I, LNAI 9651, pp. 578–590, 2016.
DOI: 10.1007/978-3-319-31753-3_46

Despite their popular usage, they suffer from serious problems such as drift, noise, selectivity and their inability to provide true concentrations [1, 2].

An electronic nose is a device used to identify, measure, and analyse chemical analytes [3]. Various applications exist for e-nose including: agriculture, disease detection, foods and beverages, air quality and environment protection, water and waste water quality control [3, 4]. Despite their wide applications, they are very susceptible to noise and drift, making them unstable. The stability issue can be resolved via periodic recalibration. Unfortunately, recalibration process is an expensive and complex task. Compensation for the need of periodic recalibration and ability to postpone the recalibration period using machine learning will need to solve the aforementioned dynamic matters. In general, the performance of e-noses is affected by drift and noise. Sensor drift is unpredictable degradation of sensor sensitivity and selectivity over time. The degradation can be due to sensor poising, sensor ageing, and environmental fluctuations such as humidity, temperature, pressure and system sampling non-specific adsorption [1, 2]. In contrast, noise is random signal deviation caused by unknown effect. Sensor noise can be due to electronic circuit error, environmental effects, sensor poising and aging. Because sensor drift is not deterministic, it is difficult, if not impossible, to distinguish it from noise and *vice versa* [5].

Based on above discussion, it is challenging to predict gas concentrations by means of a chemical measurement system. In this paper, we propose a new approach to detect gas types together with the quantification of the detected gases in a dynamic environment. The fundamental idea of our approach is to detect the gas types first, then the confidence of this prediction can be measured. After that, the quantification of the detected gases with high confidence can be carried out by using a regression approach. The primary contributions of our work are as follows: (1) An ensemble classifier of Kernel Fuzzy C-Mean with Fuzzy Support Vector Machine (FCQM-KFSVM) is trained in different chunks of the data. (2) We present a technique to estimate classifier confidence scores. (3) We focus on improving the model sensitivity to air quality exposure limits rather than attempting to predict just gas concentrations. (4) We present a new framework by integrating classification and quantification technique with confidence evaluation to address both of concept drift, identification and quantification of gases in a dynamic environment.

The rest of the paper is organized as follows. Section 2 discusses the related work. Section 3 describes the Methodology. Our proposed approaches of fuzzy based classification, confidence scores and Scalable gaussian quantification method will be introduced. Section 4 describes the experiments and evaluates the results. Finally, Sect. 5 concludes our paper.

2   Related Work

The ultimate goal of electronic e-nose outdoor air quality monitoring systems is to identify and quantify gases in the ambient air. Chemo-sensors has the ability to measure a specific gas concentration in a mixture of gases using a chemical interface. The chemo-sensor interface interacts selectively with predefined gases. These sensors have the ability to measure gas concentrations with good accuracy in a closed environment. Installing e-nose in an outdoor environment introduces many challenges such as noise, drift and high uncertainties. Several machine learning models including Self Organizing Maps (SOMs), multiple SOM, and neural networks have been used for gas identification [6–8]. Various quantification methods have been used in the literature such as Support Vector Regression (SVR), partial least squares regression (PLS), and Artificial Neural Network (ANN). Desai et al. used SVR in soft sensor applications to demonstrate the conversance and generalization capability of SVR. Feed forward neural networks (FFNNs) were used by Gulbag and Temuras [9] to quantify two gas concentrations (trichloroethylene and acetone). Linear regression methods such PCR and PLS have been used in the literature due to their ability to reduce the dimensions before fitting a regression function [10,11]. In [12] they use gaussian process to recalibrate the sensors while in our work we do not calibrate the sensors. Instead we are interested in identifying then quantifying gas concentrations levels without recalibration.

Large portions of the existing work have focused on odour classification using chemo-sensors and few attempts have been made to address the quantification part. Moreover, very few works in the literature have focused on a holistic approach that considers gas type detection and concentration measurement using e-noses for air quality monitoring system. This motivated us to introduce a new methodology which is different than other problems found in the literature. On this research we are interested in a model able to classify and quantify data instances while taking into considerations confidence scores, underestimation and overestimation rates of gas concentrations.

3 Methodology

In this section we formulate a new data mining research problem which is applicable for real world applications such as air quality monitoring. We attempt to predict air quality level with high confidence rather than predicting gas concentrations with high accuracies. Therefore, the proposed model should tolerate overestimation and be more sensitive to underestimation of high concentrations.

Problem Definition. Consider a multi-class classification problem with a set of features x produced by a gas sensor. Let $S_1, \ldots, S_t$ be a batch of examples received from e-nose sensor at time interval t. S_t consists of labeled L and unlabeled U examples, $S = L \cup U$. The labeled sets $L_t = \{(x_{1t}, y_{1t}, z_{1t}), \ldots, (x_{nt}, y_{nt}, z_{nt})\}$ are of size tn where $x_{nt} \in R^N$ and $y_{nt} \in \{1, \ldots, M\}$ where M refers to the class label (gas type) of instance x_{nt} and $m \in M$. Gas concentration for instance x_{nt} is represented by z_{nt} where $z_{nt} > 0$. The unlabeled examples of size tn are given by $U_t = \{x_{1t}, \ldots, x_{nt}\}$. Therefore, each example can be represented by $< S, Y, N >$ where $Y \in \{1, \ldots, M\}$, indicating the class label of a labeled example $x_n t \in S$; $N \in \{0,1\}$, indicating whether a given example S is labeled ($N = 1$) or unlabelled ($N = 0$). Each Instance in L is labeled, represented by $< S, Y \in M, N = 1 >$.

Consequently, each instance in U is unlabeled and can be represented by $< S, Y =?, N = 0 >$. Our objective is to train classifiers f_m using the training set $T(L)$ and a validation set $V(U)$ such that the unlabeled gas data U can be correctly classified with high confidence Φ. The classification decision process is made by applying all classifiers represented by f_p for $p \in \{1, \ldots, P\}$ to new unlabeled sample $x_{nt} \in U$ and predicting the gas type $\hat{y}_{nt}$ for which the corresponding classifier reports the highest confidence score:

$$\hat{y}_{nt} \in \underset{p \in \{1,\ldots,P\}}{\operatorname{argmax}} f_p(x_{nt}) \quad \text{where} \quad \Phi_{nt} = \underset{p \in \{1,\ldots,P\}}{\max} \Phi_p(x_{nt}). \tag{1}$$

3.1 System Architecture

Gas type detection together with quantification is an unsolved challenge. Different gases may be present in the atmosphere which will reduce sensor selectivity for differentiating different types of gases. Furthermore, as explained previously, sensors are subject to high uncertainties due to environmental changes. Moreover, chemo-sensors suffer from drift and noise. Therefore, the identification and quantification models should consider all the variabilities surrounding e-nose sensors. To address these issues we propose a three stage approach. At the first stage, we construct multiple FCQM-KFSVM classifiers to detect the gas type. At the second stage, a confidence evaluation is conducted to evaluate the prediction generated by all classifiers. Only samples with high confidence will be passed to the third stage for quantification. For each gas type (class) a corresponding training data is created for gas quantification measurement. Finally, at the third stage, the concentration of the predicted gas is estimated using FCQM-SGP. Fig. 1 illustrates the full life cycle of the proposed approach.

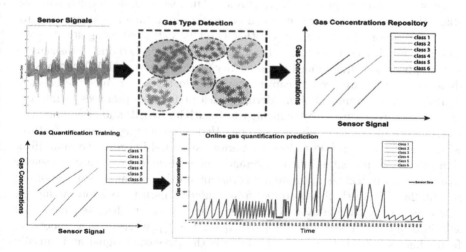

Fig. 1. The top part shows the E-nose gas detection process. The bottom part demonstrates the On-line gas quantification for e-noses.

3.2 Feature Extraction

The performance of e-nose classification is highly dependent on the effectiveness of features being used. Features that provide high discrimination power between different classes will result in a better classification performance for a learning model. The practical value of metal oxide-based gas sensors is affected due to their poor selectivity, sensitivity, stability, and slow response. Many authors have demonstrated the ability to improve sensor performance using intelligent methods for sensor feature extraction and selection [13]. Various methods have been used for feature extraction of sensor arrays, such as the Fast Fourier Transform (FFT) and Discrete Wavelet Transform (DWT) [14,15]. Transforming signals from the time domain to the frequency domain fails to preserve important transient characteristics of the original signal. Therefore, using a method that can preserve the signals transient characteristics is very important. In this study, we initially considered steady-state and **dynamic-based** feature extraction methods, which are standard method for chemo-sensory feature extraction. Steady-state features, are extracted from the steady-state response produced by a gas sensor refer to Fig. 2a. When a gas sensor is exposed to the same gas, it can produce a static change in resistance as long as the same gas concentration keeps flowing [14]. The change in resistance of a semiconductor gas sensor ΔR is the difference between the maximal resistance change and the base line as shown in this equation $\|\Delta R\| = \frac{R_{gas} - R_{air}}{R_{air}}$. R_{gas} is the sensor response to a given gas in steady-state, R_{air} is the sensor response to pure air, which is used as a baseline. The steady-state feature ΔR has two major issues. Firstly, it becomes available late in the response as shown in Fig. 2a. Secondly, it is susceptible to drift. Therefore dynamic feature enable us to capture the transient phase of the signal. Dynamic-based features are extracted from transient chemical gas sensor response (adsorption and desorption). In the adsorption stage of gas sensor responses, the resistance R increases, while in the desorption stage, the resistance R decreases, as shown in Fig. 2a. Dynamic features can provide information about the analyte that can't be extracted using steady-state methods. Furthermore, steady-state values are rarely reached due to gas sensors slow response and environmental complexity (e.g. airflow turbulence). Therefore, dynamic features are used to capture transient features to reduce the effect of dynamic real environment. Considering transient features promises to improve classification performance by reducing the sensitivity to drift and providing more discriminate information about each gas. Hence, combining transient features with steady-state can boost classifier performance and improve the quantification process. Exponential moving average ($ema\alpha$) is used to capture the sensor dynamics of the increasing/decaying transient portion of the sensor responses; see Fig. 2a. This method is borrowed from financial market stock forecasting [2]. The exponential moving average $ema\alpha$ transforms sensors discrete-time signal $x[k]$ into: $y[k] = (1 - \alpha)y[k - 1] + \alpha(r[k] - r[k - 1])$, where α is a smoothing factor between $[0, 1]$, $k = 1, 2, .., T$ and y is the forecasted signal and initially set to zero($y[1] = 0$). Various features are extracted from raw sensor responses using different smoothing parameters α e.g. ($\alpha = 0.1, \alpha = 0.01, \alpha = 0.001$)

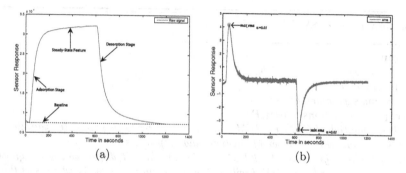

Fig. 2. (a) Raw sensor response when exposed to 223 ppmv of Ethylene. (b) Shows the exponential moving average analysis when $\alpha = 0.01$.

3.3 Classification

Because e-noses generate unpredictable and unstable sensor response, we propose a fuzzy approach to address this issue. When e-nose is exposed to mixture of gases unknown signal output can be produced. This output could belong to multiple classes with different relevance rates. Thus sensory signal output will have various levels of importance or relevance. The standard Support Vector Machine (SVM) suffers from noise sensitivity because all data points receive same treatments.Therefore, we introduce **Sample Relevance** (SR) which defines how each sample x_i contributes to a given class C_n. SR creates useful cluster representations to identify gas sensor samples' relevance to each class. We seek to obtain a high degree of SR with little redundancy between classes (no overlapping). For each training pair (x_i, y_i) an SR value s_i is calculated, the pairs with high s_i degree will effect the prediction decision more than pairs with a lower s_i degree. A Kernel-based Fuzzy C-Means clustering (KFCM) [16] algorithm is used to calculate SR which will be used as input for the fuzzy classifier. Finally, our approach address concept drift more efficiently and provide more generalization because a fuzzy ensemble is trained on different chunks of data.

Using sensor data, KFCM model will calculate SR degrees. After that, multiple classifiers are trained using KFSVM model [1,17]. Improved predictability and generalization is anticipated when using SR with FSVM ensemble. In order to find the ideal training model, a grid search is executed to find the best γ and C in $[2^{-10}, 2^{-9}, \ldots, 2^4, 2^5]$ and $[2^{-5}, 2^{-4}, \ldots, 2^9, 2^{10}]$. FCQM-KFSVM is performed using the following algorithm:

3.4 Confidence Scores

The incoming unlabelled sensor data has high prediction uncertainties due to noise and drift. Therefore performing regression directly is biased. In our proposed approach, confidence classification evaluation is applied on the prediction received from the FCQM-KFSVM ensemble. The output then is evaluated using ranking and confidence functions. Instances with high confidence are extracted

---

Algorithm 1. Fuzzy SVM Ensemble

1: **Input:**
   b_n: *Input batches* $b_1, b_2, \ldots, b_n$
   U_opt: *SR matrix*
   σ , γ *and* ε *parameters*
2: **Output:** *predictions matrix PR*
3:   *Train multiple FCQM-KFSVM classifiers and perform parameter initialization*
   for kernel function, number of clusters, and stop flag ε.
4: *Apply the SRs generated by KFCM to FSVM. An SR degree will be assigned to*
   each sensor output and FSVM will be trained using $\{x_i, y_i, s_i\}$, *where s is the SR*
   degree.
5: *Perform gas type prediction for incoming Xtn.*
6: **return** *PR.*

---

to be used by the FCQM-SGP model. The ranking function R is used to measure the ranking level for the predicted class label of the input instance:

$$R_t(j) = \left(\sum_{a=1}^{k} \frac{I(K(a) = M(j))}{N} \right) \text{ where } R_t(j) \geq 0, \sum_{j=1}^{n} R_t(j) = 1. \quad (2)$$

Where R_t is the ranking for class M for the incoming data points at time t. I is an indicator function which outputs 0 or 1.

The confidence level C_t for class M for the incoming data points at time t is defined by the following:

$$C_t(j) = \left(\sum_{k \in K | f_k = M(j)} C(f_k) \right) \text{ where } C_t(j) \geq 0, \sum_{j=1}^{n} C_t(j) = 1. \quad (3)$$

$$\Phi_t(m) = \frac{1}{2} \left(\sum_{j=1}^{n} (R_t(j) + C_t(j)) \right) \text{ where, } \Phi_t(i) \geq 0, \sum_{j=1}^{n} \Phi_t(i) = 1. \quad (4)$$

$$\hat{\Phi}_{nt} = \max_{m \in \{1, \ldots, M\}} \Phi_m(x_{nt}) \text{ where } \Phi_m(x_{nt}) \geq \Gamma. \quad (5)$$

The final confidence score for each class m is given by the confidence function Φ for incoming point x at time t (Eq. 4). As we are interested in incoming sensor data with high prediction confidence for a given class, we select the predicted gas type which has the highest confidence, exceeding a given threshold Γ. Only the data points x_{nt} with the highest confidence $\hat{\Phi}_{nt}$ are selected to be used in the gas concentration prediction.

3.5 Scalable Gaussian Process for Quantification

The ultimate goal for e-nose monitoring system is to classify and quantify the detected gas. We define the quantifier as a mapping function between the gas sensor response and the reference gas set provided in the training stage. To map gas sensor signals into concentrations, a prior distribution is used. We use a probabilistic approach inspired by Gaussian Process Regression (GPR) for gas quantification. This method, called the Scalable Gaussian process (FCQM-SGP) which was introduced by Belisle et al. [18]. The traditional GP has a very high computational cost, typically n^3 where n is the number of data points. However, FCQM-SGP significantly reduces the computational time, allowing on-line learning. The FCQM-SGP approach consists of three steps. First step is batch query processing. In this step, the query points are grouped by similarity. The number of points per query group is determined by a parameter e. Secondly, the training database is condensed by eliminating similar data, where an arbitrary condensation parameter is chosen. In this work, we used a value of 1 %. This means that when the measure of similarity between two points in the training set is smaller than 1 %, these two points are merged. Finally, for each group of query points formed in the first step, a subset of the training data created during the second step is selected to perform the GP. The points are chosen according to a distance measure, keeping only the points from the training set that are close to each query group. The number of points in the training set is determined by the parameter a (accuracy). This method is described in Algorithm 2. In this work, we use the covariance matrix introduced by Gibbs and MacKay [19]:

$$K(X, X') = \sigma_f^2 \exp\left\{ -\frac{1}{2} \sum_{j=1}^{n} \frac{(x_i - x_j')^2}{w_j} \right\} + \sigma_n^2 \delta(X, X'). \tag{6}$$

Where δ is the Kronecker delta function, σ_f^2 refers to variance of the process, n is the total number of points, X are vectors with length equal to the number of dimensions, x elements of these vectors and w is the Gaussian kernel width.

4 Experiments and Evaluation

Experimental Design. A publicly available dataset is used in this paper [2,20]. The data set was generated over three years using metal-oxide sensor array for six-gases/analytes. Different type of gases was emitted to the sensors with varying concentrations. The main objective was to detect the type of gases in the presence of drift regardless of their concentrations. It was shown by Vergara et al. that sensor drift actually occurred. A total of 13,910 samples were generated over 36 months. For more details about the dataset, the reader is referred to [2,20]. The experiments are divided into two parts, the first part is FCQM-KFSVM gas detection and the second part is FCQM-SGP gas quantification. In FCQM-KFCM the model is evaluated against different classifiers including Neural Net

---

Algorithm 2. FCQM-SGP algorithm

1: **Input:**
 X: *Training database*
 E: *Predictions database*
 θ *condensation*
 e *number of query points per group*
 a *prediction accuracy*
2: **Output:** *final quantification predictions matrix QFM*
3: *Group E, using agglomerative clustering, until e points per group.*
4: *Condense X, using parameter θ*
5: *For each group formed in step 1:*
 Select a subset of the condensed training set and perform GP.
 The size of the training subset is defined by a.
6: **return** *QFM*

---

with weight decay, NN with dropout, NN with sigmoid, and SVM. After, FCQM-KFSVM identify the gas type with high confidence the predicted data points will be used by FCQM-SGP quantification model.

To evaluate the quantification model, we tested the performance of different algorithms with different combinations of training and testing data sets. In order to create a dataset that represents a variety of scenarios and capture different sub spaces, five different training and testing proportions where created $(10, 30, 50, 70, 90)$. In 10 % proportion we use 90 % of the training set and only 10 % for testing. On the other hand 90 % indicates that we use only 10 % for training and 90 % for testing. In the 10 % proportion if we assume the data set we have contains 600 points, then 540 points will be used for training and 60 for testing. We use 10 fold cross validation to evaluate our model in each proportion. This ensures the generalization and the accuracy of the estimated error measures.It is important to mention that data points which are misclassified or has low confidence level will be ignored by FCQM-SGP.Predictions with FCQM-SGP was executed using the following parameters: $\theta = 1\%$, $e = 4$ and $a = $ medium. We compare the proposed method with Linear Regression (LR), DynaTree Linear (DT-LIN), DynaTree Constant (DT-CST), Linear SVM Regression(LSVM) and Least Squares SVM (LS-SVM). We evaluate our method in term of Classification accuracies,Root Mean Square Error (RMSE), Mean Absolute Error (MAE), Mean Bias Error (MBE), Correlation Diversity, and outliers rate.

$$RMSE = \sqrt{\frac{1}{n}\sum_{i=1}^{n}(P_i - T_i)^2} \text{ and } MAE = \frac{1}{n}\sum_{i=1}^{n}|(P_i - T_i)|$$

$$\text{Correlation Diversity}\rho = \frac{\sum_{i=1}^{n}T_iP_i - \frac{\sum_{i=1}^{n}T_i\sum_{i=1}^{n}P_i}{n}}{\sqrt{\left(\sum_{i=1}^{n}T_i^2 - \frac{(\sum_{i=1}^{n}T_i)^2}{n}\right)\left(\sum_{i=1}^{n}P_i^2 - \frac{(\sum_{i=1}^{n}P_i)^2}{n}\right)}}.$$

Results and Discussion. To evaluate the classifiers ability to deal with drift we only used batch 1 for testing and remaining are used for testing. Batch 1 is generated in the first 2 month therefore it doesn't contain drift, while batch 2–10 contain drift because it was generated after 3 month refer to [2,20]. Figure 3a show the prediction accuracies of different classifiers. Our proposed classifier FCQM-KFSVM performed better than other classifiers in batch 5 to 10 indicating it is ability to perform better in drift situations. FCQM-KFSVM archived best overall average accuracies with 58.419 %

RMSE and MAE results in Fig. 3b, c and d indicate the superiority of FCQM-SGP compared to other methods. FCQM-SGP was able to maintain good performance even when the training data set is low (10%). From MAE results in Fig. 3c and d it is clear that all methods produced inferior results compared to FCQM-SGP when the training dataset is very small (10 %). This indicates that FCQM-SGP approach is able to produce better results with the lowest number of training data points. This is very important due to the fact that labelling e-nose data is a very expensive task.

We can observe from Fig. 3c, d and f the stability of our method compared to other baseline methods. It is clear that the outliers percentage produced by the FCQM-SGP method is significantly less than the alternative methods. This is due to the fact that FCQM-SGP utilizes confidence evaluation, batch query processing, and co-clustering to achieve scalability and efficiency, as well as managing noise more efficiently than other models. Further more, drift and noise are reduced because an ensemble of fuzzy classifiers are trained on different parts of the data.

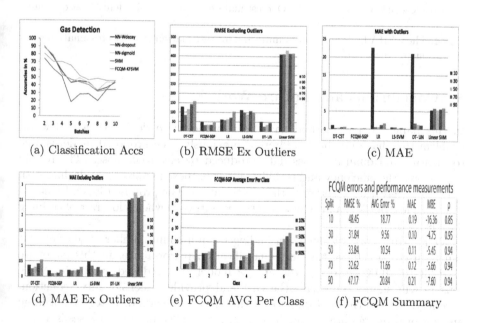

(a) Classification Accs   (b) RMSE Ex Outliers   (c) MAE

(d) MAE Ex Outliers   (e) FCQM AVG Per Class   (f) FCQM Summary

Fig. 3. The proposed model performance indicators using sixteen MOX sensors.

Air Quality Monitoring and Exposure Limits. One of the main objectives of this work is to be able to predict outdoor air quality level with high confidence. It is critical to say that we are not attempting to predict the gas concentrations accurately, rather we use the predicted gas concentrations to identify air quality level. Therefore, low concentrations are irrelevant in this sense. U.S Department of Labour, Occupational Safety & Health Administration (OSHA) has provided exposure limits for different substances [21]. Figure 4a shows the exposure limits for different gases used in our experiments.

To identify the model sensitivity to exposure limit we estimate concentration underestimation and overestimation percentage for each model. Figure 4b shows the overestimation percentage for each model. From the figure clearly FCQM-SGP model tend to have low rate of overestimation while other models tend to highly overestimate concentrations. The underestimation graph in Fig. 4c shows than DT-CST and DT-LIN has the lowest underestimation percentage when the training set is 90 % and testing is 10 %. Having said, that FCQM maintain best results when training proportions are 30 % and 50 %.

OSHA PEL permissible limits		
Gas	Class	Exposure Limits
Acetone	5	1000 PPM
Ethanol	1	1000 PPM
Toluene	6	300 PPM
Acetaldehyde	4	200 PPM
Ethylene	2	200 PPM
Ammonia	3	50 PPM

(a) Exposure Limits   (b) Overestimation Rate   (c) Underestimation Rate

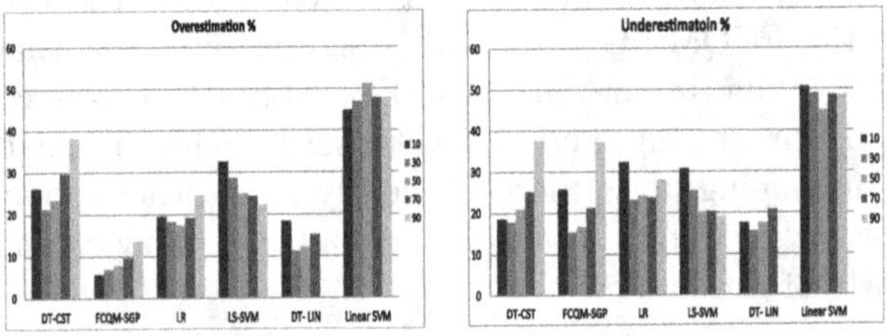

Fig. 4. The OSHA PEL Exposure Limits and overestimation and underestimation rates for various models.

5 Conclusions and Future Work

In this paper, a new method have been proposed to deal with a real world problem that have not been addressed previously. The proposed model is able to identify and quantify gases data produced by electronic noses. The results have shown the efficiency and stability of our proposed approach in dealing with e-nose gas detection together with quantification. The proposed FCQM method produced less outliers and achieved better results even when a low rate of training data is used.

Our future work will focus on finding the minimum number of sensors that can be used to achieve the best quantification performance. In this paper we used all 16 MOX sensors in the FCQM model. The reduced number of MOX sensors will significantly save the cost and provide dimensionality reduction. Secondly, despite the performance of the proposed method, it requires a large amount of training data. Manually generating labelled data for e-noses is a very expensive task. We need to work on a huge amount of labelled data to continuously update FCQM. Updating FCQM continuously with new training data using semi-supervised techniques can reduce the drift effect. Finally, we will apply the proposed model to different real world applications which have similar characteristics to air quality monitoring.

6   Reproducibility

An online implementation of our quantification model is available at: http://www.crct.polymtl.ca/SGP/run_gp.php.

References

1. Al-Maskari, S., Li, X., Liu, Q.: An effective approach to handling noise and drift in electronic noses. In: Proceedings of the 25th Australasian Database Conference on Databases Theory and Applications, ADC 2014, Brisbane, QLD, Australia, 14–16 July, pp. 223–230 (2014)
2. Vergara, A., Vembu, S., Ayhan, T., Ryan, M.A., Homer, M.L., Huerta, R.: Chemical gas sensor drift compensation using classifier ensembles. Sens. Actuators B: Chem. **166–167**, 320–329 (2012)
3. Al-Maskari, S., Saini, D., Omar, W.: Cyber infrastructure and data quality for environmental pollution control - in Oman. In: Proceedings of the 2010 DAMD International Conference on Data Analysis, Data Quality and Metada, p. 71 (2010)
4. Nimsuk, N., Nakamoto, T.: Study on the odor classification in dynamical concentration robust against humidity and temperature changes. Sens. Actuators B: Chem. **134**(1), 252–257 (2008)
5. Goodner, K.L., Dreher, J., Rouseff, R.L.: The dangers of creating false classifications due to noise in electronic nose and similar multivariate analyses. Sens. Actuators B: Chem. **80**(3), 261–266 (2001)
6. Marco, S., Pardo, A., Ortega, A., Samitier, J.: Gas identification with tin oxide sensor array and self organizing maps: adaptive correction of sensor drifts. In: Instrumentation and Measurement Technology Conference, IMTC 1997, Proceedings of the Sensing, Processing, Networking, vol. 2, pp. 904–907. IEEE, May 1997
7. Zuppa, M., Distante, C., Siciliano, P., Persaud, K.C.: Drift counteraction with multiple self-organising maps for an electronic nose. Sens. Actuators B: Chem. **98**(23), 305–317 (2004)
8. Bishop, C.M.: Neural Networks for Pattern Recognition. Oxford University Press Inc., New York (1995)
9. Gulbag, A., Temurtas, F.: A study on quantitative classification of binary gas mixture using neural networks and adaptive neuro-fuzzy inference systems. Sens. Actuators B: Chem. **115**(1), 252–262 (2006)
10. Gemperline, P.J., Long, J.R., Gregoriou, V.G.: Nonlinear multivariate calibration using principal components regression and artificial neural networks. Anal. Chem. **63**(20), 2313–2323 (1991)
11. Domanský, K., Baldwin, D.L., Grate, J.W., Hall, T.B., Li, J., Josowicz, M., Janata, J.: Development and calibration of field-effect transistor-based sensor array for measurement of hydrogen and ammonia gas mixtures in humid air. Anal. Chem. **70**(3), 473–481 (1998)

12. Geng, Z., Yang, F., Chen, X., Wu, N.: Gaussian process based modeling and experimental design for sensor calibration in drifting environments. Sens. Actuators B: Chem. **216**, 321–331 (2015)
13. Wilson, D.M., Dunman, K., Roppel, T., Kalim, R.: Rank extraction in tin-oxide sensor arrays. Sens. Actuators B: Chem. **62**(3), 199–210 (2000)
14. Llobet, E., Brezmes, J., Ionescu, R., Vilanova, X., Al-Khalifa, S., Gardner, J., Barsan, N., Correig, X.: Wavelet transform and fuzzy artmap-based pattern recognition for fast gas identification using a micro-hotplate gas sensor. Sens. Actuators B: Chem. **83**(13), 238–244 (2002). Selected Papers from TRANSDUCERS 2001 EUROSENSORS XV
15. Nakata, S., Nakamura, H., Yoshikawa, K.: New strategy for the development of a gas sensor based on the dynamic characteristics: principle and preliminary experiment. Sens. Actuators B: Chem. **8**(2), 187–189 (1992)
16. Zhang, D.Q., Chen, S.C.: A novel kernelized fuzzy c-means algorithm with application in medical image segmentation. Artif. Intell. Med. **32**(1), 37–50 (2004). Artificial Intelligence in Medicine in China
17. Nahar, V., Al-Maskari, S., Li, X., Pang, C.: Semi-supervised learning for cyberbullying detection in social networks. In: Wang, H., Sharaf, M.A. (eds.) ADC 2014. LNCS, vol. 8506, pp. 160–171. Springer, Heidelberg (2014)
18. Bélisle, E., Huang, Z., Gheribi, A.: Scalable gaussian process regression for prediction of material properties. In: Wang, H., Sharaf, M.A. (eds.) ADC 2014. LNCS, vol. 8506, pp. 38–49. Springer, Heidelberg (2014)
19. Gibbs, M.N., MacKay, D.J.C.: Efficient implementation of gaussian processes. Submitted to Statistics and Computing
20. Rodriguez-Lujan, I., Fonollosa, J., Vergara, A., Homer, M., Huerta, R.: On the calibration of sensor arrays for pattern recognition using the minimal number of experiments. Chemometr. Intell. Lab. Syst. **130**, 123–134 (2014)
21. U.S. Department of Labor, O.S.H.A.: occupational exposure limits osha annotated table (2015)

Predicting Unknown Interactions Between Known Drugs and Targets via Matrix Completion

Qing Liao[1]([✉]), Naiyang Guan[2], Chengkun Wu[2], and Qian Zhang[1]

[1] Department of Computer Science and Engineering,
Hong Kong University of Science and Technology, Hong Kong, Hong Kong
{qnature,qianzh}@cse.ust.hk
[2] College of Computer, National University of Defense Technology, Changsha, China
{chengkun_wu,ny_guan}@nudt.edu.cn

Abstract. Drug-target interactions map patterns, associations and relationships between drugs and target proteins. Identifying interactions between drug and target is critical in drug discovery, but biochemically validating these interactions are both laborious and expensive. In this paper, we propose a novel interaction profiles based method to predict potential drug-target interactions by using matrix completion. Our method first arranges the drug-target interactions in a matrix, whose entries include interaction pairs, non-interaction pairs and undetermined pairs, and finds its approximation matrix which contains the predicted values at undetermined positions. Then our method learns an approximation matrix by minimizing the distance between the drug-target interaction matrix and its approximation subject that the values in the observed positions equal to the known interactions at the corresponding positions. As a consequence, our method can directly predict new potential interactions according to the high values at the undetermined positions. We evaluated our method by comparing against five counterpart methods on "gold standard" datasets. Our method outperforms the counterparts, and achieves high AUC and F_1-score on enzyme, ion channel, GPCR, nuclear receptor and integrated datasets, respectively. We showed the intelligibility of our method by validating some predicted interactions in both DrugBank and KEGG databases.

Keywords: Drug-target interaction · Matrix completion · Drug discovery

1 Introduction

Associations between drugs and targets are essential for understanding the pharmacology of drugs and for repositioning known drug [1–4]. Capturing associations between drugs and targets using traditional biochemical experiments is a laborious and time-consuming procedure that is also very expensive [5–7]. One alternative is to compute potential associations between drugs and targets via

© Springer International Publishing Switzerland 2016
J. Bailey et al. (Eds.): PAKDD 2016, Part I, LNAI 9651, pp. 591–604, 2016.
DOI: 10.1007/978-3-319-31753-3_47

in-silico way [8,9]. Molecular docking [10–15], literature mining [16] and ligand-bases [17–19] are three common computational approaches. Docking requires information about the 3D structure of a target/protein to calculate how well each drug candidate can bind with the target, but this type of information is missing for many targets, like GPCR and ion channel [20,21]. Moreover, docking is computationally expensive, which makes it difficult to process large-scale datasets. Text mining approaches is heavily relied on domain dictionaries to deal with semantic ambiguity like aliases and synonyms in the literature [16]. Ligand-based approaches such as QSAR (Quantitative Structure Activity Relationship) compare a candidate ligand with the known ligands of a target protein to predict its bindings [17,18], and the performance of ligand-bases approach decreases when the number of known ligands is limited [19].

Recently, machine learning methods has been shown to be effective in finding the drug-target interactions based on chemical properties of drug compounds, genomic properties of targets, and interaction profiles [22–27]. Those approaches share the identical assumption that similar drugs tend to interact closely with similar target proteins [28,29]. Existing studies utilized drug/target similarity information and known interactions to capture potentially novel interactions between drugs and targets. For instance, Jacob and Vert proposed a SVM-based method [30–32] called pairwise kernel method (PKM) to generate similarity kernels over drug-target pairs [27]. Yamanishi [22] proposed the kernel regression-based method (KRM) to infer unknown drug-target interaction in a unified space called "pharmacological space". Yamanishi [23] proposed the bipartite graph inference (BLM) method, which builds a bipartite local model to predict links between drugs and targets in a bipartite graph. Gonen [24] proposed Kernelized Bayesian matrix factorization (KBMF2K) to predict interactions by projecting drug compounds and target proteins onto a unified subspace via joint Bayesian formulation. Laarhoven [25] proposed a Gaussian interaction profile (GIP) method to predict drug-target interactions by generating a Gaussian kernel from interaction profiles and similarity information among drugs and among targets. Xia [26] proposed a NetLapRLS method by incorporating Laplacian regularized least square (LapRLS) and a new kernel established from the known interaction network in a unified framework. Moreover, Wang and Zeng [33] built a restricted boltzmann machine (RBM) method to predict drug-target pairs and to describe types of predicted pairs.

Some methods utilize similarity information and partial drug-target interaction profiles to predict potential interactions between drugs and targets. However, similarity information might not be available in some cases. For example, it is extremely difficult to collect complete similarity information in large scale database like STITCH [34] and the 3D shape similarity of many proteins/targets, especially GPCRs are unavailable [21,35]. On the other side, drug similarity can be calculated based on different types of biological knowledge such as chemical structure (CS), and anatomical therapeutic chemical classification system (ATC) [32], and target similarity can also be calculated from genomic sequence (GS) [22,36] and gene ontology (GO) [37,38]. It is difficult to decide which types of similarity is the most appropriate one, as each measure has its private biochemical properties. Zheng [39] showed that the effect of the same type of similarity might vary dramatically on different datasets. For example, the GS similarity over target is very critical for predicting interactions on GPCR dataset, while it is almost useless on the nuclear receptor, ion channel and enzyme datasets. Therefore, integrating different similarity metrics is still a challenging problem.

Drug-target interaction prediction problem can infer interactions with less information. For example, Cheng [40] proposed a network-based inference (NBI) method to predict interactions by using the interaction profiles only. Moreover, Cobanoglu and Bahar *etc.* [41] assumed that all the samples obey the Gaussianly distributed probability to present probabilistic matrix factorization (PMF) to predict pairs by only using the interaction profiles. Experiments demonstrate that the prediction of NBI is more reliable than drug-based similarity inference (DBSI) and target-based similarity inference (TBSI), because choosing an improper similarity data may introduce extra noise to the model building process due to inaccurate selection of similar pairs. Although above mentioned methods exhibit a satisfactory AUC value (area under ROC curve) performance, its precision and recall are still unsatisfactory.

In this paper, we proposed a novel drug-target interaction prediction method which uses the matrix completion for prediction based on only interaction information. Our method assumes that similar drugs often interact with similar proteins and converts the interaction prediction problem into a collaborative filtering problem, which infers missing entries in the interaction matrix by using known interactions. We evaluated our method by comparing with the existing methods and ten-fold cross-validation on a "gold standard" datasets including enzyme, ion channels, G-protein-coupled receptors (GPCRs), nuclear receptor and integrated datasets. Experimental results show that our method achieves high performance in both AUC and F_1-score, and validation of predicted pairs in the latest DrugBank and KEGG databases shows that our method is intelligible.

2   Methods

2.1   Drug-Target Interaction Databases

In the drug-target interaction prediction literature, four datasets include enzyme, ion channels, GPCRs, and nuclear receptor are usually regarded as the "gold standard" dataset [22–27,39,40]. In this study, we also combined them together to generate an integrated dataset for further verification. Yamanishi [22] proposed a widely used benchmark for drug-target interactions which includes four subsets for different types of targets: enzyme, ion channels, GPCRs and nuclear receptor. These datasets were collected from curated databases including KEGG BRITE [42], BERENDA [43], SuperTarget [44] and DrugBank [45], respectively. The numbers of drugs, targets and drug-target interactions are listed in Table 1. We also generated an integrated dataset that combines all four subsets. The integrated data of these four datasets contains 5127 interactions between 989 target proteins and 791 drugs.

Table 1. Summary of the drug-target interaction datasets.

Datasets	# of drugs	# of targets	# of drug-target interactions
Enzyme	445	664	2926
Ion channel	210	204	1476
GPCR	223	95	635
Nuclear receptor	54	26	90
Integrated	791	989	5127

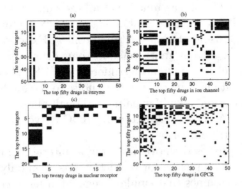

Fig. 1. Visualization of the low-rank pattern of drug-target interactions. The pattern of sorted interactions in four datasets. (a) Sorted interaction between first fifty drugs/target proteins in enzyme dataset. (b) Sorted interactions between first fifty drug/target proteins in ion channel dataset. (c) Sorted interactions between first twentieth drugs/target proteins in nuclear receptor dataset. (d) Sorted interaction between first fifty drug/target proteins in GPCR dataset.

2.2 Motivation by Data Visualization

To predict unknown interactions from the drug-target dataset, we first analyze the "gold standard" datasets. Due to drug-target interaction matrix is too sparse to difficult find some observations, we sorted the drug and target protein in descending order by their number of interactions to make figure. Figure 1 is an example to show the interaction matrix of the first fifty or twenty drugs/target proteins based on sorted interactions in four datasets. From Fig. 1(a), we can observe that many drugs have similar interactions, such as drugs #16 to #23 [KEGG:D00538, D00224, D00377, D00391, D00401, D01069, D00136, D00569], and drugs #28 to #34 [KEGG:D03778, D03781, D00947, D00963, D01397, D02441, D03776] have very similar interactions. The same observation can also be found from the target protein side in Fig. 1(a). For instance, if a target protein has interactions from drugs #28 to #34, it may also have interaction with drug #2 [KEGG:D00521], because there exists a strong relationship between drug #28 to #34 and drug #2. From the above observation, it is evident that correlation does exist between node instances in the enzyme dataset.

For instance, if a target has interactions with drug #28 to #34, then it has a high probability to interact with #2. The same phenomenon can also be observed on the other subsets. We also analyzed the most important part of the ion channel, nuclear receptor and GPCR subset and found the same observations in Fig. 1(b), (c), and (d), respectively. In Fig. 1(c), we only select the first twenty drug/target proteins from the nuclear receptor subset, as the interaction matrix size is a 56×26 matrix. From Fig. 1(c) and (d), the observation is less evident comparing with Fig. 1(a) and (b), but the same observation can still be found from target proteins #9 to #13 [KEGG:hsa3174, hsa367, hsa4306, hsa5241, hsa5465] in Fig. 1(c) and from drugs #10 to #13 [KEGG:D00136, D00139, D00180, D00225], and from target proteins #6 to #9 [KEGG:hsa1129, hsa1131, hsa1132, hsa1133] in Fig. 1(d).

According to these observations, all four subsets have some latent factors that contribute to the prediction of interactions. In other words, from the viewpoint of drugs, all drugs interact with the target proteins in a few patterns, and thus we can leverage the known interaction information to predict unknown interactions by predicting missing values of the interaction matrix. It is also true from the viewpoint of target proteins. It therefore motivates us to apply the matrix completion technique to the drug-target interaction prediction problem.

2.3   The Drug-Target Interaction Prediction Method

The original interaction dataset needs to be pre-processed in order to apply the matrix completion. In the benchmark dataset, the value 1 denotes a confirmed/annotated interaction (positive sample), while all other unknown drug-target pairs in the training data are assumed to be non-interacted (negative sample), which is denoted by the value 0. However, our method regards 0 valued entries as the ones to be predicted. Therefore, directly performing matrix completion on original dataset may lead confusion. So we fill non-interaction entries by one value a and interaction entries by another value b, where $a \neq b$ and $\{a, b\} \neq 0$. Our method first fills missing entries with value 0 in the original matrix to be recovered by using the matrix completion, and then iteratively updates the missing entries with the predicted value.

Given a drug-target interaction matrix $M \in R^{N_d \times N_t}$ involving N_d drugs and N_t targets. The set $X_d = \{d_1, d_2, ..., d_{N_d}\}$ is the drug set and the set $X_t = \{t_1, t_2, ..., t_{N_t}\}$ is the target protein set. Let $M_{(i,j)} : (i, j) \in \Omega$ denote the set of the known samples, and ω the index set of the rest known samples. We formulate the interaction prediction problem as below:

$$\min_X f_\tau(X)$$
$$s.t., P_\Omega(X) = P_\Omega(M) \tag{1}$$

where $f_\tau(X)$ is a nonlinear function of candidate solution matrix X. P_Ω is a orthogonal projector, and $P_\Omega(X)$ is equal to $X(i, j)$ if $(i, j) \in \Omega$, and $P_\Omega(X)$ is equal to zero otherwise. In matrix completion, the predicted matrix X is usually expected to be low-rank. We therefore rewrite (1) as the following problem to minimize the rank of X because nuclear norm of X is a convex surrogate of its rank:

$$\min_X \tau\|X\|_* + \tfrac{1}{2}\|X\|_F^2$$
$$s.t., P_\Omega(X) = P_\Omega(M) \tag{2}$$

where $\|X\|_*$ signifies the nuclear norm of X which is actually the sum of singular value of matrix X, and $\|X\|_F$ is the Frobernius norm of matrix X, and $\tau \geq 0$ is a thresholding which will be used in soft-thresholding operator.

According to [46], the problem (2) can be optimized by using the Lagrangian multipler method. Specially, we introduce a Lagrangian multiplier Y and get the Lagrangian function of (2) as below:

$$L(X,Y) = f_\tau(X) + <Y, P_\Omega(M) - P_\Omega(X)> \tag{3}$$

Applying the Uzawas algorithm [47] to find a saddle point of (3) until convergence.

The Uzawas algorithm first updates X with Y fixed as

$$X^k = D_\tau(Y^{k-1}) \tag{4}$$

followed by updating Y with X fixed as

$$Y^k = Y^{k-1} + \delta_k P_\Omega(M - X^K) \tag{5}$$

where $\{\delta_k\}_{k\geq 1}$ is a sequence of step size, and the soft-thresholding operator D_τ is defined as follows:

$$D_\tau(X) := UD_\tau(\Sigma)V$$
$$D_\tau(\Sigma) = diag(\{\delta_i - \tau\}_+), \tag{6}$$

Singular value decomposition (SVD) of a matrix X can obtain a sequence of positive singular values σ_i. And $diag(\{\sigma_i - \tau\}_+)$ is the positive part of $\sigma_i - \tau$, and $\sigma_i - \tau$ is equal to zero, if $\sigma_i - \tau < 0$. The physical meaning of soft-thresholding operator can be understood that it filters the data and only leaves the most important part of the dataset. Therefore, noise or redundancy information can be eliminated via the soft-thresholding operator. According to [42], the iterative method can converge to an unique solution when $0 < \delta < 2$.

The matrix completion method iteratively updates (4) and (5) until the stopping criteria is met. In this study, we choose the well-known K.K.T. conditions [48] as the stopping criteria:

$$P_\Omega(M - X^k)_F \leq \varepsilon P_\Omega(M)_F \tag{7}$$

where ϵ is the predefined tolerance, e.g., 10^{-4}.

2.4   Performance Metrics

We choose two metrics including AUC (Area under the Receiver Operating Characteristic Curve) and F_1-score to evaluate the performance of our method.

The ROC curve plots the values of TPR (true positive rate) versus FPR (false positive rate), and it is one of the well-known metrics to evaluate the performance of existing interaction prediction method in the current study. TPR and FPR are defined as:

$$TPR = \frac{TP}{TP + FN} \tag{8}$$

$$FPR = \frac{FP}{FP + TN} \tag{9}$$

where TP, FP, TN, and FN denote true positive, false positive, true negative, false negative, respectively. However, AUC is insufficient to evaluate the performance of methods on bio-dataset because most bio-datasets have a highly imbalanced class distribution between positive samples and negative samples [49]. For example, the enzyme dataset contains less than 1% interaction entries (positive samples) in the whole dataset and the remaining 99% elements are non-interaction entries (negative samples). A naive prediction method that randomly predicts all elements as non-interaction entries can achieve a small false positive rate because most elements in original dataset are non-interaction entries. On the other hand, the true positive rate can be high even though we only find one correct interaction entry, because the number of real interaction is small. That is why existing prediction methods can easily achieve a decent of AUC about 80%–99%. Practically speaking, the ability of predicting as many as potentially correct interactions can maximize the probability of validating the unverified interaction pairs via biochemical experiments. Since AUC is insufficient to evaluate performance of model, F_1-score is used as a standard information retrieval metric, to evaluate the performance of our method and its counterparts. The F_1-score is defined based on two critical metrics including Precision and Recall, i.e.

$$Precision = \frac{TP}{TP + FP} \tag{10}$$

$$Recall = \frac{TP}{TP + FN} \tag{11}$$

$$F_1 = 2 \cdot \frac{precision \cdot recall}{precision + recall} \tag{12}$$

The difference between the F_1-score and AUC is the precision term and the utilization of FPR. FPR mainly focus on the non-interaction prediction performance while the F_1-score mainly focus on the overall interaction prediction performance. In general, precision represents the interaction prediction success rate.

3   Result

This section evaluates the effectiveness of our method in terms of both AUC and F_1-score with 10-fold cross validation CV on "gold standard datasets". In this experiment, we empirically set the step size $= 1.5$ and the stopping tolerance $= 10^{-4}$. We compared our method to five representative methods including

Table 2. AUC value of 10-fold cross validation of six methods sample.

AUC	Enzyme	Ion channel	GPCR	Nuclear receptor	Integrated data
Our method	**0.9708**	**0.9778**	0.9123	0.6640	**0.9659**
NBI	0.8941	0.9284	0.8357	0.6653	0.9087
PMF	0.9109	0.9575	**0.9311**	0.8245	0.8314
GIP	0.9516	0.9761	0.9272	**0.8609**	NA
KBMF2K	0.8475	0.9111	0.8741	0.8490	NA

Table 3. F_1-score of 10-fold cross validation of six methods.

F_1 score	Enzyme	Ion channel	GPCR	Nuclear receptor	Integrated data
Our method	**0.8437**	**0.8975**	**0.7083**	0.5204	**0.8281**
NBI	0.8325	0.8233	0.6663	0.4960	0.7925
PMF	0.6556	0.8351	0.6940	0.5295	0.5536
GIP	0.7150	0.8273	0.6730	**0.6021**	NA
KBMF2K	0.6889	0.6764	0.5580	0.5381	NA

NBI [40], GIP [25], KBMF2K [24], PMF [41] and NetLapRLS [26] according to Hao's review [49]. For GIP, KBMF2K, and NetLapRLS, we used the source codes provided by the authors; for NBI and PMF, we implemented the algorithm described in their paper. Note that GIP, KBMF2K and NetLapRLS need to exploit both similarity information and interaction profiles as input to predict interactions. As the performance of NetLapRLS was verified in a slightly different way from the remaining methods, we first compared our method to NBI, GIP, KBMF2K, and then compared it against NetLapRLS separately in the next subsection.

3.1 Performance Comparison of NBI, GIP, KBMF2K, PMF and Our Method on Gold Standard Datasets

One reason to use F_1-score as the performance metric is that existing interaction datasets tend to exhibit an imbalanced distribution of positive and negative samples, as illustrated in Table 1. A large number of non-interaction entries make AUC insufficient for measuring the performance. On the contrary, the F_1-score penalizes false positives much more than ROC [49,50], and thus it can characterize the performance better in interaction prediction. Tables 2 and 3 list the AUC and F_1-scores of NBI, PMF, GIP, KBMF2K and our methods on both "gold standard" and integrated datasets, respectively. Table 2 shows that our method achieves the highest AUC value (around 0.97) on the enzyme, ion channel and the integrated datasets; and its AUC value is also good on the GPCR dataset (above 0.9). Table 3 demonstrates that our method has the highest F_1-scores on all test sets except for the nuclear receptor subsets.

In summary, our method can achieve the best performance on the Enzyme, Ion channel and integrate databases while it requires much less information comparing with the similarity-bases methods. Moreover our method has the higher AUC and F_1 value on most datasets comparing with the NBI and PMF methods.

3.2 Comparison with the NetLapRLs Method

In this experiment, we compared the performance of our method against that of the NetLapRLS method. The main difference between NetLapRLS method and other methods is that the NetLapRLS method only utilizes known inter-actions entries (positive instances) for prediction, while the remaining methods treat unknown interaction as non-interactions (negative instances) in the train-ing data. We investigate the effects of introducing negative instances into the training data on the performance of our method and NetLapRLS methods. To do this, we performed a series of performance test by randomly adding 0 %, 10 %, ..., 90 % negative samples into the training dataset. In each test, we per-formed 10-fold cross validation.

When the percentage of negative samples equals to 0 %, our method only uses positive samples to predict all unknown interaction entries like NetLapRLs.

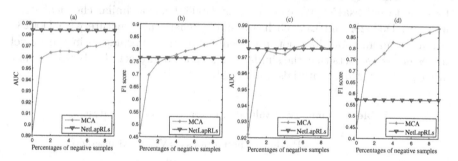

Fig. 2. AUC and F_1 of our method versus NetLapRLs on Enzyme (a), (b) and on Ion channel datasets (c), (d), respectively.

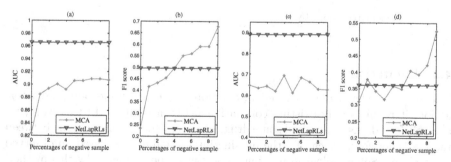

Fig. 3. AUC and F_1 of our method versus NetLapRLs on GPCRs (a), (b) and on Nuclear receptor datasets (c), (d), respectively.

We can see from Figs. 2 and 3 that when the percentage of negative samples increases, both AUC and F_1-score of our method can be gradually improved (especially for F_1-scores). More importantly, we can clearly see that even if the AUC values are stable, the F_1-score significantly improves. This also reflects the importance of including F_1-score for performance evaluation.

4   Discussion

4.1   Validated New Pairs in the Latest Databases

In order to illustrate the capability of our method in the real case, we developed a small tool [51] by Python that can automatically validate predicted links using the knowledge from DrugBank and KEGG. We decide the value of threshold when the F_1 score is the highest. If the prediction value is larger than the threshold, we regard it as candidate interaction, otherwise, non-interaction. Table 4 summaries the validation links in four datasets. According to Table 4, the percentages of new validated interactions are 16.90 %, 17.24 %, 40.78 % and 22.22 % for enzyme, ion channel, GPCR and nuclear subsets datasets, respectively. Figure 4 shows two instances of new validated drug-target interactions we truly find them in the latest database. Both drugs D00691 and D00528 interact with the same target hsa5150 and both drugs D00563 and D00283 interact with the same target hsa152. We can find that D00691 has a similar chemical structure with D00528, and D00563 has a similar chemical structure with D00283. Although our method does not apply any similarity information in the model, the result also reflects that the similar drugs tend to interact closely with similar target proteins. More materials are available at [51].

Table 4. Summary of validated interactions of four datasets.

	#Predicted interactions	#Validated interactions
Enzyme	71	12
Ion channel	58	10
GPCR	76	31
Nuclear receptor	18	4

4.2   Limitation

Due to our method only utilized interaction profiles to mine potential interaction, it requires only existing correlation between samples. In this work, we analysis four datasets and find some drugs have very similar interactions between targets. Therefore, we can fill the missing interaction entries based on observed interactions. Some biological missing value cannot be predicted by this method such as IC_{50}, EC_{50}, K_i and K_d value [52] in structure activity relationship (SAR) dataset because it does not have clear correlations between samples.

	Drug	Drug Chemical Structure	Target	Drug	Drug Chemical Structure	Target
1	D00691	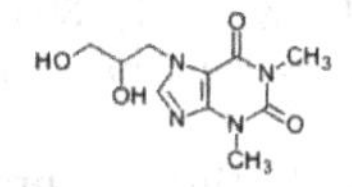	hsa5150	2	D00563	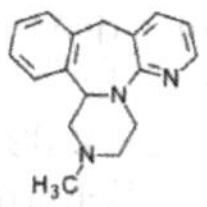 hsa152
	D00528	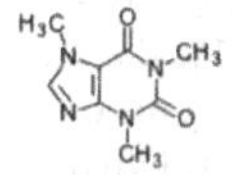			D00283	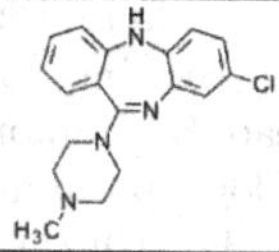

Fig. 4. Examples of drug chemical structures of the same target.

5   Conclusions

Our method is the first work to predict drug-target interactions by using the matrix completion technique [2] based on the observation that most drug-target interaction matrices are low-rank. Our method first fills missing entries with 0 in the original matrix to be recovered, and it iteratively updates the missing entries with predictive value. Moreover, in extreme case, a strong drug-drug and target-target correlation makes the interaction matrix to a low-rank one. Therefore, our method tends to adopt the lowest-rank approximate matrix as the correct solution during the iterative process.

We chose both AUC and F_1-score to evaluate the prediction performance. Five representative models: NBI, PMF, GIP, KBMF2K, NetLapRLS are used for comparative study. Among them, GIP, KBMF2K and NetLapRLS apply interaction profiles as well as similarity information of drugs/target proteins. NBI,PMF and our method predict the interaction pairs only based on interaction information. Our method outperforms other methods in terms of both AUC and F_1-score. As there is no standard similarity strategy of bio-data, our method only applies interaction profiles to predict drug-target interactions. For future work, we will extend our method by introducing similarity learning on bio datasets.

Acknowledgments. This work was supported by The National Natural Science Foundation of China (under grant No. U1435222 and No. 61502515). And this work was also supported in part by grants from 973 project 2013CB329006, RGC under the contract CERG 16212714.

References

1. Hopkins, A.L.: Drug discovery: predicting promiscuity. Nature **462**(7270), 167–168 (2009)
2. Cai, J.-F., Candès, E.J., Shen, Z.: A singular value thresholding algorithm for matrix completion. SIAM J. Optim. **20**(4), 1956–1982 (2010)

3. Ashburn, T.T., Thor, K.B.: Drug repositioning: identifying and developing new uses for existing drugs. Nat. Rev. Drug Discov. **3**(8), 673–683 (2004)
4. Dudley, J.T., Deshpande, T., Butte, A.J.: Exploiting drug-disease relationships for computational drug repositioning. Briefings Bioinform. **12**(4), 303–311 (2011)
5. Swamidass, S.J.: Mining small-molecule screens to repurpose drugs. Briefings Bioinform. **12**(4), 327–335 (2011)
6. Moriaud, F., Richard, S.B., Adcock, S.A., Chanas-Martin, L., Surgand, J.-S., Jelloul, M.B., Delfaud, F.: Identify drug repurposing candidates by mining the protein data bank. Briefings Bioinform. **12**(4), 336–340 (2011)
7. Whitebread, S., Hamon, J., Bojanic, D., Urban, L.: Keynote review: in vitro safety pharmacology profiling: an essential tool for successful drug development. Drug Discov. Today **10**(21), 1421–1433 (2005)
8. Haggarty, S.J., Koeller, K.M., Wong, J.C., Butcher, R.A., Schreiber, S.L.: Multidimensional chemical genetic analysis of diversity-oriented synthesis-derived deacetylase inhibitors using cell-based assays. Chem. Biol. **10**(5), 383–396 (2003)
9. Kuruvilla, F.G., Shamji, A.F., Sternson, S.M., Hergenrother, P.J., Schreiber, S.L.: Dissecting glucose signalling with diversity-oriented synthesis and small-molecule microarrays. Nature **416**(6881), 653–657 (2002)
10. Manly, C.J., Louise-May, S., Hammer, J.D.: The impact of informatics and computational chemistry on synthesis and screening. Drug Discov. Today **6**(21), 1101–1110 (2001)
11. Cheng, A.C., Coleman, R.G., Smyth, K.T., Cao, Q., Soulard, P., Caffrey, D.R., Salzberg, A.C., Huang, E.S.: Structure-based maximal affinity model predicts small-molecule druggability. Nat. Biotechnol. **25**(1), 71–75 (2007)
12. Rarey, M., Kramer, B., Lengauer, T., Klebe, G.: A fast flexible docking method using an incremental construction algorithm. J. Mol. Biol. **261**(3), 470–489 (1996)
13. Shoichet, B.K., Kuntz, I.D., Bodian, D.L.: Molecular docking using shape descriptors. J. Comput. Chem. **13**(3), 380–397 (1992)
14. Halperin, I., Ma, B., Wolfson, H., Nussinov, R.: Principles of docking: an overview of search algorithms and a guide to scoring functions. Proteins Struct. Funct. Bioinform. **47**(4), 409–443 (2002)
15. Shoichet, B.K., McGovern, S.L., Wei, B., Irwin, J.J.: Lead discovery using molecular docking. Curr. Opin. Chem. Biol. **6**(4), 439–446 (2002)
16. Kolb, P., Ferreira, R.S., Irwin, J.J., Shoichet, B.K.: Docking and chemoinformatic screens for new ligands and targets. Curr. Opin. Biotechnol. **20**(4), 429–436 (2009)
17. Zhu, S., Okuno, Y., Tsujimoto, G., Mamitsuka, H.: A probabilistic model for mining implicit chemical compound–generelations from literature. Bioinformatics **21**(Suppl. 2), ii245–ii251 (2005)
18. Butina, D., Segall, M.D., Frankcombe, K.: Predicting adme properties in silico: methods and models. Drug Discov. Today **7**(11), S83–S88 (2002)
19. Byvatov, E., Fechner, U., Sadowski, J., Schneider, G.: Comparison of support vector machine and artificial neural network systems for drug/nondrug classification. J. Chem. Inf. Comput. Sci. **43**(6), 1882–1889 (2003)
20. Keiser, M.J., Roth, B.L., Armbruster, B.N., Ernsberger, P., Irwin, J.J., Shoichet, B.K.: Relating protein pharmacology by ligand chemistry. Nat. Biotechnol. **25**(2), 197–206 (2007)
21. Klabunde, T., Hessler, G.: Drug design strategies for targeting g-protein-coupled receptors. Chembiochem **3**(10), 928–944 (2002)
22. Yamanishi, Y., Araki, M., Gutteridge, A., Honda, W., Kanehisa, M.: Prediction of drug-target interaction networks from the integration of chemical and genomic spaces. Bioinformatics **24**(13), i232–i240 (2008)

23. Bleakley, K., Yamanishi, Y.: Supervised prediction of drug-target interactions using bipartite local models. Bioinformatics 25(18), 2397–2403 (2009)
24. Gönen, M.: Predicting drug-target interactions from chemical and genomic kernels using bayesian matrix factorization. Bioinformatics 28(18), 2304–2310 (2012)
25. van Laarhoven, T., Nabuurs, S.B., Marchiori, E.: Gaussian interaction profile kernels for predicting drug-target interaction. Bioinformatics 27(21), 3036–3043 (2011)
26. Xia, Z., Wu, L.-Y., Zhou, X., Wong, S.T.: Semi-supervised drug-protein interaction prediction from heterogeneous biological spaces. BMC Syst. Biol. 4(Suppl. 2), S6 (2010)
27. Jacob, L., Vert, J.-P.: Protein-ligand interaction prediction: an improved chemogenomics approach. Bioinformatics 24(19), 2149–2156 (2008)
28. Klabunde, T.: Chemogenomic approaches to drug discovery: similar receptors bind similar ligands. Br. J. Pharmacol. 152(1), 5–7 (2007)
29. Schuffenhauer, A., Floersheim, P., Acklin, P., Jacoby, E.: Similarity metrics for ligands reflecting the similarity of the target proteins. J. Chem. Inf. Comput. Sci. 43(2), 391–405 (2003)
30. Nagamine, N., Sakakibara, Y.: Statistical prediction of protein-chemical interactions based on chemical structure and mass spectrometry data. Bioinformatics 23(15), 2004–2012 (2007)
31. Nagamine, N., Shirakawa, T., Minato, Y., Torii, K., Kobayashi, H., Imoto, M., Sakakibara, Y.: Integrating statistical predictions and experimental verifications for enhancing protein-chemical interaction predictions in virtual screening. PLoS Comput. Biol. 5(6), e1 000 397–e1 000 397 (2009)
32. Yabuuchi, H., Niijima, S., Takematsu, H., Ida, T., Hirokawa, T., Hara, T., Ogawa, T., Minowa, Y., Tsujimoto, G., Okuno, Y.: Analysis of multiple compound-protein interactions reveals novel bioactive molecules. Mol. Syst. Biol. 7(1), 472 (2011)
33. Wang, Y., Zeng, J.: Predicting drug-target interactions using restricted boltzmann machines. Bioinformatics 29(13), i126–i134 (2013)
34. Kuhn, M., Szklarczyk, D., Franceschini, A., von Mering, C., Jensen, L.J., Bork, P.: Stitch 3: zooming in on protein-chemical interactions. Nucleic Acids Res. 40(D1), D876–D880 (2012)
35. Ballesteros, J., Palczewski, K.: G protein-coupled receptor drug discovery: implications from the crystal structure of rhodopsin. Curr. Opin. Drug Discov. Dev. 4(5), 561 (2001)
36. Yamanishi, Y., Kotera, M., Kanehisa, M., Goto, S.: Drug-target interaction prediction from chemical, genomic and pharmacological data in an integrated framework. Bioinformatics 26(12), i246–i254 (2010)
37. Ashburner, M., Ball, C.A., Blake, J.A., Botstein, D., Butler, H., Cherry, J.M., Davis, A.P., Dolinski, K., Dwight, S.S., Eppig, J.T., et al.: Gene ontology: tool for the unification of biology. Nat. Genet. 25(1), 25–29 (2000)
38. Yu, G., Li, F., Qin, Y., Bo, X., Wu, Y., Wang, S.: Gosemsim: an r package for measuring semantic similarity among go terms and gene products. Bioinformatics 26(7), 976–978 (2010)
39. Zheng, X., Ding, H., Mamitsuka, H., Zhu, S.: Collaborative matrix factorization with multiple similarities for predicting drug-target interactions. In: Proceedings of the 19th ACM SIGKDD International Conference on Knowledge Discovery and Data Mining, pp. 1025–1033. ACM (2013)
40. Cheng, F., Liu, C., Jiang, J., Lu, W., Li, W., Liu, G., Zhou, W., Huang, J., Tang, Y.: Prediction of drug-target interactions and drug repositioning via network-based inference. PLoS Comput. Biol. 8(5), e1002503 (2012)

41. Cobanoglu, M.C., Liu, C., Hu, F., Oltvai, Z.N., Bahar, I.: Predicting drug-target interactions using probabilistic matrix factorization. J. Chem. Inf. Model. **53**(12), 3399–3409 (2013)
42. Kanehisa, M., Goto, S., Hattori, M., Aoki-Kinoshita, K.F., Itoh, M., Kawashima, S., Katayama, T., Araki, M., Hirakawa, M.: From genomics to chemical genomics: new developments in KEGG. Nucleic Acids Res. **34**(Suppl. 1), D354–D357 (2006)
43. Schomburg, I., Chang, A., Ebeling, C., Gremse, M., Heldt, C., Huhn, G., Schomburg, D.: Brenda, the enzyme database: updates and major new developments. Nucleic Acids Res. **32**(Suppl. 1), D431–D433 (2004)
44. Günther, S., Kuhn, M., Dunkel, M., Campillos, M., Senger, C., Petsalaki, E., Ahmed, J., Urdiales, E.G., Gewiess, A., Jensen, L.J., et al.: Supertarget and matador: resources for exploring drug-target relationships. Nucleic Acids Res. **36**(Suppl. 1), D919–D922 (2008)
45. Wishart, D.S., Knox, C., Guo, A.C., Cheng, D., Shrivastava, S., Tzur, D., Gautam, B., Hassanali, M.: Drugbank: a knowledgebase for drugs, drug actions and drug targets. Nucleic Acids Res. **36**(Suppl. 1), D901–D906 (2008)
46. Bertsekas, D.P.: Nonlinear programming (1999)
47. Elman, H.C., Golub, G.H.: Inexact and preconditioned uzawa algorithms for saddle point problems. SIAM J. Numer. Anal. **31**(6), 1645–1661 (1994)
48. Boyd, S., Vandenberghe, L.: Convex Optimization. Cambridge University Press, Cambridge (2009)
49. Ding, H., Takigawa, I., Mamitsuka, H., Zhu, S.: Similarity-based machine learning methods for predicting drug-target interactions: a brief review. Briefings in Bioinform., bbt056 (2013)
50. Davis, J., Goadrich, M.: The relationship between precision-recall and roc curves. In: Proceedings of the 23rd International Conference on Machine Learning, pp. 233–240. ACM (2006)
51. Wu, C., Liao, Q.: An useful tool for finding drug-target interaction in drugbank and KEGG. http://www.cse.ust.hk/~qnature/
52. Hu, Y., Bajorath, J.: Compound promiscuity: what can we learn from current data? Drug Discov. Today **18**(13), 644–650 (2013)

Author Index